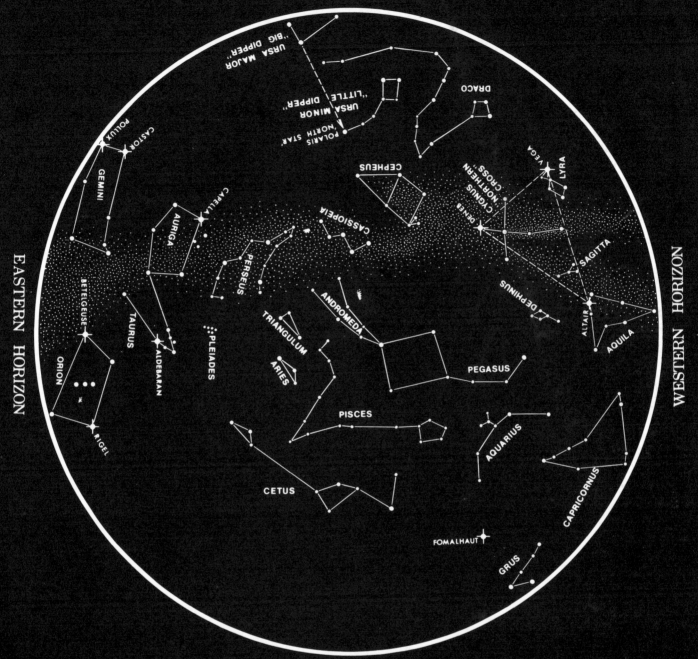

NORTHERN HORIZON

URSA MAJOR "BIG DIPPER"

URSA MINOR "LITTLE DIPPER"

POLARIS "NORTH STAR"

DRACO

CEPHEUS

VEGA

LYRA

CYGNUS "NORTHERN CROSS"

DENEB

CASSIOPEIA

SAGITTA

CAPELLA

PERSEUS

DELPHINUS

ALTAIR

AQUILA

AURIGA

ANDROMEDA

GEMINI

POLLUX

CASTOR

TRIANGULUM

PEGASUS

BETELGEUSE

TAURUS

ARIES

PLEIADES

ALDEBARAN

ORION

PISCES

AQUARIUS

RIGEL

CAPRICORNUS

CETUS

FOMALHAUT

GRUS

EASTERN HORIZON

WESTERN HORIZON

SOUTHERN HORIZON

THE NIGHT SKY IN NOVEMBER

Chart time (Local Standard Time):

10 pm...First of November
9 pm...Middle of November
8 pm...Last of November

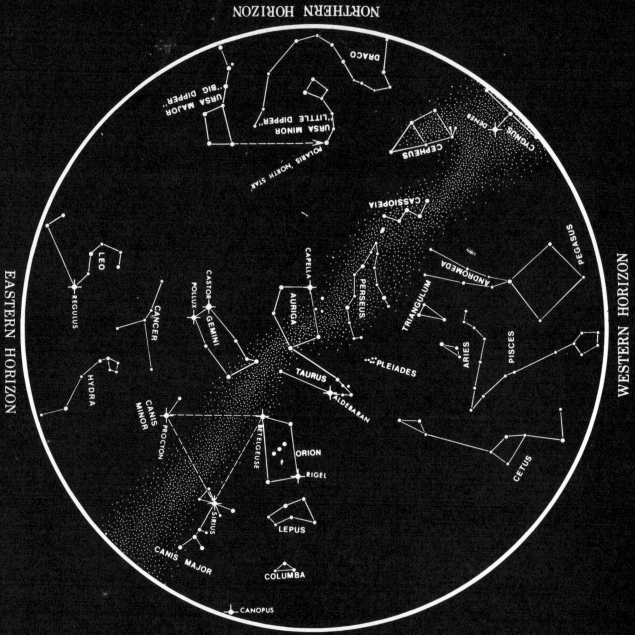

NORTHERN HORIZON

EASTERN HORIZON

WESTERN HORIZON

SOUTHERN HORIZON

THE NIGHT SKY IN JANUARY

Chart time (Local Standard Time):

10 pm...First of January
9 pm...Middle of January
8 pm...Last of January

DISCOVERING THE UNIVERSE

sixth edition

DISCOVERING THE UNIVERSE

sixth edition

NEIL F. COMINS
University of Maine

WILLIAM J. KAUFMANN III
Late of San Diego State University

W. H. FREEMAN AND COMPANY
New York

*To my mother, Pearl Mirianne Comins,
and to the memory of my father, Francis Malcolm Comins*

Senior Acquisitions Editor: Patrick Farace
Publisher: Michelle Russel Julet
Senior Development Editor: Terri Ward
Marketing Manager: Jeffrey Rucker
Associate Editor: Danielle Swearengin
Project Editor: Bradley Umbaugh
Senior Media Editor: Charlie Van Wagner
Cover and Text Designer: Blake Logan
Illustration Coordinator: Shawn Churchman
Illustrations: Fine Line Illustrations
Photo Editor: Trish Marx
Photo Researchers: Tobi Zausner and Laura Nash
Copy Editor and Indexer: Louise B. Ketz
Composition: Sheridan Sellers (W. H. Freeman and Company,
    Electronic Publishing Center) and Patrice Sheridan
Production Coordinator: Paul W. Rohloff
Manufacturing: RR Donnelley & Sons Company

Cover Image: The Hubble Space Telescope imaged this breathtaking star-forming region, 30 Doradus Nebula. It contains the most spectacular cluster of massive stars in our cosmic neighborhood of about 25 galaxies. Ultraviolet radiation and high-speed material ejected by stars in the region are triggering the collapse of gas and dust clouds resulting in the formation of new stars. (Image courtesy of NASA/N. Walborn and J. Maiz-Appelániz–Space Telescope Science Institute, Baltimore, MD/R. Barbá–La Plata Observatory, La Plata, Argentina)

Library of Congress Control Number: 2002105822

ISBN 0-7167-4450-3

Printed in the United States of America
First printing 2002

TEXT OVERVIEW

The total solar eclipse of August 11, 1999 (©2000 by Fred Espenak, MrEclipse.com)

An artist's conception of the nucleus of a coment (STScl)

An artist's conception of the explosion of Supernova 1987A (T. Goertel, STScl)

A computer simulation of colliding galaxies (Joshua E. Barnes, University of Hawaii)

CONTENTS

II ESSENTIALS FOR UNDERSTANDING THE SOLAR SYSTEM 115

PREFACE

Welcome to the cosmos. The information that has gone into this sixth edition of *Discovering the Universe* includes discoveries about the universe that were made just days before it went to press. Besides providing solid current information and some of the latest images available, this edition provides a gateway to a vast amount of astronomy information on the Web. The CD-ROM has also been greatly enhanced, and I am especially excited about the upgraded *Starry Night Backyard*™ software included on the CD. I use *Starry Night Backyard*™ in my lectures to excellent effect, and students enjoy the power and accuracy it provides for hands-on student activities.

I wrote *Discovering the Universe* with special attention to the way people learn. It evolved from my perceptions of the backgrounds, needs, and expectations of the more than 8500 students I have taught introductory astronomy to over the past twenty-three years.

Although most of the material in this book is descriptive, I do include some essential equations used in astronomy. So as not to interrupt the flow of the text, these are mostly set off in numbered boxes called An Astronomer's Toolbox along with additional explanations and practice with unfamiliar equations. Worked examples help those who are uncomfortable with mathematics follow the reasoning step by step, whatever their math background. Self-test questions reinforce this practice, and answers appear at the end of the book.

In the fourth edition of *Discovering the Universe*, I pioneered the practice of directly addressing common misconceptions about astronomy. That effort expands in this edition. **I have written this book expressly to help students identify their own misconceptions and to ponder alternatives.** Throughout the text I explain why common, incorrect beliefs are wrong.

I want students to understand and retain what they learn in this course. Memorizing for an exam is one thing; true understanding is something else altogether. Extensive research shows that it is virtually impossible for anyone to accept new information that conflicts with their established beliefs unless they first understand why their old ideas are wrong. Nearly everyone develops ideas about astronomy before taking a course in the subject. The origins of misconceptions are numerous. Many occur because the laws of nature are often so counterintuitive, defying our common sense. Others occur because we misinterpret what our senses tell us. Still others occur because sources of scientific information, such as television and the movies, are often incomplete or inaccurate. Misconceptions and deep-seated beliefs that are inconsistent with accepted scientific knowledge are inevitable as students struggle to understand our complex environment. Such incorrect beliefs do not imply that students are either naive or incapable of learning; rather they show how fertile and active the human mind is. My research into misconceptions about astronomy has led me to identify more than 1700 of them, along with their origins and common reasoning that leads to incorrect beliefs.

Presenting facts isn't enough. A book has to guide students through the facts to the ideas behind them. **Therefore, each of this book's four parts begins with an "Essentials" minichapter that provides an overview of upcoming chapters and presents essential physical concepts.** Presenting this basic material just when it is needed reinforces a student's sense of discovery.

The book begins the study of astronomy with what people have seen for millennia. As a result, students start with familiar sights and concepts, then move into realms visible only through telescopes and about which they may know less. *Discovering the Universe* has been written so that stars can be studied before or after planets, as instructors may prefer.

■ **Essentials I,** "Essentials for Understanding Astronomy," begins with naked-eye astronomy, the nature of gravity, and its effects on the planets and other bodies. Then we turn to the nature of light and how telescopes work, along with how stars and other astronomical bodies emit light.

■ **Essentials II,** "Essentials for Understanding the Solar System," begins with the Earth and Moon. We explore each of the planets and their moons, the smaller bodies in the solar system, and finally the Sun.

■ **Essentials III,** "Essentials for Understanding the Stars," describes the nature and evolutionary paths of stars, concluding with a discussion of black holes.

■ **Essentials IV,** "Essentials for Understanding the Universe," explores the Milky Way Galaxy, other galaxies, quasars, and the entire universe, including information about the possible acceleration of the universe. The book ends by exploring the age-old question of whether other life exists in the universe.

People learn in various ways. Some are visual learners; others rely more heavily on text. The text itself is designed to help students more easily understand and retain information:

■ Full-sentence headings introduce each topic. Grouped beneath a broader main heading of only a word or two or three, they also provide an outline for review.

■ Important astronomical terms appear in boldface type where they are introduced and also at the end of the chapter in a Key Words list followed by a full summary list of Key Ideas. Key words are again defined in the Glossary.

■ Review Questions and Advanced Questions at the end of each chapter stimulate students to explore the ideas in the text. Discussion Questions require even more thought, more physics and chemistry, and sometimes even outside reading. Any question relying on mathematics is preceded by an asterisk (*) and answered briefly at the back of the book. "What If …" questions will stimulate viewing the universe from new perspectives. The question sections also include Observing Projects, many for use in conjunction with *Starry Night Backyard*™ and the star charts located inside the book's covers.

■ A completely media-connected text with icons throughout the chapters allows students to connect the topic they are reading about to specific areas of the CD-ROM or the Web site or both. See "Guide to Media" for details.

ACKNOWLEDGMENTS

I am deeply grateful to the astronomers and teachers who reviewed the manuscript of this edition:

Gordon Baird, *University of Mississippi*
Peter A. Becker, *George Mason University*
Gene Byrd, *University of Alabama at Tuscaloosa*
Eugene R. Capriotti, *Michigan State University*
Michael W. Castelaz, *Pisgah Astronomical Research Institute*
David S. Chandler, *Porterville College*
Christine Clement, *University of Toronto*
Charles Curry, *University of Waterloo*
James J. D'Amario, *Harford Community College*
John Dickey, *University of Minnesota*
John D. Eggert, *Daytona Beach Community College*
Bernd Enders, *College of Marin*
Martin Gaskell, *University of Nebraska*
Siegbert Hagmann, *Kansas State University*
Chuck Higgins, *Penn State University*
Kenneth Janes, *Boston University*
Marvin D. Kemple, *Indiana University - Purdue University Indianapolis (IUPUI)*
F. W. Kleinhans, *Indiana University - Purdue University Indianapolis (IUPUI)*
John Patrick Lestrade, *Mississippi State University*
C. L. Littler, *University of North Texas*
Michael C. LoPresto, *Henry Ford Community College*

Robert Manning, *Davidson College*
P. L. Matheson, *Salt Lake Community College*
J. Scott Miller, *University of Louisville*
L. D. Mitchell, *Cambria County Area Community College*
Siobahn M. Morgan, *University of Northern Iowa*
Steven Mutz, *Scottsdale Community College*
Bob O'Connell, *College of the Redwoods*
Richard P. Olenick, *University of Dallas*
Melvyn Jay Oremland, *Pace University*
David D. Reid, *Wayne State University*
Barbara Ryden, *Ohio State University*
C. Ian Short, *Florida Atlantic University*
Larry Sessions, *Metropolitan State College of Denver*
Earl F. Skelton, *George Washington University*
George F. Smoot, *University of California at Berkeley*
Paula Szkody, *University of Washington*
Michael T. Vaughn, *Northeastern University*
William F. Welsh, *San Diego State University*
R. M. Williamon, *Emory University*
Edward L. (Ned) *Wright, University of California at Los Angeles*
Nicolle E. B. Zellner, *Rensselaer Polytechnic Institute*

Survey Reviewers
J. David Batchelor, *Community College of Southern Nevada*
Robert Frostick, *West Virginia State College*
Nathan Israeloff, *Northeastern University*
Pushpa Khare, *University of Illinois at Chicago*
M. A. K. Lodhi, *Texas Tech University*

I wish to express my warmest thanks to Terri Ward, W. H. Freeman and Company senior development editor who supervised the writing phase of the project; to Patrick Farace, senior acquisitions editor; and to Charlie Van Wagner, senior media editor for coordinating the supplements and media components. Thanks are due to the talented book team: Bradley Umbaugh, project editor; Paul W. Rohloff, production coordinator; Blake Logan, designer; Trish Marx, photo editor; Tobi Zausner and Laura Nash, photo researchers; Sheridan Sellers, composition; Shawn Churchman, illustration coordinator, and Fine Line Illustrations; Louise B. Ketz, copy editor and indexer; and Barbara Hults, proofreader.

Finally, special thanks to my wife, Suzanne, and to my sons, James and Joshua. All put up with Daddy hard at work on "the book."

Every effort has been made to make this book as error-free as possible. Nevertheless, some inaccuracies have inevitably crept in. I would appreciate hearing from you if you find an error or wish to comment on the text.

Neil F. Comins
neil.comins@umit.maine.edu

GUIDE TO DISCOVERING THE UNIVERSE

II ESSENTIALS FOR UNDERSTANDING THE SOLAR SYSTEM

IN THE CHAPTERS AHEAD YOU WILL DISCOVER

how the solar system formed from a cloud of interstellar gas and dust (Essentials II)

that the planets fall into groups with similar properties (Essentials II, Chapters 5, 6, 7)

NEW!
Essentials for Understanding ...
Brief minichapters introduce each of the four parts, presenting students with the essential background material for understanding the chapters that follow.

NEW!
Media Connections Icons
Throughout the book these icons connect illustrations, text, and questions to rich media assets available on the Web site and CD-ROM.

AIMM 3.3

MORE TO KNOW 3.1

INTERACTIVE EXERCISE 3.1

STARRY NIGHT

ANIMATION 3.3

VIDEO 3.4

WEB LINK 3.2

MOVIE MISCONCEPTIONS

Austin Powers: The Spy Who Shagged Me (New Line Cinema, 1999)
(The Everette Collection)

First he fought for the Crown.
Now he's fighting for the Family Jewels.

The year is 1999 and the international man of mystery Austin Powers and the sinister Dr. Evil are back in *Austin Powers: The Spy Who Shagged Me*. In the movie there is a scene in Doctor Evil's secret lair in which the evil doctor shows his staff how he plans to put a "laser" on the Moon and use it to destroy cities on Earth. In making his presentation, he uses a large orrery consisting of a four-foot diameter model of the Earth fixed in a basket, around which a model of the Moon is free to revolve. The United States is clearly visible on the upper half of the Earth model, meaning that the north celestial pole is above the set. The model of the Earth does not move. The evil doctor explains his plan and then pushes the Moon clockwise around the Earth until it is above the United States. What is wrong with the astronomy described so far?

Despite the error that you have just identified, the Moon follows a westward path around the Earth, as would be seen correctly by someone standing on that fixed model of the Earth in the scene. Explain why the Moon follows the correct path around the model of the Earth, despite the error above.

(Answers appear at the end of the book.)

NEW!
Movie Misconceptions
These short pieces expand on the focus of dispelling common misconceptions and reveal and debunk inaccuracies put forward in popular films.

GUIDED DISCOVERY The Color of the Sun

Different people perceive the Sun to have different colors. To many it appears white, to others yellow. Still others, who notice it at sunset, believe it to be orange or even red. But we have seen that the Sun actually gives off all colors. Moreover, the peak in the Sun's spectrum falls between blue and green. Why doesn't the Sun appear turquoise? Several factors affect our perception of its color.

Before reaching our eyes, visible sunlight passes through the Earth's atmosphere. Certain wavelengths are absorbed and reemitted by the molecules in the air, a process called *scattering*. Violet light is scattered most strongly, followed in decreasing order by blue, green, yellow, orange, and red. That means that more violet, blue, and green photons are scattered by the Earth's atmosphere than are yellow photons. The intense scattering of violet, blue, and green has the effect of shifting the peak of the Sun's intensity entering our eyes from blue-green toward yellow. The sky is blue because of this same scat...

each respond to one of three ranges of colors, which are centered on red, yellow, and blue wavelengths. None of the cones is especially sensitive to blue-green photons. By adding together the color intensities detected by the three types of cones, our brains *recreate* color. After combining all the light it can, the eye is most sensitive to the yellow-green part of the spectrum. We see blue and orange less well, and violet and red most poorly. Although the eye sees yellow and green light about equally well, the dominant color from the Sun is yellow because the air scatters green light, and our eyes are relatively insensitive to blue-green. Therefore, a casual glance at the Sun leaves the impression of a yellow object.

A longer look is extremely dangerous. **Don't try it!** Hypothetically, such a glance would leave the impression of a white Sun. The Sun's light is so intense that it would saturate the color-sensitive cones in our eyes; our brains interpret ... saturation of the cones as white.

WHAT IF . . .

Humans Had Infrared-Sensitive Eyes?

Our eyes are sensitive to less than a trillionth of 1% of the electromagnetic spectrum—what we call "visible light." But this minuscule resource provides an awe-inspiring amount of information about the universe. We interpret visible-light photons as the six colors of the rainbow—red, orange, yellow, green, blue, and violet. These colors combine to form all the others that make our visual world so rich. But the Sun actually emits photons of all wavelengths. So, what would we see if our eyes had evolved to sense another part of the spectrum?

A Darker Vision Gamma rays, X rays, and most ultraviolet radiation do not pass through the Earth's atmosphere, and the world would look dark, indeed, if our eyes were sensitive only to these wavelengths. Radio waves, in contrast, easily pass through our atmosphere. But to see the same detail from radio waves that we now see from visible-light photons, our eyes would require diameters 10,000 times larger. Each would be the size of ...

Night Vision The night sky would be a spectacular sight through infrared-sensitive eyes. Gas and dust clouds in the Milky Way absorb visible light, thus preventing the light of distant stars from getting to the Earth. However, because most infrared radiation passes through these clouds, we would be able, unaided, to see distant stars that we cannot see today. On the other hand, the white glow of the Milky Way, which is caused by the scattering of starlight by interstellar clouds, would be dimmer, because the gas and dust clouds do not scatter infrared light as much as they do visible light. (The haze created by the Milky Way would not vanish, however, because when gas and dust clouds are heated by starlight, they emit their own infrared radiation.)

Our concept of stars would be different, too. Many stars, especially young, hot ones, are surrounded by cocoons of gas and dust that emit infrared radiation. This dust is heated by the nearby stars. Instead of appearing as pinpoints, many stars would appear to be surrounded ...

 Insight into Science Keep it simple When several competing theories describe the same concepts with the same accuracy, scientists choose the simplest one. That basic tenet, formally expressed by William of Occam, in the fourteenth century, is known as **Occam's razor**. Indeed, the original form of the heliocentric cosmology was appealing not because it was more accurate, it wasn't, but because it made the same predictions within a simpler model than did the geocentric cosmology. *Remember Occam's razor.*

AN ASTRONOMER'S TOOLBOX 3-1

Photon Energies

All photons with the same energy are identical: Their energy depends solely on the photon's wavelength. To find any photon's energy we use the equation

$$E = \frac{hc}{\lambda}$$

where Planck's constant h is 6.67×10^{-34} J s, the speed of light c is 300,000 km/s, and the photon's wavelength is λ.

Example: A photon of red light has a wavelength 700 nm. What is its energy? *Note:* All distances in the following equation must be converted to the same units, such as meters.

$$F_{red} = \frac{(6.67 \times 10^{-34} \text{ J s})(300,000 \text{ km/s})}{700 \text{ nm}}$$

$$E_{red} = 2.86 \times 10^{-19} \text{J}$$

(A joule, abbreviated J, is a unit of energy.) Each photon of red light with wavelength 700 nm has an energy of 2.86×10^{-19} J.

Compare! In 1 second a 25-watt light bulb emits 25 J.

Try these: A photon has energy 4.90×10^{-19} J. Calculate its wavelength. Referring to Figure 3-1, what is this photon's color? What is the wavelength of a photon with twice this energy? What is the energy of a red photon? (*Clue:* See Figure 3-4)

End-of-Chapter Pedagogy
Includes a recap of Key Words and Ideas, Review Questions, Advanced Questions, Discussion Questions, What If... Questions, and Observing Projects, many using *Starry Night Backyard*™ from the CD-ROM.

What Do You Think? and What Did You Know?
Numbered questions at the beginning of each chapter and their end-of-chapter counterparts are based on common student misconceptions. Numbered icons within the chapter help the student locate the answers.

WHAT DO YOU THINK?

1 What is the shape of the Earth's orbit around the Sun?

2 Do the planets orbit the Sun at constant speeds?

3 Do the planets all orbit the Sun at the same speed?

4 How much force does it take to keep an object moving in a straight line at a constant speed? when measured

WHAT DID YOU KNOW?

1 *What is the shape of the Earth's orbit around the Sun?* All planets have elliptical orbits around the Sun.

2 *Do the planets orbit the Sun at constant speeds?* No. The closer a planet is to the Sun in its orbit, the faster it is moving. It moves fastest at perihelion and slowest at aphelion.

3 *Do the planets all orbit the Sun at the same speed?* No. A planet's speed depends on its average distance from the Sun. The closest planet moves fastest, the most distant planet moves slowest.

4 *How much force does it take to keep an object moving in a straight line at a constant speed?* Unless an object is subject to an outside force, like friction, it takes no force at all to keep it moving in a straight line at a constant speed.

5 *How does an object's mass differ when measured on Earth and on the Moon?* Assuming the object doesn't shed or collect pieces, its mass remains constant whether on the Earth or on the Moon. Its weight, however, is less on the Moon.

Frontiers yet to be discovered

The science related to forces and orbits described in this chapter was well established by the beginning of the nineteenth century. However, questions remained. Careful observation revealed that Newton's law of gravitation gave very slightly inaccurate predictions for the orbital path of Mercury. We will see in Chapter 13 that this problem was resolved by Albert Einstein. Nevertheless, one fundamental question from this chapter's material remains to be answered: What is "mass"? Many physicists studying the building blocks of matter (such as the elementary particles protons, neutrons, and electrons) believe that mass comes from the presence of a particle permeating the universe called the Higgs boson. When elementary particles interact with it, they gain the property we call mass. Searches for Higgs bosons are underway at high-energy particle accelerators such as CERN, near Geneva, Switzerland, and the Fermi National Accelerator in Illinois. If these experiments are successful, then we may know the origin of mass by the end of this decade.

 Further Reading on These Topics

Frontiers yet to be discovered
Sections concluding each chapter present the directions that science is exploring.

WEB/CD-ROM QUESTIONS

25 Search the Web for information about Galileo. What were his contributions to physics? Which of Galileo's new ideas were later used by Newton to construct his laws of motion? What incorrect beliefs about astronomy did Galileo hold?

26 Search the Web for information about Kepler. Before he realized that the planets move on elliptical paths, what other models of planetary motion did he consider? What was Kepler's idea of the "music of the spheres"?

27 Search the Web for information about Newton. What were some of the contributions that he made to physics other than developing his laws of motion? What contributions did he make to mathematics?

 28 **Monitoring the Retrograde Motion of Mars** Access and view the animation "The Path of Mars in 2004–2005" in Chapter 2 of the *Discovering the Universe* Web site or CD-ROM.
(a) Through which two constellations does Mars move?
(b) On approximately what date does Mars stop its direct (west to east) motion and begin its retrograde motion? *Hint:* Use the "Stop" and "Start" functions on your animation controls. (c) Over how many days does Mars move in the retrograde direction?

NEW!
Web/CD-ROM Questions
New to this edition, media questions appear at the end of each chapter. To answer them, students use the resources of the Web site or the CD-ROM.

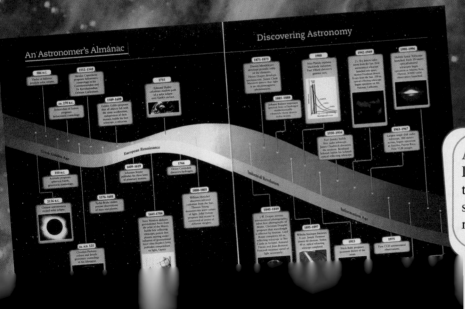

An Astronomer's Almanac
Dynamic timelines, located after the Essentials chapters, offering students a perspective on astronomy relative to other historical events.

Guide to the Media *Discovering the Universe* fully integrates the text with electronic media

One of the new features of the sixth edition is that students will use the book, Web site, and the CD-ROM as aspects of a seamless learning experience. On almost every page, new **media icons** point students to specific, relevant information on the CD-ROM and the *Discovering the Universe* Web site:

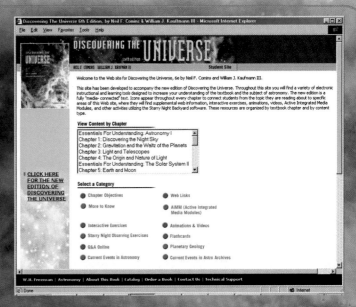

 Animations illustrating key concepts, including animated versions of selected text figures.

 Videos from space missions as well as ground-based observations.

 More to Know that expands on key text topics.

 Links to external Web sites where students can find additional information and images.

 Interactive Exercises test student understanding of selected images and diagrams in the text.

 Active Integrated Media Modules are interactive learning tools that simplify repetitive calculations and allow students to gain quantitative understanding.

 Many end-of-chapter **Observing Projects** use the award-winning *Starry Backyard Night*™ planetarium software provided on the CD-ROM.

DISCOVERING THE UNIVERSE WEB SITE

at www.whfreeman.com/dtu6e
For instructors, the site offers further information about the topics in the book, links to additional resources, and a forum for sharing ideas about using the book.

Chapter Objectives: Learning objectives help students formulate study strategies.

Q & A: The self-quizzing feature helps students prepare for exams.

Interactive Exercises: "Drag and drop" exercises that help students understand the vocabulary of astronomy in the context of illustrations from the text.

Flashcards: Electronic flashcard exercises using Key Terms and definitions from the text book.

Starry Night™ **Observation Exercises:** In-depth projects that students can perform using the *Starry Night*™ software provided on the student CD-ROM. These complement the shorter *Starry Night*™ Observing Projects in the textbook.

Current Events in Astronomy: Up-to-date links to information and resources.

Animations and Videos, both original and from sources such as NASA and ESA.

Web Links lead to Internet sites with supplemental material about astronomy that are updated regularly.

Active Integrated Media Modules (AIMM), interactive modules that take students deeper into key topics from the text, including determining distances to the stars and beyond, the Doppler Effect, the nearest stars, and relativistic redshift.

More to Know: In-depth boxes, thoroughly updated, extend topics discussed in the text.

Discovering the Universe 6.0 CD-ROM featuring *Starry Night Backyard*™

Packaged with every copy of the textbook, the *Discovering the Universe* 6.0 CD-ROM contains all the content that appears on the Web site (except for the Web Links and "Current Events in Astronomy"). It also includes the easy-to-use, award-winning *Starry Night Backyard*™ planetarium software (for both Macintosh and Windows), the ideal electronic accompaniment for any astronomy course. *Starry Night Backyard*™ observational activities are provided in each chapter of *Discovering the Universe*. Others are available on the *Discovering the Universe* Web site.

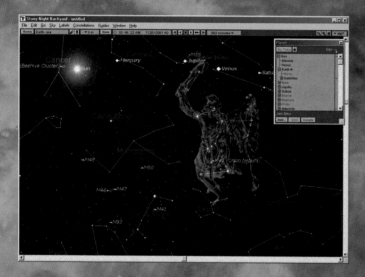

New Features include

- Observations of the sky from anywhere on Earth, any Solar System body, or any location up to 20,000 light years away.
- 1,000,000 stars along with 110 deep space objects like galaxies, star clusters and nebulae.
- Automatic updates—with each use the program checks the Web for updates and makes changes on your hard drive.
- Enhanced zoom capacity—in up to 600,000 times to view images and movies.
- Point and click navigation with tools to learn and use

Active Integrated Media Modules

Interactive modules that take students deeper into key topics from the text, including determining distances to the stars and beyond, the Doppler effect, the nearest stars, and relativistic redshift.

Small-Angle Formula

Telescope Magnification

Wien's Law

Parallax

The Drake Equation

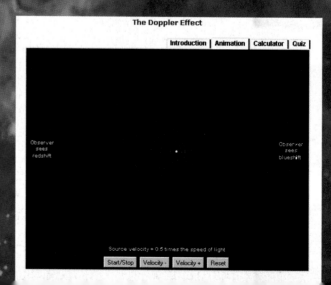

For Instructors

Instructor's CD-ROM

Supplied to all instructors who adopt *Discovering the Universe*, this CD-ROM provides instructors with the tools to create in-class presentations:

- **All text images in JPEG, PICT, and PowerPoint™ format** for use with any standard presentation software.

- All animations, videos, and AIMMs from the accompanying media available in a downloadable format.

- **Presentation Manager Pro software** lets you build classroom presentations using images and video from the *Discovering the Universe* CD-ROM, plus your own digital material (including video) imported from the Internet or other sources.

Instructor's Manual and Resource Guide, George A. Carlson, Citrus College

The Manual is provided in a Word format to make it easier to view, edit, and print. The Manual includes worked-out solutions to all end-of-chapter questions in *Discovering the Universe*. It also includes detailed chapter outlines as well as classroom-tested teaching strategies and hints. The extensive Resource Guide covers reading materials for students and instructors, audiovisual material, and discussion/paper topics.

Starry Night™ **Pro Version CD-ROM from Space.com**
Available to qualified instructors, the deluxe version of *Starry Night*™ features over 19 million celestial objects (including the complete Hubble Guide Star Catalogue), photo-realistically rendered 3D planets, astrophotographs of all 110 Messier objects, a Datamaker application for adding your own custom astronomical databases, movie-making abilities, and more.

PowerPoint™ Lecture Presentation:

Instructors adopting *Discovering the Universe* have access to PowerPoint™ files that include lecture notes for each chapter of the book, along with all the textbook images and their captions. The PowerPoint™ files can be downloaded and customized to suit the individual instructor's needs.

Overhead Transparency Set

100 full-color transparencies of key illustrations, photos

On-Line Course Materials in WebCT

Adopters using WebCT can use this service, which provides a fully loaded E-Pack that includes the instructor and student resources for *Discovering the Universe*. The files can be used as is or customized to fit specific needs. Course outlines, pre-built quizzes, links, activities, and a whole array of materials are included.

Test Bank, William J. F. Wilson and T. Alan Clark, University of Calgary

This set of more than 2000 multiple-choice questions is available on CD-ROM in both Macintosh and Windows formats. All questions are page-referenced and designated as "assignment" or "test" questions. The CD-ROM version makes it easy to print, add, edit, and re-sequence questions to suit your needs. The CD-ROM is also the access point for Diploma Online Testing.

Diploma Online Testing, from the Brownstone Research Group

Diploma makes it easy to create and administer secure exams over either the Internet or a local area network. Questions (taken from the Test Bank CD-ROM) can incorporate multimedia and interactivity. The program lets you restrict tests to specific computers or time blocks. It includes an impressive suite of gradebook and result-analysis features.

For Students

Student CD-ROM

Contains all the media from the Web site in an easy to use format as well as the award winning planetarium Software *Starry Night Backyard*™.

Starry Night ™ Observing Guide

William Wilson, T. A. Clark, University of Calgary and Marcel Bergman— Seventeen new, comprehensive lab activities for *Starry Night Backyard*™ provide even more opportunities for students to explore the cosmos. Each activity is tied directly to *Discovering the Universe* by section number.

DISCOVERING THE UNIVERSE

sixth edition

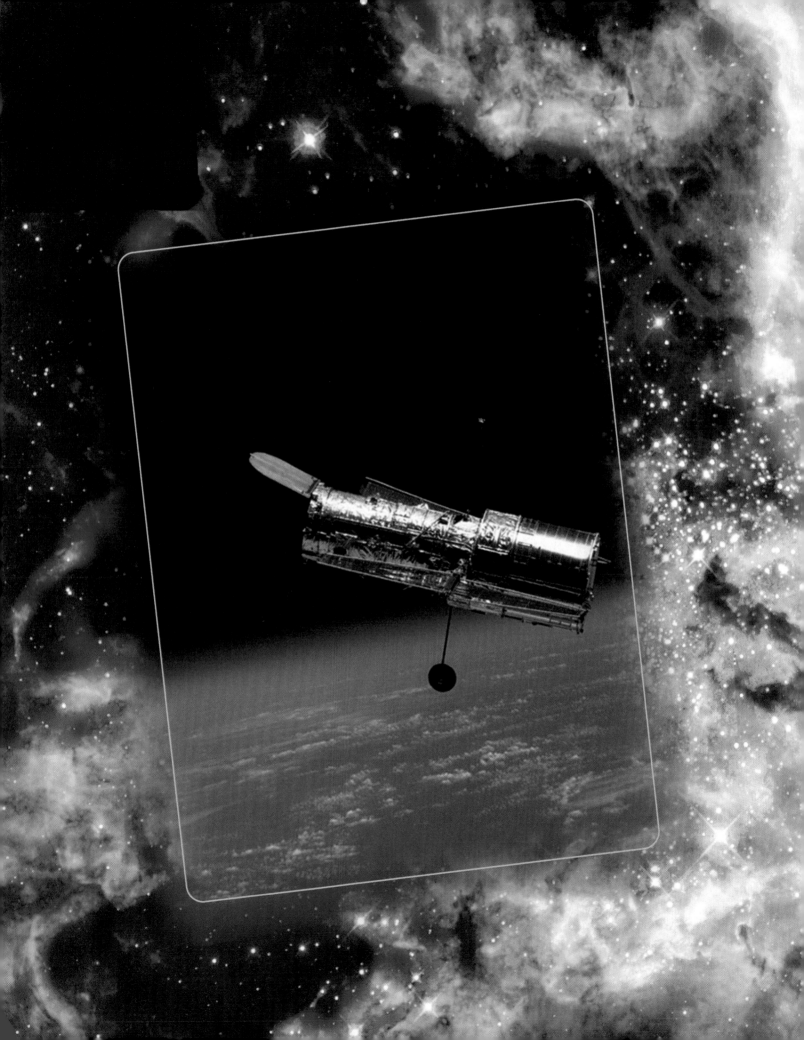

I ESSENTIALS FOR UNDERSTANDING ASTRONOMY

IN THIS SECTION OF THE BOOK YOU WILL DISCOVER

- the scientific process astronomers use to study the universe (Essentials I)

- how astronomers organize the night sky (Chapter 1)

- how the Earth's motion on its axis and its motion around the Sun cause changes on Earth and in what we see in the sky at night (Chapter 1)

- that the Moon's orbit causes eclipses and its changing phases (Chapter 1)

- Kepler's orbital laws and Newton's gravitational law (Chapter 2)

- the nature of light and how telescopes work (Chapter 3)

- the structure of atoms and what observing light from them reveals about the universe (Chapter 4)

THE HUBBLE SPACE TELESCOPE
Our quest for knowledge about the universe is epitomized by the building of such technological marvels as the Hubble Space Telescope (HST), shown here in 1997 after a servicing mission by members of the space shuttle DISCOVERY crew. HST is large enough—13.1 meters (43 feet) long and 4.3 meters (14 feet) wide—to be seen with the naked eye from the Earth's surface. (NASA)

WHAT DO YOU THINK?

1   What makes a theory scientific?

2   What do astronomers do?

3   What practical value does astronomy have?

Look for answers beside the boxed numbers in the text margins.

For thousands of years people have looked up at the sky and found themselves inspired to contemplate the nature of the universe. How was it created? Where did the Earth, Moon, and Sun come from? What are the planets and stars made of? What are our place and role in the cosmic scope of space and time?

The beauty of the star-filled night sky or a comet majestically arching across the sky (Figure I-1) or the drama of an eclipse are just a few of the things that make astronomy fascinating. But there are also practical reasons for an interest

R I V U X G

FIGURE I-1   **The Starry Sky** Daylight hides all signs of the stars and other wonders shining above the sky's azure curtain. Our thoughts about the universe overhead focus on the Sun, the Moon, and our ever-changing weather. But, ah, the night! No light show, artist's brush, or poet's words can truly capture the beauty of this breathtaking panorama. This photograph of saguaro cactus, the night sky, and Halley's Comet was taken in northern Mexico. (Courtesy of D. L. Mammana)

in the universe. The ancient Greeks knew the connection between the changing height of the noontime Sun and the different patterns of stars in the night throughout the year. This information enabled them to predict the seasons, a useful skill for farming. Many early seafaring cultures were also aware that the positions of the Moon and Sun influence the tides. This knowledge helped these people plan and navigate sailing voyages.

I-1   The universe is comprehensible

What causes these relationships between the Earth and heavenly bodies? Astronomical phenomena were first explained as the result of supernatural forces and divine intervention. The heavens were thought to be populated by demons and heroes, gods and goddesses. Yet, despite superstitious beliefs, some people have always realized that the universe is logical and comprehensible. Astronomical cycles, such as the seasons, the reappearance of stars, the tides, and day and night, led many early societies to study the patterns and motions found in the night sky. Progress in science is embodied in improved technology, such as the invention of the telescope. Improvements in technology have in turn led to further discoveries about the fundamental laws of **physics,** the science that investigates the nature of matter and energy and the relationships between them.

This section begins with the "everyday" aspects of astronomy, from naked-eye observations of the sky to the motion of the planets and the use of telescopes to widen our vistas. You will discover the process scientists use to explore natural phenomena through systematic observations and refined theories, and you will see how the knowledge gained by the scientific study of astronomy has led to an understanding of events and phenomena that our ancestors never could have imagined.

I-2   Science demands systematic observations

Astronomers gain knowledge by using the **scientific method** to observe, predict, and explain physical reality. Observations lead scientists to create a **scientific theory,** an idea or collection of ideas that proposes to explain the observed phenomenon. Scientific theories are expressed mathematically as **models.** For example, Newton's law (or theory, as such ideas are now called) of gravitation is written as an equation that predicts how bodies attract each other. This model predicts that the Sun's gravitational force makes the planets move in elliptical orbits. Gravitation is the unique attractive force between all matter in the universe. The word "gravity" is often used as shorthand for "gravitation" and both are used in this book.

A scientific theory can be independently tested and potentially disproved. Newton's law can be tested and potentially disproved by observations and thus qualifies as a scientific theory. The idea that the Earth was created in six days cannot be tested, much less disproved. It is not a scientific theory but rather a matter of faith.

Scientific theories also make testable *predictions* that can be verified using new observations and experiments. Testing is a crucial aspect of the scientific method, which requires that the theory accurately forecast the results of new observations in its realm of validity.

If a theory proves inconsistent with observations, the theory is either modified, applied only in limited circumstances, or discarded in favor of a more accurate explanation. For example, Newton's law of gravitation is entirely adequate for describing the motion of the Space Shuttle around the Earth or the Earth around the Sun, but it is inaccurate in the vicinity of a black hole, where matter is especially dense. In this realm, Newton's law of gravitation is supplanted by Einstein's theory of general relativity, which accurately describes gravitational behavior in a much wider range of conditions than Newton's law, but at the cost of greater mathematical complexity. When applied to the motion of the Earth around the Sun, general relativity gives the same results as Newton's law of gravitation.

> ## Insight into Science   Theories and beliefs   New theories are personal creations, but science is not a personal belief system. Scientific theories make predictions that can be tested independently. If everyone who performs tests of the theory's predictions gets results consistent with the theory, the theory is considered valid. In comparison, belief systems such as which sports team or political system is best are personal matters. People will always hold differing opinions about such issues.

We see an enormous number of objects in the universe, but we see them all from one vantage point in space and time. Therefore, we cannot always be certain of their true shape and form or of their history. Fortunately, many of the bodies in space are similar. By categorizing them suitably and then applying the scientific method to these groups of objects, we form theories about them and their evolutionary histories that can be tested and modified.

Most scientific theories discussed in the following chapters are supported by a broad range of observations while a few are still largely unsubstantiated. These latter theories await confirming observations that may be made in the next few years. The scientific method can be summarized in five words: observe, theorize, predict, test, modify. I urge you to watch for applications of the scientific method in action throughout this book.

The power of the scientific method was demonstrated centuries ago, during the late Renaissance, when a few courageous scientists proposed that the Earth orbits the Sun. The prevailing belief system of the sixteenth century held that the Earth was the center of everything (see Guided

FIGURE I-2   Galileo's Telescope When Galileo turned his telescope toward the sky in the early 1600s, he discovered craters on the Moon, the phases of Venus, and satellites orbiting Jupiter, and he confirmed the historical observations that there are spots on the Sun. These controversial discoveries flew in the face of conventional wisdom and threatened to undermine the teachings of organized religion of the time. (Scala/Art Resource)

Discovery: The Earth-Centered Universe). This makes sense: To the untrained eye, all the astronomical objects appear to orbit our planet!

This Earth-centered theory ran into trouble when observations of the motions of the stars, planets, Moon, and Sun were shown to be inconsistent with the predictions of the theory. Centuries of modifications to the Earth-centered theory made it truly unwieldy, and even those changes could not keep it consistent with increasingly accurate observations. In the mid-1500s, the Polish mathematician Nicolaus Copernicus resurrected a theory first proposed by Aristarchus nearly 18 centuries earlier—that the Earth orbits the Sun. Copernicus was motivated by an effort to simplify the celestial scheme. This Sun-centered theory of the known universe gained strength when the Italian scientist Galileo Galilei (Figure I-2), the first person to point a telescope toward the sky, saw the moons of Jupiter orbit that planet. This discovery flew in the face of the Earth-centered theory and fueled the search to discover the relationship between the Earth and the rest of the cosmos.

> ## Insight into Science   Shedding misconceptions
> Ideally, a scientist must be willing to discard even the most cherished theories if they fail to agree with observation and experiment. However, scientists are human; they have beliefs that they are loath to let go. Often a disproved theory, such as the belief that the Earth is at the center of the universe, dies only with its advocates.

GUIDED DISCOVERY   The Earth-Centered Universe

At the dawn of the twenty-first century, most of us find it hard to understand why anyone would believe that the Sun, planets, and stars orbit the Earth. After all, we *know* that the Earth spins on its axis; we *know* that the gravitational force from the Sun holds the planets in orbit, just as the Earth's gravitational force holds the Moon in orbit. And we *know* that the stars in the night sky all lie far past the boundaries of our solar system. These facts have become part of our understanding of the motions of the heavenly bodies.

Psychologists call this, the background information that we use to help explain things, a conceptual framework. Any conceptual framework contains all the information we take for granted. For example, when the Sun rises, moves across the sky, and sets today, we take for granted that it is the Earth's rotation that causes the Sun's apparent motion.

Our ancestors possessed a different conceptual framework than one based on science's understanding of the cosmos. They didn't know that the Earth rotates. They didn't know that the then-mysterious force that held them to the ground is the same force that attracts the Earth to the Sun and the Moon to the Earth. They didn't know that the Sun is a star, just like the fixed points of light in the sky. And they didn't know any of the other laws of physics related to motion that we take for granted.

Because they did not feel the Earth move under their feet, nor see any other indication that the Earth is in motion, our forebears sensed nothing to support the belief that the Earth moves. The obvious conclusion for one who has a prescientific conceptual framework, even today, is that the Earth stays put while objects in the heavens move around it.

This prescientific conceptual framework for understanding the motions of the heavenly bodies was strictly based on the senses, that is, people observed motions and drew "obvious," commonsense conclusions. Today, we incorporate the known and tested laws of physics in our understanding of the natural world. Many of these concepts are utterly counterintuitive, and, therefore, the conceptual frameworks we possess are less consistent with common sense than those held in the past. Studying science helps us develop intuition that is consistent with the actual workings of nature.

I-3   Astronomical distances are, well, astronomical

Astronomy is a quantitative science; its discoveries are based on facts and expressed in terms of numbers and associated units, like 1800 seconds or 8.3×10^{12} kilograms. The incredible ranges of distances, sizes, and masses in astronomy require a shorthand for large and small numbers called "powers of ten" or **scientific notation.** (Please read An Astronomer's Toolbox I-1 if you are not familiar with scientific notation.) We will also use the metric system of units, which is now standard in science. Table I-1 lists some comparisons and conversions between metric and the traditional British units of measure. Even the metric system is too cumbersome for studying such remote objects as stars, galaxies, and the farthest realms of the universe. An Astronomer's Toolbox I-2 thus introduces other important astronomical distance units.

In our everyday lives we typically deal with distances ranging from millimeters or fractions of an inch to thousands of kilometers or miles. These larger sizes are some 10^9 times bigger than the smaller ones. Astronomical distances are obviously greater, but how much so? When asked how far the closest stars (other than the Sun) are from Earth, most people give answers in the range of millions of kilometers or miles. To appreciate the range of sizes in the universe, let's briefly explore the true distances to some astronomical objects.

The Moon is typically about 384,000 km (239,000 mi) from the Earth, and the Sun's average distance is an impres-

TABLE I-1   Common Conversions Between British and Metric Units

1 inch	=	2.54 centimeters (cm)
1 cm	=	0.394 inch
1 yard	=	0.914 meter (m)
1 meter	=	1.09 yards = 39.37 inches
1 mile	=	1.61 kilometers (km)
1 km	=	0.621 mile

AN ASTRONOMER'S TOOLBOX I-1

Powers-of-Ten Notation

Astronomy is a science of extremes. As we examine various cosmic environments, we find an astonishing range of conditions, from the incredibly hot, dense centers of stars to the frigid, near-perfect vacuum of interstellar space. To describe such divergent conditions accurately, we need a wide range of both large and small numbers. Astronomers avoid such confusing terms as "a million billion billion" (1,000,000,000,000,000,000,000,000) by using a standard shorthand system. All the cumbersome zeros that accompany such a large number are consolidated into one term consisting of 10 followed by an *exponent*, which is written as a superscript and called the **power of ten**. The exponent merely indicates how many zeros you would need to write out the long form of the number. Thus,

$$10^0 = 1$$
$$10^1 = 10$$
$$10^2 = 100$$
$$10^3 = 1000$$
$$10^4 = 10,000$$

and so forth. The exponent tells you how many tens must be multiplied together to yield the desired number. For example, ten thousand can be written as 10^4 ("ten to the fourth") because $10^4 = 10 \times 10 \times 10 \times 10 = 10,000$. Similarly, 273,000 can be written as 2.73×10^5.

In scientific notation, numbers are written as a figure between 1 and 10 multiplied by the appropriate power of 10. The distance between the Earth and the Sun, for example, can be written as 1.5×10^8 km. Once you get used to it, you will find this notation more convenient than writing "150,000,000 kilometers" or "one hundred and fifty million kilometers."

This powers-of-ten system can also be applied to numbers that are less than 1 by using a minus sign in front of the exponent. A negative exponent tells you that the location of the decimal point is as follows:

$$10^0 = 1.0$$
$$10^{-1} = 0.1$$
$$10^{-2} = 0.01$$
$$10^{-3} = 0.001$$
$$10^{-4} = 0.0001$$

and so forth. For example, the diameter of a hydrogen atom is 1.1×10^{-8} cm. That is more convenient than saying "0.000000011 centimeter" or "11 billionths of a centimeter." Similarly, .000728 equals 7.28×10^{-4}.

Using the powers-of-ten shorthand, one can write large or small numbers like these compactly:

$$3,416,000 = 3.416 \times 10^6$$
$$0.000000807 = 8.07 \times 10^{-7}$$

Because powers-of-ten notation bypasses all the awkward zeros, a wide range of circumstances can be numerically described conveniently:

$$\text{one thousand} = 10^3$$
$$\text{one million} = 10^6$$
$$\text{one billion} = 10^9$$
$$\text{one trillion} = 10^{12}$$

and also

$$\text{one thousandth} = 10^{-3} = 0.001$$
$$\text{one millionth} = 10^{-6} = 0.000001$$
$$\text{one billionth} = 10^{-9} = 0.000000001$$
$$\text{one trillionth} = 10^{-12} = 0.000000000001$$

Try these questions: Write 3141000000 and .0000000031831 in scientific notation. Write 2.718282×10^{10} and 3.67879×10^{-11} in standard notation.
(Answers appear at the end of the book.)

sive 1.5×10^8 km (93 million miles) from Earth. So, if the nearest stars were millions of miles away, they would be at roughly the same distance from Earth as the Sun. They should then appear as bright as the Sun, which clearly they do not. Astronomers have measured the actual distances of the nearest stars to be 4.0×10^{13} km (25 trillion miles), and the most distant stars in our Milky Way Galaxy are

twenty-five thousand times farther away than that. Furthermore, the visible universe is populated with some fifty billion galaxies, the farthest of which are more than 10^{23} km away.

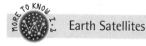 Earth Satellites

AN ASTRONOMER'S TOOLBOX I-2

Astronomical Distances

Throughout this book we will find that some of our traditional units of measure become cumbersome. It is fine to use kilometers to measure the diameters of craters on the Moon or the heights of volcanoes on Mars. However, it is as awkward to use kilometers to express distances to planets, stars, or galaxies as it is to talk about the distance from New York City to San Francisco in millimeters. Astronomers have therefore devised new units of measure.

When discussing distances across the solar system, astronomers use a unit of length called the **astronomical unit (AU)**, which is the average distance between the Earth and the Sun:

$$1 \text{ AU} \approx 1.5 \times 10^8 \text{ km} \approx 9.3 \times 10^7 \text{ miles}$$

Jupiter, for example, is an average of 5.2 times farther from the Sun than is the Earth. Thus, the distance between the Sun and Jupiter can be conveniently stated as 5.2 AU. This can be converted into kilometers or miles using the relationship above.

When talking about distances to the stars, astronomers choose between two different units of length. One is the **light-year (ly)**, which is the distance that light travels in a vacuum (in the absence of air) in one year:

$$1 \text{ ly} \approx 9.46 \times 10^{12} \text{ km} \approx 63{,}000 \text{ AU}$$

One light-year is roughly equal to six trillion miles. Proxima Centauri, the star (other than the Sun) nearest to the Earth, is just over 4.2 ly from Earth.

The second commonly used unit of length is the **parsec (pc)**, the distance at which two objects separated by 1 AU make an angle of 1 arcsecond. Imagine taking a journey far into space, beyond the orbits of the outer planets. Watching the solar system as you move away, the angle between the Sun and the Earth becomes smaller and smaller. When the Sun and Earth are side by side and you measure the angle between them as $1/3600°$ (called 1 arcsecond), you have reached a distance astronomers call 1 parsec, as shown in the figure below. The parsec turns out to be longer than the light-year. Specifically,

$$1 \text{ pc} \approx 3.09 \times 10^{13} \text{ km} \approx 3.26 \text{ ly}$$

Thus, the distance to the nearest star can be stated as 1.3 pc as well as 4.2 ly. Whether one uses light-years or parsecs is a matter of personal taste.

For even greater distances, astronomers commonly use *kiloparsecs* (kpc) and *megaparsecs* (Mpc), in which the prefixes simply mean "thousand" and "million," respectively:

$$1 \text{ kpc} = 10^3 \text{ pc}$$
$$1 \text{ Mpc} = 10^6 \text{ pc}$$

For example, the distance from Earth to the center of our Milky Way Galaxy is about 8.6 kpc, and the rich cluster of galaxies in the direction of the constellation Virgo is 20 Mpc away.

Try these questions: The nearest star (other than the Sun) is 4.22 ly away. How many miles away is it? How many kilometers? How many parsecs?

(Answers appear at the end of the book.)

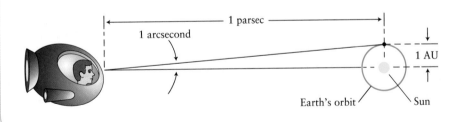

A Parsec The parsec, a unit of length commonly used by astronomers, is equal to 3.26 ly. The parsec is defined as the distance at which 1 AU perpendicular to the observer's line of sight makes an angle of 1 arcsecond.

Figure I-3 summarizes lengths in the cosmos ranging from the sizes of atomic particles to the size of the entire universe visible to us. Unlike distance measured on a ruler, the clockwise arc in this figure shows distances increasing by powers of 10. For example, going from the size of a proton (roughly 10^{-15} meters) up to the size of an atom (roughly 10^{-10} meters) takes about the same space along the arc as going from the distance between the Earth and Sun to the distance between the Earth and the nearby stars. Therefore, the largest-sized objects represented here are more than 10^{40} times larger than the smallest.

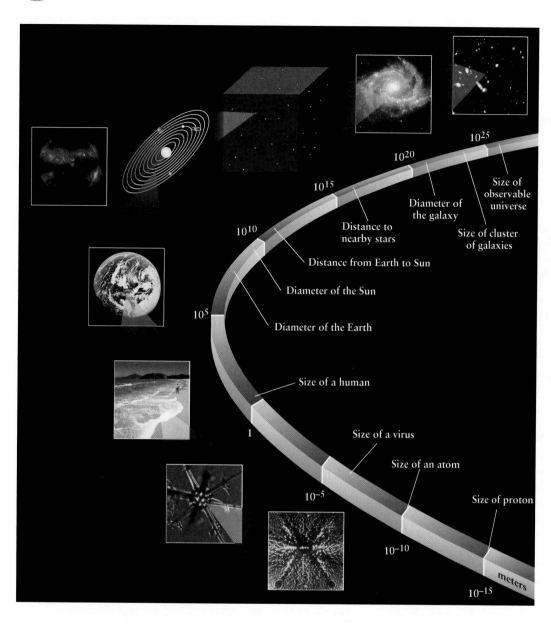

FIGURE I-3   **Examples of Powers-of-Ten Notation** The scale gives the sizes of objects in meters, ranging from subatomic particles at the bottom to the entire observable universe. Keep in mind that every 0.56 cm up along the arc represents a factor of 10 larger. (Top to bottom: R. Williams and the Hubble Deep Field Team [STScI] and NASA; AAT; L. Golub, Naval Observatory, IBM Research, NASA; Richard Bickel/Corbis; Scientific American Books)

I-4   Astronomers do much more than just make observations

The process of astronomical discovery involves more than observing and recording the heavens. The research activities of astronomers today fall into three basic categories: observing, recording, and analyzing observations; theorizing; and computer modeling. Most people think that an astronomer spends his or her time observing the sky, spending long nights directing the most powerful eyes on Earth to reveal the secrets of space. In reality, research telescopes are constantly booked, and most astronomers who get observing time—even a few weeks each year—consider themselves very lucky.

Planning observations and analyzing data take up the majority of an observational astronomer's time. Most obser-

vational astronomers use data-collecting equipment provided by research observatories, but some design and build their own specialized apparatus, which they then connect to existing telescopes.

Other astronomers, theoreticians called *astrophysicists*, never use telescopes. Rather, they hypothesize or expand on theories in an effort to explain the observations of others. Some astrophysicists construct computer models to predict outcomes based upon existing theories.

Astronomers are found in a wide variety of professional positions. In North America alone there are about 6000 astronomers. About 30 percent are employed at observatories and by various governments. Another 60 percent teach at colleges and universities. About 5 percent work in museums, planetaria, or other facilities that help educate the public, while the remaining 5 percent work in industry.

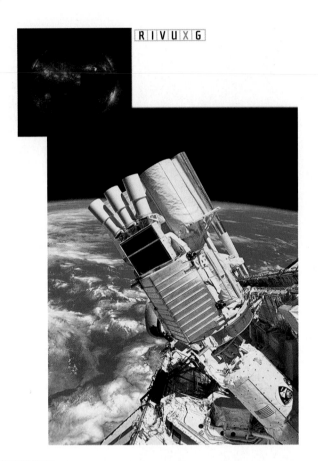

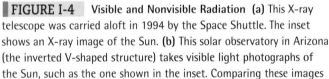

FIGURE I-4   **Visible and Nonvisible Radiation (a)** This X-ray telescope was carried aloft in 1994 by the Space Shuttle. The inset shows an X-ray image of the Sun. **(b)** This solar observatory in Arizona (the inverted V-shaped structure) takes visible light photographs of the Sun, such as the one shown in the inset. Comparing these images reveals how important observing nonvisible radiation from astronomical phenomena is to furthering our understanding of how the universe operates. (Top left and right, L. Golub Naval Observatory, IBM Research, NASA; bottom left, NASA; bottom right, NOAO)

Astronomical discovery is always exciting and sometimes totally unexpected. Unusual observations often lead astronomers and astrophysicists in new directions of research as they try to explain what they see or to reconcile apparent contradictions between theories and observations. For example, at the end of the twentieth century astronomers observed that the universe is actually expanding faster and faster, contrary to the previous belief that the universe is expanding outward more slowly all the time. This was utterly unexpected and revolutionized how astronomers view the cosmos.

In turn, the theories developed to explain the universe have led to a much deeper and richer understanding of the Earth. Many people think that astronomy deals only with the faraway and is of no significance to everyday life. But consider, as one example among many, that Newton's law of gravitation explains both the motions of the planets and why we and everything around us are held to the Earth's surface. Gravity also explains the time it takes an egg falling off a table to hit the floor and the hang time for basketball players under the net. By understanding the law of gravitation, engineers can also control the friction between your car's tires and the road or design an airplane wing to lift a jumbo jet.

I-5   Telescopes enlarge our vision of the universe

How do astronomers gather information? Beginning with Galileo, astronomers peered directly through telescopes to observe visible light from stars. Visible light is a form of energy, usually called **electromagnetic radiation,** that is emitted by stars and other objects. In 1800, William Herschel discovered infrared radiation, the first of a variety of nonvisible forms of electromagnetic radiation. These include radio waves, infrared radiation, ultraviolet rays, X rays, and gamma rays. By early in the twentieth century, all these types of electromagnetic radiation had been discovered.

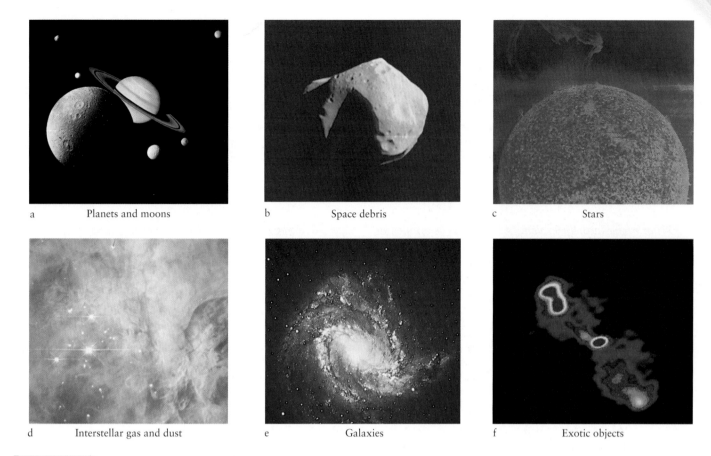

a   Planets and moons   b   Space debris   c   Stars

d   Interstellar gas and dust e   Galaxies   f   Exotic objects

FIGURE I-5   **Inventory of the Universe** Pictured here are examples of the major categories of objects that have been found throughout the universe. You will discover more about each type in the chapters that follow. All these photographs are at visible wavelengths, except the last, exotic objects, which is a radio image. (a: NOAO; b: NASA; c: NASA; d: C. R. O'Dell, Rice University; John Bally, University of Colorado; Ralph Sutherland, University of Colorado; e: R. J. Dufour, Rice University; f: F. N. Owen and J. J. Puschell/NRAO/AUI)

In the middle of the nineteenth century, scientists began recording visible-light astronomical images on film. During the past half century, astronomers have constructed telescopes and image recording systems sensitive to all forms of nonvisible electromagnetic radiation. These telescopes are providing us with ever-more-detailed images of objects in space. Every form of electromagnetic radiation has revealed new information about cosmic bodies (Figure I-4).

Whether located on Earth or in orbit above the obscuring effects of our atmosphere, telescopes give us views vastly superior to anything our eyes can see. They are crucial to our understanding of familiar objects like the Sun, and they also provide important clues about the nature of such exotic objects as neutron stars, pulsars, quasars, and black holes.

What, then, have astronomers seen of the universe? Figure I-5 inventories some of the features we will explore in this text, including planets, stars, black holes, galaxies, and quasars. Each object is constantly changing; each had an origin and each will have an end. We will study these processes, too.

The ideas of science fiction writers, such as Jules Verne and H. G. Wells, pale when compared to current reality. Ours is an age of exploration and discovery more profound than any since the voyages of Columbus and Magellan began to reveal the surface features of the Earth. We have walked on the Moon, dug into Martian soil, and landed on an asteroid. Our probes have discovered active volcanoes and barren ice fields on the satellites of Jupiter. We have visited the shimmering rings of Saturn. We have seen evidence of planets orbiting other stars.

Never before has so much been revealed in so short a span of time. As you proceed through this book, you will gain a new appreciation of the awesome power of the human mind to reach out, to explore, to observe, and to comprehend. One of the great lessons of modern astronomy is that by gaining, sharing, and passing on knowledge, we can transcend the limitations of our bodies and the brevity of human life.

I-6   Frontiers yet to be discovered

Most of our information about the universe still comes from imaging the various forms of electromagnetic radiation. But today astronomers are gathering an ever-growing body of firsthand cosmic knowledge from space debris found on Earth and the Moon, as well as from exotic particles that pepper the Earth from space. Examples of the latter include neutrinos and so-called cosmic rays (misnamed, because cosmic rays are ac-

tually particles). Astronomers are also on the brink of directly detecting the fluctuations in the fabric of space called gravitational radiation predicted by Einstein's theory of general relativity. Using these nonelectromagnetic sources of information, we may soon be able to see directly into the hearts of stars and farther back toward the beginning of time than ever before.

 Further Reading on These Topics

WHAT DID YOU THINK?

1 *What makes a theory scientific?* A theory is an idea or set of ideas proposed to explain something about the natural world. A theory is scientific if it makes objective predictions that can be objectively tested and potentially disproved.

2 *What do astronomers do?* Most astronomers spend their time analyzing observational data, building equipment to make better observations, creating theories, and carrying out computer simulations. Observational astronomers spend only a few days or weeks each year actually using research telescopes.

3 *What practical value does astronomy have?* It provides us with such information as the cause of the seasons and the nature and implications of the gravitational force for the existence of stars and other bodies. It explains why the various types of objects in the universe exist and why they behave in the ways they do. For example, astronomers have discovered (see Chapter 9) why the Sun shines and, therefore, why it has been able to keep the Earth inhabitable for billions of years. Astronomy also provides an explanation for why the universe has evolved as it has, as well as predicting the fate of the universe.

KEY WORDS

astronomical unit (AU), 8
electromagnetic radiation, 10
light-year (ly), 8
model, 1

parsec (pc), 8
physics, 1
power of ten, 7
scientific method, 1

scientific notation, 6
scientific theory, 1

KEY IDEAS

• The universe is comprehensible.

• The scientific method is a procedure for formulating theories that correctly predict how the universe behaves.

• A scientific theory must be testable, that is, capable of being disproved.

• Theories are tested and verified by observation or experimentation and result in a process that often leads to

their refinement or replacement and to the progress of science.

• Observations of the heavens have led astronomers to discover some fundamental physical laws of the universe.

• Space debris collected on Earth, as well as particles from space, such as neutrinos and cosmic rays, and an exotic energy called gravitational radiation are providing astronomers with new insights into the universe.

REVIEW QUESTIONS

The answers to all computational problems, which are preceded by an asterisk (*), appear at the end of the book.

1 Compile a list of five examples of counterintuitive astronomy ideas, such as "the Earth rotates," that we now accept as correct.

*2 Convert the following into scientific notation: (a) Earth's mass: 5,974,000,000,000,000,000,000,000 kg; (b) Sun's

mass: 1,989,000,000,000,000,000,000,000,000,000 kg; (c) Sun's radius: 696,000 km; (d) one year: 31,536,000 s.

*3 Convert: (a) 8.3 pc into light-years; (b) 6.52 ly into parsecs; (c) 8450 AU into kilometers; (d) 2.7×10^3 Mpc into kiloparsecs.

ADVANCED QUESTIONS

4   The dictionary defines *astrology* as "the study that assumes and attempts to interpret the influence of the heavenly bodies on human affairs." Based on what you know about scientific theory, is astrology a science? Why or why not? Feel free to explore astrology further if you wish before answering this question.

WHAT IF ...

5   The laws of physics changed from time to time? What coping skills would life require, if it could exist at all? What else besides the laws of physics would scientists need to comprehend in order to understand and make predictions about the universe?

6   Astronomers only collected data (made observations) about the universe and didn't create scientific theories? How would our perspective on the cosmos be different from what it is today?

7   The skies of Earth were perpetually cloudy? How might that have changed the history of our understanding of the cosmos and how might humans under such conditions eventually learn what is really "out there"?

8   Scientists remained believers in the first theory of the cosmos that they decided was correct? How might that change the dynamics by which science evolves in the face of new data that conflicts with earlier theories?

WEB/CD-ROM QUESTIONS

9   Search the Web for more information about the scientific method. Then explain in some detail what a theory and a hypothesis are in science and how they are related. What is Occam's razor (sometimes Ockham's razor) and how does it relate to the process of science? If requested by your instructor, make a flow chart of the scientific process.

OBSERVING PROJECTS

10   Install the *Starry Night Backyard*™ planetarium software from the CD-ROM in your book into your computer. The manual and answers to FAQs about *Starry Night Backyard*™ are provided at the Web link adjacent to this problem. Use *Starry Night Backyard*™ to determine when the Moon is visible today during the day and when it is visible tonight. Determine which, if any, of the following planets, all visible to the naked eye, are observable in the sky tonight: Mercury, Venus, Mars, Jupiter, and Saturn.

Hints: (1) If you are just learning to use *Starry Night Backyard*™, experiment with settings as much as necessary. Change the time to after sunset in order to see the stars or click on *Go* then *Atlas*. You can always return to your starting screen by clicking on the *Home* button in the window normally on the upper left of your screen or by clicking on *Go* then *Home*. To see the sky from your actual location on Earth, select *Set Home Location...* in the *Go* menu and click on the *Lookup...* button to find your city or town. (2) You can move the sky around by moving your mouse up until a little hand appears. Hold the mouse button (on a Windows computer the left button) as you move the mouse and you will move the sky. (3) You can change the time of day and change how rapidly time appears to pass using the Control Panel at the top of the main window. (4) Use the *Find...* command in the *Edit* menu to locate specific planets or stars by name. (5) To get more information about any object in the sky, point the cursor at the object and double-click the mouse (on a Windows computer, click the right button).

11   Observe the stars with your unaided eyes and record what you see. This is intentionally an open-ended question, so as to encourage you to think and observe as broadly as possible. For example, if Orion is up, look carefully at its sword. Reviewing the data you have written down, make a list of what *properties* of the stars you have recorded. If possible, compare your observations with those of your peers and discuss with them and with your teacher whether there are characteristics of stars that you might have missed.

AN ASTRONOMER'S ALMANAC

586 B.C.

Thales of Miletus predicts solar eclipse.

1512-1543

Nicolas Copernicus proposes heliocentric cosmology in his Commentariolus and De Revolutionibus Orbium Coelestium.

1715

Edmund Halley calculates shadow path of a solar eclipse over Earth's surface.

ca. 270 B.C.

Aristarchus of Samos proposes heliocentric cosmology.

1589-1609

Galileo Galilei proposes that all objects fall with the same acceleration, independent of their masses; builds his first telescope, a refractor.

Greek Golden Age

European Renaissance

350 B.C.

Aristotle proposes spherical Earth, geocentric cosmology.

1609-1619

Johannes Kepler publishes his three laws of planetary motion.

1766

Henry Cavendish discovers hydrogen.

2136 B.C.

Chinese astronomers record solar eclipse.

1576-1601

Tycho Brahe makes precise observations of stars and planets.

1800-1803

William Herschel discovers infrared radiation from the Sun. Thomas Young demonstrates wave nature of light. John Dalton proposes that matter is composed of atoms of different weights.

1665-1704

Isaac Newton deduces gravitational force from the orbit of the Moon; builds first reflecting telescope; proves that planets moving under influence of gravitational force obey Kepler's laws; publishes compendium on light, Optics.

ca. A.D. 125

Claudius Ptolemy refines and details geocentric cosmology in his Almagest.

DISCOVERING ASTRONOMY

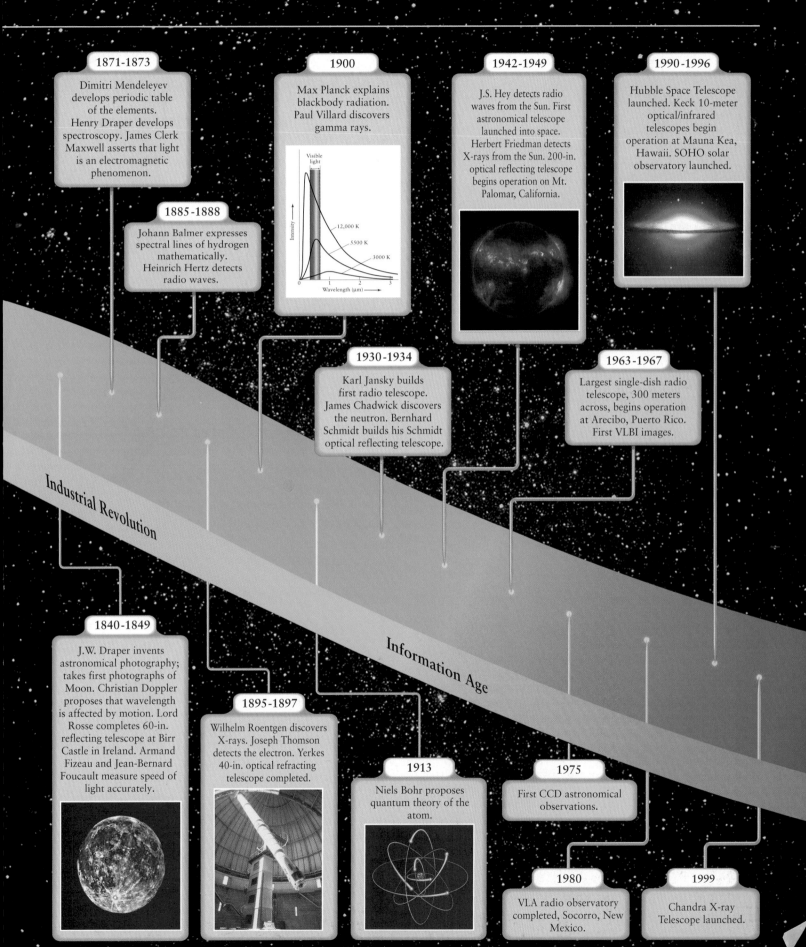

1871-1873
Dimitri Mendeleyev develops periodic table of the elements. Henry Draper develops spectroscopy. James Clerk Maxwell asserts that light is an electromagnetic phenomenon.

1900
Max Planck explains blackbody radiation. Paul Villard discovers gamma rays.

1942-1949
J.S. Hey detects radio waves from the Sun. First astronomical telescope launched into space. Herbert Friedman detects X-rays from the Sun. 200-in. optical reflecting telescope begins operation on Mt. Palomar, California.

1990-1996
Hubble Space Telescope launched. Keck 10-meter optical/infrared telescopes begin operation at Mauna Kea, Hawaii. SOHO solar observatory launched.

1885-1888
Johann Balmer expresses spectral lines of hydrogen mathematically. Heinrich Hertz detects radio waves.

1930-1934
Karl Jansky builds first radio telescope. James Chadwick discovers the neutron. Bernhard Schmidt builds his Schmidt optical reflecting telescope.

1963-1967
Largest single-dish radio telescope, 300 meters across, begins operation at Arecibo, Puerto Rico. First VLBI images.

Industrial Revolution

Information Age

1840-1849
J.W. Draper invents astronomical photography; takes first photographs of Moon. Christian Doppler proposes that wavelength is affected by motion. Lord Rosse completes 60-in. reflecting telescope at Birr Castle in Ireland. Armand Fizeau and Jean-Bernard Foucault measure speed of light accurately.

1895-1897
Wilhelm Roentgen discovers X-rays. Joseph Thomson detects the electron. Yerkes 40-in. optical refracting telescope completed.

1913
Niels Bohr proposes quantum theory of the atom.

1975
First CCD astronomical observations.

1980
VLA radio observatory completed, Socorro, New Mexico.

1999
Chandra X-ray Telescope launched.

1 DISCOVERING THE NIGHT SKY

IN THIS CHAPTER YOU WILL DISCOVER

- how astronomers map the night sky to help them locate objects in it

- that the Earth's spin on its axis causes day and night

- how the tilt of the Earth's axis of rotation and the Earth's motion around the Sun combine to create the seasons

- that the Moon's orbit around the Earth creates the phases of the Moon and lunar and solar eclipses

- how the year is defined and how the calendar was developed

Circumpolar Star Trails
This long exposure, taken from Australia's Siding Spring Mountain and aimed at the south celestial pole, shows the rotation of the sky. The building in the foreground houses the Anglo-Australian Telescope, one of the largest telescopes in the southern hemisphere. During the exposure, someone carrying a flashlight walked along the catwalk on the outside of the telescope dome. Another flashlight made the wavy trail at the ground level. (Anglo-Australian Observatory)

R I V U X G

WHAT DO YOU THINK?

1 Is the North Star—Polaris—the brightest star in the night sky?

2 Do astronomers regard constellations as simply the familiar patterns of stars in the sky first identified by ancient stargazers?

3 What causes the seasons?

4 How many zodiac constellations are there?

5 Does the Moon have a dark side that we never see from Earth?

6 Is the Moon ever visible during the daytime?

When you gaze at the sky on a clear, dark night, there seem to be millions of stars twinkling overhead. In reality, the unaided human eye can detect only about 6000 stars over the entire sky. At any one time, you can see roughly 3000 stars in dark skies, because only half of the stars are above the *horizon*, the boundary between the Earth and the sky.

 You probably have noticed patterns formed by bright stars and are probably familiar with some common names for these patterns, such as the bowl-shaped Big Dipper and broad-shouldered Orion. These recognizable patterns of stars, which we call constellations in everyday conversation, have names derived from ancient legends (Figure 1-1a).

PATTERNS OF STARS

1-1 Constellations make locating stars easy

 You can orient yourself on Earth with the help of easily recognized constellations. For instance, if you live in the northern hemisphere, you can use the Big Dipper to find the direction north. To do this, locate the Big Dipper and imagine that its bowl is resting on a table. If you see the Dipper upside down in the sky, as you frequently will, imagine the dipper resting on an upside-down

a

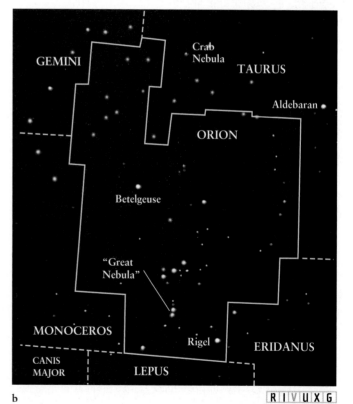

b   R I V U X G

FIGURE 1-1 **The Constellation Orion** **(a)** The pattern of stars called Orion is a prominent winter constellation. From the Northern Hemisphere, it is easily seen high above the southern horizon from December through March. You can see in this photograph that the various stars have different colors, something to watch for when you observe the night sky. **(b)** The region of the sky called Orion and parts of other nearby constellations are depicted in this photograph. All the stars inside the boundary of Orion are members of that constellation. The celestial sphere is covered by 88 constellations of differing sizes and shapes. (John Sanford/Astrostock)

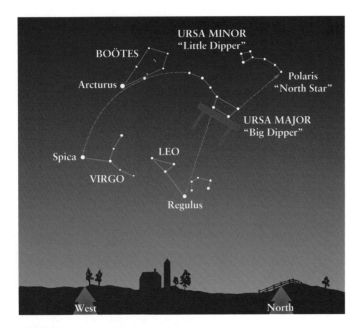

FIGURE 1-2   **The Big Dipper as a Guide**   In the northern hemisphere, the Big Dipper is an easily recognized pattern of seven bright stars. This star chart shows how the Big Dipper can be used to point out the North Star as well as the brightest stars in three other constellations. While the Big Dipper appears right side up in this drawing, at other times of the night it appears upside down. Why does this happen?

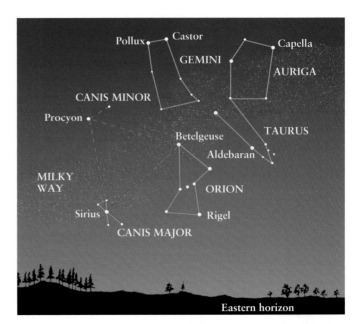

FIGURE 1-3   **The Winter Triangle**   This star chart shows the eastern sky as it appears during the evening in December. Three of the brightest stars in the sky make up the winter triangle. In addition to the constellations involved in the triangle, Gemini (the Twins), Auriga (the Charioteer), and Taurus (the Bull) are also shown.

table above it. Locate the two stars of the bowl farthest from the Big Dipper's handle. These are called the *pointer stars*. Draw a mental line through these stars leading away from the table, as shown in Figure 1-2. The first *moderately bright* star you then encounter is Polaris, also called the North Star because it is located almost directly over the Earth's north pole. So, while Polaris is not even among the 20 brightest stars (Appendix Table A-5), it is easy to locate. Whenever you face Polaris, you are facing north. East is then on your right, south is behind you, and west is on your left.

The Big Dipper example illustrates the fact that easily recognized constellations make it easy to locate other stars. The most effective way to do this is to use vivid visual connections, especially those of your own devising. For example, imagine gripping the handle of the Big Dipper and slamming its bowl through the imaginary table and onto the head of Leo (the Lion). Leo comprises the first group of bright stars your dipper encounters. As shown in Figure 1-2, the brightest star in this group is Regulus, the dot of the backward question mark that traces the lion's mane. As another example, put the Big Dipper back on its table and then follow the arc of its handle away from its bowl. The first bright star you encounter along that arc beyond the handle is Arcturus in Boötes (the Herdsman). Follow the same arc farther to the prominent bluish star Spica in Virgo (the Virgin). Spotting these stars is easy if you remember the saying "Arc to Arcturus and speed on to Spica."

During the winter months in the northern hemisphere, you can see some of the brightest stars in the sky. Many of them are in the vicinity of the "winter triangle," which connects bright stars in the constellations of Orion (the Hunter), Canis Major (the Larger Dog), and Canis Minor (the Smaller Dog), as shown in Figure 1-3. The winter triangle passes high in the sky at night during the middle of winter. It is easy to find Sirius, the brightest star in the night sky, by locating the belt of Orion and following a straight mental line from it to the left (as you face Orion). The first bright star that you encounter is Sirius.

Insight into Science   **Flexible thinking**   Part of learning science is learning to look at things from different perspectives. For example, when learning to identify the prominent constellations, be sure to learn them from different orientations (that is, with the star chart rotated at different angles), so that you can find them at different times of the night and the year.

The "summer triangle," which graces the summer sky as shown in Figure 1-4, connects the bright stars Vega in Lyra (the Lyre), Deneb in Cygnus (the Swan), and Altair in Aquila (the Eagle). A conspicuous portion of the Milky Way forms a beautiful background for these constellations, which are

GUIDED DISCOVERY   The Stars and Constellations

 Many students take an introductory astronomy course expecting to be taught the familiar stars and constellations, something most teachers just don't have time to cover. You can, however, learn them on your own using two helpful techniques for memorizing the night sky.

1. Observe easily identified constellations and use these to find nearby, lesser-known ones. For example, you may not remember where the star Aldebaran in Taurus is in the sky, but if you remember that Orion is fighting Taurus, you can easily locate Aldebaran by following a line defined by the belt of Orion to the right (away from Sirius). The first bright star you encounter is Aldebaran.

2. Make your own connections between the constellations. Have fun while you're doing it. Chances are that you will remember the phrase "slam the Big Dipper's bowl downward to hit Leo the Lion on the head" rather than "the first bright group of stars directly below the Big Dipper is Leo."

 A very efficient way to learn the constellations is to use the board and card game *Stellar 28*. In addition, astronomy computer programs such

as *Starry Night Backyard*™, included with this text, are available. *Starry Night Backyard*™ shows many things besides what constellations are up at night, including the motion, location, and phases of the planets; the location of deep sky objects, such as nebulae; and the sky as seen from any location on Earth on any date.

Another good way to familiarize yourself with the night sky is to use star charts to see which constellations are up each night. You will find a set of star charts from the *Griffith Observer* magazine at the front and back of this book. To use the charts, first select the one that best corresponds to the date and time of your observation. Take the chart outside at night and compare it directly with the sky. Hold the chart vertically and turn it so that the direction you are facing shows at the bottom. Using a flashlight with a red plastic coating over the light will make it easier to read the chart without constricting your vision. You will find stargazing a surprisingly enjoyable experience.

Stars' Technical Names

Astronomers assign all stars technical names, based on the constellations they are in. You can learn more about how these assignments are made at the adjacent More to Know link.

---

nearly overhead during the middle of summer at midnight. For more on the constellations see the above Guided Discovery Box.

Astronomers require more accuracy in locating dim objects than is possible simply by moving from constellation to constellation. They have therefore created a celestial map and applied a coordinate system to it analogous to the coordinate system of north-south latitude and east-west longitude used to navigate on the Earth. If a star's celestial coordinates are known, it can be quickly located. For such a sky map to be useful in finding stars, the stars must be fixed on it as cities are fixed on maps of the Earth.

1-2   The celestial sphere aids in navigating the sky

If you look at the night sky year after year, you will see that the stars do indeed appear fixed relative to each other. Furthermore, throughout each night the entire pattern of stars appears to rigidly orbit the Earth. We employ this Earth-based view of the heavens

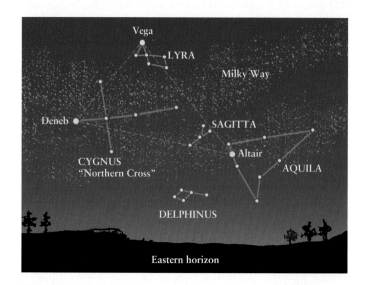

FIGURE 1-4   The Summer Triangle This star chart shows the northeastern sky as it appears in the evening in June. In addition to the three constellations involved in the summer triangle, the faint stars in the constellations of Sagitta (the Arrow) and Delphinus (the Dolphin) are also shown.

to make celestial maps by pretending that the stars are attached to the inside of an enormous hollow shell, the **celestial sphere,** with the Earth at its center (Figure 1-5).

We discussed the common usage of the word *constellation* at the beginning of this chapter. Astronomers use the word technically to describe an entire region of the sky and all the objects in that region (see Figure 1-1b). The celestial sphere is divided into 88 unequal regions, and these regions are what astronomers refer to when they use the term **constellations.** (Astronomers call the traditional star patterns *asterisms* rather than constellations when there is a danger of confusion.) Some constellations (like Ursa Major) are very large, while others (like Sagitta) are relatively small. To locate stars, we might say "Albireo in the constellation Cygnus," much as we would refer to "Ithaca in New York State."

The stars seem fixed on the celestial sphere only because of their remoteness. In reality, they are at widely varying distances from the Earth, and they do move relative to each other. But we neither see their motion nor perceive their relative distances because the stars are so far from here. You can understand this by imagining a jet plane a kilometer overhead traveling at 1000 kilometers (620 miles) an hour across the sky. Its motion is unmistakable. But a plane moving at the same speed along the distant horizon appears to be moving about a hundred times more slowly. And an object at the distance of the Sun traveling at the same speed and moving across the sky would appear to be going nearly a hundred million times more slowly than the plane overhead.

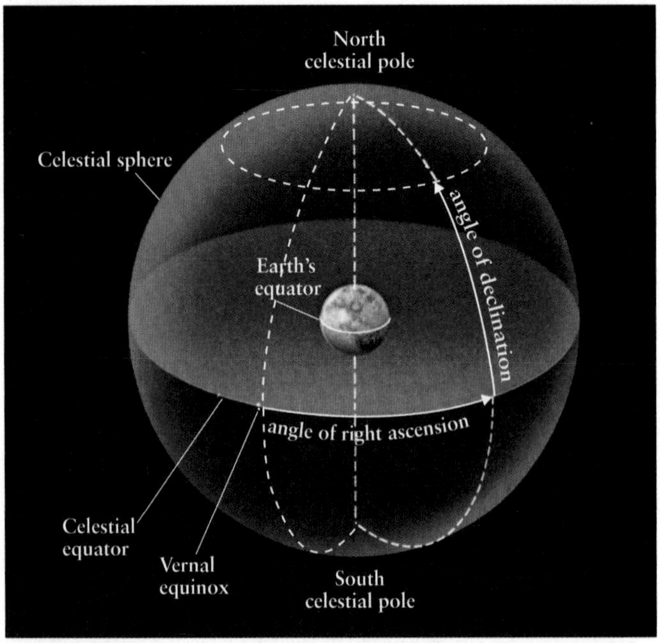

The stars (other than the Sun) are all more than 40 trillion kilometers (25 trillion miles) from us. Therefore, although the patterns of stars in the sky do change, their great distances prevent us from seeing those changes over the course of a human lifetime. Thus, as unrealistic as it is, the celestial sphere is so useful for navigating the heavens that it is used by astronomers even at the most sophisticated observatories around the world.

As shown in Figure 1-5, we can project key geographic features from Earth out into space to establish directions and bearings. If we expand the Earth's equator onto the celestial sphere, we obtain the **celestial equator.** The celestial equator divides the sky into northern and southern hemispheres, just as the Earth's equator divides the Earth into two hemispheres. We can also imagine extending the Earth's north and south poles out into space along the Earth's axis of rotation. Doing so gives us the **north celestial pole** and the **south celestial pole,** also shown in Figure 1-5. With the celestial equator and poles as reference features, astronomers denote the position of an object in the sky in much the same way that latitude and longitude are used to specify a location on Earth.

Just as we need two coordinates (latitude and longitude) to find any location on Earth, two coordinates are needed to locate any object on the celestial sphere. The equivalent to latitude on Earth is **declination** on the celestial sphere. It is measured north or south of the celestial equator. The equivalent of longitude on Earth is **right ascension** on the celestial sphere, measured around the celestial equator (see Figure 1-5).

We will see later in this chapter that the Sun moves in a circle around the celestial sphere during the course of a year. The celestial equator and the Sun's path intersect at two points. The equivalent on the celestial sphere of the Earth's prime meridian (from which degrees of longitude are measured on Earth) is where the Sun crosses the celestial equator moving northward. Angles of right ascension are measured from this point, called the *vernal equinox* (Figure 1-5).

In navigating on the celestial sphere, astronomers measure the distance between objects in terms of angles. If you are not familiar with measuring the separation between objects using this method, read An Astronomer's Toolbox 1-1.

FIGURE 1-5   The Celestial Sphere The celestial sphere is the apparent "bowl" or hollow sphere of the sky. The celestial equator and poles are projections of the Earth's equator and axis of rotation out into space. The north celestial pole is therefore located directly over the Earth's north pole, while the south celestial pole is directly above the Earth's south pole.

EARTHLY CYCLES

The everyday rhythms of the Earth and all life on it arise from three celestial motions: the Earth's rotation, which causes day and night; the Earth's revolution around the Sun, which creates the seasons and the year; and the Moon's revolution around the Earth, which creates the lunar phases,

AN ASTRONOMER'S TOOLBOX 1-1

Observational Measurements Using Angles

Astronomers have inherited many useful concepts from antiquity. Among other things, ancient mathematicians invented angles and a system of angular measure that is still used to denote the positions and apparent sizes of objects in the sky. To locate stars, for example, we do not need to know their distances from Earth (which are all different). All we need to know is the angle from one star to another in the sky, a property that remains fixed over our lifetimes.

An **arc angle**, often just called an **angle**, is the opening between two lines that meet at a point. Angular measure is a method of describing the size of an angle. The basic unit of angular measure is the **degree**, designated by the symbol °. A full circle is divided into 360°. A right angle measures 90°. As shown in the figure below, the angle between the two "pointer stars" in the Big Dipper is about 5°.

Astronomers also use angular measure to describe the apparent sizes of celestial objects. For example, imagine looking up at the full Moon. The angle covered by the Moon's diameter is nearly $1/2°$. We therefore say that the **angular diameter**, or angular size, of the Moon is $1/2°$. Alternatively, astronomers say that the Moon "subtends" an angle of $1/2°$. In this context, *subtend* means "to extend across."

The adult human hand held at arm's length provides a means of estimating angles. For example, your fist covers an angle of 10°, whereas the tip of your finger is about 1° wide. Various segments of your index finger extended to arm's length can be similarly used to estimate angles a few degrees across as shown in the figure below, right.

To talk about smaller angles, we subdivide the degree into 60 arcminutes (abbreviated 60 arcmin or 60′). An arcminute is further subdivided into 60 arcseconds (abbreviated 60 arcsec or 60″). A dime viewed face-on from a distance of 1 mile has an angular diameter of about 2 arcsec. From everyday experience, we know that an object looks big when it is nearby but small when it is far away. The angular size of an object therefore does not necessarily tell you anything about its actual physical size. For example, the fact that the Moon's angular diameter is $1/2°$ does not tell you how big the Moon really is. But if you also happen to know the distance to the Moon, then you can calculate the Moon's physical diameter. In general, the physical diameter of an object can be calculated from the equation:

$$\text{physical diameter} = \text{distance} \times \tan (\text{angular diameter})$$

where tan (angular diameter) means the tangent of the angle denoted "angular diameter." In the Moon's case, using a measured distance (Appendix Table A-3) of 384,400 km and an angular diameter of $1/2°$, we find the diameter to be roughly 3350 km. The difference between this and the exact diameter of 3476 km is due primarily to the approximate value of $1/2°$ that we have used.

Try these questions: The Sun is 1.5×10^8 km away and has a diameter of 1.4×10^6 km. How large an angle does it make in our sky? How is that relevant to the arc angle of the Moon in our sky? What arc angle would the Moon make in our sky if it were twice as far away? Half as far?

(Answers appear at the end of the book.)

The Big Dipper The angular distance between the two "pointer stars" at the front of the Big Dipper is about 5°. For comparison, the angular diameter of the Moon is about $1/2°$.

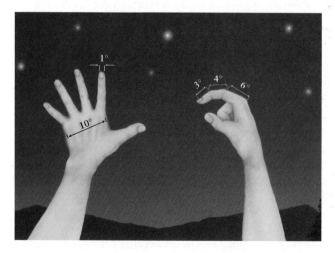

Estimating Angles with the Human Hand Various parts of the adult human hand extended to arm's length can be used to estimate angular distances and sizes in the sky.

the cycle of tides, and the spectacular phenomena we call eclipses.

Contrary to common sense, the seasons are not caused by the change in the Earth's distance from the Sun. If the seasons were caused by the changing distance from the Earth to the Sun, all parts of the Earth should have the same seasons at the same time. In fact, the northern and southern hemispheres have exactly opposite seasons. Furthermore, the Earth is closest to the Sun on January 3 of each year—the dead of winter in the northern hemisphere! We will explore the seasons further in Section 1-6.

Insight into Science Expect the unexpected Many phenomena in the universe defy commonsense explanations. The process of science requires that we question the obvious, that is, what we think we know. The fact that the changing distance from the Earth to the Sun has a negligible effect on the seasons is an excellent example.

1-3 Earth's rotation causes the stars to appear to move

The Earth spins on its axis. Such motion is called **rotation.** The Earth's 24-hour rotation causes the constellations—as well as the Sun, Moon, and planets—to appear to rise on the eastern horizon, move across the sky, and set on the western horizon. Bear in mind that the Sun and Moon rise due east and set due west only on certain days of the year. The **diurnal motion,** or daily motion, of the celestial bodies is apparent in time-exposure photographs, such as the one that opens this chapter. The Earth's rotation causes day and night because it makes the Sun appear to follow a diurnal path across the sky.

People who spend time outdoors at night are familiar with the diurnal motion of the stars. Take a friend outside on a clear, warm night to observe it for yourself. Soon after dark, find a spot away from bright lights, and note the constellations in the sky relative to some prominent landmarks near you on Earth. A few hours later, check again from the same place. You will find that the entire pattern of stars (as well as the Moon, if it is visible) has shifted. New constellations will have risen above the eastern horizon, while other constellations will have disappeared below the western horizon. If you check again just before dawn, you will find the stars that were just rising in the east when the night began are now low in the western sky.

Different constellations are visible at night during different times of the year. This occurs because the Earth orbits, or *revolves,* around the Sun. **Revolution** is the motion of any astronomical object around another astronomical object. The Earth takes one year, or about 365 1/4 days, to go once around the Sun. As a result of this motion, the darkened,

nighttime side of the Earth is turned toward different parts of the heavens at different times of the year. (We explore this further in Section 1-5.) Another result of the Earth's motion around the Sun is that every star rises approximately 4 minutes earlier each night than it did the night before.

We spoke earlier of stars rising on the eastern horizon and setting on the western horizon. Depending on your latitude, some of the stars and constellations never disappear below the horizon. Instead, they trace complete circles in the sky over the course of each night. To understand why this happens, imagine that you are standing on the Earth's north pole at night. Looking straight up, you see Polaris. Wherever you are, objects directly overhead are said to be at your **zenith.** Because the Earth is spinning around its axis directly under your feet, all the stars appear to move from left to right in horizontal rings around you. The exception is Polaris, which always remains at the north pole's zenith. As seen from the north pole, no stars rise or set (Figure 1-6). They just seem to revolve around Polaris in horizontal circles. Stars and constellations that never go below the horizon are called **circumpolar.** All stars visible from the north or south pole are circumpolar.

Now visualize yourself at the equator. All the stars appear to rise straight up in the eastern sky and set straight down in the western sky (Figure 1-7). Polaris is barely visible on the northern horizon. While Polaris never sets, all the other stars do, and therefore none of the stars are circumpolar as seen from the equator.

You may already have concluded from these two mental exercises that the angle at which the stars rise and set

FIGURE 1-6   Motion of Stars at the Poles Because the Earth rotates around its poles, stars seen from these locations appear to move in huge, horizontal circles. This is the same effect you would get by standing up in a room and spinning around; everything would appear to move in circles around you. At the north pole stars move left to right, while at the south pole they move right to left.

FIGURE 1-7   Rising of Stars at the Equator Standing on the equator, you are perpendicular to the axis around which the Earth rotates. As seen from there, the stars rise straight up on the eastern horizon and set straight down on the western horizon. This is the same effect you get when driving straight over the crest of a hill; the objects on the other side of the hill appear to move straight upward as you descend.

R I V U X G

FIGURE 1-8   Rising and Setting of Stars at Middle North Latitudes Unlike the motion of the stars at the poles (see Figure 1-6), the stars at all other latitudes do change angle above the ground throughout the night. The time-lapse photograph shows stars setting. The latitude determines the angle at which the stars rise and set. (David Miller/DMI.)

depends on your viewing latitude. Figure 1-8, for example, shows stars setting at 35° north latitude. In Orono, Maine (44° 45′ north latitude), where this book was written, stars rise at an angle of nearly 45° to the eastern horizon and set at an angle of 45° to the western horizon. Polaris is fixed at 45° above the horizon in Orono's northern sky, not at the zenith there, as it is at the north pole or on the horizon as seen from the equator. As another example, except for those stars in the upper corners, the stars whose paths are shown in the photograph on page 14 are circumpolar because they are visible all night, every night.

If you live in the northern hemisphere, Polaris is always located above your northern horizon at an angle equal to your latitude. Only the stars and constellations that pass between Polaris and the land directly below it are circumpolar. The farther north you go in the northern hemisphere, the greater the number of stars and constellations that are circumpolar. The opposite is true in the southern hemisphere.

1-4   The rate at which the Earth rotates determines the length of the day

WEB LINK 1.7   The Sun's daily motion through the sky provided our ancestors' earliest reference for time, because the Sun's location determines whether it is day or night, whether we are awake or asleep, and whether it is time for breakfast or dinner. In other words, the Sun determines the length of the **solar day,** upon which our 24-hour day is based. However, the length of the solar day

varies throughout the year. The Sun is not a perfect time-keeper, because the Earth's orbit around it is not circular, as we will study in Chapter 3. Because the Earth moves slightly more rapidly along its orbit when it is near the Sun than when it is farther away, the speed of the Sun across the sky also varies throughout the year. Using the average time interval between consecutive noontimes to determine the solar day corrects this problem, and this average defines our 24-hour day.

WEB LINK 1.8   Astronomically, noon is defined as the instant when the Sun is highest in the sky. However, at different longitudes the Sun is highest at different times. Thus, astronomical noon in New London, Connecticut, occurs slightly earlier than it does in New Haven, Connecticut. Before the advent of time zones, local time was based on astronomical noon. To travel from New London west to New Haven by train, for example, you had to know the departure time at New London, using New London time, as well as the arrival time in New Haven (say, if someone was going to meet you) in New Haven time. Such time considerations became very confusing and burdensome as society became more complex. Fortunately, in the late nineteenth century, time zones were established to remove this problem. In a **time zone,** everyone agrees to set their clocks alike. Time zones, originally developed for scheduling rail transportation, are based on the time at 0° longitude in Greenwich, England, a location called the *prime meridian.* With some variation due to geopolitical boundaries, every 15° of longitude around the globe begins a new time zone. The resulting 24 time zones are shown in Figure 1-9. Going from one time zone to the

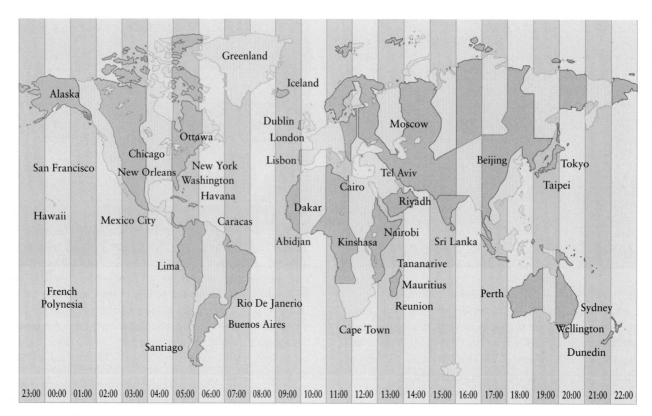

23:00 00:00 01:00 02:00 03:00 04:00 05:00 06:00 07:00 08:00 09:00 10:00 11:00 12:00 13:00 14:00 15:00 16:00 17:00 18:00 19:00 20:00 21:00 22:00

FIGURE 1-9   Time Zones of the World   For convenience, Earth's 360° circumference is divided into 24 time zones. Ideally, each time zone would run due north-south. However, political considerations make some zones irregular. Indeed, there are even a few zones only a half hour wide.

next usually requires you to change the time on your wristwatch by exactly 1 hour.

1-5   The Earth's orbit determines the length of the year and which stars are up at night

Just as the day originates in the Earth's rotation, the year is the unit of time based on the Earth's revolution about the Sun. The Earth does not take exactly 365 days to orbit the Sun, so the year is not exactly 365 days long. Ancient astronomers realized that the length of a year is approximately 365¼ days. The Roman statesman Julius Caesar established the system of leap years to account for this extra quarter of a day. By adding an extra day to the calendar every four years, Caesar hoped to ensure that seasonal astronomical events, such as the beginning of spring, would occur on the same date year after year.

Caesar's system would have worked if the year were exactly 365¼ days long and the Earth's rotation axis never changed direction. Unfortunately, this is not the case. Thus, over time, a discrepancy accumulated between Caesar's "Julian" system and actual time. Annual events began to fall on different dates each year. To straighten things out, Pope Gregory XIII reformed the Julian calendar in 1582. He began by dropping ten days (October 5, 1582, was proclaimed to be October 15, 1582), which brought the first day of spring back to March 21. Next, he modified Caesar's system of leap years. Caesar had added February 29 to every calendar year that is evenly divisible by four. For example, 1988, 1992, and 1996 were all leap years with 366 days. But this system produces an error of about three days every four centuries. To solve the problem, Pope Gregory decreed that century years would be leap years only if evenly divisible by 400. For example, the years 1700, 1800, and 1900 were not leap years under the improved Gregorian system. But the year 2000—which can be divided evenly by 400—was a leap year.

We use the Gregorian system today. It assumes that the year is 365.2425 mean solar days long, which is very close to the length of the *tropical year,* defined as the time interval from one vernal equinox to the next. In fact, the error is only one day in every 3300 years. That won't cause any problems for a long time. Two major impacts of our annual journey around the Sun are the changing stars in our night sky and the unfolding of the seasons.

Figures 1-10a and b show why different constellations are visible at night during different times of the year. When the Sun is "in" Virgo (September 18–November 1), the hemisphere containing the Sun and the constellations around Virgo are in daylight (Figure 1-10a). When the Sun is up, so are Virgo and the surrounding constellations, and so we cannot see them. During that time of year, the constellations on the other side of the celestial sphere, centered on the constellation Pisces, are in darkness. So, when the Sun is "in" Virgo, Pisces and the constellations around it are high in our sky at night.

Six months later, when the Sun is "in" Pisces, that half of the sky is filled with daylight, while Virgo and the constellations around it are high in the night sky (Figure 1-10b). These arguments apply everywhere on Earth at the same time because the Sun moves through the zodiac constellations very slowly as seen from Earth, taking a year to make one complete circuit.

1-6   The seasons result from the tilt of the Earth's rotation axis combined with its revolution around the Sun

The seasons are due to the combined effects of the Earth's revolution and the tilt of the Earth's axis of rotation relative to the plane of our orbit around the Sun. Imagine that you could see the stars even during the day, so that you could follow the Sun's apparent motion against the background constellations throughout the year. (The Sun appears to move among the stars, of course, because the Earth orbits around it.)

From day to day, the Sun would trace a straight path on the celestial sphere. You can see this with your *Starry Night Backyard*™ software by choosing *Go/Atlas*, locking on the Sun, zooming out to about 90° of the sky, setting the timestep to 3 days, and stepping along or setting the time to run continuously. You will see the stars moving relative to the Sun, meaning that the Sun is moving along the ecliptic around the celestial sphere. As you can see in Figure 1-11a, the ecliptic makes a closed circle bisecting the celestial sphere.

The term **ecliptic** has another use in astronomy. The Earth orbits the Sun in a plane also called the ecliptic. You can see that the two ecliptics exactly coincide. Imagine yourself on the Sun watching the Earth move day by day. The path of the Earth on the celestial sphere as seen from the Sun is precisely the same as the path of the Sun as seen from the Earth (Figure 1-11b).

> **Insight into Science** **Define your terms** The words "ecliptic" and "constellation" have more than one scientific meaning. Most often, however, scientists restrict words to one well-defined meaning, because more than one meaning can lead to misunderstandings. Be sure that you understand each word in astronomy and how to use it.

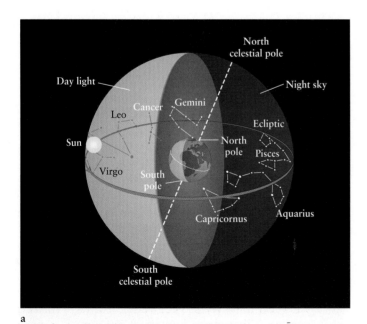

a

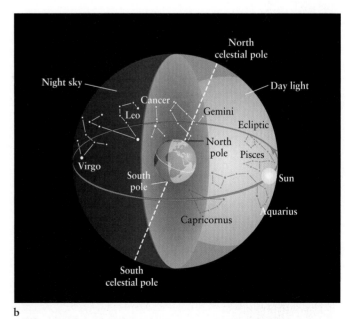

b

FIGURE 1-10   **Why different constellations are visible at different times of the year. (a)** On autumnal equinox each year, the Sun is in the constellation Virgo. As seen from Earth, that part of the sky is in daylight and we see stars only on the other half of the sky, centered around the constellation Pisces. **(b)** Six months later, the Sun is in Pisces. This side of the sky is bright, while the side centered on Virgo is now in darkness.

MOVIE MISCONCEPTIONS

First he fought for the Crown.
Now he's fighting
for the Family Jewels.

Austin Powers: The Spy Who Shagged Me (New Line Cinema, 1999)
(The Everette Collection)

The year is 1999 and the international man of mystery Austin Powers and the sinister Dr. Evil are back in *Austin Powers: The Spy Who Shagged Me*. In the movie there is a scene in Doctor Evil's secret lair in which the evil doctor shows his staff how he plans to put a "laser" on the Moon and use it to destroy cities on Earth. In making his presentation, he uses a large orrery consisting of a four-foot diameter model of the Earth fixed in a basket, around which a model of the Moon is free to revolve. The United States is clearly visible on the upper half of the Earth model, meaning that the north celestial pole is above the set. The model of the Earth does not move. The evil doctor explains his plan and then pushes the Moon clockwise around the Earth until it is above the United States. What is wrong with the astronomy described so far?

Despite the error that you have just identified, the Moon follows a westward path around the Earth, as would be seen correctly by someone standing on that fixed model of the Earth in the scene. Explain why the Moon follows the correct path around the model of the Earth, despite the error above.

(Answers appear at the end of the book.)

The celestial equator and the ecliptic are different circles on the celestial sphere because the Earth's axis is tilted $23\frac{1}{2}°$ away from a line perpendicular to the ecliptic, as shown in Figure 1-12. Except for tiny changes each year, discussed below, the Earth maintains this tilted orientation as it orbits the Sun. Polaris is above the north pole throughout the year. For half the year, the northern hemisphere is tilted toward the Sun, and, as a result, the Sun rises higher in the northern

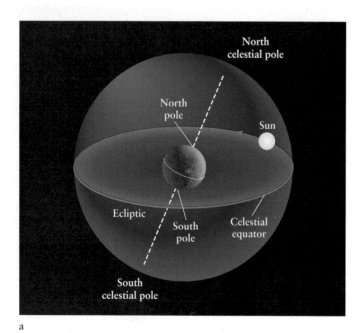

a

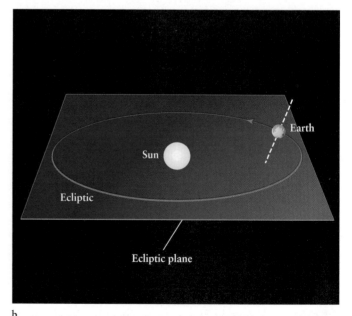

b

FIGURE 1-11   The Ecliptic   (a) The ecliptic is the apparent annual path of the Sun on the celestial sphere. (b) The ecliptic is also the plane described by the Earth's path around the Sun. The planes created by the two ecliptics exactly coincide. As in (a), the rotation axis of the Earth is shown here tilted $23\frac{1}{2}°$ from being perpendicular to the ecliptic.

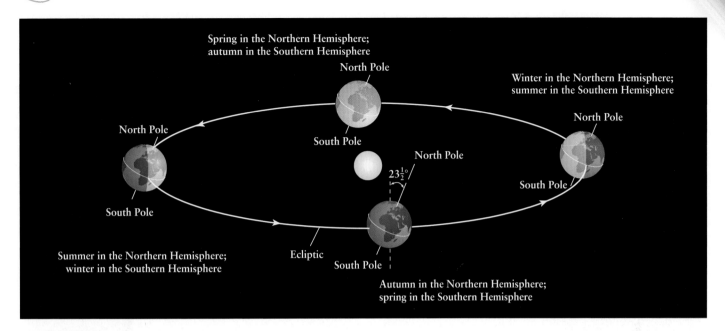

Spring in the Northern Hemisphere;
autumn in the Southern Hemisphere

North Pole

South Pole

North Pole

South Pole

North Pole

$23\frac{1}{2}°$

North Pole

Winter in the Northern Hemisphere;
summer in the Southern Hemisphere

North Pole

South Pole

Summer in the Northern Hemisphere;
winter in the Southern Hemisphere

Ecliptic

South Pole

Autumn in the Northern Hemisphere;
spring in the Southern Hemisphere

FIGURE 1-12   The Tilt of the Earth's Axis The Earth's axis of rotation is tilted $23\frac{1}{2}°$ from being perpendicular to the plane of the Earth's orbit. The Earth maintains this orientation (with its north pole aimed at the north celestial pole near the star Polaris) throughout the year as it orbits the Sun. Consequently, the amount of solar illumination and the number of daylight hours at any location on Earth vary in a regular fashion throughout the year.

hemisphere's sky than it does during the other half of the year. Equivalently, when the Sun is tilted toward the southern hemisphere, it rises higher in the southern hemisphere's sky (Figure 1-13). By the way, the seasons are not caused by the fact that one hemisphere is slightly closer to the Sun than the other—that difference in distance is much, much too small to cause any temperature difference.

 The Height of the Sun

The higher the Sun rises during the day, the more daylight hours. During these longer periods of daylight, more light and heat energy from the Sun strike that hemisphere than during the seasons when the north pole is pointed away from the Sun, and, therefore, the Sun does not rise nearly as high in the sky. Furthermore, when the Sun is higher in the sky, its energy is more concentrated on the Earth's surface (Figure 1-14). The temperature in any region on Earth is determined by the duration of daylight there and the height of the Sun in the sky.

Because of the tilt of the Earth's axis of rotation, the ecliptic and the celestial equator are inclined to each other by $23\frac{1}{2}°$, as shown in Figure 1-15. These two circles intersect at only two points, which are exactly opposite each other on the celestial sphere. Both points are called **equinoxes** (from the Latin words meaning "equal night"), because

R I V U X G

FIGURE 1-13   The Height of the Sun The maximum height of the Sun in the sky varies throughout the year because of the $23\frac{1}{2}°$ tilt of the Earth's axis. This photograph shows the height of the Sun in the sky at the same clock time at 10-day intervals throughout the year, as well as its changing east-west location. The east-west variation in the Sun's location is caused by its rising, reaching its maximum height in the sky, and setting at different places throughout the year (See Figure 1-16). (Dennis Di Cicco)

a

b

FIGURE 1-14 **The Energy Deposited by the Sun** In the northern hemisphere, the angle of the Sun above the southern horizon determines how much heat and light strike each square meter of ground. In the southern hemisphere, the Sun's angle is measured above the northern horizon. **(a)** During the summer at middle latitudes, a shaft of sunlight illuminates a nearly circular patch of ground at noon. **(b)** During the winter, the same shaft of sunlight at noon strikes the ground at a steeper angle, spreading the same amount of sunlight over a larger, oval shape. Because the sunlight's energy is diluted over a larger area, the ground receives less heat during the winter than during the summer.

when the Sun appears at either point, it is directly over the Earth's equator, and there are 12 hours of daytime and 12 hours of nighttime everywhere on Earth on that day.

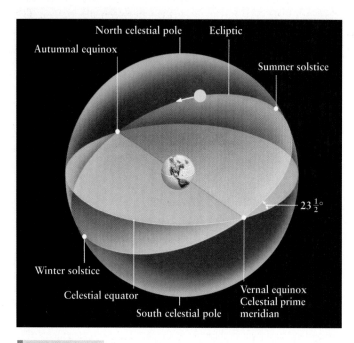

FIGURE 1-15 **The Seasons Are Coupled to Equinoxes and Solstices** The ecliptic is inclined to the celestial equator by $23\frac{1}{2}°$ because of the tilt of the Earth's axis of rotation. The ecliptic and the celestial equator intersect at two points called the equinoxes. The northernmost point on the ecliptic is called the summer solstice; the southernmost point is called the winter solstice.

The **vernal equinox** occurs around March 21, when the Sun crosses the celestial equator heading northward; the **autumnal equinox** occurs six months later, around September 22, with the Sun heading southward. The vernal equinox is the "prime meridian" of the celestial sphere, as discussed earlier. The midpoints between the vernal and autumnal equinoxes along the ecliptic are significant as well. The point on the ecliptic farthest north of the celestial equator is the **summer solstice.** It signals the day of the year in the northern hemisphere with the largest number of daylight hours (and the fewest daylight hours in the southern hemisphere). The Sun reaches this point around June 21 each year. Six months later, around December 21, the Sun is farthest south of the celestial equator at the **winter solstice,** the day of the year in the northern hemisphere with the fewest daylight hours (and the most daylight hours in the southern hemisphere).

The Sun is highest in the northern sky on the summer solstice. This marks the beginning of summer in the northern hemisphere. As the Sun moves southward, the amount of daylight decreases. The autumnal equinox marks a midpoint in the amount of heat deposited by the Sun onto the northern hemisphere and is the beginning of autumn. When the Sun reaches the winter solstice, it is lowest in the northern sky and is above the horizon for the shortest time of the year. This is the beginning of winter. Returning northward, the Sun crosses the celestial equator once again on the vernal equinox, the beginning of spring.

Figure 1-16 shows the seasonal changes in the Sun's daily path across the sky. On the first day of spring or autumn (when the Sun is at one of the equinoxes), the Sun rises directly in the east and sets directly in the west. Daytime

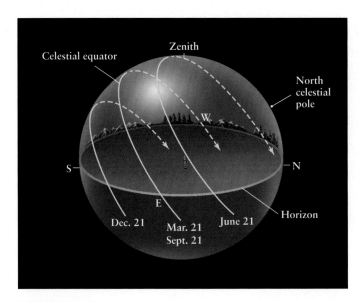

FIGURE 1-16   **The Sun's Daily Path** On the first day of spring and the first day of fall, as seen from middle latitudes in the northern hemisphere, the Sun rises precisely in the east and sets precisely in the west. During summer in the northern hemisphere, the Sun rises in the northeast and sets in the northwest. In winter, the Sun rises in the southeast and sets in the southwest.

and nighttime are of equal duration everywhere on Earth on those days only.

During the northern hemisphere's summer months, when the northern hemisphere is tilted toward the Sun, the Sun rises in the northeast and sets in the northwest. The Sun provides more than 12 hours of daylight in the northern hemisphere and passes high in the sky at noontime. At the summer solstice, the Sun is as far north as it gets, giving the greatest number of daylight hours to the northern hemisphere.

During the northern hemisphere's winter months, when the northern hemisphere is tilted away from the Sun, the Sun rises in the southeast. Daylight lasts for fewer than 12 hours, as the Sun skims low over the southern horizon and sets in the southwest. Night is longest in the northern hemisphere when the Sun is at the winter solstice.

People at different latitudes see the noontime Sun at different angles in the sky each day. The farther from the equator you are, the lower the Sun is in the sky at noon. The farther north you go, less of the Sun's heat and light energy are deposited (see Figure 1-14) and, therefore, the colder the land is. At latitudes above $66^{1}/_{2}°$ north latitude or below $66^{1}/_{2}°$ south latitude, the Sun does not rise at all during parts of their fall and winter months. During their spring and summer months, those same regions of the Earth have continuous sunlight for weeks or months, hence the name "Land of the Midnight Sun."

The Sun takes one year to complete a trip around the ecliptic. Since there are about $365^{1}/_{4}$ days in a year and 360° in a circle, the Sun appears to move along the ecliptic at a rate of slightly less than 1° per day. The constellations through which the Sun moves throughout the year are called **zodiac** constellations. We cannot see the stars of these constellations when the Sun is among them, of course, but we can plot the Sun's path on the celestial sphere to determine through which constellations it moves. One traditionally learns that there are 12 zodiac constellations, which are used by astrologers. These 12 are based on constellation boundaries used in antiquity. Those boundaries have since been redefined by astronomers, and the Sun moves through 13 modern constellations throughout the year. (The thirteenth zodiac constellation is Ophiuchus, the Serpent Holder. The Sun passes through Ophiuchus from December 1 to December 19 each year.) Table 1-1 lists all the zodiac constellations and the dates the Sun passes through them.

1-7   Precession is a slow, circular motion of the Earth's axis of rotation

As noted above, the position of the Earth's axis of rotation changes slightly with respect to the celestial sphere (that is, it "points" in a slightly different direction) each year. This change in orientation is caused by gravitational forces from the Moon and Sun. Recall from Section I-2 that gravitation is the universal force of attraction between all matter. The Earth's rotation creates an "equatorial bulge"—our planet's diameter is about 43 km (27 mi) greater at the equator than from pole to pole.

TABLE 1-1   **The 13 Constellations of the Zodiac**

Constellation	Dates of Sun's Passage Through
Pisces	March 13–April 20
Aries	April 20–May 13
Taurus	May 13–June 21
Gemini	June 21–July 20
Cancer	July 20–August 11
Leo	August 11–September 18
Virgo	September 18–November 1
Libra	November 1–November 22
Scorpius	November 22–December 1
Ophiuchus	December 1–December 19
Sagittarius	December 19–January 19
Capricorn	January 19–February 18
Aquarius	February 18–March 13

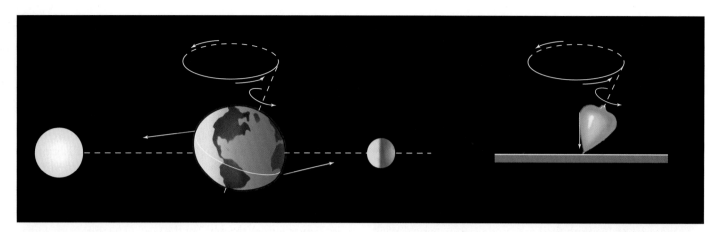

FIGURE 1-17 **Precession** The gravitational pulls of the Moon and the Sun on the Earth's equatorial bulge cause the Earth to precess. As the Earth precesses, its axis of rotation slowly traces out a circle in the sky. The situation is analogous to that of a spinning top. The top of the toy top shows the motion of Earth's north or south pole, while the point on which the top spins represents the center of the Earth. As the top spins, the Earth's gravitational pull causes the top's axis of rotation to move in a circle.

Because of the Earth's tilted axis of rotation, the Sun and Moon are usually not located directly over the Earth's equator. As a result, their gravitational attraction on the Earth tries to force the equatorial bulge to be as close to them as possible. However, the Earth does not respond to these forces from the Sun and Moon by straightening up perpendicular to them on its axis. Instead, it changes the direction in which its axis of rotation points on the celestial sphere—a motion called **precession.** This is exactly the same behavior that is exhibited by a spinning top (Figure 1-17). If the top were not spinning, gravity would pull it over on its side. But when it is spinning, the combined actions of gravity and rotation cause the top's axis of rotation to precess in a circular path. As with the toy top, the combined actions of gravity and rotation cause the Earth's axis to trace out a circle in the sky while remaining tilted about 23 1/2° away from the perpendicular. In the mid-1990s, astronomers simulated the behavior of the Earth and discovered that without a large Moon, the Earth would not keep to a 23 1/2° tilt but rather would change its tilt wildly, and even flip over!

The Earth's rate of precession is slow compared to human time scales. It takes about 26,000 years for the north celestial pole to trace out a complete circle around the sky, as shown in Figure 1-18. (The south celestial pole executes a similar circle in the southern sky.) At the present time, the Earth's north axis of rotation points within 1° of the star Polaris. In 3000 B.C., it was pointing near the star Thuban in the constellation of Draco (the Dragon). In A.D. 14,000, the pole star will be near Vega in Lyra.

As the Earth's axis of rotation precesses, its equatorial plane also moves. Because the Earth's equatorial plane defines the location of the celestial equator in the sky, the celestial equator also precesses. Because the intersections of the celestial equator and the ecliptic define the equinoxes, these key locations in the sky also shift slowly from year to year. This entire phenomenon is often called the **precession of the equinoxes.** This change was discovered by the great Greek astronomer

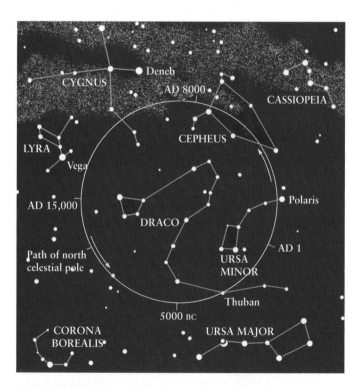

FIGURE 1-18 **The Path of the North Celestial Pole** As the Earth precesses, the north celestial pole slowly traces out a circle among the northern constellations. At the present time, the north celestial pole is near the moderately bright star Polaris, which serves as the pole star. The total precession period is about 26,000 years.

Hipparchus in the second century B.C. Today, the vernal equinox is located in the constellation Pisces (the Fishes). Two thousand years ago, it was located in Aries (the Ram). Around the year A.D. 2600, the vernal equinox will move into Aquarius (the Water Bearer).

1-8   The phases of the Moon originally inspired the concept of the month

The Moon's contribution to the Earth's precession causes changes that take millennia. Other lunar effects are noticeable every day. As the Moon orbits the Earth, we see different **lunar phases.** As with the Earth and other spherical bodies, the Sun illuminates half of the Moon at any time. The Moon's phase depends on how much of its sunlit hemisphere is exposed to our Earth-based view. When the Moon is closest to the Sun in the sky, its dark hemisphere faces the Earth. This phase, during which the Moon is at most a tiny crescent, is called the *new* Moon (Figure 1-19).

During the seven days following the new phase, more of the Moon's illuminated hemisphere becomes exposed to our view, resulting in a phase called the *waxing crescent* Moon. At the *first quarter* Moon, we see half of the illuminated hemisphere and half of the dark hemisphere. "Quarter Moon" refers to how far in its cycle the Moon has gone, rather than what fraction of the Moon appears lit by sunlight.

  During the next week still more of the illuminated hemisphere can be seen from Earth, giving us the phase called the *waxing gibbous* Moon. When the Moon arrives on the opposite side of the Earth from the Sun, we see almost all of the fully illuminated hemisphere, which is called the *full* Moon. Over the following two weeks, we see less and less of the illuminated hemisphere as the Moon continues along its orbit. This movement produces the phases called the *waning gibbous* Moon, *third quarter* Moon, and the *waning crescent* Moon.

The Moon completes a full cycle of phases in 29¹⁄₂ days. Thus, the side of the Moon facing away from Earth, its *far side,* is not always dark, and so the Moon's far side is not necessarily its "dark side."

Figure 1-19 shows the Moon at various positions in its orbit. Remember that the bright side of the Moon is on the right (west) side of the waxing Moon, while the bright side is on the left (east) side of the waning Moon. When looking at the Moon through a telescope, the best place to see details is where the shadows are longest. This occurs at the boundary between the bright and dark regions, called the **terminator.**

Figure 1-19 also shows local time around the globe, from noon, when the Sun is highest in the sky, to midnight, when it is on the opposite side of the Earth. These time markings can be used to correlate the phase and position of the Moon with the time of day. For example, at first quarter, the Moon is 90° east of the Sun in the sky; hence, moonrise occurs approxi-

---

MOVIE MISCONCEPTIONS

The Mummy Returns (Universal Pictures, 2001)
(The Everette Collection)

The movie *The Mummy Returns* is set in 1935, where our heroes Rick and Evelyn O'Connell are married, raising their nine-year-old son, Alex. In the movie, Alex puts on an ancient Egyptian bracelet which, he then learns, must be carried into a pyramid within one week or *awful things* will happen. At the same time, the mummy of the Egyptian high priest Imhotep, the mummy of the film's title, is being brought back to life. The reincarnation of his former mistress, Anck-Su-Namun is waiting for him and, projected mystically and momentarily to the pyramid, they kiss under a thin crescent moon. (This scene might be construed as a flashback to their former lives, but for the purposes of this question, we will assume it occurs at the time of the rest of the movie's events.) The boy is then kidnapped by the mummy's minions and whisked on a journey to the temple. He is followed by his father and mother, among others, traveling part of the trip in a balloon that is shown crossing the full Moon. The boy gets into the temple on time and the rest of the movie unfolds. What is wrong with the astronomy in the movie, as described above?

(Answers appear at the end of the book.)

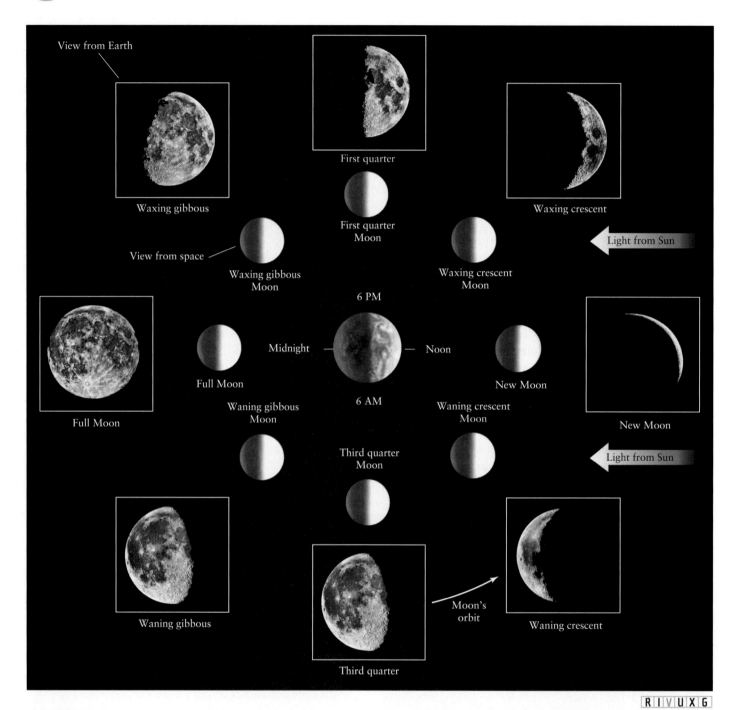

R I V U X G

FIGURE 1-19 **The Phases of the Moon** The diagram shows the Moon at eight locations on its orbit as viewed from far above the Earth's north pole. The corresponding photographs show the resulting lunar phases as seen from Earth. The waning and third quarter photographs look backwards, but they are correctly oriented as seen from Earth. Light from the Sun illuminates one-half of the Moon at all times, while the other half is dark. It takes about $29\frac{1}{2}$ days for the Moon to go through all its phases. (Yerkes Observatory & Lick Observatory)

mately at noon. At full Moon, the Moon is opposite the Sun in the sky; thus, moonrise occurs at sunset. Using this information, you can see that the Moon is visible during the daytime (Figure 1-20) for a part of most days of the year.

Since the dawn of civilization, people have sought accurate timekeeping systems. Ancient Egyptians wanted to know when the Nile would flood, and farmers everywhere needed to know when to plant crops. Migratory tribes wanted to know when the weather would change. Religious leaders scheduled observances in accordance with celestial events. Thus, astronomers have traditionally been responsible for telling time. Indeed, of the four ways in which time cycles are set, three are astronomical in origin: Time is determined by the positions of the Moon, Sun, or stars or, in our own age, by technological means, such as atomic clocks.

The approximately four weeks that the Moon takes to complete one cycle of its phases inspired our ancestors to invent the concept of a month. Astronomers find it useful to define two types of months, depending on whether the Moon's motion is measured relative to the stars or to the Sun. Neither type corresponds exactly to the months of our usual calendar, which have different lengths.

The **sidereal month** is the time it takes the Moon to complete one full orbit of 360° around the Earth (Figure 1-21). This is also the time it takes the Moon to start at one place on the celestial sphere and return to exactly the same place again. Indeed, sidereal motion of any astronomical body is motion measured with respect to the distant stars. This true orbital period for the Moon is approximately 27.3 days. The **synodic month, or lunar month,** is the time it takes the Moon to complete one 29½-day cycle of phases (that is, from new Moon to new Moon or from full Moon to full Moon) and thus is measured with respect to the Sun rather than the distant stars.

The synodic month is longer than the sidereal month, because the Earth is orbiting the Sun while the Moon goes through its phases. As shown in Figure 1-21, the Moon must travel *more* than 360° along its orbit to complete a cycle of phases (for example, from one new Moon to the next), which takes about 2.2 days longer than the sidereal month.

Both the sidereal month and synodic month vary somewhat, because the gravitational pull of the Sun on the Moon affects the Moon's speed as it orbits the Earth. The sidereal

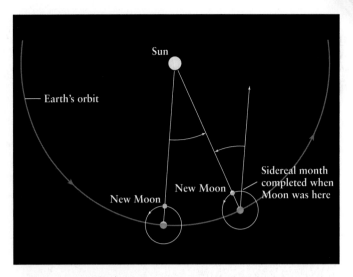

FIGURE 1-21 The Sidereal and Synodic Months The sidereal month is the time it takes the Moon to complete one revolution with respect to the background stars. However, because the Earth is constantly moving in its orbit about the Sun, the Moon must travel through more than 360° to get from one new Moon to the next. The synodic month is the time between consecutive new moons or consecutive full moons. Thus, the synodic month is slightly longer than the sidereal month.

month can vary by as much as 7 hours, while the synodic month can vary by as much as 12 hours.

The terms *synodic* and *sidereal* are also used in discussing the motion of the other bodies in the solar system. The synodic period of a planet is the time between consecutive straight alignments between the Sun, Earth, and that planet (during which time the planet also goes through a cycle of phases, as seen from the Earth). On the other hand, any orbit measured with respect to the distant stars is called "sidereal," including orbits of the planets around the Sun, as well as orbits of moons around their planets. The Earth's sidereal year is 365.2564 days. The difference between the sidereal year and the tropical year is due primarily to the Earth's precession (see Section 1-7).

ECLIPSES

1-9 Eclipses occur only when the Moon crosses the ecliptic during the new or full phase

Eclipses are among the most spectacular natural phenomena. During a **lunar eclipse,** the brilliant full Moon often darkens to a deep red, while during a solar eclipse, broad daylight is transformed into an eerie twilight, as the Sun

FIGURE 1-20 The Moon during the Day The Moon is visible at some time during daylight hours virtually every day. The time of day or night it is up in our sky depends on its phase. (Art Wolfe)

seems to be blotted from the sky. A lunar eclipse occurs when the Moon passes through the Earth's shadow. This can happen only when the Sun, Earth, and Moon are in a straight line at full Moon. A **solar eclipse** occurs when the Moon's shadow moves across the Earth's surface. As seen from Earth, the Moon moves in front of the Sun. This can happen only when the Sun, Moon, and Earth are aligned at new Moon.

At first glance, it would seem that eclipses should happen at every new and full Moon, but, in fact, they occur much less often. Eclipses occur infrequently because the Moon's orbit is tilted 5° out of the ecliptic, as shown in Figure 1-22. Because of this tilt, the new Moon and full Moon usually occur when the Moon is either above or below the plane of the Earth's orbit. In such positions, a perfect alignment between the Sun, Moon, and Earth is not possible and an eclipse cannot occur.

When the Moon crosses the plane of the ecliptic during its new or full phase, an eclipse takes place. The Moon crosses the ecliptic at what is called the **line of nodes** (see Figure 1-22). By calculating the number of times a new Moon takes place on the line of nodes, we find that at least two and no more than five solar eclipses occur each year.

Lunar eclipses occur just about as frequently as solar eclipses, but the maximum number of eclipses (solar plus lunar) possible in a year is seven.

 Maximum Number of Eclipses Annually

1-10 There are three types of lunar eclipse

The Earth's shadow has two distinct parts, as shown in Figure 1-23. The **umbra** is the part of the shadow where all direct sunlight is blocked by the Earth. The **penumbra** of the shadow is where the Earth blocks only some of the sunlight. Depending on how the Moon travels through the Earth's shadow, three kinds of lunar eclipses may occur. A **penumbral eclipse,** when the Moon passes through only the Earth's penumbra, is easy to miss. The Moon still looks full, just a little dimmer than usual or even reddish in color.

When only part of the lunar surface passes through the umbra, a bite seems to be taken out of the Moon, and we see a **partial eclipse.** When the Moon travels completely into the umbra, as sketched in Figure 1-23, we see a **total eclipse** of the Moon. Totality is the time when the entire Moon is in

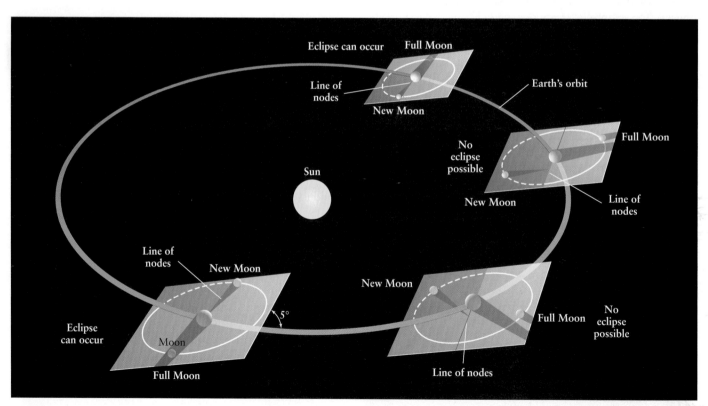

 FIGURE 1-22   **Conditions for Eclipses** The Moon must be very nearly on the ecliptic at new Moon for a solar eclipse to occur. A lunar eclipse occurs only if the Moon is very nearly on the ecliptic at full Moon. When new Moon or full Moon phases occur away from the ecliptic, no eclipse is seen, because the Moon and the Earth do not pass through each other's shadows.

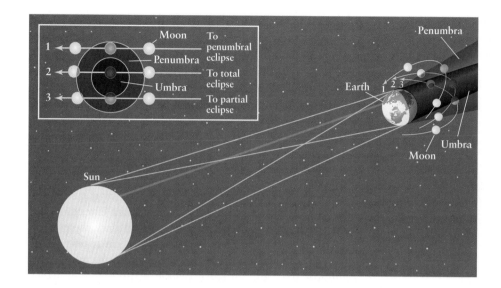

FIGURE 1-23 Three Types of Lunar Eclipses People on the nighttime side of the Earth see a lunar eclipse when the Moon moves through the Earth's shadow. The umbra is the darkest part of the shadow. In the penumbra, only part of the Sun is covered by the Earth. The inset shows the various lunar eclipses that occur, depending on the Moon's path through the Earth's shadow.

the Earth's umbra. Eclipses with the maximum duration of totality, lasting for up to 1 hour 47 minutes, occur when the Moon is closest to the Earth and travels directly through the center of the umbra. Table 1-2 lists all the total and partial lunar eclipses from late 2002 through 2006.

Even during a total eclipse, the Moon does not completely disappear. A small amount of sunlight passing through the Earth's atmosphere is bent into the Earth's umbra. The light deflected into the umbra is primarily red and orange, and thus the darkened Moon glows faintly in rust-colored hues during totality, as shown in Figure 1-24. At sunrise and sunset the Sun appears red or orange for the same reason, specifically, because at those times red and orange light are deflected from the Sun toward you. Other colors, especially violet, blue, and green, are actually deflected by the air so much more than red, orange, and yellow

that the former colors from the Sun do not enter the Earth's umbra or reach us at sunrise and sunset.

Lunar eclipses are perfectly safe to watch with the naked eye. However, solar eclipses are *never* safe to view without suitable eye protection. *Viewing the Sun directly at any time without an approved filter causes permanent eye damage.*

1-11 There are also three types of solar eclipse

Because of their different distances from Earth, the Sun and the Moon have nearly the same angular diameter as seen from Earth—about 1/2°. When the Moon completely covers the Sun, the result is a *total solar eclipse.* You must be at a location within the Moon's umbra to see a total solar eclipse.

TABLE 1-2 Lunar Eclipses, 2002–2006

Date	Visible from	Type	Duration of totality (h:min)
2002 November 20	Americas, Europe, Africa, eastern Asia	Penumbral	
2003 May 16	Americas, Europe, Africa	Total	0:53
2003 Nov 9	Americas, Europe, Africa, central Asia	Total	0:24
2004 May 4	South America, Europe, Africa, Asia, Australia	Total	1:16
2004 October 8	Americas, Europe, Africa, central Asia	Total	1:21
2005 April 24	Americas, eastern Asia, Australia	Penumbral	
2005 October 17	North America, Asia, Australia	Partial	
2006 March 14	Americas, Europe, Africa, Asia	Penumbral	
2006 September 7	Europe, Africa, Asia, Australia	Partial	

R I V U X G

FIGURE 1-24 A Total Eclipse of the Moon Notice the distinctly reddish color of the Moon in this photograph, taken by an amateur astronomer during the lunar eclipse of September 6, 1979. (M. Harms)

During those few precious moments, hot gases (the **solar corona**) surrounding the Sun can be observed and photographed (Figure 1-25). Astronomers can then learn more about the Sun's temperature, chemistry, and atmospheric activity. You can see in Figure 1-26 that only the tip of the

Moon's umbra ever reaches the Earth's surface. As the Earth turns and the Moon orbits, the tip traces an **eclipse path** across the Earth's surface. Only people within this path are treated to the spectacle of a total solar eclipse. Figure 1-26 shows the dark spot produced by the Moon's umbra on the Earth's surface during a total solar eclipse.

The Earth's rotation and the orbital motion of the Moon cause the umbra to race along the eclipse path at speeds in excess of 1700 km/h (1050 mph). Totality never lasts for more than 7½ minutes at any one location on the eclipse path, and it usually lasts for only a few moments.

The Moon's umbra is also surrounded by a penumbra. During a solar eclipse, the Moon's penumbra extends over a large portion of the Earth's surface. When only the penumbra sweeps across the Earth's surface, the Sun is only partly covered by the Moon. This circumstance results in a *partial eclipse of the Sun*. Similarly, people in the penumbra of a total eclipse see a partial eclipse.

The Moon's orbit around the Earth is not quite a perfect circle. The distance between the Earth and the Moon, which averages 384,400 km (238,900 mi), varies by a few percent as the Moon goes around the Earth. The width of the eclipse path depends primarily on the Earth-Moon distance during an eclipse. The eclipse path is widest—up to 270 km (170 mi)—when the Moon happens to be at the point in its orbit nearest the Earth. Usually, however, the path is much narrower.

If a solar eclipse occurs when the Moon is farthest from the Earth, then the Moon's umbra falls short of the Earth and

R I V U X G

FIGURE 1-25 A Total Eclipse of the Sun During a total solar eclipse, the Moon completely covers the Sun's disk, and the solar corona can be photographed. This halo of hot gases extends for millions of kilometers into space. This gorgeous image is a composite of several taken in Chisamba, Zambia, during the June 21, 2001 solar eclipse. (F. Espenak)

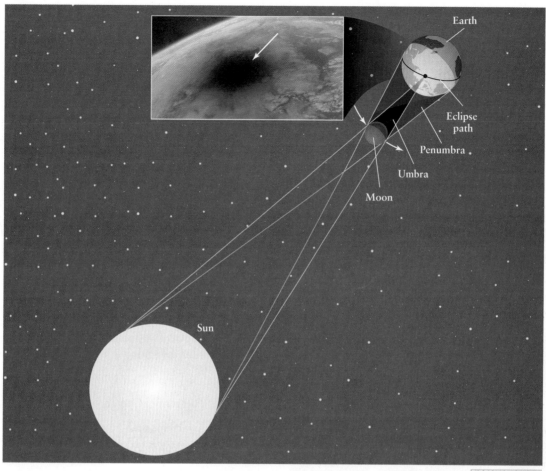

FIGURE 1-26 The Geometry of a Total Solar Eclipse During a total solar eclipse, the tip of the Moon's umbra traces an eclipse path across the Earth's surface. People inside the eclipse path see a total solar eclipse, whereas people inside the penumbra see only a partial eclipse. The photograph in this figure shows the Moon's shadow on the Earth. It was taken from the *Mir* space station during the August 11, 1999 total solar eclipse. The Moon's umbra appears as a dark spot on the eastern coast of the United States. (Jean-Pierre Haigneré, Centre National d'Etudes Spatiales, France/GSFS)

no one sees a total eclipse. From the Earth's surface, the Moon then appears too small to cover the Sun completely, and a thin ring or "annulus" of light is seen around the edge of the Moon at mid-eclipse. This type of eclipse is called an **annular eclipse** (Figure 1-27). The length of the Moon's umbra is nearly 5000 km (3100 mi) shorter than the average distance between the Moon and the Earth's surface. Thus, the Moon's shadow often fails to reach the Earth, making annular eclipses more common than total eclipses. Table 1-3 lists all the total, partial, and annular solar eclipses from late 2002 through 2006.

A total solar eclipse is a dramatic event. The sky begins to darken, the air temperature falls, and the winds increase as the Moon's umbra races toward you. All nature responds: Birds go to roost, flowers close their petals, and crickets begin to chirp as if evening had arrived. As totality approaches, the landscape is bathed in shimmering bands of light and dark as the last few rays of sunlight peek out from behind the edge of the Moon. Finally, the corona blazes forth in a star-studded daytime sky. It is an awesome sight but should only be viewed through a suitable protective filter.

1-12 Frontiers yet to be discovered

Geologists have discovered a variety of cycles of global temperature change that have occurred over the history of the Earth. Indeed, in the 1990s geologists discovered that much of the Earth once suffered a global freezing. These changes occur over tens of thousands of years, hundreds of thousands of years, and possibly longer cycles. Some of the causes of these changes have yet to be discovered. Even today the Earth's global climate is changing over the years. All the causes of this change are not yet known. For example, are our activities the only cause of global warming or

TABLE 1-3 Solar Eclipses, 2002–2006

Date	Type	Area	Notes
2002 December 4	Total	southern Africa, Australia, Indonesia	Maximum length 2:04
2003 May 31	Annular	Europe, Asia, northwest North America	Maximum length 3:37
2003 November 23	Total	Australia, southern South America	Maximum length 1:57
2004 April 19	Partial	southern Africa	
2004 October 14	Partial	northeast Asia, Hawaii, Alaska	
2005 April 8	Total	New Zealand, North and South America	Maximum length 0:42
2005 October 3	Annular	Europe, Africa, southern Asia	Maximum length 4:32
2006 March 29	Total	Africa, Europe, western Asia	Maximum length 4:07
2006 September 22	Annular	South America, western Africa	Maximum length 7:09

R | I | V | U | X | G

FIGURE 1-27   An Annular Eclipse of the Sun This composite of five exposures taken at sunrise in Costa Rica shows the progress of an annular eclipse of the Sun that occurred on December 24, 1974. Note that at mid-eclipse the edge of the Sun is visible around the Moon. (Dennis Di Cicco)

are there long-term seasonal changes that affect it and, if so, how? Another question yet to be answered is how the Earth got its $23\frac{1}{2}°$ tilt.

 Further Reading on These Topics

WHAT DID YOU KNOW?

▌1▐ *Is the North Star—Polaris—the brightest star in the night sky?* Polaris is a star of medium brightness compared with other stars visible to the naked eye.

▌2▐ *Do astronomers regard constellations as simply the familiar patterns of stars in the sky first identified by ancient stargazers?* Astronomers sometimes use the common definition of a constellation as a pattern of stars. Formally, however, a constellation is an entire region of the celestial sphere and all the stars and other objects in it. Viewed from Earth, the entire sky is covered by 88 different-sized constellations. If there is any room for confusion, astronomers refer to the patterns as asterisms.

▌3▐ *What causes the seasons?* The tilt of the Earth's rotation axis with respect to the ecliptic causes the seasons. They are not caused by the changing distance from the Earth to the Sun that results from the shape of Earth's orbit.

▌4▐ *How many zodiac constellations are there?* There are 13 zodiac constellations, the lesser-known one being Ophiuchus.

▌5▐ *Does the Moon have a dark side that we never see from Earth?* Half of the Moon is always dark. Whenever we see less than a full Moon, we are seeing part of the Moon's dark side. So, the dark side of the Moon is not the same as the far side of the Moon, which we never see from Earth.

1 6 *Is the Moon ever visible during the daytime?* The Moon is visible at some time during daylight hours almost every day of the year. Different phases are visible during different times of the day.

KEY WORDS

angle, 19
angular diameter (angular size), 19
annular eclipse, 35
arc angle, 19
autumnal equinox, 26
celestial equator, 18
celestial pole, 18
celestial sphere, 18
circumpolar stars, 20
constellation, 18
declination, 18
degree (°), 19
diurnal motion, 20
eclipse path, 34
ecliptic, 23

equinox, 25
line of nodes, 32
lunar eclipse, 31
lunar phases, 29
north celestial pole, 18
partial eclipse, 32
penumbra, 32
penumbral eclipse, 32
precession, 28
precession of the equinoxes, 28
revolution, 20
right ascension, 18
rotation, 20
sidereal month, 31
solar corona, 34

solar day, 21
solar eclipse, 32
south celestial pole, 18
summer solstice, 26
synodic month (lunar month), 31
terminator, 29
time zone, 21
total eclipse, 32
umbra, 32
vernal equinox, 26
winter solstice, 26
zenith, 20
zodiac, 27

KEY IDEAS

Patterns of Stars

• The surface of the celestial sphere is divided into 88 unequal regions called constellations.

Earthly Cycles

• The celestial sphere appears to revolve around the Earth once in each day-night cycle. In fact, it is the Earth's rotation that causes that apparent motion.

• The poles and equator of the celestial sphere are determined by extending the axis of rotation and the equatorial plane of the Earth out onto the celestial sphere.

• Earth's axis of rotation is tilted at an angle of $23\frac{1}{2}°$ from the perpendicular to the plane of the Earth's orbit. This tilt causes the seasons.

• Equinoxes and solstices are significant points along the Earth's orbit that are determined by the relationship between the Sun's path on the celestial sphere (the ecliptic) and the celestial equator.

• The Earth's axis of rotation slowly changes direction relative to the stars over thousands of years, a phenomenon called precession. Precession is caused by the gravitational pull of the Sun and Moon on the Earth's equatorial bulge.

• The length of the day is based upon the average motion of the Sun along the celestial equator, which produces the 24-hour day upon which our clocks are based.

• The phases of the Moon are caused by the relative positions of the Earth, Moon, and Sun. The Moon completes one cycle of phases in a synodic month, which averages $29\frac{1}{2}$ days.

• The Moon completes one orbit around the Earth with respect to the stars in a sidereal month, which averages 27.3 days.

Eclipses

• The shadow of an object has two parts: the umbra, where direct light from the source is completely blocked; and the penumbra, where the light source is only partially obscured.

• A lunar eclipse occurs when the Moon moves through the Earth's shadow. During a lunar eclipse, the Sun, Earth, and Moon are in alignment, and the Moon is in the plane of the ecliptic.

• A solar eclipse occurs when a strip of the Earth passes through the Moon's shadow. During a solar eclipse, the Sun, Earth, and Moon are in alignment, and the Moon is in the plane of the ecliptic.

• Depending on the relative positions of the Sun, Moon, and Earth, lunar eclipses may be penumbral, partial, or total, and solar eclipses may be annular, partial, or total.

REVIEW QUESTIONS

1 How are constellations useful to astronomers?

2 What is the celestial sphere, and why is this ancient concept still useful today?

3 What is the celestial equator, and how is it related to the Earth's equator? How are the north and south celestial poles related to the Earth's axis of rotation?

4 What is the ecliptic, and why is it tilted with respect to the celestial equator?

5 By about how many degrees does the Sun move along the ecliptic each day?

6 Through how many constellations does the Sun move every day?

7 Through how many constellations does the Sun move every year?

8 Why does the tilt of the Earth's axis relative to its orbit cause the seasons as the Earth revolves around the Sun? Draw a diagram to illustrate your answer.

9 What are the vernal and autumnal equinoxes? What are the summer and winter solstices? How are these four points related to the ecliptic and the celestial equator?

10 What is precession, and how does it affect our view of the heavens?

11 How does the daily path of the Sun across the sky change with the seasons?

12 Why is it warmer in the summer than in the winter?

13 Why is it convenient to divide the Earth into time zones?

14 Why does the Moon exhibit phases?

15 What is the difference between a sidereal month and a synodic month? Which is longer? Why?

16 What is the line of nodes, and how is it related to solar and lunar eclipses?

17 What is the difference between the umbra and the penumbra of a shadow?

18 What is a penumbral eclipse of the Moon? Why is it easy to overlook such an eclipse?

19 Which type of eclipse—lunar or solar—have most people seen? Why?

20 How is an annular eclipse of the Sun different from a total eclipse of the Sun? What causes this difference?

21 When is the next leap year?

22 At which phase(s) of the Moon does a solar eclipse occur? A lunar eclipse?

23 Is it safe to watch a solar eclipse without eye protection? A lunar eclipse?

24 During what phase is the Moon least visible in the daytime?

ADVANCED QUESTIONS

The answers to all computational problems, which are preceded by an asterisk (*), appear at the end of the book.

25 During what phase(s) does the Moon rise after sunrise and before sunset? After sunset and before sunrise? At sunset? At sunrise?

26 Why can't a person in Australia use the Big Dipper to find north?

27 Are there any stars in the sky that are not members of a constellation?

28 At what places on Earth is Polaris seen on the horizon?

29 Where do you have to be on the Earth to see the Sun at your zenith? If you stay at one such location for a full year, on how many days will the Sun pass through the zenith?

30 Where do you have to be on Earth to see the south celestial pole at your zenith? What is the maximum

possible elevation (angle) of the Sun above the horizon at that location? On what date is this maximum elevation achieved?

31 Where on the horizon does the Sun rise at the time of the vernal equinox?

32 Consult a star map of the southern hemisphere and determine which, if any, of the bright southern stars could someday become south celestial pole stars.

33 Are there stars in the sky that never set where you live? Are there stars that never rise where you live? Does your answer depend on your location on Earth? Why or why not?

34 Using a diagram, demonstrate that your latitude on Earth is equal to the altitude of the north celestial pole above your northern horizon.

35 Using a star map, determine which bright stars, if any, could someday mark the location of the vernal equinox. Give the approximate years when this should happen.

36 What is the phase of the Moon if it (a) rises at 3 A.M.? (b) sets at 9 P.M.? At what time does (c) the full Moon set? (d) the first quarter Moon rise?

37 What is the phase of the Moon if, on the first day of spring, the Moon is located at the position of (a) the vernal equinox, (b) the summer solstice, (c) the autumnal equinox, (d) the winter solstice?

*38 How many more sidereal months than synodic months are there in a year? Why?

39 How do we know that the phases of the Moon are not due to the Moon moving in the Earth's shadow?

40 Do the paths of total solar eclipses fall more frequently on oceans or on land? Explain.

41 Can one ever observe an annular eclipse of the Moon? Why or why not?

42 Which of the five images of the Sun is the first and which is the last in the sequence shown in the photograph for Figure 1-27? Justify your answer.

43 During a lunar eclipse, does the Moon enter the Earth's shadow from the east or the west? Explain your answer.

44 Do we see all of the Moon's surface from the Earth? *Hint:* Carefully examine the photographs in Figure 1-19.

45 Explain why the waning gibbous, third quarter and waning crescent photographs in Figure 1-19 are correctly oriented as seen from Earth.

46 Why is a small crescent of light often observed on the Moon when it is exactly in the new phase?

47 Make a drawing of an annular solar eclipse as seen from space. Be sure to make clear how it differs from a total solar eclipse.

*48 Assuming that the Sun makes an angle of $1/2°$ in our sky and is at a distance of 1.496×10^{11} m, what is the Sun's diameter? Divide this by 2 to find the Sun's radius and explain why this result is slightly different from the value given in Appendix Table A-6.

49 How long was the exposure for the photograph of circumpolar star trails that opens this chapter?

50 Determine which stars whose paths are shown in the photo that opens this chapter are *not* circumpolar. An easy way to write the answer is in terms of distance on the photograph from the location of the south celestial pole.

DISCUSSION QUESTIONS

51 Examine a list of the 88 constellations. Are there any constellations whose names obviously date from modern times? Where are these constellations located? Why do they not have ancient names?

52 In his novel *King Solomon's Mines*, H. Rider Haggard described a total solar eclipse that was seen in both South Africa and the British Isles. Is such an eclipse possible? Why or why not?

53 Describe how a lunar eclipse would look if the Earth had no atmosphere.

54 Examine a listing of total solar eclipses over the next several decades. What are the chances that you might be able to travel to one of the eclipse paths? Do you think you might go through your entire life without ever seeing a total eclipse of the Sun?

WHAT IF ...

55 The Moon moved about the Earth in an orbit perpendicular to the plane of the Earth's orbit? What would the cycle of lunar phases be? Would solar and lunar eclipses be possible under these circumstances?

56 The Earth's axis of rotation were tilted at a different angle? What would the seasons be like where you are now if the axis of rotation is (a) 0° and (b) 45° to its orbital plane. What would be different about the seasons and the day-night cycle if you lived at one of the Earth's poles?

57 You watched the Earth from the Moon? What would you see for Earth's (a) daily motion, (b) motion along the celestial sphere, and (c) cycle of phases.

58 The Moon didn't rotate? Describe how its surface *features* would appear from the Earth—that is, would we see all sides of it over time? Why or why not? (Ignore the change in phase when discussing its appearance.) *Carefully* study the photographs in Figure 1-19 and state whether the same features are visible at all times or whether we see different features over time. (Again, ignore the phases.) What can you conclude about whether the Moon actually rotates?

WEB/CD-ROM QUESTIONS

*59 Work through the AIMM (Active Integrated Media Module) called "Small-Angle Toolbox" in Chapter 1 of the *Discovering the Universe* CD-ROM or Web site. Use it to determine the diameters in kilometers of the Sun, Saturn, and Pluto given the following distances and angular sizes:

Object	Distance (km)	Angular size (″)
Sun	1.5×10^8	1800
Saturn	1.5×10^9	16.5
Pluto	6.3×10^9	0.06

60 Search the Web for information about the Great Nebula of Orion (also called the Orion nebula, see Figure 1-1). Can the Great Nebula be seen with the naked eye? Does it exist alone in space or is it part of a larger system of interstellar material? What has been learned by examining the Great Nebula with telescopes sensitive to infrared light?

61 Search the Web for information about the national flags of Australia, New Zealand, and Brazil, and the state flag of Alaska. What stars are depicted on these flags? Explain any similarities or differences among these flags.

62 Search the Web for the English meaning of the Japanese word *Subaru*. Make a drawing of the Subaru car's emblem and explain it.

63 Use the U.S. Naval Observatory's Web site to determine the times of sunset and sunrise on (a) your birthday and (b) the date this assignment is due. Are the times the same? Explain why or why not.

64 Search the Web for information about the next total solar eclipse. Through which major cities, if any, does the path of totality pass? What is the maximum duration of totality? Find a location where this maximum duration is observed. Will the eclipse be visible (even as a partial eclipse) from your present location?

65 Search the Web for information about the next total lunar eclipse. Will the total phase of the eclipse be visible from your present location? If not, will the penumbral phase be visible? Draw a picture of the Sun, Earth, and Moon at totality and indicate your location on the drawing of the Earth.

 66 Access the animation "The Moon's Phases" in Chapter 1 of the *Discovering the Universe* Web site or CD-ROM. This shows the Earth-Moon system as seen from a vantage point above the Earth's north pole. (a) Describe where you would be on the diagram if you are on the equator and the time were 6:00 P.M. (b) If it were 6:00 P.M. and you were standing on the Earth's equator, would a third-quarter Moon be visible? Why or why not? If it would be visible, describe its appearance.

OBSERVING PROJECTS

67 On a clear, cloud-free night, use the star charts within the covers of this book to see how many constellations of the zodiac you can identify. Which ones are easy to find? Which are difficult?

 68 Using *Starry Night Backyard*™ with all pattern lines off, see if you can identify the zodiac constellations that are visible in the sky this evening. Set the time to sometime after sunset and turn off the sky motion by clicking on the square to the left of the timestep indicator. Then "grab" the sky and move around until you recognize some constellations. Use the star charts within the covers of this book for guidance, if necessary. If you still have difficulty finding zodiac constellations, click on *Guides* and then on *The Ecliptic* to highlight the part of the sky in which the zodiac is located. If you still have trouble, go to *Constellations* and click on *Astronomical* to see the asterisms. If you need even more help, go to *Constellations* again and click on *Auto Identify*. Then as you move the little hand over a constellation, its name and classical image will appear.

Watch out for planets, which are also found in the zodiac—you may have to ignore one or more of these bright objects in picking out the constellation patterns. Moving the hand icon over an object will tell you what it is. Return to *Home* (assuming it is daylight when you do this), switch off daylight and determine which zodiac constellation the Sun is passing through today.

69 Use *Starry Night Backyard*™ to observe the diurnal motion of the sky. (a) First set *Starry Night Backyard*™ to display the sky as seen from where you live. Select *Set Home Location…* in the *Go* menu and click on the *Lookup* button to find your city or town. Then, using the hand cursor, center your field of view on the northern horizon (if you live in the northern hemisphere) or southern horizon (if you live in the southern hemisphere). In the Control Panel on the top of the main window, set the timestep to 3 minutes and click on the *Forward* button. To see the stars during the daytime, turn off daylight (select *Daylight* in the *Sky* menu). Do the stars appear to revolve clockwise or

counterclockwise? Explain this in terms of the Earth's rotation. Are any of the stars circumpolar? (b) Recenter your field of view on the southern horizon (if you live in the northern hemisphere) or the northern horizon (if you live in the southern hemisphere). Describe what you see. Are any of these stars circumpolar?

70 Use the *Starry Night Backyard*™ program to observe the Sun's motion on the celestial sphere. Select *Atlas* in the *Go* menu to see the entire celestial sphere, including the part below the horizon. Center on the Sun by using the *Find...* command in the *Edit* menu. Zoom out until you see about 90° of the celestial sphere. (a) Select *Auto Identify* in the *Constellations* menu to display the constellations at the center of the screen. In which constellation is the Sun located today? Is this the same as the astrological sign for today's date (see Table 1-1)? (b) Be sure you are locked on the Sun. In the Control Panel at the top of the main window, set the timestep to 3 days and click on the *Forward* button. Observe the Sun for a full year of simulated time. How many constellations does it pass through? Where does it cross the celestial equator? What path does it follow? Does it ever change direction?

71 Examine the star charts that are published monthly in such popular astronomy magazines as *Sky & Telescope* and *Astronomy*. How do they differ from the star charts within the covers of this book? On a clear, cloud-free night, use one of these star charts to locate the celestial equator and the ecliptic on the night sky. Note the inclination of the Milky Way to the ecliptic and celestial equator. What do your observations tell you about the orientation of the Earth and its orbit around the Sun relative to the rest of the Galaxy?

72 Observe the Moon on each clear night over the course of a month. Note the Moon's location among the constellations and record that location on a star chart that also shows the ecliptic. After a few weeks, your observations will begin to trace the Moon's orbit. Identify the orientation of the line of nodes by marking the points where the Moon's orbit and the ecliptic intersect. On what dates is the Sun near the nodes marked on your star chart? Compare these dates with the dates of the next solar and lunar eclipses.

73 Use *Starry Night Backyard*™ to study the Moon's path in the sky and eclipses. Begin by setting *Starry Night Backyard*™ to "Atlas" mode. Do this by clicking on *Go* and then *Atlas*. Find the Moon on the celestial sphere by clicking on *Edit* and then *Find*. Type in "moon" and press enter. Increase the angle of the sky you see to 15° using the button on the upper right of the screen. Click on it and then click on 15°, which will appear in a list. Note whether the Moon lies on the ecliptic. Set the motion to increment by 6 hours per timestep. Keeping the Moon locked in the center of your screen, step ahead. (a) In what direction does the Moon move against the background of stars? Does it ever change direction? (b) Change the timestep to 1 day. Determine how many days elapse between successive times the Moon is on the ecliptic. (If you don't see the green line representing the ecliptic, select *The Ecliptic* in the *Guides* menu.) (c) Move forward in time until the Moon is either full or new *and* it is on the ecliptic. What type of eclipse could occur on that day? (d) Using Tables 1-2 and 1-3, set the program to the date of the next such eclipse. Does the Moon have the same phase as in part (c)? (e) Moving the time ahead or back by 1 hour per timestep, observe the eclipse. If the Sun and Moon "miss" each other, explain why this occurred.

74 It is quite possible that a lunar eclipse will occur while you are taking this course. If so, look up on the Internet the precise time that it will happen; in the current issue of a reference from the U.S. Naval Observatory, such as the *Astronomical Almanac* or *Astronomical Phenomena*; or in such magazines as *Sky & Telescope* and *Astronomy*, which generally run articles about eclipses the month before they happen. Make arrangements to observe the next lunar eclipse. Note the times at which the Moon enters and exits the Earth's umbra.

75 Use *Starry Night Backyard*™ to determine when the Sun rises and sets today. Similarly, find out when the Moon rises and sets today. Starting with the time set to last midnight, determine during what hours the Moon is visible during daylight hours today and during what hours it is visible at night.

WHAT IF . . .

EARTH'S AXIS LAY ON THE ECLIPTIC?

Imagine the Earth tilted so that its axis of rotation lies in the plane of its orbit about the Sun (see the accompanying figure). New Earth still rotates once every 24 hours. We arbitrarily fix New Earth's north pole to point toward the star Tau Tauri, a star nearly on the ecliptic, just north of the bright star Aldebaran in the constellation Taurus (see Figure 1-3).

New Seasons Let's see how the seasons unfold on New Earth. It is March 21, the date of the spring equinox. The Sun is directly over the equator. For the next three months, the Sun rises higher in the northern hemisphere. Unlike on our Earth, the Sun does not stop moving northward when it is over $23^{1}/_{2}°$ north latitude. Rather, the Sun rises farther and farther north day by day until it appears over the north pole of New Earth at the summer solstice, around June 22. Three months later, at the autumnal equinox, the Sun again rises over the equator, and day and night have the same length everywhere.

The Sun appears over the south pole of New Earth at the winter solstice, around December 22. The Sun's apparent motion through New Earth's sky completes the seasonal cycle by moving north, appearing over the equator once again around March 21.

New Climate Day and night take on new meanings for inhabitants of New Earth. On our Earth, the regions above the Arctic Circle and below the Antarctic Circle have days or weeks of continual light in summer and days or weeks of constant darkness in winter. But on New Earth, *every place* has extended winter periods of constant darkness followed by extended summer periods of constant daylight. Spring and fall on New Earth have daily cycles of daylight and darkness, which separate the periods of continual daylight and darkness.

A New Day At the latitude of Atlanta, Georgia, (33°46′) on New Earth, the day-night cycle occurs for only seven and a half months out of the year. During the other four and a half months, there is continuous day or continuous night, coupled with harsh summers and winters. With variations, this sequence of events occurs everywhere on New Earth.

The seasonal cycle on New Earth prevents the formation of permanent polar ice caps. Polar regions experience the same tropical heating and high temperatures as the equatorial regions of our Earth. The polar regions of New Earth in winter are exceptionally cold, so seasonal polar ice caps may form. Because the southern polar cap resting on Antarctica is not permanent, the oceans, and the shorelines on the continents, are higher than those on our Earth.

If seasonal polar ice caps form, the dominant force controlling weather may shift from the jet streams that circle our Earth along lines of latitude to a pole-to-pole flow. Thermal flows created by intense heating at one location and cooling at others may replace our Earth's trade winds and other east-west winds. How else would New Earth differ from our world?

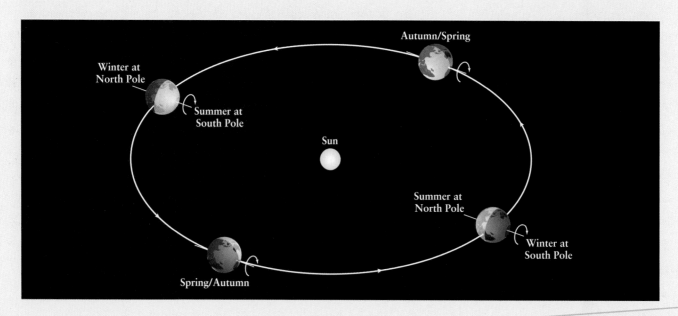

2 GRAVITATION AND THE WALTZ OF THE PLANETS

IN THIS CHAPTER YOU WILL DISCOVER

- the scientific revolution that dethroned Earth from its location at the center of the universe

- Copernicus's argument that the planets orbit the Sun

- Kepler's determination of the shapes of planetary orbits

- Galileo's first views of craters on the Moon and moons orbiting Jupiter

- how Isaac Newton used all these discoveries to formulate the basic laws of physics

- how the Sun holds the planets in their orbits

Gravity's Pull
The gravitational force of the Moon is approximately 1/6th the gravitational force of the Earth. This low gravity enabled Apollo astronauts to function well in spacesuits weighing 48 kg (106 lb). Furthermore, there is very little atmosphere on the Moon. The flag in this Apollo 17 photograph has wires running through it to make it appear that it is waving in a breeze. In an experiment run on Apollo 15, astronaut Dave Scott dropped a hammer and feather simultaneously from the same height. The lack of atmosphere led to both objects striking the moon's surface simultaneously. (NASA)

R I V U X G

WHAT DO YOU THINK?

1   What is the shape of the Earth's orbit around the Sun?

2   Do the planets orbit the Sun at constant speeds?

3   Do the planets all orbit the Sun at the same speed?

4   How much force does it take to keep an object moving in a straight line at a constant speed?

5   How does an object's mass differ when measured on the Earth and on the Moon?

The groundwork for modern science was set down by Greek mathematicians and philosophers beginning around 2500 years ago, when Pythagoras and his followers began using mathematics to describe natural phenomena. About 200 years later, Aristotle asserted that the universe is comprehensible: It is governed by regular laws. Just as important, the Greeks were also among the first to leave a written record of their ideas, so that succeeding generations could develop, criticize, and test their conclusions. This concept evolved into writing physical theories quantitatively in mathematical terms so that we can test the theories by observing nature.

Early Greek astronomers tried to explain the motion of the five then-known planets: Mercury, Venus, Mars, Jupiter, and Saturn. Most people at that time held a *geocentric* view of the universe: They assumed that the Sun, the Moon, the stars, and the planets revolve about the Earth. A theory of the overall structure and evolution of the universe is called a **cosmology**, so the prevailing cosmology was geocentric. Based on observations of the motions of heavenly bodies, the **geocentric cosmology** is so compelling that it held sway for nearly 2000 years. It was only in the face of more and more accurate observations of planet motions among the stars that its validity came into question.

ORIGINS OF A SUN-CENTERED UNIVERSE

The Greeks knew that the positions of the planets slowly shift against the background of "fixed" stars in the constellations. In fact, the word *planet* comes from a Greek term meaning "wanderer." The Greeks observed that planets do not move at uniform rates through the constellations. From night to night, as viewed in the northern hemisphere, they usually move slowly to the left (eastward) relative to the background stars. This eastward movement is called **direct motion.** Occasionally, however, a planet seems to stop and then back up for several weeks or months. This westward movement is called **retrograde motion.**

These planetary motions are much slower than the apparent daily movement of the entire sky caused by the Earth's rotation, and so they are superimposed on it. Therefore, the planets always rise in the east and set in the west, as the stars do. Both direct and retrograde motion are best detected by mapping the nightly position of a planet against the background stars over a long period. For example, Figure 2-1 shows the path of the planet Mars from July 2005 through February 2006.

Explaining the motions of the five planets in a geocentric (Earth-centered) universe was one of the main challenges facing the astronomers of antiquity. The effort resulted in an increasingly contrived model, especially in explaining retrograde motion. The mechanical description of a geocentric model of the universe (see Guided Discovery: Geocentric Cosmology) was complex, and, as observations improved, the model increasingly failed to fit the data. Because simplicity and accuracy are hallmarks of science, the complex geocentric model had to give way to a simpler, more elegant one. The ancient Greek astronomer Aristarchus proposed a more straightforward explanation of planetary motion, namely, that all the planets, including the Earth, revolve about the Sun.

The retrograde motion of Mars in the heliocentric cosmology, for example, occurs because the Earth overtakes and passes the red planet, as shown in Figure 2-2. The occasional retrograde movement of a planet is merely the result of our changing viewpoint as we orbit the Sun—an idea that is beautifully simple compared to a geocentric system with all its complex planetary motions.

 Insight into Science   Keep it simple When several competing theories describe the same concepts with the same accuracy, scientists choose the simplest one. That basic tenet, formally expressed by William of Occam, in the fourteenth century, is known as **Occam's razor.** Indeed, the original form of the heliocentric cosmology was appealing not because it was more accurate, it wasn't, but because it made the same predictions within a simpler model than did the geocentric cosmology. *Remember Occam's razor.*

2-1 Copernicus devised the first comprehensive heliocentric cosmology

Dethroning Earth from its central role in the universe was difficult for a variety of reasons. Important among them is the fact that the Earth just doesn't seem to move! Combining this with the human desire to be at the center of everything and with geocentric religious teaching kept the belief that the Earth is at the center of everything firmly seated for more than 1300 years after Aristarchus proposed

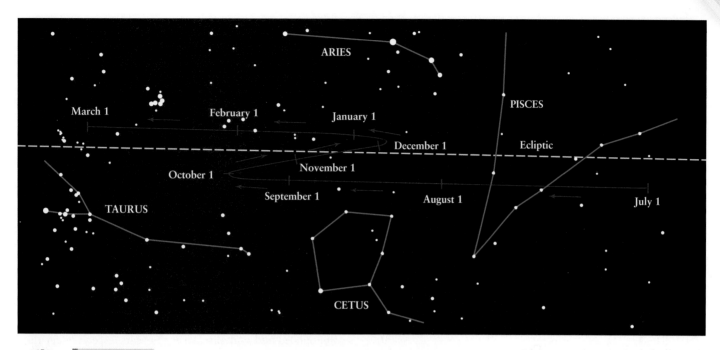

FIGURE 2-1 **The Path of Mars in 2005–2006** From July 2005 through February 2006, Mars moves across the constellations of Pisces, Aries, and Taurus. From October 1 through December 12, 2005, Mars's motion will be retrograde.

the **heliocentric** (Sun-centered) **cosmology.** The person who took on the ancient authorities was the sixteenth-century Polish lawyer, physician, mathematician, economist, canon, and artist, Nicolaus Copernicus.

 Copernicus turned his attention to astronomy in the early 1500s. He found that by assuming that everything orbits the Sun rather than the Earth, he could determine which planets are closer to the Sun than the Earth and which are farther away. Because Mercury and Venus are always observed fairly near the Sun, Copernicus correctly concluded that their orbits must lie inside that of

the Earth. The other planets visible to the naked eye—Mars, Jupiter, and Saturn—can be seen high in the sky in the middle of the night, when the Sun is far below the horizon. This can occur only if the Earth comes between the Sun and a planet. Copernicus therefore concluded that the orbits of Mars, Jupiter, and Saturn lie outside the Earth's orbit. Three more-distant planets (Uranus, Neptune, and Pluto) were discovered after the telescope was invented.

The geometrical arrangements among the Earth, another planet, and the Sun are called **configurations.** For example, when Mercury or Venus is directly between the Earth and

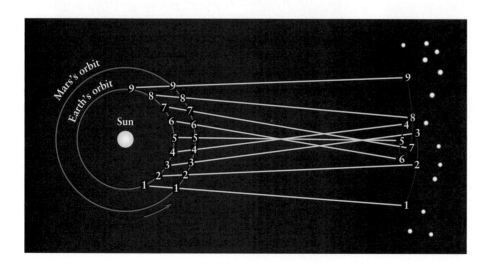

FIGURE 2-2 **A Heliocentric Explanation of Planetary Motion** The Earth travels around the Sun more rapidly than does Mars. Consequently, as the Earth overtakes and passes this slower-moving planet, Mars appears (from points 4 through 6) to move backward among the background stars for a few months.

GUIDED DISCOVERY   Geocentric Cosmology

The Greeks developed many theories to account for retrograde motion and the resulting loops that the planets trace out against the background stars. One of the most successful ideas was expounded by the last of the great ancient Greek astronomers, Ptolemy, who lived in Alexandria, Egypt, 1900 years ago. The basic concepts are sketched in the accompanying figures. Each planet is assumed to move in a small circle called an **epicycle,** the center of which moves in a larger circle called a **deferent,** whose center is offset from the Earth. As viewed from Earth, the epicycle moves eastward along the deferent, and both circles rotate in the same direction (counterclockwise).

Most of the time, the motion of the planet on its epicycle adds to the eastward motion of the epicycle on the deferent. Thus, the planet is seen to be in direct (eastward) motion against the background stars throughout most of the year. However, when the planet is on the part of its epicycle nearest the Earth, its motion along the epicycle subtracts from the motion of the epicycle along the deferent. The planet thus appears to slow and then halt its usual eastward movement among the constellations, even seeming to go backward for a few weeks or months. Using this concept of epicycles and deferents, Greek astronomers were able to explain the retrograde loops of the planets.

Using the wealth of astronomical data in the library at Alexandria, including records of planetary positions covering hundreds of years, Ptolemy deduced the sizes of the epicycles and deferents and the rates of revolution needed to produce the recorded paths of the planets. After years of arduous work, Ptolemy assembled his calculations in the *Almagest,* in which the positions and paths of the Sun, Moon, and planets were described with unprecedented accuracy. In fact, the *Almagest* was so successful that it became the astronomer's bible. For more than 1000 years, Ptolemy's cosmology endured as a useful description of the workings of the heavens.

Eventually, however, the commonsense explanation of the Earth-centered cosmology began to go awry. Errors and inaccuracies that were unnoticeable in Ptolemy's day compounded and multiplied over the years, especially errors due to precession, the slow change in the direction of the Earth's axis of rotation. Fifteenth-century astronomers made some cosmetic adjustments to the Ptolemaic system. However, the system became less and less satisfactory as more complicated and arbitrary details were added to keep it consistent with the observed motions of the planets. After Newton's time, scientists knew that orbital motion required a force acting on the body. However, nothing in Ptolemy's epicycle theory produced such a force.

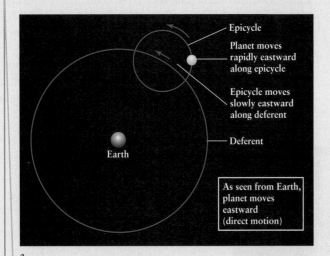

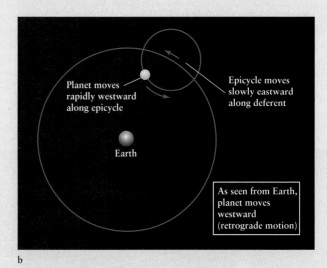

 A Geocentric Explanation of Planetary Motion Each planet revolves about an epicycle, which in turn revolves about a deferent centered approximately on the Earth. As seen from Earth, the speed of the planet on the epicycle alternately **(a)** adds to or **(b)** subtracts from the speed of the epicycle on the deferent, thus producing alternating periods of direct and retrograde motion.

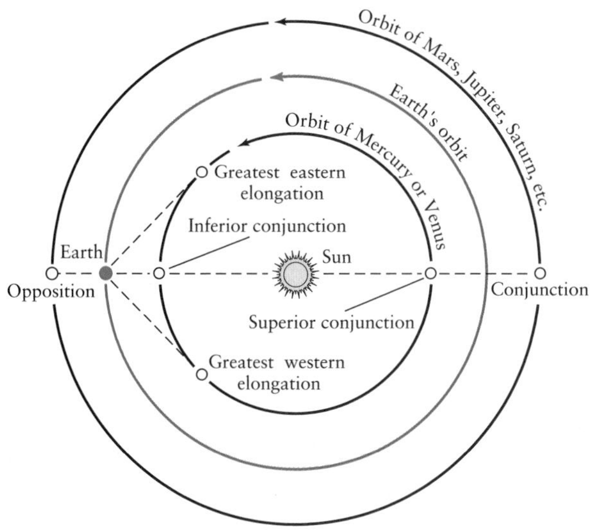

INTERACTIVE EXERCISE 2.1 **FIGURE 2-3** **Planetary Configurations** It is useful to specify key points along a planet's orbit, as shown in this figure. These points identify specific geometric arrangements between the Earth, another planet, and the Sun.

the Sun, as in Figure 2-3, we say the planet is in a configuration called an **inferior conjunction;** when either of these planets is on the opposite side of the Sun from the Earth, its configuration is called a **superior conjunction.**

The angle between the Sun and a planet as viewed from the Earth is called the planet's **elongation.** A planet's elongation varies from zero degrees to a maximum value, depending upon where we see it in its orbit around the Sun. At **greatest eastern** or **greatest western elongation,** Mercury and Venus are as far from the Sun in angle as they can be. This is about 28° for Mercury and about 47° for Venus. When either Mercury or Venus rises before the Sun, it is visible in the eastern sky as a bright "star" and is often called the "morning star." Similarly, when either of these two planets sets after the Sun, it is visible in the western sky and is then called the "evening star." Because these two planets are not always at their greatest elongations, they are often very close in angle to the Sun and therefore hard to see. This is especially true of Mercury, which is never more than 28° from the Sun, while Venus is often nearly halfway up the sky at sunrise or sunset and therefore quite noticeable at these times. Because they are so bright and sometimes appear to change color due to motion of the Earth's atmosphere, Venus and Mercury are often mistaken for UFOs. (The same motion of the air causes the road in front of your car to shimmer on a hot day.)

Planets whose orbits are larger than Earth's have different configurations. For example, when Mars is located behind the Sun, as seen from Earth, it is said to be in **conjunction.** When it is opposite the Sun in the sky, the planet is at **opposition.** It is not difficult to determine when a planet happens to be located at one of the key positions in Figure 2-3. For example, when Mars is at opposition, it appears high in the sky at midnight.

It is easy to follow a planet as it moves from one configuration to another. However, these observations alone do not tell us the planet's actual orbit, because the Earth, from which we make the observations, is also moving. Copernicus was therefore careful to distinguish between two characteristic time intervals, or *periods,* of each planet.

Recall from your study of the Moon in Chapter 1 that the sidereal period of orbit is the true orbital period. A planet's **sidereal period** is the time it takes the planet to complete one circuit around the Sun as would be measured by a fixed observer at the Sun's location watching each planet move through the background stars. The sidereal period determines the length of a year for each planet. Another useful time interval for astronomers is the **synodic period.** The synodic period is the time that elapses between two successive identical configurations as seen from the Earth. It can be from one opposition to the next, for example, or from one conjunction to the next (Figure 2-4). It tells us, for example, when to expect a planet to be closest to the Earth and, therefore, most easily studied.

Thus, nearly 500 years ago, Copernicus was able to obtain the first six entries shown in Table 2-1 (the others are contemporary results included for completeness). Copernicus was then able to devise a straightforward geometric method for determining the distances of the planets from the Sun. His answers turned out to be remarkably close to the modern values, as shown in Table 2-2. From these two tables it is apparent that the farther a planet is from the Sun, the longer it takes to complete its orbit.

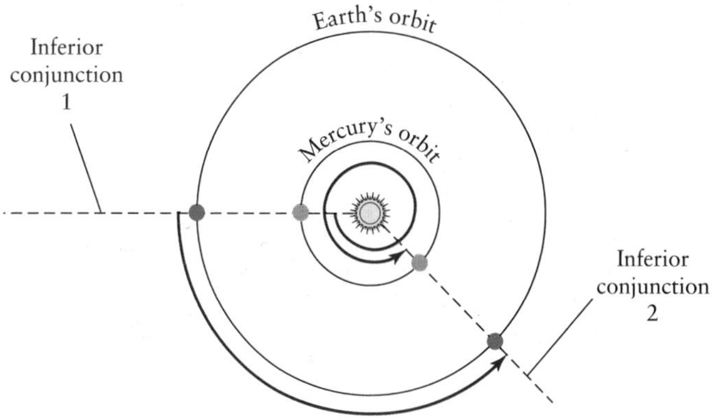

FIGURE 2-4 **Synodic Period** The time between consecutive conjunctions of the Earth and Mercury is 116 days. Typical of synodic periods for all planets, the location of the Earth is different at the beginning and end of the period. You can visualize the synodic periods of the exterior planets by putting the Earth in Mercury's place in this figure and putting one of the outer planets in the Earth's place.

TABLE 2-1   Synodic and Sidereal Periods of the Planets (in Earth Years)

	Synodic Period	Sidereal Period
Mercury	0.318 yr	0.241 yr
Venus	1.599 yr	0.616 yr
Earth	—	1.0 yr
Mars	2.136 yr	1.9 yr
Jupiter	1.092 yr	11.9 yr
Saturn	1.035 yr	29.5 yr
Uranus	1.013 yr	84.0 yr
Neptune	1.008 yr	164.8 yr
Pluto	1.005 yr	248.5 yr

TABLE 2-2   Average Distances of the Planets from the Sun

	Measurement (AU)	
	By Copernicus	Modern
Mercury	0.38	0.39
Venus	0.72	0.72
Earth	1.00	1.00
Mars	1.52	1.52
Jupiter	5.22	5.20
Saturn	9.07	9.54
Uranus	Unknown	19.19
Neptune	Unknown	30.06
Pluto	Unknown	39.53

Insight into Science   Take a fresh look   When the science is hard to visualize, try another perspective. For example, a planet's sidereal period of orbit is easy to understand from the perspective of the Sun but more complicated from the Earth. The synodic period of each planet, on the other hand, is easily determined from the Earth. As we will see, especially when we study Einstein's theories of relativity, each of these perspectives is called a *frame of reference*.

Copernicus presented his heliocentric cosmology, including supporting observations and calculations, in a book entitled *De revolutionibus orbium coelestium* (On the Revolutions of the Celestial Spheres), which was published in 1543, the year of his death. Copernicus's great insight was the simplicity of a heliocentric cosmology compared to geocentric views. However, Copernicus incorrectly assumed that the planets travel along circular paths around the Sun. As a result, his predictions were no more accurate than those of a geocentric theory!

2-2   Tycho Brahe made astronomical observations that disproved ancient ideas about the heavens

In November 1572, a bright star suddenly appeared in the constellation Cassiopeia. At first it was even brighter than Venus, but then it began to grow dim. After 18 months it faded from view.

Modern astronomers recognize this event as a supernova explosion, the violent death of a certain type of star (see Chapter 12). In the sixteenth century, however, the prevailing opinion was quite different. Teachings dating back to Aristotle and Plato argued that the heavens were permanent and unalterable. Consequently, the "new star" of 1572 could not really be a star at all, because the heavens do not change. It must instead be some sort of bright object quite near Earth, perhaps not much farther away than the clouds overhead. A 25-year-old Danish astronomer named Tycho Brahe realized that straightforward observations might reveal the distance to this object.

WEB LINK 2.3   It is everyone's common experience that when you walk from one place to another, nearby objects appear to change position against the background of more distant objects. Furthermore, the closer an object is, the more you have to change the angle at which you observe it as you move. This apparent change in position of the object with the changing position of the observer is called **parallax** (Figure 2-5).

Tycho reasoned as follows: If the new star is nearby, its position should shift against the background stars over the course of a night, as shown in Figure 2-6a. His careful observations failed to disclose any parallax, and so the new star had to be far away, farther from Earth than anyone had imagined (Figure 2-6b). Tycho summarized his findings in a small book, *De stella nova* (On the New Star), published in 1573.

Tycho's astronomical records were soon to play an important role in the development of a heliocentric cosmology. From 1576 to 1597, Tycho made comprehensive observations, measuring planetary positions with an accuracy of 1 arcminute, about as precise as is possible with the naked

FIGURE 2-5 **Parallax** Nearby objects are viewed at different angles from different places. These objects also appear to be in a different place with respect to more distant objects when viewed at the same time by observers located at different positions. Both effects are called parallax, and they are used by astronomers, surveyors, and sailors to determine distances.

eye. Arcminute (arcmin) is defined in An Astronomer's Toolbox 1-1. Upon his death in 1601, most of these invaluable records were given to his gifted assistant, Johannes Kepler.

KEPLER'S AND NEWTON'S LAWS

Until Johannes Kepler's time, astronomers had assumed that heavenly objects move in circles. For philosophical and aesthetic reasons, circles were considered the most perfect and most harmonious of all geometric shapes. However, using circular orbits failed to yield accurate predictions for the positions of the planets. For years, Kepler tried to find a shape for orbits that would fit Tycho Brahe's observations of the planets' positions against the background of distant stars. Finally, he began working with a geometric form called an **ellipse.**

2-3 Kepler's laws describe orbital shapes, changing speeds, and the lengths of planetary years

An ellipse can be drawn as shown in Figure 2-7a. Each thumbtack is at a **focus** (plural **foci**). The longest diameter across an ellipse, called the *major axis*, passes through both foci. Half of that distance is called the **semimajor axis,** whose length is usually designated by the letter *a*. In astronomy, the length of the semimajor axis is also the average distance between a planet and the Sun.

To Kepler's delight, the ellipse turned out to be the curve he had been searching for. Predictions of the locations of planets based on elliptical paths were in very close agreement with where the planets actually were. He published this discovery in 1609 in a book known today as *New*

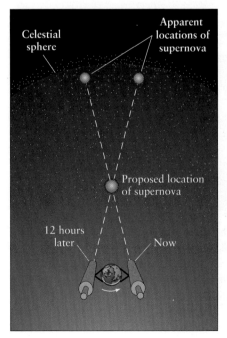

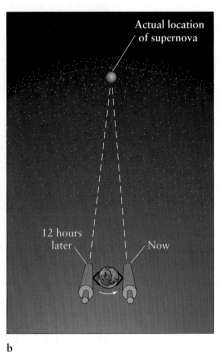

FIGURE 2-6 **The Parallax of a Nearby Object in Space** Tycho thought that the Earth doesn't rotate and that the stars revolve around it. From our modern perspective, the changing position of the supernova would be due to the Earth's rotation as shown in this figure. **(a)** Tycho Brahe argued that if an object is near the Earth, its position relative to the background stars should change over the course of a night. **(b)** Tycho failed to measure such changes for the supernova in 1572. This is illustrated in **(b)** by the two telescopes being parallel to each other. He therefore concluded that the object was far from the Earth.

a   b

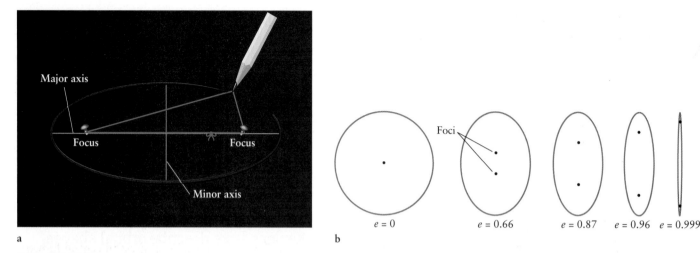

a   b

FIGURE 2-7 Ellipse **(a)** The construction of an ellipse: An ellipse can be drawn with a pencil, a loop of string, and two thumbtacks, as shown in this figure. If the string is kept taut, the pencil traces out an ellipse. The two thumbtacks are located at the two foci of the ellipse. **(b)** A series of ellipses with different eccentricities, e. Eccentricities range between 0 (circle) to just under 1.0 (almost a straight line).

Astronomy. This important discovery is now considered the first of **Kepler's laws:**

The orbit of a planet about the Sun is an ellipse with the Sun at one focus.

Ellipses have two extremes. The roundest ellipse is a circle. The most elongated ellipse approaches being a straight line. The shape of a planet's orbit around the Sun is described by its **orbital eccentricity,** designated by the letter e, which ranges from 0 (circular orbit) to just under 1.0 (nearly a straight line). Figure 2-7b shows a sequence of ellipses and their associated eccentricities. Observations have since revealed that there is no object at the second focus of each elliptical planetary orbit.

Tycho's observations also showed Kepler that planets do not move at uniform speeds along their orbits. Rather, a planet moves most rapidly when it is nearest the Sun, a point on its orbit called **perihelion.** Conversely, a planet moves most slowly when it is farthest from the Sun, a point called **aphelion.**

After much trial and error, Kepler discovered a way to describe how fast a planet moves anywhere along its orbit. This discovery, also published in *New Astronomy*, is illustrated in Figure 2-8. Suppose that it takes 30 days for a planet to go from point A to point B. During that time, the line joining the Sun and the planet sweeps out a nearly triangular area (shaded in Figure 2-8). Kepler discovered that the line joining the Sun and the planet sweeps out the same area during any other 30-day interval. In other words, if the planet also takes 30 days to go from point C to point D, then the two shaded segments in Figure 2-8 are equal in

area. Kepler's second law, also called the **law of equal areas,** can be stated thus:

A line joining a planet and the Sun sweeps out equal areas in equal intervals of time.

The physical content of Kepler's second law is that each planet's speed decreases as it moves from perihelion to aphe-

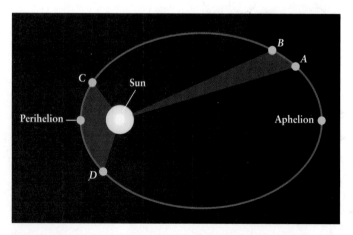

FIGURE 2-8 Kepler's First and Second Laws According to Kepler's first law, every planet travels around the Sun along an elliptical orbit with the Sun at one focus. According to his second law, the line joining the planet and the Sun sweeps out equal areas in equal intervals of time. *Note:* This drawing shows a highly elliptical orbit, with e = 0.74. Even though this is a much greater eccentricity than that of any planet in the solar system, the concept still applies to all planets and other orbiting bodies.

lion. The speed then increases as the planet moves from aphelion toward perihelion.

Kepler was also able to relate a planet's year to its distance from the Sun. This discovery, published in 1619, stands out because of its impact on future developments in astronomy. Now called *Kepler's third law*, it predicts the planet's sidereal period if we know the length of the semimajor axis of the planet's orbit:

The square of a planet's sidereal period around the Sun is directly proportional to the cube of the length of its orbit's semimajor axis.

The relationship is easiest to use if we let *P* represent the sidereal period in Earth years and *a* represent the length of the semimajor axis measured in astronomical units (as we discussed in An Astronomer's Toolbox I-2). Now we can write Kepler's third law as

$$P^2 = a^3$$

In other words, a planet closer to the Sun has a shorter year than does a planet farther from the Sun. Combining this with the second law reveals that planets closer to the Sun move more rapidly than those farther away. Using data from Tables 2-1 and 2-2, we can demonstrate Kepler's third law as shown in Table 2-3.

When Newton derived Kepler's third law using the law of gravitation, discussed below, he discovered that there is contribution to the period from the mass of the planet. However, this correction is vanishingly small for all the planets in the solar system, which is why the above equation, as shown in Table 2-3, gives such good results for the planets' orbits.

Insight into Science   Theories and explanations Scientific theories (or laws) based on observations can be useful for making predictions even if the reasons that these laws work are unknown. The explanation for Kepler's laws came decades after Kepler deduced them, in 1665, when Newton applied his mathematical expression for gravitation, the force that holds the planets in their orbits.

2-4 Galileo's discoveries strongly supported a heliocentric cosmology

While Kepler was making rapid progress in central Europe, an Italian physicist was making equally dramatic observations in southern Europe. Galileo Galilei did not invent the telescope, but he was one of the first people to point the new device toward the sky and publish his observations. He saw things that no one had ever imagined—mountains on the Moon and sunspots on the Sun. He also discovered that Venus exhibits phases.

After only a few months of observation, Galileo noticed that the apparent size of Venus as seen through his telescope was related to the planet's phase. Venus appears smallest at gibbous phase and largest at crescent phase. A geocentric cosmology could not explain why, but a heliocentric cosmology does. Galileo's observations therefore supported the conclusion that Venus orbits the Sun, not the Earth (Figure 2-9).

In 1610, Galileo also discovered four moons near Jupiter. In honor of their discoverer, these are today called the **Galilean moons** (or **satellites**, another term for moon).

TABLE 2-3   A Demonstration of Kepler's Third Law

	Sidereal period P (yr)	Semimajor axis a (AU)	P^2	=	a^3
Mercury	0.24	0.39	0.06		0.06
Venus	0.61	0.72	0.37		0.37
Earth	1.00	1.00	1.00		1.00
Mars	1.88	1.52	3.53		3.51
Jupiter	11.86	5.20	140.7		140.6
Saturn	29.46	9.54	867.9		868.3
Uranus	84.01	19.19	7,058		7,067
Neptune	164.79	30.06	27,160		27,160
Pluto	248.54	39.53	61,770		61,770

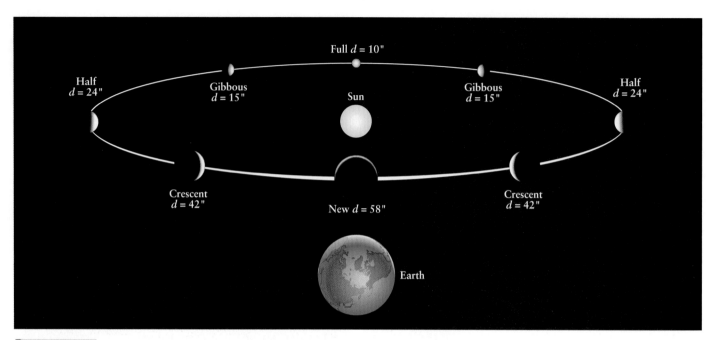

Full *d* = 10"

Half
d = 24"

Gibbous
d = 15"

Gibbous
d = 15"

Half
d = 24"

Sun

Crescent
d = 42"

Crescent
d = 42"

New *d* = 58"

Earth

FIGURE 2-9   **The Changing Appearance of Venus** This figure shows how the appearance (phase) of Venus changes as it moves along its orbit. The number below each view is the angular diameter (*d*) of the planet as seen from Earth, in arcseconds. Note that the phases correlate with the planet's angular size and its angular distance from the Sun, both as seen from Earth. These observations clearly support the idea that Venus orbits the Sun.

Galileo concluded that the moons are orbiting Jupiter because they move across from one side of the planet to the other. Confirming observations were made in 1620 (Figure 2-10). These observations all provided further proof that the Earth is not at the center of the universe. Like the Earth in orbit around the Sun, Jupiter's four moons obey Kepler's third law: The square of a moon's orbital period about Jupiter is directly proportional to the cube of its average distance from the planet.

Galileo's telescopic observations constituted the first fundamentally new astronomical data since humans began recording what they saw in the sky. In contradiction to then-prevailing opinions, these discoveries strongly supported a heliocentric view of the universe. Because Galileo's ideas could not be reconciled with certain passages in the Bible or with the writings of Aristotle and Plato, the Roman Catholic Church condemned him, and he was forced to spend his latter years under house arrest "for vehement suspicion of heresy."

A major stumbling block prevented seventeenth-century thinkers from accepting Kepler's laws and Galileo's conclusions about the heliocentric cosmology. At that time, the relationships between matter, motion, and forces were not understood. People did not know about the gravitational force of the Sun, which keeps the planets in orbit. They did not know how planets, since they had started in orbit around the Sun, could keep moving. Once anything on

Earth is put in motion, it quickly comes to rest. Why didn't the planets orbiting the Sun stop, too?

All those mysteries were soon explained by the brilliant and eccentric scientist Isaac Newton, who was born on Christmas Day in 1642, less than a year after Galileo died. In the decades that followed, Newton revolutionized science more profoundly than any person before him, and in doing so, he found physical and mathematical proofs of the heliocentric cosmology.

2-5 Newton formulated three laws that describe fundamental properties of physical reality

 Until the mid-seventeenth century, virtually all mathematical astronomy used the same approach. Astronomers from Ptolemy to Kepler worked *empirically*, that is, directly from data and observations. They adjusted their ideas and calculations until they finally came out with the right answers.

Isaac Newton introduced a new approach. He made just three assumptions, now called **Newton's laws of motion,** which he applied to all forces and bodies. He also found a formula for the force of **gravity,** the attraction between all objects due to their masses. Newton then showed that Kepler's three laws mathematically and, hence, physically, fol-

b   R I V U X G

FIGURE 2-10   Jupiter and Its Largest Moons   (a) In 1610 Galileo discovered four "stars" that move back and forth across Jupiter. He concluded that they are four moons that orbit Jupiter just as our Moon orbits the Earth. This figure shows observations made by Jesuits in 1620. (b) This photograph, taken by amateur astronomer C. Holmes, shows the four Galilean satellites alongside an overexposed image of Jupiter. Each satellite would be bright enough to be seen with the unaided eye were it not overwhelmed by the glare of Jupiter. (b: Courtesy of C. Holmes)

a

low from these laws. Using his formula, Newton accurately described the observed orbits of the Moon, comets, and other objects in the solar system. His laws also apply to the motions of all bodies on the Earth.

Newton's first law, the **law of inertia,** states that

A body remains at rest or moves in a straight line at a constant speed unless acted upon by a net outside force.

At first, this law might seem to conflict with your everyday experience. For example, if you shove a chair, it does not continue at a constant speed forever but rather it comes to rest after sliding only a short distance. From Newton's viewpoint, however, a "net outside force" does indeed act on the moving chair, namely, friction between the chair's legs and the floor. Without friction, the chair would continue in a straight path at a constant speed. A **force** changes the motion of an object.

Newton's first law tells us that there must be an outside force acting on the planets. If there were no force acting on them, they would move away from the Sun along straight-line paths at constant speeds. In other words, they would leave their curved orbits. Because this does not happen, Newton concluded that some force confines the planets to their elliptical orbits. As we shall see, that force is gravity.

Newton's second assumption describes how a force changes the motion of an object. To appreciate the concepts

of force and motion better, we must first understand the quantities that describe motion: speed, velocity, and acceleration.

Imagine an object in space. Push on the object and it begins to move. At any moment, you can describe the object's motion by specifying both its speed and direction. Speed and direction of motion together constitute the object's **velocity.** If you continue to push on the object, its speed will increase—it will accelerate.

Acceleration is the rate at which velocity changes with time. Because velocity involves both speed and direction, a slowing down, a speeding up, or a change in direction are all types of acceleration.

Suppose an object revolved about the Sun in a perfectly circular orbit. This body would have acceleration that involved only a change of direction. As this object moved along its orbit, its speed would remain constant, but its direction of motion would be continuously changing. Therefore, it would still be continuously accelerating.

Newton's second law says that the acceleration of an object is proportional to the force acting on it. In other words, the harder you push on an object, the greater the resulting acceleration. Newton's law also says that a greater mass pushed or pulled by a force accelerates more slowly than does an object of lesser mass pushed or pulled by the same force. That is why by pushing a child's wagon, you can accelerate it faster than you can accelerate a car by pushing

MOVIE MISCONCEPTIONS

Planet of the Apes (Twentieth Century Fox, 2001)
(The Kobal Collection)

irector Tim Burton rein-vents Pierre Boulle's clas-sic novel in the newest *Planet of the Apes* movie. Set in the year 2029, the film opens on a space station orbiting Saturn. The station is shown to not be rotating. We know this because the actors are standing and walking on the disk-shaped floors, occasionally looking out windows along the edge of the station. (If the station were rotating, the actors would be standing facing inward on the outer, curved edge of the structure, as was depicted in the Stanley Kubrick movie *2001: A Space Odyssey*). The actors wear normal shoes, and things lying on tables remain in place even when the station undergoes slight jostling.

A short time into the movie the hero enters a vehi-cle, puts on a helmet that rests on his jacket, and leaves the space station. It is worth noting that space helmets are designed to be part of airtight spacesuits that protect astronauts in case their spacecraft loses air pressure.

Describe the two scientific errors in *Planet of the Apes* as described in the paragraphs above.

(Answers appear at the end of the book.)

---

on it. Newton's second law can be succinctly stated as an equation. If a force acts on an object, the object will experi-ence an acceleration such that

$$\text{Force} = \text{mass} \times \text{acceleration}$$

The **mass** of an object is a measure of the total amount of material in the object, which we measure in kilograms. For example, the mass of the Sun is 2×10^{30} kg, the mass of a hydrogen atom is 1.7×10^{-27} kg, and the mass of the author of this book is 83 kg. At rest, the Sun, a hydrogen atom, and I have these same masses regardless of where we happen to be in the universe.

It is important not to confuse the concept of mass with that of weight. **Weight** is the force with which an object is pulled down while on the ground (due to gravity's attrac-tion) or, equivalently, feels inside an accelerating rocket. Force is usually expressed in pounds or newtons. For exam-ple, the force with which I am pressing down on the ground is 183 pounds.

But I weigh 183 pounds only on the Earth. I would weigh less on the Moon. Orbiting in the Space Shuttle, my apparent weight (measured by standing on a scale in the shuttle), would be zero, but my mass would be the same as when I am on Earth. An astronaut in the shuttle would still have to push me with a force to get me to move. Whenever we describe the properties of planets, stars, or galaxies, we speak of their masses, never of their weights.

Newton's final assumption, called *Newton's third law,* is the famous *law of action and reaction:*

Whenever one body exerts a force on a second body, the sec-ond body exerts an equal and opposite force on the first body.

For example, I weigh 183 pounds, and so I press down on the floor with a force of 183 pounds. Newton's third law says that the floor is also pushing up against me with an equal force of 183 pounds. (If it was less, I would fall through the floor, and if it was more, I would be lifted up-ward.) In the same way, Newton realized that because the Sun is exerting a force on each planet to keep it in orbit, each planet must also be exerting an equal and opposite force on the Sun. As each planet accelerates toward the Sun, the Sun in turn accelerates toward each planet.

The Sun is pulling the planets, so why don't they fall onto it? **Conservation of angular momentum,** a fundamental con-sequence of Newton's second law of motion, provides the answer. Angular momentum is a measure of how much energy is stored in an object due to its rotation and revolution (presented in An Astronomer's Toolbox 2-1). As the orbiting planets fall Sunward, their angular momentum provides them with motion perpendicular to that infall, meaning that plan-ets continually fall toward the Sun, but they continually miss it. Because their angular momentum is conserved, planets nei-ther spiral into the Sun or away from it. Angular momentum remains constant unless acted on by an outside torque.

Angular momentum depends on three things: how fast the body rotates or revolves, how much mass it has, and how spread out that mass is. The greater a body's angular motion or mass, or the more the mass is spread out, the greater its angular momentum. Consider, for example, a twirling ice

AN ASTRONOMER'S TOOLBOX 2-1

Energy and Momentum

Scientists identify two types of energy that are available to any object. The first, called **kinetic energy,** is associated with the object's motion. For speeds much less than the speed of light, we can write the amount of kinetic energy, KE, in an object as:

$$KE = \tfrac{1}{2}mv^2$$

where m is the object's mass (total number of particles) and v is its velocity. Kinetic energy is a measure of how much work the object can do on the outside world or, equivalently, how much work the outside world has done to put the object in motion.

Work is also a rigorously defined concept that often is at odds with our intuition. It is defined as the product of the force, F, acting on an object times the distance, d, over which the object moves in the direction of the force:

$$W = Fd$$

For example, if I exert a horizontal force of 50 newtons (a unit of force) and thereby move an object 10 meters in that direction, then I have done 50 N × 10 m = 500 joules of work. (I have used the relationship that 1 newton × 1 meter = 1 joule.)

The second type of energy is called **potential energy.** It represents how much energy is stored in an object as a result of its location in space. For example, if you hold a pencil above the ground, the pencil has potential energy that can be converted into kinetic energy by the Earth's gravitational force. How does that conversion get underway? Just let go of the pencil.

There are various kinds of potential energy, such as the potential energy stored in a battery and the potential energy stored in objects under the influence of gravity. We will focus on *gravitational potential energy.* Far from extremely massive objects, like stars, or extremely dense objects, like black holes, gravitational potential energy can be written as:

$$PE = GmM/r$$

where $G = 6.668 \times 10^{-11}$ N m²/kg², m is the mass of the object whose gravitational potential energy you are measuring, M is the mass of the object creating the gravitational potential energy, and r is the distance between the centers of masses of the object feeling the gravitational potential energy and the object creating that energy.

Near the surface of the Earth, this equation simplifies to:

$$PE = mgh$$

where $g = 9.8$ m/s² is the gravitational acceleration at the Earth's surface, and h is the height of the object above the Earth's surface.

Potential energy can be converted into kinetic energy and vice versa. By dropping the pencil, its gravitational potential energy begins decreasing while its kinetic energy begins increasing *at the same rate.* The pencil's total energy is conserved. Conversely, if you throw a pencil up in the air, the kinetic energy you give it will immediately begin decreasing, while its potential energy increases at the same rate.

Related to the motion of an object, and hence to its kinetic energy, are the concepts of linear momentum, usually just called **momentum,** and **angular momentum.** Momentum, p, is described by the equation:

$$\mathbf{p} = m\mathbf{v}$$

where v is the velocity of the object. Both p and v are in boldface to indicate that they both represent motion in *some direction* or another, as well as having some numerical value. Simple algebra reveals that kinetic energy and momentum are related by:

$$KE = p^2/2m$$

Linear momentum, then, indicates how much energy is stored in an object because of its motion in a straight line (its linear motion).

Angular momentum, L, can be expressed mathematically as:

$$L = I\omega$$

where I is the moment of inertia of an object and ω gives the speed and direction in which it is revolving or rotating. Just as the mass indicates how hard it is to change an object's straight-line motion, the moment of inertia indicates how hard it is to change the rate at which an object rotates or revolves. The moment of inertia depends on an object's mass and shape. Kinetic energy stored in angular motion can be written:

$$KE = L^2/2I$$

Newton's *first law* can also be expressed in terms of conservation of linear momentum:

A body maintains its linear momentum unless acted upon by a net external force.

Equivalently, for angular motion we can write the conservation of angular momentum:

(continued on the following page)

AN ASTRONOMER'S TOOLBOX 2-1 (continued)

A body maintains its angular momentum unless acted upon by a net external torque.

Torques are created when a force acts on an object in some direction other than toward the center of the object's angular motion, as shown in the figure. The Earth has angular momentum and keeps spinning on its rotational axis and orbiting around the Sun. Likewise, the Moon has angular momentum and keeps spinning on its rotation axis and orbiting the Earth. Virtually all objects in astronomy have angular momentum, and I think it's fair to say that conservation of angular momentum is among the most important laws in the cosmos. After all, it is what keeps the planets in orbit around the Sun, the moons in orbit around the plan-

ets, the astronomical bodies rotating at relatively constant rates, and many other rotation-related effects that we will encounter throughout this book.

Try these questions: How does tripling the linear momentum of an object change its kinetic energy? How does halving the angular momentum of an object change its kinetic energy? How much work would you do if you pushed on a desk with a force of 100 N and moved it 20 m? How much work would you do if you pushed on a desk with a force of 500 N and moved it 0 m? What two things can you vary to change the angular momentum of an object?

(Answers appear at the end of the book.)

Angular Momentum and Torque (a) When a force acts through an object's rotation axis or toward its center of mass, then the force does *not* exert a torque on the object. **(b)** When a force acts in some other direction, then it exerts a torque, causing the body's angular momentum to change. If the object can spin around a fixed axis, like a globe, then the rotation axis is the rod running through it. If the object is not held in place, then the rotation axis is in a line through a point called the object's center of mass. The center of mass of any object is the point that follows an elliptical, parabolic, or hyperbolic path in a gravitational field. All other points in the spinning object wobble as it moves.

skater. She rotates with a constant mass, practically free of outside forces. When she wishes to rotate more rapidly, she decreases the spread of her mass distribution by pulling her arms in closer to her body. According to the conservation of angular momentum, as the spread of mass decreases, the rotation rate must increase. In astronomy, we encounter many instances of the same law, as giant objects, like stars, contract.

We have now reconstructed the central relationships between matter and motion. Scientific belief in the heliocentric cosmology still requires a force to hold the planets in orbit around the Sun and the moons in orbit around the planets. Newton identified that, too.

2-6 Newton's description of gravity accounts for Kepler's laws

Isaac Newton did not invent the idea of gravity. An educated seventeenth-century person would understand that some force pulls things down to the ground. It was Newton, however, who gave us a precise description of the action of gravity, or *gravitation*, as it is more properly called. Using his

first law, Newton proved mathematically that the force acting on each of the planets is directed toward the Sun. This discovery led him to suspect that the nature of the force pulling a falling apple straight down to the ground is the same as the nature of the force on the planets that is always aimed straight at the Sun.

Newton succeeded in formulating a mathematical model describing the behavior of the gravitational force that keeps the planets in their orbits (presented in An Astronomer's Toolbox 2-2). Newton's **universal law of gravitation** states:

Two bodies attract each other with a force that is directly proportional to the product of their masses and inversely proportional to the square of the distance between them.

In other words, gravitational force decreases with distance: Move twice as far away from a body and you feel only one-quarter of the force from it that you felt before.

Using his law of gravity, Newton found that he could mathematically explain Kepler's three laws. For example, whereas Kepler discovered by trial and error that the period of orbit, *P*, and average distance between the Sun and planet, *a*,

GUIDED DISCOVERY   Astronomy's Foundation Builders

In the two centuries between 1500 and 1700, human understanding of the motion of celestial bodies and the nature of the gravitational force that keeps them in orbit surged forward as never before. Theories related to this subject were developed by brilliant thinkers, whose work established and verified the heliocentric model of the solar system and the role of gravity.

Nicolaus Copernicus (1473–1543) Copernicus was born in Torun, Poland, the youngest of four children. He pursued his higher education in Italy, where he received a doctorate in canon law and studied medicine. Copernicus developed a heliocentric theory of the known universe, and published his work in 1543 under the title *De revolu-*

(E. Lessing/Art Resource)

tionibus orbium coelestium. Copernicus became ill and died just after receiving the first copy.

Tycho Brahe (1546–1601) and **Johannes Kepler (1571–1630)** Tycho, depicted within a portrait of Kepler, was born to nobility in the Danish city of Knudstrup, which is now part of Sweden. At age 20 he lost part of his nose in a duel and wore a metal replacement thereafter. In 1576 the Danish king Frederick II built Tycho an astronomical observatory that Tycho named Uraniborg. He rejected Copernicus's heliocentric theory

(Painting by Jean-Leon Huens, courtesy of National Geographic Society)

and the Ptolemaic geocentric system and devised a halfway theory called the "Tychonic system." According to Tycho's theory, the Earth is stationary, with the Sun and Moon revolving around it, while all the other planets revolve around the Sun. Tycho died in 1601.

Kepler was educated in Germany, where he spent three years studying mathematics, philosophy, and theology. In 1596, Kepler published a booklet in which he attempted to mathematically predict the planetary orbits.

Although his theory was altogether wrong, its boldness and originality attracted the attention of Tycho Brahe, whose staff Kepler joined in 1600. Kepler deduced his three laws from Tycho's observations.

Galileo Galilei (1564–1642) Born in Pisa, Italy, Galileo studied medicine and philosophy at the University of Pisa. He abandoned medicine in favor of mathematics. He held the chair of mathematics at the University of Padua and eventually returned to the University of Pisa as a professor of mathematics. There

(Art Resource)

Galileo formulated his famous law of falling bodies: All objects fall with the same acceleration regardless of their weight. In 1609 he constructed a telescope and made a host of discoveries that contradicted the teachings of Aristotle and the Roman Catholic Church. He summed up his life's work on motion, acceleration, and gravity in the book *Dialogues Concerning Two New Sciences,* published in 1632.

Isaac Newton (1642–1727) Although he delighted in constructing mechanical devices—sundials, model windmills, a water clock, and a mechanical carriage—Newton showed no exceptional academic ability at Cambridge University, where he received a bachelor's degree in 1665. While

(National Portrait Gallery, London)

pursuing experiments in optics, Newton constructed a reflecting telescope and also discovered that white light is actually a mixture of all colors. His major work on forces and gravitation was the tome *Philosophiae Naturalis Principia Mathematica,* which appeared in 1687. In 1704, Newton published his second great treatise, *Opticks,* in which he described his experiments and theories about light and color. Upon his death in 1727, Newton was buried in Westminster Abbey, the first scientist to be so honored.

---

are related by $P^2 = a^3$, Newton demonstrated mathematically that this equation (corrected with a tiny contribution due to the mass of the planet) follows from his law of gravitation.

Newton's version of Kepler's third law can be easily recast to predict the orbits of any objects under the influence of a gravitational attraction. Indeed, all three of Kepler's laws apply to all orbiting bodies. This includes moons orbit-

ing planets, artificial satellites orbiting the Earth, and even two stars revolving about each other. Throughout this book, we will see that Kepler's three laws have a wide range of practical applications.

Newton also discovered that some objects have nonelliptical orbits around the Sun. His equations led him to conclude that the orbits of some objects are **parabolas** and

AN ASTRONOMER'S TOOLBOX 2-2

Gravitational Force

From Newton's law of gravitation, if two objects having masses m_1 and m_2 are separated by a distance r, then the gravitational force F between them is

$$F = G(m_1 m_2 / r^2)$$

In this formula, G is the **universal constant of gravitation,** whose value has been determined from laboratory experiments:

$$G = 6.668 \times 10^{-11} \text{ N m}^2 \text{ kg}^{-2}$$

where N is the unit of force, a newton.

The equation $F = G(m_1 m_2 / r^2)$ gives, for example, the force from the Sun on the Earth and, equivalently, from the Earth on the Sun. If m_1 is the mass of the Earth (6.0×10^{24} kg), m_2 is the mass of the Sun (2.0×10^{30} kg), and r is the distance from the Earth to the Sun (1.5×10^{11} m),

$$F = 3.6 \times 10^{22} \text{ N}$$

This number can then be used in Newton's second law, $F = ma$, to find the acceleration of the Earth due to the Sun. This yields:

$$a_{Earth} = F/m_1 = 6.0 \times 10^{-3} \text{ m/s}^2$$

Newton's third law says that the Earth exerts the same force on the Sun, so the Sun's acceleration due to the Earth's gravitational force is

$$a_{Sun} = F/m_2 = 1.8 \times 10^{-8} \text{ m/s}^2$$

In other words, the Earth pulls on the Sun, causing the Sun to move toward it. Because of the Sun's greater mass, however, the amount that the Sun accelerates the Earth is more than 300,000 times the amount that the Earth accelerates the Sun.

Try these questions: The Earth's radius is 6.4×10^6 m and 1 kg = 2.2 lb. What is the force that the Earth exerts on you? What is the force that you exert on the Earth? What is the Earth's acceleration on you? What would the Sun's force be on the Earth, if our planet were twice as far from the Sun as it is? How does that force compare to the force from the Sun at our present location?

(Answers appear at the end of the book.)

---

hyperbolas (Figure 2-11). For example, comets hurtling toward the Sun from the depths of space often follow parabolic or hyperbolic orbits.

Newton's ideas turned out to be applicable in an incredibly wide range of situations. The orbits of the planets and their satellites could now be calculated with unprecedented precision. Using Newton's laws, mathematicians proved that the Earth's axis of rotation must precess because of the gravitational pull of the Moon and the Sun on the Earth's equatorial bulge (recall Figure 1-17).

In addition, Newton's laws and mathematical techniques were used to predict new phenomena. Newton's friend, Edmund Halley, was intrigued by historical records of a comet that was sighted about every

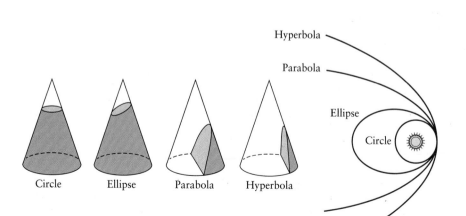

FIGURE 2-11 Conic Sections
A conic section is any one of a family of curves obtained by slicing a cone with a plane, as shown in this figure. The orbit of one body about another can be an ellipse, a parabola, or a hyperbola. Circular orbits are possible because a circle is just an ellipse for which both foci are at the same point.

FIGURE 2-12 **Halley's Comet** Halley's Comet orbits the Sun with an average period of about 76 years. During the twentieth century, the comet passed near the Sun twice—once in 1910 and again, shown here, in 1986. The comet will pass close to the Sun again in 2061. While dim in 1986, it nevertheless spread more than 5° across the sky, or 10 times the diameter of the Moon. (Science Photo Library)

76 years. Using Newton's methods, Halley worked out the details of the comet's orbit and predicted its return in 1758. It was first sighted on Christmas night of that year, and to this day the comet bears Halley's name (Figure 2-12).

Perhaps the most dramatic confirmation of Newton's ideas was their role in the discovery of the eighth planet in our solar system. The seventh planet, Uranus, had been discovered accidentally by William Herschel in 1781 during a telescopic survey of the sky. Fifty years later, however, it was clear that Uranus was not following the orbit predicted by Newton's laws. Two mathematicians, John Couch Adams in England and Urbain-Jean-Joseph Leverrier in France, independently calculated that the deviations of Uranus from its predicted orbit could be explained by the gravitational pull of a then—unknown, more distant planet. Each man predicted that the planet would be found at a certain location in the constellation of Aquarius in September 1846. A telescopic search on September 23, 1846, revealed Neptune less than 1° from its calculated position. Although sighted with a telescope, Neptune was really discovered with pencil and paper (Figure 2-13).

> **Insight into Science**   **Quantify predictions** Mathematics provides a language that enables science to make quantitative predictions that can be checked by anyone. For example, we have seen in this chapter how Kepler's third law and Newton's universal law of gravitation correctly predict the motion of objects under the influence of the Sun's gravitational attraction.

Over the years, Newton's ideas were successfully used to predict and explain motion here on Earth and throughout the universe. Even today, as we send astronauts into Earth orbit and send probes to the outer planets, Newton's equations are used to calculate the orbits and trajectories of the spacecraft.

It is a testament to Newton's genius that his three laws were precisely the basic ideas needed to understand the motions of the planets. Newton brought a new dimension of elegance and sophistication to our understanding of the workings of the universe.

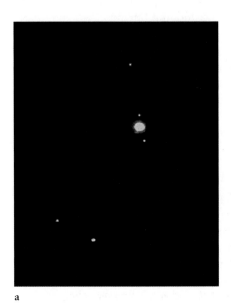

a

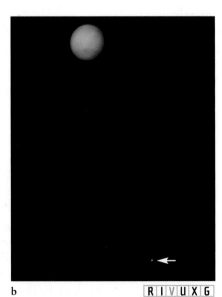

b   R I V U X G

FIGURE 2-13 **Uranus and Neptune** The discovery of Neptune was a major triumph for Newton's laws. In an effort to explain why Uranus (shown on the left, with two of its moons) deviated from its predicted orbit, astronomers predicted the existence of Neptune (shown on the right, with one of its moons indicated by an arrow). Uranus and Neptune are nearly the same size; both have diameters about 4 times that of Earth. (a: John Chumack/Photo Researchers; b: NASA)

Frontiers yet to be discovered

The science related to forces and orbits described in this chapter was well established by the beginning of the nineteenth century. However, questions remained. Careful observation revealed that Newton's law of gravitation gave very slightly inaccurate predictions for the orbital path of Mercury. We will see in Chapter 13 that this problem was resolved by Albert Einstein. Nevertheless, one fundamental question from this chapter's material remains to be answered: What is "mass"? Many physicists studying the building blocks of matter (such as the elementary particles protons, neutrons, and electrons) believe that mass comes from the presence of a particle permeating the universe called the Higgs boson. When elementary particles interact with it, they gain the property we call mass. Searches for Higgs bosons are underway at high-energy particle accelerators such as CERN, near Geneva, Switzerland, and the Fermi National Accelerator in Illinois. If these experiments are successful, then we may know the origin of mass by the end of this decade.

 Further Reading on These Topics

WHAT DID YOU KNOW?

|1 *What is the shape of the Earth's orbit around the Sun?* All planets have elliptical orbits around the Sun.

|2 *Do the planets orbit the Sun at constant speeds?* No. The closer a planet is to the Sun in its orbit, the faster it is moving. It moves fastest at perihelion and slowest at aphelion.

|3 *Do the planets all orbit the Sun at the same speed?* No. A planet's speed depends on its average distance from the Sun. The closest planet moves fastest, the most distant planet moves slowest.

|4 *How much force does it take to keep an object moving in a straight line at a constant speed?* Unless an object is subject to an outside force, like friction, it takes no force at all to keep it moving in a straight line at a constant speed.

|5 *How does an object's mass differ when measured on Earth and on the Moon?* Assuming the object doesn't shed or collect pieces, its mass remains constant whether on the Earth or on the Moon. Its weight, however, is less on the Moon.

KEY WORDS

acceleration, 53
angular momentum, 55
aphelion, 40
configuration (of a planet), 45
conjunction, 47
conservation of angular momentum, 54
cosmology, 44
deferent, 46
direct motion, 44
ellipse, 49
elongation, 47
epicycle, 46
focus (of an ellipse), 49
force, 53
Galilean moons (satellites), 51

geocentric cosmology, 44
gravity, 52
greatest elongation, 47
heliocentric cosmology, 44
hyperbola, 57
inferior conjunction, 47
Kepler's laws, 50
kinetic energy, 55
law of equal areas, 50
law of inertia, 53
mass, 54
momentum, 55
Newton's laws of motion, 52
Occam's razor, 44
opposition, 47

orbital eccentricity, 50
parabola, 57
parallax, 48
perihelion, 50
potential energy, 55
retrograde motion, 44
semimajor axis (of an ellipse), 49
sidereal period, 47
superior conjunction, 47
synodic period, 57
universal constant of gravitation, 58
universal law of gravitation, 56
velocity, 53
weight, 54
work, 55

KEY IDEAS

• The ancient Greeks laid the groundwork for progress in science. Early Greek astronomers devised a geocentric cosmology, which placed the Earth at the center of the universe.

Origins of a Sun-Centered Universe

• Copernicus's heliocentric (Sun-centered) theory simplified the general explanation of planetary motions compared to the geocentric theory.

• In a heliocentric cosmology, the Earth is but one of several planets that orbit the Sun.

• The sidereal orbital period of a planet is measured with respect to the stars. It determines the length of the planet's year. Its synodic period is measured with respect to the Sun as seen from the moving Earth (for example, from one opposition to the next).

Kepler's and Newton's Laws

• Ellipses describe the paths of the planets around the Sun much more accurately than do circles. Kepler's three laws give important details about elliptical orbits.

• The invention of the telescope led Galileo to new discoveries, such as the phases of Venus and the moons of Jupiter, that supported a heliocentric view of the universe.

• Newton based his explanation of the universe on three assumptions now called Newton's laws of motion. These laws and his universal law of gravitation can be used to deduce Kepler's laws and to describe planetary motions with extreme accuracy.

• The mass of an object is a measure of the amount of matter in the object; its weight is a measure of the force with which the gravity of some other object pulls on the object's mass.

• In general, the path of one astronomical object about another, such as that of a comet about the Sun, is an ellipse, a parabola, or a hyperbola.

REVIEW QUESTIONS

1 How did Copernicus explain the retrograde motions of the planets?

2 Which planets can never be seen at opposition? Which planets can never be seen at inferior conjunction?

3 At what configuration (superior conjunction, greatest eastern elongation, etc.) would it be best to observe Mercury or Venus with an Earth-based telescope? At what configuration would it be best to observe Mars, Jupiter, or Saturn? Explain your answers.

4 What are the synodic and sidereal periods of a planet?

5 What are Kepler's three laws? Why are they important?

6 In what ways did the astronomical observations of Galileo support a heliocentric cosmology?

7 How did Newton's approach to understanding planetary motions differ from that of his predecessors?

8 What is the difference between mass and weight?

9 Why was the discovery of Neptune a major confirmation of Newton's universal law of gravitation?

10 Why does an astronaut have to exert a force on a weightless object to move it?

ADVANCED QUESTIONS

The answers to all computational problems, which are preceded by an asterisk (*), appear at the end of the book.

11 Is it possible for an object in the solar system to have a synodic period of exactly one year? Explain your answer.

12 Explain qualitatively (in words) the systematic decrease in the synodic periods of the planets from Mars outward, as shown in Table 2-1.

*13 A line joining the sun and an asteroid was found to sweep out 5.2 square astronomical units of space in 1994. How much area was swept out in 1995? in five years?

*14 A comet moves in a highly elongated orbit about the Sun with a period of 1000 years. What is the length of the semimajor axis of the comet's orbit? What is the farthest the comet can get from the Sun?

*15 The orbit of a spacecraft about the Sun has a perihelion distance of 0.5 AU and an aphelion distance of 3.5 AU. What is the spacecraft's orbital period?

*16 Look up orbital data for the largest moons of Jupiter on the Internet, in the current issue of a reference from the U.S. Naval Observatory, such as the *Astronomical Almanac* or *Astronomical Phenomena*, or in such magazines as *Sky & Telescope* and *Astronomy*. Demonstrate that these orbits obey Kepler's third law.

17 Make diagrams of Jupiter's phases as seen from Earth and as seen from Saturn.

18 In what direction (left or right, eastward or westward) across the celestial sphere do the planets normally appear to move as seen from Australia? In what direction is retrograde motion as seen from there?

DISCUSSION QUESTIONS

19 Which planet would you expect to exhibit the greatest variation in apparent brightness as seen from earth? Explain your answer.

20 Use two thumbtacks (or pieces of tape), a loop of string, and a pencil to draw several ellipses. Describe how the shapes of the ellipses vary as you change the distance between the thumbtacks.

WHAT IF ...

*21 The Earth were 2 AU from the Sun? What would the length of the year be? Assuming such physical properties as rotation rate were as they are today, what else would be different here? What if the Earth was ¹/₂ AU from the Sun?

*22 The Earth was moved to a distance of 10 AU from the Sun? How much stronger or weaker would the Sun's gravitational pull be on Earth?

*23 The Earth had twice its present mass? Assume that all other properties of the Earth and its orbit remain the same.

What would be the acceleration of the New Earth due to the Sun compared to the present acceleration of the Earth from the Sun? *Hint:* Try combining $F = m_1 a$ and the force equation in An Astronomer's Toolbox 2-2, where m_1 is the mass of the Earth in both equations. Since the acceleration determines the period of the planet's orbit, how would the year on the more massive Earth compare to a year today?

24 The Sun suddenly disappeared? What would the Earth's path in space be in response to such an event? Describe how the Earth would change as a result and how humans might survive on a sunless planet.

WEB/CD-ROM QUESTIONS

25 Search the Web for information about Galileo. What were his contributions to physics? Which of Galileo's new ideas were later used by Newton to construct his laws of motion? What incorrect beliefs about astronomy did Galileo hold?

26 Search the Web for information about Kepler. Before he realized that the planets move on elliptical paths, what other models of planetary motion did he consider? What was Kepler's idea of the "music of the spheres"?

27 Search the Web for information about Newton. What were some of the contributions that he made to physics

other than developing his laws of motion? What contributions did he make to mathematics?

 28 **Monitoring the Retrograde Motion of Mars** Access and view the animation "The Path of Mars in 2004–2005" in Chapter 2 of the *Discovering the Universe* Web site or CD-ROM. (a) Through which two constellations does Mars move? (b) On approximately what date does Mars stop its direct (west to east) motion and begin its retrograde motion? *Hint:* Use the "Stop" and "Start" functions on your animation controls. (c) Over how many days does Mars move in the retrograde direction?

OBSERVING PROJECTS

 29 It is quite probable that within a few weeks of your reading this chapter one of the planets will be in opposition or at greatest eastern elongation, making it readily visible in the evening sky. Using the *Starry Night Backyard*™ computer program, the Internet, or consulting a reference book, such as the current issue of the *Astronomical Almanac* or the pamphlet *Astronomical Phenomena* (both published by the U.S. government), select a planet that is at or near such a configuration. To use *Starry Night Backyard*™, set the time for this evening, right click on the screen and make sure "show planets" is checked. Press the Labels button at the top and make sure "planets/Sun" is checked. Then grab the screen with the mouse and search the sky for planets. If none are up, change the date by, say, three days and search again. Repeat until you have found visible planets up at night. Plan to make your observations at that time. At opposition, would you predict that planets move rapidly or slowly from night to night against the background stars? Verify your

predictions by observing the planet once a week for a month, recording your observations on a star chart. How can you determine whether the change in position that you observe represents rapid or slow motion across the celestial sphere?

30 If Jupiter is visible in the evening sky, observe it with a small telescope on five consecutive clear nights. Record the position of the four Galilean satellites by making nightly drawings, just as the Jesuit priests did in 1620 (see Figure 2-10). From your drawings, can you tell which moon orbits closest to Jupiter and which orbits farthest? Are there nights when you saw fewer than four of the Galilean moons? What happened to the other moons on those nights?

 31 Use the *Starry Night Backyard*™ software to observe the moons of Jupiter. Set for Atlas mode (*Go:Atlas*). Center and lock on Jupiter (*Edit:Find:Jupiter*) and set the angular size of the sky you are examining to 30′ using the size button on the upper right of your screen. Lock on Jupiter (right click on Jupiter,

click on Centre/Lock). (a) Note the positions of the moons. Step forward in increments of six hours and draw the positions of the moons at each timestep. (b) From your drawings, can you tell which moon orbits closest to Jupiter and which orbits farthest away? Explain your reasoning. (c) Determine the periods of orbits of these moons. (d) Are there timesteps when you see fewer than four Galilean moons? What happened to the other moons at those times?

32 If Venus is visible in the evening sky, observe the planet with a small telescope once a week for a month. On each night, make a drawing of the phase that you see. Can you determine from your drawings if the planet is nearer or farther from the Earth than the Sun is? Do your drawings show any changes in the phase from one week to the next? If so, can you deduce if Venus is coming toward us or moving away from us?

 33 Use your *Starry Night Backyard*™ software to observe the phases of Venus. Set for Atlas mode (*Go:Atlas*). Center and lock on Venus (*Edit:Find: Venus*). Set the angular size of the sky to 7′. (a) Draw the current shape (phase) of Venus. Adjust the timestep to 30 days, make a single timestep, and again draw Venus, to scale. Make a total of 20 timesteps and drawings. (b) From your drawings, determine when the planet is nearer or farther from the Earth than the Sun is. (c) Deduce from your drawings when Venus is coming toward us or is moving away from us. (d) Explain why Venus goes through this particular cycle of phases.

34 Perform observing project 32 using Mars instead of Venus. If you have done project 32, compare your results for the two planets. Why are the cycles of phases as seen from Earth different for the two planets?

3 LIGHT AND TELESCOPES

IN THIS CHAPTER YOU WILL DISCOVER

- the nature of visible light and other types of electromagnetic radiation

- how telescopes collect and focus light

- the limitation of telescopes

- why different types of telescopes are used for different types of research

- how astronomers are developing new generations of high-technology telescopes

- how astronomers use the entire spectrum of electromagnetic radiation to observe the stars and other astronomical objects and events

WEB LINK 3.1

Mauna Kea Observatories
Astronomers prefer to build ground-based observatories on isolated mountaintops far from city lights, and where the air is dry, stable, and cloud-free. Mauna Kea Observatories, on the island of Hawaii, is home for a dozen optical and infrared telescopes. The twin Keck Telescopes are visible on the right of the photograph.
(Richard J. Wainscoat, University of Hawaii)

WHAT DO YOU THINK?

|1| What is light?

|2| Which type of electromagnetic radiation is most dangerous to life?

|3| What is the main purpose of a telescope?

|4| Why do stars twinkle?

|5| What type(s) of electromagnetic radiation can telescopes currently detect?

The telescope is the single most important tool of astronomy. Using a telescope, we see objects in space far more brightly and clearly and at greater distances than we can with the naked eye. Telescopes have played a major role in revealing the universe since Galileo viewed the craters on the Moon four centuries ago.

Refracting telescopes, which use large lenses to collect incoming starlight, were popular with astronomers 150 years ago. Modern astronomers strongly prefer reflecting telescopes, which gather light with large curved mirrors. In either case, the main purpose of a telescope is to collect as much light as possible. Astronomers attach a variety of equipment to telescopes to record incoming starlight.

THE NATURE OF LIGHT

Visible light is the form of electromagnetic energy to which our eyes are sensitive. But what exactly is visible light? How is it produced? What is it made of? How does it move through space? Scientists have struggled with these questions for the past 400 years.

3-1 Newton discovered that white light is not a fundamental color and debated whether light is particles or waves

From the time of Aristotle until the late seventeenth century, most people believed that white was a fundamental color of light. The colors of the rainbow (or, equivalently, the colors created by light passing through a prism) were believed to be added somehow as white light went from one medium through another. Isaac Newton performed experiments during the late 1600s that provided our first insights into the nature of light and disproved these earlier beliefs. Newton started by passing a beam of sunlight through a glass prism, which spread the light out into the colors of the rainbow, as shown in Figure 3-1a.

This rainbow, called a **spectrum** (plural; **spectra**), suggested to Newton that white light is actually a mixture of all the primary colors and that these colors separate out by changing direction or *refracting* by different amounts while going through the prism. He then selected a single color and sent it through a second prism (Figure 3-1b). The light emerging from this prism was the same color, heading in a different direction. After discovering that individual colors could not be further separated, Newton recombined all the colors, thereby re-creating white light.

So different colors are different entities. But what? The two major issues were determining whether they were particles or waves and how rapidly light traveled.

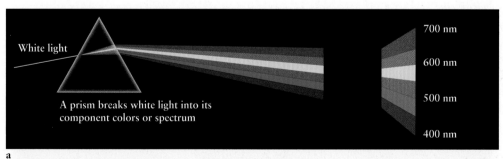

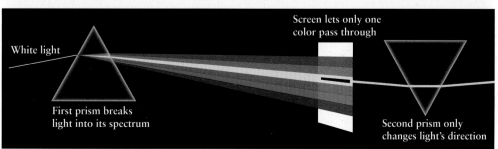

FIGURE 3-1 **A Prism and a Spectrum** (a) When a beam of white light passes through a glass prism, the light is separated or refracted into a rainbow-colored band called a spectrum. The numbers on the right side of the spectrum indicate wavelengths in nanometers (10^{-9} m). (b) Drawing of Newton's experiment showing that glass does not add to the color of light, only changes its direction.

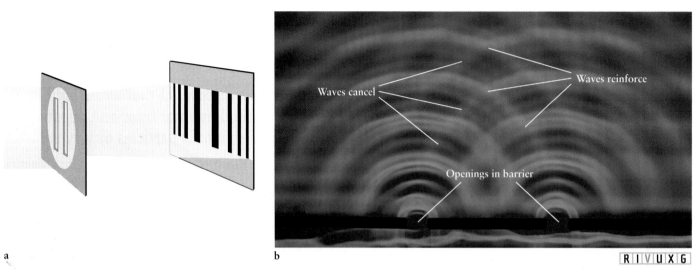

a   b

R I V U X G

█ FIGURE 3-2 Electromagnetic Radiation Travels as Waves **(a)** Thomas Young's interference experiment shows that light of a single color passing through a barrier with two slits creates alternating light and dark patterns on a screen. **(b)** The intensity of light on the screen is analogous to the height of water waves that strike the shore after passing through a barrier with two openings. In certain places, ripples from both openings reinforce each other to produce extra high waves. At intermediate locations along the shoreline, ripples and troughs cancel each other, producing still water. (© Eric Schrempp/Photo Researchers)

In the mid-1600s, the Dutch scientist Christiaan Huygens proposed that light travels in the form of waves. Isaac Newton, on the other hand, performed many experiments in optics that convinced him that light is composed of tiny particles of energy.

The English physicist Thomas Young demonstrated in 1801 that light is composed of waves. Young showed that the shadows of objects in light of a single color are not crisp and sharp. Instead, the boundary between illuminated and shaded areas is overlaid with patterns of closely spaced dark and light bands. These patterns are similar to the patterns produced by water waves passing through a barrier in the ocean (Figure 3-2). Young showed that a wavelike description of light could explain the results of his experiment but Newton's particle theory could not. Analyses of similar experiments soon produced overwhelming evidence for the wavelike behavior of light.

Further insight into the wave character of light came from calculations by the Scottish physicist James Clerk Maxwell in the 1860s. Maxwell succeeded in describing all the basic properties of electricity and magnetism in four equations. By combining these equations, Maxwell demonstrated that electrical and magnetic effects should travel through space together in the form of waves. Maxwell's suggestion that some of these waves are observed as light was soon confirmed by a variety of experiments. Because of its electric and magnetic properties, visible light is a form of *electromagnetic radiation* (Figure 3-3).

Newton showed that sunlight is actually composed of all the colors of the rainbow. Young and others showed that light travels as waves. What makes the colors of the rainbow distinct from each other? The answer is surprisingly simple: The only difference between different colors is the distance between two successive wave crests in the light wave. This dis-

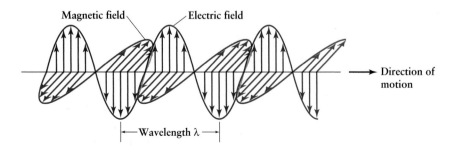

Magnetic field   Electric field

Direction of motion

├── Wavelength λ ──┤

█ FIGURE 3-3 Electromagnetic Radiation All forms of electromagnetic radiation (radio waves, infrared radiation, visible light, ultraviolet radiation, X rays, and gamma rays) consist of oscillating electric and magnetic fields perpendicular to each other that move through empty space at a speed of 3×10^5 km/s. The distance between two successive crests is called the wavelength of the light.

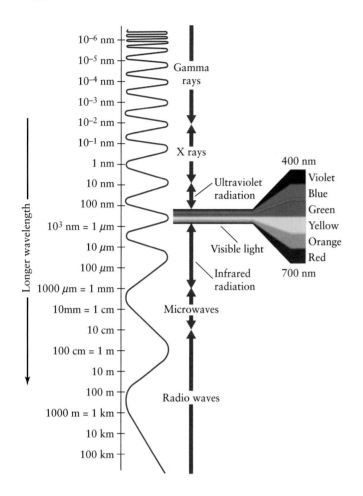

FIGURE 3-4 **The Electromagnetic Spectrum** The full array of all types of electromagnetic radiation is called the electromagnetic spectrum. It extends from the shortest-wavelength gamma rays to the longest-wavelength radio waves. Visible light forms only a tiny portion of the full electromagnetic spectrum. Note that 1 μm (one micrometer) is 10^{-6} meters.

tance is called the **wavelength** of the light, usually designated by the lowercase Greek letter λ (lambda; see Figure 3-3). Maxwell's calculations predicted that light waves all travel at the same speed in empty space, regardless of wavelength.

The wavelengths of all colors are extremely small, less than a thousandth of a millimeter. To express these tiny distances conveniently, scientists use a unit of length called the *nanometer* (nm), where 1 nm = 10^{-9} m. Experiments demonstrated that visible light has wavelengths covering the range from about 400 nm for the shortest wavelength of violet light to about 700 nm for the longest wavelength of red light. Intermediate colors of the rainbow fall between these wavelengths (see Figure 3-4). The complete spectrum of colors from the longest wavelength to the shortest is red, orange, yellow, green, blue, and violet.

3-2 Light travels at a finite speed

Light of any color travels incredibly quickly, far faster than sound. This is why we see lightning before we hear the accompanying thunderclap. But does light travel instantaneously from one place to another, or does it move with a measurable speed?

It is interesting to see how the discovery of the Galilean moons set the stage for another discovery, the finite speed of light. The first evidence for the finite speed of light came in 1675, when Ole Rømer, a Danish astronomer, carefully timed eclipses of Jupiter's moons (Figure 3-5). Rømer discovered that the moment at which a moon enters Jupiter's shadow depends on the distance between the Earth and Jupiter. When Jupiter is in opposition, that is, when Jupiter and the Earth are on the same side of the Sun, the Earth-Jupiter distance is relatively short compared to when Jupiter is near conjunction. At opposition, Rømer found that eclipses occur slightly earlier than expected, while they occur slightly later than predicted by Kepler's laws when Jupiter is near conjunction.

Rømer correctly concluded that light takes more time to travel longer distances across space. The greater the distance to Jupiter, the longer the image of an eclipse takes to reach our eyes. From his timing measurements, Rømer concluded that it takes $16\frac{1}{2}$ minutes for visible light to traverse the diameter of the Earth's orbit (2 AU). Incidentally, Rømer's interpretation of the data makes sense only if both the Earth and Jupiter orbit the Sun. His experiment therefore supported the heliocentric cosmology but not the geocentric one.

Rømer's subsequent calculation of the speed of light was off by 25 percent, due to a highly inaccurate value for

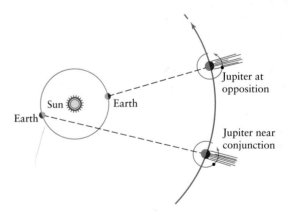

FIGURE 3-5 **Measuring the Speed of Light** An eclipse of one of Jupiter's moons seen at opposition (Earth and Jupiter on the same side of the Sun) appears to occur earlier than an eclipse seen near conjunction (Earth and Jupiter on opposite sides of the Sun). The differences in apparent times of the eclipses are due to the extra time it takes light to travel the additional distance when the planets are near conjunction. The actual time of the eclipse is determined using Kepler's laws.

AN ASTRONOMER'S TOOLBOX 3-1

Photon Energies

All photons with the same energy are identical: Their energy depends solely on the photon's wavelength. To find any photon's energy we use the equation

$$E = \frac{hc}{\lambda}$$

where Planck's constant h is 6.67×10^{-34} J s, the speed of light c is 300,000 km/s, and the photon's wavelength is λ.

Example: A photon of red light has a wavelength 700 nm. What is its energy? *Note:* All distances in the following equation must be converted to the same units, such as kilometers.

$$E_{red} = \frac{(6.67 \times 10^{-34} \text{ J s})(300,000 \text{ km/s})}{700 \text{ nm}}$$

$$E_{red} = 2.86 \times 10^{-19} \text{J}$$

(A joule, abbreviated J, is a unit of energy.) Each photon of red light with wavelength 700 nm has an energy of 2.86×10^{-19} J.

Compare! In 1 second a 25-watt light bulb emits 25 J.

Try these questions: A photon has energy 4.90×10^{-19} J. Calculate its wavelength in nanometers. Referring to Figure 3-1, what is this photon's color? What is the wavelength of a photon with twice this energy? What is the energy of a green photon? (*Clue:* See Figure 3-1.)

(Answers appear at the end of the book.)

---

the AU existing at that time. The first accurate laboratory measurements of the speed of visible light were performed in the mid-1800s. The constant speed of light in a vacuum, usually designated by the letter *c,* is now known to be 299,792.458 km/s, which we generally round to

$$c = 3.0 \times 10^5 \text{ km/s} = 1.86 \times 10^5 \text{ mi/s}$$

(Standard abbreviations for units of speed, such as km/s for kilometers per second and mi/s for miles per second will be used throughout the rest of this book.) Light traveling through air, water, glass, or any other substance always moves more slowly than it does in a vacuum. Indeed, in 1999, scientists were able to get light slowed to 61 km/hr (38 mph) and in 2001 they were able to momentarily stop it altogether.

The value *c* is a fundamental property of the universe. It appears in equations that describe, among other things, atoms, gravity, electricity, magnetism, and time. Furthermore, according to Einstein's special theory of relativity (Chapter 13), nothing can travel faster than the speed of light in a vacuum.

3-3   Einstein showed that light sometimes behaves as particles and sometimes as waves

In 1905, just as scientists were becoming comfortable with the wave nature of light, Albert Einstein proposed that light sometimes behaves as particles! Einstein used this idea to explain the photoelectric effect: Shorter wavelengths of light can knock some electrons off the surfaces of metals while longer wavelengths of light cannot, no matter how intense the beam of long-wavelength light. Einstein knew that electrons are bound onto a metal's surface by electric forces and that it takes energy to overcome those forces. Because some colors (or, equivalently, wavelengths) can remove the electrons and others cannot, the electrons must receive different amounts of energy from different colors of light. But how? Einstein proposed that light travels as waves enclosed in discrete packets called **photons,** and that photons with different wavelengths have different amounts of energy. *Specifically, the shorter the wavelength, the higher a photon's energy.*

$$\text{Photon energy} = \frac{\text{Planck's constant} \times \text{speed of light}}{\text{wavelength}}$$

where Planck's constant (named for the German physicist Max Planck) has the value 6.67×10^{-34} J s, where J is the unit of energy called a joule (see An Astronomer's Toolbox 3-1). Einstein's concept of light means that it can act both as waves and as particles.

All photons with the same wavelength are identical to each other, and, therefore, every photon of a given wavelength carries the same amount of energy as every other photon of that wavelength. The energy delivered by a photon is either enough to rip an electron off the surface of the metal or it is not. There is no middle ground, as there would be if light of one color came in packages with different numbers of waves so that photons of the same color could carry different amounts of energy. If this were the case, some photons of a certain color would have enough energy to free an electron, while other photons of the same color would not. Experimentally, this never happens. The concept of photons as just described has now been thoroughly proved, and astronomers incorporate it in their model of light.

3-4 Light is only one type of electromagnetic radiation

Visible light has a narrow range of wavelengths, from about 400 to 700 nm. But Maxwell's equations placed no length restrictions on the wavelengths of electromagnetic radiation. Researchers therefore began to look for forms of this radiation outside the range of wavelengths to which the cells of the human retina respond.

Around 1800 the British astronomer William Herschel discovered **infrared radiation** in an experiment with a prism. When he held a thermometer just beyond the red end of the visible spectrum, the thermometer registered a temperature increase, indicating that it was being heated by an invisible form of energy. Infrared radiation was identified as electromagnetic radiation with wavelengths slightly longer than red light. In experiments with electric sparks in 1888, the German physicist Heinrich Hertz first succeeded in producing electromagnetic radiation a few centimeters in wavelength, now known as **radio waves.** In 1895 Wilhelm Roentgen invented a machine that produces electromagnetic radiation with wavelengths shorter than 10 nm, now called **X rays.** Modern versions of Roentgen's machine are found in medical and dental offices and airport security checkpoints. Over the years, forms of radiation have been discovered with many other wavelengths.

We now know that visible light occupies only a tiny fraction of the full range of possible wavelengths, collectively called the **electromagnetic spectrum.** As shown in Figure 3-4, the electromagnetic spectrum stretches from the longest-wavelength radio waves, through infrared radiation, visible light, ultraviolet radiation, and X rays, to the shortest-wavelength photons, gamma rays. On the long-wavelength side of the visible spectrum, infrared radiation covers the range from about 700 nm to 1 mm. Astronomers interested in infrared radiation often express wavelength in *micrometers* or *microns* (abbreviated μm), where 1 μm = 1000 nm = 10^{-6} m. From roughly 1 mm to 10 cm is the range of microwaves, which are sometimes considered to be infrared radiation and sometimes radio waves.

At wavelengths shorter than those of visible light, **ultraviolet (UV) radiation** extends from about 400 nm down to 10 nm. Next are X rays, with wavelengths between about 10 and 0.01 nm, and beyond them are **gamma rays.** It should be noted that these boundaries are approximate and are primarily used as convenient divisions in the electromagnetic spectrum, which is actually continuous.

These various types of electromagnetic radiation share many basic properties. For example, they are all photons, they all travel at the speed of light, and they all sometimes behave as particles and sometimes as waves. But because of their different wavelengths (and therefore different energies), they interact very differently with matter. Your body tissues are nearly transparent to X rays but not to visible light; your eyes respond to visible light but not to infrared radiation; your radio detects radio waves but not ultraviolet radiation.

The Earth's atmosphere is relatively transparent to both visible light and radio waves, meaning that both pass through to reach ground-based telescopes sensitive to these forms of electromagnetic radiation. Astronomers say that the atmosphere has *windows* for these parts of the electromagnetic spectrum (Figure 3-6). Infrared radiation has a

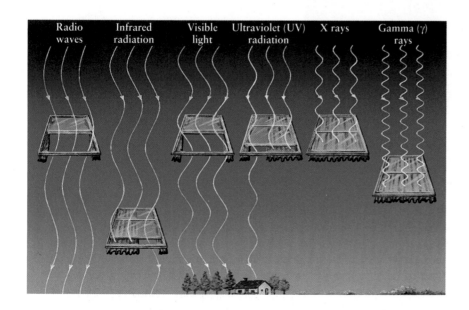

FIGURE 3-6 Windows Through the Atmosphere The Earth's atmosphere allows different types of electromagnetic radiation to penetrate into it in varying amounts. Visible light and radio waves reach all the way to the Earth's surface. Some infrared and ultraviolet radiation also can reach the ground. The other types of radiation are absorbed or reflected by the gases in the air at different characteristic altitudes. While the atmosphere does not have actual "windows," astronomers use the term to characterize the passage of radiation through it.

limited window through the atmosphere. We detect this radiation as heat.

Likewise, the longest-wavelength ultraviolet radiation, called UV-A, has a limited window. This radiation causes tanning and sunburns. Ozone (O_3) in the Earth's atmosphere screens out intermediate-wavelength ultraviolet radiation. This *ozone layer* is being depleted by human-made chemicals, such as chlorofluorocarbons (CFCs). As a result, more intermediate-wavelength ultraviolet, called UV-B, is reaching the Earth's surface, and these highly energetic photons damage living tissue, causing skin cancer and glaucoma, among other diseases.

The Earth's atmosphere is completely opaque to the other types of electromagnetic radiation, meaning that they do not reach the Earth's surface. (This is a good thing, because short-wavelength ultraviolet radiation called UV-C and X rays and gamma rays are devastating to living tissue. Gamma rays are the deadliest.) Observations of these wavelengths must be performed high in the atmosphere or, ideally, from space.

Detecting electromagnetic radiation is the essence of observational astronomy. Until recently, photons were the only sources of detailed astronomical information that we had. However, in the past three decades, astronomers have built detectors of nonelectromagnetic energies and particles from space. Chapters 9 and 13 discuss two of these devices —neutrino detectors and gravity wave detectors.

OPTICS AND TELESCOPES

Since the time of Galileo, astronomers have been designing instruments to collect more light than the human eye can collect on its own. Collecting more light enables us to see things more brightly, in more detail, and at a greater distance. There are just two basic types of telescopes—those that collect light through lenses and those that collect it from mirrors. Let's begin with the earliest telescopes, which used lenses.

3-5   A refracting telescope uses a lens to concentrate incoming light

Although light travels at about 300,000 km/s in a vacuum, it moves more slowly through a dense substance such as glass. As light enters the glass, it slows abruptly, much like a person walking from a boardwalk onto a sandy beach: Her pace suddenly slows as she steps from the smooth pavement onto the sand. And just as the person stepping back onto the boardwalk easily resumes her original pace, light exiting a piece of glass resumes its original speed.

As a result of changing speed, light can also change direction as it passes from one transparent medium into another— a phenomenon called **refraction**. You see refraction every day when looking through windows. Imagine a stream of photons from a star entering a window, as shown in Figure 3-7a.

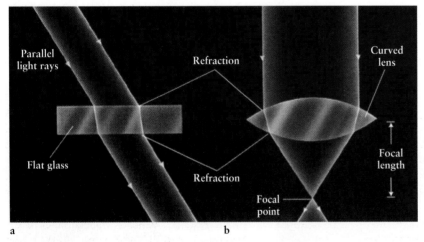

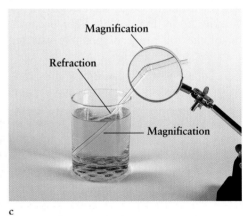

a   b   c

■ FIGURE 3-7   Refraction Through Uniform and Variable Thickness Glasses **(a)** Refraction is the change in direction of a light ray when it passes into or out of a transparent medium such as glass. When light rays pass through a flat piece of glass, the two refractions bend the rays in opposite directions. There is no overall change in the direction in which the light travels. **(b)** If the glass is in the shape of a suitable convex lens, parallel light rays converge to a focus at a special point called the focal point. The distance from the lens to the focal point is called the focal length of the lens. **(c)** A light ray entering a denser medium, like going from air into water or glass, is bent or refracted to an angle more perpendicular to the surface than the angle at which it was originally traveling. If the glass is uniformly thick, then the light leaving the glass is refracted back to the direction it had before entering the glass. If the glass has curved surfaces (that is, a convex lens), then parallel light rays converge to a focus. Such focusing leads to magnification, as discussed in the text. The straw as seen through the side of the liquid is magnified and offset from the straw above the liquid because the liquid is given a curved shape by the side of the glass. The straw as seen through the top of the liquid is refracted, but does not appear magnified because the top is flat. (c: Ray Moller/Dorling Kindersley)

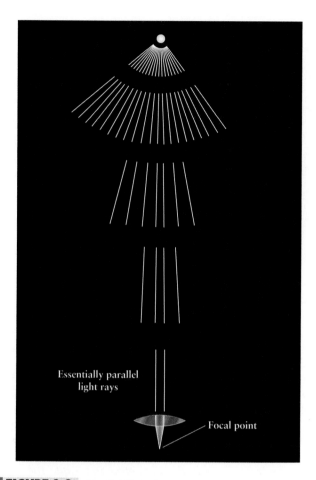

FIGURE 3-8 **Parallel Light Rays from Distant Objects** As light travels away from any object, the light rays, all moving in straight lines, separate. By the time light has traveled trillions of miles, only the light rays moving in parallel tracks are still near each other.

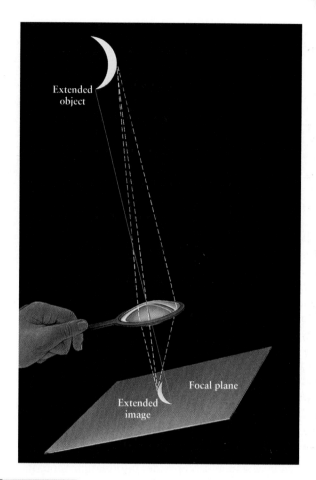

FIGURE 3-9 **Extended Objects Create a Focal Plane** Light from objects larger than points in the sky does not all converge to the focal point of a lens. Rather, the object creates an image at the focal length in what is called the focal plane.

Astronomers call such photon flows *light rays*. As a light ray goes from the air into the glass, the light ray's direction changes so that it is more perpendicular to the surface of the glass than it was before entering. Once inside the glass, the light ray travels in a straight line. Upon emerging from the other side, the light ray bends once again, resuming its original direction and speed. The net effect is only a slight, uniform displacement of the objects beyond the glass.

A telescope lens uses refraction to collect light. Unlike a window, lenses have surfaces of varying thickness (Figure 3-7b). These curved surfaces force the light rays to emerge from the lens in different directions than they had before entering the lens. Different rays striking the top surface of the lens at different places are refracted by different amounts: If the lens is shaped correctly, parallel light rays start converging once they enter the glass. As the light emerges from the glass, it converges further.

Light rays meet at a point called the **focal point** of the lens. Its distance from the lens is the **focal length** (Figure 3-7b). Actually, only light rays that were *parallel* before en-

tering the lens meet at the focal point. This occurs for light from sources that are extremely far away, like the stars (Figure 3-8 shows why). If the object is close enough to be more than just a dot as seen through the telescope, all the light from it does not converge at the focal point but rather focuses along a surface, the **focal plane** (Figure 3-9). The Moon and planets are examples of objects with images that extend over the focal plane.

A **refracting telescope,** or **refractor** (Figure 3-10), is an arrangement of two lenses used to gather light. The resulting image is a brighter, and often a bigger and clearer, view of the object. The lens at the top of the telescope, called the **objective lens,** has a large diameter and long focal length. Its purpose is to collect as much light as possible. The lens at the bottom of the telescope, the **eyepiece lens,** is smaller and has a short focal length. It restraightens the light rays after they have passed through the focal point or focal plane of the objective lens, making them parallel once again. Because the light is now more concentrated, objects are brighter as seen through a telescope.

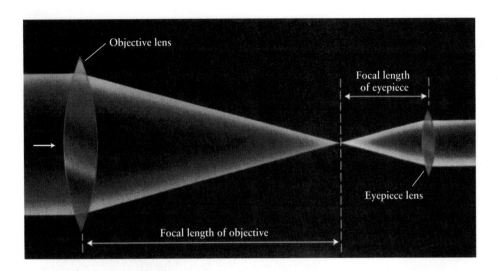

 FIGURE 3-10 Essentials of a Refracting Telescope A refracting telescope consists of a large, long-focal-length objective lens that collects and focuses light rays and a small, short-focal-length eyepiece lens that restraightens the light rays. The lenses work together to brighten, resolve, and magnify the image formed at the focal plane of the objective lens.

3-6 Telescopes brighten, resolve, and magnify

A telescope's most important function is to provide the astronomer with as many photons from the object as possible. The more photons, the more information the astronomer can extract—even if the object still appears as a pinpoint. For example, given enough photons, astronomers can measure the relative intensities of different wavelengths emitted by a star, which provides information about its temperature, chemical composition, age, and motion.

The observed brightness of any object depends on the total number of photons collected from it, which in turn depends on the area of the telescope's objective lens. A large objective lens intercepts and focuses more light than does a small objective lens. A large objective lens can therefore produce brighter images and detect fainter stars than a small objective lens can for an equal exposure time (Figure 3-11). No wonder telescopes are sometimes called "photon buckets."

> **Insight into Science** Costs and Benefits Science now relies heavily on technology to conduct experiments or make observations. The cost of cutting-edge astronomical observations may run to hundreds of millions of dollars. The return on such investments is a better understanding of how the universe works, how we can harness its capabilities, and our place in it.

The **light-gathering power** of a telescope is directly related to the area of the telescope's objective lens (or primary mirror; see Section 3-8). Recall that the area and diameter of a circle are related by

$$\text{Area} = \frac{\pi d^2}{4}$$

where d is the diameter of the lens and π (pi) is about 3.14.

Consequently, a lens with twice the diameter of another lens has 4 times the area of the smaller lens and therefore collects 4 times as much light as the smaller one. For example, a 36-cm-diameter lens has 4 times the area of an 18-cm-diameter lens. Therefore, the 36-cm telescope has 4 times the light-gathering power of a telescope half its size. The general rule is: *Double the diameter, quadruple the light-gathering power.*

The second most important function of any telescope is to reveal more details of the extended objects under study. A large telescope increases the sharpness of the image and the degree of detail that can be seen. **Angular resolution** measures the clarity of images (Figure 3-12). Poor angular resolution causes images of galaxies or planets to be fuzzy and blurred. A telescope with good angular resolution produces images that are sharp.

The angular resolution of a telescope is measured as the arc angle between two adjacent stars whose images can just barely be distinguished by that telescope. The smaller the angle, the sharper the image. Large, modern telescopes, like the Keck telescopes in Hawaii, which we will discuss later in this chapter, have angular resolutions better than 0.1 arcsec. As a general rule, *double the diameter, double the detail that can be seen.*

The final function of a telescope—often the least important one—is to make objects appear larger. This property is called **magnification**. The magnification of a refracting telescope is equal to the focal length of the objective lens divided by the focal length of the eyepiece lens:

$$\text{Magnification} = \frac{\text{focal length of the objective}}{\text{focal length of the eyepiece}}$$

RIVUXG

FIGURE 3-11   **Light-Gathering Power**   Because a large lens intercepts more starlight than a small lens, a large lens produces a brighter image. The same principle applies to telescopes that collect light using a primary mirror rather than an objective lens. The two photographs of the Andromeda galaxy were taken through telescopes with different diameters and were exposed for equal lengths of time at equal magnification. (AURA)

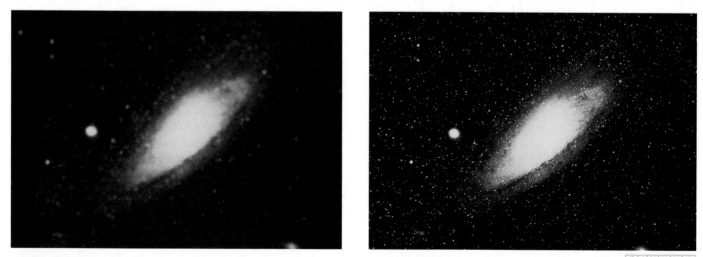

RIVUXG

FIGURE 3-12   **Resolution**   The larger the diameter of a telescope's objective lens or primary mirror, the greater the detail the telescope can resolve. These two images of the Andromeda galaxy, taken through telescopes with different diameters, show this difference. Increasing the exposure time of the smaller diameter telescope (left), will only brighten the image, not improve the resolution. (AURA)

MOVIE MISCONCEPTIONS

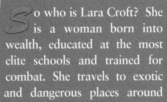

Lara Croft: Tomb Raider (Paramount Pictures, 2001)
(Claro Cortez/Reuters)

So who is Lara Croft? She is a woman born into wealth, educated at the most elite schools and trained for combat. She travels to exotic and dangerous places around the world in search of adventure. In the movie *Lara Croft: Tomb Raider*, based on the popular video game of the same name, Lara faces her greatest challenge yet. The basic plot of this film is that every 5000 years all the plan-

ets align. This means that from Earth they all appear together as a single point in the sky. During an early scene in the movie, Lara Croft is shown lying down under an indoor telescope in a well-lit room. She is allegedly watching the planets align. She sees them as disks moving across her telescope's line of sight.

There are at least 4 scientific errors in the movie, as described in just this paragraph. What are they?
(Answers appear at the end of the book.)

For example, if the objective of a telescope has a focal length of 100 cm and the eyepiece has a focal length of 0.5 cm, then the magnifying power of the telescope is

$$\text{Magnification} = \frac{100 \text{ cm}}{0.5 \text{ cm}} = 200$$

This property is usually expressed as 200×.

There is a limit to the magnification of any telescope. Try to magnify beyond that limit and the image becomes distorted. As a rule, *double the objective lens's diameter, double the telescope's maximum magnification.*

3-7 Refracting telescopes have drawbacks

For convenience in manufacturing, objective lenses for refractors have spherical surfaces. However, a spherical lens distorts images, a problem called **spherical aberration.** This is only the first drawback of using objective lenses in large research telescopes.

The amount of refraction that light undergoes varies with color (recall Figure 3-1). Because different colors have different focal lengths, any lens distorts the colors of images (Figure 3-13), a problem called **chromatic aberration.** This problem is corrected with a *compound lens* composed of two different types of glass with different refractive properties. Because compound lenses in telescopes and cameras bring all the wavelengths into focus at the same focal length, they are called **achromatic lenses** (Figure 3-14).

A third problem, *sagging* of large lenses, became apparent in the nineteenth century as larger, heavier lenses were

developed. Despite their apparent rigidity, the shapes of glass structures, like extremely large lenses, distort under the pull of the Earth's gravity. This distortion occurs because the lens can only be supported around its edge, which is very thin. When a lens is larger than about a meter in diameter, its thick center weighs it down and causes the glass to deform. As a result, the image is distorted. Moreover, as the telescope *tracks,* or follows, a star, the orientation of the telescope changes and, along with it, the amount of distortion. There is no way of knowing the exact distortion that will occur. As a result, sagging causes the images of stars to move during a single observation, causing them to appear blurry on film or other recording medium. The largest diameter telescope objective lenses ever built are in the one-meter class. They do not distort their image by sagging, but larger glass lenses would do so, and hence larger refracting telescopes were never constructed.

Unwanted refractions are a fourth problem. Lenses are ground from large, thick disks of glass that are formed by pouring molten glass into a mold. As the liquid glass cools, gas in it becomes trapped, creating air bubbles in the solidified disk. When the lens is then ground into shape, any air bubbles inside it create unwanted and unpredictable extra refractions that blur the images. Part of a lensmaker's expertise was to choose a volume of glass for grinding with as few air bubbles as possible.

A fifth problem is that glass is *opaque* to certain ranges of wavelengths. These wavelengths of light cannot pass through glass lenses. Even visible light is dimmed substantially in passing through the glass lens at the front of a refractor, and ultraviolet radiation is largely absorbed by the glass.

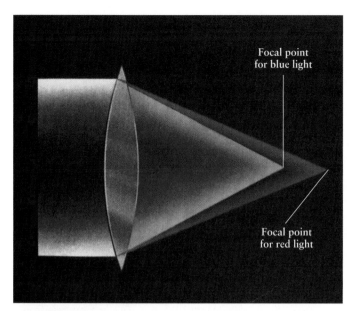

FIGURE 3-13 Chromatic Aberration Spreads Colors Out
Light of different colors passing through a lens is refracted by different amounts. This effect is called chromatic aberration. As a result, different-colored objects have different focal lengths. Consequently, the images seen through uncorrected lenses are blurred.

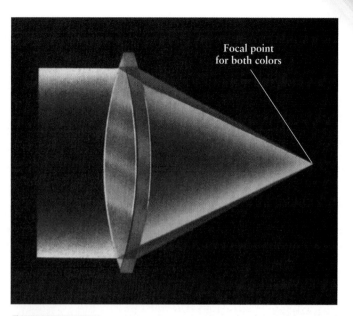

FIGURE 3-14 An Achromatic Lens Corrects for Chromatic Aberration A second lens (not the eyepiece) that refracts colors by different amounts than does the objective lens can bring all colors into focus at the same focal length.

We have already mentioned that spherical lenses produce smeared images (Figure 3-15a). Spherical aberration is overcome by focusing light through complex aspheric (nonspherical) surfaces (Figure 3-15b), but grinding aspheric lenses is so difficult that for practical purposes, objective lenses always have spherical surfaces. Spherical aberration can also be minimized by making the objective lens so thin that its spherical surfaces nearly coincide with the appropriate aspheric surfaces. A thin lens, however, has an extremely long focal length, which is why research refracting telescopes are many meters long (Figure 3-16).

Nineteenth-century master opticians devoted their lives to overcoming the problems inherent in refracting telescopes, and several magnificent refractors were constructed in the late 1800s. The largest, completed in 1897 and located at the Yerkes Observatory near Chicago (see Figure 3-16), has an objective lens 102 cm (40 in.) in diameter with a focal length of $19^{1}/_{3}$ m (63.5 ft). The second largest refracting telescope, located at Lick Observatory near San Jose, California, has an objective lens of 91 cm (36 in.) in diameter. No major refracting telescopes were constructed in the twentieth century.

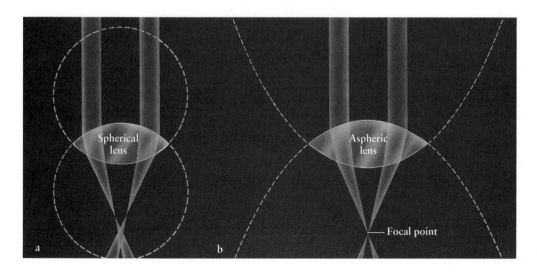

FIGURE 3-15 Spherical Aberration and Aspheric Lenses (a) Different parts of a spherical lens refract light to different focal points. (b) The ideal shape for a lens is a complex aspherical shape. Such a lens focuses all the light passing through it at the same focal length.

FIGURE 3-16   The Largest Refracting Telescope This giant refracting telescope, built in the late 1800s, is housed at Yerkes Observatory near Chicago. The objective lens is 102 cm (40 in.) in diameter, and the telescope tube is 19¹/₃ m (63¹/₂ ft) long. (Yerkes Observatory)

image of the distant object is formed, is the focal length of the mirror. The magnifying power of such a reflecting telescope is calculated in the same way as for a refractor: The focal length of the large or **primary mirror** is divided by the focal length of the eyepiece.

To view the image, Newton placed a small, flat mirror at a 45° angle in front of the focal point, as sketched in Figure 3-19a. This **secondary mirror** reflects the light rays to one side of the telescope, and the astronomer views the image through an eyepiece lens. A telescope with this optical design is still called a **Newtonian reflector.**

Newtonian telescopes are very popular with amateur astronomers, because they are convenient to use while the observer is standing up. However, they are not used in research observatories because they are lopsided. If astronomers attach their often heavy and bulky research equipment onto its side, the telescope sags and distorts the image in unpredictable ways.

Three basic designs exist for the reflecting telescopes used in research. In the first design, the astronomer actually sits at the undeflected focal point, directly in front of the primary mirror, and controls the light-collecting equipment located there. This arrangement is called a **prime focus** (Figure 3-19b). All telescopes must be kept at the same temperature as the outdoors to prevent a temperature difference from expanding or contracting the telescope, thereby distorting the image. It is therefore often a chilling experience to ride throughout a mountaintop winter night in the open prime focus "observing cage."

3-8   Reflecting telescopes use mirrors to concentrate incoming starlight

Prompted by the problem of chromatic aberration, early in the eighteenth century Isaac Newton set about to replace the objective lens with a curved mirror to collect light. Called **reflecting telescopes** or **reflectors,** telescopes with mirrors easily overcome most of the problems inherent in refracting telescopes.

A mirror collects light using **reflection** rather than refraction. In Figure 3-17, the ray of light strikes a flat mirror, and we imagine a perpendicular line coming out of the mirror at that point. According to the principle of reflection, the angle between the arriving light ray and the perpendicular (dashed line) is always equal to the angle between the reflected light ray and the perpendicular. The rule is the same regardless of the light's wavelength or the mirror's shape.

Using this principle, Newton realized that a concave, parabolic mirror will cause parallel light rays to converge to a focal point, as shown in Figure 3-18b. The distance between the reflecting surface and the focal point, where the

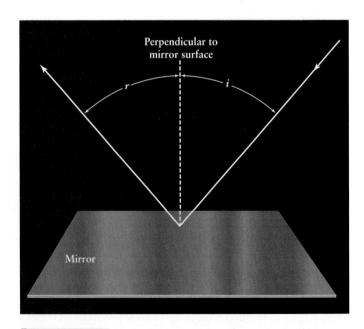

FIGURE 3-17   Reflection The angle at which a beam of light strikes a mirror (the angle of incidence i) is always equal to the angle at which the beam is reflected from the mirror (the angle of reflection r).

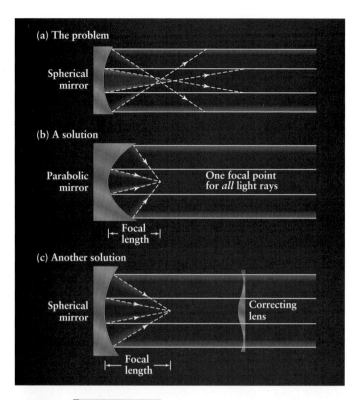

FIGURE 3-18 Spherical Aberration (a) Different parts of a spherically concave mirror reflect light to slightly different focal points. This effect, spherical aberration, causes image blurring. This problem can be overcome by (b) using a parabolic mirror or (c) using a Schmidt corrector plate (lens) in front of the telescope.

In a second popular design, the **Cassegrain focus,** a hole is drilled directly through the center of the primary mirror. A convex secondary mirror placed between the primary mirror and its focal point reflects the light rays back through the hole (Figure 3-19c). This curved secondary mirror extends the telescope's focal length. Light-gathering equipment is bolted to the bottom of the telescope, and the light is brought into focus in it. This design has an advantage over Newtonian telescopes, in that the attached equipment is balanced and does not distort the telescope frame and, hence, the image.

The third design is handy for long and bulky optical equipment that cannot be mounted directly on the telescope or for observations that benefit from extremely long focal lengths. In this design (Figure 3-19d), a series of mirrors channels the light rays away from the telescope to a remote focal point called the **coudé focus** (named after a French word meaning "bent like an elbow").

Reflecting telescopes have numerous advantages over refracting ones. First, they avoid the problem of chromatic aberration, because, unlike household mirrors, all telescope mirrors have coated top surfaces, and light never enters the glass at all. Second, the mirrors, which can weigh tons, do not warp at all, because they can be rigidly supported from underneath. Third, a mirror maker needs only to find a surface, rather than an entire volume, that is free of bubbles. Finally, because light does not enter the glass, the problem of a lens's opacity to different wavelengths never arises.

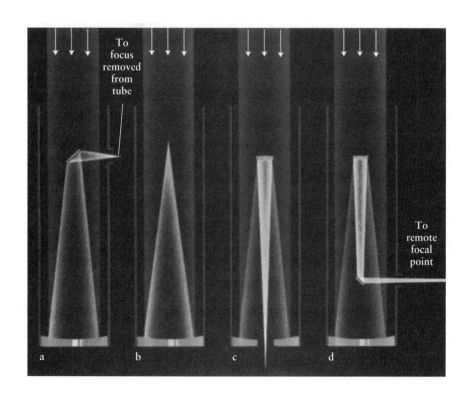

FIGURE 3-19 Reflecting Telescopes Four of the most popular optical designs for reflecting telescopes: (a) Newtonian focus (popular among amateur astronomers); and the three major designs used by researchers—(b) prime focus, (c) Cassegrain focus, and (d) coudé focus.

3-9 Reflecting telescopes also have limitations

The mirrored surfaces of research telescopes are polished so smoothly that the highest bumps are less than 1/500 the thickness of a human hair. Nonetheless, reflecting telescopes are not perfect; there are several prices to pay for the advantages reflectors offer over refractors. Two of the most important are *blocked light* and *spherical aberration*. Let's consider each problem briefly.

You have probably noticed that the secondary mirror of a reflector blocks some incoming light, one unavoidable price that astronomers must pay. Typically, a secondary mirror prevents about 10 percent of the incoming light from reaching the primary mirror. This problem is addressed by constructing primary mirrors with sufficiently large surface areas to compensate for the loss of light. You might think as well that because light is missing from the center of the telescope due to blockage by the secondary mirror, a corresponding central "hole" appears in the images. However, this problem does not occur, because light from all parts of each object enters all parts of the telescope (Figure 3-20).

To make a reflector, an optician traditionally grinds and polishes a large slab of glass into a concave spherical surface. However, light entering a spherical telescope mirror at different distances from the mirror's center comes into focus at different focal lengths, as occurs for a spherical lens, and images taken with all such telescopes appear blurry (see Figure 3-18a).

Spherical aberration in reflecting telescopes with spherical primaries may be overcome by including a thin correcting lens, called a **Schmidt corrector plate**, at the top of the telescope (see Figure 3-18c). The light coming into the telescope is refracted by the plate just enough to compensate for spherical aberration and to bring all the light into focus at the same focal length. These correctors have the added benefit of focusing light from a larger angle in the sky than would be in focus without the plate. A Schmidt corrector plate enables astronomers to map large areas of the sky with relatively few photographs at moderately high magnification. In other words, the Schmidt corrector plate acts like a wide-angle lens on a camera. However, the plate does not allow for as much magnification as a telescope with a parabolic mirror.

While parabolic primary mirrors have been meticulously ground over the past century, the advent of computer-controlled grinding and rotating furnaces (Figure 3-21), in which the liquid glass actually spins into a parabolic shape, has now made it economical to cast parabolic mirrors with diameters of several meters.

3-10 Earth's atmosphere hinders astronomical research

Besides the problems and restrictions inherent in the optics of telescopes, the Earth's atmosphere also introduces difficulties.

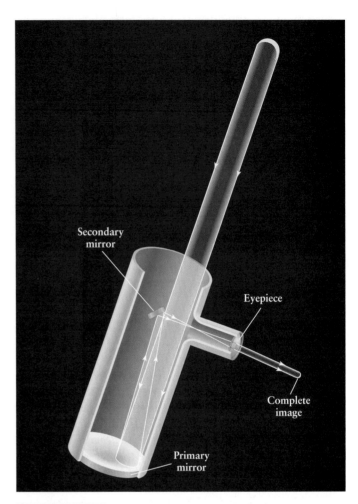

FIGURE 3-20 **The Secondary Mirror Does Not Create a Hole in the Image** Because the light rays from distant objects are parallel, light from the entire object reflects off all parts of the mirror. Therefore, every part of the object sends photons to the eyepiece. This figure shows the reconstruction of the entire Moon from light passing through just part of this telescope. The same drawing applies everywhere on the primary mirror that is not blocked by the secondary mirror.

The problems arise because the air is turbulent and filled with impurities. You have probably seen turbulence while driving in a car on a hot day, when the road ahead appears to shimmer. Blobs of air, heated by the Earth, move upward to create this effect. Light passing through such a blob is refracted, because each hot air mass has a different density than the cooler air around it. Hence, because each blob behaves like a lens, images of objects beyond it appear distorted.

The atmosphere over our heads is similarly moving and changing density, and the starlight passing through it is similarly refracted. Because the air density changes rapidly, the resulting changes in refraction make the stars appear to change brightness and position rapidly, an effect we see as **twinkling**. When photographed through large telescopes,

a

b

FIGURE 3-21 Rotating Furnace for Making Parabolic Telescope Mirrors **(a)** To make each 8.4-meter primary mirror for the Large Binocular Telescope II on Mount Graham, in Arizona, 40,000 pounds of glass are loaded into a rotating furnace and heated to 1450 K (2150°F).

This image shows glass fragments loaded into the cylindrical furnace. **(b)** After melting, spinning, and cooling, the mirror's parabolic surface is ready for final smoothing and coating with a highly reflective material. (a: Roger Ressmeyer/Corbis; b: The University of Arizona, Steward Observatory)

twinkling smears out a star's image, causing it to look like a disk rather than a pinpoint of light (Figure 3-22a). Astronomers use the expression *"seeing"* to describe how much twinkling is occurring and, therefore, how smeared out are the images from their telescopes.

The angular diameter of a star's smeared-out image, called the **seeing disk,** is a realistic measure of the best possible resolution. The size of the seeing disk varies from one observatory site to another. At Mount Palomar in California, the seeing disk is roughly one arcsec (1″). The best con-

ditions on Earth, with a seeing disk of 0.2″, have been reported at the observatory on the 14,000-ft summit of Mauna Kea, the tallest volcano on the island of Hawaii (pictured in the photo that opens this chapter).

Without the effects of the Earth's atmosphere, stars do not twinkle. As a result, photographs taken from telescopes in space reveal stars as much finer points (Figure 3-22b) and more detail for extended objects, such as planets and galaxies. The Hubble Space Telescope, as we will see often in this book, achieves magnificent resolution.

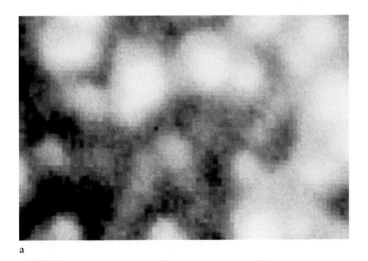

a

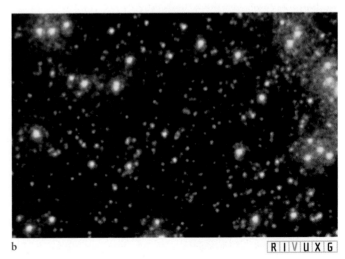

b

R I V U X G

FIGURE 3-22 Effects of Twinkling The same star field photographed with **(a)** a ground-based telescope, which is subject to twinkling, and **(b)** the Hubble Space Telescope, which is free from the effects of twinkling. (NASA/ESA)

Light pollution from cities poses another problem for Earth-based telescopes (Figure 3-23). Keep in mind that the larger the primary mirror, the more light is gathered and therefore the more information astronomers can obtain. The 5-m (200 in.) telescope at the Palomar Observatory between San Diego and Los Angeles, California, was the first truly great large telescope, providing astronomers with invaluable insights into the universe for decades. However, light pollution from the two cities now fills the night sky, seriously reducing the ability of that telescope to collect light from the stars. Not surprisingly, the best observing sites in the world are high on mountaintops—above smog, water vapor, and clouds—and far from city lights.

3-11   The Hubble Space Telescope provides stunning details about the universe

For decades, astronomers dreamed of observatories in space. Such facilities would eliminate the image distortion created by twinkling and by poor atmospheric transparency due to pollution, volcanic debris, and water vapor. They could operate 24 hours a day and over a wide range of wavelengths—from the infrared through the visible range and far out into the ultraviolet part of the spectrum. NASA has plans for four such Great Observatories, the first of which is the Hubble Space Telescope (HST), which was carried aloft by the Space Shuttle in April 1990.

Soon after HST was placed in orbit, astronomers discovered that the telescope's 2.4-m primary mirror lacked the proper curvature, which caused its star images to be surrounded by a hazy glow. During a repair mission in December 1993, astronauts installed corrective optics that eliminated the problem (Figure 3-24). It was fully upgraded in 2002. Now HST has a resolution of better than 0.1″, which is better than can be obtained by telescopes on the Earth's surface without the use of advanced technology (see section 3-12).

The observations taken by HST have staggered the imaginations of astronomers. It has made new observations related to the planets in our solar system, planetary systems forming around other stars, the distances to other galaxies, black holes, quasars, the formation of the ear-

FIGURE 3-23 Light Pollution These two images of Tucson, Arizona, were taken from the Kitt Peak National Observatory, which is 38 linear miles away. They show the dramatic growth in ground light output between 1959 (top) and 1989 (bottom). Since 1972, light pollution, a problem for many observatories around the world, has been at least partially controlled by a series of local ordinances. (NOAO/AURA/NSF/Galaxy)

FIGURE 3-24 The Hubble Space Telescope (HST) This photograph of HST hovering above the Space Shuttle's cargo bay was taken in 1993, at completion of the first servicing mission. During its 20-year lifetime, HST is studying the heavens at wavelengths from the infrared through the ultraviolet. (NASA)

liest galaxies, and the age of the universe, among many other things. This success, coupled with rapidly improving technology, has spurred scientists and engineers to begin developing the Next Generation Space Telescope called NGST to replace Hubble. NGST is scheduled for launch in 2009.

3-12 Advanced technology is spawning a new generation of superb telescopes

The clarity of images taken by the Hubble Space Telescope may suggest that ground-based observational astronomy is a dying practice. However, two exciting ground-based techniques, called active optics and adaptive optics, promise to match the quality of Hubble—or better it!

Active optics finds the best orientation for the primary mirror in response to changes in temperature and the shape of the telescope mount. It adjusts the mirror every few seconds to help keep the telescope aimed at its target. With active optics, the New Technology Telescope in Chile and the Keck telescopes in Hawaii routinely achieve resolutions as fine as 0.3″ when the resolution is much worse for telescopes without active optics at the same sites.

Adaptive optics uses sensors to determine the amount of twinkling created by atmospheric turbulence. The stellar motion is neutralized by computer-activated, motorized supports that actually reshape either the primary mirror or a smaller mirror installed farther down the optical path of the telescope. Adaptive optics effectively eliminates atmospheric distortion and produces remarkably sharp images (Figure 3-25). Many large ground-based telescopes now use adaptive optics on at least some observations, resulting in images comparable to those from HST.

Until the 1980s, telescopes with primary mirrors of between 2 and 6 m were the largest and most powerful in the world. Now, new technologies in mirror building and computer control allow us to construct much larger telescopes. There are at least 49 reflectors around the world today with primary mirrors measuring 2 m or

a   b   c   R I V U X G

FIGURE 3-25 Images from Earth and Space (a) Image of Saturn from an Earth-based telescope without adaptive optics. (b) Image of Saturn from an Earth-based telescope with adaptive optics. (c) Image of Saturn from the Hubble Space Telescope, which does not incorporate adaptive optics technology. (a & b: Air Force Research Laboratory, Starfire Optical Range, Kirtland AFB, NM; c: Reta Beebe/New Mexico State University, D. Gilmore, L. Bergeron/STScI, and NASA)

FIGURE 3-26 The 10-m Keck Telescopes Located on the (hopefully) dormant Mauna Kea volcano in Hawaii (see chapter opening photograph), these huge twin telescopes each consist of 36 hexagonal mirrors measuring 1.8 m (5.9 ft) across. Each Keck telescope has the light-gathering, resolving, and magnifying ability of a single mirror 10 m in diameter. *Inset:* View down the Keck I telescope. The hexagonal apparatus near the top of the photograph shows the housing for the 1.4 m secondary mirror. (W. M. Keck Observatory, Courtesy of Richard J. Wainscoat)

more in diameter. This number is up by 11 in only 3 years. Among these are ten with mirrors between 8 and 10 m in diameter, with more than a dozen other very large telescopes under construction.

Because the cost of building very large mirrors is enormous, astronomers have devised less expensive ways to collect the same amount of light. One approach is to make a large mirror out of smaller pieces, fitted together like floor tiles. The largest examples of this segmented-mirror technique are the 10-m (400 in.) Keck I and Keck II telescopes on the summit of Mauna Kea in Hawaii. Thirty-six hexagonal mirrors are mounted side by side in each telescope to collect the same amount of light as a single primary mirror 10 m in diameter (Figure 3-26). Another method combines images from different telescopes. For example, used together to observe the same object, the Keck telescopes have the resolving power of a single 85-meter telescope.

3-13 Storing and analyzing light from space is key to understanding the cosmos

The invention of photography during the nineteenth century was a boon for astronomy. By taking a long exposure with a camera mounted at the focal point of a telescope, an astronomer could record extremely faint features that could not be seen just by looking through the telescope. The reason is that our eyes clear the images in them several times a second, whereas film adds up the intensity of all the photons affecting its emulsion. Brighter, higher-resolution images therefore reveal greater detail in galaxies, star clusters, and planets than we could otherwise see. Moreover, we can collect even more light by taking long exposures—hour-long exposures are quite routine.

Astronomers have long known, however, that a photographic plate is an inefficient detector of light because it depends on a chemical reaction to produce an image. Typically, only 2% of the light striking a photographic plate triggers a reaction in the photosensitive material of the plate. Thus, roughly 98% of the light falling onto a photographic plate is wasted.

Technology has changed all that. We have now replaced photographic film with highly efficient electronic light detectors called **charge-coupled devices (CCDs)**. Each CCD is roughly the size of a large postage stamp (Figure 3-27). A CCD is divided into an array

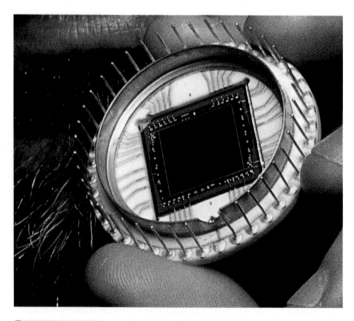

FIGURE 3-27 A Charge-Coupled Device (CCD) This tiny silicon square contains 16,777,216 light-sensitive pixels that store images in a one-piece CCD array. Electronic circuits transfer the data to a waiting computer. (IRFA, University of Hawaii, and Roger Ressmeyer. ©1993 Corbis)

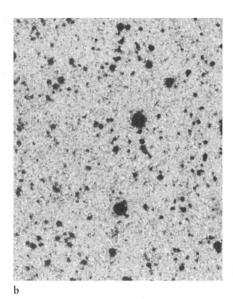

a   b   c   R I V U X G

FIGURE 3-28   Ordinary Photography Versus CCD Images
These three views of the same part of the sky, each taken with the same 4-m telescope, compare CCDs to photographic plates. **(a)** A negative print (black stars and white sky) of a photographic image.

(b) A negative CCD image. Notice that many faint stars and galaxies virtually invisible in the ordinary photograph can be seen clearly in this CCD image. **(c)** This (positive) color view was produced by combining a series of CCD images taken through colored filters. (Patrick Seitzer, NOAO)

of small, light-sensitive squares called picture elements or, more commonly, **pixels.** For example, one of the latest CCDs has more than 16 million pixels arranged in 4096 rows by 4096 columns. This is about 1000 times more pixels per square centimeter than in a typical computer screen.

When an image from a telescope is focused on the CCD, an electric charge builds up in each pixel in direct proportion to the intensity of the light falling on that pixel. When the exposure is finished, the charge on each pixel is read into a computer. The computer then transfers the image onto a television monitor. CCDs commonly respond to 70% of the light falling on them; their resolution is better than that of film, and they respond more uniformly to light of different colors. Figure 3-28 shows one photograph and two CCD images of the same region of the sky, all taken with the same telescope. Notice that many details visible in the CCD images are absent in the ordinary photograph.

RADIO ASTRONOMY—AND BEYOND

Until the 1930s, all information that astronomers gathered from the universe was based on visible light. Scientists began to wonder if objects in the universe might also emit radio waves, infrared and ultraviolet radiation, and perhaps even X rays or gamma rays. Little did anyone realize back then the enormous range of objects and activities in the universe that emit one or more of these radiations without giving off any detectable visible light!

3-14 A radio telescope uses a large concave dish to reflect radio waves to a focal point

The first evidence of nonvisible radiation from outer space came from the work of a young radio engineer, Karl Jansky, working at Bell Laboratories. Using long antennas, Jansky was investigating the sources of radio static that affect short-wavelength radiotelephone communication. In 1932 he realized that a certain kind of radio noise is strongest when the constellation Sagittarius is high in the sky. Because the center of our Galaxy is located in the direction of Sagittarius, Jansky concluded that he was detecting radio waves from elsewhere in the Galaxy.

> **Insight into Science**   Think "Outside the Box"
> Observations and experiments often require scientists to make connections between seemingly unrelated concepts. Jansky's proposal that some radio waves originate in space is an example of a scientist connecting apparently disparate scientific fields—astronomy and radio engineering.

Radio telescopes record radio signals from the sky. At first, astronomers were not enthusiastic about detecting radio noise from space, in part because of the poor angular resolution of early radio telescopes. The angular resolution

GUIDED DISCOVERY   Buying a Telescope

If you feel elated discovering treasures, you may want to consider viewing the night sky through powerful binoculars or a telescope. Both types of instruments enable you to find many beautiful, exciting objects in space. Binoculars are nice because you can quickly scan the heavens to see craters on the Moon, star clusters, the Great Nebula of Orion, and even the Andromeda galaxy. Telescopes open new vistas and even enable amateur astronomers to discover new comets, among other things, adding to our scientific database and immortalizing the discoverers' names.

There are four basic types of telescope mounts that steer the telescopes around the sky: The Fork Equatorial Mount, the German Equatorial Mount, the Altitude-Azimuth (Alt-Azimuth) Mount, and the Dobsonian Mount (see figure below).

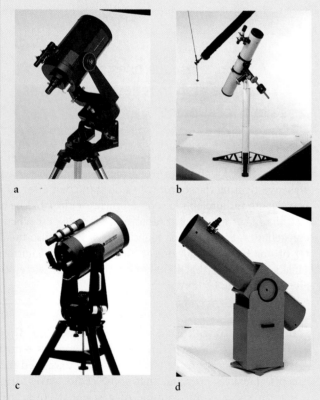

a   b

c   d

Buying a Telescope **(a)** Fork Equatorial Mount, **(b)** German Equatorial Mount, **(c)** Alt-Azimuth Mount, **(d)** Dobsonian Mount. (a, b, & d: Andy Crawford/Dorling Kindersley; c: Celestron Images)

The table on the next page presents the advantages and disadvantages of each mount.

Reflecting and refracting telescopes both have pros and cons, as discussed in the text. Perhaps the most significant difference is in their lengths. Newtonian reflectors tend to be longer and more bulky than any other type. Nevertheless, Newtonian reflectors are the least expensive and simplest to build. As a result, they are the telescope of choice for the Dobsonian mounts. If you plan to take photographs, a Cassegrain telescope is the best instrument because it remains balanced as the telescope tracks objects across the sky.

If you want to see especially large areas of the sky, you may want to purchase a telescope with a *Schmidt corrector plate*. The most common wide-field telescopes are of the Schmidt-Cassegrain design. These telescopes have a Schmidt corrector plate on top and a Cassegrain mirror on the bottom with their eyepieces located underneath.

If you build or buy a telescope with the eyepiece positioned so that you have to crawl under it or climb a ladder to see through it, you will find the experience less satisfying than if you can observe in a comfortable position. If the eyepiece is not easily accessible, you can buy a *diagonal mirror* (a right-angle mirror that goes between the telescope body and the eyepiece) to help correct this problem.

You will also want to get a few different *eyepieces* (three is a good number to start with), so that you can look at large areas of the sky under low magnification and details of small areas under high magnification. As discussed in the text, the magnification of a telescope depends on the focal lengths of the objective lens and the eyepiece. The objective lens or primary mirror is fixed in the telescope. However, all telescopes come with removable eyepieces. Change the eyepiece to another of different focal length, and you change the telescope's magnification. Keep in mind, however, that increasing the magnification decreases the area of the sky that you see.

Unless your telescope is computerized, it is also essential to have a *finder scope*, which is a small, low-magnification, very-wide-field telescope, with crosshairs, attached directly onto your main telescope. If the finder and your main scope are well aligned, you can quickly zero in on an object.

If you plan to take photographs with a CCD camera or other instrument on your telescope or to show the cosmos to large groups of people, you will need a *tracking motor* to enable the telescope to follow the stars automatically. Tracking motors require their own power from batteries, an adaptor for a car cigarette lighter, or a 110-V outlet. The motor should also be able to run at high speed so you can slew (turn) the telescope rapidly from one object to the next. For photography, you will also need a *camera adapter*.

Viewing the Sun is very exciting, especially when you can observe sunspots. If you want to observe the Sun, it is essential that you buy a good *Sun filter.* **Never, ever look at the Sun directly, either through a telescope or with your naked eye. Doing so for even a second can lead to partial or total blindness!**

Finally, you will need a flashlight with a red plastic film over the lens or one with a red LED light. Because red light does not cause the pupils of your eyes to contract, they will remain dilated (wide open) while you use the red flashlight to inspect your equipment and your star charts.

Telescope Mounts

Type of Telescope	Pros	Cons
Fork Equatorial	Can track objects in the sky with a clock drive Useful for taking CCD or film photographs Not too heavy or cumbersome	Eyepiece is in an uncomfortable position near celestial poles
German Equatorial	Can track objects in the sky with a clock drive Useful for taking CCD or film photographs Easy to change direction telescope points	Hard at first to learn to set up Takes a long time to set up Very heavy compared to other types of mounts
Alt-Azimuth	Compact—easy to set up, store, and carry Eyepiece is convenient for viewing	Must be computerized in order to track objects in the sky
Dobsonian	Least expensive for a given diameter primary mirror Easy to set up Easy to use	Eyepiece is often in inconvenient position Cannot track objects in the sky or take photographs without specialized equipment

FIGURE 3-29   **A Radio Telescope** Recall that the secondary mirror or prime focus on most telescopes block incoming light or other radiation. This new radio telescope at the National Radio Astronomy Observatory in Green Bank, West Virginia, has its prime focus hardware located off-center from the telescope's 100 m reflector. By using this new design, there is no such loss of signal. Such configurations are also common on microwave dishes used to receive satellite TV transmissions. (National Radio Astronomy Observatory)

of any telescope decreases as the wavelength increases. In other words, the longer the wavelength, the fuzzier the picture. Because radio radiation has very long wavelengths, astronomers first thought that radio telescopes could only produce blurry, indistinct images.

VIDEO 3.1 Radio telescopes each have a large, reflecting, concave dish that acts exactly like a mirror in an optical telescope (Figure 3-29). A small antenna tuned to the desired wavelength is located either at the prime focus or the Cassegrain focus. The incoming signal is relayed to amplifiers and recording instruments. Very large radio telescopes create sharper radio images because, as with optical telescopes, the bigger the dish, the better the angular resolution. For this reason, most modern radio telescopes have dishes more than 25 m in diameter. Nevertheless, even the largest radio dish in existence (305-m diameter in Arecibo, Puerto Rico) cannot come close to the resolution of the best optical telescopes. For example, a 6-m optical telescope has 2000 times better resolution than a 6-m radio telescope detecting radio waves of 1-mm wavelength.

To overcome the limitation on resolution set by telescope diameter, a clever technique enables radio astronomers to produce radio images often better than those available in the optical part of the spectrum. Called **interferometry**, the idea is to combine the data received simultaneously by two or more telescopes. The telescopes can be kilometers or even continents apart. The radio signals received by all the dishes are made to "interfere," or blend together, and with suitable

computer-aided processing the combined image of the source is sharp and clear. The results are impressive: The resolution of such a system is equivalent to that of one gigantic dish with a diameter equal to the distance between the farthest-most telescopes in the array.

Interferometry, exploited for the first time in the late 1940s, gave astronomers their first detailed views of "radio objects" in the sky. More recently, radio telescopes separated by thousands of kilometers have been linked together in **very-long-baseline interferometry (VLBI)**, producing images that are much sharper and clearer than even those from optical telescopes. The best angular resolution on Earth is obtained by combining radio data from telescopes on opposite sides of the Earth. In that case, features as small as 0.00001″ can be distinguished at radio wavelengths—10,000 times sharper than the best views obtainable from single optical telescopes. Radio telescopes are also being put into space and used in even longer-baseline interferometers. Interferometry is so successful that it has recently been applied successfully to optical and infrared telescopes.

One of the most complex systems of radio telescopes began operating in 1980 on the Plains of San Agustin near Socorro, New Mexico. Called the Very Large Array (VLA), it consists of 27 concave dishes, each 26 m (85 ft) in diameter. The 27 telescopes are positioned along the three arms of a gigantic Y that can span a distance of 36 km (22 mi). Working together, they can create radio images with 0.1″ resolution. Figure 3-30 shows the VLA. This system produces

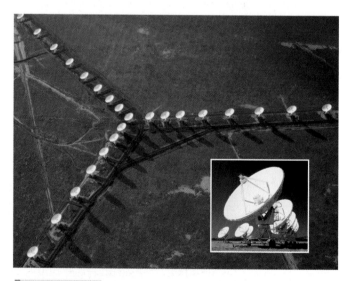

FIGURE 3-30   **The Very Large Array (VLA)** The 27 radio telescopes of the VLA system are arranged along the arms of a Y in central New Mexico. The telescopes can be moved so that the array can detect either wide areas of the sky (when they are close together, as in this photograph) or small areas with higher resolution (when they are farther apart). The inset shows the traditional secondary mirror assembly in the center of each of these antennas. (Jim Sugar/Corbis; *inset:* David Nunuk/Science Photo Library/Photo Researchers)

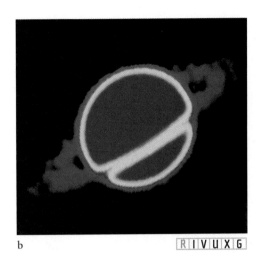

a   R I V U X G   b   R I V U X G

FIGURE 3-31 Optical and Radio Views of Saturn **(a)** This picture was taken by a camera on board a spacecraft as it approached Saturn. The view was produced by sunlight scattered from the planet's cloudtops and rings. **(b)** This false-color picture, taken by the VLA, shows radio emission from Saturn at a wavelength of 2 cm (a: NASA; b: ©1982 Associated Universities, Inc. under contract with the National Science Foundation/VLA observations by Imke de Pater, J. R. Dickel)

radio views of the sky with resolutions comparable to that of the very best optical telescopes.

To make radio images more comprehensible, radio astronomers often use false colors or gray scales to display their radio views of astronomical objects. An example of the use of false colors is shown in Figure 3-31. The most intense radio emission is shown in red, the least intense in blue. Intermediate colors of the rainbow represent intermediate levels of radio intensity. Black indicates that there is no detectable radio radiation. Astronomers working at other nonvisible-wavelength ranges also frequently use false-color techniques to display views obtained from their instruments.

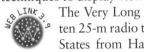 The Very Long Baseline Array (VLBA) consists of ten 25-m radio telescopes located across the United States from Hawaii to New Hampshire. With a maximum baseline of 8000 km, VLBA has a resolving power of 0.001″. It is making major contributions to astronomy, such as the discovery in 1994 of a black hole in a nearby galaxy (discussed in Chapter 13). The 1997 addition of a radio telescope in space tripled the maximum baseline for radio telescope arrays.

3-15 Infrared and ultraviolet telescopes also use reflectors to collect their electromagnetic radiation

As the success of radio astronomy mounted, astronomers started exploring the possibility of making observations at other nonvisible wavelengths. The next two parts of the spectrum to be explored were infrared and ultraviolet. As with radio telescopes, infrared and ultraviolet telescopes are all reflectors. Because water vapor is the main absorber of infrared radiation from space, locating infrared observatories at sites of low humidity can overcome much of the atmosphere's hindrance. For example, the summit of Mauna Kea on Hawaii is exceptionally dry (most of the moisture in the air is

below the height of this volcano), and infrared observations are the primary function of NASA's 3-m Infrared Telescope Facility (IRTF) there.

 The best way of avoiding water vapor is to place a telescope in orbit around the Earth. The 1983 Infrared Astronomical Satellite (IRAS), the HST NICMOS (Near Infrared Camera and Multi-Object Spectrometer), the 1995 Infrared Space Observatory (ISO), and the 2002 Space Infrared Telescope Facility (SIRTF, Figure 3-32) have done much to reveal the full richness and variety of the infrared sky. SIRTF is the infrared equivalent to HST. These are among NASA's Great Observatories. Infrared astronomy has provided astronomers with the ability to take the temperatures of asteroids, see the otherwise invisible surface features on Saturn's cloud-enshrouded moon Titan, the bands of dust in our Galaxy, the dust disks around nearby stars, and the distant galaxies that emit most of their radiation at infrared wavelengths, among many other things. These infrared observatories have located more than a quarter million infrared sources in the sky. Most of these features are invisible to optical telescopes.

The best observations at ultraviolet wavelengths are also made from space, because the Earth's atmosphere absorbs much of this radiation. During the early 1970s, both Apollo and Skylab astronauts used small telescopes above the Earth's atmosphere to give us some of our first views of the ultraviolet sky. Small rockets have also been used to place ultraviolet cameras briefly above the Earth's atmosphere. A typical ultraviolet view is shown in Figure 3-33, along with a corresponding infrared view from IRAS, a view in visible light, and a star chart.

Some of the finest early ultraviolet astronomy was accomplished by the International Ultraviolet Explorer (IUE), which was launched in 1978 and functioned until 1996. The satellite was built around a Cassegrain telescope with a 45-cm (18-in.) mirror and a total focal length of 6.74 m (22 ft). Observations covered the ultraviolet range from 116 to

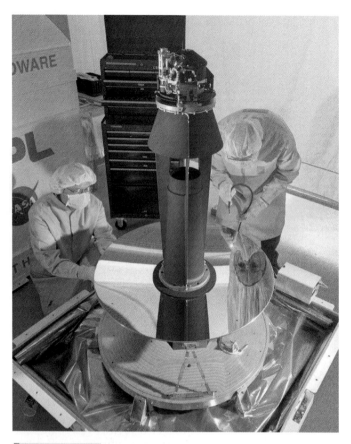

FIGURE 3-32 Mirror Assembly for the Space Infrared Telescope Facility (SIRTF) Launched in 2002, this Great Observatory is scheduled to observe the infrared cosmos for five years. It will take images and spectra of planets, comets, gas, and dust around other stars and in interstellar space, galaxies, and the large-scale distribution of matter in the universe. (Balz/SIRTF Science Center)

320 nm. The Space Shuttle was transformed into an orbiting observatory twice in the 1990s, carrying aloft and then returning to Earth three ultraviolet telescopes. In 1992 the Extreme Ultraviolet Explorer (EUVE) was launched. It is sensitive to photons with wavelengths between 7 and 76 nm, at the short end of the ultraviolet spectrum. As with infrared observations, ultraviolet images reveal sights previously invisible and often unexpected. Many objects in space emit ultraviolet radiation that astronomers can use to study the chemistries of these cosmic bodies. Therefore, astronomers launched the Far Ultraviolet Spectroscopic Explorer (FUSE). It is providing us with information about such things as how much deuterium (hydrogen nuclei with one neutron) was created when the universe formed and the chemical evolution of galaxies.

3-16 X-ray and gamma ray telescopes cannot use normal reflectors to gather information

Because neither X rays nor gamma rays penetrate the Earth's atmosphere, observations at these extremely short wavelengths must be made from space. Astronomers got their first look at the X-ray sky during brief rocket flights in the late 1940s. Several small satellites launched during the early 1970s viewed the entire X-ray and gamma-ray sky, revealing hundreds of previously unknown short-wavelength sources, including several black holes (see Chapter 13). X-ray telescopes have also been carried on Space Shuttle missions (see Figure I-4a).

Because of their high energies, X rays penetrate even highly polished surfaces that they meet head on. Therefore,

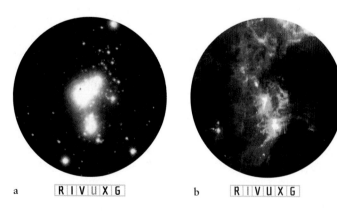

a R I V U X G   b R I V U X G

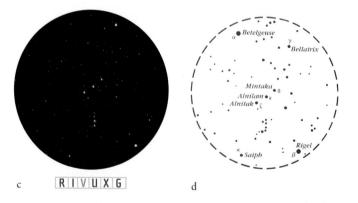

c R I V U X G   d

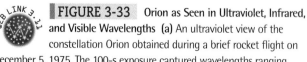

FIGURE 3-33 Orion as Seen in Ultraviolet, Infrared, and Visible Wavelengths (a) An ultraviolet view of the constellation Orion obtained during a brief rocket flight on December 5, 1975. The 100-s exposure captured wavelengths ranging from 125 to 200 nm. (b) A false-color view from the Infrared

Astronomical Satellite uses color to display specific ranges of infrared wavelengths: red indicates long-wavelength radiation; green, intermediate-wavelength radiation; and blue, short-wavelength radiation. For comparison, (c) an ordinary optical photograph and (d) a star chart are included. (a: G. R. Carruthers, NRL; b: NASA; c: R. C. Mitchell, Central Washington University)

normal reflecting mirrors cannot be used to focus them. However, when barely skimming or grazing a surface, X rays can be reflected and thereby focused. Figure 3-34 shows the design of "grazing incidence" X-ray telescopes.

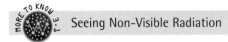
Seeing Non-Visible Radiation

The Chandra X-ray Observatory, another NASA Great Observatory, was launched in 1999 and provides astronomers with the highest resolution X-ray images available. To date, thousands of X-ray sources have also been discovered all across the sky. Among these are planetary atmospheres, stars, stellar remnants, vast clouds of intergalactic gas, jets of gas emitted by galaxies, black holes, quasars, clusters of galaxies, and a diffuse X-ray glow that fills the universe.

The electromagnetic radiation with the shortest wavelengths and the most energy are gamma rays. In 1991 the Compton Gamma Ray Observatory was carried aloft by the Space Shuttle. Named in honor of Arthur Holly Compton, an American physicist who made important discoveries about gamma rays, this orbiting observatory carried four instruments that performed a variety of observations, giving us tantalizing views of the gamma-ray sky. Gamma rays are too powerful even for grazing incidence telescopes. Therefore, astronomers have devised other methods of detecting them and determining where they came from. These include absorbing them in crystals; allowing them to pass through tiny holes called collimators whose directions are well determined; and using chambers in which the gamma rays transform into electrons and positrons (positively charged electrons), leaving a track whose direction can be determined. These techniques are nowhere near as precise as those used in other parts of the spectrum, and the best resolution gamma ray instruments are only accurate to about 5″. In summary, we can now observe in virtually all parts of the electromagnetic spectrum (Figure 3-35).

3-17 Frontiers yet to be discovered

Before the twentieth century, astronomers were like the blind person trying to describe the elephant. Our ancestors did not have the technology that could enable them to see the big picture of the universe. However, we are beginning to see it, and as you will learn in the chapters that follow, our understanding of the cosmos is therefore increasing dramatically. A vast amount of observational information remains to be gathered. Indeed, literally every planet, moon, piece of interplanetary debris, star, stellar remnant, gas cloud, galaxy, quasar, cluster of galaxies, and supercluster of galaxies has a story to tell. Observational astronomy is so new an activity that we are still making new and often unexpected discoveries almost daily. Perhaps the greatest challenge to observational astronomy is the

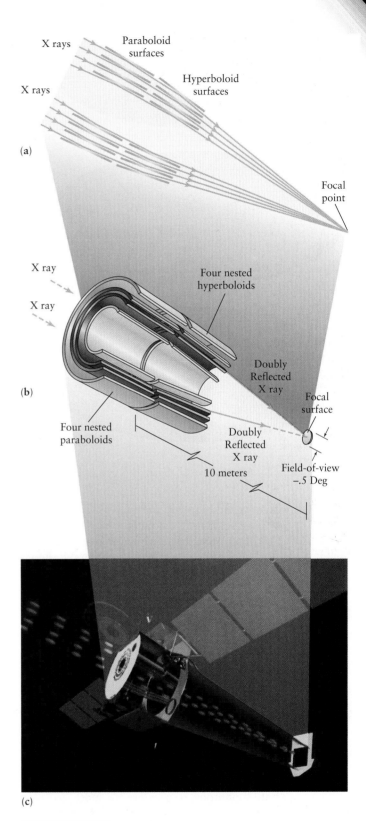

(a)

X rays
Paraboloid surfaces
Hyperboloid surfaces
Focal point

(b)

X ray
X ray
Four nested hyperboloids
Doubly Reflected X ray
Focal surface
Four nested paraboloids
Doubly Reflected X ray
10 meters
Field-of-view −.5 Deg

(c)

FIGURE 3-34 Grazing Incidence X-ray Telescopes X rays penetrate objects they strike head on. To focus them, they have to be gently nudged by skimming off cylindrical "mirrors." The shapes of the mirrors optimize the focus. The bottom diagram shows how X rays are focused in the Chandra X-ray Telescope. (a & b: NASA/CXC/SAO; c: NASA/ Chandra X-ray Observatory Center/Smithsonian Astrophysical Observatory)

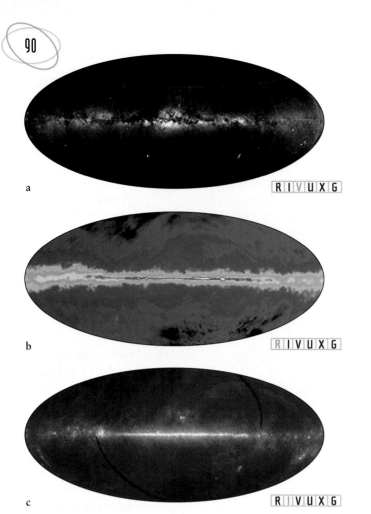

a RIVUXG

b RIVUXG

c RIVUXG

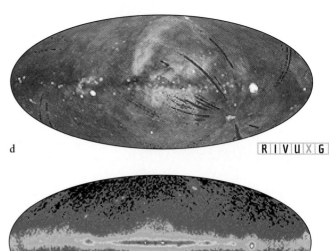

d RIVUXG

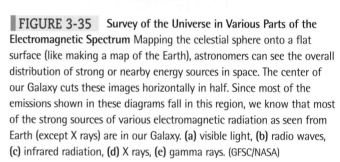

e RIVUXG

FIGURE 3-35 Survey of the Universe in Various Parts of the **Electromagnetic Spectrum** Mapping the celestial sphere onto a flat surface (like making a map of the Earth), astronomers can see the overall distribution of strong or nearby energy sources in space. The center of our Galaxy cuts these images horizontally in half. Since most of the emissions shown in these diagrams fall in this region, we know that most of the strong sources of various electromagnetic radiation as seen from Earth (except X rays) are in our Galaxy. **(a)** visible light, **(b)** radio waves, **(c)** infrared radiation, **(d)** X rays, **(e)** gamma rays. (GFSC/NASA)

fact that when we add up all the currently observable matter in the universe, it only amounts to about 5 percent of all the "stuff" that must exist. We know that there is so much more out there because of its observed gravitational effects. It helps keep galaxies from flying apart, but we cannot yet see it. There is still a lot to discover about the elephant.

 Further Reading on These Topics

WHAT DID YOU THINK?

1 *What is light?* Light, more properly "visible light," is one form of electromagnetic radiation. All electromagnetic radiation (radio waves, infrared radiation, visible light, ultraviolet radiation, X rays, and gamma rays) has both wave and particle properties.

2 *What type of electromagnetic radiation is most dangerous to life?* Gamma rays have the highest energies of all photons, so they are the most dangerous to life. However, ultraviolet radiation from the Sun is the most common everyday form of dangerous electromagnetic radiation we encounter.

3 *What is the main purpose of a telescope?* A telescope is designed primarily to collect as much light as possible.

4 *Why do stars twinkle?* Rapid changes in the density of the Earth's atmosphere cause passing starlight to change direction, making stars appear to twinkle.

5 *What type(s) of electromagnetic radiation can telescopes currently detect?* Telescopes have been built that can observe the entire electromagnetic spectrum.

KEY WORDS

achromatic lens, 74
active optics, 81

adaptive optics, 81
angular resolution, 72

Cassegrain focus, 77
charge-coupled device (CCD), 82

KEY IDEAS

The Nature of Light

• Photons, compact units of vibrating electric and magnetic fields, all carry energy through space at the same speed, "the speed of light" (300,000 km/s in a vacuum, slower in any medium).

• Radio waves, infrared, visible light, ultraviolet radiation, X rays, and gamma rays are all forms of electromagnetic radiation.

• Visible light occupies only a small portion of the electromagnetic spectrum.

• The wavelength of a visible light photon is associated with its color. Wavelengths of visible light range from about 400 nm for violet light to 700 nm for red light.

• Infrared radiation and radio waves have wavelengths longer than those of visible light. Ultraviolet radiation, X rays, and gamma rays have wavelengths that are shorter.

Optics and Telescopes

• A telescope's most important function is to gather as much light as possible. Its second function is to reveal the observed object in as much detail as possible. Often the least important function of a telescope is to magnify objects.

• Refracting telescopes, or refractors, produce images by bending light rays as they pass through glass lenses. Glass impurity, opacity to certain wavelengths, and structural difficulties make it inadvisable to build extremely large refractors.

• Reflecting telescopes, or reflectors, produce images by reflecting light rays from concave mirrors to a focal point or focal plane. Reflectors are not subject to many of the problems that limit the usefulness of refractors. Telescopes that employ advanced technologies, such as active or adaptive optics, produce extremely sharp images.

• Charge-coupled devices (CCDs) are used at a telescope's focal point to record images.

• Earth-based telescopes are being built with active and adaptive optics. These advanced technologies yield resolving power comparable to the Hubble Space Telescope.

Radio Astronomy—and Beyond

• Radio telescopes have large reflecting antennas (dishes) that are used to focus radio waves.

• Very sharp radio images are produced with arrays of radio telescopes linked together in a technique called interferometry.

• The Earth's atmosphere is transparent to most visible light and radio waves, along with some infrared and ultraviolet, arriving from space, but it absorbs much of the electromagnetic radiation at other wavelengths.

• For observations at other wavelengths, astronomers depend upon telescopes carried above the atmosphere by rockets and satellites. Satellite-based observatories are giving us a wealth of new information about the universe and permit coordinated observation of the sky at all wavelengths.

REVIEW QUESTIONS

The answers to all computational problems, which are preceded by an asterisk (*), appear at the end of the book.

1 Describe refraction and reflection. How do these processes enable astronomers to build telescopes?

2 Give everyday examples of refraction and reflection.

3 Describe a refracting telescope by doing Interactive Exercise 3-1 and transcribe the drawing and correct labels to paper, if requested.

*4 How much more light does a 3-m-diameter telescope collect than a 1-m-diameter telescope?

5 With the aid of a diagram, describe a reflecting telescope. Describe four different ways in which an astronomer can access the light collected by reflecting telescopes.

6 Explain some of the advantages of reflecting telescopes over refracting telescopes.

7 What are the three major functions of a telescope?

8 What is meant by the angular resolution of a telescope?

9 What limits the ability of the 5-m telescope on Mount Palomar to collect starlight?

10 Why will many of the very large telescopes of the future make use of multiple mirrors or ultrathin mirrors?

11 What is meant by adaptive optics? What problem does adaptive optics overcome?

12 Compare an optical reflecting telescope and a radio telescope. What do they have in common? How are they different?

13 Why can radio astronomers observe at any time during the day, whereas optical astronomers are mostly limited to observing at night?

14 Why must astronomers use satellites and Earth-orbiting observatories to study the heavens at X-ray or gamma-ray wavelengths?

15 Why did Rømer's observations of the eclipses of Jupiter's moons support the heliocentric, but not the geocentric, cosmogony?

ADVANCED QUESTIONS

16 Advertisements for telescopes frequently give a magnification for the instrument. Is this a good criterion for evaluating telescopes? Explain your answer.

*17 The observing cage in which an astronomer sits at the prime focus of the 5-m telescope on Mount Palomar is about 1 m in diameter. What fraction of the incoming starlight is blocked by the cage? *Hint:* The area of a circle of diameter d is $\pi d^2/4$, where $\pi \approx 3.14$.

*18 Compare the light-gathering power of the Palomar 5-m telescope to that of the fully dark-adapted human eye, which has a pupil diameter of about 5 mm.

19 Show by means of a diagram why the image formed by a simple refracting telescope is "upside down."

*20 Suppose your Newtonian reflector has a mirror with a diameter of 20 cm and a focal length of 2 m. What magnification do you get with eyepieces whose focal lengths are (a) 9 mm, (b) 20 mm, and (c) 55 mm?

21 Why does no major observatory have a Newtonian reflector as its primary instrument, whereas Newtonian reflectors are extremely popular among amateur astronomers?

DISCUSSION QUESTIONS

22 Discuss the advantages and disadvantages of using a small optical telescope in Earth orbit versus a large optical telescope on a mountaintop.

23 If you were in charge of selecting a site for a new observatory, what factors would you consider?

WHAT IF ...

24 Telescopes were first invented today? What objects or areas of the sky would you recommend that astronomers explore first? Why?

25 An observatory were established on the Moon? List the advantages and disadvantages for astronomy.

26 We had eyes sensitive to radio waves? How would we be different and how would our visual perceptions of the world be different?

27 Humans were unable to detect any electromagnetic radiation? How would that change our lives and what alternatives might evolve (indeed have, for some species) to provide information about distant objects?

WEB/CD-ROM QUESTIONS

28 Several telescope manufacturers build telescopes with a design called a "Schmidt-Cassegrain." These use a correcting lens in an arrangement like that shown in Figure 3-18c. Consult advertisements on the Web to see the appearance of these telescopes and find out their cost. Why are they very popular among amateur astronomers?

29 Projects are under way to build large optical reflectors in South Africa (SALT, the Southern African Large Telescope) and in the Canary Islands (GTC, the Gran Telescopio Canarias). Search for current information about these telescopes on the Web. What will be the sizes of the primary mirrors? Are the telescopes designed for imaging, for spectroscopy, or both? Will they observe only at visible wavelengths? In what ways do they complement or surpass existing telescopes?

30 Access the Active Integrated Media Module "Telescope Magnification" in Chapter 3 of the *Discovering the Universe* Web site or CD-ROM. A common telescope found in department stores is a 3-inch (76-mm) diameter refractor with a $f_{obj} = 750$ mm that boasts a magnification of 300 times. Use the magnification calculator to determine the magnifications that are achieved by using each of the following commonly found eyepieces on that telescope: Eyepiece A with focal length 40 mm; Eyepiece B with focal length 25 mm; Eyepiece C with focal length 12 mm; and Eyepiece D with focal length 2.5 mm. Which eyepiece was used in the advertisement and why was that one chosen?

OBSERVING PROJECTS

31 During the daytime, obtain a telescope and several eyepieces of differing focal lengths. If you can determine the telescope's focal length (often printed on it), calculate and record the magnifying power of each eyepiece. Focus the telescope on some familiar object, such as a distant lamppost or tree. **DO NOT FOCUS ON THE SUN! Looking directly at the Sun through a telescope will cause blindness.** Describe the image you see through the telescope. Is it upside down? How does the image move as you slowly and gently shift the telescope left and right or up and down? Examine the distant objects under different magnifications. How do the field of view and the quality of the image change as you go from low magnification to high magnification?

32 On a dark, clear, moonless night, can you see the Milky Way from where you live? If so, briefly describe its appearance. If not, what is interfering with your ability to see it?

33 Determine what fraction of stars visible to the naked eye you can see from your location. On the next dark, clear night, observe the night sky and then run *Starry Night Backyard*™. Press the "sky" button on top and then choose and record which of the following most accurately represents your actual night sky: No Sky Pollution/Small City Light Pollution/Large City Light Pollution. Right mouse click and then choose centre/lock with the little hand near the center of the screen. Set the sky pollution to match your sky. If you have used either sky pollution setting, the goal of this project is to estimate what fraction of the visible stars you are not seeing. If you are using the No Sky Pollution setting, the goal is to see what fraction of the stars the inner-city viewers are missing. Using a roughly 3 inch/8 cm-square section of the sky on the screen (about the size of a 1.44 MB floppy disk), count and record the number of stars with your present setting. Now set the sky to either No Sky Pollution in the first case or Large City Light Pollution in the second and count the stars. What fraction of the stars were you seeing in the first case? What fraction of the stars are large city viewers missing in the second case?

34 Observe the stars when the Moon is either full, new, or in a quarter phase. Record the phase of the Moon and note, qualitatively, whether you see many more stars than those just visible in the well-known asterisms such as the Big Dipper or Orion. You are taking this information so you can compare it with observations on a later date. Repeat these observations on a clear night about a week later, again noting the phase of the Moon and the numbers of visible stars. Compare the numbers of stars on the two nights. On which night did you see more stars? Why? During what three phases of the Moon did you expect that astronomers most like to make their observations?

35 On a clear night, view the Moon, a planet, and a star through a telescope using eyepieces of various focal lengths. (Use your *Starry Night Backyard*™ program or consult a source on the Internet or such magazines as *Sky & Telescope* or *Astronomy* to determine the phase of the Moon and the locations of the planets.) How do the images change as you view with increasing magnification? Do they degrade at any point?

36 Many towns and cities have amateur astronomy clubs. Attend a "star party" hosted by your local club. People bring their telescopes to these gatherings and are delighted to show you their instruments and take you on a telescopic tour of the heavens. Such experiences can lead to a very enjoyable, lifelong hobby.

WHAT IF . . .

HUMANS HAD INFRARED-SENSITIVE EYES?

Our eyes are sensitive to less than a trillionth of 1% of the electromagnetic spectrum—what we call "visible light." But this minuscule resource provides an awe-inspiring amount of information about the universe. We interpret visible-light photons as the six colors of the rainbow—red, orange, yellow, green, blue, and violet. These colors combine to form all the others that make our visual world so rich. But the Sun actually emits photons of all wavelengths. So, what would happen if our eyes had evolved to sense another part of the spectrum?

A Darker Vision Gamma rays, X rays, and most ultraviolet radiation do not pass through the Earth's atmosphere, and the world would look dark, indeed, if our eyes were sensitive only to these wavelengths. Radio waves, in contrast, easily pass through our atmosphere. But to see the same detail from radio waves that we now see from visible-light photons, our eyes would require diameters 10,000 times larger. Each would be the size of a baseball field!

What about infrared radiation? While not all incoming infrared photons get through our atmosphere, short-wavelength ("near") infrared radiation passes easily through air. Depending on wavelength, the Sun emits between one-half and a ten billionth as many infrared photons as optical photons. Fortunately, most of these are in the near infrared.

Heat-Sensitive Vision To see infrared photons, human eyes would need to be only 5 to 10 times larger. Some snakes have evolved infrared vision. Portable infrared "night vision" cameras and goggles are available to us humans. Because everything that emits heat emits infrared photons, infrared sight would be very useful. And not everything we see with infrared sight would be due just to reflected sunlight—hotter objects would be intrinsically brighter than cool ones. For example, seeing infrared would allow us to observe changes in a person's emotional state. Someone who is excited or angry often has more blood near the skin, and thus gives off more infrared (heat) than normal. Conversely, someone who is scared has less blood near the skin and thus emits less heat.

Night Vision The night sky would be a spectacular sight through infrared-sensitive eyes. Gas and dust clouds in the Milky Way absorb visible light, thus preventing the light of distant stars from getting to the Earth. However, because most infrared radiation passes through these clouds, we would be able, unaided, to see distant stars that we cannot see today. On the other hand, the white glow of the Milky Way, which is caused by the scattering of starlight by interstellar clouds, would be dimmer, because the gas and dust clouds do not scatter infrared light as much as they do visible light. (The haze created by the Milky Way would not vanish, however, because when gas and dust clouds are heated by starlight, they emit their own infrared radiation.)

Our concept of stars would be different, too. Many stars, especially young, hot ones, are surrounded by cocoons of gas and dust that emit infrared radiation. This dust is heated by the nearby stars. Instead of appearing as pinpoints, many stars would appear to be surrounded by wild strokes of color, and we would have an Impressionist sky.

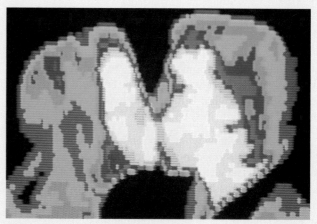

R I V U X G

The infrared (heat) from this kissing couple has been converted into visible light colors so that we can interpret the invisible radiation. The hottest regions are white, with successively cooler areas shown in yellow, orange, red, green, sky blue, dark blue, and violet. (D. Montrose/Custom Medical Stock Photo)

4 THE ORIGIN AND NATURE OF LIGHT

IN THIS CHAPTER YOU WILL DISCOVER

- the origins of electromagnetic radiation

- the structure of atoms

- that stars with different surface temperatures emit different intensities of electromagnetic radiation

- that astronomers can determine the chemical compositions of stars and interstellar clouds by studying the wavelengths of electromagnetic radiation that they absorb or emit

- how to tell whether an object in space is moving toward or away from Earth

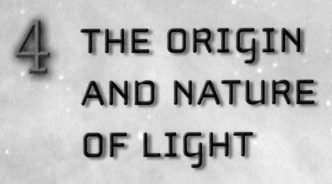

Stellar Spectra
Starlight passing through a prism or diffraction grating spreads into its component colors. From such spectra we can learn an incredible amount about stars, including their masses, surface temperatures, diameters, chemical compositions, and motions toward or away from us. This image, called a spectrogram, shows the spectrum of the Sun sliced and stacked to fit on this page. In it you can see thousands of absorption lines. As you study astronomy, note the distinct differences between the intensities of the various colors emitted by different stars. (NOAO)

R I V U X G

WHAT DO YOU THINK?

1 How hot is a "red hot" object compared to objects glowing other colors?

2 What color is the Sun?

3 Can carbon dating be used to determine the age of space debris found on Earth?

Seeing objects in space is one thing, understanding their physical properties—temperature, chemical composition, mass, size, and motion—is something else altogether. Because almost all our knowledge about space comes from electromagnetic radiation, we now delve deeper into the nature of light. By understanding how light and matter interact, astronomers glean an amazing amount of information about the heavenly bodies.

BLACKBODY RADIATION

We begin our study of the interaction between light and matter with a perfectly familiar property of matter—temperature.

4-1   An object's peak color shifts to shorter wavelengths as it is heated

Imagine that you have taken hold of an iron rod in a completely darkened room. You cannot see the rod, but you can feel its warmth in your hands. The rod is emitting enough infrared radiation (heat) for you to feel, but too few visible photons for your eyes to see. Now imagine that there is also a propane torch in the room. You light it and heat the iron rod for several seconds, until it begins to glow (Figure 4-1a).

The rod's first visible color is red—it glows "red hot." Heated a little more, the rod appears orange and brighter (Figure 4-1b). Heated more, it appears yellow and brighter still (Figure 4-1c). After even more heating, the rod appears white-hot and brighter yet. If it doesn't melt, further heating will make the rod appear blue and even brighter. This "thought experiment," confirmed by the photographs in Figure 4-1, shows how the light any object emits changes with its temperature:

1. As an object heats up, it gets brighter because it emits more electromagnetic radiation.

2. The dominant color or wavelength of the emitted radiation changes with temperature. A cool object emits most of its energy at long wavelengths, such as infrared or red. A hotter object emits most of its energy at shorter wavelengths, such as blue, violet, or even ultraviolet.

Although the iron rod appears to emit a single color depending on its temperature, it actually gives off all wavelengths of electromagnetic radiation. So do stars, rocks, animals, and most other things in nature. However, a single wavelength of emitted radiation, denoted λ_{max}, dominates. When the object is cool, like a rock or animal, λ_{max} is a radio or infrared wavelength. When it is hot enough, λ_{max} is in the range of visible light, giving a hot object its characteristic color. Sufficiently hot stars emit λ_{max} in the ultraviolet part of the electromagnetic spectrum.

Iron bars and stars are good approximations to an important class of theoretical objects that astrophysicists call **blackbodies**. A blackbody absorbs all the electromagnetic radiation that strikes it. None of the incoming radiation is

a   b   c   R I V U X G

FIGURE 4-1 **Heating a Bar of Iron** This sequence of photographs shows the changing appearance of a piece of iron as it is heated. As the temperature increases, the amount of energy radiated by the bar increases. The apparent color of the bar also changes because, as the temperature goes up, the dominant wavelength of light emitted by the bar decreases. (© 1984 Richard Megna Fundamental Photogaphs)

reflected or scattered off its surface. The incoming radiation heats up the blackbody, which then reemits the energy it has absorbed, but not with the same intensities at each wavelength as it received.

 Intensity measures how much energy a blackbody emits per second from any area on its surface. The intensities vary with the wavelength or color, and studies of blackbodies reveal that *the intensity of each wavelength emitted by a blackbody is determined solely by its temperature.* Profiles of the intensity of blackbody radiation at different wavelengths, called **blackbody curves,** are shown for three temperatures in Figure 4-2. (Temperatures throughout this book are expressed in Kelvins. If you are not familiar with the Kelvin temperature scale, review Toolbox A-1, "Temperature Scales," at the back of the book.)

Unlike our iron rod, stars produce their own electromagnetic radiation rather than just absorbing and reradiating light from an outside source. Even so, stars behave like blackbodies, and the self-generated radiation they emit closely follows the idealized blackbody curves. We can, therefore, measure a real star's surface temperature from the location of the peak of its intensity curve. To do that, we need to make our observations quantitative, that is, to express them mathematically.

4-2   The intensities of different emitted colors reveal a star's temperature

 Wien's law is the mathematical relationship between the location of the color peak for each curve in Figure 4-2 and that blackbody's temperature. In 1893 the German physicist Wilhelm Wien found that

The dominant wavelength of radiation emitted by a blackbody is inversely proportional to its temperature.

In other words, the hotter any object becomes, the shorter its λ_{max}, and vice versa.

Wien's law proves very useful in computing the surface temperature of a star, because all we need to know is the dominant wavelength of the star's electromagnetic radiation—we don't need to know the star's size, distance, or any other physical property.

A blackbody curve describes in detail the electromagnetic radiation a star emits. In 1879 the Austrian physicist Josef Stefan observed that

An object emits energy at a rate proportional to the fourth power of its temperature in Kelvins.

In other words, if you double the temperature of an object (for example, from 500 to 1000 K), the energy emitted from each square meter of the object's surface each second in-

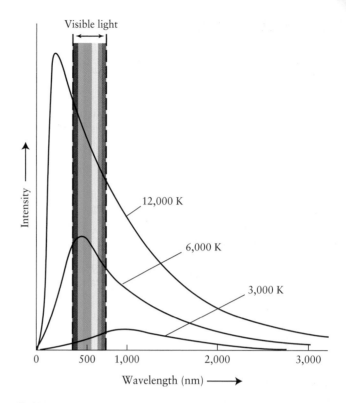

FIGURE 4-2   **Blackbody Curves** Three representative blackbody curves are shown here. Each curve shows the intensity of radiation at every wavelength emitted by a blackbody at a particular temperature. Note that hotter objects emit more of every wavelength and that the peak of emitted radiation changes with temperature.

creases by a factor of 2^4, or 16 times. If you triple the temperature (for example, from 500 to 1500 K), the rate at which energy is emitted increases by a factor of 3^4, or 81 times. Stefan's experimental results were put on firm theoretical ground by Ludwig Boltzmann in 1885. The intensity-temperature relationship for blackbodies is named the **Stefan-Boltzmann law** in their honor.

The Stefan-Boltzmann law and Wien's law are powerful tools for understanding the stars. If you would like to learn how to use these two radiation laws to make predictions, see An Astronomer's Toolbox 4-1. These laws describe only two basic properties of *blackbody radiation,* namely, the relative rate at which it is emitted and the location of its peak. Because blackbodies give off all wavelengths of electromagnetic radiation, the blackbody curves, such as those already shown in Figure 4-2, give a more complete picture. These curves show the intensities of electromagnetic radiation emitted at different wavelengths by a blackbody with a given temperature. As we will see in the next section, the *differences* between an ideal blackbody curve and the curves seen from actual stars reveal stellar chemistries and motions of stars toward or away from us, among other things.

AN ASTRONOMER'S TOOLBOX 4-1

The Radiation Laws

Wien's law can be stated as a simple equation. If λ_{max}, the wavelength of maximum intensity, is measured in meters, then

$$\lambda_{max} = \frac{2.9 \times 10^{-3}}{T}$$

where T is the temperature of the blackbody measured in Kelvins.

Example: The Sun's maximum intensity is at a wavelength of about 500 nm, or 5×10^{-7} m. From Wien's law, we can calculate the Sun's surface temperature as

$$T = \frac{2.9 \times 10^{-3}}{5 \times 10^{-7}} = 5800 \text{ K}$$

The total energy emitted from each square meter of an object's surface each second is called the **energy flux**, F. In this context, flux means "rate of flow." Using this concept, we can write the Stefan-Boltzmann law as

$$F = \sigma T^4$$

where T is the temperature of the object in Kelvins and σ (lowercase Greek sigma) is a number called the Stefan-Boltzmann constant:

$$\sigma = 5.67 \times 10^{-8} \text{J}/(\text{m}^2 \cdot \text{K}^4 \cdot \text{s})$$

where J is the energy unit joules, m is the unit meters, K is degrees Kelvin, and s is the unit seconds.

Because F is the energy emitted per second from each square meter of an object, multiplying F by the surface area, $4\pi r^2$, where r is the object's radius, yields the total energy emitted by the object each second. Denoted L, and called **luminosity**, we have

$$L = F \cdot 4\pi r^2 = \sigma T^4 4\pi r^2$$

Luminosity is the total energy emitted per second by an entire object.

Compare! The Stefan-Boltzmann law has several applications. Suppose you observe two stars of equal size, one with a surface temperature of 10,000 K and the other with the same surface temperature as the Sun (5800 K). You can use the Stefan-Boltzmann law to determine how much brighter the hotter star is:

$$\frac{F_{hotter}}{F_{colder}} = \frac{\sigma \times 10,000^4}{\sigma \times 5800^4} \approx 8.8$$

So, the hotter star emits 8.8 times as much energy from each square meter of its surface than does the cooler star. Because the stars are the same size, the hotter one is therefore 8.8 times brighter than the cooler one.

Try these questions: The color yellow is centered around 550 nm. If the Sun were actually yellow, what would be its surface temperature? To find out why it is not yellow, see Guided Discovery: The Color of the Sun on page 100. What is the peak wavelength in nanometers given off by a blackbody at room temperature of 300 K? Referring to Figure 3-4, what part of the electromagnetic spectrum is that in?

(Answers appear at the end of the book.)

Figure 4-3 shows how the intensity of sunlight varies with wavelength. The blackbody curve for a body with a temperature of 5800 K is also plotted in Figure 4-3. Note how the observed intensity curve for the Sun closely follows the ideal blackbody curve at most wavelengths. Because the observed intensity curves for most stars and the idealized blackbody curves are so closely correlated, the laws of blackbody radiation can be applied to starlight. The peak of the intensity curve for the Sun is at a wavelength of about 500 nm, which is in the blue-green part of the visible spectrum.

By the end of the nineteenth century, physicists realized that they had reached an impasse in understanding electromagnetic radiation. While intensities and peak wavelengths were understood, all attempts to explain the characteris-

tic shapes of blackbody curves had failed. A breakthrough finally came in 1900, when the German physicist Max Planck derived a mathematical formula for the blackbody curves. He assumed that electromagnetic radiation is emitted in separate packets of energy. Light, a wave, behaves like a beam of particles. It was a remarkable result verified in 1905 by Albert Einstein, who called these particles of light *photons.*

We can now explain our thought experiment with the iron bar in terms of the properties of photons. Recall from Chapter 3 that the energy carried by a photon of light is inversely proportional to its wavelength. In other words, long-wavelength photons, such as radio waves, carry little energy. Short-wavelength photons, like X rays and gamma

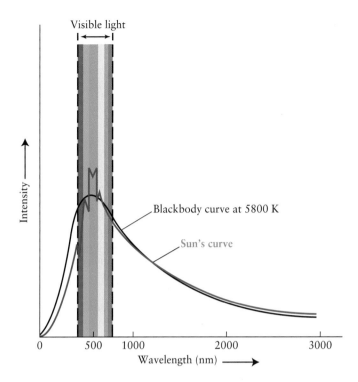

FIGURE 4-3 **The Sun as a Blackbody** This graph compares the intensity of sunlight (in red) over a wide range of wavelengths with the intensity of radiation from a blackbody at a temperature of 5800 K. The Earth's atmosphere scatters shorter-wavelength photons more than it scatters longer-wavelength ones. As seen from Earth, this scattering lowers the intensity of sunlight and shifts the blackbody peak to the right from where it actually occurs. To avoid these effects, the measurements of the Sun's intensity in this figure were made in space. The Sun mimics a blackbody remarkably well.

rays, carry much more energy. The relationship between the energy of a photon and its wavelength is called **Planck's law.** (You may want to refer back to An Astronomer's Toolbox 3-1.)

Together, Planck's law and Wien's law relate the temperature of an object to the energy of the photons it emits (Table 4-1). A cool object emits primarily long-wavelength photons that carry little energy, while a hot object gives off mostly short-wavelength photons that carry much more energy. In later chapters, we will find these ideas invaluable for understanding how stars of various temperatures interact with gas and dust in space.

DISCOVERING SPECTRA

In 1814, the German optician Joseph von Fraunhofer repeated Newton's classic experiment of shining a beam of sunlight through a prism (recall Figure 3-1). But Fraunhofer magnified the resulting rainbow-colored spectrum. He discovered that the solar spectrum contains hundreds of fine dark lines, which became known as **spectral lines.** Fraunhofer counted more than 600 such lines, and today physicists have detected more than 30,000 of them. Thousands of spectral lines are visible in the photograph of the Sun's spectrum shown in the figure opening this chapter.

By the mid-1800s, chemists discovered that they could produce spectral lines in the laboratory. Around 1857, the German chemist Robert Bunsen invented a special gas burner that produces a clean, colorless flame. Certain chemicals are easy to identify by the distinctive colors emitted when bits of the chemical are sprinkled into the flame of a Bunsen burner.

TABLE 4-1   Some Properties of Electromagnetic Radiation

	Wavelength (nm)	Photon energy (eV)*	Blackbody temperature (K)
Radio	$>10^7$	$<10^{-4}$	<0.03
Microwave**	10^7 to 4×10^5	10^{-4} to 3×10^{-3}	0.03 to 30
Infrared	4×10^5 to 7×10^2	3×10^{-3} to 2	30 to 4100
Visible	7×10^2 to 4×10^2	2 to 3	4100 to 7300
Ultraviolet	4×10^2 to 10^1	3 to 10^3	7300 to 3×10^6
X ray	10^1 to 10^{-2}	10^3 to 10^5	3×10^6 to 3×10^8
Gamma ray	$<10^{-2}$	$>10^5$	$>3 \times 10^8$

Note: > means greater than, < means less than.
*1 eV = 1.6×10^{-19} J.
**Microwaves, listed here separately, are often classified as radio waves or infrared radiation.

GUIDED DISCOVERY   The Color of the Sun

Different people perceive the Sun to have different colors. To many it appears white, to others yellow. Still others, who notice it at sunset, believe it to be orange or even red. But we have seen that the Sun actually gives off all colors. Moreover, the peak in the Sun's spectrum falls between blue and green. Why doesn't the Sun appear turquoise? Several factors affect our perception of its color.

Before reaching our eyes, visible sunlight passes through the Earth's atmosphere. Certain wavelengths are absorbed and reemitted by the molecules in the air, a process called *scattering*. Violet light is scattered most strongly, followed in decreasing order by blue, green, yellow, orange, and red. That means that more violet, blue, and green photons are scattered by the Earth's atmosphere than are yellow photons. The intense scattering of violet, blue, and green has the effect of shifting the peak of the Sun's intensity entering our eyes from blue-green toward yellow. (The sky is blue because of the strong scattering of blue light—it isn't violet because the Sun emits many fewer violet photons than blue ones.)

The perception of a yellow Sun is further enhanced by our eyes themselves. Our eyes do not see all colors equally well. Rather, the light-sensitive cones in our eyes each respond to one of three ranges of colors, which are centered on red, yellow, and blue wavelengths. None of the cones is especially sensitive to blue-green photons. By adding together the color intensities detected by the three types of cones, our brains *recreate* color. After combining all the light it can, the eye is most sensitive to the yellow-green part of the spectrum. We see blue and orange less well, and violet and red most poorly. Although the eye sees yellow and green light about equally well, the dominant color from the Sun is yellow because the air scatters green light, and our eyes are relatively insensitive to blue-green. Therefore, a casual glance at the Sun leaves the impression of a yellow object.

A longer look is extremely dangerous. **Don't try it!** Hypothetically, such a glance would leave the impression of a white Sun. The Sun's light is so intense that it would saturate the color-sensitive cones in our eyes; our brains interpret such saturation of the cones as white.

At sunrise and sunset we see an orange or red Sun. This occurs because close to the horizon the Sun's violet, blue, green, and even yellow photons are strongly scattered by the thick layer of atmosphere through which they travel, leaving the Sun looking redder and redder as it sets.

Bunsen's colleague, Gustav Kirchhoff, suggested that light from the colored flames could best be studied by passing it through a prism (Figure 4-4). The chemists promptly discovered that the spectrum from a flame consists of a pattern of thin, bright spectral lines against a dark background (not a blackbody spectrum!). They next found that *each chemical element produces its own characteristic pattern of spectral lines.* Thus was born in 1859 the technique of **spectral analysis,** the identification of chemical substances by their spectral lines.

4-3   Each chemical element produces its own unique set of spectral lines

A chemical **element** is a fundamental substance because it cannot be broken down into more basic units and still retain its properties. By the mid-1800s, chemists had already identified such familiar elements as hydrogen, oxygen, carbon, iron, gold, and silver. Spectral analysis promptly led to the discovery of additional elements, many of which are quite rare.

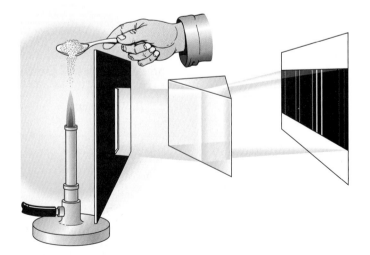

FIGURE 4-4   The Kirchhoff-Bunsen Experiment In the mid-1850s, Kirchhoff and Bunsen discovered that when a chemical substance is heated and vaporized, the resulting spectrum exhibits a series of bright spectral lines. In addition, they found that each chemical element produces its own characteristic pattern of spectral lines.

After Bunsen and Kirchhoff recorded the prominent spectral lines of all the known elements, they began to discover other spectral lines in mineral samples. In 1860, for example, they found a new line in the blue portion of the spectrum of mineral water. After chemically isolating the previously unknown element responsible for the line, they named it cesium (from the Latin *caesius*, meaning gray-blue). The next year a new spectral line in the red portion of the spectrum of a mineral sample led to the discovery of the element rubidium (from *rubidus*, for red).

During a solar eclipse in 1868, astronomers found a new spectral line in the light coming from the upper atmosphere of the Sun when the main body of the Sun was hidden by the Moon. This line was attributed to a new element, which was named helium (from the Greek *helios*, meaning Sun). Helium was not actually discovered on Earth until 1895, when it was identified in gases obtained from a uranium compound.

A list of the chemical elements is most conveniently displayed in the form of a **periodic table** (Figure 4-5). Each element has a unique atomic number (described in Section 4-5), and the elements are arranged in the periodic table by their atomic numbers. With a few exceptions, this sequence also corresponds to increasing average mass of the atoms of the elements. Thus, hydrogen (symbol H) with atomic number 1 is the lightest element. Iron (Fe) has atomic number 26 and is a moderately heavy element.

All the elements in a single vertical column of the periodic table have similar chemical properties. For example, the elements in the far right column are all gases at room temperature and normal air pressure, and they rarely react chemically with other elements.

In addition to the 92 naturally occurring elements, Figure 4-5 lists the artificially produced elements. All the human-made elements are heavier than uranium (U), and all are highly radioactive, meaning that they spontaneously decay into lighter elements shortly after being created in the laboratory.

> **Insight into Science** Seek relationships Finding how properties relate often provides invaluable insights into how new things work. The periodic table, for example, enables scientists to determine the properties of similar chemical elements. Keep an eye out for other such relationships throughout this book.

Because each chemical element produces its own unique pattern of spectral lines, scientists can determine the chemical composition of a remote astronomical object by identifying

Periodic Table of the Elements

1 H Hydrogen																	2 He Helium
3 Li Lithium	4 Be Beryllium											5 B Boron	6 C Carbon	7 N Nitrogen	8 O Oxygen	9 F Fluorine	10 Ne Neon
11 Na Sodium	12 Mg Magnesium											13 Al Aluminum	14 Si Silicon	15 P Phosphorus	16 S Sulfur	17 Cl Chlorine	18 Ar Argon
19 K Potassium	20 Ca Calcium	21 Sc Scandium	22 Ti Titanium	23 V Vanadium	24 Cr Chromium	25 Mn Manganese	26 Fe Iron	27 Co Cobalt	28 Ni Nickel	29 Cu Copper	30 Zn Zinc	31 Ga Gallium	32 Ge Germanium	33 As Arsenic	34 Se Selenium	35 Br Bromine	36 Kr Kryton
37 Rb Rubidium	38 Sr Strontium	39 Y Yttrium	40 Zr Zirconium	41 Nb Niobium	42 Mo Molybdenum	43 Tc Technetium	44 Ru Ruthenium	45 Rh Rhodium	46 Pd Palladium	47 Ag Silver	48 Cd Cadmium	49 In Indium	50 Sn Tin	51 Sb Antimony	52 Te Tellurium	53 I Iodine	54 Xe Xenon
55 Cs Cesium	56 Ba Barium	57 La Lanthanum	72 Hf Hafnium	73 Ta Tantaium	74 W Tungsten	75 Re Rhenium	76 Os Osmium	77 Ir Iridium	78 Pt Platinum	79 Au Gold	80 Hg Mercury	81 Tl Thallium	82 Pb Lead	83 Bi Bismuth	84 Po Polonium	85 At Astatine	86 Rn Radon
87 Fr Francium	88 Ra Radium	89 Ac Actinium	104 Rf Rutherfordium	105 Db Dubnium	106 Sg Seaborgium	107 Bh Bohrium	108 Hs Hassium	109 Mt Meitnerium	110	111	112		114		116		

58 Ce Cerium	59 Pr Praseodymium	60 Nd Neodymium	61 Pm Promethium	62 Sm Samarium	63 Eu Europium	64 Gd Gadolinium	65 Tb Terbium	66 Dy Dysprosium	67 Ho Holmium	68 Er Erbium	69 Tm Thulium	70 Yb Ytterbium	71 Lu Lutetium
90 Th Thorium	91 Pa Protactinium	92 U Uranium	93 Np Neptunium	94 Pu Plutonium	95 Am Americium	96 Cm Curium	97 Bk Berkelium	98 Cf Californium	99 Es Einsteinium	100 Fm Fermium	101 Md Mendelevium	102 No Nobelium	103 Lr Lawrencium

FIGURE 4-5 Periodic Table of the Elements The periodic table is a convenient listing of the elements arranged according to their atomic numbers and chemical properties. Due to naming disagreements, elements 110–112, 114, and 116 do not yet have universally accepted names. Elements 113, 115, 117, and beyond have not yet been discovered.

FIGURE 4-6　Iron in the Sun's Atmosphere　The upper spectrum is a portion of the Sun's spectrum from 420 to 430 nm. Numerous dark spectral lines are visible. The lower spectrum is a corresponding portion of the spectrum of vaporized iron. Several bright spectral lines can be seen against the black background. The fact that the iron lines coincide with some of the solar lines proves that there is some iron (albeit a very tiny amount) in the Sun's atmosphere. (Carnegie Observatories)

the lines in its spectrum. For example, Figure 4-6 shows a portion of the Sun's spectrum along with the spectrum of an iron sample taken here on Earth. No other chemical has iron's particular pattern of spectral lines. It is iron's own distinctive "fingerprint." Because the spectral lines of iron also appear in the Sun's spectrum, we can reasonably conclude that the Sun's atmosphere contains some vaporized iron.

Bunsen and Kirchhoff collaborated in designing and constructing the first **spectroscope.** This device consists of a prism and several lenses that magnify the spectrum so that it can be closely examined. After photography was invented, scientists preferred to produce a permanent photographic record of spectra. A device for photographing a spectrum is called a **spectrograph,** and this instrument is among the astronomer's most important tools.

In its basic form, a spectrograph consists of a slit, two lenses, and a prism arranged to focus the spectrum of an astronomical object onto a CCD (see Figures 3-27 or 3-28), as shown in Figure 4-7. This optical device typically mounts at the focal point of a telescope. The image of the object to be examined is focused on the slit. After the spectrum of a star or galaxy has been photographed, the CCD collects light from a known source, such as a gas composed of helium, neon, and argon, that is focused on the slit. This "comparison spectrum" is placed next to the spectrum of the object, as in Figure 4-6. Because the wavelengths of the spectral lines in the comparison spectrum are already known from laboratory experiments, these lines can be used to identify and measure the wavelengths of the lines in the spectrum of the star or galaxy under study.

There are drawbacks to this type of spectrograph. A prism does not spread colors evenly: The blue and violet portions of the spectrum are spread out more than the red portion. In addition, because the blue and violet wavelengths must pass through more glass than the red wavelengths (see Figure 3-1), light is absorbed unevenly across the spectrum. Indeed, a glass prism is opaque to ultraviolet wavelengths.

A better device for breaking starlight into the colors of the rainbow is a **diffraction grating,** a piece of glass on which thousands of closely spaced lines are cut. Some of the finest diffraction gratings have as many as 10,000 lines per centimeter. The spacing of the lines must be very regular. Light rays reflected from different parts of the diffraction grating interfere with each other to produce a spectrum. Figure 4-8 shows the design of a modern diffraction grating spectrograph.

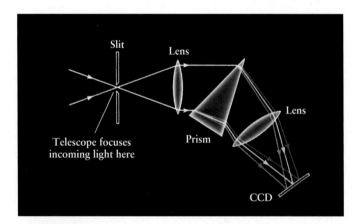

FIGURE 4-7　A Prism Spectrograph　This optical device uses a prism to break up the light from an object into a spectrum. The lenses focus that spectrum onto a CCD.

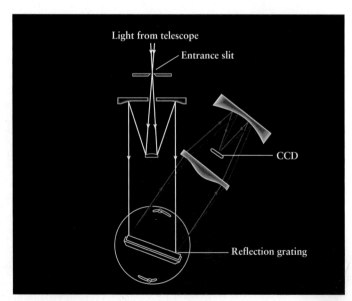

FIGURE 4-8　A Grating Spectrograph　This optical device uses a grating to break up the light from an object into a spectrum. An arrangement of lenses and mirrors focuses that spectrum onto a CCD.

The spectral data are converted by computer to a graph that plots light intensity against wavelength. Dark lines in the rainbow-colored spectrum appear as depressions or valleys on the graph, while bright lines in the spectrum appear as peaks. For example, Figure 4-9 shows both a picture and a plot of a spectrum for hydrogen in which five dark spectral lines appear.

4-4   The brightnesses of spectral lines depend on conditions in the spectrum's source

During his pioneering experiments with spectra, Kirchhoff sometimes saw dark spectral lines, called *absorption lines,* among the colors of the rainbow. In other experiments, he saw bright spectral lines, called *emission lines,* against an otherwise dark background (see Figure 4-4). By the early 1860s, Kirchhoff had discovered the conditions under which these different types of spectra are observed. His description is summarized today as **Kirchhoff's laws:**

Law 1 A hot object or a hot, dense gas produces a **continuous spectrum** (also called a **continuum**)—a complete rainbow of colors without any spectral lines. This is a blackbody spectrum.

Law 2 A hot, rarefied gas produces an **emission line spectrum**—a series of bright spectral lines against a dark background.

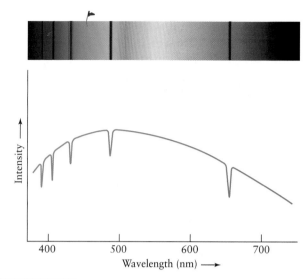

FIGURE 4-9   Spectrum of Hydrogen Gas (a) When a CCD is placed at the focus of a spectrograph, a rainbow-colored spectrum is recorded. (b) The spectrum is converted by computer into a graph of intensity versus wavelength. Note that the dark spectral lines appear as dips in the intensity-versus-wavelength curve.

Law 3 A cool gas in front of a continuous source of light produces an **absorption line spectrum**—a series of dark spectral lines among the colors of the rainbow.

To test your understanding of the above, you may want to try Interactive Exercise 4-1 before examining Figure 4-10, which summarizes how absorption and emission lines are formed.

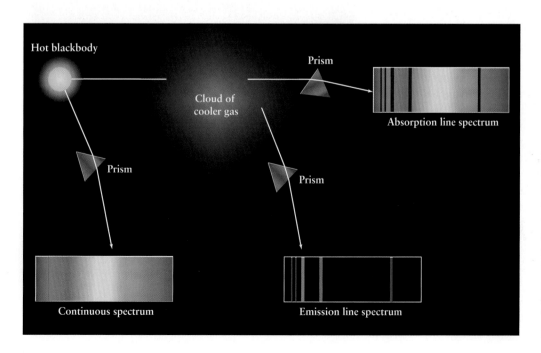

Continuous spectrum

Emission line spectrum

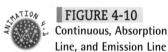

FIGURE 4-10

Continuous, Absorption Line, and Emission Line Spectra This schematic diagram summarizes how different types of spectra are produced. A hot, glowing object emits a continuous spectrum. If this source of light is viewed through a cool gas, dark absorption lines appear in the resulting spectrum. When the same gas is viewed against a cold, dark background, its spectrum consists of just bright emission lines. The animation link shows the details of the activity drawn here.

Absorption lines are seen if the background is hotter than the gas. Emission lines are seen if the background is cooler. Note that the bright lines in the emission spectrum of a particular gas occur at exactly the same wavelengths as the dark lines in the absorption spectrum of that gas.

Consider, for example, the spectrum of the Sun. We know that the Sun's surface emits a continuous, blackbody spectrum, but here on Earth many absorption lines are seen in it (see the figure that opens this chapter). Kirchhoff's third law explains why. There must be a cooler gas between the surface of the Sun and the Earth. In fact, there are two: the Sun's lower atmosphere and the Earth's entire atmosphere. With these observations in hand, we turn now to exploring why the different spectra occur.

ATOMS AND SPECTRA

An atom is the smallest particle of a chemical element that still has the properties of that element. At the time of Kirchhoff's discoveries, scientists knew that all matter is composed of atoms, but they did not know their structures. Scientists saw that atoms of a gas somehow extract light of specific wavelengths from white light that passes through the gas, leaving dark absorption lines, and they concluded that the atoms then radiate light of precisely the same wavelengths—the bright emission lines (see Figure 4-6). But traditional theories of electromagnetism could not explain this phenomenon. The answer came early in the twentieth century with the development of quantum mechanics and nuclear physics.

4-5 An atom consists of a small, dense nucleus surrounded by electrons

The first important clue about the internal structure of atoms came in 1910 from Ernest Rutherford, a chemist and physicist from New Zealand. Rutherford and his colleagues at the University of Manchester in England were investigating the recently discovered phenomenon of radioactivity. Over time, a radioactive element naturally and spontaneously transforms into another element by emitting particles. Certain radioactive elements, such as uranium and radium, were known to emit such particles with considerable speed. It seemed plausible that a beam of these high-speed particles could penetrate a thin sheet of gold. Rutherford and his associates found that almost all the particles passed through the gold sheet with little or no deflection. To the surprise of the experimenters, however, an occasional particle bounced right back. It must have struck something very dense indeed.

Rutherford was quick to realize the implications of his experiment. He correctly reasoned that most of the mass of an atom is concentrated in a compact lump that occupies only a small part of the atom's volume. Most of the radioac-

tive particles pass freely through the nearly empty space that makes up most of the atom, but a few particles happen to strike the dense mass at the center and rebound.

Rutherford proposed a new model for the structure of an atom. According to this model, a massive, positively charged **nucleus** at the center of the atom is orbited by tiny, negatively charged *electrons* (Figure 4-11). Rutherford concluded that at least 99.95% of the mass of an atom is concentrated in the nucleus, whose diameter is only about one ten-thousandth the diameter of the atom.

Further research revealed that the nucleus of an atom contains two types of particles: *protons* and *neutrons*. A proton has almost the same mass as a neutron, and each proton or neutron is about 2000 times more massive than an electron. A proton has a positive electric charge and a neutron has no charge. Because like charges repel each other, protons do not naturally stay bound together; they try to move as far away from each other as possible. Neutrons help keep the protons bound together. Conversely, opposite charges attract, and this attraction keeps electrons in orbit around the nucleus, held there by the positively charged protons.

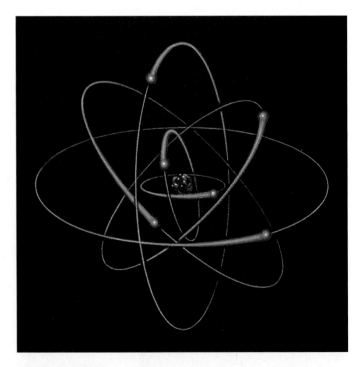

FIGURE 4-11 Rutherford's Model of the Atom Electrons orbit the atom's nucleus, which contains most of the atom's mass. The nucleus contains two types of particles: protons and neutrons. Because the nucleus and the electrons are so small compared to the distance between them, atoms are mostly empty space. The electrons, traveling at a few percent of the speed of light, make the nuclei appear to be surrounded by clouds.

AN ASTRONOMER'S TOOLBOX 4-2

Radioactivity and the Ages of Objects

 Radioactive Age-Dating

The isotopes of many elements are radioactive, meaning that the elements spontaneously transform into other elements. Each radioactive isotope has a distinctive *half-life*, the time it takes half of the initial concentration of the isotope to transform into another element. After two half-lives, a radioactive isotope is reduced to $1/2 \times 1/2$ or $1/4$ of its initial concentration (see the accompanying figure). Among the most important radioactive elements for determining the age of objects in astronomy is the isotope of uranium with 146 neutrons, ^{238}U. The half-life of ^{238}U decaying into lead is 4.5 billion years.

To determine the time since an object, such as a piece of space debris discovered on Earth, solidified, scientists estimate how much lead it had when it formed. Then they measure the amounts of uranium and lead it contains now. Subtracting the amount of original lead, they use the amount of uranium and lead with the graph to determine how long ago the object formed.

Example: Suppose a piece of space debris discovered on Earth was determined to have equal amounts of lead and uranium. How long ago did this debris form? Assuming that it originally had no lead, we see from the chart that a 1-to-1 mix of lead to uranium occurs 4.5 billion years after the object formed. This is one half-life of uranium.

Compare! This process works with any radioactive isotope. However, some isotopes have such short half-lives that they are not useful in astronomy. For example, the well-known carbon dating used to determine the ages of ancient artifacts on Earth is of little use in astronomy because ^{14}C has a half-life of only 5730 years. Because

so much of it has decayed away by then, carbon dating is useful for fewer than 100,000 years, usually a period over which little of astronomical importance occurs.

Try these questions: What fraction of a kilogram of radioactive material remains after 3 half-lives have passed? Using the figure below, estimate how much Uranium will remain after $6^3/4$ billion years. Approximately how many half-lives of ^{14}C pass in 100,000 years?

(Answers appear at the end of the book.)

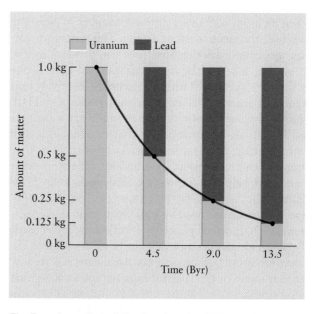

The Transformation of Uranium into Lead This figure shows the rate that 1 kg of uranium decays into lead, as described in the text. It contains $1/8$ kg of uranium after 13.5 billion years.

The number of protons in an atom's nucleus determines what element that atom is. Each element is assigned an **atomic number** that equals the number of protons it contains. Of the naturally occurring elements, a hydrogen nucleus always has 1 proton, a helium nucleus always has 2, and so forth, up to uranium with 92 protons in its nucleus (see Figure 4-5).

In contrast, the number of neutrons in the nuclei of different atoms of the same element may vary. For example, oxygen, the eighth element in the periodic table, with an atomic number of 8, always has eight protons, but it may

have eight, nine, or ten neutrons. These three slightly different kinds of oxygen are called **isotopes.** The isotope of oxygen with eight neutrons is by far the most abundant variety.

Many elements have isotopes that are radioactive. For example, carbon with six neutrons, ^{12}C, is stable, while carbon with eight neutrons, ^{14}C, is unstable. ^{14}C decays into nitrogen with seven neutrons, ^{14}N. To learn how radioactive decay is used to determine the ages of different objects, see An Astronomer's Toolbox 4-2.

Normally, the number of electrons orbiting an atom is equal to the number of protons in the nucleus, thus making

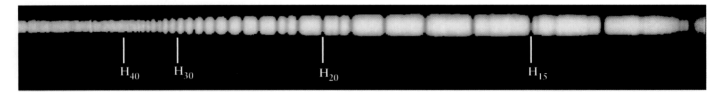

FIGURE 4-12 **Balmer Lines in the Spectrum of a Star** This portion of the spectrum of a star called HD 193182 shows more than two dozen Balmer lines. HD are the initials of Henry Draper. Draper and succeeding astronomers cataloged this and many other stars. The series converges at 364.56 nm, just to the left of H_{40}. This star's spectrum also contains the first 12 Balmer lines (H_α through H_{12}), but they are not visible in this particular spectrogram. (Carnegie Observatories)

the atom electrically neutral. Astronomers denote neutral atoms by writing the atomic symbol followed by the roman numeral I. For example, neutral hydrogen is written as H I and neutral iron is Fe I.

When an atom contains a different number of electrons than protons, the atom is called an **ion.** The process of creating an ion is called **ionization.** Ions are denoted by the atomic symbol followed by a roman numeral that is one greater than the number of missing electrons. Positively ionized hydrogen (missing its one electron) is denoted H II, while positively ionized iron with seven electrons missing is denoted Fe VIII. It is worth noting that negative ions also exist, where nuclei have more electrons orbiting than they have protons. These are found, for example, in the Sun's outer layers and they help account for the Sun's blackbody curve (see Figure 4-3).

One way to create positive ions from neutral atoms is *photoionization,* in which photons of sufficiently high energy literally rip an electron completely out of orbit and into the space between atoms. Ionization occurs in stars, as seen by the fact that many of the spectral lines for stars correspond to ionized atoms.

Atoms can also share electrons and, by doing so, remain bound together. Such groups are called **molecules.** They are the essential building blocks of all complex structures, including life.

4-6 Spectral lines occur when an electron jumps from one energy level to another

WEB LINK 4.3 The challenge of reconciling Rutherford's atomic model with the observations of spectral analysis was undertaken by the young Danish physicist Niels Bohr, who joined Rutherford's group at Manchester in 1911. Bohr began by trying to understand the structure of hydrogen, the lightest of the elements. The simplest hydrogen atom consists of a single electron and a single proton. Hydrogen has a visible spectrum consisting of a pattern of lines that begins at 656.28 nm and ends at 364.56 nm. The longest-

wavelength spectral line is called H_α, the second H_β, the third H_γ and so forth, ending with the shortest-wavelength line, H_∞, at 364.56 nm. (The first few lines of the series are identified by Greek-letter subscripts; the remainder are identified by numerical subscripts.) The closer you get to 364.56 nm, the more spectral lines you see.

This spectral pattern had been described mathematically in 1885 by Johann Jakob Balmer, a German schoolteacher. By trial and error, Balmer discovered a formula for calculating the wavelengths of the hydrogen lines. Because of his discovery, the spectral lines of hydrogen at visible wavelengths are called *Balmer lines,* and the entire pattern from H_α to H_∞ is called the *Balmer series.* The spectrum of the star shown in Figure 4-12 exhibits more than two dozen Balmer lines, from H_{13} through H_{40}.

Bohr's goal was to mathematically derive Balmer's formula from basic laws of physics. He began by assuming that the electron in a hydrogen atom moves around the nucleus only in certain specific orbits. As shown in Figure 4-13, it is customary to label these orbits $n = 1$, $n = 2$, $n = 3$, and so on. They are called the *Bohr orbits.*

Bohr argued that for an electron to jump, or to make a **transition,** from one orbit to another, the hydrogen atom must gain or lose a specific amount of energy. An electron jump from an inner orbit to an outer orbit requires energy; an electron jump from an outer orbit to an inner one releases energy. The energy gained or released by the atom when the electron changes orbits is the difference in energy between these two orbits.

According to Planck and Einstein, the packet of energy gained or released is a photon, whose energy is inversely proportional to its wavelength. Using these ideas, Bohr mathematically derived the formula that Balmer had discovered by trial and error. Furthermore, Bohr's discovery elucidated the meaning of the Balmer series: All the Balmer lines are produced by electron transitions between the second Bohr orbit ($n = 2$) and higher orbits ($n = 3$, 4, 5, and so forth). As an example, Figure 4-14 shows the electron transition that gives rise to the H_α spectral line, which has a wavelength of 656.28 nm.

In addition to giving the wavelengths of the Balmer series, Bohr's formula correctly predicts the wavelengths of other series of spectral lines that occur at nonvisible wavelengths. For example, electron transitions between the lowest Bohr orbit ($n = 1$) and all higher orbits also produce spectral lines. These transitions create the *Lyman series*, which is entirely in the ultraviolet wavelengths. At infrared wavelengths is the *Paschen series*, which arises out of transitions to and from the third Bohr orbit ($n = 3$). Additional series exist at still longer wavelengths.

Bohr's ideas also help explain Kirchhoff's laws. Each spectral line corresponds to one specific transition between the orbits of the electrons of a particular element. An absorption line is created when an electron jumps from an inner orbit to an outer orbit, extracting the required photon from an outside source of energy, such as the continuous spectrum of a hot, glowing object. An emission line is produced when an electron transitions to a lower orbit and emits a photon.

Today's view of the atom owes much to the Bohr model, but it is enhanced in certain ways. The modern theory of atoms is called **quantum mechanics,** a branch of physics dealing with photons and subatomic particles that was first developed during the 1920s. As a result of this work, physicists have discarded the concept that electrons are solid particles with planetlike orbits about the nucleus. Instead, electrons are now known to have both wave and particle properties; they are said to occupy certain allowed **energy levels** in the atom.

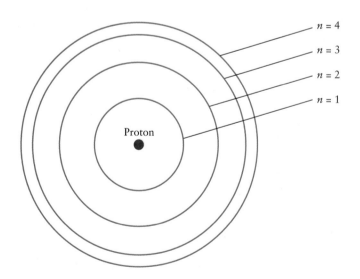

FIGURE 4-13   **Bohr Model of the Hydrogen Atom**   According to Bohr's model of the atom, an electron circles the nucleus only in allowed orbits $n = 1, 2, 3$, and so on. The first four Bohr orbits are shown here.

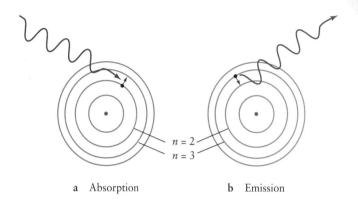

a   Absorption   b   Emission

FIGURE 4-14   **The Absorption and Emission of an H$_\alpha$ Photon**   This schematic diagram, drawn according to the Bohr model of the atom, shows what happens when a hydrogen atom absorbs or emits an H$_\alpha$ photon, which has a wavelength of 656.28 nm. **(a)** The photon is absorbed by the atom as the electron jumps from orbit $n = 2$ up to orbit $n = 3$. **(b)** The photon is emitted by the atom as the electron falls from orbit $n = 3$ down to orbit $n = 2$.

An extremely useful way of displaying the structure of an atom is an energy level diagram, such as that shown in Figure 4-15 for hydrogen. The lowest energy level, called the **ground state,** corresponds to the $n = 1$ Bohr orbit. An electron can jump from the ground state up to the $n = 2$ level only if the atom absorbs a photon of wavelength 121.6 nm. The energy of a photon is often expressed in electron volts (abbreviated eV). The 121.6 nm photon has an energy of 10.19 eV, so the energy level $n = 2$ is shown on the diagram as having an energy 10.19 eV above the energy of the ground state, which is usually assigned a value of 0 eV. Similarly, the $n = 3$ level is 12.07 eV above the ground state and so forth, up to the $n = \infty$ level at 13.6 eV. These states with more energy than the ground state are collectively called *excited states*.

If any electron in a hydrogen atom absorbs a photon with an energy greater than 13.6 eV, that electron will be knocked completely out of the atom. This is the process of photoionization mentioned above.

4-7   Spectral lines shift due to the relative motion between the source and the observer

Christian Doppler, a professor of mathematics in Prague, pointed out in 1842 that wavelength is affected by motion. As shown for the observer on

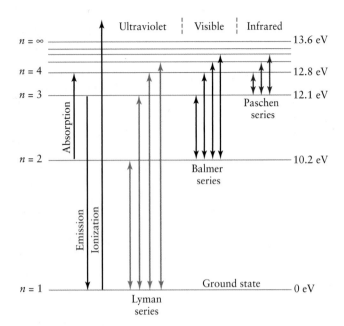

FIGURE 4-15 **Energy Level Diagram of Hydrogen** The structure of the hydrogen atom is conveniently displayed in a diagram showing the energy levels above the ground state. A variety of electron jumps, or transitions, are shown, including those that produce some of the most prominent lines in the hydrogen spectrum. For another perspective on the same information, you can try the linked Interactive Exercise.

the left in Figure 4-16, the wavelengths of electromagnetic radiation from an approaching source are compressed. The circles represent waves emitted from consecutive wave peaks as the source moves along. Because each successive wave is emitted from a position slightly closer to you, you see a shorter wavelength than you would if the source were stationary. All the spectral lines in the spectrum of an approaching source, such as a star, are therefore shifted toward the short-wavelength (blue) end of the spectrum, regardless of its distance. This phenomenon is called a **blueshift.**

Conversely, electromagnetic waves from a receding source are stretched out. You see a longer wavelength than you would if the source were stationary. All the spectral lines in the spectrum of a receding source, regardless of its distance, are shifted toward the longer-wavelength (red) end of the spectrum, producing a **redshift.** In either case, the effect of relative motion on wavelength is called the **Doppler shift.** This effect varies directly with approaching or receding speed: An object approaching twice as fast as another has its colors (or wavelengths) blueshifted twice as much when the approach speeds are small compared to the speed of light. An Astronomer's Toolbox 4-3 gives a numerical example.

The speed determined from the Doppler shift is called the **radial velocity,** because the motion is along our line of sight, or along the "radius" drawn from Earth to the star. Of course, the star may well have motion perpendicular to our line of sight, across the celestial sphere. This **proper motion** does not affect the perceived wavelength and cannot be determined by Doppler shift. Proper motion is determined by measuring a star's motion relative to background stars (Figure 4-17). The star with the greatest proper motion as seen from Earth is Barnard's star (Figure 4-18).

Proper motions are so small that they can be measured only for relatively nearby stars in our Galaxy. However, the radial velocity of virtually every object in space can be determined regardless of their distances from us. For example, Doppler shift measurements of the spectra of hot gases on the Sun's surface reveal that they rise and fall. Doppler shift measurements of stars in double star systems give crucial data about the speeds of the stars orbiting around each other. Doppler measurements of the spectra of distant galaxies enable us to determine the rate at which the entire universe is expanding. In later chapters, we will refer to the Doppler shift whenever we need to convert an observed wavelength shift into a speed.

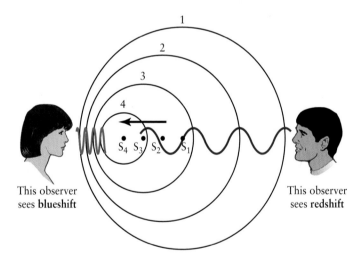

FIGURE 4-16 **The Doppler Effect** Wavelength is affected by motion between the light source and an observer. A source of light is moving toward the left. The four circles (numbered 1 through 4) indicate the location of light waves that were emitted by the moving source when it was at points S_1 through S_4, respectively. Note that the waves are compressed in front of the source but stretched out behind it. Consequently, wavelengths appear shortened (blueshifted) if the source is moving toward the observer and lengthened (redshifted) if the source is moving away from the observer. Motion perpendicular to an observer's line of sight does not affect wavelength.

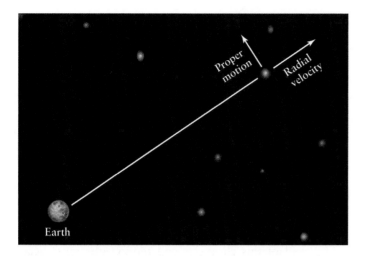

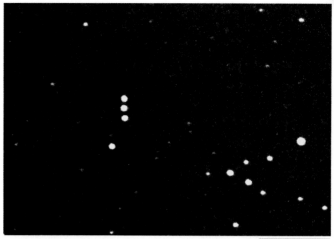

FIGURE 4-17 Radial and Proper Motions of a Star
The speed of a star toward or away from the Earth is its radial velocity. This motion creates a Doppler shift in the star's spectrum. The motion of the star across the sky, perpendicular to our line of sight, is called proper motion. Proper motion does not affect the star's spectrum.

FIGURE 4-18 Barnard's Star This is a series of three superimposed photographs, taken over a four-year period. The line of three dots shows the proper motion of Barnard's star during that time. In addition to having the largest known proper motion (10.3" per year), Barnard's star is one of the closest stars to Earth. (John Sanford/Science Photo Library/Photo Researchers)

AN ASTRONOMER'S TOOLBOX 4-3

The Doppler Shift

Suppose that λ_0 is the wavelength of a spectral line from a stationary source. This is the wavelength that a reference book would list or a laboratory experiment would yield. If the source is moving, this particular spectral line is shifted to a different wavelength λ. The size of the wavelength shift is usually written as $\Delta\lambda$ (delta lambda), where $\Delta\lambda = \lambda - \lambda_0$. This $\Delta\lambda$ is the difference between the wavelength that you actually observe in the spectrum of a star or galaxy and the wavelength listed in reference books.

Doppler proved that the wavelength shift is governed by the simple equation

$$\Delta\lambda/\lambda_0 = v/c$$

where v is the speed of the source measured along the line of sight between the source and the observer. As usual, c is the speed of light (3×10^5 km/s).

Example: The spectral lines of hydrogen appear in the spectrum of the bright star Vega as shown in the figure. The prominent hydrogen line H_α has a normal wavelength of 656.285 nm, but in Vega's spectrum the line is located at 656.255 nm. The wavelength shift is −0.030 nm, so the star is approaching us with a speed of −14 km/s. The minus sign indicates that the star is moving toward us.

Try these questions: How fast would a star be moving toward us if all its wavelengths shifted to nine-tenths of their rest wavelengths? If a star is moving away from us at 100 km/s, what is the change, $\Delta\lambda/\lambda$? If a star is moving toward us at 0.1 c, what is the change, $\Delta\lambda/\lambda$, in its wavelengths?

(Answers appear at the end of the book.)

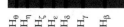

H_θ H_η H_ζ H_ε H_δ H_γ H_β H_α

The Spectrum of Vega Several Balmer lines are seen in this photograph of the spectrum of Vega, the brightest star in the constellation Lyra (the Lyre). All the spectral lines are shifted equally and very slightly toward the blue side of the spectrum, indicating that Vega is approaching us. (NOAO)

Insight into Science   Remote science Astronomical objects are so remote and their activities are often so complex that astrophysicists must use a tremendous amount of physics to interpret observations. For example, a single spectrum can contain information about stars, extrasolar planets, interstellar gas, the Earth's motion and atmosphere, and the performance of the observing telescope and the equipment attached to it. All of these factors must be understood theoretically and accounted for.

With the work of people like Planck, Einstein, Rutherford, Bohr, and Doppler, the interchange between astronomy and physics came full circle. Modern physics was born when Newton set out to understand the motions of the planets. Two and a half centuries later, physicists in their laboratories discovered the basic properties of electromagnetic radiation and the structures of atoms. As we will see in later chapters, the fruits of their labors have important implications for astronomy even today.

4-8   Frontiers Yet to be Discovered

Spectra provide much of what we know about the cosmos. Spectra of objects in space continue to provide new insights into the chemical composition of the planets, moons, local space debris, stars, interstellar gas and dust, and galaxies, among other things. Furthermore, as we obtain more accurate Doppler measurements of objects in our Galaxy, in other nearby galaxies, and, indeed, of entire galaxies, we are better able to determine the rate at which the objects in the universe are moving relative to each other. The more we understand of this chemistry and motion, the more accurately we will be able to explain the evolution of the solar system, stars, galaxies, and the universe.

 Further Reading on These Topics

WHAT DID YOU KNOW?

1 *How hot is a "red hot" object compared to objects glowing other colors?* Of all objects that glow visibly from heat stored or generated inside them, those that glow red are the coolest. Those that glow blue are the hottest.

2 *What color is the Sun?* The Sun emits all wavelengths of electromagnetic radiation. The colors it emits most intensely are in the blue-green part of the spectrum. Since the human eye is less sensitive to blue-green than to yellow and the Earth's atmosphere scatters blue-green wavelengths more readily than longer wavelengths, we see the Sun as yellow.

3 *Can carbon dating be used to determine the age of space debris found on Earth?* No. Carbon dating is only reliable for objects that formed within the past 100,000 years. Space debris found on Earth formed more than 4.5 billion years ago.

KEY WORDS

absorption line spectrum, 103
atomic number, 105
blackbody, 96
blackbody curve, 97
blueshift, 108
continuous spectrum, 103
diffraction grating, 102
Doppler shift, 108
element, 100
emission line spectrum, 103
energy flux, 98

energy level, 107
ground state, 107
ion, 106
ionization, 106
isotope, 105
Kirchhoff's laws, 103
luminosity, 98
molecule, 106
nucleus (of an atom), 104
periodic table, 101
Planck's law, 99

proper motion, 108
quantum mechanics, 107
radial velocity, 108
redshift, 108
spectral analysis, 100
spectral lines, 99
spectrograph, 102
spectroscope, 102
Stefan-Boltzmann law, 97
transition (of an electron), 106
Wien's law, 97

KEY IDEAS

• By studying the wavelengths of electromagnetic radiation emitted and absorbed by an astronomical object, astronomers can learn the object's temperature, chemical composition, and movement through space.

Blackbody Radiation

• A blackbody is a hypothetical object that perfectly absorbs electromagnetic radiation at all wavelengths. Because a blackbody does not reflect any electromagnetic radiation from outside sources, the radiation that it does emit depends only on its temperature. Stars closely approximate blackbodies.

• Wien's law states that the dominant wavelength of radiation emitted by a blackbody is inversely proportional to its temperature. The intensities of radiation emitted at various wavelengths by a blackbody at a given temperature are shown as a blackbody curve.

• The Stefan-Boltzmann law relates the temperature of a blackbody to the rate at which it radiates energy.

Discovering Spectra

• Spectroscopy—the study of electromagnetic spectra—provides important information about the chemical composition of remote astronomical objects.

• Kirchhoff's three laws of spectral analysis describe the conditions under which absorption lines, emission lines, and a continuous spectrum can be observed. Spectral lines serve as distinctive "fingerprints" for the chemical elements and chemical compounds comprising a light source.

Atoms and Spectra

• An atom consists of a small, dense nucleus (composed of protons and neutrons) surrounded by electrons that can occupy only certain energy levels.

• The spectral lines of a particular element correspond to the various electron transitions between allowed energy levels of that element. When an electron shifts from one energy level to another, a photon of the appropriate energy (and hence a specific wavelength) is absorbed or emitted by the atom.

• The spectrum of hydrogen at visible wavelengths consists of the Balmer series, which arises from electron transitions between the second energy level of the hydrogen atom and higher levels.

• When an atom loses or gains one or more electrons it is said to be charged, or ionized. One way for it to lose an electron is for the electron to absorb an energetic photon and thereby fly free from its atom.

• The spectral lines of an approaching light source are shifted toward shorter wavelengths (a blueshift); the spectral lines of a receding light source are shifted toward longer wavelengths (a redshift).

• The equation describing the Doppler effect states that the size of a wavelength shift is proportional to the radial velocity between the light source and the observer.

REVIEW QUESTIONS

1 What is a blackbody? What does it mean to say that a star behaves almost like a blackbody? If stars behave like blackbodies, why are they not black?

2 What is Wien's law? How could you use it to determine the temperature of a star's surface?

3 What is the Stefan-Boltzmann law? Why are astronomers interested in it?

4 Using Wien's law and the Stefan-Boltzmann law, explain the changes in color and intensity that are observed as the temperature of a hot, glowing object increases.

5 Describe the experimental evidence that supports the Bohr model of the atom.

6 Explain how the spectrum of hydrogen is related to the structure of the hydrogen atom.

7 Why do different elements have different patterns of lines in their spectra?

8 What is the Doppler shift, and why is it important to astronomers?

9 Explain why the Doppler shift tells us only about the motion directly along the line of sight between a light source and an observer.

ADVANCED QUESTIONS

The answers to all computational problems, which are preceded by an asterisk (*), appear at the end of the book.

*10 Approximately how many times around the world could a beam of light travel in 1 second?

*11 The bright star Regulus in the constellation of Leo (the Lion) has a surface temperature of 12,200 K. Approximately what is the dominant wavelength (λ_{max}) of the light it emits?

*12 The bright star Procyon in the constellation of Canis Minor (the Little Dog) emits the greatest intensity of radiation at a wavelength λ_{max} = 445 nm. Approximately what is the surface temperature of the star in Kelvins?

*13 The wavelength of H_β in the spectrum of the star Megrez in the Big Dipper is 486.112 nm. Laboratory measurements demonstrate that the normal wavelength of this spectral line is 486.133 nm. Is the star coming toward us or moving away from us? At what speed?

*14 In the spectrum of the bright star Rigel, H_α has a wavelength of 656.331 nm. Is the star coming toward us or moving away from us? How fast?

*15 Imagine driving down a street toward a traffic light. How fast would you have to go so that the red light would appear green?

DISCUSSION QUESTIONS

16 Compare the technique of identifying chemicals by their spectral line patterns with that of identifying people by their fingerprints.

17 Suppose you look up at the night sky and observe some of the brightest stars with your naked eye. Is there any way of telling which stars are hotter and which are cooler? Explain.

18 How can we exclude the Earth's atmosphere as the source of the iron absorption lines in Figure 4-6?

WHAT IF ...

*19 The Sun were twice its actual diameter, but still had the same surface temperature? At what wavelength would that new Sun emit its radiation most intensely? How many times brighter than the present Sun would the new Sun be? How might things on Earth be different under the bigger Sun?

20 All the nearby stars were observed to have redshifted spectra? What conclusions could we draw? What other measurement would be useful to have for each star and why?

21 No stars had any Doppler shift? What would that say about stellar motions relative to the solar system? Is that possible? *Clue:* Consider Newton's law of gravitation from Chapter 3.

22 The spectrum of the Sun (chapter-opening figure) had several absorption lines missing when taken from above the Earth's atmosphere compared to its spectrum taken at the Earth's surface?

WEB/CD-ROM QUESTIONS

*23 **Measuring Stellar Temperatures** Access the Active Integrated Media Module "Blackbody Curves" in Chapter 4 of the *Discovering the Universe* Web site or CD-ROM. (a) Use the module to determine the range of temperatures over which a star's peak wavelength is in the visible spectrum. (b) Determine if any of the following stars have a peak wavelength in the visible spectrum: Rigel, T = 14,000 K; Deneb, T = 9500 K; Arcturus, T = 4500 K; Vega, T = 11,500 K; Betelgeuse, T = 3100 K.

OBSERVING PROJECTS

24 As soon as weather conditions permit, observe a rainbow. Confirm that the colors are in the same order as shown in Figure 3-1a.

25 Obtain a glass prism or a diffraction grating, available from science museum stores, catalogs, or your college physics department. Look through the prism or grating at various light sources, such as an ordinary incandescent lightbulb, a neon sign, and a mercury vapor street lamp. ***Do not look at the Sun! Looking directly at the Sun causes blindness.*** Do you have any trouble seeing spectra? What do you have to do to see a spectrum? Describe the differences in the spectra of the various light sources you observed.

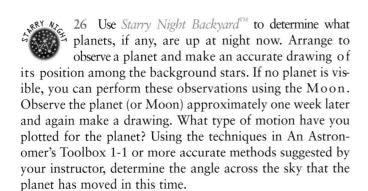

26 Use *Starry Night Backyard*™ to determine what planets, if any, are up at night now. Arrange to observe a planet and make an accurate drawing of its position among the background stars. If no planet is visible, you can perform these observations using the Moon. Observe the planet (or Moon) approximately one week later and again make a drawing. What type of motion have you plotted for the planet? Using the techniques in An Astronomer's Toolbox 1-1 or more accurate methods suggested by your instructor, determine the angle across the sky that the planet has moved in this time.

27 Use *Starry Night Backyard*™ to examine three distant celestial objects. First view the entire sky (*Go:Atlas*). Then search for objects (i), (ii), (iii) listed below (*Edit:Find:object name*). Based on its appearance, state whether it is likely to have a continuum spectrum, an absorption line spectrum, or an emission line spectrum and explain your reasoning. (i) Trifid Nebula, (ii) M31, the Andromeda galaxy, (iii) the Moon. (*Hint:* Recall that moon is simply reflected sunlight.) Your teacher may assign additional objects to find.

II ESSENTIALS FOR UNDERSTANDING THE SOLAR SYSTEM

IN THE CHAPTERS AHEAD YOU WILL DISCOVER

how the solar system formed from a cloud of interstellar gas and dust (Essentials II)

that the planets fall into groups with similar properties (Essentials II, Chapters 5, 6, 7)

how astronomers characterize each planet's "personality" by its size, chemical composition, rotation rate, atmosphere, and features (Essentials II, Chapters 5, 6, 7)

that the moons throughout the solar system are extremely varied (Chapters 5, 6, 7)

that the solar system also contains debris—asteroids, meteoroids, and comets—left over from its formation (Chapter 8)

why the Sun shines (Chapter 9)

THE PLANETS OF OUR SOLAR SYSTEM
This photomontage shows the Sun and the nine planets of our solar system. At the top (from left to right) are Mercury, Venus, Earth, and Mars. At the bottom (from left to right) are Jupiter, Saturn, Uranus, Neptune, and Pluto. (The images are not reproduced to the same scale. See Figure II-10 for a scale drawing.) (Calvin J. Hamilton)

R I V U X G

WHAT DO YOU THINK?

1   How many stars are there in the solar system?

2   Was the solar system created as a direct result of the formation of the universe?

3   How long has the Earth existed?

4   Is Pluto always the farthest planet from the Sun?

5   What typical shape(s) do moons have?

For as long as people have looked up at the heavens, they have wondered about the Sun, the Moon, and the planets visible to the naked eye. Telescopes revealed many new features of these objects, as well as the existence of other planets and moons, along with asteroids, meteoroids, and comets, all orbiting the Sun. The Sun and all the bodies that orbit it make up our **solar system.**

The solar system has only one star, the Sun. The stars you see in the night sky, plus billions of stars too far away to see with the naked eye, are members of a larger system called the *Milky Way Galaxy.* The solar system is also part of the Milky Way, a topic we discuss in Chapter 14.

Did the solar system form all at once, or did it come together by serendipity? How were its varied building blocks of rock, metal, ice, and gas created? How has the solar system changed since its formation, and what does its history tell us about the planets we see today? Within the past few decades, telescopes and space probes, along with the theories of modern science, have finally provided answers to these age-old questions. Our new wealth of information is giving us a rapidly growing understanding of our nearest neighbors in space.

II-1   The solar system formed from a cloud of cold gas and dust

Astronomers believe that the solar system formed from a vast cloud of interstellar gas and dust, called the **solar nebula.** This cloud condensed 4.6 billion years ago under the influence of its own gravitational force to form the planets, the Sun, and other bodies. We can determine where the matter comprising the solar nebula came from by examining the chemical elements found in the bodies of the solar system and in the stars.

By studying the spectra of stars and interstellar clouds, we know that hydrogen and helium are by far the most abundant elements in the universe. Together these elements account for 98% of the mass of all the material in existence—all the other elements combined account for only 2%. However, the Earth's mass contains less than 0.15% hydro-

gen and helium. Somehow the early solar system was enriched with the heavier elements we see today—oxygen, silicon, aluminum, iron, carbon, calcium, and others. Although these elements are rare in the universe as a whole, they are commonplace on Earth, and they are found to varying extents in all the other objects in the solar system. Our picture of the emerging solar nebula must explain these differences.

There is a good reason for the overwhelming abundance of hydrogen and helium throughout the universe. Most astronomers believe that the universe formed between 14 and 15 billion years ago with a violent event called the *Big Bang.* Only the lightest elements—hydrogen, helium, and a tiny amount of lithium—emerged when the cosmos was formed. The first stars, composed only of these three elements, condensed out of this primeval matter, probably within a few hundred million years after the Big Bang.

The difference in chemical composition between the early universe and the solar system shows that the solar system did not form as a *direct* result of the Big Bang. The solar system came billions of years later, and the heavier elements came to the solar nebula from stars that existed and exploded before the solar system formed.

Stars use various elements, such as hydrogen, helium, and carbon, to create energy. In this process, called *fusion,* lighter elements are transformed into heavier ones, such as hydrogen becoming helium or helium becoming carbon. We call fusion a *thermonuclear reaction* because the tremendous heat generated in the centers of stars is what enables nuclei of lighter elements there to combine, or fuse, into heavier elements. We explore details of the creation of this heavier matter in Chapters 11 and 12.

Near the ends of their lives, most stars cast matter out into space to form clouds of interstellar gas and dust. Often a star's outer layers are gradually expelled (Figure II-1a and b), but some stars end their lives with spectacular detonations called *supernovae* (Figure II-1c), which blow the stars apart. Either way, the space between the stars becomes filled with gas containing mostly hydrogen and helium—but now also enriched with heavy elements created in the stars. The solar nebula was a fragment of a large cloud of such gas and dust.

Sometimes, the explosive force of a supernova compresses fragments of interstellar gas and dust clouds that had been ejected from earlier generations of stars. These fragments are pulled inward on themselves by the mutual gravitational attraction of their gas and dust particles, thereby creating new star and planetary systems (Figure II-2). The collapsing matter is primarily hydrogen and helium, but the stars that ejected it had converted some of it to the other 90 naturally formed elements. These include the building blocks of life and this is believed to be how the solar system formed—the debris left over from a previous generation of stars. *We are made of star dust!*

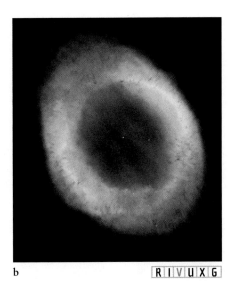

a | R I V U X G b | R I V U X G c | R I V U X G

FIGURE II-1 **How Stars Lose Mass** **(a)** The brightest star in Scorpius, Antares, is nearing the end of its existence. Strong winds from its surface are expelling large quantities of gas and dust, creating this nebula reminiscent of an Impressionist painting. The scattering of starlight off this material makes it appear especially bright, even at a distance of 604 ly. **(b)** The Ring Nebula is just over 2000 ly from Earth. The central star shed its outer layers of gas and dust in an expanding spherical shell now about $1/2$ ly across. The relatively gentle emission of this matter is called a planetary nebula, discussed in Chapter 12. **(c)** A supernova is the most powerful known mechanism for a star to shed mass. This is the Crab Nebula. Although it is about 6000 ly from Earth, it was visible during the day for three weeks during A.D. 1054. (a: David Malin/Anglo-Australian Observatory; b: NASA; c: Malin/Pasachoff/Caltech)

R I V U X G

FIGURE II-2 **A Dusty Region of Star Formation** These young stars in the constellation Orion are still surrounded by much of the gas and dust from which they formed. The bluish, wispy nebulosity is caused by starlight reflecting off abundant interstellar dust grains. These grains are made of heavy elements (such as carbon, silicon, and iron) produced by earlier generations of stars. Astronomers hypothesize that the solar system formed from such a cloud of gas and dust. (©1980, 1984 Anglo-Australian Observatory)

II-2 Gravity and heat shaped the young solar system

To find out more about the solar system's origins, we look for clues in its pieces. Especially valuable is the interplanetary debris we know as asteroids, meteoroids, and comets. Some of these bodies are believed to be unchanged remnants from the formation of the solar system. Other clues to the origin of our solar system have recently been found in distant stars, where planetary systems like our own are developing.

Initially, it was very cold inside the solar nebula—well below the freezing point of water. The nebula began with a diameter of at least 100 AU and a total mass about 2 to 3 times the mass of the Sun. Ice and ice-coated dust grains composed of heavy elements were scattered abundantly across this vast volume. Hydrogen and helium gas, accompanied by this ice and dust, fell toward the center of the solar nebula under the influence of the gravitational attraction of other gas in the nebula. As a result, the density and pressure at the center of the nebula began to increase, producing a concentration of matter called the **protosun**.

As the nebula continued to contract, atoms at its center collided with one another with increasing speed and frequency. Such collisions created heat, causing the temperature deep inside the solar nebula to soar. Therefore, the first heat in the solar system came from colliding gas, not from fusion, which began later.

a

b

c

d

e

f

ANIMATION II-1

FIGURE II-3   The Formation of the Solar System

This sequence of drawings shows six stages in the formation of the solar system. (a) A slowly rotating cloud of interstellar gas and dust begins to contract because of its own gravity. (b) A central condensation, the protosun, forms as the cloud flattens and rotates faster. (c) A flattened disk of gas and dust surrounds the protosun, which has begun to shine. (d) The Sun's rising temperature removes the gas from the inner regions, leaving dust and larger debris revolving in place. (e) The planets have established dominance in their regions of the solar system. (f) The solar system as it appears today.

The planets and other bodies orbiting the protosun formed because the solar nebula was swirling (rotating). As a result of this rotation, the nebula had angular momentum (see An Astronomer's Toolbox 2-1). Conservation of angular momentum prevented some of the nebula from falling inward. Without this initial rotation, everything in the solar nebula would have fallen straight into the protosun, leaving no matter behind to form the planets. Mathematical studies show that the combined effects of gravity and rotation transform even an irregular cloud fragment (which the solar nebula probably was) into a rotating disk with a warm center and cold edges, as shown in Figure II-3. The disk shape explains why the orbits of the planets are all nearly in the same plane.

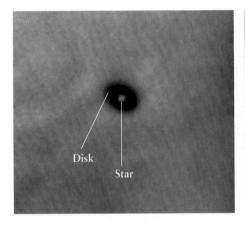

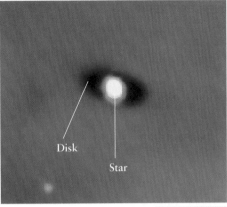

Disk

Star

Disk

Star

 FIGURE II-4 Young Circumstellar Disks of Matter These Hubble Space Telescope images show circumstellar disks surrounding two stars 1500 light-years away from Earth in the Orion Nebula. These stars, both less than 10 million years old, are believed to be surrounded by protoplanetary gas and dust. (STScI, M. J. McCaughrean/MPIA, C. R. O'Dell/ Rice University, NASA)

While we cannot see our solar system as it was back then, astronomers have found disks of gas and dust surrounding other young stars, like those shown in Figure II-4. Called **protoplanetary disks** or **proplyds,** these systems are believed to be undergoing the same initial stage of evolution that we are describing here for our solar system.

Temperatures around the young protosun soon began to climb. The rising temperatures vaporized all the common icy substances in the inner region of the solar nebula and pushed light gases like hydrogen and helium outward. In this inner region, where the Earth orbits today, mostly heavier elements remained. These heavy elements were eventually to come together and form the four inner planets: Mercury, Venus, Earth, and Mars.

II-3 Collisions in the early solar system led to the formation of planets

The formation of the inner planets was the result of innumerable impacts of small, rocky particles in the solar system's protoplanetary disk. Initially, neighboring dust grains (Figure II-5) and pebbles in the solar nebula collided and stuck together. Then, over a period of a few million years, these accumulations of dust and pebbles coalesced into larger objects called **planetesimals,** with diameters of about 10 km.

During the next stage, the planetesimals collided. Because of their mutual gravitational attraction, they coalesced into still larger objects called **protoplanets.** This accumulation of material is called **accretion.** Computer simulations of the inner solar system based on Newton's laws (see Section 2-5) suggest that accretion continues for roughly 100 million years and should lead to the formation of fewer than half a dozen planets (Figure II-6). The agreement between predictions of the number of planets and the number in our inner solar system is encouraging. However, observations indicate that the planets form more quickly.

Computer Simulations

Insight into Science **Computer Aid Analysis** Many equations are so complex that they require computer analysis to enable us to understand their implications. An example is the physics of the formation of the planets and the Sun. Observations then lead to refined computer models.

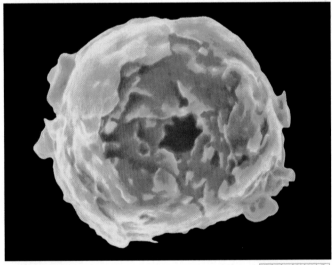

FIGURE II-5 Space Dust Collected by a high-flying aircraft, this piece of interplanetary dust is believed to be typical of the state of matter in the protoplanetary disk. It is about 20 μ (0.02 mm) across. Such particles stuck together to form larger and larger chunks of matter, eventually creating such objects as the planets, moons, and asteroids. (Scott Messenger/University of Washington, St. Louis/Science Photo Library/Photo Researchers)

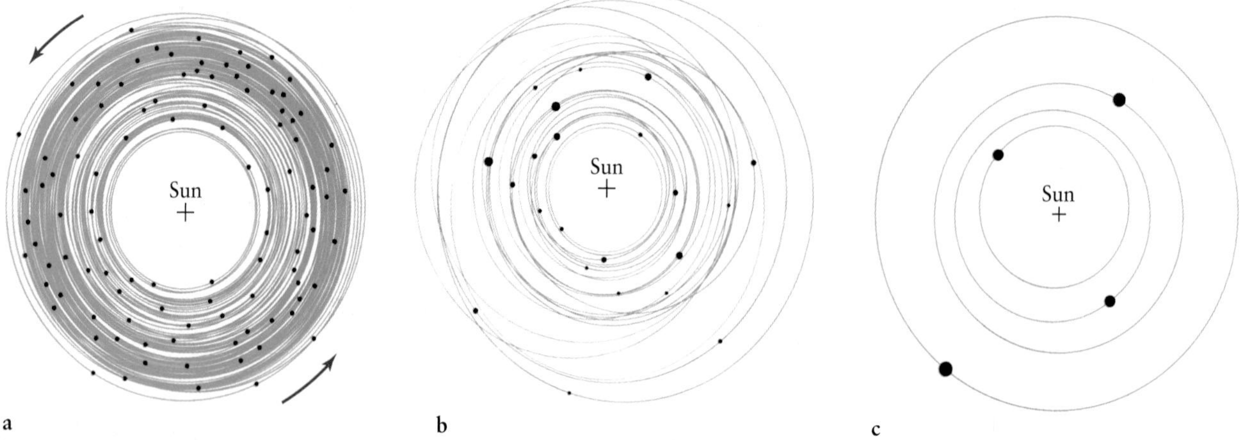

a   b   c

FIGURE II-6   Accretion of the Inner Planets These three drawings show the results of a computer simulation of the formation of the inner planets. **(a)** The simulation begins with 100 planetesimals. **(b)** After 30 million years, these planetesimals have coalesced into 22 protoplanets. **(c)** This final view is for an elapsed time of 441 million years, but the formation of the inner planets in this simulation is almost complete after only 100 million years.

While the inner regions of the solar system were heating up, temperatures in the outer regions of the solar nebula remained quite cool. Ice and ice-coated dust, along with hydrogen and helium gas, were able to survive in these cooler regions. The large outer planets—Jupiter, Saturn, Uranus, and Neptune—also probably formed from the accretion of planetesimals. For each, a rocky core even bigger than the Earth apparently served as a "seed," and for about a million years the core became coated with additional rock and gas. When the gas-rich envelope became as massive as the rocky core, the gas began to accumulate at a runaway pace. Thereafter, the envelope pulled in all the gas in its vicinity, creating a huge hydrogen-rich shell surrounding a core of rocky material.

Because of their gaseous shells, the outer planets today include abundant quantities of hydrogen, helium, methane, ammonia, and water. Many of the moons of the outer planets are also partially composed of these low-density substances.

During the millions of years that the planets were forming, so, too, was the Sun. During this time the temperature and pressure at the center of the contracting protosun continued to climb. Finally, the center of the protosun became hot enough to ignite thermonuclear reactions. Now, hydrogen fused into helium in its core, thereby releasing huge amounts of energy, and the Sun was born. Sunlike stars take approximately 100 million years to form from a nebula and settle down, which means that the Sun became a full-fledged star at about the same time as the accretion of the inner planets was being completed. Indeed, radiation from the Sun prevented new planetesimal formation, thereby limiting the sizes of the planets.

According to radioactive dating of the oldest debris ever discovered from space, the solar system was roughly in the form we know today just under 4.6 billion years ago. While the planets were in place, only a few of the dozens of moons were in orbit, and innumerable pieces of rocky and icy debris were on collision courses with the planets. Indeed, evidence indicates that our Moon was created by such a collision (see Chapter 5).

Once formed, our Moon's surface was scarred by numerous impacts (Figure II-7). These scars, called **craters,** are also found on those planets and moons that do not have appreciable atmospheres or geological activity that would otherwise erase these features. Indeed, the Moon's virtually airless environment has preserved important information about the early history of the solar system. (Note that we capitalize the word *moon* only when we are referring to the Earth's Moon.)

Radioactive dating (see An Astronomer's Toolbox 4-2) of Moon rocks brought back by the Apollo astronauts indicates that the rate of impacts declined dramatically about 3.8 billion years ago. Since that time, impact cratering has proceeded at a very low rate. Thus, most of the craters on the Moon and planets were formed during the first 800 million years of the solar system's history, as the young planets swept up rocky debris left over from the solar nebula. Conditions developed on the young Earth, especially the presence of liquid water, so that biological evolution could begin to eventually occur here. There is growing evidence that liquid water existed on Mars and may still exist inside it and inside Jupiter's moon Europa, raising the hotly debated question of whether simple life evolved on those worlds, too.

R I V U X G

FIGURE II-7   **Our Moon** This photograph, taken by astronauts in 1972, shows thousands of the craters produced by impacts of rocky debris left over from the formation of the solar system. Age-dating of lunar rocks brought back by the astronauts indicates that the Moon is about 4.5 billion years old. Most of the lunar craters were formed during the Moon's first 700 million years of existence, when the rate of bombardment was much greater than it is now. (NASA)

II-4   Comparisons among the nine planets show distinct similarities and significant differences

The nine planets are Mercury, Venus, Earth, Mars, Jupiter, Saturn, Uranus, Neptune, and Pluto. As shown in Figure II-8, the orbits of the four inner planets—Mercury, Venus, Earth, and Mars—are crowded close to the Sun. In contrast, the orbits of the four large outer planets—Jupiter, Saturn, Uranus, and Neptune—are widely spaced at greater distances from the Sun. Pluto usually orbits at the far fringes of the part of the solar system inhabited by planets.

Kepler's laws (see Chapter 2) showed us that all the planets have elliptical orbits. Even so, most of their orbits are nearly circular. The exceptions are Mercury and Pluto, whose orbits are noticeably elliptical. In fact, Pluto's orbit sometimes takes it nearer to the Sun than its neighbor, Neptune. In 1979, Pluto passed inside Neptune's orbit. Neptune was the most distant planet from the Sun until February 1999, when Pluto's elliptical orbit once again took it farther out than Neptune (Figure II-9). Pluto will be farther from the Sun than Neptune for about the next 230 years.

All the planetary orbits except that of Pluto lie in planes close to the ecliptic. Pluto's orbit has a conspicuous tilt (called an *orbital inclination*) compared with the orbits of

14

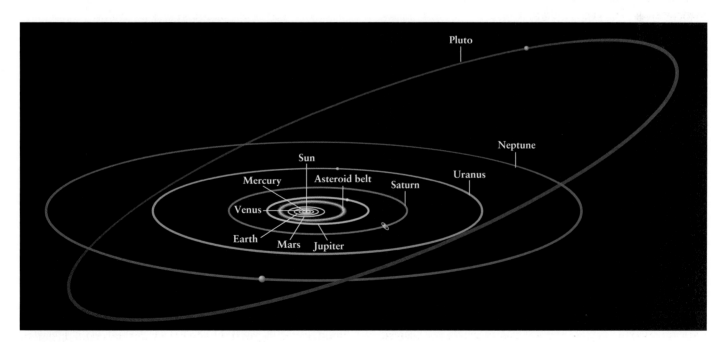

FIGURE II-8   **The Solar System** This scale drawing shows the distribution of planetary orbits around the Sun. The four inner planets are located close to the Sun; the five outer planets orbit at much greater distances from the Sun. The viewpoint of this figure makes the orbits appear much more oval-shaped than they really are. Seen from above the disk of the solar system, most of the orbits appear nearly circular, as shown for Neptune in Figure II-9. All orbits are counterclockwise because the view is from above the Earth's north pole.

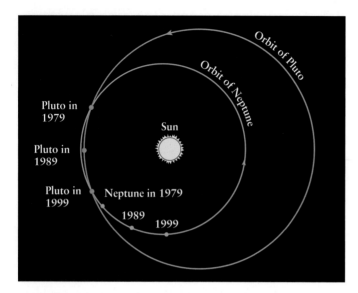

FIGURE II-9 **Pluto's Orbit Is Highly Elliptical** Pluto has the most elliptical orbit of any planet. Indeed, it spends part of the time closer to the Sun than to Neptune. It last moved inside Neptune's orbit in 1979 and moved back outside in 1999. Note that Pluto and Neptune will never collide because their orbits are synchronized by their gravitational interaction to keep them apart.

TABLE II-1   Orbital Characteristics of the Planets

| | Average distance from Sun | | Orbital period |
	(AU)	(10⁶ km)	(yr)
Mercury	0.39	58	0.24
Venus	0.72	108	0.62
Earth	1.00	150	1.00
Mars	1.52	228	1.88
Jupiter	5.20	778	11.86
Saturn	9.54	1427	29.46
Uranus	19.19	2871	84.01
Neptune	30.06	4497	164.79
Pluto	39.53	5914	248.54

the other planets (see Figure II-8). Table II-1 lists some orbital characteristics of the nine planets.

Size is the most obvious physical difference among the planets. The smallest of the four giant outer planets, Neptune, is nearly 4 times larger in diameter than the Earth, the largest of the four inner planets. First place among the giant planets goes to Jupiter, whose diameter is about 11 times bigger than that of Earth. Pluto is the smallest of all the planets, and is even smaller than the seven largest moons, including our own. Figure II-10 shows the Sun and the plan-

ets drawn to the same scale. The diameters of the planets are given in Table II-2.

Mass is another characteristic that distinguishes the inner and outer planets. The four inner planets have low masses compared with the giant planets. Again, first place goes to Jupiter, whose mass is 318 times greater than Earth's (see Table II-2).

Size and mass can be combined in a useful way to provide information about the chemical composition of a planet (or any other object, for that matter). Matter composed of heavy elements, like iron or lead, has more particles compressed into the same volume than matter composed of light elements, like hydrogen, helium, or carbon. Therefore, objects composed primarily of heavier elements have a greater **average density** than objects composed primarily of lighter elements. Average density is given by the equation

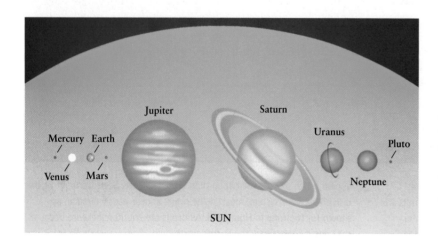

FIGURE II-10 **The Sun and the Planets** This drawing shows the nine planets in front of the disk of the Sun, with all ten bodies drawn to the same scale. The four planets orbiting nearest the Sun (Mercury, Venus, Earth, and Mars) are small and made of rock and metal. The next two planets (Jupiter and Saturn) are large and composed primarily of hydrogen and helium. Uranus and Neptune are also large but contain much water, as well. Pluto has roughly equal amounts of rock and ice.

TABLE II-2 Physical Characteristics of the Planets

	Diameter		Mass		Average density
	(km)	(Earth = 1)	(kg)	(Earth = 1)	(kg/m³)
Mercury	4,878	0.38	3.3×10^{23}	0.06	5430
Venus	12,100	0.95	4.9×10^{24}	0.81	5250
Earth	12,756	1.00	6.0×10^{24}	1.00	5520
Mars	6,786	0.53	6.4×10^{23}	0.11	3950
Jupiter	142,984	11.21	1.9×10^{27}	317.94	1330
Saturn	120,536	9.45	5.7×10^{26}	95.18	690
Uranus	51,118	4.01	8.7×10^{25}	14.53	1290
Neptune	49,528	3.88	1.0×10^{26}	17.14	1640
Pluto	2,300	0.18	1.3×10^{22}	0.002	2030

$$\text{Average density} = \frac{\text{total mass}}{\text{total volume}}$$

The chemical composition (kinds of elements) of any object therefore determines how much space the object occupies per unit mass. For example, a kilogram of iron takes up less space (is denser) than a kilogram of water, even though both have the same mass. Average density is expressed in kilograms per cubic meter. To help us grasp this concept, we compare the average densities of the planets to something familiar, namely water, which has an average density of 1000 kg/m³.

The four inner planets have high average densities compared to water (see Table II-2). In particular, the average density of the Earth is 5520 kg/m³. Because the density of typical surface rock is only about 3000 kg/m³, the Earth must contain a large amount of material inside it that is denser than surface rock.

In sharp contrast, the outer planets have relatively low average densities. For example, Saturn has an average density less than that of water. Their low densities support the belief that the giant outer planets contain significant amounts of the light elements hydrogen and helium.

Pluto again is an oddity. Although it is smaller than any of the inner planets, its average density falls between those of the Earthlike and the giant outer planets. Pluto is probably composed of a mixture of rock and ice, because its average density is between that of rock and ice.

Further information about the chemical composition of bodies in the solar system (among other places) can be obtained from their spectra. For solar system objects, this radi-

ation is primarily sunlight scattered off the surface or clouds surrounding each object. As we saw in Chapter 4, the spectra provide us with details of an object's chemical composition. Spectra, for example, tell us about the chemical composition of atmospheres in the solar system.

 Spectra

Insight into Science Astronomical Measurement
The laws of physics allow us to infer things we cannot measure directly. For example, we derive information about the chemical compositions of planets from their masses and volumes. Kepler's laws yield the mass of each planet from the periods of their moons' orbits, while the measured diameters of the planets yield their volumes. As shown in the equation above, dividing mass by volume yields the average density, which tells us about their chemistries.

The planets can be placed roughly into four groups. The four inner planets form the first group, followed by Jupiter and Saturn as the second, Uranus and Neptune as the third, and Pluto as the fourth. A group of one? In fact, Pluto actually appears to be the largest of myriad small, distant bodies in the solar system called *Kuiper belt objects*. Indeed, at least 485 such bodies orbiting the Sun beyond Pluto have been identified, as we will discuss in Chapter 8.

From photographs, spectra, and average densities, we can draw conclusions about the compositions of all the planets.

The four inner ones, composed primarily of rock and metal, are called **terrestrial planets** because they resemble the Earth (in Latin, *terra*). Mountains, craters, canyons, and volcanoes are common on their hard, rocky surfaces.

Jupiter and Saturn are composed primarily of hydrogen and helium. Uranus and Neptune have extremely large quantities of water as well as much hydrogen and helium, and as noted above, Pluto is probably a mix of rock and ice. Clouds enshroud Jupiter, Saturn, Uranus, and Neptune. These four were traditionally called the *Jovian planets* (the Roman god Jupiter was also called Jove). However, this single grouping hides significant differences in the sizes and chemistries of Jupiter and Saturn, on the one hand, and Uranus and Neptune on the other, so we will explore them as two groups in Chapter 7. The montage of photographs that opens this section shows the distinctive appearances of the planets.

The different surfaces (or the upper cloud layers) of the planets reflect different amounts of light back out into space. The fraction of incoming light returning directly into space is given by each body's **albedo.** An object that reflects no light has an albedo of 0.0; for example, powdered charcoal has an albedo of nearly 0.0. An object that reflects all the light incident upon it (a high-quality mirror comes close) has an albedo of 1.0. The albedo multiplied by 100 gives the percentage of light directly scattered off that body. The three inner planets with solid surfaces exposed to space (Mercury, Earth, and Mars) have albedos of 0.37 or less. The five planets surrounded by clouds have albedos of 0.47 or more. While Pluto's albedo of 0.5 is relatively high, we do not yet know much about its surface.

Every planet except Mercury and Venus has moons (sometimes called *natural satellites*). There are at least 88 moons in the solar system, and more are still being discovered. Unlike our Moon, most moons are irregularly shaped, more like potatoes than spheres. We will discover that there is as much variety among moons as there is among planets.

II-5   Minor debris from the formation of the solar system still exists

In addition to the nine planets and their moons, many other objects orbit the Sun. Between the orbits of Mars and Jupiter are believed to be millions of such bodies, called **asteroids,** most of them smaller than a kilometer across. This region is called the **asteroid belt.** Asteroids are composed primarily of metal and rock. The largest asteroid, Ceres, has a diameter of about 900 km. The next largest, Pallas and Vesta, are each about 500 km in diameter. Still smaller ones are increasingly numerous. There are many thousands of kilometer-sized asteroids. A close-up picture of the asteroid Gaspra is shown in Figure II-11. In addition to the asteroids between Mars and Jupiter, other asteroids have highly ellip-

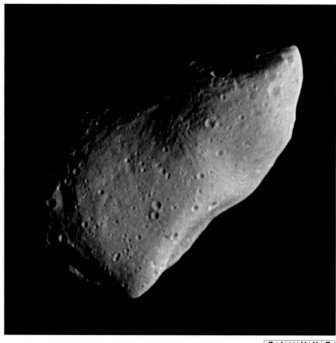

FIGURE II-11   An Asteroid This picture of the asteroid Gaspra was taken in 1991 by the Galileo spacecraft on its way to Jupiter. The asteroid measures $12 \times 20 \times 11$ km. Millions of similar chunks of rock orbit the Sun between the orbits of Mars and Jupiter. (NASA)

tical orbits that take them across the paths of some of the planets. Some asteroids have their own moons, such as asteroid Ida, which is orbited by heavily cratered Dactyl.

 Pieces of rocky and metallic debris even smaller than asteroids are called **meteoroids.** Typical meteoroids are boulder-sized or smaller and apparently exist throughout the disk of the solar system, although most are in the asteroid belt. Chemical studies show that many meteoroids are small fragments of asteroids that broke off when larger bodies collided.

 Quite far from the Sun, beyond the orbit of Pluto, are hundreds of billions of chunks of ice called **comets.** Some comets orbit in a doughnut-shaped region centered on the ecliptic, the Kuiper belt, while others are believed to be distributed in a sphere around the Sun. Many comets have highly elongated orbits that occasionally bring them close to the Sun. When this happens, the Sun's radiation vaporizes some of the comet's ices, producing long, flowing tails (Figure II-12).

Icy debris is also believed to have been observed orbiting other stars. Studying this material is one method astronomers use in their quest to locate planets outside the solar system.

GUIDED DISCOVERY   The Formation of the Solar System

4.7 billion years ago
* Solar nebula begins collapsing.
* Protosun starts growing from in-falling gas.

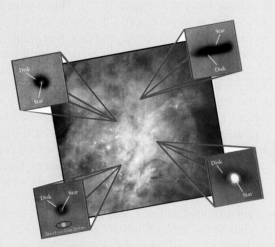

4.6 billion years ago
* Disk of matter forms around the young protosun.
* As regions of the disk clump and collide, planets begin to form.

4.51 billion years ago
* Formation of meteoroids, asteroids, comets, and planets essentially complete.
* Sun starts nuclear fusion and brightens.
* Sun's energy causes remaining gas and dust to dissipate into interstellar space.
* Meteoroids, asteroids, and comets continue to strike planets and the Sun.

3.1 billion years ago
* Most planetesimals have become parts of planets or the Sun.
* Frequent major impacts end.

R I V U X G

FIGURE II-12 **A Comet** The solid part of a typical comet is an irregularly shaped chunk of ice and rocky debris 10 km in diameter. When a comet passes near the Sun, solar radiation vaporizes some of the comet's ices, and the resulting gases and dust form one or two tails millions of kilometers long. This photograph shows Comet Kohoutek in 1974. (NASA)

II-6 Planets orbiting other stars have been discovered

Astronomers are discovering planets outside our solar system around an ever-growing number of stars. More than 80 such planets have been located. These discoveries began with visual images showing disks of material orbiting a few stars, such as Beta Pictoris. The disk around Beta Pictoris is warped, a sign that planets may exist there as well. The planets would tug on different areas of the disk differently, thereby creating the warp (Figure II-13). This star and the material orbiting it formed only 20 million years ago. In 2001, astronomers also detected evidence that millions of comets have formed in this system. By studying such systems as Beta Pictoris, we hope to better understand the formation of the solar system, including addressing the question of whether the Earth acquired its water from comets in the solar system.

Planets outside the solar system are called *extrasolar planets*. Observing them in orbit around other stars is challenging, because even high-albedo planets reflect less than a millionth as much light as a star emits. So far, extrasolar planets have been too dim or too close to their parent stars to be seen easily. Hence, scientists look for planets by indirect observation, primarily by detecting their gravitational tugs on the stars they orbit.

Some planets are discovered by measuring the radial velocity (defined in Chapter 4) of their stars, as measured by the Doppler shifts of the stars' spectra. This motion toward or away from Earth is created by the gravitational pull of the

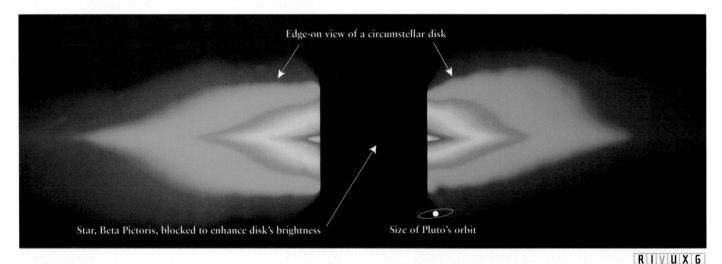

R I V U X G

FIGURE II-13 **A Circumstellar Disk of Matter** This photograph shows a disk of material orbiting the star Beta Pictoris (blocked out in this image) 50 ly from Earth. Twenty million years old, this disk is believed to be composed primarily of iceberglike bodies orbiting the star. Notice that the edge-on disk is warped. This is believed to be due to the gravitational tug of at least one planet orbiting Beta Pictoris inside the inner edge of the disk (in the central region that is blocked out). (STScI, A. Schultz/CSC/STScI and NASA)

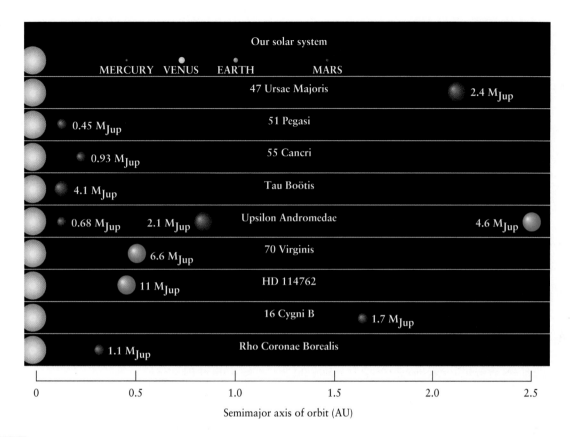

Semimajor axis of orbit (AU)

FIGURE II-14   **Planets Around Normal Stars** This figure shows the separation between extrasolar planets and nine Sunlike stars. The corresponding star names are given in the middle of each line. Our solar system, at the top, shows how close many of these large extrasolar planets are to their stars. (Adapted from G. Marcy and R. P. Butler)

planet. The Doppler shift changes cyclically, and the length of time of one cycle is the period of the planet's orbit.

Other planets are discovered from variations in a star's proper motion, or motion among the background stars (see Chapter 4). A planet's attraction causes its star to deviate from motion in a straight line. This *astrometric* method of discovery looks for just such a wobble.

Finally, planets are located by changes in the brightness of their stars. As a planet passes between us and its star, it partially eclipses the star. Using this method in 2001, astronomers were able to study the atmosphere of a planet. As a gas giant passed in front of the star HD 209458, sodium in the planet's atmosphere absorbed certain wavelengths of starlight. This absorption was observed using the *Hubble Space Telescope.*

The astrometric method also tells us the mass of a planet. The two lowest-mass planets known with confidence have 3 or 4 times the mass of the Earth, and one may have been observed with a mass only slightly greater than that of our Moon. The vast majority, however, range

between a half and a few times the mass of Jupiter, and many orbit surprisingly close to their stars (Figure II-14). From our model of how our solar system formed, these planets should have formed much farther from their stars than they are at present. Nevertheless, the solar system is one of several star systems that do not have Jupiter-sized planets located within an AU of their star.

In 1999, astronomers began discovering star systems with more than one planet. This is done by observing some stars wobbling in ways too complicated to be caused by a single planet. In a multiple planet system, each planet contributes a tug to the star. By combining the effects of two or more planets, the observed pattern of the star's motion (Doppler or astrometric) can be recovered. The first star discovered with multiple planets was Upsilon Andromedae (Figure II-15).

A dramatic consequence of the inward spiraling of planets was discovered in 2001, when the remnants of at least one planet were discovered in the atmosphere of a star that contains at least two other planets. The star's atmosphere

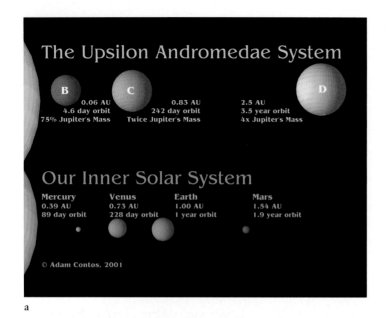

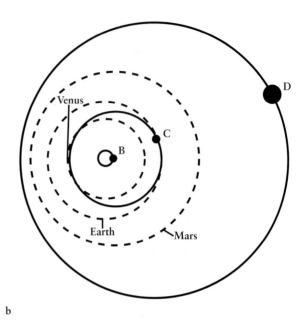

a   b

FIGURE II-15   **A Star with Three Planets** (a) The star Upsilon Andromedae is believed to have at least three planets, discovered by measuring the complex Doppler shift of the star. This star system is located 44 ly from Earth and the planets all have masses similar to that of Jupiter. (b) The orbital paths of the planets, labeled B, C, and D, along with the orbits of Venus, Earth, and Mars, drawn in for comparison. (Adam Contos, Harvard-Smithsonian Center for Astrophysics)

contained a rare form of lithium that is found in planets, but which is destroyed in stars within 30 million years after they form. The presence of this isotope, ^{6}Li, means that at least one planet spiraled so close to the star that it was vaporized.

Extrasolar planets orbit a breathtaking variety of stars, including Earth-sized, burned-out hulks of stars much more massive than the Sun. Some known planets, however, do orbit stars much like our own. The vast majority of extrasolar planets have highly elliptical orbits, with eccentricities up to at least to $e = 0.71$ (see Section 2-3). Star systems with massive planets on highly eccentric orbits are unlikely to have life-sustaining earthlike planets. This is because the changing gravitational force from the massive planet is likely to prevent smaller planets from staying in stable orbits.

All the *known* extrasolar planets probably do not support life as we know it. In the first place, like our giant planets, most of them lack solid surfaces and they are composed primarily of hydrogen and helium. The terrestrial-type planets among them are orbiting only tiny stellar remnants. These stars exploded long ago with such titanic force that any life on their planets would have been annihilated. Furthermore, most extrasolar planets are so close to their stars as to be too hot for organic compounds to remain stable. Life outside the solar system may yet exist. We just have to find the right planets.

We are most likely to discover extraterrestrial life on extrasolar planets similar to Earth orbiting at about 1 AU from stars similar to the Sun. This is because stars supporting life have to last long enough for life to evolve and the planets must be at distances at which water can exist as a liquid. And the search is on, thanks to the power of adaptive optics and space telescopes. Planets are brighter at infrared wavelengths, compared to their stars, than they are at visible wavelengths. Will an infrared telescope therefore point us to the first traces of distant life?

II-7   Frontiers yet to be discovered

Despite their relative closeness to us, there are still countless objects to be discovered and studied in the solar system. The most poorly understood bodies are the Kuiper belt objects. But, as we will see in the coming chapters, virtually every other object in the solar system has secrets to be uncovered. Our understanding of the formation of the solar system also has the potential of being greatly advanced by further study of proplyds and planets orbiting other stars.

We begin a detailed exploration of the solar system by examining the two bodies we know best—the Earth and its Moon. By understanding them, we will be better able to make some sense of the remarkably alien neighboring worlds we encounter thereafter.

WHAT DID YOU THINK?

1 *How many stars are there in the solar system?* The solar system has only one star, the Sun.

2 *Was the solar system created as a direct result of the formation of the universe?* No. All matter and energy were created by the Big Bang. However, much of the material that exists in our solar system was processed inside stars that evolved before the solar system existed. The solar system formed billions of years after the Big Bang occurred.

3 *How long has the Earth existed?* The Earth formed along with the rest of the solar system 4.6 billion years ago.

4 *Is Pluto always the farthest planet from the Sun?* No. From 1979 to 1999, Neptune was the farthest planet from the Sun. This was because Pluto's orbit is highly eccentric, bringing that planet inside Neptune's orbit for about 20 years once every 250 years.

5 *What typical shape(s) do moons have?* While some moons are spherical, most look roughly like potatoes.

KEY WORDS

accretion, 119
albedo, 124
asteroid, 124
asteroid belt, 124
average density, 122

comet, 124
crater, 120
meteoroid, 124
planetesimal, 119
protoplanet, 119

protoplanetary disks (proplyds), 119
protosun, 117
solar nebula, 116
solar system, 116
terrestrial planet, 124

KEY IDEAS

• Hydrogen, helium, and traces of lithium, the three lightest elements, were formed shortly after the creation of the universe. The heavier elements were produced much later by stars and cast into space when the stars died. By mass, 98% of the matter in the universe is hydrogen and helium.

• The solar system formed 4.6 billion years ago from a swirling, disk-shaped cloud of gas, ice, and dust called the solar nebula.

• The four inner planets formed through the accretion of dust particles into planetesimals and then into larger protoplanets. The four large outer planets probably formed through the runaway accretion of gas onto rocky protoplanetary cores.

• The Sun formed at the center of the solar nebula. After about 100 million years, the temperature at the protosun's center was high enough to ignite thermonuclear reactions. For 800 million years after the Sun formed, impacts of asteroidlike objects on the young planets dominated the history of the solar system.

• The four inner planets of the solar system share many characteristics and are distinctly different from the four giant outer planets and from Pluto.

• The four inner, terrestrial planets are relatively small, have high average densities, and are composed primarily of rock and metal.

• Jupiter and Saturn have large diameters and low densities and are composed primarily of hydrogen and helium. Uranus and Neptune have large quantities of water as well as much hydrogen and helium. Pluto, the smallest of the nine planets, is of intermediate density owing to its rock-ice composition.

• Asteroids are rocky and metallic debris in the solar system larger than about a kilometer in diameter. Meteoroids are smaller pieces of such debris. Comets are debris that contain both ice and rock.

• Astronomers have observed disks of gas and dust orbiting young stars.

• At least 80 extrasolar planets have been discovered orbiting other stars.

REVIEW QUESTIONS

1 Why was Pluto recently closer to the Sun than Neptune?

2 Why do astronomers believe that the solar nebula was rotating?

3 To test your understanding of the formation of the solar system do Interactive Exercise II-1 on the CD-ROM or Web site. Explain what is happening in each figure. You can print out your results, if required.

4 Describe three methods that exist for discovering extrasolar planets.

ADVANCED QUESTIONS

5 What two properties of a planet must be known in order for its average density to be determined? How are these properties determined?

6 How can Neptune have more mass than Uranus but a smaller diameter? *Hint:* See Table II-2.

WHAT IF ...

7 The Earth had formed around one of the first-generation stars created in the universe? What chemical elements would the Earth then be composed of? Could we humans exist on that version of Earth? Why or why not?

8 The Earth had a highly elliptical orbit, like Pluto? What would be different about the Earth and life on it?

9 The accretion process of planet formation was still going on? How would life in general on Earth be different and how might we be different?

10 The solar system passed through a cloud of gas and dust that was beginning to collapse to form a new star and planet system? What might happen to the Earth and how would the passage affect the appearance of the sky?

WEB/CD-ROM QUESTIONS

11 Search the Web for information about recent observations of protoplanetary disks. What insights about the formation of the solar system have astronomers gained from these observations? Explain the evidence astronomers have, from these observations, that planets are forming in these disks.

12 In 2000, extrasolar planets with masses comparable to that of Saturn were first detected around the stars HD 16141 (also called 79 Ceti) and HD 46375. Search the Web for information about these planets. How do their masses compare to previously discovered extrasolar planets? Do these planets move around their stars in the same kind of orbit as Saturn follows around the Sun? If not, explain the differences.

13 Search the Web for information about the planet orbiting the star HD 209458. This planet has been detected using two methods. What are they? What have astronomers learned about this planet?

14 **Determining Terrestrial Planet Orbital Periods.** Access the animation "Planetary Orbits" in Essentials II of the *Discovering the Universe* Web site or CD-ROM. Focus on the motions of the inner planets during the last half of the animation. Using the stop and start buttons, determine how many days it takes Mars, Venus, and Mercury to orbit the Sun once if it takes Earth approximately 365 days.

OBSERVING PROJECTS

15 Planetarium software can help you get an overview of our solar system. We will learn more about it in the next chapters. For now, however, open your *Starry Night Backyard*™ program and put it into Atlas mode (*Go/Atlas*). Turn on the planets (Sky: Planets/Sun).Describe and comment on the brightnesses of the various planets relative to the stars. Now turn on daylight (Sky: Daylight) and by changing the time determine which, if any, planets are visible in the night sky this month where you live.

16 There are many young stars still embedded in the clouds of gas and dust from which they formed. In the winter evening sky in the northern hemisphere, for example, is the famous Orion Nebula (Figure II-2). In the summer night sky in the northern hemisphere, the Lagoon, Omega, and Trifid nebulae can be observed. Examine some of these nebulae with a telescope. Describe their appearances. Try to determine which stars in your field of view are actually associated with the nebulosity.

You can use the *Starry Night Backyard*™ program to help you find these nebulae. Also, the table below gives their coordinates (right ascension and declination; see Section 1-2).

Nebula	Right ascension	Declination
Lagoon	18^h 03.8^m	$-24°$ $23'$
Omega	18 20.8	-16 11
Trifid	18 02.3	-23 02
Orion	5 35.4	-5 27

17 Use the *Starry Night Backyard*™ program to observe magnified images of at least four of the planets. First go into Atlas mode (*Go/Atlas*) and then put up the planet list (Sky/Planet List). In the Planet palette, double click on the name of the planet you wish to view. When it appears, right click on it and press *magnify*. Describe the planet's appearance. From what you observe, state whether you are looking at the planet's surface or a cloud cover and justify your assertion.

AN ASTRONOMER'S ALMANAC

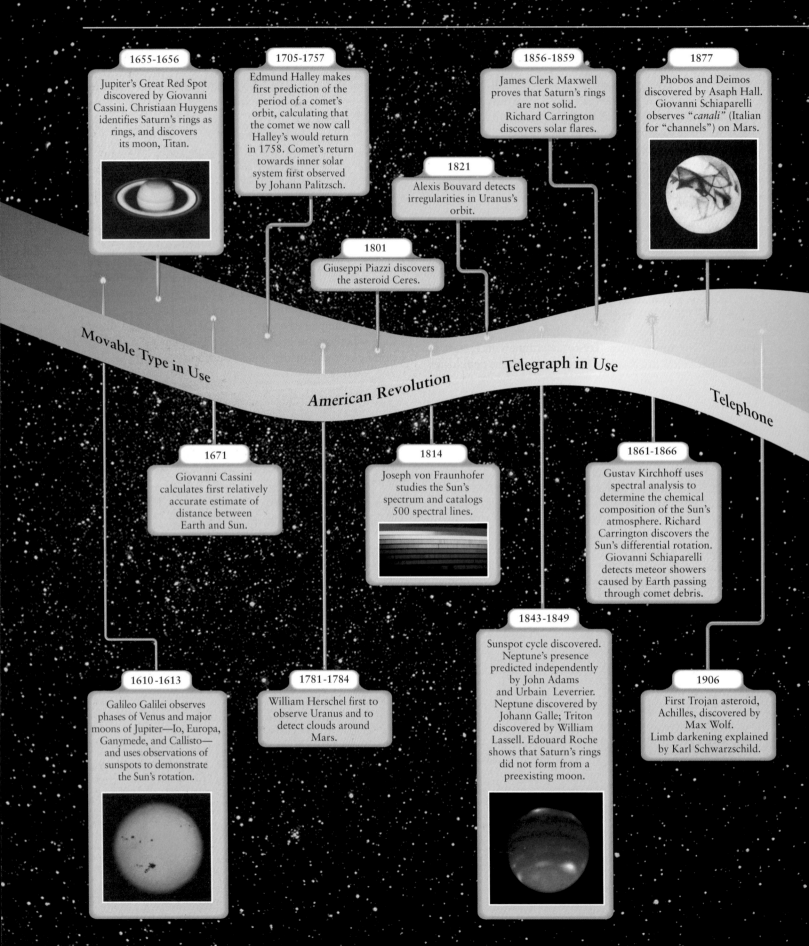

1655-1656
Jupiter's Great Red Spot discovered by Giovanni Cassini. Christiaan Huygens identifies Saturn's rings as rings, and discovers its moon, Titan.

1705-1757
Edmund Halley makes first prediction of the period of a comet's orbit, calculating that the comet we now call Halley's would return in 1758. Comet's return towards inner solar system first observed by Johann Palitzsch.

1856-1859
James Clerk Maxwell proves that Saturn's rings are not solid. Richard Carrington discovers solar flares.

1877
Phobos and Deimos discovered by Asaph Hall. Giovanni Schiaparelli observes *"canali"* (Italian for "channels") on Mars.

1821
Alexis Bouvard detects irregularities in Uranus's orbit.

1801
Giuseppi Piazzi discovers the asteroid Ceres.

Movable Type in Use

Telegraph in Use

American Revolution

Telephone

1671
Giovanni Cassini calculates first relatively accurate estimate of distance between Earth and Sun.

1814
Joseph von Fraunhofer studies the Sun's spectrum and catalogs 500 spectral lines.

1861-1866
Gustav Kirchhoff uses spectral analysis to determine the chemical composition of the Sun's atmosphere. Richard Carrington discovers the Sun's differential rotation. Giovanni Schiaparelli detects meteor showers caused by Earth passing through comet debris.

1843-1849
Sunspot cycle discovered. Neptune's presence predicted independently by John Adams and Urbain Leverrier. Neptune discovered by Johann Galle; Triton discovered by William Lassell. Edouard Roche shows that Saturn's rings did not form from a preexisting moon.

1610-1613
Galileo Galilei observes phases of Venus and major moons of Jupiter—Io, Europa, Ganymede, and Callisto—and uses observations of sunspots to demonstrate the Sun's rotation.

1781-1784
William Herschel first to observe Uranus and to detect clouds around Mars.

1906
First Trojan asteroid, Achilles, discovered by Max Wolf. Limb darkening explained by Karl Schwarzschild.

THE SOLAR SYSTEM

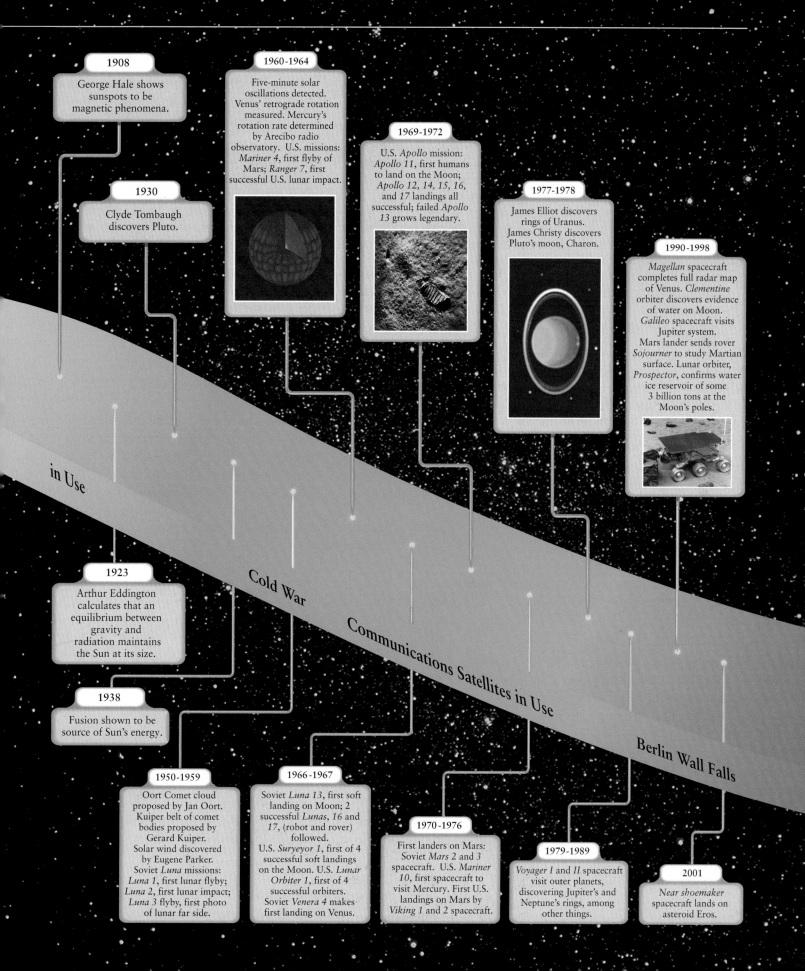

1908

George Hale shows sunspots to be magnetic phenomena.

1930

Clyde Tombaugh discovers Pluto.

1960-1964

Five-minute solar oscillations detected. Venus' retrograde rotation measured. Mercury's rotation rate determined by Arecibo radio observatory. U.S. missions: *Mariner 4*, first flyby of Mars; *Ranger 7*, first successful U.S. lunar impact.

1969-1972

U.S. *Apollo* mission: *Apollo 11*, first humans to land on the Moon; *Apollo 12, 14, 15, 16,* and *17* landings all successful; failed *Apollo 13* grows legendary.

1977-1978

James Elliot discovers rings of Uranus. James Christy discovers Pluto's moon, Charon.

1990-1998

Magellan spacecraft completes full radar map of Venus. *Clementine* orbiter discovers evidence of water on Moon. *Galileo* spacecraft visits Jupiter system. Mars lander sends rover *Sojourner* to study Martian surface. Lunar orbiter, *Prospector*, confirms water ice reservoir of some 3 billion tons at the Moon's poles.

in Use

1923

Arthur Eddington calculates that an equilibrium between gravity and radiation maintains the Sun at its size.

Cold War

Communications Satellites in Use

Berlin Wall Falls

1938

Fusion shown to be source of Sun's energy.

1950-1959

Oort Comet cloud proposed by Jan Oort. Kuiper belt of comet bodies proposed by Gerard Kuiper. Solar wind discovered by Eugene Parker. Soviet *Luna* missions: *Luna 1*, first lunar flyby; *Luna 2*, first lunar impact; *Luna 3* flyby, first photo of lunar far side.

1966-1967

Soviet *Luna 13*, first soft landing on Moon; 2 successful *Lunas*, 16 and 17, (robot and rover) followed. U.S. *Suryeyor 1*, first of 4 successful soft landings on the Moon. U.S. *Lunar Orbiter 1*, first of 4 successful orbiters. Soviet *Venera 4* makes first landing on Venus.

1970-1976

First landers on Mars: Soviet *Mars 2* and *3* spacecraft. U.S. *Mariner 10*, first spacecraft to visit Mercury. First U.S. landings on Mars by *Viking 1* and 2 spacecraft.

1979-1989

Voyager I and *II* spacecraft visit outer planets, discovering Jupiter's and Neptune's rings, among other things.

2001

Near shoemaker spacecraft lands on asteroid Eros.

5 EARTH AND MOON

IN THIS CHAPTER YOU WILL DISCOVER

- why the Earth provides such an ideal environment for life

- that the Earth's insides are churning, its surface is constantly in motion, and its atmosphere has evolved dramatically

- how Earth's magnetic field helps protect life from high-energy particles from space

- that the Moon's surface has been pounded by countless impacts from space debris

- how the Sun and Moon cause the Earth's tides

- that both the Earth and Moon have two major types of surface features

- that water has apparently been discovered at the Moon's poles

The Earth and the Moon
Earth is a dynamic planet whose surface is covered mostly with water and whose nitrogen-oxygen atmosphere supports life. As you will see, the Moon is a barren, desolate world. Because the Moon has no liquid water and virtually no atmosphere, lunar rocks have not been subjected to weathering by wind and water. They thus preserve information about the early history of the solar system. (GSFC/NASA)

R I V U X G

WHAT DO YOU THINK?

1   Will the Earth's ozone layer, which is now being depleted, naturally replenish itself?

2   Who was the first person to walk on the Moon, and on what Apollo space mission did this event occur?

3   Do we see all parts of the Moon's surface at some time throughout the lunar cycle of phases?

4   Does the Moon rotate and, if so, how fast?

5   What causes the ocean tides?

6   When does the spring tide occur?

uppose that you arrive in our solar system from a distant star system, looking for inhabitable worlds. Your spacecraft approaches the inner solar system on the opposite side of the Sun from the Earth. You encounter Mars, with its thin, chilled, unbreathable atmosphere and barren desert landscapes. Not very promising. Moving much closer to the Sun, the next planet you happen to encounter, Venus, is enshrouded in corrosive clouds hiding a menacingly hot surface. Finally, moving back out you spy Earth, orbiting the Sun between these two relatively forbidding planets—one too cold, the other too hot. Ever-changing white clouds pirouette above the browns and blues of Earth's continents and oceans. Its close companion, the Moon, pockmarked by countless craters, causes those oceans to move up and down in a ceaseless rhythm. Is this Earth a world that your funding agency back home can use to justify more astronomical research?

EARTH: A DYNAMIC, VITAL WORLD

A space traveler might first study the sunlight reflected from the Earth's shifting clouds and its surface. The Earth has an average albedo of 0.37, meaning it scatters about 37% of the sunlight it receives right back into space. That scattered light would tell an interstellar traveler that the Earth is indeed an appealing world to investigate, because it indicates large quantities of liquid water as well as dry land. Water, the nearly universal solvent in which life probably first formed, covers about 71% of the Earth's surface.

Beneath the clouds, our visitor would find a geologically active world. Earthquakes shake many regions of it. Volcanoes pour huge quantities of molten rock from inside the Earth onto the surface, and gas from within it vents into the atmosphere. Some mountains are still rising, while others are wearing away. Water flow erodes topsoil and carves river valleys. Rain and snow help rid the atmosphere of dust particles. And life teems virtually everywhere, making the Earth unique in the solar system. Figure 5-1 details Earth's important

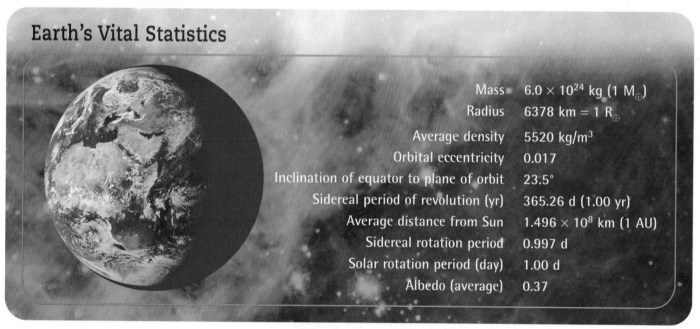

Earth's Vital Statistics

Mass	6.0×10^{24} kg (1 $M_\oplus$)
Radius	6378 km = 1 $R_\oplus$
Average density	5520 kg/m³
Orbital eccentricity	0.017
Inclination of equator to plane of orbit	23.5°
Sidereal period of revolution (yr)	365.26 d (1.00 yr)
Average distance from Sun	1.496×10^8 km (1 AU)
Sidereal rotation period	0.997 d
Solar rotation period (day)	1.00 d
Albedo (average)	0.37

FIGURE 5-1   **A View of Earth's Surface**   An oasis in the forbidding void of space, the Earth is a world of unsurpassed beauty and variety. Ever-changing cloud patterns drift through its skies. More than two-thirds of its surface is covered with oceans. This liquid water, in combination with a huge variety of chemicals from its lands, led to the formation and evolution of life over most of the planet's surface. (NASA)

R I V U X G

physical and orbital properties. The symbol ⊕ in Figure 5-1 and hereafter is astronomer's shorthand for "Earth."

5-1 The Earth's atmosphere has evolved over billions of years

Essential to the existence of complex life is an atmosphere containing oxygen molecules. The Earth's early atmosphere contained no oxygen, and we begin our survey of the Earth with a study of its atmosphere. The Earth's atmosphere is unique among the planets. The air we breathe is predominantly a 4-to-1 mixture of nitrogen to oxygen, two gases that are found only in small amounts in the atmospheres of other planets. This is, however, the third atmosphere to envelop the Earth.

The first one, composed of trace remnants of hydrogen and helium left over from the formation of the solar system, did not last very long. These gases are too light to stay near the Earth (consider what happens to helium balloons). Heated by sunlight, these gases quickly dissipated into space. The gases of the second atmosphere came from inside the Earth through volcanoes and fissures. It was composed primarily of carbon dioxide and water along with some nitrogen. In fact, there was roughly 100 times as much gas in this second atmosphere as there is today. Carbon dioxide and water store a lot of heat, helping to create what is called the *greenhouse effect,* about which we will say more shortly. The high concentration of these gases early in the Earth's life stored more heat than the atmosphere does today. The Earth back then would have been a serious hothouse had the Sun reached its full intensity. Evidence indicates it had not and so, without the high concentrations of carbon dioxide and water in the air, the young Earth would have actually been frozen solid.

Oceans began forming when water precipitated out of this atmosphere, as well as from impacts of water-rich space debris. The oceans eventually absorbed about half the carbon dioxide in that second atmosphere. (Water holds a lot of carbon dioxide, as you can see by the amount of carbon dioxide fizz in a can of soda.) As life evolved in the oceans, much of that dissolved carbon dioxide was transformed into the shells of many creatures, and then, as the animals died, their carbon-rich shells sank to the ocean bottoms. Shells piled up and compressed the shells underneath them into rock such as limestone.

Early plant life in the oceans and then on the shores removed most of the remaining carbon dioxide in the air by converting it into oxygen and the nutrients that the plants needed to survive. The most efficient of such conversion mechanisms is photosynthesis, which today helps maintain the balance between carbon dioxide and oxygen in the air. What remained in the air was nitrogen and water vapor.

The first oxygen that returned to the air, a by-product of plant life activity, did not stay there for long. Oxygen is highly reactive, and early atmospheric oxygen thus combined quickly with many elements on the Earth's surface, notably iron, to form new compounds. (In Chapter 6, we will see similar iron compounds on Mars.)

Eventually, after all the minerals that could combine with oxygen had done so, the atmosphere began to fill with this gas. With the carbon dioxide gone, the air became the nitrogen-oxygen mixture that we breathe today. While nitrogen had been a tiny fraction of that earlier atmosphere, it now dominates the thinner air of today.

Because it has mass, the air is pulled down toward the Earth by gravity, thereby creating a pressure on the surface. We define pressure as a force acting over some area:

$$\text{Pressure} = \frac{\text{force}}{\text{area}}$$

The average atmospheric pressure at sea level is 14.7 pounds per square inch, or 1 atmosphere (atm). This pressure is due to the weight of the air pushing down. The atmospheric pressure decreases smoothly with increasing altitude, falling by roughly half with every 5.5 km of altitude. Seventy-five percent of the mass of the atmosphere lies within 11 km (approximately 7 mi, or 36,000 ft) of the Earth's surface.

All Earth's weather—clouds, rain, sleet, and snow—occurs in this lowest region of the atmosphere, called the **troposphere.** Commercial jets generally fly at the top of the troposphere to avoid having to plow through the denser air below. If the Earth were the size of a typical classroom globe, the troposphere would only be as thick as a sheet of paper wrapped around its surface.

Starting at the Earth's surface, atmospheric temperature initially decreases upward, reaching a minimum of 218 K (−67°F) at the top of the troposphere. Above this level is the region called the **stratosphere,** which extends from 11 to 50 km (approximately 7 to 31 mi) above the Earth's surface. This is the realm of the **ozone layer.**

Insight into Science What's in a Word The importance of word meanings in science cannot be overstressed. The word "layer" in the term "ozone layer" suggests a narrow region, perhaps a few meters thick. Not so. The ozone layer extends in altitude over some 40 kilometers of the atmosphere!

 Ozone molecules (three oxygen atoms, O_3, bound together) are created in the stratosphere when ultraviolet radiation from the Sun occasionally splits O_2 up there into two separate oxygen atoms. Each resulting oxygen atom then combines with different O_2 molecules to create ozone. Once created, ozone efficiently absorbs solar ultraviolet rays, thereby heating the air in this

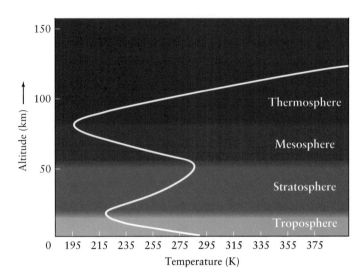

FIGURE 5-2 Temperature Profile of Earth's Atmosphere The atmospheric temperature changes with altitude because of the way sunlight interacts with various gases at different heights.

layer while preventing most of this lethal radiation from reaching the Earth's surface. The temperature therefore increases upward through the stratosphere to about 285 K (50°F) at its top (Figure 5-2).

Not much ozone actually exists in the ozone layer. If all of it was compressed to the density of the air we breathe, it would be a layer only a few millimeters (about one-eighth of an inch) thick. It is enough, however, to protect us from the Sun's ultraviolet radiation. Because the Earth's early atmosphere lacked ozone, ultraviolet radiation penetrated to the planet's surface, where it provided much of the energy necessary for life to evolve in the oceans. After life developed, however, the formation of the ozone layer by the same ultraviolet radiation was essential in lessening the amount of ultraviolet radiation that reached the surface, thereby allowing life-forms to leave the oceans and survive on land.

Today, the ozone layer is still vital, because the energy it absorbs would otherwise cause skin cancer and various eye diseases. In sufficiently high doses, ultraviolet radiation even damages the human immune system. The ozone layer is being depleted today by human-made chemicals, such as chlorofluorocarbons (CFCs). At present, ozone levels in the mid-latitudes of the northern hemisphere are decreasing by about 4% per decade. Much more rapid and dramatic drops in the ozone layer over the Earth's poles, especially over the south pole, have been observed for more than two decades. These seasonal *ozone holes* typically extend over 24 million square kilometers, or 4.6% of the Earth's surface area.

Fortunately, as explained above, sunlight creates ozone. If current international efforts aimed at slowing the depletion rate succeed, the ozone layer will naturally replenish itself, and the ozone holes will probably stop occurring.

Note in Figure 5-2 that above the stratosphere lies the **mesosphere.** Atmospheric temperature again declines as one moves up through the mesosphere, reaching a minimum of about 200 K (–103°F) at an altitude of about 80 km (50 mi). This minimum marks the bottom of the **thermosphere,** above which the Sun's ultraviolet light ionizes atoms, producing charged particles that reflect radio waves. You can tune in distant AM radio stations because their transmissions bounce off this region and return to Earth. On the other hand, FM radio stations use much higher frequencies that pass through this layer without bouncing back. That is why you must be quite near an FM station to receive it.

5-2 Plate tectonics produce major changes on the Earth's surface

Land masses protrude through Earth's oceans. These are regions of the Earth's outermost layer, or **crust,** and are composed of relatively low-density rock that floats on denser material below it. By studying the various kinds of rocks at a particular location, geologists can deduce the history of that site. For example, whether an area was once covered by an ancient sea or flooded by lava from volcanoes is readily apparent from the kinds of rocks that are present.

Beneath the Earth's oceans, the crust has distinct segments, indicating that the continents are separate bodies. We all believe, or at least hope, that the land under our feet will remain secure and unchanged. But the segmented appearance of the Earth's crust reveals that the planet is constantly changing. Earthquakes and volcanoes hint at activity hidden below the Earth's crust.

Anyone who carefully examines the shapes of Earth's continents (see Figure 5-5a) might conclude that landmasses move. Eastern South America, for example, looks like it would fit snugly against western Africa. Between 1912 and 1915, the German scientist Alfred Wegener published his observations on the remarkable fit between landmasses on either side of the Atlantic Ocean, an idea proposed by Isaac Newton two centuries earlier. Wegener was inspired to advocate the theory of **continental drift**—that the continents on either side of the Atlantic Ocean have drifted apart.

Wegener argued that a single, gigantic supercontinent called Pangaea (meaning "all lands") began to break up and drift apart some 200 million years ago. Initially, most geologists ridiculed Wegener's ideas. Although it was generally accepted that the continents do float on denser rock beneath them, few geologists could accept that entire continents move across the Earth's face at speeds as great as several centimeters per year. Wegener and other "continental drifters" could not explain what forces would shove the massive continents around or how less-dense continental rock had gotten through the dense rock of the ocean floor.

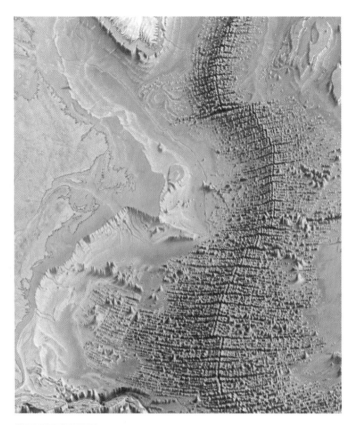

FIGURE 5-3 **The Mid-Atlantic Ridge** This artist's rendition of the floor of the North Atlantic Ocean shows an unusual mountain range in the middle of the ocean floor. Called the Mid-Atlantic Ridge, these mountains are created by lava seeping up from the Earth's interior along a rift that extends from Iceland to Antarctica.

Then, in the mid-1950s, long, underwater mountain ranges were discovered. The Mid-Atlantic Ridge, for example, stretches all the way from Iceland to Antarctica (Figure 5-3). Careful examination revealed that molten rock from the Earth's interior is being forced upward there, as the ocean floor slides apart on either side of the ridge. This **seafloor spreading** is in fact pushing the Americas away from Europe and Africa at a speed of roughly 3 cm per year.

Insight into Science From Proof to Acceptance
Scientific theories may take years or decades of testing and replication before they are accepted. Newton speculated about continental drift in the late seventeenth century. Wegener had proposed plate tectonics by 1915. His theory was not verified until the 1950s and was not accepted until the late 1960s.

Seafloor spreading provided just the physical mechanism that had been missing from the theory of continental drift. It

established Wegener's theory of crustal motion, which was developed into the modern theory of **plate tectonics.**

The Earth's surface is made up of about a dozen major tectonic plates that move relative to each other. In recent years, geologists have uncovered evidence that points to a whole succession of supercontinents that broke apart and reassembled. Pangaea is only the most recent supercontinent in this cycle, which repeats on average every 500 million years (Figure 5-4).

a 200 million years ago

b 180 million years ago

c Today

FIGURE 5-4 **The Supercontinent Pangaea** (a) The continents of today are pieces of what was once a bigger, united body called Pangaea. (b) Geologists now argue that Pangaea must have first split into two smaller supercontinents, which they call Laurasia and Gondwanaland. (c) These bodies later separated into the continents of today. Gondwanaland split into Africa, South America, Australia, and Antarctica, while Laurasia divided to become North America and Eurasia.

The boundaries between plates are the sites of some of the most impressive geological activity on our planet. Earthquakes tend to occur at the boundaries of the Earth's crustal plates, where the plates are colliding (*convergent boundaries*), separating (*divergent boundaries*), or grinding against each other (*transform fault boundaries*). The boundaries between tectonic plates (Figure 5-5a) stand out clearly when the epicenters of earthquakes are plotted on a map. The San Andreas Fault, running along the West Coast of the United States, is an example of a fault where the North American and Pacific plates are rubbing against each other. The Red Sea and Gulf of Suez are expanding as Africa and the Middle East move apart (Figure 5-5b). Great mountain ranges, such as the Himalayas, are thrust up by ongoing collisions between the Indian-Australian and Eurasian plates (Figure 5-5c). During such collisions, one plate moves downward into the Earth (*subducts*) under the other. Where old crust subducts, we find deep trenches, such as the ones off the coasts of Japan and Chile.

But what powers all this activity? Some enormous energy source has caused shifting among landmasses for billions of years. The answer to that question leads us to a scientific model of the Earth's interior.

5-3 Earth's interior consists of a rocky mantle and an iron-rich core

Geologists have calculated that the Earth was entirely molten soon after its formation, about 4.6 billion years ago. The violent impact of space debris, along with energy released by the breakup of radioactive elements, heated and kept the young Earth molten. A similar process of atoms breaking apart, called *nuclear fission*, creates heat used to generate electricity in nuclear power plants.

Iron and the other dense elements sank toward the center of the young, liquid Earth, just as a rock sinks in a pond. At the same time, less dense materials were forced upward toward the surface (Figure 5-6a). This process, called **planetary differentiation**, produced a layered structure within the Earth: a very dense central **core** surrounded by a **mantle** of less dense minerals, which in turn was surrounded by a thin crust of relatively light minerals (Figure 5-6b).

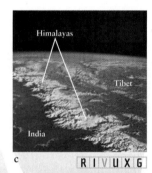

▌FIGURE 5-5 The Earth's Major Tectonic Plates (a) The Earth's surface is divided into a dozen or so rigid plates that move relative to each other. The boundaries of the plates are the scenes of violent seismic and geologic activity, such as earthquakes, volcanoes, rising mountain ranges, and sinking seafloors. The arrows indicate whether plates are moving apart (← →), together (→ ←), or sliding past one another (⇆).

 (b) The Separation of Two Plates. The plates that carry Egypt and Saudi Arabia are moving apart, leaving the trench that contains the Red Sea. This view, taken by astronauts in 1966, shows the northern Red Sea on the right.

 (c) The Collision of Two Plates. The plates that carry India and China are colliding. As a result, the Himalayas are being thrust upward. In this photograph, taken by astronauts in 1968. Mount Everest is one of the snow-covered peaks near the center. (a: Digital image by Peter W. Sloss, NOAA-NESDIS-NGDC; b: *Gemini 12*, NASA; c: *Apollo 7*, NASA)

a

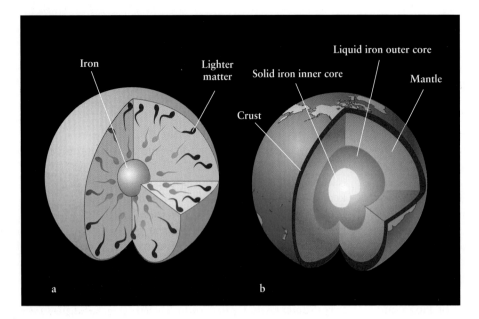

FIGURE 5-6 Cutaway Model of the Earth The early Earth was probably a homogeneous mixture of elements with no continents or oceans. (a) While molten, iron sank to the center and light material floated upward to form a crust. (b) As a result, Earth has a dense iron core and a crust of light rock, with a mantle of intermediate density between them.

Differentiation explains why most of the rocks you find on the ground are composed of lower-density elements like silicon and aluminum. The iron, gold, lead, and other denser elements found on the Earth today actually had to return to the surface via volcanoes and other lava flows. Despite this return of heavier elements to the surface, the average density of crustal rocks is 3000 kg/m^3, considerably less than the average density of the Earth as a whole (5520 kg/m^3).

The temperature in the Earth rises steadily from about 290 K on the surface to nearly 5000 K at the center. Radioactivity in the Earth's interior helps replenish heat lost through the planet's surface. In addition to temperature, pressure also increases with increasing depth below the Earth's surface. The deeper you go, the more mass presses down on you. These factors shape the Earth's inner layers.

You might think that temperatures of 5000 K (9000°F) would melt virtually any substance. The mantle, in particular, is composed largely of minerals rich in iron and magnesium, both of which have melting points of only slightly more than 1250 K at the Earth's surface. However, the melting point of anything also depends on the pressure to which it is subjected: The higher the pressure, the higher the melting point. The high pressures within the Earth (1.4 million atm at the bottom of the mantle) preserve a solid mantle down to a depth of about 2900 km and allow for a solid iron core in the center of the planet, as we will see shortly.

Geologists have deduced basic properties of the Earth's interior by studying the response of the planet to earthquakes. Earthquakes produce a variety of **seismic waves,** vibrations that travel through the Earth either as ripples like ocean waves or by compressing matter like sound waves. Geologists use sensitive **seismographs** to detect and record these vibratory motions. The varying density and composition of the Earth's interior affects the direction of seismic waves traveling through the Earth. By studying the deflection of these waves, geologists have been able to determine properties of the Earth's interior.

 Waves in and on the Earth

By 1906, the year of the great San Francisco earthquake, analysis of seismic recordings led to the discovery that the Earth has a molten iron core with a radius of about 3500 km (2150 mi). For comparison, the overall radius of our planet is 6378 km (3963 mi). More careful measurements in the 1930s revealed that the molten core actually surrounds a solid core with a radius of about 1250 km (775 mi). The cores are composed of roughly 80% iron and 20% lighter elements such as silicon. The interior of our planet therefore has a curious structure: a liquid region sandwiched between a solid inner core and a solid mantle (see Figure 5-6b).

A sophisticated computer model of the Earth's interior predicts that the solid inner core is rotating faster than the rest of the planet. Seismic evidence supports the model that the core goes around about 2 degrees more per year than the surface. Refined models and observations like these are finally revealing the secrets of the Earth's interior.

The tremendous heat and molten rock inside the Earth provide the energy that drives the plate motion. That heat, from the formation of the planet and from the decay of radioactive elements, affects the Earth's surface today. This heat transport is accomplished by **convection.** In this process a liquid or gas is heated and as a result, it expands, becomes buoyant, and therefore rises, carrying the heat it received with it. At a suitable height it gives off the heat, cools, condenses, and sinks back down. You witness convection every time you see simmering soup. Heated from below, blobs of hot soup move upward. The rising soup gives off its heat at its surface.

Heated from below, parts of the Earth's upper mantle are hot enough to have an oozing, plastic flow (like stretching Silly Putty). As sketched in Figure 5-7, molten rock, or *magma*, seeps upward along cracks or rifts in the ocean floor, where convection currents below them are carrying plates apart. Crustal rock sinks back down into the Earth where plates collide. The continents ride on top of the plates as they are pushed around by the convection currents circulating beneath them.

Supercontinents like Pangaea apparently sow the seeds of their own destruction by blocking the flow of heat from the Earth's interior. As soon as a supercontinent forms, temperatures beneath it rise. As heat accumulates, the supercontinent domes upward and cracks. Overheated molten rock wells up to fill the resulting rifts, which continue to widen as pieces of the fragmenting supercontinent are pushed apart.

The Earth's molten, iron-rich interior affects the Earth's exterior in another important way, one that has fewer obvious signs than tectonic plate motion. It is the creation of a magnetic field that extends through the Earth's surface and out into space. As we will see next, that magnetic field results from the convection of molten metal inside the Earth combined with the Earth's rotation.

5-4   The Earth's magnetic field shields us from the solar wind

If you have ever used a compass, you have seen the effect of the Earth's magnetic field. This planetary field is quite similar to the field that surrounds a bar magnet (Figure 5-8a). Magnetic fields are created by electrical charges in motion. Thus, the motion of your compass is caused by electrical charges moving inside the Earth. The wires in your house also carry moving charges called electric currents that, in turn, create magnetic fields of their own.

 The Origin of the Earth's Magnetic Field

Many geologists believe that convection of molten iron in the Earth's outer core combined with our planet's rotation creates electric currents, which in turn create the Earth's magnetic field. The details of this so-called **dynamo theory** of Earth's magnetic field are still being developed.

Magnetic fields nearly always form complete loops (see Figure 5-8b). The magnetic fields near the Earth's south rotation pole loop out tens of thousands of kilometers into space and then return near the Earth's north rotation pole. The places where the magnetic fields pierce the Earth's surface most intensely are called the north and south magnetic poles. These poles move daily. Evidence in solidified lava also reveals that the Earth's magnetic field completely reverses or flips on an irregular schedule ranging from tens of thousands to hundreds of thousands of years. The last reversal was about 600,000 years ago. Today, the magnetic and rotation poles of the Earth are about 11.3° apart; other planets have much larger angles between their magnetic and rotation poles. Figure 5-9 is a scale drawing of the magnetic fields around the Earth, which comprise the Earth's **magnetosphere**.

Our planet's magnetic field protects the Earth's surface from bombardment by energetic particles from space. Most of these particles originate in the Sun (see Chapter 9) and are called the **solar wind**. The solar wind is an erratic flow of electrically charged particles away from the Sun's upper atmosphere. Near the Earth, the particles in the solar wind move at speeds of approximately 400 km/s, or about a million miles per hour.

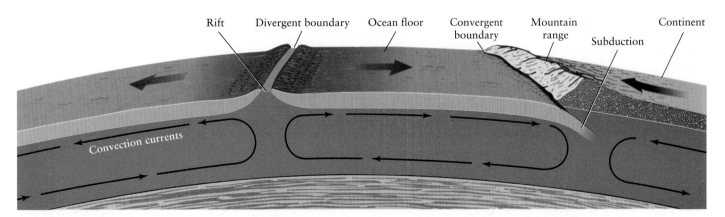

Rift   Divergent boundary   Ocean floor   Convergent boundary   Mountain range   Subduction   Continent

Convection currents

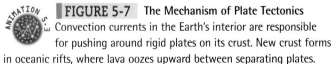
FIGURE 5-7   The Mechanism of Plate Tectonics Convection currents in the Earth's interior are responsible for pushing around rigid plates on its crust. New crust forms in oceanic rifts, where lava oozes upward between separating plates. Mountain ranges and deep oceanic trenches are formed where plates collide. Note that not all tectonic plates move together or apart—some scrape against each other.

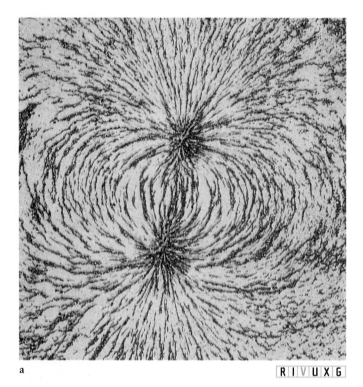

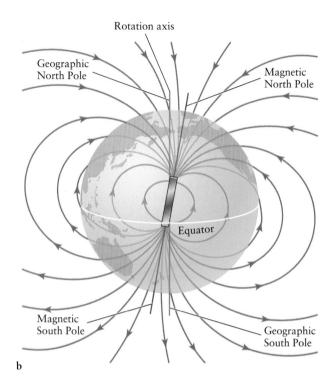

a   R I V U X G   b

FIGURE 5-8   **The Earth's Magnetic Field** (a) The magnetic field of a bar magnet is revealed by the alignment of iron filings on paper. (b) Generated in the Earth's molten, metallic core, the Earth's magnetic field extends far into space. Note that the field is not aligned with the Earth's rotation axis. By convention, the magnetic pole near the Earth's north rotation axis is called the magnetic north pole even though it is actually the south pole on a magnet! We will see similar misalignments and flipped magnetic fields when we study other planets. (a: Jules Bucher/Photo Researchers)

Magnetic fields can change the direction of moving charged particles. The solar wind particles heading toward the Earth are deflected by our planet's magnetic field. The field is strong enough to trap these charged particles in two huge, doughnut-shaped rings called the **Van Allen radiation belts,** which surround the Earth (see Figure 5-9). These belts, discovered in 1958 during the flight of the first successful U.S. Earth-orbiting satellite, were named after the physicist James Van Allen, who insisted that the satellite carry a Geiger counter to detect charged particles.

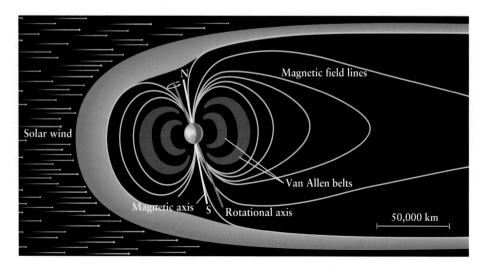

FIGURE 5-9   **Earth's Magnetosphere** A slice through the Earth's magnetic field, which surrounds the entire planet, carves out a cavity in space that excludes charged particles ejected from the Sun, called the solar wind. The Earth is located in this cavity. Most of the particles of the solar wind are deflected around the Earth by the fields in a turbulent region colored blue in this drawing. Because of the strength of the Earth's magnetic field, our planet can trap charged particles in two huge, doughnut-shaped rings called the Van Allen belts (in red).

Insight into Science **Pace of Discovery** Discoveries such as the Van Allen belts remind us of how much remains to be learned, even about our own planet. Scientists continue to make such fundamental discoveries for many reasons. The technology necessary to make certain discoveries, for example, may just have been developed. Or our understanding of the laws of physics may be incomplete. Sometimes the equations are so complex that their implications are not yet clear. Progress in science requires unquenchable curiosity and unfailing open-mindedness.

VIDEO 5.2 Sometimes the Van Allen belts overload with particles. The particles then leak through the magnetic fields near the poles and cascade down into the Earth's upper atmosphere, usually in a ring-shaped pattern. As these high-speed, charged particles collide with gases in the upper atmosphere, the gases fluoresce (give off light) like the gases in a neon sign. The result is a beautiful, shimmering display called the **northern lights (aurora borealis)** or the **southern lights (aurora australis)**, depending on the hemisphere in which the phenomenon is observed (Figure 5-10).

Occasionally a violent event on the Sun's surface called a **coronal mass ejection** sends a burst of protons and electrons straight through the Van Allen belts and into the atmosphere. These events deplete the ozone layer and cause spectacular auroral displays (see Figure 5-10) that can often be seen over a wide range of latitudes and longitudes. Such events also disturb radio transmissions and can damage communications satellites and electrical transmission lines.

THE MOON AND TIDES

Earth's only natural satellite provides one of the most dramatic sights in the nighttime sky. The Moon is so large and so nearby that some of its surface features are readily visible to the naked eye. Even without a telescope, you can easily see dark gray and light gray areas that cover vast expanses of the Moon (Figure 5-11). Appearances can be deceiving, however, as our ancestors interpreted what they saw of the Moon's surface as flat and some asserted that it is covered by large seas of water (maria). Our modern understanding of the lunar surface only began with the telescopic observations of Galileo.

In Galileo's day, most people believed that the Moon had a smooth face. Galileo's telescope revealed mountains towering above its barren surface. Was the Moon's geological history as active as that of the Earth? What are the Moon's surface features telling us about its evolution? Our exploration of the Moon starts with a study of its surface.

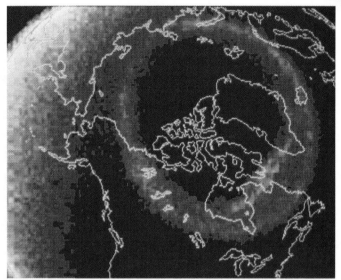

a   R I V U X G

b   R I V U X G

c   R I V U X G

WEB LINK 5.6 **FIGURE 5-10** **The Northern Lights (Aurora Borealis)** A deluge of charged particles from the Sun can overload the Van Allen belts and cascade toward the Earth, producing auroras that can be seen over a wide range of latitudes. Auroras typically occur 100 to 400 km above the Earth's surface. **(a)** View of aurora from the deep space satellite *Dynamics Explorer 1*. **(b)** View of aurora from the Space Shuttle. **(c)** This aurora seen from Earth is green because of the glow of oxygen atoms. (a: L. A. Frank and J. D. Crave, The University of Iowa; b: NASA; c: Paul A. Souders/Corbis)

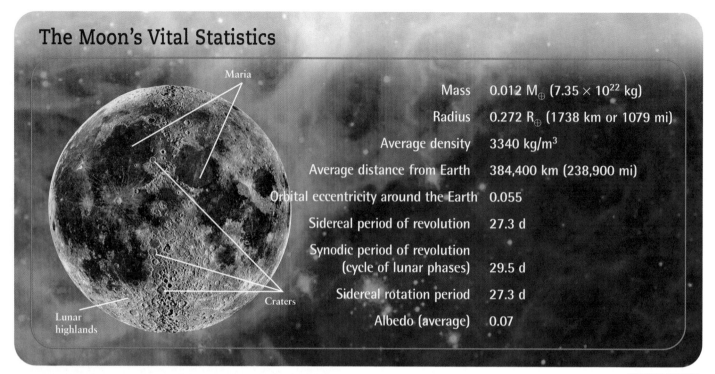

The Moon's Vital Statistics

Mass	0.012 $M_\oplus$ (7.35 × 10²² kg)
Radius	0.272 $R_\oplus$ (1738 km or 1079 mi)
Average density	3340 kg/m³
Average distance from Earth	384,400 km (238,900 mi)
Orbital eccentricity around the Earth	0.055
Sidereal period of revolution	27.3 d
Synodic period of revolution (cycle of lunar phases)	29.5 d
Sidereal rotation period	27.3 d
Albedo (average)	0.07

Maria · Craters · Lunar highlands

FIGURE 5-11   The Moon Our Moon is one of seven large satellites in the solar system. The Moon's diameter of 3476 km (2160 mi) is slightly less than the distance from New York to San Francisco. This photograph is a composite of first quarter and last quarter views, in which long shadows enhance the surface features. (Lick Observatory)

R I V U X G

5-5 The Moon's surface is covered with craters, plains, and mountains

Perhaps the most familiar and characteristic features on the Moon are its craters (Figure 5-12). Through an Earth-based telescope, some 30,000 craters, with diameters ranging from 1 km to more than 100 km, are visible. Following a tradition established in the seventeenth century, the most prominent craters are named after scientists, mathematicians, and philosophers, such as Kepler, Copernicus, Pythagoras, Plato, and Aristotle. Close-up photographs from lunar orbit reveal millions of craters too small to be seen with Earth-based telescopes. Indeed, extreme close-up photos of the Moon's surface reveal countless microscopic craters (Figure 5-13).

Earth, by comparison, has only about 200 known impact craters. Many craters on Earth have been drawn inside our planet by plate tectonic motion and obliterated. Many other craters of a few kilometers in diameter and smaller have been worn away (eroded) by weathering effects of wind and water. The Earth's atmosphere vaporized countless space rocks that would otherwise have created small craters here. Moreover, those that did hit had been slowed down by the atmosphere so that they produced smaller craters than they would have otherwise. We will discuss Earth's craters further in Chapter 8.

Virtually all lunar craters, both large and small, are the result of bombardment by meteoritic material. Nearly all these craters are circular, indicating that they were not merely gouged out by high-speed rocks. Instead, the high speed (upwards of 180,000 km/hr or 112,000 mi/hr) collisions with the Moon's surface vaporized the rapidly moving debris, often with the power of a large nuclear bomb. The resulting explosions of hot gas and debris produced the round-rimmed craters we observe today.

Meteoritic impact causes material from the crater site to be ejected onto the surrounding surface. This pulverized rock is called an **ejecta blanket.** You can see some of the more recent ejecta blankets, which are lighter colored than the older ones (see Figure 5-11). The blanket darkens with time as its surface roughens from impacts by the solar wind. The lightness of its ejecta is one clue astronomers use to determine how long ago a crater formed.

Craters larger than about 20 km also form central peaks (see Figure 5-12). These occur because the impact compresses the crater floor so much that afterward the crater rebounds and pushes the peak upward. As the peak goes up, the crater walls collapse and form terraces.

Besides the craters, the most obvious characteristic of the Moon visible from Earth is that its surface is various shades of gray. Most prominent are the large, dark gray plains called

R I V U X G

FIGURE 5-12   Details of a Lunar Crater This photograph, taken from lunar orbit by astronauts in 1969, shows a typical view of the Moon's heavily cratered far side, the hemisphere we never see from Earth. The large crater near the middle of the picture is approximately 80 km (50 mi) in diameter. Note the crater's central peak, collapsed crater wall, and the numerous tiny craters that pockmark the surrounding lunar surface. (NASA)

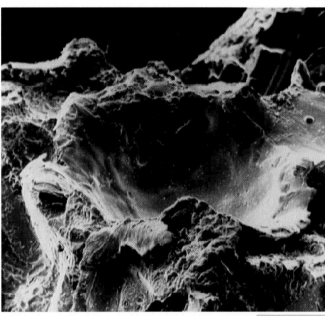

R I V U X G

FIGURE 5-13   A Microscopic Lunar Crater This photograph made with a microscope shows tiny microcraters less than 1 mm across on a piece of moon rock. (NASA)

maria (pronounced "mar-ee-uh"). The singular form of this term, **mare** (pronounced "mar-ay"), means "sea" in Latin and was introduced in the seventeenth century when observers using early telescopes thought that these features were large bodies of water. We know now that no water exists in liquid form on our satellite. Nevertheless, we retain these poetic, fanciful names, from Mare Tranquillitatis (Sea of Tranquillity) and Mare Nubium (Sea of Clouds) to Mare Nectaris (Sea of Nectar) and Mare Serenitatis (Sea of Serenity). As we will discuss shortly, maria are basins on the Moon that were filled in with lava after most cratering ended.

The largest of the maria is Mare Imbrium (Sea of Showers). It is roughly circular and measures 1100 km (700 mi) in diameter. Although the maria seem quite smooth in telescopic views from the Earth, close-up photographs from lunar orbit reveal that they contain small craters and occasional cracks called **rilles** (Figure 5-14). Rilles are the collapsed roofs of underground lava rivers. When the flowing lava solidified, it shrank and the roofs fell in on the spaces created.

As we will discuss further in Section 5-8, the same side of the Moon always faces the Earth. One of the surprises stemming from the earliest lunar exploration is that only one small mare graces the Moon's far side (Figure 5-15). Except for that lone mare, craters cover the entire far side of the Moon.

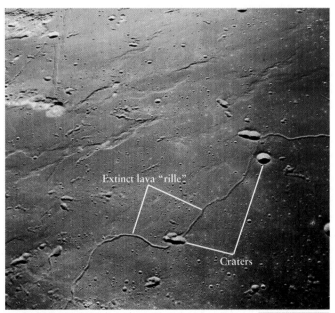

R I V U X G

FIGURE 5-14   Details of a Lunar Mare Close-up views of the lunar surface reveal numerous tiny craters and rilles on the maria. Astronauts in lunar orbit took this photograph of Mare Tranquillitatis (Sea of Tranquillity) in 1969 while searching for potential landing sites for the first manned landing. (NASA)

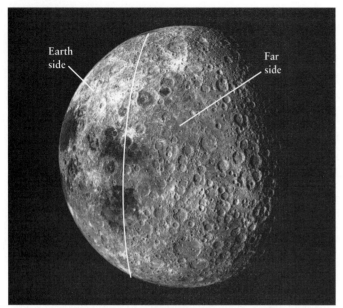

Earth side

Far side

WEB LINK 5.9   VIDEO 5.3

FIGURE 5-15 Comparing the Near and Far Sides of the Moon On the left side of this photograph are three maria visible from the Earth. In this view, taken by the crew of *Apollo 16,* the far side of the Moon is more uniformly and more heavily cratered than the side facing the Earth. The curve indicates the separation between the near and far sides of the Moon. (NASA)

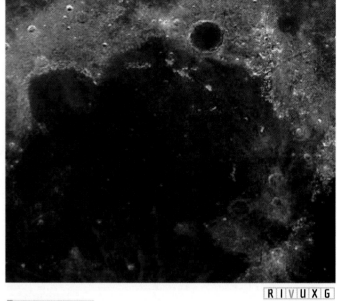

FIGURE 5-16 Mare Imbrium and the Surrounding Highlands Mare Imbrium, the largest of the 14 dark plains that dominate the Earth-facing side of the Moon, is ringed by lighter-colored highlands strewn with craters and towering mountains. The highlands were created by asteroid impacts pushing land together. (NASA)

Detailed measurements by astronauts in lunar orbit demonstrated that the maria on the Moon's Earth-facing side are 2 to 5 km below the average lunar elevation and the one on the far side is 4 to 5 km above the average lunar elevation. The flat, low-lying, dark gray maria cover only 17% of the lunar surface. The remaining 83% is composed of light gray, heavily cratered, mountainous regions called **highlands** (Figure 5-16).

Lunar mountain ranges that rim the vast maria basins suggest ancient impacts far more violent than those that produced typical craters. One such mountain range can be seen around the edges of Mare Imbrium (see Figure 5-16). Although they take their names from famous terrestrial ranges, such as the Alps, the Apennines, and the Carpathians, the lunar mountains were not formed through plate tectonics as mountains on Earth were. Analyses of Moon rocks and observations from lunar orbit imply that violent impacts of huge asteroids thrust up the lunar highlands.

5-6   Visits to the Moon yielded invaluable information about its history

VIDEO 5.4   WEB LINK 5.10

Reaching the Moon became a national obsession for the United States in the early 1960s. A series of successful space mis-

sions by the former Soviet Union spurred President John F. Kennedy's goal of landing a person on the Moon by the end of that decade. The Apollo missions focused our technologies, and, in the end, the drama of space travel and the compelling views from space of the oasis that is our Earth helped bring all of humanity a little closer together. Amid all the excitement, a prodigious amount of scientific research about the Moon was accomplished during the Apollo program.

WEB LINK 5.11

The first of six manned lunar landings, *Apollo 11,* set down on Mare Tranquillitatis on July 20, 1969. Astronaut Neil Armstrong was the first human to set foot on the Moon. In the same year, *Apollo 12* also landed in a mare. Human voyages into space encountered a setback when *Apollo 13* experienced a nearly fatal explosion en route to the Moon. Fortunately, a titanic effort by the astronauts and ground personnel brought it safely home. *Apollo 14* through *Apollo 17* took on progressively more challenging lunar terrain. The period of human visitation to our Moon lasted less than four years, culminating in 1972, when *Apollo 17* landed amid rugged mountains just east of Mare Serenitatis (Figure 5-17).

Astronauts discovered that the lunar surface is covered with a layer of fine powder and rock fragments. This layer, ranging in thickness from 1 to 20 m, is called the **regolith** (from the Greek, meaning "blanket of stone"; Figure 5-18). It formed during 4.5 billion years of relentless micro-meteorite

 FIGURE 5-17   An Apollo Astronaut on the Moon
Apollo 17 astronaut Harrison Schmitt enters the Taurus-Littow Valley on the Moon. Here an enormous boulder had once slid down a mountain, fracturing on the way. This final Apollo mission landed in the most rugged terrain of any Apollo flight. (NASA)

FIGURE 5-18   The Regolith   The Moon's surface is covered by a layer of powdered rock that was created by billions of years of bombardment by space debris. Called the regolith, this surface material sticks together like wet sand, as illustrated by this *Apollo 11* astronaut's bootprint. (NASA)

bombardment that pulverized the surface rock. We do not refer to this layer as *soil*, because the term "soil" suggests the presence of decayed biological matter, which is not found on the Moon's surface. The rough, powdered regolith absorbs most of the light incident on it. Therefore, the Moon's albedo is only 0.07, meaning that it scatters only 7% of the light striking it.

A major goal of the Apollo program was to study the geology of the Moon and help determine its history. The astronauts brought back a total of 382 kg (842 lb) of lunar rocks, which have provided important information about the early history of the Moon. By carefully measuring trace amounts of radioactive elements, geologists determined that lunar rock from the maria is solidified lava that dates back only 3.1 to 3.8 billion years. This indicates that the maria developed about a billion years after the Moon formed and that the maria were formed from molten rock. Indeed, these Moon rocks are composed mostly of the same minerals that are found in volcanic rocks on Hawaii and Iceland. The rock of these low-lying lunar plains is called **mare basalt** (Figure 5-19). About 17% of the Moon's surface rock is basalt.

In contrast to the dark mare basalt, the lunar highlands are covered with a light-colored rock called **anorthosite** (Figure 5-20). On Earth, anorthositic rock is found only in very old mountain ranges, such as the Adirondacks in the eastern United States. Compared to the mare basalts, which

FIGURE 5-19   Mare Basalt   This 1.53-kg (3.38-lb) specimen of mare basalt was brought back by *Apollo 15* astronauts in 1971. Small holes that cover about a third of its surface suggest that gas dissolved in the lava from which this rock solidified. When the lava reached the airless lunar surface, bubbles formed as the pressure dropped and the gas expanded. Some of the bubbles were frozen in place as the rock cooled. (NASA)

R I V U X G

FIGURE 5-20 **Anorthosite** The lunar highlands are covered with this ancient type of rock, which is believed to be the material of the original lunar crust. This sample's dimensions are 18 × 16 × 7 cm. While this sample is medium gray, other anorthosites retrieved from the Moon have been white, while others are darker gray than this one. It was brought back by *Apollo 16* astronauts. (NASA)

R I V U X G

FIGURE 5-21 **Impact Breccias** These rocks are created from shattered debris fused together under high temperature and pressure. Such conditions prevail immediately following impacts of space debris on the Moon's surface. (NASA)

have more of the heavier elements like iron, manganese, and titanium, anorthosite is rich in calcium and aluminum. Anorthosite is therefore less dense than basalt. Many highland rocks brought back to Earth are **impact breccias** (Figure 5-21), which are composites of different rock fused together as a result of meteorite impacts.

 Radioactive Age-Dating

In contrast to rocks from maria, typical anorthositic specimens from the highlands are between 4.0 and 4.3 billion years old. These ancient specimens of highland rock probably represent samples of the Moon's original crust (Figure 5-22).

Since spacecraft began orbiting the Moon, scientists have noted that their orbits do not follow the expected elliptical paths. Every once in a while, usually over a circular mare, the spacecraft momentarily dip Moonward, as they are pulled by a higher than normal concentration of mass near the Moon's surface. These *mass concentrations*, or **mascons**, are believed to be relatively dense rock that filled the mare after the impact of an iron meteorite. New mascons continue to be discovered, including 7 found by the *Lunar Prospector* spacecraft in 1998.

In 1994 the lunar-orbiting *Clementine* spacecraft made an incredible discovery: the first evidence of frozen water on the Moon. Located in craters near the Moon's south pole, this ice is not in danger of being evaporated by sunlight, because, like sunlight at Earth's poles, the Sun deposits very little energy at the lunar poles. Indeed, using radar, astronomers have mapped out areas

near the poles that never receive sunlight, ideal for ice to remain for eons (see Figure 5-22). In 1998 the *Lunar Prospector* confirmed *Clementine*'s discovery and also found evidence for ice at the Moon's north pole. The ice probably was deposited on the Moon during comet and asteroid impacts. Current estimates are that at least 3 billion tons of water are encased there.

The presence of water ice greatly simplifies the prospect of human settlement of the Moon, because its ice can provide us not only with liquid water but also with oxygen and rocket fuel. The Space Shuttle's primary rocket engine is fueled by liquid oxygen and liquid hydrogen. Separating the Moon's water into liquid oxygen and liquid hydrogen would create the same fuel. An existing water reservoir makes the prospect of establishing a lunar colony much less expensive and more feasible than lugging all that water from Earth or dragging a water-rich comet to the Moon.

Astronauts found no traces of life on the Moon. Life as we know it requires liquid water, of which the Moon has none. The Moon has an atmosphere of sodium gas so thin as to be considered an excellent vacuum compared to the density of the air we breathe. An atmosphere is held around a planet or moon by that body's gravitational force. If the atmospheric gases are too hot (moving too fast) for the gravitational force to hold them, they will leak into space. Our Moon has so little gravitational attraction (about one-sixth what we feel at the Earth's surface) that it is unable to retain its tenuous atmosphere. The atmosphere is renewed by sunlight striking the Moon, releasing more gas from its rocks. The Earth, in contrast, has enough mass to retain such gases as nitrogen, oxygen, carbon dioxide, water, and argon.

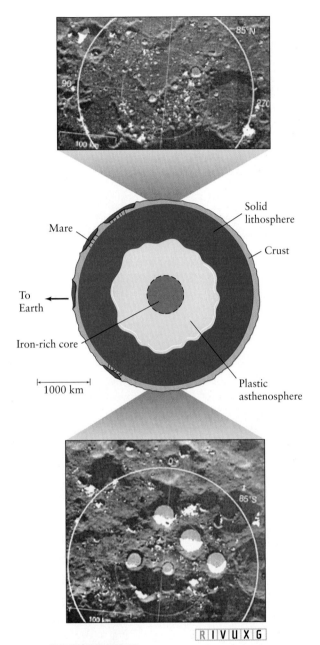

R I V U X G

FIGURE 5-22 **The Moon's Interior** Based on seismic experiments left on the Moon by Apollo astronauts and the studies done by *Lunar Prospector,* we know that the Moon has a crust, mantle, and core. The lunar crust has an average thickness of about 60 km on the Earth-facing side but about 100 km on the far side. The crust and solid upper mantle extend to about 800 km, where the nonrigid inner mantle begins. The Moon's core has a radius of between 300 and 425 km. Although the main features of the Moon's interior are analogous to those of the Earth, the proportions and details are quite different. The *Clementine* spacecraft revealed that the south pole has a significant basin. The crust was apparently stripped away by an impact. **Top inset:** A radar image of the Moon's north polar region. The areas in white and gray are *cold traps* where the Sun never shines. **Bottom inset:** A radar image of the Moon's south polar region, also showing cold traps. (top & bottom inset: NASA/Galaxy)

The slow rotation rate and a small iron core, whose presence was confirmed by *Lunar Prospector,* suggest that unlike the Earth, the Moon should not have a strong magnetic field. Spacecraft orbiting the Moon have confirmed this. However, the *Lunar Prospector* has discovered local areas where the magnetic fields are strong enough to keep the solar wind away from the Moon's surface. These anomalous regions have yet to be explained.

5-7 The Moon probably formed from debris cast into space when a huge asteroid struck the young Earth

Before the Apollo program, scientists debated three different theories about the origin of the Moon. The **fission theory** holds that the Moon was pulled out from a rapidly rotating proto-Earth. This theory does not explain why the Moon has virtually no water incorporated in its rocks, as revealed by the bone-dry samples brought back by Apollo astronauts. The **capture theory** posits that the Moon was formed elsewhere in the solar system and then drawn into orbit about the Earth by gravitational forces. However, it is physically very difficult for a planet to capture a large moon. Furthermore, because bodies formed in different places have different overall chemical compositions, this theory does not explain the similar geochemistries of the Moon and the Earth's surface. The **cocreation theory** proposes that the Earth and the Moon were formed near each other at the same time. However, this theory fails to explain why, compared to the Earth, the Moon has less of the denser elements, such as iron, and why certain types of rocks found on the Moon are not found on Earth.

A fourth theory, first proposed in the 1980s, applies the established fact of age-old collisions in the solar system to the formation of the Moon. Called the **collision-ejection theory,** it proposes that the newly formed Earth was struck at an angle by a Mars-sized asteroid that literally splashed some of the Earth's surface layers into orbit around the young planet. Evidence from Moon rocks places this event at around 4.5 billion years ago, within the first 100 million years of the Earth's existence. A computer simulation of this cataclysm is shown in Figure 5-23. This orbiting matter became a short-lived disk that eventually condensed into the Moon, just as planetesimals condensed to form the Earth.

The collision-ejection theory is consistent with many of the known facts about the Moon. For example, rock vaporized by the impact would have been depleted of volatile (easily evaporated) elements and water, leaving the parched Moon rocks we now know. Also, the Moon has little heavy iron-rich matter because that material had sunk deep into the Earth before the asteroid struck. Furthermore, most of the debris from the collision would lie near the plane of the ecliptic as long as the orbit of the impacting asteroid had

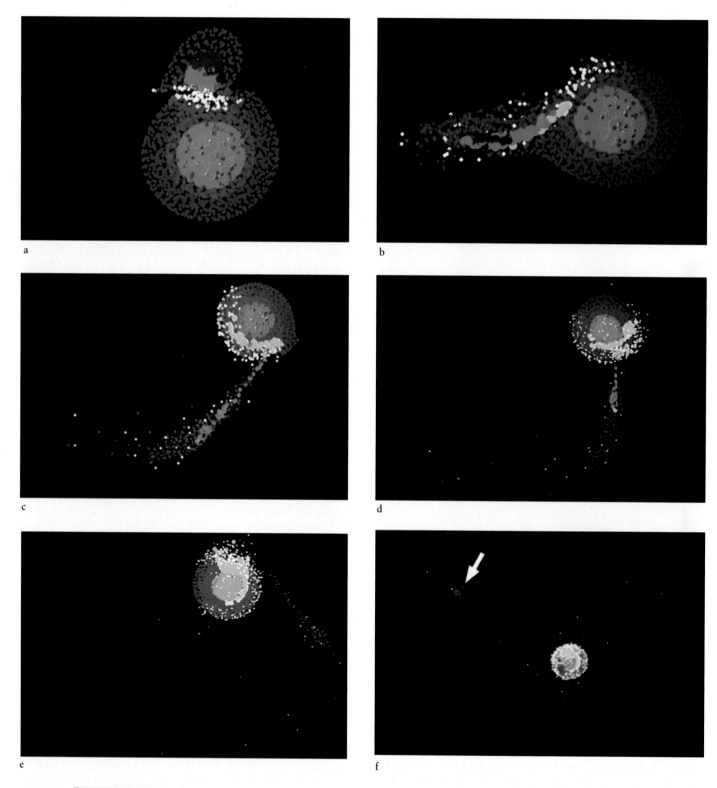

FIGURE 5-23 **The Moon's Creation** This computer simulation models the creation of the Moon from material ejected by the impact of a large asteroid with the young Earth. To follow the ejected material as it moves away from the Earth, successive views (a–f) pan out to show increasingly larger volumes of space. Blue and green indicate iron from the cores of the Earth and the asteroid; red, orange, and yellow indicate rocky mantle material. In this simulation, the impact ejects both mantle and core material, but most of the dense iron falls back onto the Earth. The surviving ejected rocky matter coalesces to form the Moon (indicated by arrow). (W. Benz)

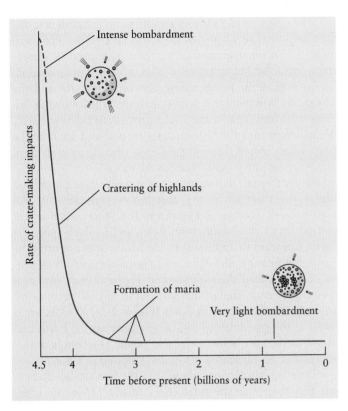

FIGURE 5-24 **The Rate of Crater Formation** The first 700 million years of the Moon's history were dominated by frequent crater-making impacts. Near the end of this intense bombardment, several large impacts gouged out the mare basins. For the past 3.8 billion years, the impact rate has been relatively low.

been near that plane. Also, the impact of an object large enough to create the Moon could have tipped the Earth's axis of rotation and so inaugurated the seasons.

The surface of the newborn Moon probably stayed molten for millennia, due to heat released during the impact of rock fragments falling onto the young satellite and the decay of short-lived radioactive isotopes. As the Moon gradually cooled, low-density lava floating on the Moon's surface began to solidify into the anorthositic crust that exists today in the highlands.

By correlating the ages of Moon rocks with the density of craters at the sites where they were collected, geologists have found that the rate of impacts on the Moon changed over the ages. The ancient, heavily cratered lunar highlands are evidence of intense bombardment that dominated the Moon's early history. As shown in Figure 5-24, this barrage extended from 4.5 billion years ago, when the Moon's surface first solidified, until about 3.8 billion years ago. This period of impacts was punctuated by a severe pounding 4 billion years ago, part of a solar-system-wide reign of terror.

The frequency of impacts gradually tapered off as meteoroids and planetesimals were swept up by the newly formed planets. Recorded among the final scars at the end of this crater-making era are the impacts of more than a dozen objects, each measuring at least 100 km across. As these huge rocks pounded the young Moon, they blasted gaping craters in what would later become the maria. Meanwhile, heat from the decay of long-lived radioactive elements like uranium and thorium began to melt the inside of the Moon again. Then, from 3.8 to 3.1 billion years ago, great floods of molten rock gushed up from the lunar interior through the weak spots in the crust created by the large impacts. Lava filled the impact basins and created the maria we see today. Because most cratering was over by then, maria have few craters. The reason that the far side of the Moon has only one maria apparently stems from the fact that the Moon's crust on the side facing Earth is thinner than on the far side. It was therefore easier for the molten rock inside the Moon to escape through the large craters on our side than those on the far side.

Relatively little has happened on the Moon since those ancient times. A few fresh large craters have been formed, but the astronauts visited a world that has remained largely unchanged for more than 3 billion years. Unlike the Earth, which is still quite geologically active, the Moon is considered to be geologically dead.

5-8 Gravitational force and the orbits of the Moon and Earth produce the tides on Earth and also force the Moon to rotate at the same rate that it revolves around the Earth

If you observe the Moon from Earth over many nights, you always see the same maria and highlands. Even as the Moon goes through its phases, the same surface features face the Earth (see Figure 1-19). Only from spacecraft have we seen the "far side."

The far side of the Moon is often, and incorrectly, called the "dark side" of the Moon. Whenever we view less than a full Moon, some sunlight is falling on its far side and we are seeing part of its dark side. Indeed, throughout each cycle of lunar phases all parts of the Moon get equal amounts of sunlight.

Although the same side of the Moon always faces the Earth, the Moon does rotate. In fact, it has **synchronous rotation:** It rotates on its axis at the same rate that it orbits the Earth (Figure 5-25). To see why the Moon must be rotating, hold your arm out with your palm facing you. Your palm represents the side of the Moon facing the Earth. Standing still, swing your arm horizontally

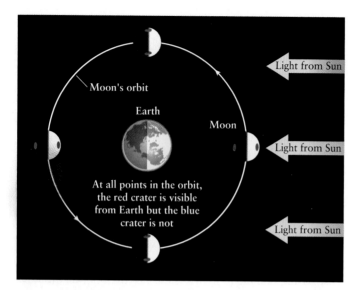

FIGURE 5-25 **Synchronous Rotation of the Moon** This figure shows the motion of the Moon around the Earth as seen from above the Earth's north polar region. In order for the Moon to keep the same side facing the Earth as it orbits our planet, the Moon must rotate on its axis at precisely the same rate that it revolves around the Earth.

always keeping the palm facing you. The only way you can do this is if your hand turns (rotates) at exactly the same rate that it swings (revolves).

The origin of the Moon's synchronous rotation is in the **tidal force** that acts between the Earth and Moon. This tidal force is also what creates ocean tides on Earth. To understand how the Moon came to have synchronous rotation, we investigate the forces acting between the Earth and the Moon. To "kill two birds with one stone," the following discussion focuses on how the Moon creates the tides on Earth. We will then turn this information around and see how the Earth once created tides on the Moon that changed its original rotation rate to synchronous rotation.

Gravitational attraction is what keeps the Earth and Moon together, while the angular momentum (see An Astronomer's Toolbox 2-1) given to the Moon when it was formed keeps the Moon from being pulled back onto the Earth. Contrary to appearances, the Moon does not orbit the Earth, but rather both bodies are in motion around their center of mass called their *barycenter* (Figure 5-26a), just as a spinning wrench sliding across a table has its parts move around a similar common point (Figure 5-26b). Tides are a result of the gravitational tug-of-war between the Earth and Moon as the two bodies orbit this point. The details of this interaction are presented in Guided Discovery: Tides.

Detailed calculations show that the Sun's tidal effect is just less than half the tidal effect of the Moon. In other words, the Moon generates about $2/3$ of the tidal activity on Earth, while the Sun generates most of the rest.

When the Sun, Moon, and Earth are aligned, the Sun and Moon pull in the same direction, and therefore the tides are highest. These **spring tides** (so-called not because they

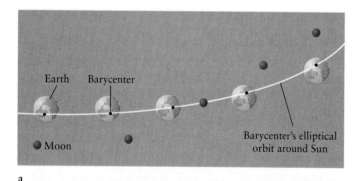

a

FIGURE 5-26 **Gravitational and Centrifugal Forces Create the Tides** (a) The paths of the Earth and Moon as their barycenter follows an elliptical orbit around the Sun. Their motions create centrifugal forces on each body that always point away from the other body. (b) This time-lapse image shows a spinning wrench sliding across a table. Note that its center of mass (the red "plus" sign) moves in a straight line, while its other parts follow curved paths. (b: Berenice Abbott/Photo Researchers)

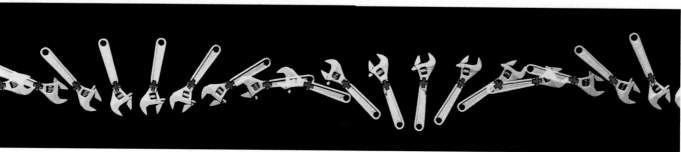

b

GUIDED DISCOVERY Tides

Since tides would occur even if the Earth were not rotating, we will ignore the Earth's rotation until after the basic tidal concepts are introduced. Likewise, we will add in the Sun's tidal effect later.

Consider the motion of the Earth and Moon around their barycenter. The barycenter is located 1068 miles below the Earth's surface on the line between the centers of the Earth and Moon (Figure 5-26a). As a result of its motion around the barycenter, every point on Earth feels a force away from the Moon. Two dancers or skaters holding each other and whirling around feel a similar force. Calculations show that the strength of the outward force acting on the Earth, labeled F_{cent} in Figure 5-27a, is the same everywhere on the planet.

Recall from Chapter 2 (An Astronomer's Toolbox 2-2) that the gravitational force from one body on another decreases with distance. Therefore, the Moon's gravitational attraction is greatest at the point on the Earth's surface closest to the Moon. This attraction is greater than the force felt at, say, the center of the Earth, which in turn is greater than the force felt on the opposite side of the Earth from the Moon. The gravitational force in various places is labeled F_{grav} in Figure 5-27a.

Because forces in the same or opposite directions simply add or subtract, respectively, we can add up the two forces F_{cent} and F_{grav} at any point on the Earth due to the presence of the Moon. The resulting net force at various places is labeled F_{tide} in Figure 5-27a. This is the tidal force acting at these places. For example, at the point on the Earth directly below the Moon, the tidal force is very strong and points toward the Moon. At the center of the Earth the two forces cancel each other completely. At the point on the opposite side of the Earth from the Moon, the tidal force is equal in strength to the tidal force directly under the Moon, but directed away from the Moon. The strength of the tidal force at various places is summarized in Figure 5-27b.

Consider the point on the Earth closest to the Moon. The strength of the tidal force, F_{tide}, seems to imply that the Moon's gravitational force lifts the water there to create the high tide. However, the force of gravitation from the Moon is not strong enough to raise that water substantially. Rather, the high tide closest to the Moon occurs because ocean waters from nearly halfway around to the opposite side of the Earth are pulled Moonward by the gravitational force acting on them. This water slides toward the Moon and fills the ocean directly under it. Likewise, the net outward force acting on the opposite side of the Earth from the Moon pushes the waters that are just over halfway around from the Moon to the side directly away from it, creating a high tide of equal magnitude on the opposite side of the Earth from the Moon. Where the water has been drained away in this process, low tides occur.

Let's spin the Earth back up. As the Earth rotates, the geographical locations of the high and low tides continually change. Figure 5-28 indicates that high tides occur when the Moon is high in the sky or below our feet, while low tides occur when the Moon is near either horizon. This means that there are two cycles of high and low tides at most places on Earth each day. These cycles do not span exactly 24 hours, because during that time the Moon is moving around the Earth, changing the time at which high or low tide occurs at any location. As a result, the time between consecutive high tides is 24 h 50 min.

Tides are complicated by the fact that the oceans are not uniformly deep and that the shores around continents and islands have a variety of shapes. As a result, there are some places where the oceans have only a single cycle of tides each day. Some places have tides that change negligibly while some channels, such as the Bay of Fundy between the United States and Canada, and the Bristol Channel between Wales and England, have much higher tides than normal.

The Earth's rotation is nearly 30 times faster than the Moon's revolution around the Earth. Therefore, the rotating Earth drags the high tides around with it. As a result, the high tide that should be directly between the centers of the Earth and Moon is dragged 10° ahead of the Moon in its orbit around the Earth (Figure 5-28). This high tide stays at roughly this position because the Moon's gravitational force pulls it westward while it is being dragged eastward by the Earth's rotation and the two effects balance each other out. A similar argument applies to the tide on the opposite side of the Earth.

We can repeat this analysis for the Earth-Sun system and get exactly the same qualitative results.

only occur in the spring—they occur in all seasons—but, from the German, *springen*, meaning to "spring up") occur at every new and full Moon (Figure 5-28a). Note that it doesn't matter whether the Sun and Moon are on the same or opposite sides of the Earth when creating the spring tides. At first quarter and third quarter, the Sun and Moon form a right angle as seen from the Earth. The gravitational forces they exert compete with each other, so the tidal distortion is

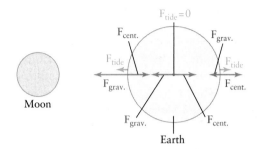

a

b

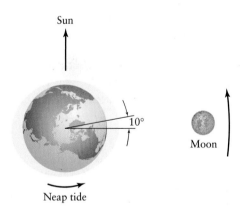

b

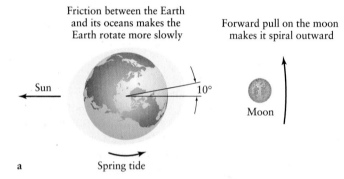

FIGURE 5-27   Tidal Forces **(a)** The Moon induces tidal forces F_{tide}, on the Earth because of the force, F_{cent}, created by the orbital motion of the two bodies combined with the changing strength of the Moon's gravitational force, F_{grav}, over the Earth's surface. The magnitude and direction of each arrow represents the strength and direction of each force. **(b)** Water slides along the Earth to create the tides. Ignoring the Earth's rotation and continents until Figure 5-28, this figure shows how two high tides are created on the Earth by the Moon's gravitational pull. The Sun has a weaker, but otherwise identical, effect.

FIGURE 5-28   Tides on the Earth The gravitational and centrifugal forces of the Moon and Sun deform the oceans. Due to Earth's rapid rotation, the high tide closest to the Moon leads it around the Earth by about 10°. **(a)** The greatest deformation (spring tides) occurs when the Sun, Earth, and Moon are aligned with the Sun and Moon either on the same or opposite sides of the Earth. **(b)** The least deformation (neap tides) occurs when the Sun, Earth, and Moon form a right angle.

the least pronounced. The smaller tidal shifts on these days are called **neap tides** (Figure 5-28b).

Finally, we can explain the Moon's synchronous rotation: When the Moon was young, it was molten and Earth's gravity created huge tides of molten rock up to 60 ft high there. The Earth's tidal force acted to keep the Moon's high tide directly between the Moon's center and the center of the Earth. Friction created by the tidal motion changed the Moon's rotation rate. As a result, the Moon's rotation rate became the same as its revolution rate. When it solidified, the Moon was locked into this synchronous rotation.

5-9   The Moon is moving away from the Earth

A century ago, Sir George Darwin (son of evolutionist Charles Darwin) proposed that the Moon is moving away from the Earth. To understand why, note in Figure 5-28 that the high tide nearest the Moon is actually 10° ahead of the Moon in its orbit around the Earth. As discussed in the Guided Discovery, this offset occurs because friction between the ocean and the ocean floor causes the rapidly spinning Earth to drag the tidal bulge ahead of the line between the center of the Earth and center of the Moon. Gravitational pull from the high ocean tide nearest the Moon pulls the Moon ahead in its orbit, giving it energy and thereby forcing it to spiral outward.

To test this concept, the *Apollo 11, 14,* and *15* astronauts placed on the Moon a set of reflectors, similar to the orange and red ones found on cars, to help measure the Moon-Earth distance. Pulses of laser light were fired at the Moon from the Earth (Figure 5-29), and the time it took the light to reach the Moon, bounce off the reflector array, and return to the Earth was carefully recorded. Knowing the speed of light and the time it takes for the light pulse to make

R I V U X G

FIGURE 5-29   **Lunar Ranging** Beams of laser light are fired through three telescopes at the Observatoire de la Côte d'Azur, France. The light is then reflected back by the corner reflectors placed on the Moon by Apollo astronauts. From the time it takes the light to reach the Moon and return to Earth, astronomers can determine the distance to the Moon to within 2 cm. (Observatoire de la Côte d'Azur)

the circuit, astronomers can calculate the distance to the Moon with an error of only a few millimeters. From such measurements over time, astronomers have established that the Moon is spiraling away from the Earth at a rate of 3.8 cm per year.

This means, of course, that the Moon used to be closer to the Earth. Although the initial distance of the Moon when it coalesced is not known, its present rate of recession tells us that it was at least half its present distance; more accurate calculations suggest one-tenth. At that distance, the tides on the Earth were a thousand times higher than they are today. (What effects would those tides have had on the geology of the young Earth?)

Where does the Moon get the energy necessary to spiral away from the Earth? The answer, of course, is from the Earth itself. As the tides move westward to try to stay under the Moon, they encounter the eastward-moving landmasses, which disrupt the water's motion. Because the oceans move westward, while the planet rotates eastward, the oceans actually push on the continents opposite to the Earth's direction of rotation. The result is that the tides are slowing the Earth's rotation. The angular momentum lost by the Earth is gained by the Moon. At present, the day is slowing down by about one thousandth of a second each century. Geologists calculate that when the Earth first formed, the day was only between 5 and 6 hours long.

Will the Moon ever leave the Earth completely as a result of these dynamics? In fact, it won't. Rather, the Earth's rotation will slow down until it is in synchronous rotation with respect to the Moon. Thereafter, the Moon will remain at a fixed distance over one side of the Earth. However, billions of years before that happens, the Sun is going to self-destruct, so it doesn't really matter!

5-10 Frontiers yet to be discovered

The Earth and Moon still hold many mysteries. We have yet to understand the different rotation rate of the Earth's core from its surface, as well as the details of the dynamo effect that generates the Earth's magnetic field. We do not know how fast the Earth was rotating initially. We also need to better understand the impact of humans on the Earth and its atmosphere.

The Moon still inspires many questions and hides many secrets. The six American manned missions and three Soviet soft, robotic lunar landings have barely scratched the Moon's surface, bringing back samples from only nine locations. We still know very little about the Moon's far side. Is there really water buried at the poles? Is the Moon's interior molten? Does the Moon really have an iron-rich core? How old are the youngest lunar rocks? Did lava flows occur over western Oceanus Procellarum only two billion years ago, as crater densities there suggest? Is the Moon really geologically dead, or does it just appear dead because our examination of the lunar surface has been so cursory? Such questions can be answered only by returning to the Moon.

WHAT DID YOU KNOW?

1   *Will the Earth's ozone layer, which is now being depleted, naturally replenish itself?* Yes. Ozone is created continuously from normal oxygen molecules by their interaction with the Sun's ultraviolet radiation.

2   *Who was the first person to walk on the Moon, and on what Apollo space mission did this event occur?* Neil Armstrong was the first person to set foot on the Moon. He and Buzz Aldrin flew on the *Apollo 11* spacecraft piloted by Michael Collins. Armstrong and Aldrin set down the *Eagle* lander on the Moon on July 20, 1969.

3   *Do we see all parts of the Moon's surface at some time throughout the lunar cycle of phases?* No. Because the Moon's rotation around Earth is synchronous, we always see the same side. The far side of the Moon has been seen only from spacecraft that pass or orbit it.

14 *Does the Moon rotate and, if so, how fast?* The Moon rotates at the same rate that it revolves around the Earth. If the Moon did not rotate, then, as it revolved, we would see its entire surface from the Earth, which we do not.

15 *What causes the ocean tides?* The tides are created by centrifugal and gravitational forces, primarily from the Moon and, to a lesser extent, from the Sun.

16 *When does the spring tide occur?* Spring tides occur twice monthly, during each full and new Moon.

KEY WORDS

KEY IDEAS

Earth: A Dynamic, Vital World

• The Earth's atmosphere is about four-fifths nitrogen and one-fifth oxygen. This abundance of oxygen is due to the biological processes of life-forms on the planet.

• The Earth's atmosphere is divided into layers named the troposphere, stratosphere, mesosphere, and thermosphere. Ozone molecules in the stratosphere absorb ultraviolet light rays.

• The outermost layer, or crust, of the Earth offers clues to the history of our planet.

• The Earth's surface is divided into huge plates that move over the upper mantle. Movements of these plates, a process called plate tectonics, are caused by convection in the mantle, upwelling of molten material along cracks in the ocean floor produces seafloor spreading. Plate tectonics is responsible for most of the major features of the Earth's surface, including mountain ranges, volcanoes, and the shapes of the continents and oceans.

• Study of seismic waves (vibrations produced by earthquakes) shows that the Earth has a small, solid inner core surrounded by a liquid outer core. The outer core is surrounded by the dense mantle, which in turn is surrounded by the thin, low-density crust. The Earth's inner and outer cores are composed primarily of iron with some nickel mixed in. The mantle is composed of iron-rich minerals.

• The Earth's magnetic field produces a magnetosphere that surrounds the planet and blocks the solar wind.

• Some charged particles from the solar wind are trapped in two huge, doughnut-shaped rings called the Van Allen radiation belts. A deluge of particles from a coronal mass ejection by the Sun can initiate an auroral display.

The Moon and Tides

• The Moon has light-colored, heavily cratered highlands and dark-colored, smooth-surfaced maria.

• Many lunar rock samples are solidified lava formed largely of minerals also found in Earth rocks.

• Anorthositic rock in the lunar highlands was formed between 4.0 and 4.3 billion years ago, whereas the mare basalts solidified between 3.1 and 3.8 billion years ago. The Moon's surface has undergone very little geological change over the past 3 billion years.

• Impacts have been the only significant "weathering" agent on the Moon; the Moon's regolith (pulverized rock layer) was formed by meteoritic action. Lunar rocks brought back to Earth contain no water and are depleted of volatile elements.

• Frozen water has been discovered at the Moon's poles.

• The collision-ejection theory of the Moon's origin holds that the young Earth was struck by a huge asteroid, and debris from this collision coalesced to form the Moon.

• The Moon was molten in its early stages, and the anorthositic crust solidified from low-density magma that floated to the lunar surface. The mare basins were created later by the impact of planetesimals and were then filled with lava from the lunar interior.

• Gravitational and centrifugal interactions between the Earth and the Moon produce tides in the oceans of the Earth and set the Moon in synchronous rotation. The Moon is moving away from the Earth, and consequently, the Earth's rotation rate is decreasing.

REVIEW QUESTIONS

1 Why does the Earth's albedo change daily? seasonally?

2 Why is the Earth's surface not riddled with craters as is that of the Moon?

3 Describe the structure of the Earth's atmosphere.

4 Describe the process of plate tectonics. Give specific examples of geographic features created by plate tectonics.

 5 To review the Earth's tectonic plates, do Interactive Exercise 5-1 and print out or sketch the completed diagram.

 6 To review the mechanism of plate tectonics, do Interactive Exercise 5-2 and print out or sketch the completed diagram.

7 How do we know about the Earth's interior, given that the deepest wells and mines extend only a few kilometers into its crust?

8 Describe the interior structure of the Earth.

9 Why is the center of the Earth not molten?

10 Describe the Earth's magnetosphere.

 11 To review the Earth's magnetosphere, do Interactive Exercise 5-3 and print out or sketch the completed diagram.

12 What are the Van Allen belts?

13 What kind of features can you see on the Moon with a small telescope?

14 On the basis of lunar rocks brought back by the astronauts, explain why the maria are dark-colored, but the lunar highlands are light-colored.

15 Why are there so few craters on the maria?

 16 To review the crater formation history in the solar system, do Interactive Exercise 5-4 and print out or sketch the completed diagram.

17 Briefly describe the main differences and similarities between Moon rocks and Earth rocks.

18 How do we know that the maria were formed after the lunar highlands?

19 What is a tidal force? How do tidal forces produce tides in the Earth's oceans?

20 What is the difference between spring tides and neap tides?

21 Why do most scientists support the collision-ejection theory for the Moon's formation?

ADVANCED QUESTIONS

22 Explain how the outward flow of energy from the Earth's interior drives the process of plate tectonics.

23 Why do some geologists believe that Pangaea was the most recent in a succession of supercontinents?

24 Why are active volcanoes, such as Mount St. Helens, usually located in mountain ranges along the boundaries of tectonic plates?

25 Why is more lunar detail visible through a telescope when the Moon is near quarter phase than when it is at full phase?

26 In *The Tragedy of Pudd'nhead Wilson*, Mark Twain wrote, "Everyone is a Moon, and has a dark side which he never shows to anybody." What, if anything, is wrong with Twain's astronomy?

27 Why are the Moon rocks retrieved by astronauts so much older than typical Earth rocks, even though both worlds formed at nearly the same time?

28 Some people who supported the fission theory proposed that the Pacific Ocean basin is the scar left when the Moon pulled away from the Earth. Explain why this idea is wrong.

29   Apollo astronauts left seismometers on the Moon that radioed seismic data back to Earth. The data showed that moonquakes occur more frequently when the Moon is at perigee (closest to the Earth) than at other locations along its orbit. Give an explanation for this finding.

DISCUSSION QUESTIONS

32   If the Earth did not have a magnetic field, do you think auroras would be more common or less common than they are today? Explain.

33   Comment on the idea that without the Moon's presence, life would have developed far more slowly.

34   Identify and compare the advantages and disadvantages of lunar exploration by astronauts as opposed to mobile, unmanned instrument packages and robots.

35   When was the last earthquake near your hometown? How far is your hometown from a plate boundary? What

WHAT IF ...

37   The Earth had two moons? Assume one is at our Moon's distance and the other is at half that distance. Describe the motion of the two moons in the sky and how they might appear to us. What would be different here on Earth?

38   The Moon orbited the Earth in the opposite direction from the Earth's rotation, rather than in the same direction? What would be different about the Earth and life on it? Assume that such a counterrotating Moon is at the same distance as our Moon.

39   Our Moon were one-tenth of its actual distance from the Earth? What would be different about the Earth and life on it? *Hint:* The heights of tides vary as $1/r^3$, where r is the distance between the centers of the Earth and the Moon.

WEB/CD-ROM QUESTIONS

43   Use the Web to learn more about plate tectonics. What global changes might accompany the formation and breakup of a supercontinent? How might these changes affect the evolution of life? What life-forms dominated the Earth when Pangaea existed some 200 million years ago, and also when fragments of the preceding supercontinent were as dispersed as today's continents?

44   Search the Web for information about "Pangaea Ultima," a supercontinent that may form in the distant future. When is it expected to form? How will it compare to Pangaea of 200 million years ago (Figure 5-4)?

45   Use the Web to determine the status of the Antarctic and Arctic ozone holes. How has the situation changed

30   Why do you think that no Apollo missions landed on the far side of the Moon?

31   How might studying albedo help astronomers locate inhabitable worlds orbiting other stars?

kinds of topography (for example, mountains, plains, seashore) dominate the geography of your hometown area? Does that topography and the frequency of earthquakes seem to be consistent with your hometown's proximity to a plate boundary?

36   The ice on the Moon is believed to be mixed with rock just under the Moon's surface. How might that water be economically extracted and purified?

40   The Moon, located where it actually is, were as massive as the Earth? What would be different about the Earth and life on it and about the Moon?

41   The Earth's interior were entirely solid, rather than partly molten? What would be different about the Earth and life on it?

42   The Earth were now in synchronous rotation with the Moon? That is, suppose that the Earth rotated at the same rate that the Moon orbits the Earth. What would be different about the Earth and life on it?

over the past few years? Explain why most scientists who study this issue blame chemicals called CFCs for the existence of the ozone holes.

46   In 1989 representatives of many countries signed a treaty called the Montreal Protocols to protect the ozone layer. Use the Web to learn of the current status of the treaty. How many nations have signed it? Has the treaty been amended? If so, how? List when various substances that destroy the ozone layer are scheduled to be phased out.

47   Several unmanned missions to the Moon were under development as of this writing. These include LUNAR-A and SELENE (Institute of Space and Astronautical Science, Japan) and SMART-1 (European Space Agency). Search

the Web for current information about these and any other upcoming lunar missions you can find. When is each scheduled to be launched? What *new* investigations are planned for each mission? What existing scientific issues may these missions resolve?

48 Refinements to the mass and impact angle of the planetesimal that struck the Earth and created the Moon have recently been made. Search the Web for this information and explain the justification for these new parameters.

49 Some astronomers have observed changes in brightness and color on the Moon, called Lunar Transient Phenomena. Search the Web and explain what these events are and what their origins are believed to be.

OBSERVING PROJECTS

  Observing Tips and Tools: You can learn a lot by observing the Moon through binoculars. Note that the Moon will appear right-side-up through binoculars, but inverted through a telescope. So, if you use a map of the Moon to aid your observations, you will need to take this into account. Inexpensive Moon maps are available, such as *Moon Map* published by *Sky &Telescope* magazine. To determine the lunar phase (when you cannot examine the Moon directly), you can usually find it on most calendars; by checking the weather page in your newspaper; by consulting the current issue of *Sky & Telescope* or *Astronomy* magazine; by using your *Starry Night Backyard*™ program; or on the Web.

50 Use a telescope or binoculars to observe the Moon. Compare the texture of the lunar surface you see on the maria with that of the lunar highlands. How does the visibility of details vary with distance from the terminator (the boundary between day and night on the Moon)? Why?

51 If you live near the ocean, observe the tides to see how the times of high and low tides are correlated with the position of the Moon in the sky.

52 Observe the Moon through a telescope or binoculars every few nights over a period of two weeks between new Moon and full Moon. Make sketches of various surface features, such as craters, mountain ranges, and maria. How does the appearance of these features change with the Moon's phase? Which features are most easily seen at a low angle of illumination (near the terminator)? Which features show up best with the Sun nearly overhead as seen from the Moon (far from the terminator)?

 53 Use the *Starry Night Backyard*™ program to view the Earth as it appears from the Moon (*Go/Earth* (from Moon)). Set the timestep to 12 hours and click the ▶ (flow time) button. (a) Describe the phases of the Earth. (b) Observe the surface features of the Earth that you see at each timestep. Do they change? If so, explain why—note that there is no significant change in the face of the Moon as seen from Earth.

 54 Use *Starry Night Backyard*™ to observe the changing appearance of the Moon. Set for today, turn on Atlas mode (*Go/Atlas*) and then find the Moon (*Edit/Find/Moon*). What phase is it in? Set the time-step to 6 hours and click the ▶ (flow time) button. (a) Describe how the phase of the Moon changes.b) Look carefully at the features near the left-hand and right-hand edges (limbs) of the Moon. Are these features always exactly in the same places *relative to the limb*? They should be if the Moon's synchronous rotation were the only motion it had relative to the Earth. If not, explain. (*Hint:* a Web search on the word"libration"[not libation] may help.) (c) Does the apparent sizeof the Moon remain constant? If not, explain what this tells us about the shape of the Moon's orbit around the Earth.

WHAT IF . . .

THE MOON DIDN'T EXIST?

The Moon—full and bright or crescent and mysterious—is a fixture in our sky and in our cultures. Throughout history, people have woven myths about the Moon and its effects on everything from childbirth to stock market activity. Countless romances have begun under a full Moon, and entire nations have dedicated themselves to reaching it.

What if the Moon never existed? Would life even have developed here? If so, how would it be different? Would a self-aware species like ourselves have evolved?

Tumbling Earth Rapidly rotating terrestrial planets without large moons are unstable and calculations show that they periodically tumble—the north pole becomes the south pole and vice versa! If the Earth had no Moon, this process would create enormous stresses on the planet's surface leading to catastrophic earthquakes, volcanic activity, tsunamis, as well as severe atmosphere and magnetic field changes. Our Moon stabilizes the Earth's rotation, preventing such flips. It would be incomparably harder for complex life to persist on a tumbling world than it is on ours.

Delayed Origins It would have been much more difficult for life to start evolving on a Moonless Earth. In the first place, the minerals from which life developed in our Earth's young oceans were swept down there in large part by gigantic tides, a thousand times higher than the tides of today, created by the Moon when it was some 10 times closer to Earth than at present. The newly formed Earth, spinning more than 4 times faster than at present, caused these tides to move miles inland and back out to sea every 2 hours or so.

A Moonless Earth would still have tides, created by the Sun. At most, however, these tides would never be more than one-third as high as Earth's *present* tides. Minerals would wash into the oceans incredibly slowly from the flow of rivers. As a result, it is likely that it would have taken much longer, perhaps hundreds of millions of years longer, for enough minerals to be dissolved for life to firmly establish itself.

Harsh Conditions Animal life's transition from oceans to land would also be much harder because of continuous winds, between 50 and 150 miles per hour, created by the planet's rapid rotation. The resulting waves (wind causes waves) would be enormous and perpetual. Fish with legs near shores would almost certainly be pounded to a pulp rather than being able to sedately walk onto land and then back into the water, as apparently happened on Earth.

But we see on Earth how life develops in the most incredible forms and places. It seems plausible, then, that sea life would find a way to make the transfer to dry land and continue to evolve there.

Allowing then for diverse terrestrial life on the Moonless Earth, what would be different about that life compared to life on our Earth? For one thing, creatures evolving there would have to withstand the perpetual pummeling from winds and the debris they carry. Turtlelike shells are one solution.

Obviously, there would be no eclipses or moonlight—all clear nights would be equally dark and star-filled. Therefore, Earthlike nocturnal animals would be less successful at hunting, foraging, and traveling. Instead, these animals might evolve more enhanced senses to compensate for their inability to see visible light well at night.

Rush Hour Clearly, naked apes do not seem a likely bet on a Moonless Earth, nor do birds battling the ever-present winds. But given enough time, complex, even self-aware life likely would evolve. After all, we evolved because of the challenges faced by our ancestors, and Earth without a Moon would clearly provide challenges of its own.

Finally, the physiology of life on Earth evolved based on a 24-hour day. This is most evident in our biological clocks, or circadian rhythms, which are the internal mechanisms that regulate sleeping and waking, eating, and other cyclic activities. Faced with a 6-hour day, these circadian rhythms would be hopelessly out of sync with the natural world. To function on such a world, all its creatures would have to evolve biological clocks based on 6 hours, which certainly could have occurred. Considering all we have to do now, imagine what life would be like with only 6 hours in each day!

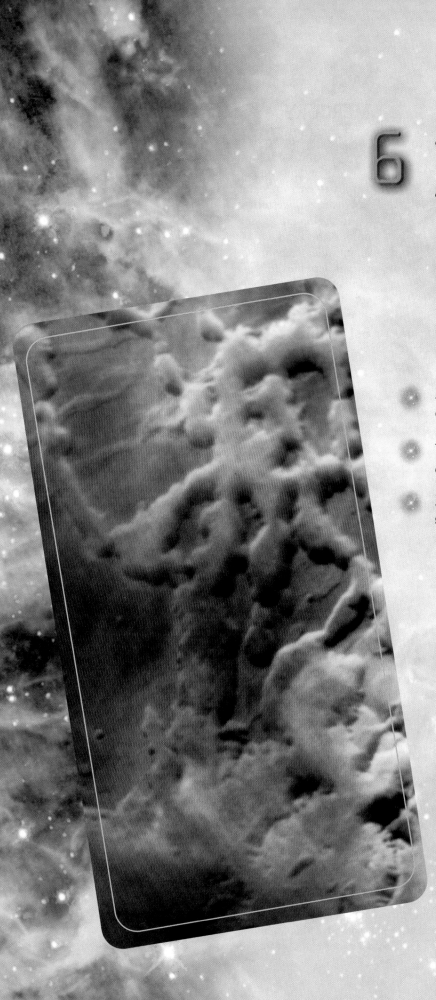

6 THE OTHER TERRESTRIAL PLANETS

IN THIS CHAPTER YOU WILL DISCOVER

- Mercury, a Sun-scorched planet with a heavily cratered surface and an iron core

- Venus, perpetually shrouded in thick, poisonous clouds and mostly covered by gently rolling hills

- Mars, the inspiration for scientific and popular speculation about extraterrestrial life

Canyon Fog
Delicate fog created by the scant water vapor in Mars's atmosphere fills canyons of Labyrinthus Noctis at sunrise and sunset. (NASA/JPL)

R I V U X G

WHAT DO YOU THINK?

1   Is the temperature on Mercury, the closest planet to the Sun, higher than the temperature on Earth?

2   What planet is most similar to Earth?

3   What is the composition of the clouds surrounding Venus?

4   Does Mars have liquid water on its surface today?

5   Is life known to exist on Mars today?

Far from the Sun are planets so alien that comparisons to Earth are virtually meaningless. Closer to home, however, a space traveler from Earth might feel like an ordinary tourist. The craters on Mercury call to mind the Moon. The clouds of Venus look almost comforting, which belies their true chemical composition. A feature on Mars, the Valles Marineris, distinctly brings to mind the Grand Canyon. How can a riverbed exist on such an arid world?

As we begin our tour of the inner solar system, it will help to keep asking: While they look like things back home, are they really the same? See the table "The Inner Planets: A Comparison."

MERCURY

 Whenever it rises before the Sun, Mercury heralds the ascending Sun in the brightening eastern sky as a "morning star." When Mercury rises after the Sun, it is seen only for a short time after sunset, hovering low over the western horizon as an "evening star." At its greatest elongation (see Figure 2-3), Mercury is sometimes among the brightest objects in the sky, with an apparent magnitude of about −0.4. Near conjunction, however, Mercury is very hard to see because its apparent magnitude rises to over 5.5 and it is viewed close to the Sun.

6-1   Photographs from *Mariner 10* reveal Mercury's lunarlike surface

Telescopic photographs of Mercury taken from Earth reveal almost nothing about the planet's surface. It was only in 1974, when *Mariner 10* coasted past the planet, that astronomers got their first glimpse of Mercury's inhospitable but familiar-looking surface. Figure 6-1 lists Mercury's properties. *Mariner 10* was sent on a remarkable orbit that enabled it to pass Mercury three times. Nevertheless, it photographed only 40% of Mercury's surface, and we still do not know what the other 60% looks like.

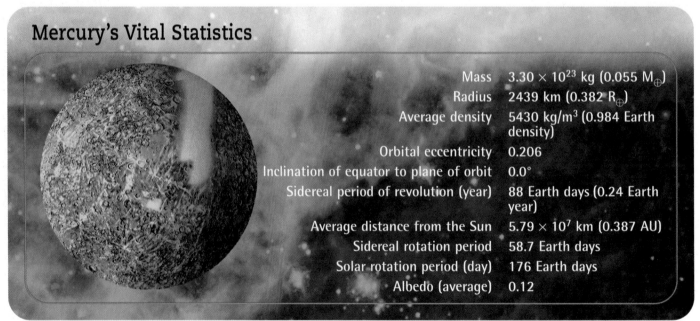

Mercury's Vital Statistics

Mass	3.30×10^{23} kg (0.055 $M_\oplus$)
Radius	2439 km (0.382 $R_\oplus$)
Average density	5430 kg/m³ (0.984 Earth density)
Orbital eccentricity	0.206
Inclination of equator to plane of orbit	0.0°
Sidereal period of revolution (year)	88 Earth days (0.24 Earth year)
Average distance from the Sun	5.79×10^7 km (0.387 AU)
Sidereal rotation period	58.7 Earth days
Solar rotation period (day)	176 Earth days
Albedo (average)	0.12

R I V U X G

FIGURE 6-1 *Mariner 10* **Mosaic of Mercury** This image of Mercury is a composite of dozens of photos taken by the *Mariner 10* spacecraft. The blank region was not imaged by the spacecraft, nor was the opposite side of the planet. Although most of these images were recorded at distances of about 200,000 km (125,000 mi) from Mercury, their resolution is still vastly superior to that of the best Earth-based images. (Astrogeology Team, U.S. Geological Survey)

The Inner Planets: A Comparison

	Interior	Surface	Temperature	Atmosphere	Magnetic Field
Mercury	Not known whether liquid or solid	Heavy cratering, scarps	700 K by day, 100 K by night	H, He, K, P, O; transient and tenuous	0.1 times Earth's field
Venus	At least partially molten	Light cratering, mostly volcanic plains, gently rolling hills, some volcanos	750 K	Mostly CO_2, N; 98 times denser than Earth's	None detected
Earth	Solid inner core, molten outer core, mantle	Very little cratering, continents and land at ocean floors, weathering, volcanos, global tectonic plates	200–315 K	Mostly N_2, O_2	Strong global field
Mars	Probably solid core	Moderate cratering, weathering, dormant volcanos, huge canyons	160–280 K	Mostly CO_2, N_2; 0.006 times as dense as Earth's	Weak, local fields

For detailed numerical comparisons between planets, see Tables A-1 and A-2.

Earth | Mercury | Mars | Venus

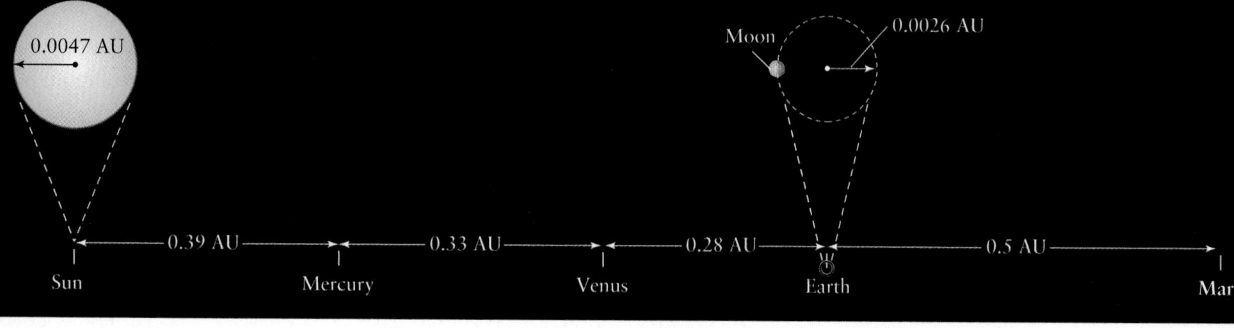

0.0047 AU · Moon · 0.0026 AU

Sun — 0.39 AU — Mercury — 0.33 AU — Venus — 0.28 AU — Earth — 0.5 AU — Mars

RIVUXG

FIGURE 6-2 **Mercury and Our Moon** Mercury (left) and our Moon are shown here to the same scale. Mercury's diameter is 4878 km and the Moon's is 3476 km. For comparison, the distance from New York to Los Angeles is 3944 km (2451 mi). Mercury's surface is more uniformly cratered than that of the Moon. Daytime temperatures at the equator on Mercury reach 700 K (800°F), hot enough to melt lead or tin. (NASA, UCO/Lick Observatory)

 Mariner 10 and the Exploration of Mercury

Astronomers conclude that most of the craters on both Mercury and the Moon were produced by impacts in the 800 million years after these bodies condensed from the solar nebula. As we saw in Chapter 5, the strongest evidence for this belief comes from analyzing and dating Moon rocks. Debris remaining after the planets were formed pounded these young worlds, gouging out most of the craters we see today. Like the Moon, Mercury has an exceptionally low albedo of 0.12 (12% reflection). It is very bright as seen from Earth only because the sunlight scattering from Mercury is so intense.

Although first impressions of Mercury evoke a lunar landscape, closer scrutiny reveals distinctly nonlunar characteristics. Lunar craters in the highlands (Figure 6-2) are densely packed, with one crater overlapping the next. Furthermore, there are very few craters on the maria. Mercury has neither maria nor other craterless regions. There are broad plains sparsely filled with relatively small craters. Elsewhere, the surface is covered with more numerous, overlapping craters. Figure 6-3 shows a typical close-up view of Mercury.

As we learned in Chapter 5, the lunar maria were produced by extensive lava flows that occurred between 3.1 and 3.8 billion years ago. Ancient lava flows also probably formed the Mercurian plains. As large meteorites punctured the planet's thin, newly formed crust, lava welled up from the molten interior to flood low-lying areas. The existence of craters pitting Mercury's plains suggests that these plains formed just over 3.8 billion years ago. Recall from Figure 5-24 that this was near the end of the era of heavy bombardment. Mercury's plains are therefore older than most of the lunar maria, leaving more time for cratering to eradicate marialike features.

The most impressive feature discovered by *Mariner 10* was a huge circular region called the Caloris Basin (Figure 6-4), which measures 1300 km (810 mi) in diameter and lies along the *terminator* (the border between day and night) in the figure. It is surrounded by a 2-km-high ring of mountains, beyond which are relatively smooth plains. Like the lunar maria, the Caloris Basin was probably gouged out by the impact of a large meteorite that penetrated the planet's crust. Because relatively few craters pockmark the lava flows that filled the basin, the Caloris impact must have occurred toward the end of the major crater-making period.

The Caloris impact was a tumultuous event that shook the entire planet. Indeed, the collision seems to have affected the side of Mercury directly opposite the Caloris Basin. That area (Figure 6-5) has a jumbled, hilly region covering nearly

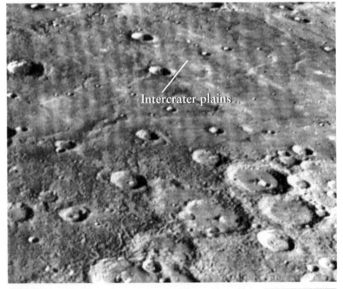

FIGURE 6-3 **Mercury's Craters and Plains** This view of Mercury's northern hemisphere was taken by *Mariner 10* as it sped past the planet in 1974. Numerous craters on the bottom half of the image and broad intercrater plains on the top half cover an area 480 km (300 mi) wide. (NASA)

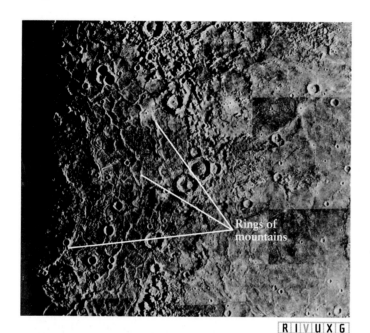

FIGURE 6-4   The Caloris Basin   *Mariner 10* sent back this view of a huge impact basin on Mercury's equator. Only about half of the Caloris Basin appears because it happened to lie on the terminator when the spacecraft sped past the planet. Although the center of the impact basin is hidden in the shadows (just beyond the left side of the picture), several semicircular rings of mountains reveal its extent. (NASA)

FIGURE 6-5   Unusual, Hilly Terrain   What look like tiny, fine-grained wrinkles on this picture are actually closely spaced hills, part of a jumbled terrain that covers nearly 500,000 km² on the opposite side from the Caloris Basin. The large, smooth-floored crater, Petrarch, near the center of this photograph has a diameter of 170 km (106 mi). This impact crater was produced more recently than the Caloris Basin. (NASA)

half a million square kilometers, about twice the size of Wyoming. The hills, which appear as tiny wrinkles that cover most of the photograph, are about 5 to 10 km wide and between 100 and 1800 m high. Geologists believe that energy from the Caloris impact traveled through the planet and became focused, like light through a lens, as it passed through Mercury. As this concentrated energy reached the far surface of the planet, jumbled hills were pushed up. A similar pair of phenomena to Caloris and the jumbled terrain on the other side of Mercury can be seen in our Moon's Mare Orientale and chaotic hills on the opposite side. Finding such a similar pair supports the theory that the impact and the jumbled surface are connected.

Mariner 10 also revealed gently rolling plains and numerous, long cliffs, called **scarps,** meandering across Mercury's surface (Figure 6-6). The scarps are believed to have developed as the planet cooled. Almost everything that cools contracts. Therefore, as Mercury's mantle and molten iron core cooled and contracted, its surface moved inward. Because it was solid, Mercury's crust could not collapse uniformly. Instead, it wrinkled as it contracted, forming the scarps. These features and the lack of recent volcanic activity suggest that the planet's interior is solid to a significant depth. Otherwise, lava would have leaked out as the scarps formed.

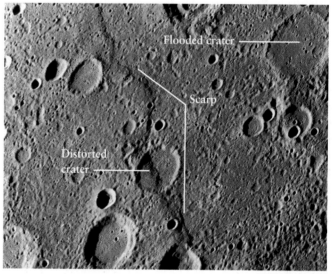

FIGURE 6-6   Scarps on Mercury   A long, meandering cliff called Santa Maria Rupes runs from north to south across this *Mariner 10* image of a region near Mercury's equator. This cliff, called a scarp by geologists, is more than 1 km high and runs for several hundred kilometers. Note how the crater in the center of the image was distorted vertically when the scarp formed. (NASA)

6-2 Mercury's temperature range is the most extreme in the solar system

By sending radio waves to Mercury and examining the reflected signal, astronomers in 1992 made an extraordinary discovery—water ice near Mercury's poles in craters that are permanently in shadow, as also found on our Moon. Dozens of circular regions, presumably craters, send back signals with characteristics distinct to ice. The origin of this ice, whether from comet impacts, from gases rising from inside the planet and then freezing, or both sources, remains to be determined.

Mercury's mass is quite low, only 5.5% that of the Earth. Like our Moon, the force of gravity is too weak on Mercury to hold a permanent atmosphere, but trace amounts of five different gases have been detected around Mercury. Scientists believe the Sun is the source of the hydrogen and helium gas near Mercury, while sodium and potassium gas escape from rocks inside the planet (a process called *outgassing,* which also occurs on the Earth). Oxygen observed in Mercury's atmosphere may come from polar ice that is slowly evaporating. All these gases drift into space and are continually being replenished in the atmosphere from their respective resources.

Because of Mercury's slow rotation and minimal atmosphere, the differences in temperature between day and night are far more noticeable there than on the Earth. At noon on Mercury, the surface temperature is 700 K (800°F). At the terminator, where day meets night, the temperature is about 425 K (305°F), and on the night side, the temperature falls as low as 100 K (–280°F)! This range of temperature, 600 K (1080°F), on Mercury contrasts tremendously with the typical change in temperature of 11 K (20°F) between day and night on Earth.

Insight into Science **Model-Building** In modeling real situations, scientists consider several different effects simultaneously. Omit any crucial property and you get inaccurate results. For example, think of how scientists might calculate how long ice can remain at Mercury's poles. They must take into account the planet's distance from the Sun, the tilt of its axis of rotation, its rotation rate, surface features, whether the ice is exposed or mixed with other material, and the chemical composition of Mercury.

6-3 Mercury has an iron core and a magnetic field like Earth

Mercury's average density of 5430 kg/m³ is quite similar to Earth's (5520 kg/m³). As we saw in Chapter 5, typical rocks on Earth's surface have a density of only about 3000 kg/m³ because they are composed primarily of lightweight elements. The high average densities of both Mercury and our planet are caused by their iron cores.

Because Mercury is less dense than the Earth, you might conclude that Mercury has a lower percentage of iron, a heavy element, than our planet. In fact, it is the Earth's greater mass pressing inward that makes it denser. This inward force compresses the iron core, making our planet denser than it would otherwise be.

Mercury is actually the most iron-rich planet in the solar system. Figure 6-7 shows a scale drawing of its interior, where an iron core fills 42% of the planet's volume. Surrounding the core is a 600-km-thick rocky mantle. For comparison, the Earth's iron core occupies only 17% of the planet's volume. How much of Mercury's core is molten is still unknown.

Events early in Mercury's history must somehow account for its high iron content. We know that the inner regions of the primordial solar nebula were incredibly hot. Perhaps only iron-rich minerals were able to withstand the solar heat there, and these subsequently formed iron-rich Mercury. According to another theory, an especially intense outflow of particles from the young Sun stripped Mercury of its low-density mantle shortly after the Sun formed. A third possibility is that, during the final stages of planet formation, Mercury was struck by a large planetesimal. Computer simulations show that this cataclysmic collision would have ejected much of the lighter mantle (Figure 6-8).

6-4 Mercury's rotation and revolution are coupled

 When the Earth was young, its gravitational force created tides on the Moon, thereby forcing the Moon into synchronous rotation. The Sun created

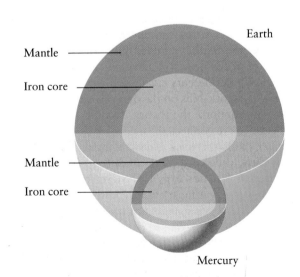

FIGURE 6-7 **The Interiors of Mercury and Earth** Mercury has the highest percentage of iron of any planet in the solar system. Its iron core occupies an exceptionally large fraction of its interior.

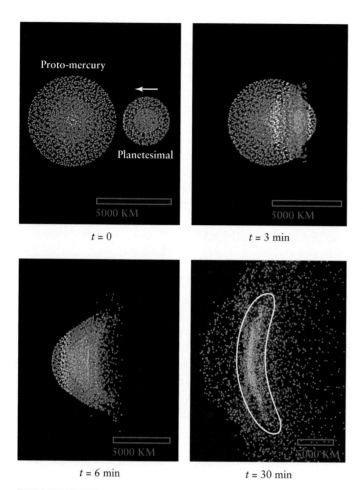

Proto-mercury

Planetesimal

5000 KM

$t = 0$

5000 KM

$t = 3$ min

5000 KM

$t = 6$ min

5000 KM

$t = 30$ min

FIGURE 6-8 **The Stripping of Mercury's Mantle** To account for Mercury's high iron content, one theory proposes that a collision with a massive asteroid stripped Mercury of most of its rocky mantle. These four images show a computer simulation of a head-on collision between proto-Mercury and a body one-sixth its mass. Both worlds are shattered by the impact, which vaporizes much of their rocky mantles. Mercury eventually reforms from the iron-rich debris left behind (material inside banana-shaped region). The rest of the material (outside the banana) left the vicinity of the planet. (W. Benz, A. G. W. Cameron, and W. Slattery)

a significant tidal bulge on Mercury that locked into place when the planet solidified. This is analogous to what the Earth would look like if the oceans suddenly froze solid. Furthermore, Mercury is so close to the Sun (average separation: 0.387 AU) that the Sun's gravitational force also influenced Mercury's rotation rate. However, Mercury's highly eccentric orbit ($e = 0.21$) prevented the planet from being locked into synchronous orbit. Instead, Mercury has what is called a **3-to-2 spin-orbit coupling.** This means that Mercury undergoes three sidereal rotations (rotations measured with respect to the distant stars, not the Sun), while undergoing two revolutions around the Sun (Figure 6-9).

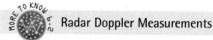

Radar Doppler Measurements

Why not synchronous rotation? Basically Mercury's orbit is too eccentric to allow it. Planets rotate at constant rates, and so for the Sun to always be up in a planet's sky, the planet would have to have a nearly circular orbit around it. Our Moon, for example, has synchronous rotation around the Earth with an orbital eccentricity of 0.055. As seen from the Moon, therefore, the Earth remains essentially fixed in the sky. However, Kepler's second law (see section 2-3) dictates

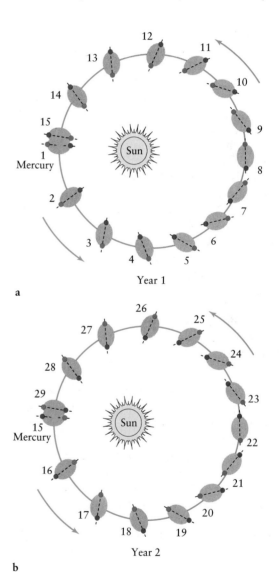

a

Year 1

b

Year 2

FIGURE 6-9 **3-to-2 Spin-Orbit Coupling** Mercury undergoes three sidereal rotations every two years. You can see this by following the red (or blue) dot from time 1 to time 29 and observe that the dot points to the right 3 times during this interval, which corresponds to two Mercurian years. Therefore, a sidereal day on Mercury is 58.7 Earth days long. During the same interval, however, the Sun is at noon as seen from the red dot's location only twice: at time 1 and time 29. Therefore, a solar day is 176 Earth days long.

that the Sun moves faster across Mercury's sky at perihelion than at aphelion. Therefore, from Mercury's surface, the Sun is seen to rise and set.

Although the strength of the Sun's tidal force on Mercury was unable to put it into synchronous rotation as the Earth did to the Moon, the Sun's tidal force did change Mercury's original rotation rate. Today, one of only two regions on Mercury (with slight variations) always faces the Sun at the planet's perihelion. In other words, at perihelion, only the regions of Mercury with the high tidal bulges are pointing toward the Sun (see Figure 6-9). This is a result of the 3-to-2 spin-orbit coupling mentioned above.

The motion of the Sun across Mercury's sky is unique in the solar system. First, a solar day there (noon to noon) is 176 Earth days long, twice the length of a year! Furthermore, standing at the location of high tide on Mercury near perihelion, you would see the Sun start to rise in the east, stop high in the sky, move back toward the east, stop again, and then continue its westward journey. Similarly, at perihelion, you could watch the Sun set, rise in the same place, and then set again a day or two later.

As we saw in Chapter 5, electric currents flowing in a planet's liquid iron core create a planetwide magnetic field. The Earth creates its electric current by rotating once a day. Mercury rotates 59 times more slowly (see Figure 6-1). This is hardly fast enough to generate a magnetic field, yet *Mariner 10* did find a weak magnetic field on Mercury. The origin of its present magnetic field remains a mystery. Be-

sides Pluto, Mercury is the planet from which we have the least observational data. Therefore, several missions to Mercury are planned for the coming decade.

VENUS

Venus and Earth have almost the same mass, the same diameter, and the same average density. Indeed, if Venus were located at the same distance from the Sun as is the Earth, then it, too, might well have evolved life. However, Venus is 30% closer to the Sun than the Earth is, and this one difference between the two planets leads to a host of others, making Venus inhospitable to life.

6-5 The surface of Venus is completely hidden beneath a permanent cloud cover

At nearly twice the distance from the Sun as Mercury, Venus is often easy to view without interference from the Sun's glare. At its greatest elongation, Venus is seen high above the western horizon after sunset, where, like Mercury, it is often called the "evening star." High in the eastern sky before sunrise, it is called the "morning star."

Venus is so easy to identify because it is often one of the brightest objects in the night sky. Only the Sun and the Moon

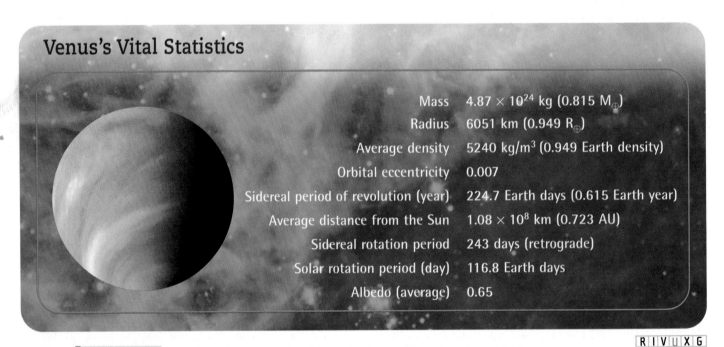

Venus's Vital Statistics

Mass	4.87×10^{24} kg ($0.815\ M_\oplus$)
Radius	6051 km ($0.949\ R_\oplus$)
Average density	5240 kg/m^3 (0.949 Earth density)
Orbital eccentricity	0.007
Sidereal period of revolution (year)	224.7 Earth days (0.615 Earth year)
Average distance from the Sun	1.08×10^8 km (0.723 AU)
Sidereal rotation period	243 days (retrograde)
Solar rotation period (day)	116.8 Earth days
Albedo (average)	0.65

R I V U X G

FIGURE 6-10 **Venus's Vital Statistics** Venus's thick cloud cover efficiently traps heat from the Sun, resulting in a surface temperature even hotter than that on Mercury. Unlike Earth's clouds, which are made of water droplets, Venus's clouds are very dry and contain droplets of concentrated sulfuric acid. This ultraviolet image was taken by the *Pioneer Venus Orbiter* in 1979. (NASA)

outshine Venus at its greatest brilliance. Venus is often mistaken for a UFO, because when it appears low on the horizon, its bright light is strongly refracted by the Earth's atmosphere, making it appear to rapidly change color and position.

Unlike Mercury, Venus is intrinsically bright (Figure 6-10), because it is completely surrounded by light-colored, highly reflective clouds. Visible light telescopes cannot penetrate this thick, unbroken layer of clouds. Until 1962, we did not even know how fast Venus rotates. In the 1960s, however, both the United States and the former Soviet Union began sending probes to Venus. The Americans sent fragile, lightweight spacecraft into orbit near Venus. The Soviets, who had more powerful rockets, sent massive vehicles directly into the Venusian atmosphere.

Building spacecraft that could survive the descent proved to be more frustrating than anyone had expected. Finally, in 1970, a Soviet probe managed to transmit data for a few seconds directly from the Venusian surface. Soviet missions, which continued until 1983, measured a surface temperature of 750 K (900°F) and a pressure of 90 atm. This is the same pressure you would feel if you were swimming 0.82 km (2700 ft) underwater on Earth.

In contrast to Earth's present nitrogen and oxygen–rich atmosphere, Venus's thick atmosphere is 96% carbon dioxide, with the remaining 4% being mostly nitrogen. This atmosphere is remarkably similar to the Earth's early carbon dioxide–rich atmosphere. These gases were vented from inside Venus through volcanoes and other openings. Unlike the Earth, however, Venus has no liquid oceans to dissolve the carbon dioxide, so it remains in the atmosphere there today.

Soviet spacecraft also discovered that Venus's clouds are confined to a 20-km-thick layer located 48 to 68 km above the planet's surface. Above and below the clouds are 20-km-thick layers of haze. Beneath the lower level of haze, the Venusian atmosphere is clear all the way down to the surface. Standing on the surface of Venus, you would experience a perpetually cloudy day.

Unlike the clouds on Earth, which appear white from above, the cloudtops of Venus appear yellowish or yellow-orange to the human eye. These colors are typical of sulfur and its compounds. Indeed, spacecraft found substantial amounts of sulfur dust in Venus's upper atmosphere and sulfur dioxide and hydrogen sulfide at lower elevations. The clouds are composed of droplets of concentrated sulfuric acid! Because of the tremendous atmospheric pressure at Venus's surface, the droplets remain suspended as a thick mist, rather than falling as rain.

All of the major chemical compounds spewed into our air by Earth's volcanoes have also been detected in Venus's atmosphere. Among these molecules are large quantities of sulfur compounds. Because many of these substances are very short-lived, they must be constantly replenished by new eruptions. In fact, the abundance of sulfur compounds in

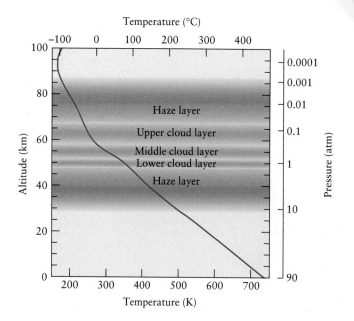

FIGURE 6-11 **Temperature and Pressure in the Venusian Atmosphere** The pressure at the Venusian surface is a crushing 90 atm (1296 lb/in.²). Above the surface, atmospheric pressure decreases smoothly with increasing altitude. The temperature of Venus's atmosphere increases smoothly from a minimum of about 173 K (–150°F) at an altitude of 100 km to a maximum of nearly 750 K (900°F) on the ground.

Venus's atmosphere does vary. These data suggest the possibility that the sulfurous compounds in Venus's atmosphere come from active volcanoes. This question is very much open, however, because of the question of lightning on Venus. Lightning is a frequent event around active volcanoes here on Earth. Astronomers have searched for lightning on Venus's atmosphere, but whether it has been detected there remains highly controversial.

The results of the early probing of Venus are summarized in Figure 6-11. Both pressure and temperature decrease smoothly with increasing altitude. From the changes in temperature and pressure above a planet's surface, we can understand the structure of its atmosphere.

6-6 The greenhouse effect heats Venus's surface

At first, no one could believe reports that the surface temperature on Venus was higher than the surface temperature on Mercury, which, after all, is closer to the Sun. After some initial skepticism astronomers quickly found a straightforward explanation—the **greenhouse effect** (Figure 6-12).

Perhaps you have had the experience of parking your car in the sunshine on a warm summer day. You roll up the

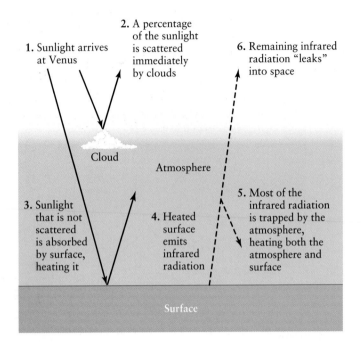

1. Sunlight arrives at Venus

2. A percentage of the sunlight is scattered immediately by clouds

6. Remaining infrared radiation "leaks" into space

Cloud

Atmosphere

3. Sunlight that is not scattered is absorbed by surface, heating it

4. Heated surface emits infrared radiation

5. Most of the infrared radiation is trapped by the atmosphere, heating both the atmosphere and surface

Surface

FIGURE 6-12   **The Greenhouse Effect** A portion of the sunlight reaching a planet penetrates through the atmosphere and heats the planet's surface. The heated surface emits infrared radiation, much of which is absorbed by water vapor and carbon dioxide. The trapped radiation helps raise the average temperatures of the surface and atmosphere. Some infrared radiation does penetrate the atmosphere and leaks into space. In a state of equilibrium, the rate at which the planet loses energy to space in this way is equal to the rate at which it absorbs energy from the Sun.

windows, lock the doors, and go on an errand. After a few hours, you return to discover that the interior of your automobile has become stiflingly hot, typically 20 K (36°F) warmer than the outside air temperature.

What happened to make your car so warm? First, sunlight entered your car through the windows. This radiation was absorbed by the dashboard and the upholstery, raising their temperatures. Because they become warm, your dashboard and upholstery emit infrared radiation, which you detect as heat and to which your car windows are opaque. This energy is therefore trapped inside your car and absorbed by the air and interior surfaces. As more sunlight comes through the windows and is trapped, the temperature continues to rise. The same trapping of sunlight warms an actual greenhouse.

Carbon dioxide is responsible for a similar warming of Venus's atmosphere. Like your car windows, carbon dioxide is transparent to visible light but opaque to infrared radiation. Although most of the visible sunlight striking the Venusian cloudtops is reflected back into space, enough light reaches the Venusian surface to heat it. The warmed surface

in turn emits infrared radiation, which cannot escape through Venus's carbon dioxide–rich atmosphere. This trapped radiation produces the high temperatures found on Venus.

Without the greenhouse effect, the surface of Venus would have a noontime temperature of 465 K. Because of it, that temperature is actually a sweltering 750 K, hotter than the hottest spot on Mercury! Furthermore, the thick atmosphere keeps the night side of Venus at nearly the same temperature, unlike the night side of Mercury, where the temperature drops precipitously. This high temperature prevents liquid water from existing on Venus; its surface is dry.

On Earth, atmospheric water and carbon dioxide both contribute to the greenhouse effect in our air. Warmed by sunlight, the Earth emits infrared radiation, some of which is absorbed by water vapor and carbon dioxide gas, causing the air temperature to rise. The small amounts of these gases in the Earth's atmosphere produce a comparatively gentle greenhouse effect. The heating due to the greenhouse effect is, however, increasing on Earth as our atmosphere acquires more carbon dioxide from burning fossil fuels and other sources. The air retains this gas due to large-scale clear-cutting of forests, which decreases the number of plants that can convert this gas into other carbon compounds and oxygen.

6-7   Venus is covered with gently rolling hills, two "continents," and numerous volcanoes

Soviet spacecraft that landed on Venus provided us with intriguing close-up images of the planet's arid surface. Figure 6-13 is a view of Venus's regolith taken in 1981. Russian scientists believe that this region was covered with a thin layer of lava that contracted and fractured upon cooling to create the rounded, interlocking shapes seen in the photograph. Indeed, measurements by several spacecraft indicate that Venusian rock is quite similar to lava rocks called basalt, which are common on Earth and the Moon.

By far the best images of the Venusian surface came from the highly successful *Magellan* spacecraft that arrived at Venus in 1990. In orbit about the planet, *Magellan* sent radar signals through the clouds surrounding Venus. It mapped Venus using a radar altimeter that bounced microwaves off the ground directly below the spacecraft. By measuring the time delay of the radar echo, scientists determined the heights and depths of Venus's hills and valleys. As a result, astronomers have been able to construct a three-dimensional map of the planet. The resolution of this map is about 75 m. You could see a football stadium on Venus if there were any! (None were detected.)

Venus is remarkably flat compared to the Earth. More than 80% of Venus's surface is covered with volcanic plains and gently rolling hills created by numerous lava flows. Figure 6-14 is a computer-generated view showing a typical Venusian landscape.

FIGURE 6-13 The Venusian Surface (a) This color photograph, taken by a Soviet spacecraft, shows rocks that appear orange because the light was filtered through the thick, sulfur-rich clouds. (b) By comparing the apparent color of the spacecraft to the color it was known to be, computers can correct for the sulfurous light. The actual color of the rocks is gray. In this view, the rocky plates covering the ground may be fractured segments of a thin layer of lava. The toothed wheel in each image is part of the landing mechanism that keeps the spherical spacecraft from rolling. (a: C. M. Pieters and the USSR Academy of Sciences)

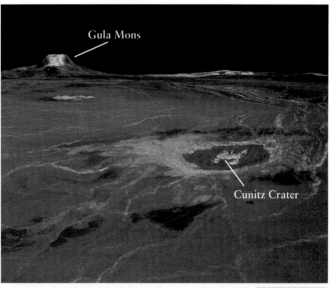

FIGURE 6-14 A Venusian Landscape Most of Venus is covered with plains and gently rolling hills produced by numerous lava flows. The crater in the foreground is 48 km (30 mi) in diameter. The volcano near the horizon is 3 km (1.9 mi) high. Note the cracks in the Venusian surface toward the right side of the image. (NASA/JPL)

Radar images revealed two large highlands, or "continents," rising well above the generally level surface of the planet (Figure 6-15). The continent in the northern hemisphere is Ishtar Terra, named after the Babylonian goddess of love and approximately the same size as Australia. Ishtar Terra is dominated by a high plateau ringed by towering mountains. The highest mountain is Maxwell Montes, whose summit rises to an altitude of 11 km above the average surface. For comparison, Mount Everest on Earth rises 9 km above sea level.

The larger Venusian continent, Aphrodite Terra (named after the Greek goddess called Venus by the Romans) is a vast belt of highlands just south of the equator. Aphrodite is 16,000 km (10,000 mi) in length and 2000 km (1200 mi)

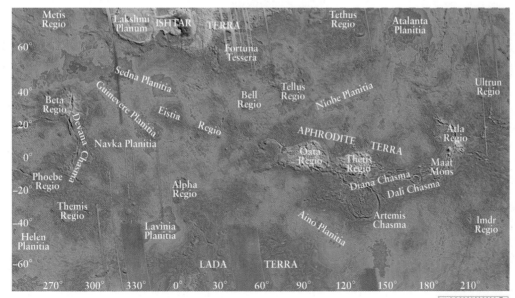

FIGURE 6-15 A Map of Venus This false-color radar map of Venus, analogous to a topographical map of Earth, shows the large-scale surface features of the planet. The equator extends across the middle of the map. Color indicates elevation—red for highest, followed by orange, yellow, green and blue for lowest. The planet's highest mountain is Maxwell Montes on Ishtar Terra. Scorpion-shaped Aphrodite Terra, a continentlike highland, contains several spectacular volcanoes. Do not confuse the blue and green for oceans and land. (Peter Ford/MIT, NASA/JPL)

wide, giving it an area about one-half that of Africa. The global view of the surface of Venus in Figure 6-16 shows that most of Aphrodite Terra is covered by vast networks of faults and fractures.

While more than 1600 major volcanoes and volcanic features were discovered on Venus, fewer than a thousand impact craters have been observed there (Figure 6-17). This compares to the hundreds of thousands seen on the Moon and Mercury. Today, of course, Venus's thick atmosphere heats, and thereby vaporizes, much of the infalling debris that would otherwise create craters. Because its atmosphere is thicker than Earth's, Venus vaporizes such space junk falling toward it more efficiently than Earth does. However, there should have been a period shortly after Venus formed and before its atmosphere developed (from gases escaping from the planet's interior) when impacts were common.

The low number and random distribution of craters on Venus lead to the belief that the planet's surface is periodically erased and replaced. Because geologists have found no large-scale tectonic plate motion on Venus to refresh the surface, the current theory explaining the lack of extensive cratering is that the entire surface of the planet melts! This would occur if the crust is very thick compared to the Earth's crust. A crust 300 km thick (Earth's crust is everywhere less than 65 km thick) would insulate the interior so much that, heated by radioactive elements in it, the mantle could become hot enough to melt the crust on occasion.

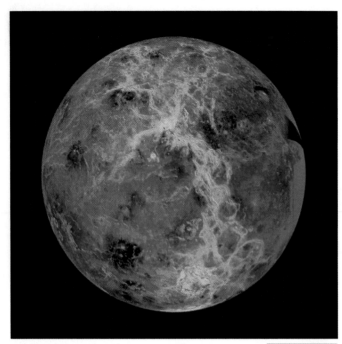

R I V U X G

FIGURE 6-16 A "Global" View of Venus A computer was used to map numerous *Magellan* images onto a simulated globe. Color is used to enhance small-scale structures. Extensive lava flows and lava plains cover about 80% of Venus's relatively flat surface. (NASA)

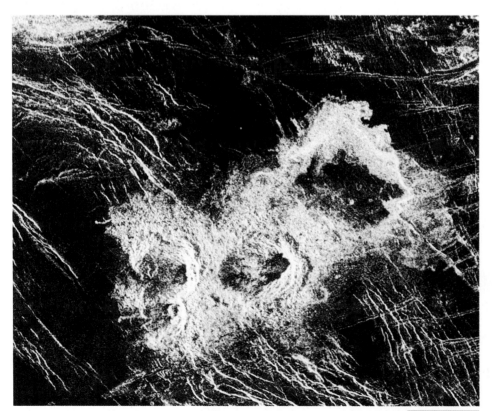

FIGURE 6-17 Craters on Venus Impact craters on Venus tend to occur in clusters, which suggests that they are formed from large, single pieces of infalling debris that are broken up in the atmosphere. Shown here is the triple-impact Crater Stein. (NASA/JPL)

R I V U X G

It has been proposed that every 700 million years or so, the entire surface of Venus liquefies until the pent-up heat escapes, a new solid crust forms, and the process begins anew.

The sulfur content of the air and the traces of active volcanoes on the surface are compelling indicators that, like the Earth, Venus has a molten interior. Because the average density of Venus is similar to that of Earth, its core is predominantly iron. Currents in the molten iron should generate a magnetic field. However, none of the spacecraft sent there has detected one. This is plausible only if Venus rotates exceptionally slowly, which it does (see Figure 6-10). In fact, the planet takes 116.8 Earth days to get from one sunrise to the next (if you could see the Sun from Venus's surface).

Unlike the Earth, Venus has **retrograde rotation.** This means that the direction of Venus's orbit around the Sun (counterclockwise as seen from space far above the Earth's north pole) is *opposite* the direction of its rotation (clockwise as seen from the same vantage point). In other words, sunrise on Venus occurs in the west. Venus's rotation axis is tilted more than 177°, compared to the Earth's 23^1/$_2$° tilt. Because Venus's axis is within 3° of being perpendicular to the plane of its orbit around the Sun, the planet has no seasons. Although we do not know the cause of Venus's retrograde rotation, one likely explanation is that a monumental impact altered the rotation early in the planet's existence.

6-8 *Magellan's* last gasp revealed irregularities in Venus's upper atmosphere

Magellan finished mapping the surface of Venus in 1994. Instead of keeping it in orbit, NASA astronomers decided to give it one last mission. They fired braking rockets on the spacecraft, sending it spiraling down toward the planet's surface. By doing so, it was hoped that *Magellan* would transmit useful data about Venus's dense atmosphere before the spacecraft vaporized. On October 12, 1994, *Magellan* stopped transmitting, but before it did so, it sent back tantalizing information.

Magellan indicated that the density of Venus's atmosphere is lower at some altitudes and higher at other altitudes than was expected. This information will be useful in helping astronomers plan future missions to the cloud-shrouded planet.

MARS

Mars is the only planet whose surface features can be seen through Earth-based telescopes. Its distinctive rust-colored hue makes it stand out in the night sky (Figure 6-18). When Mars is near opposition, even telescopes for home use reveal its seasonal changes. Dark markings on the Martian

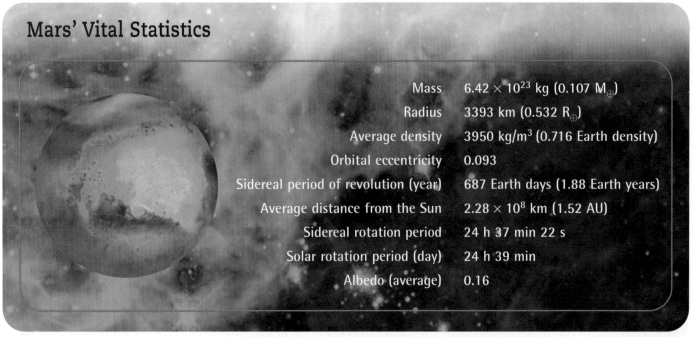

Mars' Vital Statistics

Mass	6.42×10^{23} kg (0.107 M$_\oplus$)
Radius	3393 km (0.532 R$_\oplus$)
Average density	3950 kg/m^3 (0.716 Earth density)
Orbital eccentricity	0.093
Sidereal period of revolution (year)	687 Earth days (1.88 Earth years)
Average distance from the Sun	2.28×10^8 km (1.52 AU)
Sidereal rotation period	24 h 37 min 22 s
Solar rotation period (day)	24 h 39 min
Albedo (average)	0.16

FIGURE 6-18   Mars Viewed from the Earth This photograph of Mars was taken by the Hubble Space Telescope in 2001, when the Earth-Mars distance was 68 million km (43 million miles). (NASA)

R I V U X G

surface can be seen to vary, and prominent polar caps shrink noticeably during the spring and summer months (Figure 6-19).

6-9 Earth-based observations originally suggested that Mars might harbor life

The Dutch physicist Christiaan Huygens made the first reliable observations of Mars in 1659. Using a telescope of his own design, Huygens identified a prominent, dark surface feature that reemerged about every 24 hours, suggesting a rate of rotation similar to that of Earth. Huygens's observations soon led to speculation about life on Mars because the planet seemed so similar to Earth.

In 1877, Giovanni Virginio Schiaparelli, an Italian astronomer, reported seeing 40 lines crisscrossing the Martian surface. He called these dark features *canali*, an Italian term meaning "channels." It was soon mistranslated into English as *canals*, implying the existence on Mars of intelligent creatures capable of engineering feats. This speculation led Percival Lowell, who came from a wealthy Boston family, to finance a major new observatory near Flagstaff, Arizona. By the end of the nineteenth century, Lowell had allegedly observed 160 Martian "canals."

It soon became fashionable to speculate that the Martian canals formed an enormous, planetwide irrigation network to transport water from melting polar caps to vegetation near the equator. (The seasonal changes on Mars's dark surface markings can be mistaken for vegetation.) In view of the planet's reddish, desertlike appearance, Mars was thought to be a dying planet whose inhabitants must go to great lengths to irrigate their farmlands. No doubt the Martians would readily abandon their arid ancestral home-

land and invade the Earth for its abundant resources. Hundreds of science fiction stories and dozens of monster movies owe their existence to the *canali* of Schiaparelli.

> **Insight into Science** **Perception and Reality** Our perceptual tendency to see patterns, even where they do not exist, makes it easy for us to leap to conclusions. Scientific images and photographs, however, require objective analysis.

On October 30, 1938, Orson Welles presented a radio program announcing the invasion of Earth by Martians. His broadcast, live from New York City, was inspired by the 1898 book *War of the Worlds* by H. G. Wells. (You might ask your grandparents if they heard the show.) So realistic was the performance that it caused panic throughout the New York–New Jersey area. Let's see how close to the mark H. G. Wells was when it comes to life on Mars.

6-10 Early space probes to Mars found no canals but did find some controversial features

 Spacecraft journeying to Mars since the 1970s have sent back pictures that clearly show numerous flat-bottomed craters, volcanoes, and canyons, but not one canal of a size consistent with those allegedly seen from Earth (Figure 6-20). Schiaparelli's *canali* were optical illusions, and H. G. Wells was completely off the mark. We have observed no cities or roads or other signs of civilization on the red planet. There is no evidence that intelligent Martian life has ever existed.

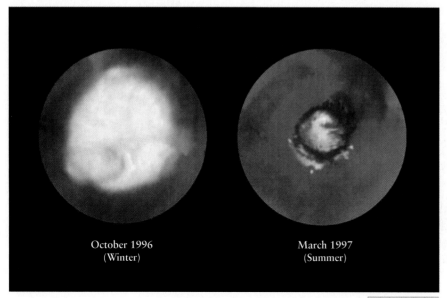

October 1996 (Winter)   March 1997 (Summer)

FIGURE 6-19 **Changing Seasons on Mars** During the Martian winter, the temperature drops so low that carbon dioxide freezes out of the Martian atmosphere. A thin coating of carbon dioxide frost covers a broad region around Mars's north pole. During summer in the northern hemisphere, the range of this north polar carbon dioxide cap decreases dramatically. In the summer a ring of dark sand dunes is exposed around Mars's north pole. (S. Lee/J. Bell/ M. Wolff/Space Science Institute/NASA)

R I V U X G

R I V U X G

FIGURE 6-20   Martian Terrain This high-altitude photograph shows a variety of the features on Mars, including broad, towering volcanoes (left) on the highland called Tharsis Bulge, impact craters (upper right), and vast, windswept plains. An enormous canyon system called Valles Marineris crosses horizontally just below the center of the image. (A. S. McEwen, USGS)

Insight into Science   Fact and Fiction Science fiction may foreshadow real scientific discoveries. (Can you think of some examples?) But science also reins in the fantasies of the science fiction writer. A century ago, millions of people, including top scientists, believed that technologically advanced life existed on Mars.

Astronomers have photographed several surface features that at first glance could have been crafted by intelligent lifeforms. In 1976, a hundred years after Schiaparelli's alleged discovery of canali, the *Viking Orbiter 1* spacecraft photographed a feature that appeared to be a humanlike face (Figure 6-21a). However, when the *Mars Orbiter* photographed the surface in greater detail and from a different angle in 1998 (Figure 6-21b), it found no facial features. If anything, the same spot looked more like a giant shoeprint.

Scientists universally believe that the face and other alleged patterns were created by shadows on wind-blown hills. Other features, in the shape of pyramids, are consistent with the winds eroding softer rock around harder rock pushed upward from inside Mars billions of years ago. The same thing happens on Earth.

Insight into Science   Keep It Simple Was the Martian "face" made by intelligent life or by wind-sculpted hills? When two alternative interpretations present themselves, scientists apply the principle of Occam's razor and always choose the one requiring the fewer unproven assumptions.

6-11 Probes to Mars found craters, volcanoes, and canyons

The partially eroded, flat bottoms of the Martian craters probably result from dust storms that frequently rage across the planet's surface. Over the past two centuries, astronomers have seen faint surface markings disappear under a reddish-orange haze as thin Martian winds have stirred up finely powdered dust from the Martian regolith. Some storms obscure the entire planet as happened in 2001. Over the ages, deposits

a

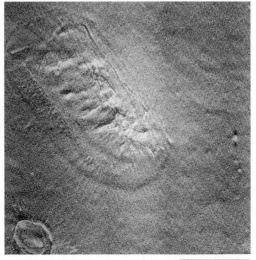

b

R I V U X G

FIGURE 6-21

In the Eye of the Beholder These two images of the same site on Mars, taken 22 years apart, show how the apparent face in **(a)** changed to a more "natural-looking" feature in **(b)**. This transformation was due to weathering of the site, improved camera technology, and the change in angle at which the photograph was taken. (a: NSSDC/NASA; b: NSSDC/NASA and Dr. Michael H. Carr)

GUIDED DISCOVERY The Inner Solar System

In this section, we will learn more about the appearances of the inner planets as seen from Earth using your *Starry Night Backyard*™ program.

Mercury Find Mercury (Edit/Find/Mercury). The first step is to see that Mercury is always close to the horizon when it is visible in the night sky.

1. *If Mercury is below the horizon:* Choose the "Reset Time" option when the "Obscured by Horizon" warning comes up. You will then see Mercury in the eastern sky.

Now, if the Sun is not yet up, Mercury is the "morning star." Step forward in time by 3 minute intervals and watch the Sun rise and Mercury disappear near the horizon. Go to section 3.

On the other hand, if the Sun is already up, it will be to the west (to the right) of Mercury. Set the time to 6 P.M., relocate Mercury, and run the time forward or backward until the Sun has set in the west but Mercury is still up. Mercury is now the "evening star," located near the western horizon. Step forward in time by 3 minute intervals and watch Mercury set. Go to section 3.

2. *If Mercury is already up in the sky:* If Mercury is west (to the right) of the Sun: We want to set up the screen with Mercury as the "morning star." To do this, re-set the time to 5 a.m., relocate Mercury, and run time forward or backward until Mercury is above the eastern horizon and the Sun is below the horizon. This shows Mercury as the "morning star." Now, run time forward at 3 minutes per timestep, and you will see the white dot of Mercury fade. Now go to section 3.

If Mercury is east (to the left) of the Sun in the sky, then set the time to 6 p.m., relocate Mercury, and run the time forward or backward until the Sun has set but Mercury is still up. Mercury is now the "evening star," located near the western horizon. Step forward in time by 3 minute intervals and watch Mercury set.

3. *Now we explore Mercury's orbit.* Go to Atlas mode (Go/Atlas) and again find Mercury. This time you will see it highly magnified. Zoom away using the magnifying glass icon with the minus sign until the angle of sky is about 70°. Make sure you are centered and locked on Mercury (grab Mercury, right click, and then left click on Centre/Lock). Set the timestep to 1 day and press the right pointing triangle. Describe and explain the motion of the Sun relative to Mercury. You can learn more about Mercury as seen from Earth by trying Observing Project 49 at the end of this chapter.

Venus Repeat the same procedures for Venus as we have described for Mercury. What similarities and differences do you see?

Mars Return to Atlas mode, set the date to 6/12/2003, time to midnight, and lock on Mars. What is Mars's phase? Zoom away until you have the Sun on the screen or you have gone the maximum distance (about 100° angle on the screen). Determine how far (in angle) the Sun is away from Mars. If the Sun is not visible, you can make this measurement by "dragging" the celestial sphere around and measuring angles from the grid. Does the Sun ever get this far in angle from Mercury or Venus, as seen from Earth? Explain.

---

of dust have filled in the crater bottoms, but 3 billion years of sporadic storms have not wiped out the craters; the Martian atmosphere is too thin to carry enough of this extremely fine-grained powder to eradicate them completely.

In 1971 the *Mariner 9* spacecraft went into orbit around Mars and began sending back detailed, close-up pictures. The lack of present tectonic plates there soon became evident. However, in 1998, the *Mars Global Surveyor* discovered patterns in the surface magnetic fields of Mars like those found where Earth's tectonic plates separate and lock the Earth's magnetic field in the rock. Geologists propose that when Mars was young, its internal "dynamo" (see section 5-4) created strong magnetic fields and that tectonic plates helped craft its surface, while locking traces of its changing magnetic field in the molten rock. However, the planet quickly cooled, its global magnetic field vanished, and tectonic-plate apparently ceased nearly 4 billion years ago.

Mariner 9 showed the existence of enormous volcanoes and vast canyons (Figure 6-22). The largest Martian vol-

cano, Olympus Mons (Figure 6-23), covers an area as big as the state of Missouri and rises 26 km (16 mi) above the surrounding plains—nearly 3 times the height of Mount Everest. The highest volcano on Earth, Mauna Loa in the Hawaiian Islands, has a summit only 17 km above the ocean floor.

On Earth, the Hawaiian Islands are only the most recent additions to a long chain of volcanoes. They resulted from **hot-spot volcanism**, a process by which molten rock rises to the surface from a fixed hot region far below. The Pacific tectonic plate is slowly moving northwest at a rate of several centimeters per year. As a result, new volcanoes are created above the hot spot, while older ones move off and become extinct, eventually disappearing beneath the ocean.

Because Mars's surface does not appear to be moving due to the lack of tectonic plate motion, one hot spot can keep pumping lava upward through the same vent for millions of years. One result is Olympus Mons—a single giant volcano rather than a long chain of smaller ones. The vol-

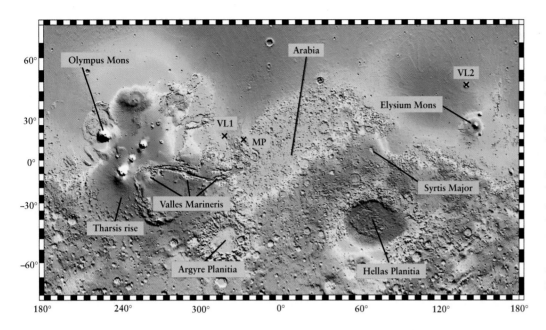

FIGURE 6-22
The Topography of Mars The color coding on this map of Mars shows elevations above (positive numbers) or below (negative numbers) the planet's average radius. To produce this map, an instrument on board *Mars Global Surveyor* fired pulses of laser light at the planet's surface, then measured how long it took each reflected pulse to return to the spacecraft. The *Viking Lander 1* (VL1), *Viking Lander 2* (VL2), and *Mars Pathfinder* (MP) landing sites are each marked with an x. (MOLA Science Team, NASA/GSFC)

cano's summit has collapsed to form a volcanic crater, called a **caldera,** large enough to contain the state of Rhode Island (see Figure 6-23). Calculations indicate that Mars's interior should be solid, so we do not expect to see any further volcanic activity there.

Since arriving in September, 1997, the *Mars Global Surveyor* has provided more information about Mars than all previous missions to Mars combined. Detailed altitude measurements reveal that Mars has distinctly different northern and southern hemispheres (see Figure 6-22). For reasons we do not yet understand, the average elevation of the northern hemisphere is about 5 km (3 mi) lower than that of the southern hemisphere. Hence,

they are referred to as Vastitas Borealis (the **northern vastness** or **northern lowlands**) and the **southern highlands.** Most of the volcanoes on Mars are in the northern hemisphere, but most of the impact craters are in the southern hemisphere. This suggests that the northern vastness has been resurfaced by some process that eradicated ancient lowland craters.

Between the two hemispheres, the earlier *Mariner 9* discovered a vast canyon running roughly parallel to the Martian equator (see Figures 6-20 and 6-24). In honor of the *Mariner* spacecraft that revealed so much of the Martian surface, this enormous chasm has been designated Valles Marineris.

Valles Marineris stretches over 4000 km, about one-fifth the circumference of Mars. The Martian canyons are up to 6 km (4 mi) deep and 190 km (120 mi) wide. Valles Marineris begins with heavily fractured terrain in the west and ends with ancient cratered terrain in the east. If this canyon were located on Earth, it would stretch from New York to Los Angeles. Geologists believe that Valles Marineris is a large crack that formed as the planet cooled. It was enhanced by nearby rising crust to its west and widened further by erosion. Some of the eastern parts of this system appear to have been formed almost entirely by water flow, like Earth's Grand Canyon.

By observing the orbit of *Mars Global Surveyor*, astronomers have concluded that the Martian crust is about 40 km thick under the northern lowlands and about 70 km thick under the southern highlands. However, they find that the boundary between the thin and thick crusts does not line up with the boundary between high and low terrain. Current models of Mars suggest that it has a core about 3400 km in diameter. However, a detailed understanding of Mars's interior waits for the placing of seismic detectors on the planet's surface.

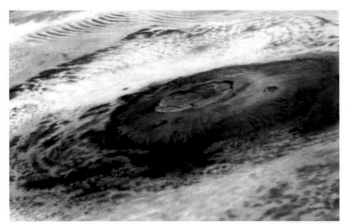

R I V U X G

FIGURE 6-23   The Olympus Caldera This view of the summit of Olympus Mons is based on a mosaic of six pictures taken by one of the *Viking* orbiters. The caldera consists of overlapping, volcanic craters and measures about 70 km across. The volcano is wreathed in mid-morning clouds brought upslope by cool air currents. The cloudtops are about 8 km below the volcano's peak. (NASA)

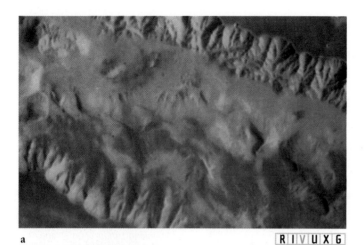

a R I V U X G

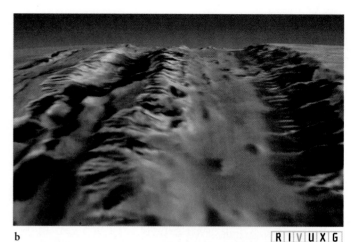

b R I V U X G

FIGURE 6-24 A Segment of Valles Marineris (a) This mosaic of *Viking Orbiter* photographs shows a segment of Valles Marineris. The canyon is about 100 km (60 mi) wide in this region. The canyon floor has two major levels. The northern (upper)

canyon floor is 8 km (5 mi) beneath the surrounding plateau, whereas the southern canyon floor is only 5 km (3 mi) below the plateau. **(b)** This westward-looking view of the Valles Marineris was created using altitude data from *Mars Global Surveyor.* (a: NASA; b: MOLA Science Team/NASA)

6-12 Surface features indicate that water once flowed on Mars

Despite disproving the theory that Mars has broad canals, Mars-orbiting spacecraft did reveal many features that look like dried-up riverbeds (Figure 6-25), lakes, and other water-

related features. Some of the riverbeds include intricate branched patterns and delicate channels meandering among flat-bottomed craters. Rivers on Earth invariably follow similarly winding courses (Figure 6-26).

The Martian riverbeds were totally unexpected because liquid water cannot exist today on Mars. Water is liquid

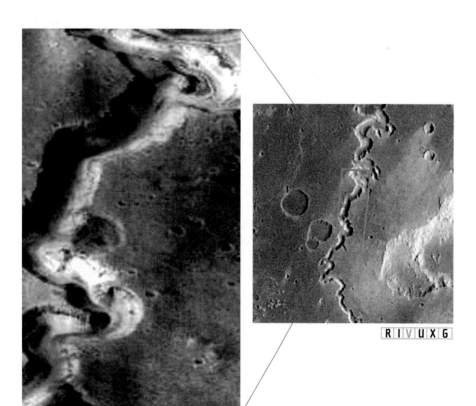

R I V U X G

FIGURE 6-25   Ancient River Channel on Mars Winding canyons on Mars, such as the one in this *Viking Orbiter I* image (on the right) appear to be at least partly due to sustained water flow. This belief is supported by the terraces seen on the canyon walls in the high resolution *Mars Orbiter* image (inset). Long periods of water flow require that the planet's atmosphere was once thicker and its climate more Earthlike. (NASA)

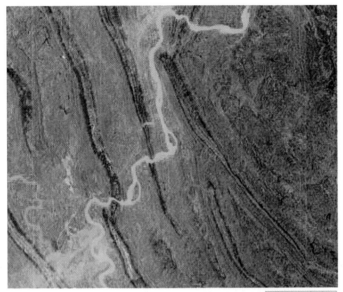

FIGURE 6-26   **River Flow on Earth**   This image of the Yangzi river near Chongqing, China, as is typical of rivers on Earth, shows the same snakelike curve as the river channels on Mars. (TMSC/NASA)

over a limited range of temperature and pressure. At low temperatures, water becomes ice; at high temperatures, it becomes steam. The atmospheric pressure also affects the state of water. If the pressure is very low, molecules easily escape from the liquid's surface, causing the water to vaporize. Because the pressure of Mars's atmosphere is only 0.6% that of the Earth's atmosphere, any liquid water on Mars today would furiously boil and rapidly evaporate into the

thin Martian air. If the gravitational attraction of any planet (or moon) to various gases in its atmosphere is too low, these gases will eventually escape into space. Calculations reveal that water escapes from Mars's atmosphere.

Further evidence that water once flowed on Mars comes from so-called *SNC meteorites* found on Earth (Figure 6-27a). These space rocks are believed to have once been pieces of Mars because their chemistries are consistent with those of rocks studied on Mars's surface and because they contain trace gases in amounts found only in the current Martian atmosphere. The meteorites were ejected into space during especially powerful impacts on that planet's surface. They also contain water-soaked clay, which is not expected to be found on any objects in the solar system besides Mars, Earth, and perhaps Europa. At least 22 Mars meteorites have been identified.

Where did the water on Mars come from and go to? Some of it was (and still is) frozen in the Martian polar caps, although their seasonal variations today are created by the freezing and evaporating of dry ice (carbon dioxide). The temperature at Mars's poles, typically 160 K (−170°F), keeps the water there permanently frozen, like the permafrost found in northern Asia and on Antarctica. Other frozen water probably lies elsewhere under the Martian surface. Heat from meteoritic impacts or from volcanic activity occasionally melted this layer of permafrost on Mars. The ground then collapsed, and millions of tons of rock pushed the water to the surface. These flash floods could account for the riverbeds we see there today.

In 2001, astronomers reported evidence that volcanic activity may still be causing episodic flows of both water and lava on Mars's surface. *Mars Global Surveyor* images

a

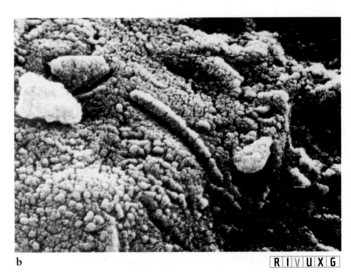

b

FIGURE 6-27   **A Piece of Mars on Earth**   Mars's thin atmosphere does little to protect it from impacts. Some of the debris ejected from impact craters there apparently traveled to Earth. **(a)** This SNC meteorite was recovered in Antarctica. It shows strong evidence of having been exposed to liquid water on Mars, perhaps for hundreds of years. **(b)** These may be the fossil remains of primitive bacterial life on Mars. (a: NASA; b: NASA/JPL)

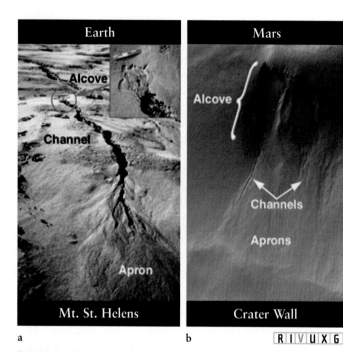

FIGURE 6-28 Water Flow on Earth and Mars (a) When underground water flows on the Earth, the surface can become unstable and slide (creating what geologists call an *alcove*). The water reaching the surface runs down, gouging out *channels* and carrying debris that creates the *apron*. (b) This image from the *Mars Global Surveyor* shows the same features, suggesting that liquid water has come from under Mars's surface. When this event occurred is not yet known. (JPL/Malin Space Science Systems/NASA)

show channels apparently carved by water flowing down the walls of pits or craters (Figure 6-28). What makes these observations especially intriguing is that they appear to be geologically young, indicating that liquid water may still exist under the Martian surface.

Evidence that water existed as recently as a few million years ago (and possibly still today) just meters below Mars's equator also comes in the form of clusters of cones on the planet's surface. These cones are created when lava flows over water-rich terrain. The lava vaporizes the water, which eventually bursts through the lava, creating the cones. Similar clusters of cones on Earth created by this mechanism are found in Iceland. We present further evidence for water on Mars in section 6-14.

6-13 Martian air is thin and often filled with dust

As on Venus, some 95% of Mars's thin atmosphere is composed of carbon dioxide. The remaining 5% consists of nitrogen, argon, and some traces of oxygen. The planet's

gravitational force is just strong enough to hold all its gases but as noted above, it is too weak to prevent most water vapor from evaporating into space. Consequently, the concentration of water vapor in Mars's atmosphere is 30 times lower than the concentration above the Earth. If all the water vapor could somehow be squeezed out of the Martian atmosphere, it would not fill even one of the five Great Lakes of North America.

Mars experiences Earthlike seasons because of a striking coincidence, first noted in the late 1700s by the German-born English astronomer William Herschel. Just as the Earth's equatorial plane is tilted $23\frac{1}{2}°$ from the plane of its orbit, Mars's equator makes an angle of about 25° with its orbit. However, the Martian seasons last nearly twice as long as Earth's, because Mars takes nearly two Earth years to orbit the Sun.

Unlike our blue sky, Mars's atmosphere is often pastel red, sometimes turning shades of pink and russet (Figure 6-29). All these colors are due to the fine dust blown from the planet's desertlike surface during windstorms. The dust is iron oxide, familiar here on Earth as rust. Mars's sky also changes color because the amount of dust in the air varies with the season. During the winter, carbon dioxide ice adheres to the dust particles and drags them to the ground. This helps clear and lighten the air. In the summer months, the carbon dioxide is not frozen, and the dust blown by surface winds remains aloft longer.

The sky color also changes over periods of many years. In 1995, for example, the amount of dust was observed to have dropped dramatically compared to that observed in the 1970s. The reason for such long-term changes is still under investigation.

6-14 Surface exploration reveals an antiseptic surface and even more signs of water

The discovery in the 1960s that water once existed on the surface of Mars rekindled speculation about Martian life. Although it was clear that Mars has neither civilizations nor fields of plants, microbial life-forms still seemed possible. Searching for signs of organic matter was one of the main objectives of the ambitious and highly successful *Viking* missions.

The two *Viking* spacecraft were launched during the summer of 1975. Each spacecraft consisted of two modules—an orbiter and a lander. Almost a year later, both *Viking* landers set down on rocky plains north of the Martian equator (see Figure 6-22).

The landers confirmed the long-held suspicion that the red color of the planet is due to large quantities of iron in its soil. Despite the high iron content of its crust, Mars has a lower average density ($3950 kg/m^3$) than that of other ter-

R I V U X G

FIGURE 6-29 **Mars's Rust-Colored Sky** Taken by the *Mars Pathfinder*, this stunning photograph of another world shows the *Sojourner* rover snuggled against a rock named Moe on the Ares Vallis. At the top of the image, the pink color of the Martian sky is evident. One of the "Twin Peaks" can be seen in the background, about 1 km away. (NASA)

restrial planets (more than 5000 kg/m^3 for Mercury, Venus, and Earth). Mars must therefore contain overall a lower percentage of iron than these other planets.

Each *Viking* lander was able to dig into the Martian regolith and retrieve rock samples for analysis (Figure 6-30). Bits of the regolith were observed to cling to a magnet mounted on the scoop, indicating that the regolith contains iron (consistent with the assertion above that Mars is rich in iron oxides). Indeed, further analysis showed the rocks at both sites to be rich in iron, silicon, and sulfur. The Martian regolith can best be described as an iron-rich clay.

The *Viking* landers each carried a compact biological laboratory designed to test for microorganisms in the Martian soil. Three biological experiments were conducted, each based on the idea that living things alter their environment: They eat, they breathe, and they give off waste products. In each experiment, a sample of the Martian regolith was placed in a closed container, with or without a nutrient substance. The container was then examined for any changes in its contents.

The first data returned by the *Viking* biological experiments caused great excitement. In almost every case, rapid and extensive changes were detected inside the sealed containers. However, further analysis showed that these changes were due solely to nonbiological chemical processes. Apparently, the Martian regolith is rich in chemicals that effervesce (fizz) when moistened. A large amount of oxygen is apparently tied up in the regolith in the form of unstable chemicals called *peroxides* and *superoxides*, which break down in the presence of water to release oxygen gas.

The chemical reactivity of the Martian regolith probably comes from ultraviolet radiation that beats down on the planet's surface. Ultraviolet photons easily break apart molecules of carbon dioxide (CO_2) and water vapor (H_2O) by

Trenches

R I V U X G

VIDEO 6.6

FIGURE 6-30 **Digging in the Martian Regolith** *Viking's* mechanical arm with its small scoop protrudes from the right side of this view of the *Viking 1* landing site. Several small trenches dug by the scoop in the Martian regolith appear near the left side of the picture. (NASA)

knocking off oxygen atoms, which then become loosely attached to chemicals in the regolith. Ultraviolet photons also produce ozone (O_3) and hydrogen peroxide (H_2O_2), which become incorporated in the regolith. In all these cases, the loosely attached oxygen atoms make the regolith extremely reactive.

Here on Earth, hydrogen peroxide is commonly used as an antiseptic. When you pour this liquid on a wound, it fizzes and froths as the loosely attached oxygen atoms chemically combine with organic material and destroy germs. The *Viking* landers *may* have failed to detect any organic compounds on Mars, because the superoxides and peroxides in the Martian regolith make it antiseptic today.

Okay, if there is no sign of life on Mars now, are there signs of ancient life from the days when water flowed? Curiously, the best evidence may lie on Earth. When cut open, several of the Martian meteorites (section 6-12) have shown microscopic features that could be fossils of Martian bacterial life and their excretions (see Figure 6-27b). These features, no larger than 500 nm (1/100th the diameter of a human hair), are 30 times smaller than bacteria found on Earth. Detailed analysis of the meteorites in 2001 revealed the presence of several organically created features, including tiny spheres found with some of Earth's bacteria, and magnetite crystals, which are also used by some bacteria on Earth as compasses to find food. It is worth noting that claims of fossils from Mars in the 1960s have been disproved. While the present meteorite evidence is not yet conclusive evidence that life existed there, it has withstood over five years of rigorous analysis and the studies continue.

The 1997 visit by *Pathfinder* and its little rover, *Sojourner* (see Figure 6-29), to the Ares Vallis has revealed evidence of several floods at the mouth of a dried flood plain. The evidence includes layers of sediment, clumps of rock and sand stuck together like similar groupings created by water on Earth, rounded rocks worn down as they were dragged by the water, and rocks aligned by water flow. The rover performed some 20 chemical analyses of rocks in the landing area. Results indicate chemical compositions consistent with those of the SNC meteorites, supporting the belief that these latter bodies came from Mars. The soil is a rusty color, laden with sulfur. The source of this sulfur is a mystery, as is the large amount of silicon-rich silica discovered on rocks such as Barnacle Bill.

Further evidence for the existence of standing water on Mars comes from layered terrain, similar to that found in Earth's Grand Canyon, located in several places on the red planet (Figure 6-31). The most plausible explanation of these formations is layers of sediment laid down by large bodies of water.

The temperature at *Pathfinder's* (see Figure 6-22) site ranged from 197 K (−105° F) just before dawn to 263 K (14°F) at the warmest time of day. During the afternoons, heat from the planet's surface warmed the air and created

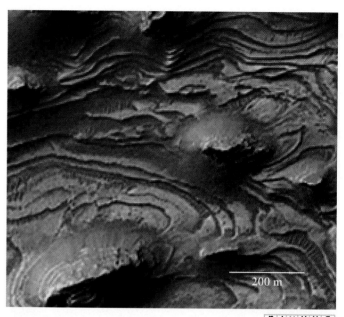

R I V U X G

FIGURE 6-31   Ancient Oceans and Lakes on Mars This *Mars Global Surveyor* images of a portion of *Valles Marineris* reveals terrain with "stair step layers." Such terrain is likely to have been created by sedimentation at the bottom of an ancient body of water. (Malin Space Science Systems/JPL/NASA)

whirlwinds called **dust devils**. A similar phenomenon occurs in dry or desert terrain on Earth. Martian dust devils reach altitudes of 6 km (20,000 ft). All three landers detected drops in air pressure as dust devils swept past. Dust devils are large enough to be seen by orbiting spacecraft (Figure 6-32)

The *Mars Global Surveyor's* images have greatly strengthened the belief that Mars once had liquid water, as seen in Figure 6-33. The total amount of water that existed on Mars's surface is not known. By studying flood channels that exist there, geologists estimate that there was enough water to cover the planet to a depth of 500 meters (1500 ft). For comparison, Earth has enough water to cover our planet to a depth of 2700 meters (8900 ft), assuming that the Earth's surface was everywhere a uniform height. The evidence is very strong now that Mars once had lakes, rivers, and perhaps even oceans of liquid water. There is observational evidence of ancient shoreline encircling the northern lowlands of Mars. The belief that the northern lowlands were an ocean is supported by the analysis of a 1.2 billion-year-old meteorite from Mars. It has remnants of salt believed to have been deposited there when the Martian ocean dried up.

The *Mars Global Surveyor* has taken visible light images of Mars's surface, seeing objects as small as 1.5 m across. Its laser has mapped the planet to a resolution of 330 m along

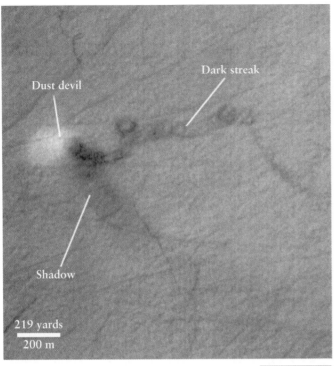

R I V U X G

FIGURE 6-32 **Martian Dust Devil** This *Mars Global Surveyor* image shows a dust devil as seen from almost directly above. This tower of swirling air and dust casts a long shadow in the afternoon Sun. The dust devil had been moving from right to left before the picture was taken, leaving a dark curlicue-shaped trail in its wake. The area shown is about 1.5 × 1.7 km (about 1 mile on a side). (NASA/JPL/Malin Space Science Systems)

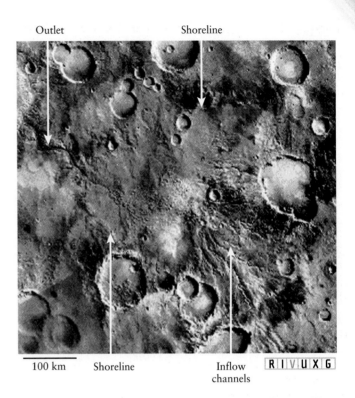

R I V U X G

FIGURE 6-33 **Evidence of Water on Mars** A dry Martian lake, photographed by the *Mars Global Surveyor* with a resolution of 1.5 m. An excellent example of how geology and astronomy overlap, the features of this dry lake are consistent with those found on lakebeds on Earth. (NASA/JPL)

the surface and 1 m in height (see Figure 6-22). *Surveyor* has also mapped heat emitted from Mars and mineral deposits on its surface, along with the planet's strange magnetic fields, discussed earlier. This remotely sensed information is invaluable in helping space scientists decide where to explore Mars's surface in upcoming missions such as the *Mars Odyssey* spacecraft, which is now collecting information about the planet's surface composition, radiation environment, and underground water.

6-15 Mars's two moons look more like potatoes than spheres

Two tiny moons orbit close to Mars's surface. Phobos (meaning "fear") and Deimos ("panic") are so small that they were not discovered until 1877. Potato-shaped Phobos is the inner and larger of the two (Figure 6-34a). It rises in the west and races across the sky in only 5¹/₂ hours, as seen from Mars's equator. It is heavily cratered, and observations

by *Mars Global Surveyor* indicate that its surface has been transformed into dust at least 1 meter thick by countless tiny impacts over the eons. Football-shaped Deimos is less cratered than Phobos (Figure 6-34b). As seen from Mars, Deimos rises in the east and takes about three Earth days to creep from one horizon to the other.

Phobos and Deimos were not formed like our Moon, by splashing off Mars. Rather, they are captured planetesimals. Both moons are in synchronous rotation as they orbit the red planet.

> **Insight into Science** **Imagine the Moon** The definitions we use for words based on our everyday experience often fail us in astronomy. For example, the word "moon" usually creates an image of a spherical body, like our Moon. In reality, most of the moons in the solar system are unsymmetrical, like Phobos and Deimos.

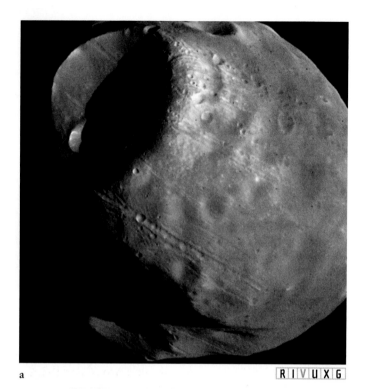

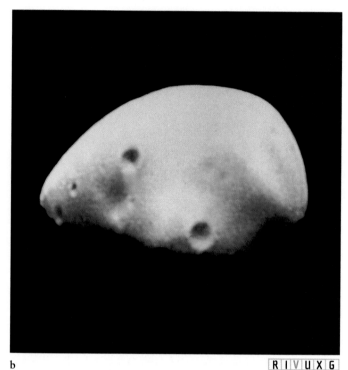

a   R I V U X G   b   R I V U X G

 FIGURE 6-34   **Phobos and Deimos** (a) Phobos, the larger of Mars's two moons, is potato-shaped and measures approximately 28 × 23 × 20 km. It is dominated by crater Stickney, seen on the left. (b) Deimos is less cratered than Phobos and measures roughly 16 × 12 × 10 km. (a & b: NASA)

6-16 Frontiers yet to be discovered

All three other terrestrial planets have much to reveal. We have yet to see all of Mercury's surface. What is the chemical composition of its surface rocks? What is Mercury's cooling history and how much of its core is molten? Is there really ice at its poles? If so, how did it get there? What will its internal structure reveal?

Does Venus have active volcanoes? If not, what supplies its atmosphere with sulfur compounds? Can we find observational evidence of what caused Venus's rotation axis to flip over? Likewise, can we find further evidence that its surface periodically undergoes significant recovering?

Perhaps one of the most significant discoveries that lies in store in our solar system is confirmation that life once existed on Mars and if there actually is liquid water under the red planet's surface, whether there is still life there. If so, how far did life evolve? What are the similarities and differences between such life and life on Earth? Significant similarities might imply a common origin. Furthermore, what is the surface water history of that world? What causes its local magnetic fields? What does its interior look like? How did its axis get tilted? Where did its moons come from? It is likely that we will have answers to many of these questions in the coming few decades.

   Further Reading on These Topics

WHAT DID YOU KNOW?

|1| *Is the temperature on Mercury, the closest planet to the Sun, higher than the temperature on Earth?* The temperature on the daytime side of Mercury is much higher than on Earth, but the temperature on the nighttime side of Mercury is much lower than on Earth, because Mercury rotates so slowly and has little atmosphere to retain heat.

|2| *What planet is most similar to Earth?* Venus is most similar to Earth in size, chemistry, and distance from the Sun. Mars is more similar to Earth (than Venus) only in its length of day and in having seasons.

|3| *What is the composition of the clouds surrounding Venus?* The clouds are made primarily of sulfuric acid.

I 4 *Does Mars have liquid water on its surface today?* No, but there are strong indications that it had liquid water on its surface in the past.

I 5 *Is life known to exist on Mars today?* No current life has yet been discovered on Mars.

KEY WORDS

3-to-2 spin-orbit coupling, 167
caldera, 177
dust devil, 182
greenhouse effect, 169

hot-spot volcanism, 176
northern vastness (northern lowlands), 177

retrograde rotation, 173
scarp, 165
southern highlands, 177

KEY IDEAS

All four inner planets are composed primarily of rock and metal, and thus they are classified as terrestrial.

Mercury

• Even at its greatest orbital elongations, Mercury can be seen from Earth only briefly after sunset or before sunrise.

• The *Mariner 10* spacecraft passed near Mercury in the mid-1970s, providing pictures of its surface. The Mercurian surface is pocked with craters like the Moon's, but extensive, smooth plains lie between these craters. Long cliffs meander across the surface of Mercury. These scarps probably formed as the planet cooled, solidified, and shrank.

• The long-ago impact of a large object formed the huge Caloris Basin on Mercury and shoved up jumbled hills on the opposite side of the planet.

• Mercury has an iron core much like that of the Earth.

Venus

• Venus is similar to the Earth in size, mass, and average density, but it is covered by unbroken, highly reflective clouds that conceal its other features from Earth-based observers.

• While most of Venus's atmosphere is carbon dioxide, its dense clouds contain droplets of concentrated sulfuric acid mixed with yellowish sulfur dust. Active volcanoes on Venus may be a constant source of this sulfurous veil.

• Venus's exceptionally high temperature is caused by the greenhouse effect, as the dense carbon dioxide atmosphere traps and retains heat emitted by the planet. The surface pressure on Venus is 90 atm, and the surface temperature is 750 K. Both temperature and pressure decrease as altitude increases.

• The surface of Venus is surprisingly flat, mostly covered with gently rolling hills. There are two major "continents" and several large volcanoes. The surface of Venus shows evidence of local tectonic activity but not the large-scale motions that play a major role in continually reshaping the Earth's surface.

Mars

• Earth-based observers found that the Martian solar day is nearly the same as that of the Earth, that Mars has polar caps that expand and shrink with the seasons, and that the Martian surface undergoes seasonal color changes.

• A century ago observers reported networks of linear features that many perceived as canals. These observations led to speculation about self-aware life on Mars.

• The Martian surface has many flat-bottomed craters, several huge volcanoes, a vast equatorial canyon, and dried-up riverbeds—but no canals formed by intelligent life. Flash-flood features and dry riverbeds on the Martian surface indicate that large amounts of water once flowed there.

• Liquid water would quickly boil away in Mars's thin present-day atmosphere, but the planet's polar caps contain some frozen water, and a layer of permafrost may exist beneath the regolith.

• The Martian atmosphere is composed mostly of carbon dioxide. The surface pressure is less than 0.01 atm.

• Chemical reactions in the regolith together with ultraviolet radiation from the Sun apparently act to sterilize the Martian surface.

• Mars has no global magnetic fields, but local fields pierce its surface in at least nine places.

• Features that may be fossil remains of bacteria have been found in several meteorites that are believed to have come from Mars.

• Mars has two potato-shaped moons, the captured planetesimals Phobos and Deimos. Both are in synchronous rotation with Mars.

REVIEW QUESTIONS

1 Why is Mercury so difficult to observe? When is the best time to see the planet? *Hint:* The Guided Discovery box, "The Inner Solar System," on page 176 can help.

2 Compare the surfaces of Mercury and our Moon. How are they similar? How are they different?

3 Compare the interiors of Mercury and Earth. How are they similar? How are they different?

 4 To better understand the interiors of Mercury and Earth, do Interactive Exercise 6-1 on the Web or CD-ROM. You can print out the result, if requested.

5 What are the longest features found on Mercury? Why are the examples of this feature probably much older than tectonic features on the Earth?

6 Briefly describe at least one theory explaining why Mercury has such a large iron core.

7 Astronomers often refer to Venus as the Earth's twin. What physical properties do the two planets have in common? In what ways are the two planets dissimilar?

8 Why is it hotter on Venus than on Mercury?

9 What is the greenhouse effect? What role does it play in the atmospheres of Venus and the Earth?

10 What evidence exists for active volcanoes on Venus?

11 Describe the Venusian surface. What kinds of geological features would you see if you could travel around the planet?

12 Why do astronomers believe that Venus's surface was not molded by the kind of tectonic activity that shaped the Earth's surface?

13 Why is Mars red?

14 When is the best time to observe Mars from Earth? *Hint:* The Guided Discovery box, "The Inner Solar System," on page 176 can help.

15 Compare the cratered regions of Mercury, the Moon, and Mars. Assuming that the craters on all three worlds originally had equally sharp rims, what can you conclude about the environmental histories of these worlds?

16 How would you tell which craters on Mars were formed by meteoritic impacts and which by volcanic activity?

17 Compare the volcanoes of Venus, Earth, and Mars. Do you think hot-spot volcanism is or was active on all three worlds? Explain.

18 What geologic features indicate plate tectonic activity once occurred on Mars? What features created by tectonic activity on Earth are not found on Mars?

 19 To better understand the surface features of Mars, do Interactive Exercise 6-2 on the Web or CD-ROM. You can print out your results, if requested.

20 What is the current knowledge concerning life on Mars? Do you think that today Mars is as barren and sterile as the Moon? Why or why not?

 21 To compare the surfaces of Mercury, Venus, and Mars, do Interactive Exercise 6-3 on the Web or CD-ROM. You can print out your results, if requested.

 22 What evidence have astronomer's accumulated that liquid water once existed in large quantitites on Mars's surface? What evidence is there that water is still there, under the surface?

ADVANCED QUESTIONS

23 What evidence do we have that the surface features on Mercury were not formed during recent geologic history?

24 Venus takes 440 days to move from greatest western elongation to greatest eastern elongation, but it needs only 144 days to go from greatest eastern elongation to greatest western elongation. With the aid of a diagram, explain why.

25 As seen from Earth, the brightness of Venus changes as it moves along its orbit. Describe the main factors that determine Venus's variations in brightness as seen from Earth. *Hint:* See the discussion of Venus in Chapter 2.

26 How might Venus's cloud cover change if all of Venus's volcanic activity suddenly stopped? How might these changes affect the overall Venusian environment?

27 Compare Venus's continents with those on the Earth. What do they have in common? How are they different?

28 Explain why Mars has the longest synodic period of all the planets, although its sidereal period is only 687 days.

29 With carbon dioxide accounting for about 95% of the atmospheres of both Mars and Venus, why do you think there is little greenhouse effect on Mars today?

30 Could the polar regions of Mars reasonably be expected to harbor life-forms, even though the Martian regolith is sterile at the *Viking* lander sites?

DISCUSSION QUESTIONS

31 If you were planning a return mission to Mercury, what features and observations would be of particular interest to you and why?

32 If you were designing a space vehicle to land on Venus, what special features would be necessary? In what ways would this mission and landing craft differ from a spacecraft designed for a similar mission to Mercury?

33 Suppose someone told you that the *Viking* mission failed to detect life on Mars simply because the tests were

designed to detect terrestrial life-forms, not Martian life-forms. How would you respond?

34 Compare the scientific opportunities for long-term exploration offered by the Moon and Mars. What difficulties would there be in establishing a permanent base or colony on each of these two worlds?

35 Imagine you are an astronaut living at a base on Mars. Describe your day's activities, what you see, the weather, the spacesuit you are wearing, and so on.

WHAT IF ...

36 Mercury had synchronous rotation? How would the temperatures on such a planet be different than they are today? Where would humans set up camp on such a world?

37 Venus had the same rotation rate, surface, and atmosphere as the Earth? In what ways would life there be different than it is here? How would their perceptions of the cosmos be different from ours?

38 Mars had the same mass, surface, and atmosphere as the Earth? In what ways would life there be different than

it is here? How would their perceptions of the cosmos be different from ours?

39 Mars rotated once every 20 days rather than once every 1.026 days? What would be different?

40 The carbon dioxide content of the Earth's atmosphere increased? What would happen to the Earth? This is not a completely hypothetical question, because carbon dioxide levels are increasing today.

WEB/CD-ROM QUESTIONS

41 Search the Web for the latest information about upcoming space missions to Mercury. When are they scheduled to be launched and when are they scheduled to arrive at Mercury? What scientific experiments will they carry? What scientific issues are these instruments intended to resolve?

 42 Elongations of Mercury Access the animation "Elongations of Mercury" in Chapter 6 of the *Discovering the Universe* Web site or CD-ROM. (a) Note the dates of the greatest eastern and western elongations in the animation. Which time interval is greater: from a greatest eastern elongation to a greatest western elongation or vice versa? (b) Based on what you observe in the animation, draw a diagram to explain your answer to the question in (a).

43 Search the Web for the latest information about proposed future missions to Venus. What scientific experiments will they carry? What scientific issues are these instruments intended to resolve?

44 Surface Temperature of Venus Access the Active Integrated Media Module "Wien's Law" in Chapter 4 of the *Discovering the Universe* Web site or CD-ROM. (a) Using the Wien's law calculator, determine Venus's approximate temperature if it emits blackbody radiation with a peak wavelength of 3866nm. (b) By trial and error, find the wavelength of maximum emission for a surface temperature of 750 K (for present-day Venus) and a surface temperature of 850 K (as it might become if its

greenhouse gas density increases). In what part of the electromagnetic spectrum do these wavelengths lie?

45 Search the Web for the latest information about upcoming space missions to Mars, including the European Space Agency's *Mars Express* and the Japanese orbiter *Nozomi*. When are they scheduled to be launched and when are they scheduled to arrive at Mars? What scientific experiments will they carry? What scientific issues are these instruments intended to resolve?

46 In 1999, two NASA spacecraft—*Mars Climate Orbiter* and *Mars Polar Orbiter*—failed to reach their destinations. Search the Web for information on these missions. What were their scientific goals? How and why did the missions fail? Which current and future missions, if any, are intended to replace these missions?

47 Search the Web for information about possible manned missions to Mars. How long would such a mission take? How expensive would they be? What are some advantages and disadvantages of a manned mission compared to an unmanned one?

48 Conjunctions of Mars Access and view the animation "The Orbits of Earth and Mars" in Chapter 6 of the *Discovering the Universe* Web site or CD-ROM. (a) The animation highlights three dates when Mars is in opposition, so that the Earth lies directly between Mars and the Sun. By using the "Stop" and

"Play" buttons in the animation, find two times during the animation when Mars is in *conjunction*, so that the Sun lies directly between Mars and the Earth (see Figure 2-3). For each conjunction, make a drawing showing the positions of the Sun, the Earth, and Mars, and record the

month and year when the conjunction occurs. (b) When Mars is in conjunction, at approximately what time of day does it rise as seen from Earth? At what time of day does it set? Is Mars suitably placed for telescopic observations when it is in conjunction?

OBSERVING PROJECTS

  49 Refer to the accompanying table to determine the dates of the next two or three greatest elongations of Mercury. Consult such magazines as *Sky & Telescope* and *Astronomy*, or use your *Starry Night Backyard*™ planetarium software to see if any of these greatest elongations is going to be especially favorable for viewing the planet. If so, make plans to be one of those rare individuals who has actually seen the innermost planet of the solar system. Set aside several evenings (or mornings) around the date of the favorable elongation to reduce the chances of being "clouded out." Select an observing site that has a clear, unobstructed view of the horizon where the Sun sets (or rises). Make arrangements to have a telescope at your disposal. Search for the planet on the dates you have selected and make a drawing of its appearance through your telescope.

Greatest Elongations of Mercury

Evening	2002 Sep 1	27.2°E
Morning	2002 Oct 13	18.1°W
Evening	2002 Dec 26	19.9°E
Morning	2003 Feb 4	25.4°W
Evening	2003 Apr 16	19.8°E
Morning	2003 Jun 3	24.4°W
Evening	2003 Aug 14	27.4°E
Morning	2003 Sep 26	17.9°W
Evening	2003 Dec 9	20.9°E
Morning	2004 Jan 17	23.9°W
Evening	2004 Mar 29	18.9°E
Morning	2004 May 14	26.0°W
Evening	2004 Jul 27	27.1°E
Morning	2004 Sep 9	18.0°W
Evening	2004 Nov 21	22.2°E
Morning	2004 Dec 29	22.4°W
Evening	2005 Mar 12	18.3°E
Morning	2005 Apr 26	27.2°W
Evening	2005 Jul 9	26.3°E
Morning	2005 Aug 23	18.4°W
Evening	2005 Nov 3	23.5°E
Morning	2005 Dec 12	21.1°W

50 Refer to the following table in the next column to see if Venus is currently near a greatest elongation. If so, view the planet through a telescope. Make a sketch of the planet's appearance. From your sketch, can you determine whether Venus is closer to us or farther from us than the Sun?

Greatest Elongations of Venus

Evening	2002 Aug 22	46.0°E
Morning	2003 Jan 11	47.0°W
Evening	2004 Mar 29	46.0°E
Morning	2004 Aug 17	45.8°W
Evening	2005 Nov 3	47.1°E

51 Observe Venus through a telescope once a week for a month and make a sketch of the planet's appearance on each occasion. From your sketches, can you determine whether Venus is approaching us or moving away from us?

 52 Consult such magazines as *Sky & Telescope* and *Astronomy*, or use your *Starry Night Backyard*™ planetarium software to determine Mars's location among the constellations. If Mars is suitably placed for observation, arrange to view the planet through a telescope. Draw a picture of what you see. What magnifying power seems to give you the best image? Can you distinguish any surface features? Can you see a polar cap or dark markings? If not, can you offer an explanation for Mars's bland appearance?

 53 Using your *Starry Night Backyard*™ software, try the activities in this chapter's Guided Discovery box, "The Inner Solar System," to follow Mercury's orbit as it nears the Sun. With the timestep still set at 1 sidereal day, describe the motion of Mercury relative to the equatorial coordinate grid (that is, the celestial sphere). What can you say about the maximum and minimum angles on the celestial sphere between Mercury and the Sun? Periodically stop the motion, zoom in on Mercury, and determine its phase. Can you find a pattern of phases?

 54 The Guided Discovery box on page 176 will introduce you to Mars's motion. In *Starry Night Backyard*™ go into Atlas mode (*Go/Atlas*), find Mars (*Edit/Find/Mars*), set the sky at 60°, and lock on Mars. Run the program with a timestep of 2 sidereal days. Notice that the equatorial coordinate system usually flows smoothly past Mars. When the stars behind Mars begin to slow down, stop the motion by pressing the ■ icon. Unlock Mars by moving it to the left quarter of the screen. Then run the program again. Describe how Mars now moves relative to the coordinate system. Stop the program, run it backward through this entire period, and stop it. Set the timestep at 1 sidereal day and rerun it forward in time, this time using the single-step feature. Make a day-by-day drawing of Mars's motion during this time. What is the name of this motion?

7 THE OUTER PLANETS

IN THIS CHAPTER YOU WILL DISCOVER

- Jupiter, an active, vibrant, multicolored world more massive than all the other planets combined

- Jupiter's diverse system of moons

- Saturn, with its spectacular system of thin, flat rings and numerous moons

- Uranus and Neptune: similar to each other but quite different from Jupiter and Saturn

- tiny Pluto and its moon, Charon, orbit each other in synchronous rotation

Auroras on Saturn
Glowing auroras crown the poles of Saturn. This image of Saturn is entirely in the ultraviolet. These auroras are just one of the myriad features of the outer planets that we are now able to view with current, powerful telescopes and from spacecraft traveling into the realm of the planets.
(J. Trauger/JPL and NASA)

R I V U X G

WHAT DO YOU THINK?

1   Is Jupiter a "failed star" or almost a star?

2   What is Jupiter's Great Red Spot?

3   Does Jupiter have continents and oceans?

4   Is Saturn the only planet with rings?

5   Are the rings of Saturn solid ribbons?

It's easy to imagine a trip to Mars, exploring its vast canyons and icy polar regions. Even Venus and Mercury are best understood by carefully comparing them to the Earth and the Moon. However, we find few similarities between the Earth and the outer planets. Granted, the auroras, shown in the photograph of Saturn opening this chapter, are similar in shape and origin to Earth's auroras as seen from space (see Figure 5-10a). However, Jupiter, Saturn, Uranus, and Neptune are so much larger, are rotating so much faster, and have such different chemical compositions from our world that we will seem indeed to have left the Earth far behind. Pluto is so small that the tidal force of its moon, Charon, has set the planet into synchronous rotation—the same side of Pluto always faces Charon—unlike the Earth, which spins some 30 times faster than our Moon orbits us.

Nothing on Earth suggests the swirling red and brown clouds of Jupiter, the ever-changing ring system of Saturn or the blue-green clouds of Uranus and Neptune. (See the table "The Outer Planets: A Comparison.") We begin our exploration of the outer planets with Jupiter, an alien world of unsurpassed splendor.

JUPITER

Even viewed through a small telescope on Earth, Jupiter's appearance sets it apart from the terrestrial worlds (Figure 7-1). As seen from the *Voyager 1* and *Voyager 2* spacecraft, which flew by Jupiter within months of each other in 1979, from the *Galileo* spacecraft over the last few years, and from the Hubble Space Telescope today, Jupiter's multicolored bands create a world of breathtaking beauty. Figure 7-2 lists Jupiter's properties.

7-1   Jupiter's rotation helps create colorful, global weather patterns

1   More than 1300 Earths could be packed into the volume of Jupiter, the largest planet in the solar system. Using the

R I V U X G

FIGURE 7-1   **Jupiter as Seen from Earth** Many features of Jupiter's clouds are visible from Earth, including multicolored parallel regions and several light and dark spots. The Great Red Spot appears in the lower, central region of the Jovian atmosphere. (S. Larson)

orbital periods of its moons and Kepler's laws, astronomers have determined that Jupiter is 318 times more massive than Earth. Jupiter has more than 2½ times as much mass as all the other planets combined. Nevertheless, it would have to be 75 times more massive still before it could generate energy as the Sun does and therefore be classified as a star.

What we see of Jupiter are the tops of clouds that permanently cover it (Figure 7-2). Because it rotates about once every 10 hours—the fastest of any planet—Jupiter's clouds are in perpetual motion and are confined to narrow bands. In contrast, winds on slower-rotating Earth wander over vast ranges of latitude (compare Figure 5-1).

Jupiter's cloud bands provide a backdrop for turbulent swirling cloud patterns, as well as for *white ovals* and *brown ovals*, which are rotating storms similar in structure to hurricanes on Earth (Figure 7-3). The white ovals are observed to be cool clouds higher than the average clouds in Jupiter's atmosphere. Conversely, the brown ovals are warmer and lower clouds. They are holes in the normal cloud layer. On Jupiter, the various oval features last from hours to centuries. In 1998, two 50-year-old storms on Jupiter were observed to merge into a single storm as large across as the Earth's diameter. Computers show us how the cloud features on Jupiter would look if the planet's atmosphere were unwrapped like a piece of paper (Figure 7-4). You can see ripples, plumes, and light-colored wisps in these images.

Jupiter's Vital Statistics

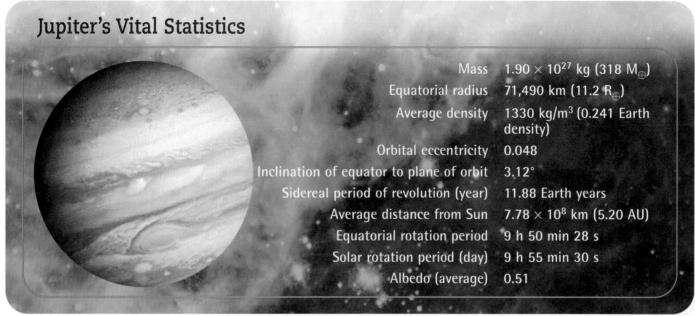

Mass	1.90×10^{27} kg (318 $M_\oplus$)
Equatorial radius	71,490 km (11.2 $R_\oplus$)
Average density	1330 kg/m³ (0.241 Earth density)
Orbital eccentricity	0.048
Inclination of equator to plane of orbit	3.12°
Sidereal period of revolution (year)	11.88 Earth years
Average distance from Sun	7.78×10^8 km (5.20 AU)
Equatorial rotation period	9 h 50 min 28 s
Solar rotation period (day)	9 h 55 min 30 s
Albedo (average)	0.51

R I V U X G

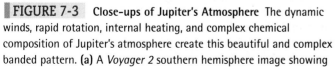

FIGURE 7-2 **Jupiter as Seen from a Spacecraft** This view was sent back from *Voyager 1* in 1979. Features as small as 600 km across can be seen in the turbulent cloudtops of this giant planet. The complex cloud motions surrounding the Great Red Spot are clearly visible. (NASA)

a R I V U X G

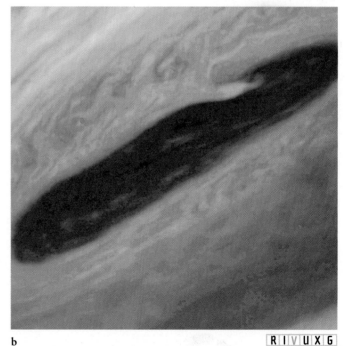

b R I V U X G

FIGURE 7-3 **Close-ups of Jupiter's Atmosphere** The dynamic winds, rapid rotation, internal heating, and complex chemical composition of Jupiter's atmosphere create this beautiful and complex banded pattern. **(a)** A *Voyager 2* southern hemisphere image showing a white oval that has existed for over 40 years. **(b)** A *Voyager 2* northern hemisphere image showing a brown oval. The white feature overlapping the oval is a high cloud. (a & b: NASA)

The Outer Planets: A Comparison

	Interior	Surface	Rings	Atmosphere	Magnetic Field
Jupiter	Terrestrial core, liquid metallic hydrogen shell, liquid hydrogen mantle	No solid surface, atmosphere gradually thickens to liquid state, belt and zone structure, hurricanelike features	Yes	Primarily H, He	19,000 × Earth's total field at its cloud layer, 14 × stronger than Earth's surface field
Saturn	Similar to Jupiter, with bigger terrestrial core and less metallic hydrogen	No solid surface, less distinct belt and zone structure than Jupiter	Yes	Primarily H, He	570 × Earth's total field at its cloud layer, $^2/_3$ × Earth's surface field
Uranus	Terrestrial core, liquid water shell, liquid hydrogen and helium mantle	No solid surface, weak belt and zone system, hurricanelike features, color from methane absorption of red, orange, yellow	Yes	Primarily H, He, some CH_4	50 × Earth's total field at its cloud layer, 0.7 × Earth's surface field
Neptune	Similar to Uranus	Like Uranus	Yes	Primarily H, He, some CH_4	35 × Earth's total field at its cloud layer, 0.4 × Earth's surface field
Pluto	Unknown	Apparently rock and ice	No	Unknown	Unknown

For detailed numerical comparisons between planets, see Appendix Tables A-1 and A-2.
*To see the orientations of these magnetic fields relative to the rotation axes of the planets, see Figure 7-31.

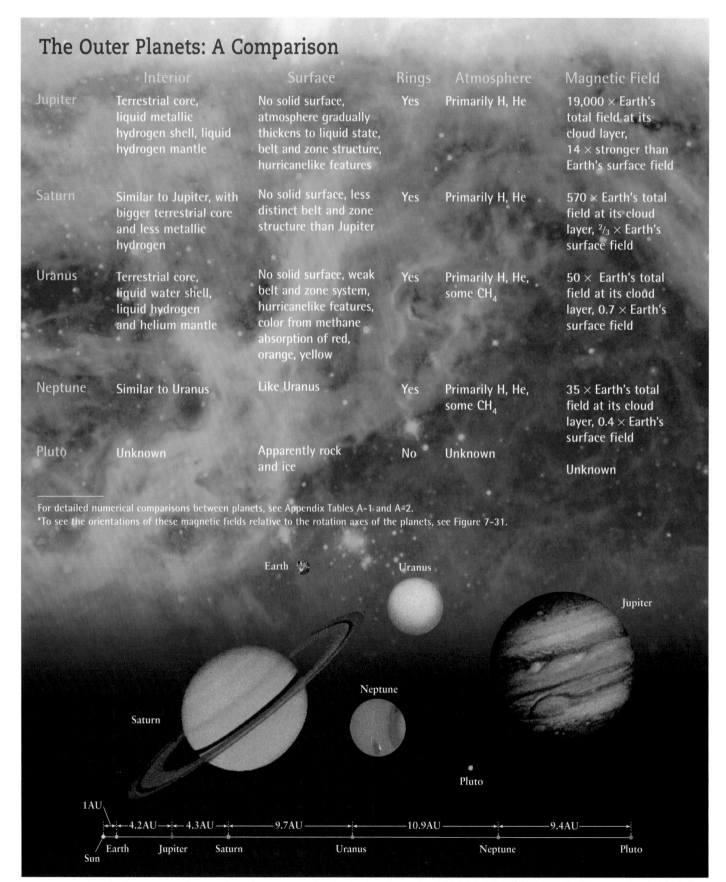

Earth   Uranus

Jupiter

Saturn   Neptune

Pluto

1AU
|← 4.2AU →|← 4.3AU →|←— 9.7AU —→|←——— 10.9AU ———→|←—— 9.4AU ——→|

Sun   Earth   Jupiter   Saturn   Uranus   Neptune   Pluto

(Stephen P. Meszaros/NASA, M. Buie, K. Horne, and D. Tholen)

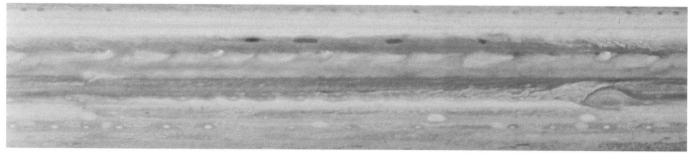

a *Voyager 1 view*

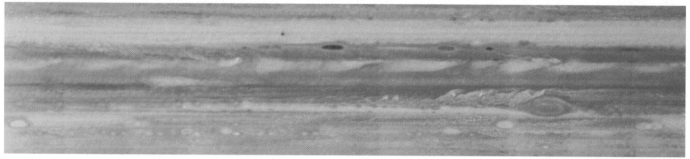

b *Voyager 2 view*   R I V U X G

❚ FIGURE 7-4   Jupiter's Northern and Southern Hemispheres
Computer processing shows the entire Jovian atmosphere from
(a) *Voyager 1* and **(b)** *Voyager 2.* A computer was used to "unwrap"
the planet's atmosphere. Notice in both views the dark ovals in the
northern hemisphere and the white ovals in the southern hemisphere.
The banded structure is absent near the poles. Notice, too, that the Great
Red Spot moved westward while the white ovals moved eastward during
the four months between the two *Voyager* flybys. (NASA)

One rust-colored oval feature called the **Great Red Spot**
(Figure 7-5) is so large that it can be seen through a small telescope. It changes dimensions, and at present it is about
25,000 km long by 12,000 km wide, large enough so that
two Earths could easily fit side by side inside it. The Great
Red Spot was first observed around 1665, either by English
scientist Robert Hooke or Italian astronomer Giovanni Cassini. Because earlier telescopes were unlikely to have been
able to see it, the Great Red Spot could well have formed
long before that time. Heat welling upward from inside
Jupiter has maintained this storm for more than three centuries. Consider what life would be like for us if the Earth
sustained storms for such long periods.

In 1690, Cassini noticed that the speeds of Jupiter's
clouds vary with latitude, an effect called **differential rotation.** Near the poles, the rotation period of Jupiter's atmosphere, 9 h 55 min 30 s, is 5 minutes longer than at the equator. Furthermore, clouds at different latitudes circulate in
opposite directions—some eastward, some westward. At
their boundaries the clouds rub against each other, creating
beautiful swirling patterns (see Figure 7-3). The interactions
of clouds at different latitudes also help provide stability for
storms like the Great Red Spot.

Spectra from Earth-based telescopes and from the
Galileo probe sent into Jupiter's upper atmosphere in 1995
give more detail about Jupiter's atmosphere. More than

R I V U X G

WEB LINK 7-4   VIDEO 7-2

❚ FIGURE 7-5   The Great Red Spot This
image of the Great Red Spot, taken by the
Galileo spacecraft in 1996, has been color
enhanced to show more details of the dynamic activity taking place in
and around this giant storm. The counterclockwise circulation of gas
in the Great Red Spot takes about six days to make one rotation. The
clouds that encounter it are forced to pass around it, and when other
oval features are near it, the entire system becomes particularly
turbulent, like the batter in a two-bladed blender. (NASA/JPL)

86% of its atoms are hydrogen and 13% are helium. The remainder consist of molecular compounds such as methane (CH_4), ammonia (NH_3), and water vapor (H_2O). Keeping in mind that different elements have different masses, we can convert these percentages of atoms into the masses of various substances in Jupiter's atmosphere: 75% hydrogen, 24% helium, and 1% other substances. Because the interior contains more heavy elements than the surface, the overall mass distribution in Jupiter is calculated to be 71% hydrogen, 24% helium, and 5% all heavier elements.

The descent of the probe from the *Galileo* spacecraft into Jupiter's atmosphere revealed wind speeds of up to 600 km/h, higher-than-expected air density and temperature, and lower-than-expected concentrations of water, helium, neon, carbon, oxygen, and sulfur. This probably occurred because the probe descended into a particularly arid region of the atmosphere called a *hot spot,* akin to the air over a desert on Earth.

Observations from spacecraft visiting the Jovian system and the scientific model of Jupiter's atmosphere developed to explain these observations indicate that Jupiter has three major cloud layers (Figure 7-6). Apparently, because it descended through a hot spot, the *Galileo* probe failed to detect them.

The uppermost Jovian cloud layer is composed of crystals of frozen ammonia. Since these crystals and the frozen water in Jupiter's clouds are white, what chemicals create the subtle tones of brown, red, and orange? The answer is as yet unknown. Some scientists think that sulfur compounds, which can assume many different colors depending on their temperature, play an important role. Others think that phosphorus might be involved, especially in the Great Red Spot.

Astronomers first determined Jupiter's overall chemical composition from its average density—only 1330 kg/m^3—which implies that Jupiter is composed primarily of light-weight elements such as hydrogen and helium surrounding a relatively small core of metal and rock. (Recall the discussion of Jupiter's formation in Essentials II.) Jupiter's surface and mantle are entirely liquid. Because the vast majority of Jupiter's mass is composed of hydrogen, we cannot easily distinguish the planet's atmosphere from its "surface." However, at a distance of 150 km below the cloudtops, the pressure (about 10 atm) is enough to liquefy hydrogen. It was just at this distance below the cloudtops that the *Galileo* probe failed, presumably crushed by the high pressure of the atmosphere.

In introducing the solar system, we also noted that a young planet heats up as it coalesces. After it forms, radioactive elements continue to heat its interior. On Earth, this heat leaks out of the surface through volcanoes and other vents. Jupiter loses heat everywhere on its surface, because, unlike the Earth, it has no landmasses to block the heat loss.

As the heat from within Jupiter warms its liquid mantle, blobs of heated hydrogen and helium move upward. When these blobs reach the cloudtops, they give off their heat and descend back into the interior. (The same process, *convection,* drives the motion of the Earth's mantle and its tectonic plates, as well as soup simmering on a stove.) Jupiter's rapid, differential rotation draws the convecting gases into bands

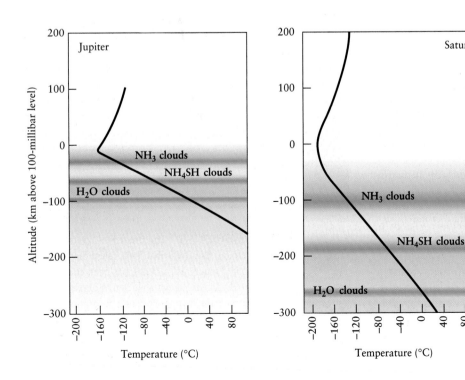

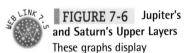

FIGURE 7-6 Jupiter's and Saturn's Upper Layers These graphs display temperature and pressure profiles of Jupiter's and Saturn's upper regions, as deduced from measurements at radio and infrared wavelengths. Three major cloud layers are shown in each, along with the colors that predominate at various depths. Data from the *Galileo* spacecraft indicate that Jupiter's cloud layers are not found at all locations around the planet; there are some relatively clear, cloud-free areas.

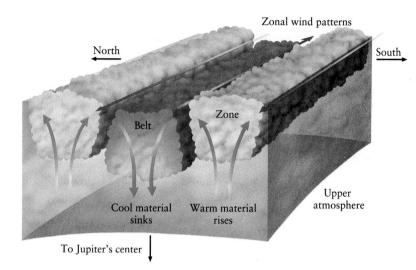

North

Zonal wind patterns

South

Belt

Zone

Cool material sinks

Warm material rises

Upper atmosphere

To Jupiter's center

FIGURE 7-7 Why Jupiter Has Belts and Zones The light-colored zones and dark-colored belts in Jupiter's atmosphere are regions of rising and descending gases, respectively. In the zones, gases warmed by heat from Jupiter's interior rise upward and cool, forming high-altitude clouds. In the belts, cooled gases descend and undergo an increase in temperature; the cloud layers seen there are at lower altitudes than in the zones. Jupiter's rapid differential rotation shapes these regions of rising and descending gas into bands parallel to the planet's equator. Differential rotation also causes the wind velocities at the boundaries between belts and zones to be predominantly to the east or west.

around the planet. Even through a small telescope, you can see on Jupiter dark, reddish bands called **belts** alternating with light-colored bands called **zones**. Jupiter's belts and zones result from the combined actions of the planet's convection and rapid rotation. The light zones are regions of hotter, rising gas, while the dark belts are regions of cooler, descending gas (Figure 7-7).

7-2 Jupiter's interior has four distinct regions

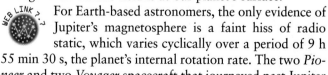

Because Jupiter is mostly hydrogen and helium, its average density is less than one-quarter that of the Earth. Yet the gravitational force created by its enormous bulk compresses and heats its interior so much that 20,000 km below the cloudtops the pressure is 3 million atmospheres. Below this depth, the pressure is high enough to transform hydrogen into **liquid metallic hydrogen.** Under such conditions hydrogen is a metal, meaning that it is an excellent conductor of electricity, like the copper wiring in a house. Electric currents running through this rotating, metallic region of Jupiter generate a powerful planetary magnetic field. At the cloud level of Jupiter, this field is 14 times stronger than Earth's field is at our planet's surface.

For Earth-based astronomers, the only evidence of Jupiter's magnetosphere is a faint hiss of radio static, which varies cyclically over a period of 9 h 55 min 30 s, the planet's internal rotation rate. The two *Pioneer* and two *Voyager* spacecraft that journeyed past Jupiter in the 1970s revealed the awesome dimensions of Jupiter's magnetosphere. The volume it engulfs is nearly 30 million kilometers across. It envelops the orbits of many of its moons. If Jupiter's magnetosphere were visible from Earth, it would cover an area in the sky 16 times larger than the full Moon.

Despite its preponderance of hydrogen and helium, Jupiter formed around a terrestrial (rock and metal) protoplanet. This core is only 4% of Jupiter's mass, which amounts to nearly 13 times the mass of the entire Earth. As discussed in Essentials II, this core formed early in Jupiter's life. Along with the rock and metal, there were almost certainly quantities of frozen water, carbon dioxide, methane, and ammonia. When astronomers talk about "ice," they generically refer to any or all of these compounds.

The tremendous crushing weight of the bulk of Jupiter above the core—equal to the mass of 305 Earths—compresses the terrestrial core down to a sphere only 20,000 km in diameter. By comparison, Earth's diameter is 12,756 km. At the same time, the pressure forced the lighter ices out of the rock and metal, thereby forming a shell of these "ices" between the solid core and the liquid metallic hydrogen layer. Calculations reveal that the temperature and pressure inside Jupiter should make these "ices" liquid there! The pressure at Jupiter's very center is calculated to be about 70 million atmospheres, and the temperature there is about 25,000 K, nearly 4 times hotter than the surface of the Sun.

7-3 Cometary fragments were observed to strike Jupiter

On July 7, 1992, a comet passed so close to Jupiter that the planet's gravitational tidal force ripped it into at least 21 pieces. The debris from this comet was first observed in March 1993 by comet hunters Gene and Carolyn Shoemaker and David Levy. (Because it was the ninth comet they had found together, it was named Shoemaker-Levy 9 in their honor.) Shoemaker-Levy 9 was an unusual comet that

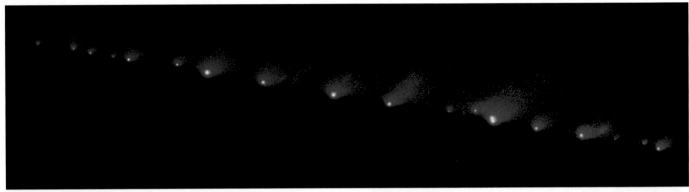

R I V U X G

 WEB LINK 7.8

FIGURE 7-8   Comet Shoemaker-Levy 9 Debris Approaching Jupiter   The comet was torn apart by Jupiter's gravitational force on July 7, 1992, fracturing into at least 21 pieces. Returning debris, shown here in May 1994, struck Jupiter between July 16 and July 22, 1994. (H. A. Weaver, T. E. Smith, STScI and NASA)

actually orbited Jupiter, rather than just orbiting the Sun. Calculations of the comet's orbit showed that the pieces would return to strike Jupiter between July 16 and July 22, 1994 (Figure 7-8).

Recall from Essentials II that impacts were extremely common in the first 800 million years of the solar system's existence. From more recent times, two chains of impact craters have been discovered on Earth, consistent with pieces of comets having hit our planet within the past 300 million years. However, it is very uncommon for pieces of space debris as large as several kilometers in diameter to collide with planets today. Therefore, the discovery that Shoemaker-Levy 9 would hit Jupiter created great excitement in the astronomical community. Seeing how a planet and a comet respond to such an impact would allow astronomers to deduce information about the planet's atmosphere and interior and also about the striking body's properties.

The impacts occurred as predicted, with most of Earth's major telescopes—as well as those on several spacecraft—watching closely (Figure 7-9). At least 20 fragments from Shoemaker-Levy 9 struck Jupiter, and 15 of them had detectable impact sites. The impacts resulted in fireballs some 10 km in diameter with temperatures of 7500 K, which is hotter than the surface of the Sun. Indeed, the largest fragment gave off as much energy as 600 million megatons of TNT, far more than the energy that could be released by all the nuclear weapons remaining on Earth, combined. Impacts were followed by crescent-shaped ejecta containing a variety of chemical compounds. Ripples or waves spread out from the impact sites through Jupiter's clouds in splotches that lasted for months.

The observations suggest that the pieces of comet did not penetrate very far into Jupiter's upper cloud layer. This fact, in turn, suggests that the pieces were not much larger than a kilometer in diameter. The ejecta from each impact included a dark plume that rose high into Jupiter's atmosphere. The darkness was apparently due to carbon compounds vaporized from the comet bodies. Also detected

R I V U X G   R I V U X G

FIGURE 7-9   Three Impact Sites of Comet Shoemaker-Levy 9 on Jupiter   Shown here are visible (left) and ultraviolet (right) images of Jupiter taken by the Hubble Space Telescope after three pieces of Comet Shoemaker-Levy 9 struck the planet. The astronomers had expected white remnants (the color of condensing ammonia or water vapor); the darkness of the impact sites may have come from carbon compounds in the comet debris. Note the auroras in the ultraviolet image. Auroras and lightning are common on Jupiter, due in part to the planet's strong magnetic fields and dynamic cloud motions. (NASA)

from the comet were water, sulfur compounds, silicon, magnesium, and iron. By watching how rapidly the impact debris traveled above the clouds, astronomers were able to determine that the winds some 250 km above the cloud layer move at speeds of 3600 km/h (more than 2200 mph).

JUPITER'S MOONS AND RINGS

Jupiter hosts at least 28 moons. Galileo was the first person to observe the four largest moons, in 1610, seen through his meager telescope as pinpoints of light. He called them the "Medicean stars" to attract the attention of the Medicis, rulers of Florence and wealthy patrons of the arts and sciences. To Galileo, the moons provided evidence supporting the then-controversial Copernican cosmology; at that time, Western theologians asserted that all cosmic bodies orbited the Earth. The fact that the Medicean stars orbited Jupiter raised grave concerns in some circles.

 To the modern astronomer, these moons are four extraordinary worlds, different both from the rocky terrestrial planets and from hydrogen-rich Jupiter. Now called collectively the **Galilean moons** or **Galilean satellites,** they are named after the mythical lovers and companions of the Greek god Zeus (called Jupiter by the early Romans). From the closest moon outward, they are Io, Europa, Ganymede, and Callisto.

These four worlds were photographed extensively by the *Voyager 1* and *Voyager 2* flybys and by the *Galileo* spacecraft (Figure 7-10). The two inner Galilean satellites, Io and Europa, are approximately the same size as our Moon. The two outer satellites, Ganymede and Callisto, are comparable in size to Mercury. Figure 7-10 presents comparative information about these six bodies.

7-4 Io's surface is sculpted by volcanic activity

Sulfury Io is among the most exotic moons in our solar system (Figure 7-11). With a density of 3570 kg/m³, it is neither terrestrial nor "Jovian" in chemical composition. Its density is similar to that of Earth's surface, suggesting that most of Io is rock, rather than denser metals or lighter water. It zooms through its orbit of Jupiter once every 1.8 days. Like our Moon, Io is in synchronous rotation with its planet. Like the Earth, Io has a sizable iron core—it extends halfway out from the moon's center.

Images of Io reveal giant plumes rich in sulfur dioxide emitted by geysers, similar to Old Faithful on Earth, such as the two labeled in Figure 7-11a. Most of this ejected material falls back onto Io's surface; the rest is moving fast enough to escape into space. Although they are geysers, the plumes are actually emitted through volcanoes (Figure 7-11b), which also emit basaltic lava flows rich in magnesium and iron. Io's volcanoes are named after gods and goddesses associated with fire in Greek, Norse, Hawaiian, and other mythologies. Io also has numerous black "dots" on its surface, which apparently are dormant volcanic vents. Old lava flows radiate from many of these locations, which are typically 10 to 50 km in diameter and cover 5% of Io's surface. Observations suggest that Io has about 300 active volcanoes emitting 10 trillion tons of matter each year in plumes up to 500 km high. That is enough material to resurface Io to a depth of 1 meter each century.

> **Insight into Science**   **What's in a Name?** Following up on the preconceptions that common words create (see the Insight into Science "Imagine the Moon" in Chapter 6), objects with familiar names often have different characteristics than we expect. We tend to envision moons as inert, airless, lifeless, dry places similar to our Moon. As we will see throughout this chapter, starting with Io, some moons have active volcanoes, atmospheres, and probably vast, underground, liquid water oceans.

Just before their discovery, the existence of active volcanoes on Io was predicted from analysis of the gravitational forces to which that moon is subjected. As it rapidly orbits Jupiter, Io repeatedly passes between Jupiter and one or another of the other Galilean satellites. These moons pull Io farther from Jupiter, changing the tidal forces acting on it from the planet. As the distance between Io and Jupiter varies, the resulting tidal stresses alternately squeeze and flex the moon. In turn, this constant tidal stressing heats Io's interior through friction, generating as much energy inside Io as the detonation of 2400 tons of TNT every second. Gas and molten rock eventually make their way to the moon's surface, where they are ejected.

Satellite instruments have identified sulfur and sulfur dioxide in the material erupting from Io's volcanoes. Sulfur is normally bright yellow. If heated and suddenly cooled, however, it forms molecules that assume a range of colors, from orange and red to black, which accounts for Io's tremendous range of colors (see Figure 7-11). Sulfur dioxide (SO_2) is an acrid gas commonly discharged from volcanic vents here on Earth and, apparently, on Venus. When eruptions on Io release this gas into the cold vacuum of space, it crystallizes into white flakes, which fall onto the surface and account for the moon's whitish deposits.

The *Galileo* spacecraft detected an atmosphere around Io. Composed of oxygen, sulfur, and sulfur dioxide, it is only one-billionth as dense as the air we breathe. Io's atmosphere can sometimes be seen to glow blue, red, or green, depending on the gases involved. Gases ejected from Io's volcanoes have also been observed extending out into space and forming a doughnut-shaped region around Jupiter called the *Io torus,* about which we will say more shortly.

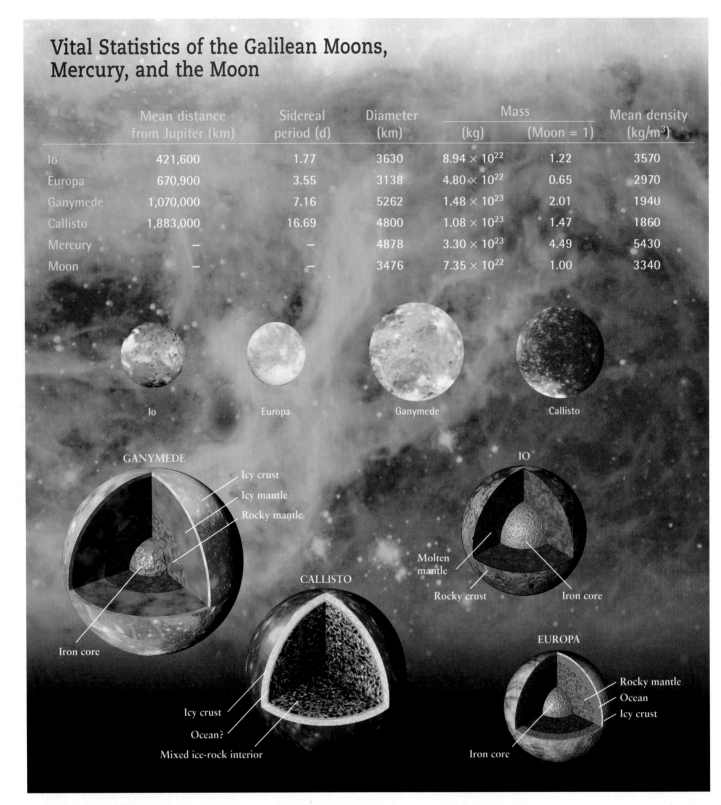

Vital Statistics of the Galilean Moons, Mercury, and the Moon

	Mean distance from Jupiter (km)	Sidereal period (d)	Diameter (km)	Mass (kg)	Mass (Moon = 1)	Mean density (kg/m³)
Io	421,600	1.77	3630	8.94×10^{22}	1.22	3570
Europa	670,900	3.55	3138	4.80×10^{22}	0.65	2970
Ganymede	1,070,000	7.16	5262	1.48×10^{23}	2.01	1940
Callisto	1,883,000	16.69	4800	1.08×10^{23}	1.47	1860
Mercury	–	–	4878	3.30×10^{23}	4.49	5430
Moon	–	–	3476	7.35×10^{22}	1.00	3340

Io   Europa   Ganymede   Callisto

GANYMEDE
Icy crust
Icy mantle
Rocky mantle
Iron core

IO
Molten mantle
Rocky crust
Iron core

CALLISTO
Icy crust
Ocean?
Mixed ice-rock interior

EUROPA
Rocky mantle
Ocean
Icy crust
Iron core

FIGURE 7-10 The Galilean Satellites The four Galilean satellites are shown here to the same scale. Io and Europa have diameters and densities comparable to our Moon and are composed primarily of rocky material. Ganymede and Callisto are roughly as big as Mercury, but their low average densities indicate that each contains a thick layer of water and ice. The cross-sectional diagrams of the interiors of the four Galilean moons show the probable internal structures of the moons based on their average densities and on information from the *Galileo* mission. (NASA and NASA/JPL)

WEB LINK 7.9

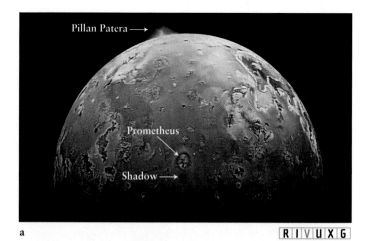

Pillan Patera ⟶

Prometheus

Shadow ⟶

a   RIVUXG

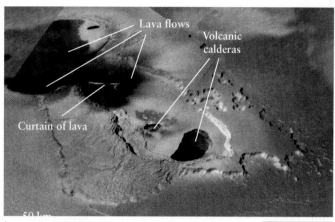

Lava flows

Volcanic
calderas

Curtain of lava

50 km

b   RIVUXG

FIGURE 7-11   **Io** **(a)** This close-up view
was taken by the *Galileo* spacecraft in 1997.
Scientists believe that the range of colors results
from surface deposits of sulfur ejected from Io's numerous volcanoes.
The plume on the top, from the active Pillan Patera volcano, rises more
than 190 km above the moon's surface. The plume from the volcano

Prometheus rises up 100 km, casting a visible shadow. Prometheus
has been active in every image taken of Io since the *Voyager* flybys of
1979. **(b)** Photographed in 1999 and then 2000 (shown here), the
ongoing lava flow from this volcanic eruption at Tvashtar Catena has
considerably altered this region of Io's surface. (a: NASA/JPL; b: University
of Arizona/JPL/NASA)

7-5   Europa apparently harbors liquid water below its surface

Images of Europa's ice and rock surface from *Voyager 2* and
the *Galileo* spacecraft suggest that Jupiter's second-closest
Galilean moon contains liquid water (Figure 7-12). Europa
orbits Jupiter every 3½ days and, like Io, is in synchronous
rotation. The changing gravitational tug of Io creates stress
inside Europa similar to the magma-generating distortion
that Io undergoes. The stress may create enough heat inside
Europa to keep the water just a few kilometers below the
moon's ice and rock surface in a liquid state.

Strong evidence for liquid water inside Europa comes
from the ice floes seen in Figure 7-12. The surface features
are similar to those seen in the Arctic region on Earth. The
movement of the surface features creating the floes, along
with swirls, strips, and ridges, appears to be driven by circulating water underneath the moon's surface, creating tectonic plate motion and tidal flexing. Also indicative of
water is the moon's reddish color. The coloring may be due
to salt deposits left after liquid water rose to the surface and
evaporated.

Galileo spacecraft images suggest that some of Europa's
features have moved within the past few million years, and

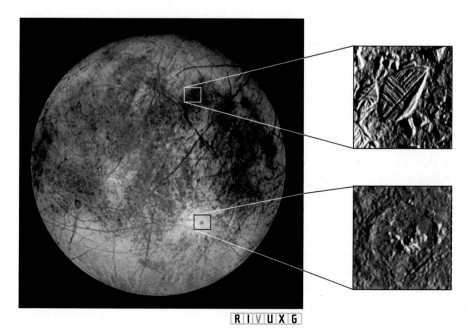

RIVUXG

FIGURE 7-12   **Europa** Imaged
by the *Galileo* spacecraft, Europa's ice
surface is covered by numerous streaks
and cracks that give the satellite a fractured
appearance. The streaks are typically 20 to 40 km
wide. The upper inset shows the thin, disturbed
icy crust of Europa, colored by dust from the
impact crater Pwyll. The grooves and ridges in this
image are typically 100 m across. The lower inset
shows details of Pwyll Crater, 50 km in diameter.
(Jet Propulsion Laboratory/NASA; upper inset: Planetary
Image Research Laboratory/University of Arizona/JPL/
NASA; lower inset: NASA)

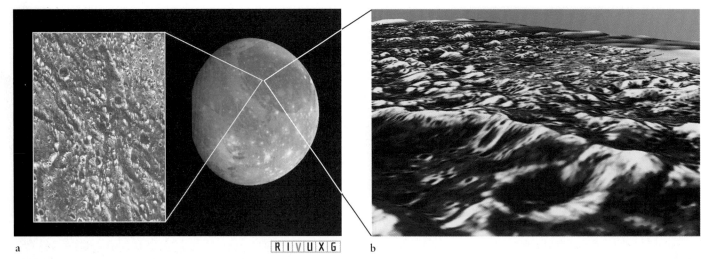

a   R I V U X G   b

FIGURE 7-13 **Ganymede** This side of Ganymede is dominated by a huge, dark, circular region called Galileo Regio, which is the largest remnant of Ganymede's ancient crust. Darker areas of the moon are older; lighter areas are younger, tectonically deformed regions. The white areas in and around some craters indicate the presence of water ice. Inset **(a)** is about 46 × 64 km (29 × 38 mi). Note in this *Galileo* spacecraft image the deep furrows in the moon's icy crust that probably resulted from crustal movement due to impacts and tectonic plate motion. The perspective in inset **(b)** was created by combining information from two *Galileo* images. (NASA/JPL)

perhaps are still in motion. Indeed, the chaotic surface revealed by *Galileo* is interpreted as having formed as a result of volcanism on Europa, strengthening the belief that this moon still has a liquid water layer. Replenishment of the surface by tectonic plate motion would explain why only a few small impact craters, such as the crater Pwyll (see Figure 7-12), have survived.

Europa's average density of 2970 kg/m³ is slightly less than Io's. A quarter of its mass may be water. It also has a metallic core of much higher density. In 1995 astronomers discovered an extremely thin atmosphere containing molecular oxygen surrounding Europa. The density of this gas is about 10^{-11} times the density of the air we breathe. The oxygen may come from water molecules broken up on the moon's surface by ultraviolet radiation from the Sun.

7-6 Ganymede is larger than Mercury

Ganymede is the largest satellite in the solar system (Figure 7-13). Its diameter is larger than Mercury's, although its density of 1940 kg/m³ is much less than that of Mercury. It also has a permanent magnetic field that is twice as strong as Mercury's field. Ganymede orbits Jupiter in synchronous rotation once every 7.2 days. Like its neighbor Europa, Ganymede has an iron-rich core, a rocky mantle, a liquid water ocean, a thin atmosphere, and a covering of dirty ice.

The existence of the ocean is implied by the discovery of a second, changing magnetic field around Ganymede that is generated by Jupiter's magnetic field. As Ganymede orbits Jupiter, the planet's powerful magnetic field creates an electrical current inside the moon, which in turn creates Ganymede's varying magnetic field. (The same effect is used to create electrical currents in electrical power stations here on Earth.) The best explanation of why the current flows inside Ganymede is that liquid salt water exists there; salt water is a good conductor of electricity. This implies, of course, the presence of a liquid ocean. Furthermore, salts have been observed on Ganymede's surface. They were apparently carried upward and deposited there as water leaked out and froze.

Like our Moon, Ganymede has two very different kinds of terrain. Dark, polygon-shaped regions are its oldest surface features, as judged by their numerous craters. Light-colored, heavily grooved terrain is found between the dark, angular islands. These lighter regions are much less cratered and therefore younger. Ganymede's grooved terrain consists of parallel mountain ridges up to 1 km high and spaced 10 to 15 km apart. These features suggest that the process of plate tectonics may have dominated Ganymede's early history. But unlike Europa, where tectonic activity still occurs today, tectonics on Ganymede bogged down 3 billion years ago as the satellite's crust froze solid.

Another mechanism that may have created Ganymede's large-scale features is a bizarre property of water. Unlike most liquids, which shrink upon solidifying, water expands when it freezes. Seeping up through cracks in Ganymede's original crust, water thus forced apart fragments of that crust. This process could have produced jagged, dark islands of old crust separated by bands of younger, light-colored, heavily grooved ice. Topping off Ganymede's varied features is the discovery that auroras occur there.

7-7 Callisto bears the scars of a huge asteroid impact

Callisto is Jupiter's outermost Galilean moon. It orbits Jupiter in 16.7 days, and, like the other Galilean moons, Callisto's rotation is synchronous. Callisto is 91% as big and 96% as dense as Ganymede. It has a thin atmosphere of hydrogen and carbon dioxide. Like Ganymede, Callisto apparently harbors a substantial liquid water ocean. Its presence is again inferred by Callisto's changing magnetic field. The heat that keeps the ocean liquid apparently comes from energy released by radioactive decay inside the moon.

While numerous large impact craters are scattered over Callisto's dark, ancient, icy crust, it has very few craters smaller than 100 m across (Figure 7-14). Astronomers speculate that the smaller craters have disintegrated. Unlike Ganymede and Europa, Callisto has no younger, grooved terrain. The absence of grooved terrain suggests that tectonic activity never began there: The satellite simply froze too rapidly. It is bitterly cold on Callisto's surface. *Voyager* instruments measured a noontime temperature of 155 K (−180°F), and the nighttime temperature plunges to 80 K (−315°F).

Callisto carries the cold, hard evidence of what happens when one astronomical body strikes another. *Voyager 1* photographed the huge impact basin, Valhalla, on Callisto (see Figure 7-14). An asteroid-sized object produced Valhalla Basin, which is located on Callisto's Jupiter-facing hemisphere. Like throwing a rock into a calm lake, ripples ran out

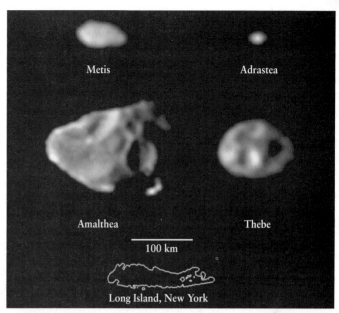

R I V U X G

 FIGURE 7-15 Irregularly Shaped Inner Moons The four known inner moons of Jupiter are significantly different than the Galilean satellites. They are roughly oval-shaped bodies. While craters have not yet been resolved on Adrastea and Metis, their irregular shapes strongly suggest that they are cratered. All four moons are named for characters in mythology relating to Jupiter (in Greek mythology, Zeus). (NASA/JPL, Cornell University)

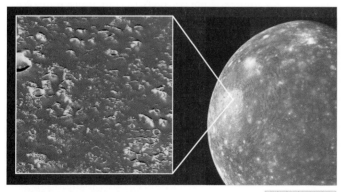

R I V U X G

 FIGURE 7-14 Callisto The outermost Galilean satellite is almost exactly the same size as Mercury. Numerous craters pockmark Callisto's icy surface. Note the series of faint, concentric rings that cover the left third of the image. These rings outline a huge impact basin called Valhalla, which dominates the Jupiter-facing hemisphere of this frozen, geologically inactive world. The inset, a high resolution *Galileo* image, shows a portion of Valhalla. Most of the very smallest craters in this close-up view have been completely obliterated, and the terrain between the craters has been blanketed by dark dusty material. The area shown is about 11 km (7 mi) across. (Left: Arizona State University/JPL/NASA; right: JPL/NASA)

from the impact site along Callisto's surface, cracking the surface and freezing into place for eternity. The largest remnant rings surrounding the impact crater have diameters of 3000 km. In 2001, the *Galileo* spacecraft revealed spires 80–100 m high on Callisto. These are also believed to have been created by an impact, perhaps the same one. The probable interiors of the Galilean moons are shown in Figure 7-10.

7-8 Other debris orbits Jupiter as smaller moons and ringlets

Besides the four Galilean moons, Jupiter has at least 24 other moons, a set of tenuous ringlets, and the doughnut-shaped Io torus of electrically charged gas particles. The non-Galilean moons are all irregular in shape and smaller than 150 km in diameter. Four of these moons are inside Io's orbit (Figure 7-15); all the other confirmed and suspected moons are outside Callisto's orbit. The Galilean moons along with the smaller moons closer to Jupiter and six of the outer moons orbit in the same direction that Jupiter rotates (**prograde orbits**). The remaining outer moons revolve in the opposite direction (**retrograde orbits**). The outer ones appear to be

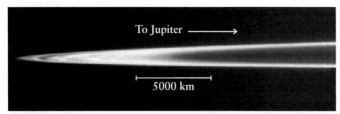

To Jupiter ⟶

5000 km

a   R I V U X G

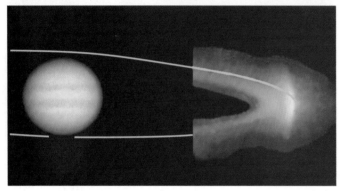

b   R I V U X G

FIGURE 7-16 Jupiter's Ring and Torus (a) A portion of Jupiter's faint ring system, photographed by *Voyager 2.* The ring is probably composed of tiny rock fragments. The brightest portion of the ring is about 6000 km wide. The outer edge of the ring is sharply defined, but the inner edge is somewhat fuzzy. A tenuous sheet of material extends from the ring's inner edge all the way down to the planet's cloudtops. **(b)** Io's torus (also called Io's plasma torus because the gas particles in it are charged—a plasma). This infrared image was obtained from an Earth-based telescope. The colors indicate temperature: Purple shows emission from hot sulfur ions; green shows emission from cooler sulfur ions. These ions were ejected from Io's volcanoes. An artist has added the line showing the full range of the Io torus, which fills a doughnut-shaped volume around Jupiter. (a: NASA/JPL, Cornell University; b: Courtesy of J. Trauger)

captured asteroids, while the inner ones are probably smaller pieces broken off a larger body.

As astronomers predicted, the cameras aboard *Voyager 1* discovered ringlets around Jupiter. There are three, the brightest of which is seen in Figure 7-16a, a *Voyager 2* image. They are the darkest, simplest set of rings in the solar system. They consist of very fine dust particles that are continuously kicked out of orbit by radiation from Jupiter and the Sun. Therefore, these rings are being replenished by material from Io and the other moons that has been knocked free by impacts from tiny pieces of interplanetary space debris. As we will see, Jupiter is only the first of four planets with rings.

A doughnut-shaped region of electrically charged gas particles, a *plasma*, orbits Jupiter in the same orbit as Io

(Figure 7-16b). Called the Io torus, it consists of sulfur and oxygen ions (charged atoms) along with free electrons. These particles were ejected by Io's geysers and they are held in orbit by Jupiter's strong magnetic field. Guided by the field, some of this matter spirals toward Jupiter, thereby creating the aurora seen there (see Figure 7-9).

SATURN

Saturn, with its ethereal rings, presents the most spectacular image of all the planets (Figure 7-17). Giant Saturn has 95 times as much mass as the Earth, making it second in mass and size to Jupiter. Like Jupiter, Saturn has a thick, active atmosphere composed predominantly of hydrogen. It also has a strong magnetic field. Ultraviolet images from the Hubble Space Telescope reveal auroras around Saturn (see Chapter 7 opener) like those seen around Jupiter (see Figure 7-9). Saturn's vital statistics are listed in Figure 7-17.

7-9   Saturn's surface and interior are similar to those of Jupiter

Partly obscured by the thick, hazy atmosphere above them, Saturn's clouds lack the colorful contrast visible on Jupiter. Nevertheless, photographs do show faint stripes in Saturn's atmosphere similar to Jupiter's belts and zones (Figure 7-18). Changing features there show that Saturn's atmosphere, too, has differential rotation—ranging from 10 hours and 14 minutes at the equator to 10 hours and 40 minutes at high latitudes. As on Jupiter, some of the belts and zones move eastward, while others move westward. Although Saturn lacks a long-lived spot like Jupiter's Great Red Spot, it does have storms, including a major new one discovered by the Hubble Space Telescope in November 1994 (Figure 7-19).

Saturn's atmosphere is composed of the same basic gases as Jupiter. However, because its mass is lower than Jupiter's, Saturn's gravitational force on its atmosphere is less. Therefore, Saturn's atmosphere is more spread out than that of its larger neighbor (see Figure 7-6). Astronomers infer that Saturn's interior structure resembles Jupiter's. A layer of molecular hydrogen just below the clouds surrounds a mantle of liquid metallic hydrogen, liquid "ices," and a solid, terrestrial core (Figure 7-20).

Because of its smaller mass, Saturn's interior is also less compressed than Jupiter's. Saturn's rocky core is larger, and the pressure there is insufficient to convert as much hydrogen into a liquid metal. Saturn's rocky core is about 32,000 km in diameter, while its layer of liquid metallic hydrogen is 12,000 km thick (see Figure 7-20). To alchemists, Saturn was associated with the extremely dense element lead. This is wonderfully ironic in that at 690 kg/m^3, Saturn is the least dense body in the entire solar system.

Saturn's Vital Statistics

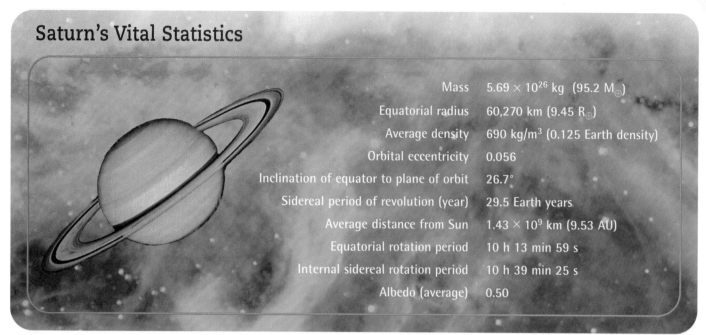

Mass	5.69×10^{26} kg (95.2 $M_\oplus$)
Equatorial radius	60,270 km (9.45 $R_\oplus$)
Average density	690 kg/m³ (0.125 Earth density)
Orbital eccentricity	0.056
Inclination of equator to plane of orbit	26.7°
Sidereal period of revolution (year)	29.5 Earth years
Average distance from Sun	1.43×10^9 km (9.53 AU)
Equatorial rotation period	10 h 13 min 59 s
Internal sidereal rotation period	10 h 39 min 25 s
Albedo (average)	0.50

R I V U X G

FIGURE 7-17 **Saturn** *Voyager 2* sent back this image when the spacecraft was 34 million kilometers from Saturn. Note that you can see the planet through the gaps in the rings. This creates the illusion that the gaps are empty. (NASA)

R I V U X G

FIGURE 7-18 **Belts and Zones on Saturn** *Voyager 1* took this view of Saturn's cloudtops at a distance of 1.8 million kilometers. Note that there is little swirling structure and substantially less contrast between belts and zones on Saturn than there is on Jupiter. (NASA)

R I V U X G

FIGURE 7-19 **A New Storm on Saturn** An arrowhead-shaped storm was discovered on Saturn in July 1994. Located near Saturn's equator, it stretches 12,700 km across, making its diameter roughly the same as that of the Earth. (Reta Beebe, New Mexico State University/ D. Gilmore and L. Bergeron, STScI/NASA)

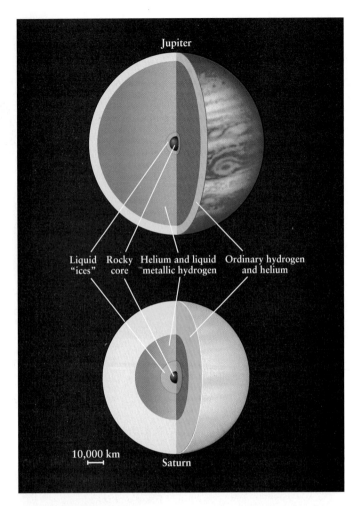

Jupiter

Liquid "ices" | Rocky core | Helium and liquid metallic hydrogen | Ordinary hydrogen and helium

10,000 km

Saturn

FIGURE 7-20 Cutaways of the Interiors of Jupiter and Saturn The interiors of both Jupiter and Saturn are believed to have four regions: a terrestrial core, a liquid "ice" shell, a metallic hydrogen shell, and a normal liquid hydrogen mantle. Their atmospheres are thin layers above the normal hydrogen, which boils upward, creating the belts and zones.

7-10 Saturn's spectacular rings are composed of fragments of ice and ice-coated rock

Even when viewed through a telescope from the vicinity of the Earth, Saturn's magnificent rings are among the most spectacular objects in the solar system. They are tilted 27° from Saturn's plane of orbit, so sometimes they are nearly edge-on to us, making them virtually impossible to see (Figure 7-21). Even when the rings are at their maximum tilt from our perspective, Saturn is so far away that our best Earth-mounted telescopes can reveal only their largest features. But even these have turned out to hold surprises. In 1675, Giovanni Cassini discovered one remark-

able feature—a dark division in the rings. This 5000-km-wide gap, called the **Cassini division**, separates the dimmer **A ring** from the brighter **B ring,** which lies closer to the planet. By the mid-1800s, astronomers using improved telescopes detected a faint **C ring** just inside the B ring and barely visible in Figure 7-19 (see Figure 7.22a for more detail). The Cassini division exists because the gravitational force from Saturn's moon Mimas combines with the gravitational force from the planet to keep the region clear of debris. Whenever matter drifts into the Cassini division, Mimas (orbiting at a different rate than the matter in the Cassini division) periodically exerts a force on this matter, thereby forcing it out of the division. This effect is called a **resonance** and is similar to what happens when you push someone on a swing at the right time and so enable them to go higher and higher.

A second gap exists in the outer portion of the A ring (again, barely visible in Figure 7-19), named the **Encke division** (Figure 7-22a), after the German astronomer Johann Franz Encke, who allegedly saw it in 1838. (Many astronomers have argued that Encke's report was erroneous, because his telescope was inadequate to resolve such a narrow gap.) The first undisputed observation of the 270-km-wide division was made by the American astronomer James Keeler in the late 1880s, with the newly constructed 36-in. refractor at the Lick Observatory in California. Unlike the Cassini division, the Encke division is kept clear because a small moon, Pan, orbits within it.

Pictures from the *Voyager* spacecraft show that Saturn's rings are remarkably thin—less than 2 km thick according to recent estimates. This is amazing when you consider that the total ring system has a width (from inner edge to outer edge) of more than 89,000 km.

Because Saturn's rings are very bright (albedo = 0.80), the particles that form them must be highly reflective. Astronomers had long suspected that the rings consist of ice and ice-coated rocks. Spectra taken by Earth-based observatories and by the *Voyager* spacecraft have confirmed this suspicion. Because the temperature of the rings ranges from 93 K (−290°F) in the sunshine to less than 73 K (−330°F) in Saturn's shadow, frozen water is in no danger of melting or evaporating from the rings.

Saturn's rings are slightly salmon-colored. This coloring suggests that they contain traces of organic molecules, which often have similar hues. It appears that the rings have gained this material by being bombarded with debris from the outer solar system.

To determine the size of the particles in Saturn's rings, *Voyager* scientists measured the brightness of different rings from many angles as the spacecraft flew past the planet. They also measured changes in radio signals received from the spacecraft as it passed behind the rings. The largest particles in Saturn's rings are roughly 10 m across, although snowball-sized particles about 10 cm in diameter are more abundant than larger pieces. High-resolution images from *Voyager* revealed that the ring structure seen from Earth

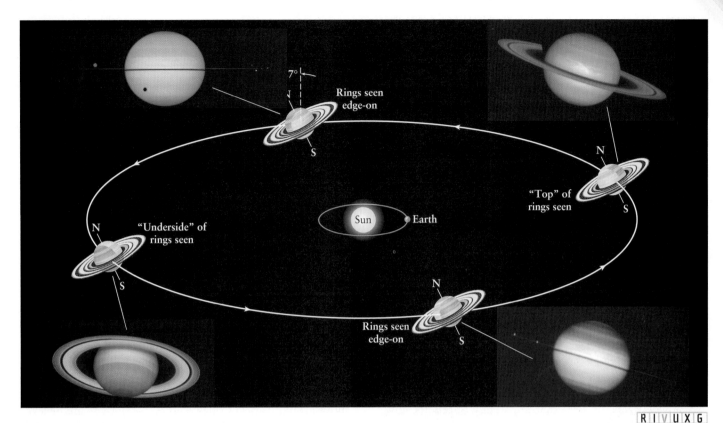

R I V U X G

FIGURE 7-21   **Saturn from the Earth** Saturn's rings are aligned with its equator, which is tilted 27° from the plane of Saturn's orbit around the Sun. Therefore, Earth-based observers see the rings at various angles as Saturn moves around its orbit. The plane of Saturn's rings and equator keeps the same orientation in space as the planet goes around its orbit, just as the Earth does as it orbits the Sun. The accompanying Earth-based photographs show how the rings seem to disappear entirely about every 15 years. (Top left: E. Karkoschka/U. of Arizona Lunar and Planetary Lab and NASA; bottom left: AURA/STScI/NASA; bottom right and top right: A. Bosh/Lowell Obs. and NASA)

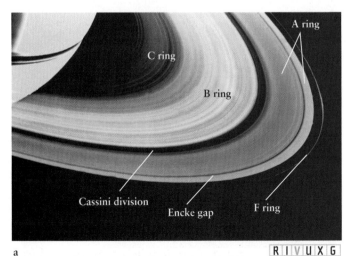

a   R I V U X G

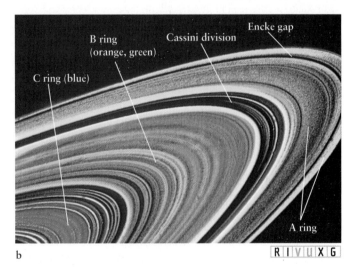

b   R I V U X G

FIGURE 7-22   **Numerous Thin Ringlets Constitute Saturn's Rings** **(a)** This *Voyager 1* image hints that Saturn's rings contain numerous ringlets. Note the shadow of the rings on the planet and the gap in the shadow from Cassini's division. **(b)** Details of Saturn's rings are visible in this photograph sent back by *Voyager 2*. It shows that the rings are actually composed of thousands of closely spaced ringlets. The colors are exaggerated by computer processing to show the different ringlets more clearly. (NASA/JPL)

actually consists of hundreds upon hundreds of closely spaced thin bands, or **ringlets,** of particles (Figure 7-22b). Furthermore, myriad dust-sized particles fill the Cassini and Encke divisions.

The *Voyager* cameras also sent back the first high-quality pictures of the **F ring,** a thin set of ringlets just beyond the outer edge of the A ring. Two tiny satellites following orbits on either side of the F ring serve to keep the ring intact (Figure 7-23). The outer of the two satellites orbits Saturn at a slower speed than do the ice particles in the ring. As the ring particles pass near it, they receive a tiny, backward gravitational tug, which slows them down, causing them to fall into orbits a bit closer to Saturn. Meanwhile, the inner satellite orbits the planet faster than the F ring particles. Its gravitational force pulls them forward and nudges the particles into a higher orbit. The combined effect of these two satellites is to focus the icy particles into a well-defined, narrow band about 100 km wide.

Because of their confining influence, these two moons, Prometheus and Pandora, are called **shepherd satellites** or **shepherd moons.** Among the most curious features of the F ring is that the ringlets are sometimes braided (Figure 7-24) and sometimes separate.

Saturn's shell of liquid metallic hydrogen produces a planetwide magnetic field that apparently affects its rings. Saturn's slower rotation and much smaller volume of liquid metallic hydrogen produce a surface magnetic field only about $2/3$ as strong as Earth's surface field. Data from spacecraft show that Saturn's magnetosphere contains radiation

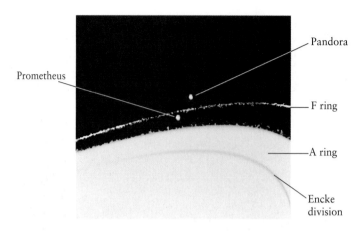

R I V U X G

FIGURE 7-23 The F Ring and Its Two Shepherds Two tiny satellites, Prometheus and Pandora, each measuring about 50 km across, orbit Saturn on either side of the F ring. The gravitational effects of these two shepherd satellites confine the particles in the F ring to a band about 100 km wide. (NASA)

belts similar to those of Earth. Furthermore, dark **spokes** move around Saturn's rings (Figure 7-25); these are believed to be created by the magnetic field, which lifts charged particles out of the plane in which the rings orbit. Spreading the particles out decreases the light scattered from them and therefore makes the rings appear darker.

R I V U X G

FIGURE 7-24 Braided F Ring This photograph from *Voyager 1* shows several strands, each measuring roughly 10 km across, that comprise the F ring. The total width of the F ring is about 100 km. (NASA)

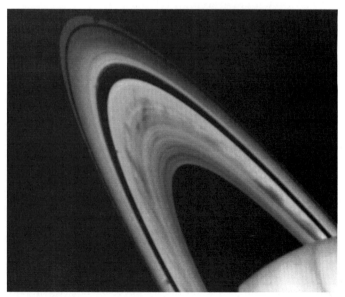

R I V U X G

FIGURE 7-25 Spokes in Saturn's Rings Believed to be caused by Saturn's magnetic field temporarily lifting particles out of the ring plane, these dark regions move around the rings like the spokes on a rotating wheel. (NASA)

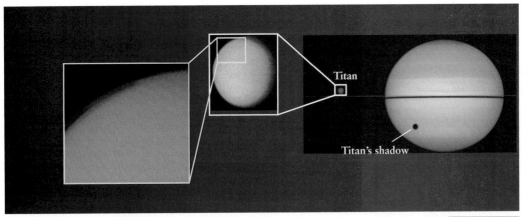

FIGURE 7-26

Titan These views of Titan were taken by *Voyager 2.* Very few features are visible in the thick, unbroken haze that surrounds this large satellite. The main haze layer is located nearly 300 km above Titan's surface. (Left: JPL/NASA; center: NASA; right: Erich Karkoschka, LPL/STScI/NASA)

R I V U X G

7-11 Titan has a thick, opaque atmosphere rich in nitrogen, methane, and other hydrocarbons

Only 7 of Saturn's 30 known moons are spherical. The rest are oblong, suggesting that they are captured asteroids. The 12 moons discovered in 2001 move in clumps, indicating that they may be pieces of a larger moon that was broken up by impacts. Saturn's largest moon, Titan, is second in size to Ganymede among the moons of the solar system, and the only moon to have a dense atmosphere. About 10 times more gas lies above each square meter of Titan's surface than lies above each square meter of the Earth.

Christiaan Huygens discovered Titan in 1655, the same year he proposed that Saturn has rings. By the early 1900s, several scientists had begun to suspect that Titan might have an atmosphere, because it is cool enough and massive enough to retain heavy gases.

Because of its atmosphere, Titan was a primary target for the *Voyager* missions. To everyone's disappointment, the *Voyagers* spent hour after precious hour sending back featureless images (Figure 7-26). Unexpectedly, Titan's thick cloud cover completely blocked any view of its surface. The same dense haze allows little sunlight to penetrate—that moon's surface must be a dark, gloomy place.

Voyager data indicated that roughly 90% of Titan's atmosphere is nitrogen. Most of this nitrogen probably formed from the breakdown of ammonia (NH_3) by the Sun's ultraviolet radiation into hydrogen and nitrogen atoms. Because Titan's gravity is too weak to retain hydrogen, this gas has escaped into space, leaving behind ample nitrogen. The unbreathable atmosphere is about 4 times as dense as Earth's.

The second most abundant gas on Titan is methane, a major component of natural gas. Sunlight interacting with methane induces chemical reactions that produce a variety of other carbon-hydrogen compounds, or **hydrocarbons.** For example, spacecraft have detected small amounts of ethane

(C_2H_6), acetylene (C_2H_2), ethylene (C_2H_4), and propane (C_3H_8) in Titan's atmosphere.

Ethane, the most abundant of these compounds, condenses into droplets as it is produced and falls to Titan's surface to form a liquid. Enough ethane may exist to create rivers, lakes, and even oceans on Titan (Figure 7-27). Nitrogen combines with these hydrocarbons to produce other compounds. Although one of these compounds, hydrogen cyanide (HCN), is a poison, some of the others are the building blocks of life's organic molecules.

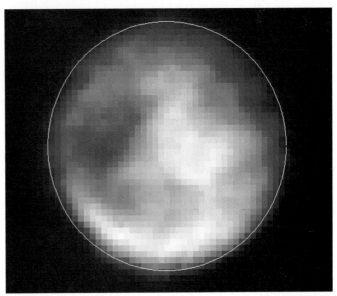

R I V U X G

FIGURE 7-27 **Surface Features on Titan** The Keck I Telescope, using an infrared camera, was able to peer through Titan's haze (see Figure 7-26). The dark region, emitting little infrared, may be a sea of liquid methane or more complex hydrocarbons, while the brighter regions may be rock and water ice. The white circle indicates Titan's circumference. (Lawrence Livermore National Laboratory)

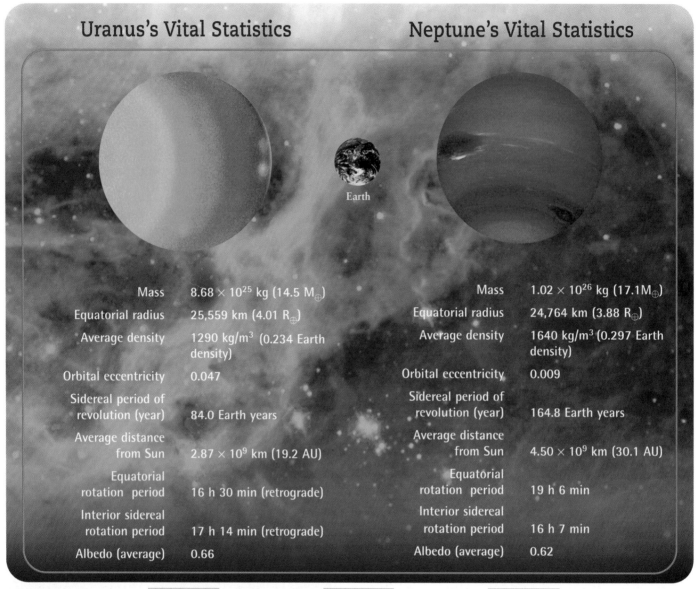

Uranus's Vital Statistics

Mass	8.68×10^{25} kg (14.5 $M_\oplus$)
Equatorial radius	25,559 km (4.01 $R_\oplus$)
Average density	1290 kg/m^3 (0.234 Earth density)
Orbital eccentricity	0.047
Sidereal period of revolution (year)	84.0 Earth years
Average distance from Sun	2.87×10^9 km (19.2 AU)
Equatorial rotation period	16 h 30 min (retrograde)
Interior sidereal rotation period	17 h 14 min (retrograde)
Albedo (average)	0.66

Neptune's Vital Statistics

Mass	1.02×10^{26} kg (17.1$M_\oplus$)
Equatorial radius	24,764 km (3.88 $R_\oplus$)
Average density	1640 kg/m^3 (0.297 Earth density)
Orbital eccentricity	0.009
Sidereal period of revolution (year)	164.8 Earth years
Average distance from Sun	4.50×10^9 km (30.1 AU)
Equatorial rotation period	19 h 6 min
Interior sidereal rotation period	16 h 7 min
Albedo (average)	0.62

Earth

R I V U X G   R I V U X G   R I V U X G

FIGURE 7-28 **Uranus, Earth, and Neptune** These images of Uranus, the Earth, and Neptune are to the same scale. Uranus and Neptune are quite similar in mass, size, and chemical composition. Both planets are surrounded by thin, dark rings, quite unlike Saturn's, which are broad and bright. The clouds on the right of Uranus are each the size of Europe. (NASA)

Some molecules can join together in long, repeating molecular chains to form substances called **polymers,** of which plastics are the best known example. Many of the hydrocarbons and carbon-nitrogen compounds in Titan's atmosphere can form such polymers. Scientists hypothesize that droplets of lighter polymers remain suspended in Titan's atmosphere to form a mist, while heavier polymer particles settle down onto Titan's surface. Using the Keck I telescope in 1999, astronomers observed infrared emissions from Titan's surface. The bright areas in Figure 7-27 are inferred to be islands or continents of rock and ice, while the dark areas are believed to be ethane oceans.

We have little reason to suspect that life exists on Titan, however; its surface temperature of 95 K (−288°F) is prohibitively cold. Nevertheless, a more detailed study of the chemistry of Titan may shed light on the origins of life on Earth.

Just as we have recently revisited Jupiter, we are sending back a spacecraft to Saturn. Launched in October 1997, spacecraft *Cassini* is scheduled to go into orbit in July 2004 on a four-year mission to study the ringed planet and the bodies that orbit it. Like *Galileo* at Jupiter, *Cassini* will deploy a probe, named *Huygens,* that will descend into Saturn's atmosphere.

URANUS

Uranus and its largest moons are so far from the Sun (19.2 AU) that from Earth they appear to be a small cluster of stars. No wonder they seem fixed in the heavens: A Uranian year equals 84 Earth years. Since its discovery in 1781, Uranus has only orbited the Sun just over 2½ times.

7-12   Uranus sports a hazy atmosphere and clouds

Until early 1996, observations of Uranus, the fourth most massive planet in our solar system, revealed few notable features in the visible part of the spectrum. Even the 1986 visit of *Voyager 2* to Uranus showed a remarkably featureless world. It took the Hubble Space Telescope's infrared camera to find what *Voyager*'s visible light camera could not: Uranus has a system of belts

and zones. Its hydrogen atmosphere has traces of methane along with a high-altitude haze under which are clear air and ever-changing methane clouds that dwarf the typical cumulus clouds we see on Earth. Uranus's clouds are towering and huge, each typically as large as Europe (Figure 7-28). Along with the rest of the atmosphere, the clouds go around the planet once every 16½ hours.

Uranus contains 14½ times as much mass as the Earth and is 4 times bigger in diameter (see Figure 7-28). Its outer layers are composed predominantly of gaseous hydrogen and helium. The temperature in the upper atmosphere of the planet is so low (about 73 K, or −330°F) that the methane and water there condense to form clouds of ice crystals. Because methane freezes at a lower temperature than water, methane forms higher clouds over Uranus. Methane efficiently absorbs red light, giving Uranus its blue-green color.

Earth-based observations show that Uranus rotates once every 17 hours 14 minutes on an axis of rotation that lies very nearly in the plane of the planet's orbit. Uranus's rotation axis is inclined 98° from a line perpendicular to its plane of orbit. Therefore, it is one of only three planets with retrograde rotation (Venus and Pluto are the other two). As Uranus orbits the Sun, its north and south poles alternately point almost directly toward or directly away from the Sun, producing exaggerated seasons (Figure 7-29). In the

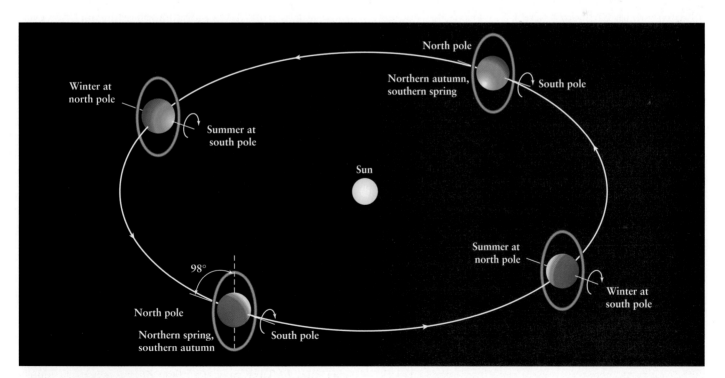

FIGURE 7-29   **Exaggerated Seasons on Uranus** Uranus's axis of rotation is tilted so steeply that it lies nearly in the plane of its orbit. Seasonal changes on Uranus are thus greatly exaggerated. For example, during

midsummer at Uranus's south pole, the Sun appears nearly overhead for many Earth years, while the planet's northern regions are subjected to a long, continuous winter night. Half an orbit later, the seasons are reversed.

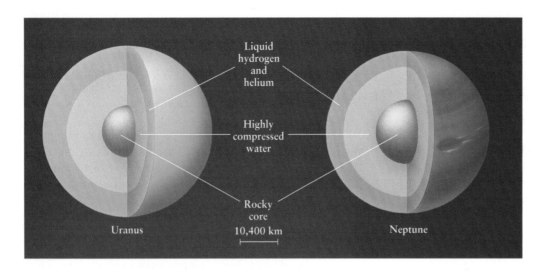

FIGURE 7-30 Cutaways of the Interiors of Uranus and Neptune The interiors of both Uranus and Neptune are believed to have three regions: a terrestrial core surrounded by a liquid water mantle, which is surrounded in turn by liquid hydrogen and helium. Their atmospheres are thin layers at the top of their hydrogen and helium layers.

summertime, near Uranus's north pole, the Sun is almost directly overhead for many Earth years, at which time southern latitudes are subjected to a continuous, frigid winter night. Forty-two Earth years later, the situation is reversed. For more on this, see the essay "What If ... Earth's Axis Lay on the Ecliptic?" on page 42.

From *Voyager* photographs, planetary scientists have concluded that each of the five largest Uranian moons has probably had at least one shattering impact. A catastrophic collision with an Earth-sized object may also have knocked Uranus on its side, as we see it today.

From its mass and density (1290 kg/m³), astronomers conclude that Uranus's interior has three layers. The outer 30% of the planet is liquid hydrogen and helium, the next 40% inward is highly compressed liquid water (with some methane and ammonia), and the inner 30% is a rocky core

(Figure 7-30). Indirect evidence for the water layer comes from the apparent deficiency of ammonia on Uranus. This gas dissolves easily in water, so an ocean would explain the scarcity of ammonia in the planet's atmosphere.

Voyager 2 passed through the magnetosphere of Uranus, revealing that the planet's surface magnetic field is about ³/₄ that of the Earth. That strength is reasonable, considering the planet's mass and rotation rate, but everything else about the magnetic field is extraordinary. It is remarkably tilted—59% from its axis of rotation—and does not even pass through the center of the planet (Figure 7-31).

Because of the large angle between the magnetic field of Uranus and its rotation axis, the magnetosphere of Uranus wobbles considerably as the planet rotates. Such a rapidly changing magnetic field will help us in explaining pulsars, a type of star we will study in Chapter 12.

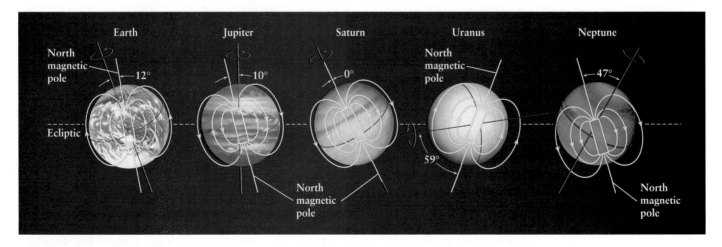

FIGURE 7-31 The Magnetic Fields of Five Planets This drawing shows how the magnetic fields of Earth, Jupiter, Saturn, Uranus, and Neptune are tilted relative to their rotation axes. Note that the magnetic fields of Uranus and Neptune are offset from the centers of the planets and steeply inclined to their rotation axes. Jupiter, Saturn, and Neptune have north magnetic poles on the hemisphere where Earth has its south magnetic pole.

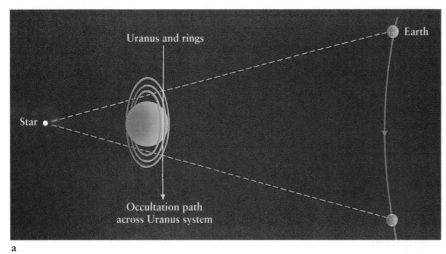

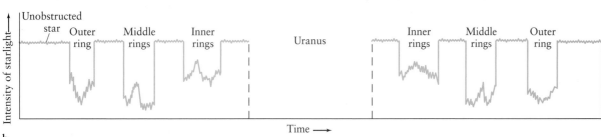

a

b

WEB LINK 7.22   ANIMATION 7.3

FIGURE 7-32
Discovery of the Rings of Uranus **(a)** Light from a star is reduced as the rings move in front of it. **(b)** With sensitive light detectors, astronomers can detect the variation in light intensity. Such dimming led to the discovery of Uranus's rings. Of course, the star vanishes completely when Uranus occults it.

7-13   A system of rings and satellites revolves around Uranus

Nine of Uranus's thin, dark rings were discovered accidentally in 1977 when Uranus passed in front of a star. The star's light was momentarily blocked by each ring, thereby revealing their existence to astronomers (Figure 7-32). Blocking the light of a more distant object, such as the star here, by something between it and us, such as Uranus's rings, is called an **occultation**. A picture taken while *Voyager* was in Uranus's shadow revealed more thin rings (Figure 7-33).

Most of Uranus's moons, like its rings, orbit in the plane of the planet's equator. Five of these satellites, ranging in diameter from 480 to nearly 1600 km, were known before

R I V U X G

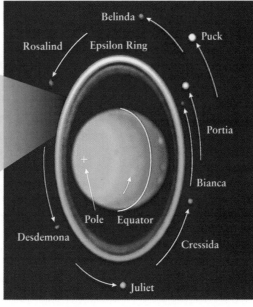

FIGURE 7-33   The Rings and Moons of Uranus This image of Uranus, its rings, and eight of its moons was taken by the Hubble Space Telescope. Inset: Close-up of part of the ring system taken by *Voyager 2* when the spacecraft was in Uranus's shadow looks back toward the Sun. Numerous fine dust particles between the main rings gleam in the sunlight. Uranus's rings are much darker than Saturn's, and this long exposure revealed many very thin rings and dust lanes. The short streaks are star images blurred because of the spacecraft's motion during the exposure. **(Inset: NASA)**

FIGURE 7-34 **Miranda** The patchwork appearance of Miranda in this mosaic of *Voyager 2* images suggests that this satellite consists of huge chunks of rock and ice that came back together after an ancient, shattering impact by an asteroid or a neighboring Uranian moon. The curious banded features that cover much of Miranda are parallel valleys and ridges that may have formed as dense, rocky material sank toward the satellite's core. At the very bottom of the image—where a "bite" seems to have been taken out of the satellite—is a range of enormous cliffs that jut upward as high as 20 km, twice the height of Mount Everest. (NASA)

the *Voyager* mission. However, *Voyager*'s cameras discovered ten additional satellites, each fewer than 50 km across. Still others have since been observed from Earth. Several of these tiny, irregularly shaped moons are shepherd satellites whose gravitational pull confines the particles within the thin rings that circle Uranus.

The smallest of Uranus's five main satellites, Miranda, is the most fascinating and bizarre of its 21 known moons. Unusual wrinkled and banded features cover Miranda's surface (Figure 7-34). Its highly varied terrain suggests that it was once seriously disturbed. Perhaps a shattering impact temporarily broke it into several pieces that then recoalesced, or perhaps severe tidal heating, as we saw on Io, moved large pieces of its surface.

Miranda's core originally consisted of dense rock, while its outer layers were mostly ice. If a powerful impact did occur, blocks of debris broken off from Miranda drifted back together through mutual gravitational attraction. Recolliding with that moon, they formed a chaotic mix of rock and ice.

In this scenario, the landscape we see today on Miranda is the result of huge, dense rocks trying to settle toward the satellite's center, forcing blocks of less dense ice upward toward the surface.

NEPTUNE

Neptune is physically similar to Uranus (review Figures 7-28 and 7-30). Neptune has 17.1 times the Earth's mass, 3.88 times Earth's diameter, and a density of 1640 kg/m^3. Unlike Uranus, however, cloud features can readily be discerned on Neptune. Its whitish, cirruslike clouds consist of methane ice crystals. The methane absorbs red light, leaving the planet's belts and zones with a banded, bluish appearance (Figure 7-35). Like Jupiter, the atmosphere of Neptune also experiences differential rotation. The winds on Neptune blow as fast as 2000 km/h—among the fastest in the solar system.

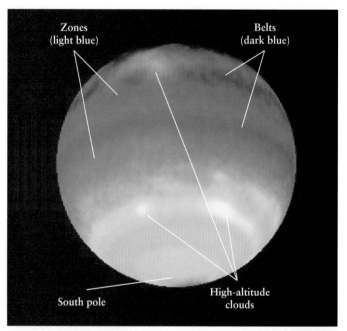

FIGURE 7-35 **Neptune's Banded Structure.** Several Hubble Space Telescope images at different wavelengths were combined to create this enhanced-color view of Neptune. The dark blue and light blue areas are the belts and zones, respectively. The dark belt running across the middle of the image lies just south of Neptune's equator. White areas are high-altitude clouds, presumably of methane ice. The very highest clouds are shown in yellow-red, as seen at the very top of the image. The green belt near the south pole is a region where the atmosphere absorbs blue light, perhaps indicating some differences in chemical composition. (Lawrence Sromovsky, University of Wisconsin-Madison and STScI/NASA)

7-14 Neptune was discovered because it had to be there

Neptune's discovery is storied because it illustrates a scientific prediction leading to an expected discovery. In 1781 the British astronomer William Herschel discovered Uranus. Its position was carefully plotted, and by the 1840s, it was clear that even considering the gravitational effects of all the known bodies in the solar system, Uranus was not following the path predicted by Newton's and Kepler's laws. Either the theories behind these laws were wrong, or there had to be another, yet-to-be-discovered body in the solar system pulling on Uranus.

Independent, nearly simultaneous calculations by an English mathematician, John Adams, and a French astronomer, Urbain Leverrier, predicted the same location for the alleged planet. That planet, Neptune, was located in 1846 by the German astronomer Johann Galle, within a degree or two of where it had to be to have the observed influence on Uranus.

In August 1989, nearly 150 years after its discovery, *Voyager 2* arrived at Neptune to cap one of NASA's most ambitious and successful space missions. Scientists were overjoyed at the detailed, close-up pictures and wealth of data about Neptune sent back to Earth by the spacecraft.

> **Insight into Science** Process and Progress Scientific theories make testable predictions (see Chapter 1). Based on the details of Uranus's orbit around the Sun, Newton's law of gravitation predicted that Uranus's orbit was being affected by the gravitational attraction of another planet. The law also predicted where that planet was located, leading to the discovery of Neptune.

At the time *Voyager 2* passed it, a giant storm raged in Neptune's atmosphere. Called the **Great Dark Spot**, it was about half as large as Jupiter's Great Red Spot. The Great Dark Spot (Figure 7-36) was located at about the same latitude on Neptune and occupied a similar proportion of Neptune's surface as the Great Red Spot does on Jupiter. Although these similarities suggested that similar mechanisms created the spots, the Hubble Space Telescope in 1994 showed that the Great Dark Spot had disappeared. Then, in April 1995, another storm developed in the opposite hemisphere.

Neptune's interior is believed to be very similar in composition and structure to that of Uranus: a rocky core surrounded by ammonia- and methane-laden water (see Figure 7-30). Neptune's surface magnetic field is about 40 percent that of the Earth. Also, as with Uranus, Neptune's magnetic axis, the line connecting its north and south magnetic poles, is tilted sharply from its rotation axis. In this case, the tilt is 47%. Again like Uranus, Neptune's magnetic axis does not pass through the center of the planet (see Figure 7-31).

We saw in section 7-2 that the magnetic fields of Jupiter and Saturn are believed to be generated by the motions of their liquid metallic hydrogen. However, Uranus and Neptune lack this material, and their magnetic fields have a different origin. These fields are believed to exist because molecules such as ammonia dissolved in their water layers lose electrons (become ionized). These ions, moving with the planets' rotating, fluid interiors, create the same dynamo effect that produces the magnetic field.

7-15 Neptune has rings and has captured most of its moons

Like Uranus, Neptune is surrounded by a system of thin, dark rings (Figure 7-37). It is so cold at these distances from the Sun that both planets' ring particles retain methane ice. Scientists speculate that eons of radiation damage have converted this methane ice into darkish carbon compounds, thus accounting for the low reflectivity of the rings.

Neptune has eight known moons. Seven have irregular shapes and highly elliptical orbits, which suggest that Neptune captured them. Triton, discovered in 1846, is spherical and was quickly observed to have a nearly circular, retrograde orbit around Neptune. It is difficult to imagine how a

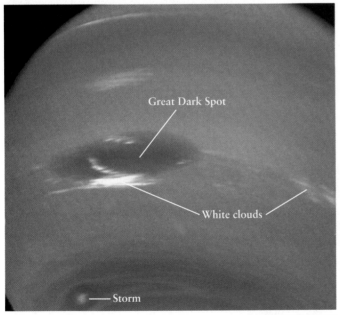

Great Dark Spot

White clouds

Storm

R I V U X G

FIGURE 7-36 **Neptune** This view from *Voyager 2* looks down on the southern hemisphere of Neptune. The Great Dark Spot, whose diameter at the time was about the same size as the Earth's diameter, is near the center of this picture. It has since vanished. Note the white, wispy methane clouds. (NASA/JPL)

VIDEO 7.10

RIVUXG

FIGURE 7-37   Neptune's Rings   Two main rings are easily seen in this view alongside overexposed edges of Neptune. Careful examination also reveals a faint inner ring. A fainter-still sheet of particles, whose outer edge is located between the two main rings, extends inward toward the planet. (NASA)

satellite and planet could form together but rotate in opposite directions. Indeed, only a few of the small outer satellites of Jupiter and Saturn have retrograde orbits, and these bodies are probably captured asteroids. Some scientists have therefore suggested that Triton may have been captured 3 or 4 billion years ago by Neptune's gravity.

Upon being captured, Triton was most likely in a highly elliptical orbit. However, the tidal force the moon creates on Neptune's liquid surface would have made Triton's orbit more circular. Conversely, the tidal force created on Triton by Neptune due to the moon's changing distance to Neptune would have caused the moon to stretch and flex, providing enough energy to melt much of the satellite's interior and obliterate Triton's original surface features, including craters. Triton's south polar region is shown in Figure 7-38. Note that very few craters are visible. Calculations based on these observations indicate that Triton's present surface is about 100 million years old.

Triton does exhibit some surface features seen on other icy worlds, such as long cracks resembling those on Europa and Ganymede. Other features unique to Triton are quite puzzling. For example, the top half of Figure 7-38 reveals a wrinkled terrain that resembles the skin of a cantaloupe.

Triton also has a few frozen lakes like the one shown in Figure 7-39. Some scientists have speculated that these lakelike features are the calderas of extinct ice volcanoes. A mixture of methane, ammonia, and water, which can have a melting point far below that of pure water, could have formed a kind of cold lava on Triton.

Voyager instruments measured a surface temperature of 36 K (−395°F), making Triton the coldest world that our probes have ever visited. Nevertheless, *Voyager* cameras did glimpse two towering plumes of gas extending up to 8 km above the satellite's surface. These are apparently plumes of nitrogen gas warmed by interior radioactive decay and escaping through vents or fissures.

In the same way that our Moon raises tides on Earth, Triton raises tides on Neptune. Whereas the tides on Earth cause our Moon to spiral outward, the tides on Neptune cause Triton (in its retrograde orbit) to spiral inward. Within the next quarter of a billion years, Triton will reach the **Roche limit,** the distance at which a planet creates tides on its moon's solid surface high enough to pull its moon apart. Pieces of Triton will then literally float into space until the entire moon is demolished! By destroying Triton, Neptune

Canteloupe terrain

South polar region   Dark, wind-blown deposits

RIVUXG

FIGURE 7-38   Triton's South Polar Cap   Approximately a dozen high-resolution *Voyager 2* images were combined to produce this view of Triton's southern hemisphere. The pinkish polar cap is probably made of nitrogen frost. A notable scarcity of craters suggests that Triton's surface was either melted or flooded by icy lava after the era of bombardment that characterized the early history of the solar system. (NASA)

FIGURE 7-39 **A Frozen Lake on Triton** Scientists believe that the feature in the center of this image is a basin filled with water ice. The flooded basin is about 200 km across. (NASA)

will create a new ring system that will be much more substantial than its present one.

MORE TO KNOW 7.1 Roche Limit

PLUTO AND BEYOND

In 1930, the astronomer Clyde W. Tombaugh was searching for a giant planet that was allegedly affecting Neptune's orbit. Instead, Tombaugh discovered tiny Pluto (Figure 7-40), a planet far too small to have a noticeable gravitational effect on Neptune. Figure 7-41 lists Pluto's data. Refined observations show that Neptune's orbit is not being affected by another giant planet. No such planet has ever been found.

Tombaugh recognized Pluto as a planet because it moved among the background stars from night to night, but he had no idea how strange its orbit is compared with those of the other planets. As discussed in Essentials II (see Figure II-8), Pluto's orbit is so elliptical that it is sometimes closer to the Sun than Neptune, as it was from 1979 to 1999. It is now farther away from the Sun than Neptune and will continue to be so for about the next 230 years. Pluto's orbit is tilted with respect to the plane of the ecliptic more than any other planet.

7-16 Pluto and its moon Charon are about the same size

Pluto was little understood for half a century until astronomers noticed that its image sometimes appears oblong (Figure 7-42). This observation led to the discovery in 1978 of Pluto's only known moon, Charon (pronounced KAR-en, after the mythical boatman who ferried souls across the River Styx to Hades, the domain ruled by Pluto). From 1985 through 1990, the orbit of Charon was oriented so that Earth-based observers could watch it eclipse Pluto. Astronomers used observations of these eclipses to determine that Pluto is only twice as broad as its satellite: its diameter is 2380 km and Charon's is 1190 km.

The average distance between Charon and Pluto is less than 1/20 the distance between the Earth and our Moon.

R I V U X G

FIGURE 7-40 **Discovery of Pluto** Pluto was discovered in 1930 by searching for a dim, starlike object that slowly moves against the background stars. These two photographs were taken one day apart. (Lick Observatory)

WEB LINK 7.26

Pluto's Vital Statistics

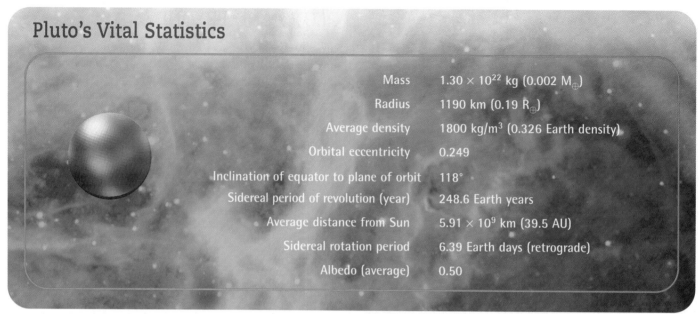

Mass	1.30×10^{22} kg (0.002 $M_{\oplus}$)
Radius	1190 km (0.19 $R_{\oplus}$)
Average density	1800 kg/m^3 (0.326 Earth density)
Orbital eccentricity	0.249
Inclination of equator to plane of orbit	118°
Sidereal period of revolution (year)	248.6 Earth years
Average distance from Sun	5.91×10^9 km (39.5 AU)
Sidereal rotation period	6.39 Earth days (retrograde)
Albedo (average)	0.50

R I V U X G

FIGURE 7-41   **Pluto and Its Vital Statistics** This Hubble Space Telescope image of Pluto shows little detail but indicates that the major features of Pluto's surface each cover large amounts of its surface area. (Alan Stern, Southwest Research Institute, Marc Buie, Lowell Observatory, NASA/ESA)

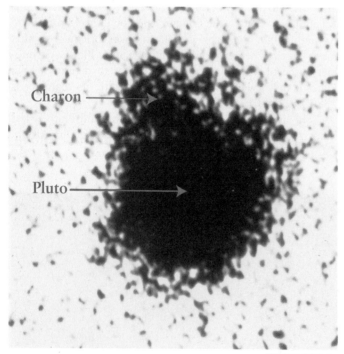

R I V U X G

FIGURE 7-42   **Discovery of Charon** Long ignored as just a defect in the photographic emulsion, the bump on the upper left side of this image of Pluto led astronomer James Christy to discover the moon Charon. (U.S. Naval Observatory)

Furthermore, Pluto always keeps the same side facing Charon. This is the only case in which a moon and a planet both have synchronous rotation with respect to each other. As seen from the satellite-facing side of Pluto, Charon neither rises nor sets, but instead hovers in the sky, perpetually suspended above the horizon.

The exceptional similarities between Pluto and Charon suggest that this binary system may have formed when Pluto collided with a body of similar size. Perhaps chunks of matter were stripped from this second body, leaving behind a mass, now called Charon, that was vulnerable to capture by Pluto's gravity. Alternatively, perhaps Pluto's gravity captured Charon into orbit during a close encounter between the two worlds.

Both of these are unlikely scenarios. For either to be feasible, many Pluto-sized objects must have existed in the outer regions of the young solar system. One astronomer estimates that there must have been at least a thousand Pluto-sized bodies in order for a collision or close encounter between two of them to have occurred at least once since the solar system formed 4.6 billion years ago.

The best pictures of Pluto and Charon were taken by the Hubble Space Telescope (Figure 7-43). Like Neptune's moon Triton, these worlds are probably composed of nearly equal amounts of rock and ice. The surface of Pluto shows more large-scale features than any object in the solar system other than the Earth. Examination of Hubble Space Tele-

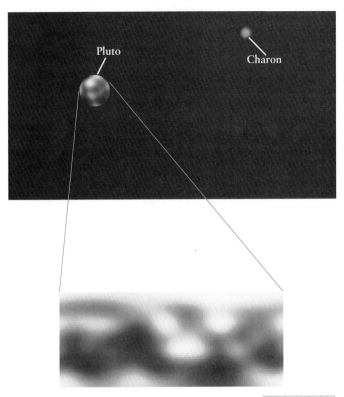

R I V U X G

FIGURE 7-43   Pluto and Charon This picture of Pluto and its moon, Charon, is a composite of Hubble Space Telescope images. It shows the greatest detail we have of Pluto's surface. Pluto and Charon are separated here by only 19,700 km. **Inset:** A map covering 85% of Pluto's surface. The contrast between dark and light strongly suggests regions covered with ice and regions covered with rocky material. (Pluto: Alan Stern/SwRI, Mar Cuie/Lowell Observatory, NASA, & ESA; Charon: R. Albrecht, ESA/ESO Space Telescope European Coordinating Facility, and NASA; inset: A. Stern/SwRI, M. Buie/Lowell Observatories, NASA, and ESA)

scope images has revealed some 12 distinct regions, and studies of Pluto's surface temperature show that different regions vary by up to 25 K from each other.

Pluto's spectrum shows that its surface contains frozen nitrogen, methane, and carbon monoxide. The planet is also observed to have a very thin atmosphere of nitrogen and

carbon monoxide when it is closest to the Sun. In contrast, Charon's surface appears to be covered predominantly with water ice.

Ever since Pluto was discovered, astronomers have searched for other objects in our solar system. Neptune's captured moons support the idea of other bodies at the outskirts of the solar system, and astronomers have begun discovering them by the hundreds. These distant masses of ice and rock, which we will discuss further in Chapter 8, lie in a region now called the Kuiper belt, which is believed to extend out 500 AU from the Sun.

Insight into Science   Paradigm Shift Usually it takes overwhelming evidence contradicting established beliefs before scientists begin looking at old ideas in completely new ways. Such changes in fundamental beliefs are sometimes called *paradigm shifts*. The acceptance of tectonic plate motion is one such change. Until the past few years, astronomers took it for granted that Pluto is a planet. Perhaps another paradigm shift in the near future will reclassify tiny Pluto as the largest object in the Kuiper belt.

7-17   Frontiers yet to be discovered

The outer planets hold countless new insights into the formation and evolution of the solar system. Considering that virtually all present extrasolar planets are Jupiter-like gas giants, our outer planets all have a lot to tell us about planets throughout our Galaxy. Why did the *Galileo* probe fail to detect the atmospheric structures believed to exist on Jupiter? How has the Great Red Spot persisted for so long? Is there life in the oceans of the outer Galilean moons? What, exactly, do the ring particles around each of the planets look like? How long ago did Saturn's ring system form? How long will it last? Are the spokes in Saturn's rings really caused by its magnetic field, and, if so, why do the spokes appear as they do? What caused Miranda's surface to become so profoundly disturbed? How did the Pluto-Charon system form? These are but a few of the questions about the outer planets that will be answered during this century.

Further Reading on These Topics

WHAT DID YOU KNOW?

1 *Is Jupiter a "failed star" or almost a star?* No. Jupiter has 75 times too little mass to shine as a star.

2 *What is Jupiter's Great Red Spot?* The Great Red Spot is a long-lived, oval cloud circulation similar to a hurricane on Earth.

3 *Does Jupiter have continents and oceans?* No. Jupiter is surrounded by a thick atmosphere primarily of hydrogen and helium that becomes liquid as one moves inward. The only solid matter in Jupiter is its core.

4 *Is Saturn the only planet with rings?* No. Four planets (Jupiter, Saturn, Uranus, and Neptune) have rings.

5 *Are the rings of Saturn solid ribbons?* Saturn's rings are all composed of thin, closely spaced ringlets consisting of particles of ice and ice-coated rocks. If they were solid ribbons, Saturn's gravitational tidal force would tear them apart.

KEY WORDS

KEY IDEAS

Jupiter and Saturn

• Jupiter is by far the largest and most massive planet in the solar system.

• Jupiter and Saturn probably have rocky cores surrounded by a thick layer of liquid metallic hydrogen and an outer layer of ordinary liquid hydrogen. Both planets have an overall chemical composition very similar to that of the Sun.

• The visible features of Jupiter exist in the outermost 100 km of its atmosphere. Saturn has similar features, but they are much fainter. Three cloud layers exist in the upper atmospheres of both Jupiter and Saturn. Because Saturn's cloud layers extend through a greater range of altitudes, the colors of the Saturnian atmosphere appear muted.

• The colored ovals visible in the Jovian atmosphere represent gigantic storms, some of which (such as the Great Red Spot) are stable and persist for years or even centuries.

• Jupiter and Saturn have strong magnetic fields created by electric currents in the metallic hydrogen layer.

• Four large satellites orbit Jupiter. The two inner Galilean moons, Io and Europa, are roughly the same size as our Moon. The two outer moons, Ganymede and Callisto, are approximately the size of Mercury.

• Io is covered with a colorful layer of sulfur compounds deposited by frequent explosive eruptions from volcanic vents. Europa is covered with a smooth layer of frozen water crisscrossed by an intricate pattern of long cracks.

• The heavily cratered surface of Ganymede is composed of frozen water with large polygons of dark, ancient crust separated by regions of heavily grooved, lighter-colored, younger terrain. Callisto has a heavily cratered ancient crust of frozen water.

• Saturn is circled by a system of thin, broad rings lying in the plane of the planet's equator. Each major ring is composed of a great many narrow ringlets consisting of numerous fragments of ice and ice-coated rock. Jupiter has a much less substantial ring system.

Uranus and Neptune

• Uranus and Neptune are quite similar in appearance, mass, size, and chemical composition. Each has a rocky core probably surrounded by a dense, watery mantle; the axes of their magnetic fields are steeply inclined to their axes of rotation; and both planets are surrounded by systems of thin, dark rings.

• Uranus is unique in that its axis of rotation lies nearly in the plane of its orbit, producing greatly exaggerated seasons on the planet.

• Uranus has five moderate-sized satellites, the most bizarre of which is Miranda.

• The largest satellite of Neptune, Triton, is an icy world with a tenuous nitrogen atmosphere. Triton moves in a retrograde orbit that suggests it was captured into orbit by Neptune's gravity, and its orbit is spiraling down toward Neptune.

✦ Pluto and Beyond

• Pluto, the smallest planet in the solar system, and its satellite, Charon, are icy worlds that may well resemble Triton.

• Other objects orbit the Sun beyond the orbit of Pluto. Hundreds of these Kuiper belt objects have been observed.

REVIEW QUESTIONS

1   Describe the appearance of Jupiter's atmosphere. Which features are long-lived and which are fleeting?

 2   To test your knowledge of Jupiter's belt and zone structure, do Interactive Exercise 7-1 on the CD-ROM or Web. You can print out your results, if required.

3   What causes the belts and zones in Jupiter's atmosphere?

 4   To test your knowledge of Jupiter's internal structure, do Interactive Exercise 7-2 on the CD-ROM or Web. You can print out your results, if required.

5   What is liquid metallic hydrogen? Which planets appear to contain this substance? What produces this form of hydrogen?

6   Compare and contrast the surface features of the four Galilean satellites, discussing their geologic activity and their evolution.

 7   To test your knowledge of the Galilean Moons, do Interactive Exercise 7-3 on the CD-ROM or Web. You can print out your results, if required.

8   What energy source powers Io's volcanoes?

9   Why are numerous impact craters found on Ganymede and Callisto but not on Io or Europa?

10   Describe the structure of Saturn's rings. What are they made of?

 11   To test your knowledge of Saturn's rings, do Interactive Exercise 7-4 on the CD-ROM or Web. You can print out your results, if required.

12   Why do features in Saturn's atmosphere appear to be much fainter and "washed out" compared with features in Jupiter's atmosphere?

13   Explain how shepherd satellites operate. Is "shepherd satellite" an appropriate term for these objects? Explain.

14   Describe Titan's atmosphere. What effect has sunlight presumably had on Titan's atmosphere?

15   Describe the seasons on Uranus. Why are the Uranian seasons different from those on any other planet?

16   Briefly describe the evidence supporting the idea that Uranus was struck by a large planetlike object several billion years ago.

17   Why are Uranus and Neptune distinctly bluer than Jupiter and Saturn?

18   Compare the ring systems of Saturn and Uranus. Why were Uranus's rings unnoticed until the 1970s?

19   How do the orientations of Uranus's and Neptune's magnetic axes differ from those of the other planets?

 20   To test your knowledge of planetary magnetic fields, do Interactive Exercise 7-5 on the CD-ROM or Web. You can print out your results, if required.

21   Suppose you were standing on Pluto. Describe the motions of Charon relative to the horizon. Under what circumstances would you never see Charon?

22   Describe the circumstantial evidence supporting the idea that Pluto is one of a thousand similar icy worlds that once occupied the outer regions of the solar system.

23   Explain why Triton will never collide with Neptune, even though Triton is spiraling toward that planet.

24   The discovery of Pluto's moon, Charon, could have been made decades before it actually was. What caused the delay?

ADVANCED QUESTIONS

25   Consult the Internet or such magazines as *Sky & Telescope* or *Astronomy* to determine what space missions are now under way. What data and pictures have they sent back that update information presented in this chapter?

26   Long before the *Voyager* flybys, Earth-based astronomers reported that Io appeared brighter than usual for a few hours after emerging from Jupiter's shadow. From what we know about the material ejected from Io's volcanoes, explain this brief brightening of Io.

27 Compare and contrast Valhalla on Callisto with the Caloris Basin on Mercury.

28 Can we infer from naked-eye observations that Saturn is the most distant of the planets visible without a telescope? Explain.

29 As seen by Earth-based observers, the intervals between successive edge-on presentations of Saturn's rings alternate between 13 years 9 months and 15 years 9 months. Why are these two intervals not equal?

30 Compare and contrast the internal structures of Jupiter and Saturn with the internal structures of Uranus and Neptune. Can you propose an explanation for why the differences between these two pairs of planets occurred?

31 Neptune has the third largest mass of all the planets, but Uranus has the third largest diameter. Reconcile these two facts.

DISCUSSION QUESTIONS

32 Suppose that you were planning a mission to Jupiter employing an airplanelike vehicle that would spend many days, even months, flying through the Jovian clouds. What observations, measurements, and analyses should this aircraft make? What dangers might it encounter, and what design problems would you have to overcome?

33 Discuss the possibility that Europa, Ganymede, or Callisto might harbor some sort of marine life.

34 Suppose you were planning separate missions to each of Jupiter's Galilean moons. What questions would you want these missions to answer, and what kinds of data would you want your spacecraft to send back? Given the

different environments on the four satellites, how would the designs of the four spacecraft differ?

35 NASA and the Jet Propulsion Laboratory have tentative plans to place spacecraft in orbit about Uranus and Neptune in this century. What kinds of data should be collected and what questions would you like to see answered by these missions?

36 Would you expect the surfaces of Pluto and Charon to be heavily cratered? Explain.

WHAT IF ...

37 Jupiter, at its present location, were a star? What would Earth be like? *Hint:* Recall that to be a star, Jupiter would have to have 75 times more mass than it has today.

38 Jupiter had formed at one-third its present distance of 5.2 AU from the Sun? What would Earth be like?

39 Io were struck by another object of similar size? *Hint:* You can create a variety of different scenarios by

imagining the impacting body striking from different directions and with different speeds and at different angles.

40 Jupiter were orbiting in the opposite direction that it actually is? What effects might this have on the other planets? Would this change affect Earth? If so, how?

WEB/CD-ROM QUESTIONS

 41 **Moving Weather Systems on Jupiter** Access and view the video "The Great Red Spot" in Chapter 7 of the *Discovering the Universe* Web site or CD-ROM. (a) Near the bottom of the video window you will see a white oval moving from left to right. By stepping through the video one frame at a time, estimate how long it takes this oval to move a distance equal to its horizontal dimension. (*Hint:* You can keep track of time by noticing how many frames it takes a

feature in the Great Red Spot, at the center of the video window, to move in a complete circle around the center of the spot. The actual time for this feature to complete a circle is about six days.) (b) The horizontal dimension of the white oval is about 4000 km. At what approximate speed (in km/hr) does the white oval move? (Speed = distance/time).

42 Search the Web, especially the Web sites at NASA's Jet Propulsion Laboratory and the European Space Agency,

for information about the current status of the *Cassini* mission. When will *Cassini* arrive at Saturn? What are the current plans for its tour of Saturn's satellites? What ideas are being considered for the *Cassini* extended mission, to begin in 2008?

43 In 2000, astronomers reported the discovery of several new moons of Saturn. Search the Web for information about these. How did astronomers discover them? Have the observations been confirmed? How large are these moons? What sort of orbits do they follow?

44 The Rotation Rate of Saturn Access and view the video "Saturn from the Hubble Space Telescope" in Chapter 7 of the *Discovering the Universe* Web site or CD-ROM. The total time that actually elapses in this video is 42.6 hours. Using this information, identify and follow an atmospheric feature and determine the rotation period of Saturn.

45 The discovery of Charon, Figure 7-42, was made by an astronomer at the U.S. Naval Observatory. Search the Web to find out why the U.S. Navy carries out work in astronomy.

OBSERVING PROJECTS

46 Consult such magazines as *Sky & Telescope* and *Astronomy* or your *Starry Night Backyard*™ software to determine whether Jupiter is currently visible in the night sky. If so, make arrangements to view the planet through a telescope. What magnifying power seems to give you the best view? Draw a picture of what you see. Can you see any belts and zones? How many? Can you see the Great Red Spot?

47 Make arrangements to view Jupiter's Great Red Spot through a telescope. Consult the *Sky & Telescope* Web site, which lists the times when the center of the Great Red Spot passes across Jupiter's central region, as seen from Earth. The Great Red Spot is well placed for viewing for 50 minutes before and after this time. You will need a refractor with an objective lens of at least 15 cm (6 in.) diameter or a reflector with an objective of at least 20 cm (8 in.) diameter. Using a pale blue or green filter can increase the color contrast and make the spot more visible. For other useful hints, see the article, "Tracking Jupiter's Great Red Spot" by Alan MacRobert (*Sky & Telescope,* September 1997). Sketch Jupiter based on your observations.

48 If it is visible at night, observe Jupiter through a pair of binoculars or telescope. Can you see all four Galilean moons? Make a drawing of what you observe. To identify the moons, use *Starry Night Backyard*™. Start by going to Atlas mode (*Go/Atlas*). Then locate Jupiter (*Edit/Find/Jupiter*). Zoom out to 30' angle. You should be able to see all the moons labeled in their present positions (be sure that the time is *not* changing). Compare the locations of the moons on the screen with your observations to determine which ones you are seeing.

49 Use *Starry Night Backyard*™ to observe the motion of the Galilean moons of Jupiter. First go to Atlas mode (*Go/Atlas*). Lock on Jupiter (*Edit/Find/Jupiter*) and zoom out to 30' angle. Set the timestep to 1 hour and click the ▶ button. You will see the four Galilean moons orbiting Jupiter. (a) Are all four moons ever on the same side of Jupiter? (b) Observe the moons passing in front of and behind Jupiter (zoom in as needed). Explain how your observations tell you that all four satellites orbit Jupiter in the same direction.

50 Consult such magazines as *Sky & Telescope* and *Astronomy* or your *Starry Night Backyard*™ software to determine whether Saturn is currently visible in the night sky. If so, view Saturn through a small telescope. Make a sketch of what you see. Estimate the angle at which the rings are tilted to your line of sight. Can you see the Cassini division? Can you see any belts or zones in Saturn's clouds? Do you observe a faint, starlike object near Saturn that might be Titan? What observations could you perform to test whether the starlike object is a Saturnian satellite?

51 Use the *Starry Night Backyard*™ program to observe the changing appearance of Saturn. First go to Atlas mode (*Go/Atlas*) and then find Saturn (*Edit/Find/Saturn*). Zoom in until Saturn nearly fills the screen. Set the timestep to 1 year. Use the single-step time control buttons to observe the changing aspect or orientation of the rings. During which of the next 30 years will we see the rings edge-on?

52 Determine whether Uranus or Neptune is currently visible in the night sky. If so, make arrangements to view them through a telescope. To help you find these planets, use your *Starry Night Backyard*™ software or the star chart published each January in *Sky & Telescope* showing the paths of Uranus and Neptune against the background stars. For more detailed images than you can get through a telescope, locate these planets with your *Starry Night Backyard*™ software and zoom to high resolution. Also, locate Triton with the software.

53 Use the *Starry Night Backyard*™ program to observe Pluto and Charon. First go to Atlas mode (*Go/Atlas*) and find Pluto (*Edit/Find/Pluto*). Zoom in as close as possible. Select Planet List in the Window menu and click on the triangle to the left of the name Pluto. Then, in the "Orbit" column on the right-hand side of the Planet List, click to the right of the name Charon. A mark will appear in this column and Charon's orbit will appear in the main window. In the Control Panel, set the timestep to 3 hours. (a) Use the single-step time button to step through enough time to determine the period of Charon's orbit. How does your answer compare with Pluto's sidereal rotation period given in Figure 7-41? Explain. (b) What is the apparent shape of Charon's orbit around Pluto? How can you reconcile this with the fact that Charon's actual orbit is nearly a perfect circle?

WHAT IF . . .

WE LIVED ON A METAL-POOR EARTH?

Earth provides a wealth of building blocks necessary for the development and evolution of complex life-forms. More than 80 elements on or near Earth's surface combine in countless ways essential for its diversity of flora and fauna. Especially important for life on Earth are metals such as iron. The human body typically contains more than 3 grams of iron, mostly in the form of hemoglobin that helps transport oxygen through the bloodstream. A slight iron imbalance leads to anemia or toxicity. And without abundant metals, most of our technologies would never have developed.

But what if the solar system formed from an interstellar gas cloud containing fewer metals and high-mass elements, that is, a solar system with less iron, nickel, copper, and other metals crucial for life as we know it?

Metal-Rich versus Metal-Poor   Earth formed with almost 6×10^{21} metric tons of star matter, almost one-third of it iron. If the solar system formed from interstellar gas containing relatively few heavy elements, Earth would contain a much lower fraction of such elements as uranium, lead, iron, and nickel. The Earth would be comprised of a correspondingly higher fraction of lower-mass elements, such as carbon, nitrogen, oxygen, silicon, and aluminum. That version of Earth—let's call it Lithia—would be profoundly different from our planet.

The Environment of Lithia   Without heavy elements like iron, Lithia's density, gravity, and magnetic fields are much lower than Earth's. Let's assume that the density of Lithia is similar to that of the Moon. (We'll assume that the Moon doesn't change in its characteristics or its distance from Lithia.)

If Lithia has the same radius as Earth, the force of gravity on Lithia's surface is only 60% that on Earth. That is, you would weigh 40% less on Lithia. Mobile life-forms that evolve on the metal-poor planet require much less muscle strength to get about and less bone density to resist the planet's gravity.

With few radioactive elements to provide heat, the core of Lithia cooled off quickly compared with the core of Earth. As a result, Lithia has much less heat flowing upward through its mantle and crust, so there most likely is no plate tectonic activity. Lithia also has less volcanic activity. But without plate motion to spread the lava flows, volcanoes will grow to higher elevations than on Earth. The lack of crustal motions also means that pockets of high-density material will not rise from below, forming the metal ores mined on Earth.

The Moon's orbit and the tides it induces are different, too. With Lithia having 60% of Earth's mass, the Moon at its present distance orbits once every 36 days instead of once every 29.5 days. Thus, the cycle of lunar phases is longer, but the cycle of lunar-induced tides is slightly shorter. Surprisingly, the tides have roughly the same height, because the tidal effects depend only on the Moon's gravity and not Earth's, and the distance between Earth and Lithia and their satellite is the same.

Life on Lithia   The low abundance of iron and the resulting lack of a planetary magnetic field might preclude the development of complex life-forms on Lithia. And yet, the incredible diversity of life we find on Earth suggests that evolution on Lithia might well occur. If advanced life-forms evolve on Lithia before the Sun uses up its nuclear fuel and becomes a red giant, the chemical differences between Lithia and Earth dictate that they will differ profoundly from the complex life we find on Earth.

Without the heat of radioactive elements to keep Lithia's core molten for billions of years and with fewer metals, Lithia will have a much weaker magnetic field—if any at all—than Earth. Inhabitants can expect auroras to grace the night sky continuously. On the other hand, high-energy particles and the radiation they create in the atmosphere will bombard any life-forms on the surface continuously and will also adversely affect the ozone layer. A larger amount of the Sun's harmful ultraviolet radiation will reach the surface. Without more protection, life as we know it could not survive.

A Barren Landscape With a thin atmosphere and an abundance of lighter elements such as silicon (silicon dioxide is sand), Lithia might resemble a barren desert here on Earth. (Photri)

8 VAGABONDS OF THE SOLAR SYSTEM

IN THIS CHAPTER YOU WILL DISCOVER

- asteroids and meteoroids, pieces of interplanetary rock and metal

- comets, bodies containing large amounts of ice and rocky debris

- space debris that falls through the Earth's atmosphere

- that impacts from space 250 million and 65 million years ago caused mass extinctions of life on Earth

- that a wayward asteroid could again threaten life on Earth

WEB LINK 8.1

Comet Hyakutake
A comet is almost always named after the person who first sees it. The Japanese amateur astronomer Yuji Hyakutake first observed the comet named for him using binoculars on the morning of January 30, 1996. This photograph was taken two months later, in April 1996, when the comet extended more than 30° across the sky—more than 60 times the diameter of the full Moon as seen from Earth. Comet Hyakutake passed within 0.1 AU (15 million kilometers or 9 million miles) of Earth. (Jerry Lodriguss)

R I V U X G

WHAT DO YOU THINK?

|1| Are the asteroids a planet that was somehow destroyed?

|2| How far apart are the asteroids on average?

|3| Why do comets have tails?

|4| In which direction does a comet tail point?

|5| What is a shooting star?

The formation of the solar system was not a tidy affair. Planetesimals collided by the billions to form the planets we see today and their moons. For nearly a billion years, pieces of smaller debris struck them often, scarring and shaping their surfaces. This era of frequent collisions ended some 3.8 billion years ago, but innumerable pieces of debris still orbits the Sun, and dramatic impacts continue today. Large, rocky bodies occasionally pass startlingly close to the Earth—closer even than the Moon. Myriad small ones penetrate the atmosphere daily. Comets weave glorious trails across the sky. Icy bodies such as these may once have provided Earth with water and other material essential to the evolution of life.

These leftovers, the vagabonds of the solar system, are the asteroids, meteoroids, and comets. We have already seen pieces of Comet Shoemaker-Levy 9 plunge into Jupiter. You may have spotted Comet Hale-Bopp during the summer of 1997. What do the characteristics of these interplanetary travelers reveal about the composition of the solar system and about how life began here?

ASTEROIDS

We know that the solar system formed from a rotating disk of gas and dust. The matter that had too much angular momentum to fall onto the protosun coalesced at varying distances into planetesimals. Many of these chunks of rock and metal eventually collided, forming the planets and larger moons of the solar system. Others were captured whole by various planets as small, irregularly shaped moons, like Phobos and Deimos in orbit around Mars. However, many planetesimals still orbit the Sun today in splendid isolation. These are the **asteroids,** sometimes called *minor planets.*

8-1 Most asteroids orbit the Sun between Mars and Jupiter

On New Year's Day 1801, the Sicilian astronomer Giuseppe Piazzi was carefully mapping faint stars in the constellation of Taurus. He noticed a dim, previously uncharted "star" that shifted its position slightly over the next several nights. Uranus had been discovered that way by William Herschel just 20 years before. Was Piazzi's object another planet? The answer is no. His lucky sighting was too small to qualify as a full-fledged planet. Rather, he was the first to discover an asteroid.

Later that year, the orbit of this object was determined to lie between Mars and Jupiter. At Piazzi's request, the object was named Ceres (pronounced see-reez), after the patron goddess of Sicily. Ceres is spherical like the planets (Figure 8-1), but its diameter is a scant 940 km, only one-quarter the diameter of our Moon.

In 1802, the German astronomer Heinrich Olbers discovered another faint, starlike object that moved against the background stars. He called it Pallas, after the Greek goddess of wisdom. Like Ceres, Pallas orbits the Sun in a low eccentricity (nearly circular) orbit between the orbits of Mars and Jupiter. Pallas is even dimmer and smaller than Ceres, with a diameter of only 600 km.

Only two more of these minor planets—Juno and Vesta—were found until the mid-1800s, when telescopes improved. Astronomers then began to stumble across many more asteroids orbiting the Sun at distances from 2 and 3½ AU, between the orbits of Mars and Jupiter. This region of the solar system is now called the **asteroid belt** (Figure 8-2). Asteroids whose orbits lie entirely within this region are called **belt asteroids.**

R I V U X G

FIGURE 8-1 Comparison of Ceres with the Moon and the Earth Ceres, the Moon, and the Earth are shown here to scale. Ceres, shown in this infrared photo (Earth and Moon appear in visible light) is the largest asteroid but is so small that it is not considered a planet. Because it does not orbit a body other than the Sun, it is also not classified as a moon. (NASA)

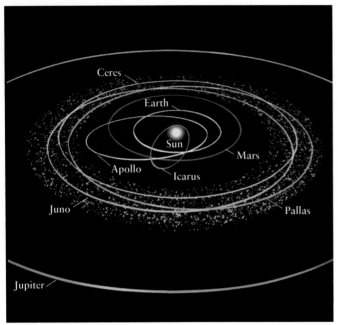

a

b

FIGURE 8-2 **Asteroid Orbits** **(a)** An artist's rendition of some asteroids and their orbits. Most asteroids orbit the Sun in a 1$^{1}/_{2}$-AU-wide belt between the orbits of Mars and Jupiter. The orbits of belt asteroids Ceres, Pallas, and Juno are indicated to scale. Some asteroids, such as Apollo and Icarus, have highly eccentric paths that cross Earth's orbit. Others, called the Trojan asteroids, follow the same orbit as Jupiter. **(b)** Actual positions of all known asteroids at Jupiter's orbit or closer. The locations of the belt asteroids are indicated by green dots. Objects passing closer than 1.3 AU to the Sun are shown by red circles. Objects observed at more than one opposition are indicated by filled circles, objects seen at only one opposition are indicated by outline circles. Jupiter's Trojan asteroids are deep blue squares. Comets are filled and unfilled light-blue squares. (b: Minor Planet Center)

The next real breakthrough came in 1891, when the German astronomer Max Wolf applied photographic techniques to the search for asteroids. A total of 300 asteroids had been found up to that time, each painstakingly discovered by scrutinizing the skies for faint, uncharted "stars" whose positions shifted slowly from one night to the next. With the advent of astrophotography, however, the floodgates were opened. Astronomers could simply aim a camera-equipped telescope at the stars and take long exposures. If an asteroid happened to be in the field of view, it left a distinctive trail on the photographic plate (Figure 8-3). Using this technique, Wolf alone discovered 228 asteroids.

Insight into Science **Confirming Observations**
New findings must be confirmed or replicated by other competent scientists before discoveries and observations are accepted by the scientific community. Therefore, asteroid observations need to be repeated and repeated in order to determine exact orbits and to eliminate the possibility that a sighting is of a known asteroid or other object.

R I V U X G

FIGURE 8-3 **Discovering Asteroids** In 1998, the Hubble Space Telescope found this asteroid while observing in the constellation Centaurus. The exposure, tracking stars, shows the asteroid as a 19 arcsecond streak. This asteroid is about 2 km in diameter and was located about 140 million km (87 million miles) from Earth. (R. Evans and K. Stapelfeldt, Jet Propulsion Laboratory and NASA)

Ceres, the largest asteroid, alone accounts for about 30% of the mass of all the known asteroids combined. Only three asteroids—Ceres, Pallas, and Vesta—have diameters greater than 300 km. Records of all objects smaller than planets in our solar system are kept by the Minor Planet Center in Cambridge, Massachusetts. The center reports that as of May 2002, 39,462 asteroids had been confirmed to exist, with more than 125,000 other observations awaiting confirmation as new asteroids. The number of asteroids increases dramatically with *decreasing size*: Only 30 asteroids have diameters between 200 and 300 km; 200 more are bigger than 100 km across; there are estimated to be millions that are less than 1 km across.

You may have heard the common belief that the asteroids were once a single planet that was somehow destroyed. It is much more likely that Jupiter's gravitational force pulling by differing amounts on the planetesimals in the region of the asteroid belt made it virtually impossible for large numbers of asteroids to meet, coalesce, and form a single object. (We consider Jupiter's gravitational effect in more detail in the next section.)

In fact, if all the asteroids had once been part of a single body, it would have had a diameter of only 1500 km, or 12% of the Earth's diameter. This is less than two-thirds the diameter of Pluto and half the diameter of our Moon. Because Pluto just barely makes it into the category of a planet, a single body containing the entire asteroid mass would not qualify as a planet.

8-2 Jupiter's gravity creates gaps in the asteroid belt

In 1867, the American astronomer Daniel Kirkwood called attention to gaps in the asteroid belt. These features, called **Kirkwood gaps,** show the influence of Jupiter's gravitational attraction, as best seen in a graph of asteroid orbital periods, like the one in Figure 8-4. Note the gaps at simple fractions ($\frac{1}{3}$, $\frac{2}{5}$, $\frac{3}{7}$, and $\frac{1}{2}$) of Jupiter's orbital period. The gravitational effect of Jupiter creating gaps in the asteroid belt resembles the gravitational *resonance* effect of Mimas creating the Cassini division in Saturn's rings (see Chapter 7).

    Resonances

Although it is likely that the asteroid belt contains millions of asteroids, their typical separation is a staggering 10 million kilometers. This is quite unlike the image that has been created by innumerable science fiction movies of asteroids so close together that you must dodge them as you fly past.

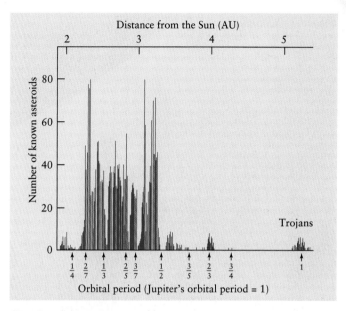

FIGURE 8-4   **The Kirkwood Gaps** This graph displays the number of asteroids at various distances from the Sun. Notice that few orbital periods of asteroids correspond to such simple fractions as $\frac{1}{3}$, $\frac{2}{5}$, $\frac{3}{7}$, and $\frac{1}{2}$ of Jupiter's orbital period. Repeated alignments with Jupiter have deflected asteroids away from these orbits. The Trojan asteroids accompany Jupiter as it orbits the Sun.

Insight into Science   **Get Real** Scientists continually apply the "laws of nature" to new situations, problems, observations, and experiments, even to popular culture. For example, applying Newton's law of gravity to the asteroids reveals that they could never swarm, as science fiction movies suggest. At those close quarters, their gravity would cause them either to collide or to pass so close together that they would actually fly rapidly and permanently apart.

Despite the large average separation between asteroids, the gravitational influences of Mars and Jupiter have sent some asteroids caroming into each other at various times over the 4.6 billion years that the solar system has existed. In 1918, the Japanese astronomer Kiyotsugu Hirayama drew attention to groups of asteroids that share nearly identical orbits. These are fragments of parent asteroids.

A collision between kilometer-sized asteroids must be an awe-inspiring event. Typical collision velocities are estimated to be 3600 to 18,000 km/h (2000 to 11,000 mph), which is more than sufficient to shatter rock. In some collisions, the resulting fragments may not have enough speed to escape from each other's gravitational attraction, and they reassemble. Alternatively, several large fragments may end up orbiting or in contact with each other. The asteroid

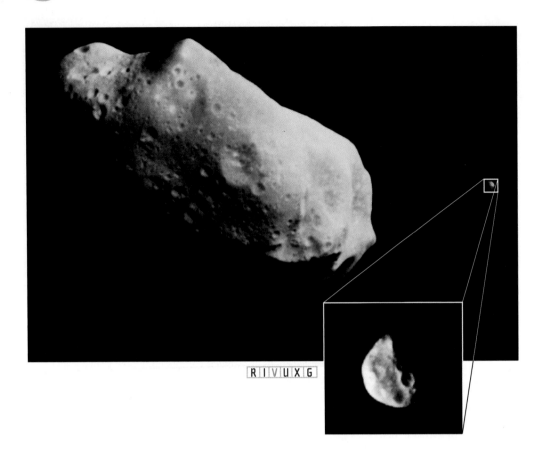

R I V U X G

FIGURE 8-5 Ida and Its Moon The 55-km-long rocky asteroid Ida, shown here with its moon Dactyl, is about twice the size of the younger asteroid Gaspra (see Figure II-11). **Inset:** Dactyl is also heavily cratered. (NASA)

Toutatis appears to be composed of two comparably sized pieces connected to each other, as does the asteroid Castalia.

In the early 1990s, the Jupiter-bound *Galileo* spacecraft passed near two asteroids—Gaspra (see Figure II-11) and Ida (Figure 8-5)—and sent back close-up views. Both asteroids are probably fragments of larger parent bodies that were broken apart by catastrophic collisions. Because Ida's surface is more heavily cratered than Gaspra's, Ida is much older.

  At least two asteroids have their own moons: Ida, accompanied by Dactyl, and Dionysus, whose moon has yet to be named. Dactyl is a pockmarked asteroid some 1.5 km in diameter that orbits Ida at a distance of 100 km. Dionysus's moon appears to be 0.5 km across, orbiting only a few kilometers from the larger body.

8-3 Asteroids exist outside the asteroid belt

While Jupiter's gravitational pull clears out certain orbits within the asteroid belt, it actually captures asteroids at two locations in the path of its own orbit. The gravitational forces of the Sun and Jupiter work together to hold asteroids in orbit at these locations, called **stable Lagrange points,** in honor of the French mathematician Joseph Lagrange, whose calculations explained them. One Lagrange point is located 60° ahead of Jupiter, and the other is 60° behind, as shown in Figure 8-6.

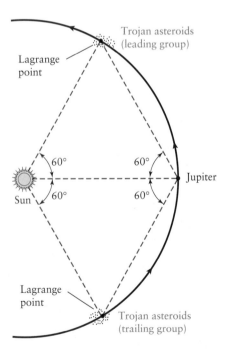

FIGURE 8-6 Jupiter's Trojan Asteroids Groups of asteroids orbit at the two stable Lagrange points along Jupiter's orbit, trapped by the combined gravitational forces of Jupiter and the Sun. Asteroids at these locations are named after Homeric heroes of the Trojan War. For more details of the location of Jupiter's Trojan asteroids, see Figure 8-2b.

The asteroids trapped at Jupiter's Lagrange points are called **Trojan asteroids,** each named after a hero of the Trojan War. As of May 2002, 1215 Trojan asteroids orbiting Jupiter have been catalogued (see Figure 8-2b). Closer to home, six asteroids have been discovered at one of Mars's Lagrange points, and asteroid number 3753 (Cruithne) orbits at one of Earth's Lagrange points.

Some asteroids have highly elliptical orbits that bring them into the inner regions of the solar system (see Figure 8-2). Others have similarly elliptical orbits that extend from the asteroid belt out beyond the farthest reaches of Pluto's orbit. The *Amor asteroids,* cross Mars's orbit, while the **Apollo asteroids** even cross Earth's orbit.

The Apollo asteroid 433 Eros passed within 23 million kilometers of our planet in 1931. On October 30, 1937, the asteroid Hermes passed within 900,000 km of Earth—only a little more than twice the distance to the Moon. On June 14, 1968, asteroid Icarus passed Earth at a distance of only 6 million kilometers. In 1972, space debris was observed to skip off Earth's atmosphere and retreat back into space.

There were several close calls recently. On December 9, 1994, asteroid 1994 XM1 passed within 105,000 km of the Earth, and on January 8, 2002, an asteroid passed within 375,000 km. The former asteroid is about 10 m across—the size of a small bus (Figure 8-7). At least 1690 *Earth-crossing* asteroids are known.

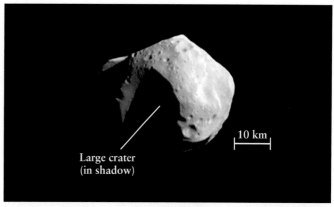

R I V U X G

FIGURE 8-8 Asteroid Mathilde Reflecting only half as much light as a charcoal briquette, Mathilde is half as dense as typical, stony asteroids. Slightly larger than Ida, irregularly shaped Mathilde measures $66 \times 48 \times 46$ km, rotates once every 17.4 d, and has a mass equivalent to 110 trillion tons. The part of the asteroid shown is about 59×47 km. The large crater in shadow is about 20 km across. (Johns Hopkins University, Applied Physics Laboratory)

During these close encounters, astronomers can examine the details of asteroids. For example, an asteroid's brightness often varies as it rotates because different surface features scatter different amounts of light. Such data show that typical rotation periods for asteroids are between 5 and 20 hours, although one asteroid, labeled 1998 KY26, rotates about once every 10.7 minutes. This is the fastest-rotating object known in the solar system.

The proximity of asteroids and the promise of learning more about these ancient and extremely varied members of our solar system prompted NASA to send the *Near Earth Asteroid Rendezvous (NEAR) Shoemaker* spacecraft to visit asteroids Mathilde (Figure 8-8) and Eros. *NEAR Shoemaker* revealed that Mathilde is only 1.3 times denser than water and has an albedo of 0.04, making it darker than charcoal. This heavily cratered, carbon-rich body therefore has only half the density of the other asteroids astronomers have studied, such as Ida. Eros is about 3 times as dense as water and rotates once every $5^{1}/_{4}$ hours. In February 2000, *NEAR Shoemaker* went into orbit around Eros, and for a year the spacecraft's cameras and other sensors sent back a wealth of information about the asteroid.

NEAR Shoemaker showed that Eros (Figure 8-9) is a solid chunk of rock and metal. Analyzing Eros's spectra reveals that it is probably much the same as it was when it coalesced 4.6 billion years ago. This means that it was never hot enough to differentiate (separate rock from metal). Infrared observations reveal that like our Moon, Eros has a regolith. It also has several substantial craters and is strewn

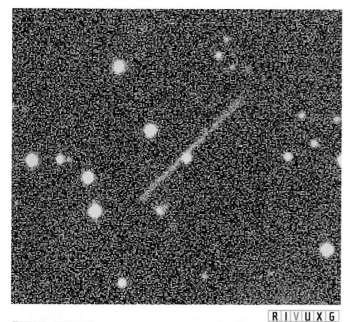

R I V U X G

FIGURE 8-7 Asteroid 1994 XM1 This image was obtained on December 9, 1994, shortly before the asteroid arrived in the Earth's vicinity. When it passed by the Earth just 12 hours later, asteroid 1994 XM1 was less than half the distance from the Earth to the Moon. (Jim Scotti, Spacematch on Kitt Peak)

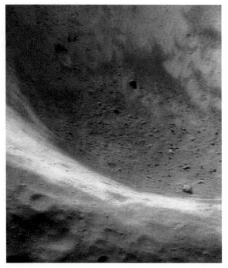

a   b   c   R I V U X G

FIGURE 8-9 Eros **(a)** The *Near Earth Asteroid Rendezvous* (*NEAR*) *Shoemaker* spacecraft took this image of asteroid Eros in February 1999. The top of the figure is the asteroid's north polar region. Eros's dimensions are 33 × 13 × 13 km (21 × 8 × 8 mi) and it rotates every 5^1/$_4$ hours. Its density is 2700 kg/m^3, close to the average density of the Earth's crust and twice as dense as asteroid Mathilde. **(b)** Looking into the large crater near the top of (a), which is 5.3 km (3.3 miles) across. **(c)** The penultimate image taken by *NEAR Shoemaker* before it was gently landed on Eros. Taken from an altitude of 250 m (820 ft), the image is only 12 meters across. You can see rocks and boulders buried to different depths in the regolith.

(a, b & c: Johns Hopkins Applied Physics Laboratory)

with boulders (Figure 8-9c). On February 12, 2001, NASA engineers landed *NEAR Shoemaker* on Eros so gently that the spacecraft continued to transmit data after landing. Details of Eros as small as 1.4 meters across were imaged by *NEAR Shoemaker*.

Most of the Apollo asteroids will eventually strike a moon or planet—even the Earth. Some asteroids will end up heading straight into the Sun. However, the chance of the Earth being hit by an asteroid in the next few thousand years is remote. We will examine this possibility at the end of the chapter.

The Sun is not the only star with an asteroid belt. In 2001, astronomers discovered that the star Zeta Leporis, 70 light-years from Earth in the constellation Lepus (the hare), has a disk of debris that appears to contain asteroids. This star system is less than 0.5 billion years old and astronomers hope it will provide insights into the early evolution of the asteroids and other objects in the disk of gas and dust surrounding the early Sun.

COMETS

While asteroids consist primarily of rock and metal, other pieces of space debris are composed of frozen water, along with rock, metal, and ices of other compounds. We have seen in earlier chapters that water ice, along with carbon dioxide, methane, and ammonia ices, was locked up in planets and moons. Ices in the young solar system also condensed with roughly equal amounts of small rocky and metallic debris into bodies that still remain in orbit around the Sun. These dirty icebergs in space are the **comets**. To better determine their composition, the *Stardust* probe is on its way to Comet Wild 2, where it will collect samples and return to Earth in 2006.

8-4 Comets come from far out in the solar system

In the first few hundred millions years of the solar system's existence, comets formed in its outer reaches at roughly the distances of Saturn, Uranus and Neptune. In that region, water was plentiful and the temperature low enough for the ices to condense into chunks several kilometers across. In 2001, astronomers were able to measure the temperature at which ammonia ice formed in the comet Linear, thereby determining that this comet formed between the orbits of Saturn and Uranus. Then, gravitational tugs from Uranus and Neptune flung the comets in every direction.

Today, the solar system is believed to contain two reservoirs of comets. Most comets that eventually return to the inner solar system and develop the long tails that we usually associate with them are believed to come from a doughnut-shaped region beyond the orbit of Pluto. This **Kuiper belt**, named after the American astronomer Gerard Kuiper, who first proposed its existence in 1951, is centered on the plane of the ecliptic and extends out some 500 AU from the Sun (Figure 8-10).

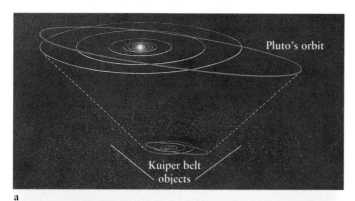

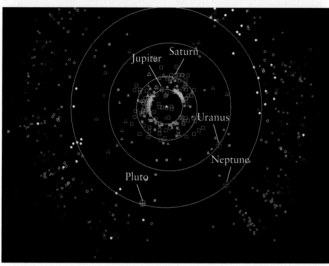

a

b

FIGURE 8-10   **The Kuiper Belt** **(a)** The Kuiper belt of comets
spreads from Pluto out 500 AU from the Sun, as depicted in this
artist's conception. Most of the estimated 200 million belt comets are
believed to orbit in or near the plane of the ecliptic. More than 530
of these objects have been located; the largest is 320 km across. Some
astronomers believe that Pluto and Charon are members of the Kuiper
belt. **(b)** The current positions of the minor bodies in the outer solar
system are shown in this diagram. Unusual high eccentricity objects
are shown as cyan triangles. Objects roaming among the outer
planets, called Centaur objects, are orange triangles. Plutinos are
white circles. Miscellaneous objects are magenta circles and "classical"
or "main-belt" objects as red circles. Objects observed at only one
opposition are denoted by open symbols, objects with multiple-
opposition orbits are denoted by filled symbols. Comets are filled and
unfilled light-blue squares. (b: Minor Planet Center)

More than 530 Kuiper belt objects have been observed
(Figure 8-11). At least one of them, named 1998 WW31,
has a moon of its own (Figure 8-11c). Another, 2000 WR106,
is between one quarter and one half the size of Pluto. At
least 40 Kuiper belt objects orbit the Sun in the same region
as Pluto. Because these latter bodies are smaller than Pluto,
they are called *Plutinos*. Given the number and locations of
these objects, astronomers estimate that the Kuiper belt contains at least 200 million comets.

a   b   c   R I V U X G

FIGURE 8-11   **Kuiper Belt Objects** **(a) and (b)** These 1993 images show the discovery (white
lines) of one of more than 530 known Kuiper belt objects. These two images of Kuiper belt object
1993 SC were taken 4.6 hours apart, during which time the object moved from **(a)** to **(b)** against the
background stars. **(c)** The Kuiper belt object 1998 WW31 and its moon (lower left). (a & b: Alan
Fitzsimmons, Queen's University of Belfast; c: C. Veillet/CFHT)

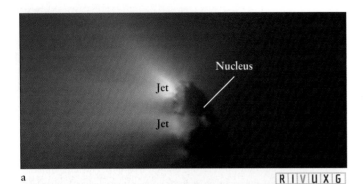

a   R I V U X G

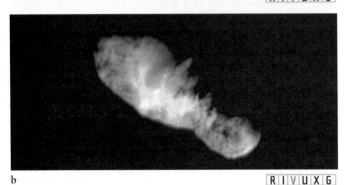

b   R I V U X G

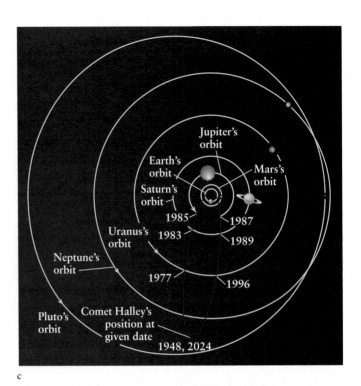

c

FIGURE 8-12   Comet Nuclei (a) The nucleus of Comet Halley. This image, taken by the *Giotto* spacecraft, shows the potato-shaped nucleus of the comet. Its dark nucleus measures 15 km in its longest dimension and about 8 km in its shortest. The Sun illuminates the comet from the left. The numerous bright areas on the nucleus are icy outcroppings that reflect more sunlight than surrounding areas of the comet. Two jets of gas can be seen emanating from the left side of the nucleus. **(b)** The nucleus of Comet Borrelly. Taken by *Deep Space 1*, this image has resolution of 45 m (150 ft). The nucleus is 8 km (5 mi) long, and the image was taken from 3417 km (about 2000 mi) away. **(c)** Comet Halley's orbit. Its highly eccentric orbit takes Halley out beyond Neptune with a period of about 76 years. (a: Max-Planck-Institut für Aeronomie; b: Deep Space 1 Team, JPL, NASA)

The vast majority of the several billion comets estimated to exist are believed to lie even farther from the Sun. Unlike the Kuiper belt comets and the rest of the solar system, these comets are believed to have a spherical distribution around the Sun called the **Oort cloud,** named after the Dutch astronomer Jan Oort, who first proposed its existence in the 1950s. Astronomers calculate that the Oort cloud extends out at least 50,000 AU, one-fifth the distance to the nearest stars. Most of these comets have orbits so nearly circular that they never even get as close as Pluto is to the Sun. However, occasionally a passing star's gravitational force nudges a distant comet toward the inner solar system. As a result of its inward plunge, comets from the Oort comet cloud, like Hale-Bopp and Hyakutake, have highly elliptical orbits. Often comets also have orbits that are highly tilted, even perpendicular, to the plane of the ecliptic.

Because the comets in the Kuiper belt and Oort cloud are far from the Sun, they are completely frozen. Solid comet bodies, called **nuclei** (*singular* **nucleus**), are typically 20 kilometers across. The first pictures of a comet's nucleus were obtained when a fleet of spacecraft flew past Comet Halley

in 1986 (Figure 8-12a). Halley's potato-shaped nucleus is darker than coal, probably because of carbon-rich compounds left behind after its ice evaporated. In 2001, the *Deep Space 1* spacecraft took the highest resolution image ever of a comet nucleus (Figure 8-12b). The ends of the comet are very rugged, while the center has long, rolling plains, from which jets of gas appear to originate.

As a comet nucleus comes within 20 AU of the Sun, solar radiation begins to vaporize the ices on its surface. The liberated gases form an atmosphere, or **coma,** around the nucleus. Because the coma scatters sunlight, it appears as a fuzzy, luminous ball. The largest coma ever measured was more than a million kilometers across—nearly as large as the Sun. Not visible to the human eye is the **hydrogen envelope,** a sphere of tenuous gas surrounding the comet's nucleus and measuring as much as 20 million kilometers in diameter (Figure 8-13).

Of course, the most visible and inspiring features of comets are their long, flowing, diaphanous **tails** (Figure 8-14). Comet tails develop from coma gases and dust pushed outward from the Sun. This means that comet tails do not trail behind the nucleus, as the exhaust from a jet plane does in the

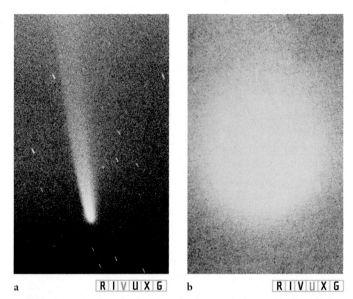

a ⬛ R I V U X G　b ⬛ R I V U X G

■ FIGURE 8-13　**Comet Kohoutek and Its Hydrogen Envelope**
Comet Kohoutek as seen in visible light **(a)** and to the same scale at
ultraviolet wavelengths **(b)**. The ultraviolet picture reveals a huge
hydrogen cloud surrounding the comet's nucleus. (a & b: Johns Hopkins
University and Naval Research Laboratory)

⬛ R I V U X G

■ FIGURE 8-14　**Comet West**　Astronomer Richard M. West first
noticed this comet on a photograph taken with a telescope in 1975.
After passing near the Sun, Comet West became one of the brightest
comets of the 1970s. This photograph shows the comet in the
predawn sky in March 1976. (Hans Vehrenberg)

Earth's atmosphere. Rather, *at the comet's nucleus, the tails
always point away from the Sun* (Figure 8-15), regardless of
the direction of the comet's motion. The implication that
something from the Sun was "blowing" the comet's gases
radially outward led Ludwig Biermann to predict the existence of the solar wind (see Chapters 5 and 9). This stream of
particles from the Sun was actually discovered in 1962, a full
decade later, by instruments on the spacecraft *Mariner 2*.

> **Insight into Science**　**Working in Reverse**　Science
> often advances by working from observations and
> experiments that require scientific explanations. For
> example, seeing comet tails and how they behave led
> Biermann to wonder what caused them. The simplest
> physical explanation is that matter from each comet
> body is being evaporated and pushed upon by some-
> thing from the sun, the solar wind. Remember Occam's
> razor.

The Sun's radiation usually produces two comet tails: a
gas (or ion) tail and a **dust tail** (Figure 8-16). Positively
charged ions (atoms missing one or more electrons) from the
coma are swept directly away from the Sun by the solar
wind to form the gas tail. This tail often appears blue,
because of blue light emitted by carbon monoxide ions in
it. Ions typically leave the coma at speeds of 1.4 million

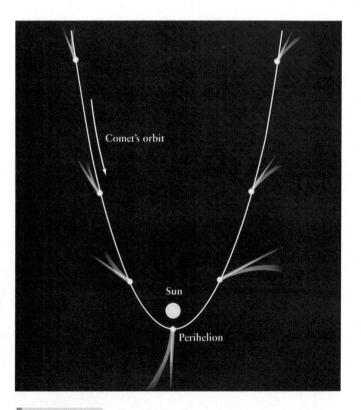

■ FIGURE 8-15　**The Orbit and Tails of a Comet**　The solar wind
and sunlight blow a comet's dust particles and ionized atoms away from
the Sun. Consequently, comets' tails always point away from the Sun.

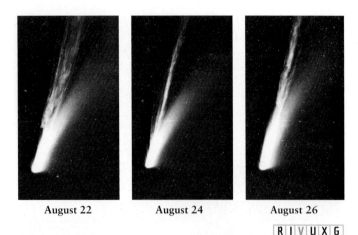

| August 22 | August 24 | August 26 |

R I V U X G

FIGURE 8-16   The Two Tails of Comet Mrkos Comet Mrkos dominated the evening sky in August 1957. These three views, taken at two-day intervals, show dramatic changes in the comet's gas tail. In contrast, the slightly curved dust tail remained fuzzy and featureless. (Palomar Observatory)

kilometers per hour. The relatively straight gas tail can change dramatically from night to night (see Figure 8-16).

The dust tail is formed when photons strike dust particles that have been freed from the comet's evaporating nucleus. Light exerts pressure on any object that absorbs or scatters it. While this pressure, called **radiation pressure** or

photon pressure, is quite weak, fine-grained dust particles in a comet's coma are sufficiently light to be blown away from the comet, thus producing a dust tail. The dust tail often is the color of sunlight. The dust particles are massive enough not to flow straight away from the Sun, rather, the dust tail arches in a path that lies between the gas tail and the direction from which the comet came (Figures 8-15 and 8-17).

Figure 8-17 outlines the structure of a comet. Some, like the comet shown in Figure 8-18, have a large, bright coma but short, stubby tails. Others have an inconspicuous coma but one or more tails of astonishing length. The gas tail in Figure 8-19 stretches more than 150 million kilometers (1 AU) in length.

8-5   Comets do not last forever

Discovering Comets

Astronomers, many of them amateurs, typically discover at least a dozen new comets each year. Falling Sunward from the Kuiper belt or Oort cloud, most are **long-period comets,** which have such eccentric orbits that they leave the inner solar system after one pass by the Sun and typically take 1 million to 30 million years to return.

However, sometimes a comet passes so close to a planet that the planet's gravitational force changes the comet's orbit, slowing it down and trapping it in the inner solar sys-

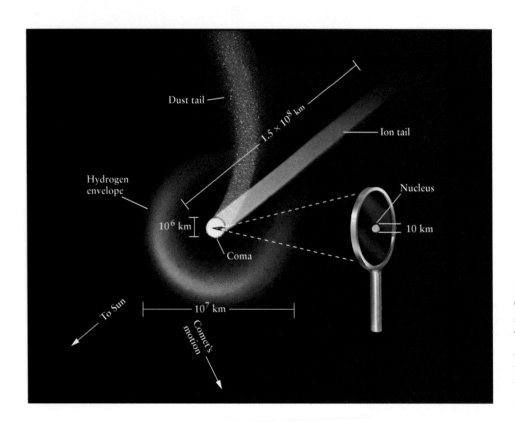

FIGURE 8-17   The Structure of a Comet The solid part of a typical comet (the nucleus) is roughly 10 km in diameter. The coma can be as large as 10^5 to 10^6 km across, and the hydrogen envelope is typically 10^7 km in diameter. A comet's tail can be enormous—as long as 1 AU. (This drawing is not to scale.)

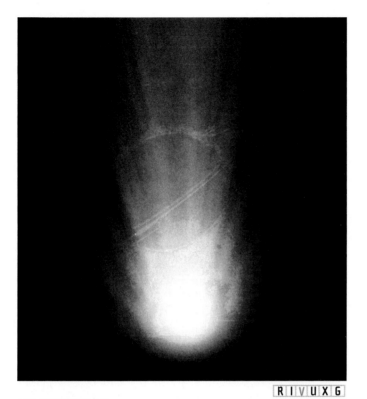

R I V U X G

FIGURE 8-18 The Head of Comet Brooks This comet, named after its discoverer, had an exceptionally large, bright coma. It dominated the night skies in October 1911. (UCO/Lick Observatory)

R I V U X G

FIGURE 8-19 The Tail of Comet Ikeya-Seki Named after its codiscoverers in Japan, this comet dominated the predawn skies in late October 1965. Although its coma was tiny, its tail spanned 1 AU. (C. R. Lynds/Kitt Peak National Observatory/Tucson, AZ)

tem (Figure 8-20). The comet then becomes a **short-period comet,** orbiting the Sun in fewer than 200 years. Like Comet Halley, short-period comets appear again and again at predictable intervals. If you missed Halley's comet in its 1985 visit, your next chance to see it is in 2061.

Comets cannot survive an infinite number of passages near the Sun. A typical comet is estimated to lose between 1/60th and 1/100th of its mass at each perihelion (closest approach to the Sun). Therefore, a typical comet survives at most 100 close passes to the Sun before its ices evaporate completely. It is likely, then, that Comet Halley, which has been seen at least 27 times, will survive for fewer than 5600 more years. Figure 8-21 shows the nucleus of a comet breaking up shortly after it passed perihelion. Soon thereafter, its remaining dust and rock fragments spread out in a loose collection of debris that continues to circle the Sun along the comet's orbital path.

A comet can also be torn apart when it comes too close to a planet, or it can be destroyed completely by striking a planet, a moon, or the Sun. A spectacular example was Comet Shoemaker-Levy 9. As discussed in Chapter 7, that comet fragmented under the tidal force from Jupiter in 1992. Two years later, with the world's astronomers watching carefully, the pieces returned and struck the planet (see Figures 7-8 and 7-9).

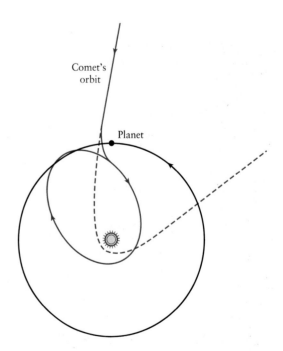

FIGURE 8-20 Transforming a Long-Period Comet into a Short-Period Comet The gravitational force of a planet can change a comet's orbit. Initially, on highly elliptical orbits, comets are sometimes deflected into more circular paths that keep them in the inner solar system.

The comet's disintegration provided an important clue to its structure. For Shoemaker-Levy 9 to break up in the first place tells astronomers that its nucleus must have been held together very weakly. It may actually have been composed of separate pieces that had stuck together until Jupiter's tidal force pulled them apart. Since then, astronomers have observed at least three other comets break up, in 1999, 2000 (Comet LINEAR) and 2001. These observations suggest that this structure may be typical of other comets as well.

Comets occasionally lose mass quickly by ejecting it in bursts. Several times, starting in September 1995, astronomers observed Comet Hale-Bopp eject 7 to 10 times more mass than usual (Figure 8-22). Astronomers believe that this event resulted from surface ice and dust being heated by the Sun and then being rapidly ejected from a local region on the comet's surface called a *vent*. Hale-Bopp rotates with a period of about 12 hours. The ejected matter therefore formed a pinwheel-shaped distribution spiraling from the comet's nucleus at a speed of 109 km/hr (68 mph). While the ejected matter did not represent a significant fraction of the comet's mass, its light-reflecting dust made it look like a huge comet fragment.

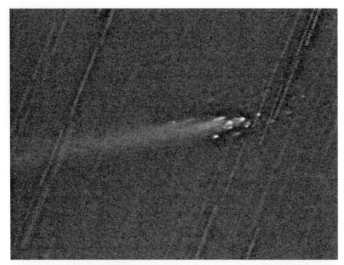

R I V U X G

FIGURE 8-21   The Fragmentation of Comet LINEAR On July 26, 2000, Comet LINEAR passed its perihelion 0.74 AU from the Sun. Within days, the comet was observed to have fragmented, like comets Shoemaker-Levy 9 and West (among many others) before it. (European Southern Observatory)

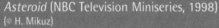

MOVIE MISCONCEPTIONS

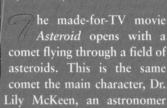

Asteroid (NBC Television Miniseries, 1998)
(© H. Mikuz)

The made-for-TV movie *Asteroid* opens with a comet flying through a field of asteroids. This is the same comet the main character, Dr. Lily McKeen, an astronomer from Colorado, has been studying as it makes its trip through the solar system. Dr. McKeen discovers that as the comet passed through the asteroid belt it struck and dislodged two asteroids from their harmless orbits, sending them Earthward. The small one will create an explosion more powerful than the atomic bomb dropped on Hiroshima and the bigger one could send the Earth into another ice age. Dr. McKeen decides it is time to notify Washington of the impending catastrophe.

In the movie, the comet was completely unscathed by the encounters and continues on its original orbit. There are several scientific inaccuracies with this scene. Describe the errors. *Hint:* consider, among other things, the sizes and chemical compositions of comets and asteroids.

If this type of interaction were possible, then the likelihood of asteroids coming earthward would be quite high, because each year several comets actually pass through the asteroid belt. Movie plots like the one presented in *Asteroid* create fear that the Earth is in imminent danger of a life-threatening impact. Such concerns can be used to justify spending billions of dollars for a system to detect these bodies and defend the Earth from them. Is this concern justified? You may find a search of the Web useful in answering this question, but be careful to check the credentials of the sources you use and cite.

Near the end of the movie *Asteroid*, there is a scene showing the audience the original comet flying briskly across the sky with its tail trailing behind it. What two aspects of this scene are unrealistic?

(Answers appear at the end of the book.)

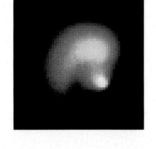

WEB LINK 8.13
FIGURE 8-22 Comet **Hale-Bopp** Discovered on July 23, 1995, this comet was at its breathtaking best in mid-1997. Did you see it? **Inset:** Jets of gas and debris were observed shooting out from Comet Hale-Bopp several times. This image shows the comet nucleus (lower bright region), an ejected piece of the comet's surface (upper bright region), and a spiral tail. The ejected piece eventually disintegrated, following the same spiral pattern as the tail. (Tony and Daphne Hallas, Astrophotos)

R I V U X G

The Sun is not the only star around which comets orbit. A star labeled CW Leonis, about 500 light-years from Earth, is presently enlarging in size and vaporizing billions of comets around it. The spectra of the water vapor from these bodies has been observed.

METEOROIDS, METEORS, AND METEORITES

WEB LINK 8.14
As noted earlier, asteroids occasionally collide with each other, sending fragments into interplanetary space. These smaller pieces, along with rocky and metallic fragments from evaporating comets, and debris that never coalesced with larger bodies, are still strewn throughout the solar system. Indeed, by studying the chemistry of various asteroids, astronomers are now beginning to identify which asteroids were the origins for others. For example, asteroid Braille (named in honor of Louis Braille, developer of the alphabet for the blind), is a piece of debris 2 km long blasted off the 500-km-diameter asteroid Vesta.

8-6   Small rocky debris peppers the solar system

Meteoroids are rocky and metallic debris smaller than asteroids scattered throughout the solar system. Although no official size standard distinguishes the two, meteoroids are no more than a few tens of meters across, and the vast majority are smaller than a millimeter.

Meteoroids are often pulled by gravity into the Earth's atmosphere. Air friction generates so much heat that a meteoroid's outer layer begins to vaporize. (The same process creates the heat on the hull of the Space Shuttle as it reenters our atmosphere.) As the meteoroid penetrates farther into the air, leaving behind a trail of dusty gas, it becomes a **meteor** (Figure 8-23). Common names for these dramatic streaks of light flashing across the sky include *shooting stars*, *fireballs*, and *bolides*. Fireballs are meteors at least as bright as Venus; bolides are bright meteors that explode in the air. Therefore, "shooting stars" are not stars of any kind, nor are they dying stars.

8-7   Impact craters and meteor showers mark remnants of space debris on Earth

Most meteors vaporize completely before they can strike the Earth. Their dust settles to the ground, often carried by raindrops. (This is not the source of acid rain, however, which comes from natural and human-made gases ejected into the atmosphere.) Any part of a meteor that survives its fiery descent to Earth may leave an **impact crater**. Weather and

R I V U X G

FIGURE 8-23   Meteors A streak of light is produced when
a piece of interplanetary rock or dust strikes the Earth's atmosphere
at high speed. Seen through a fisheye lens, this image of the 2001
Leonids meteor shower shows at least a doxen meteors. (Juraj Toth/
Comenius U. Bratislava/Modra Observatory/NASA)

FIGURE 8-24   The Barringer Crater An iron
meteoroid measuring 50 m across struck the ground in
Arizona 50,000 years ago. The result was this beautifully
symmetric impact crater. (D. J. Roddy & K. Zeller/USGS)

water erosion are wearing away all of the 200 or so impact
craters now known on the Earth, and thousands more have
long since been drawn into the Earth by the motion of its
tectonic plates. Indeed, those known today are all less than
500 million years old because of forces reshaping the Earth's
surface.

One of Earth's best-preserved impact craters is the
famous Barringer (or Meteor) Crater near Winslow, Arizona
(Figure 8-24). Measuring 1.2 km across and 200 m deep, it
formed 50,000 years ago when an iron-rich meteoroid some
50 m across (about half the length of a football field) struck
the ground at 40,000 km/h (25,000 mph). The blast was like
the detonation of a 20-megaton hydrogen bomb.

On a typical clear night, you can expect to see a meteor
about every 10 minutes. However, at predictable times
throughout each year, the Earth is inundated with them.
These **meteor showers** occur when the Earth moves through
the orbit of debris left behind by a comet (Figures 8-23, 8-25,
and 8-26).

Some 30 meteor showers can be seen each year.
Because the meteors in each shower appear to come from a
fixed region of the sky, they are named after the constellation from which the meteors appear to radiate (Figure
8-26). For example, meteors in the Perseid shower appear
to originate in the constellation Perseus. More than one
meteor can be seen each minute at the peaks of such prodi-

Comet debris

Earth

FIGURE 8-25   The Origin of Meteor Showers As comets
dissipate, they leave debris behind that spreads out along their orbit.
When the Earth plows through such material, many meteors can be
seen emanating from the same place within a very short time—a
meteor shower. As shown in this diagram, many comets have high
orbital inclinations.

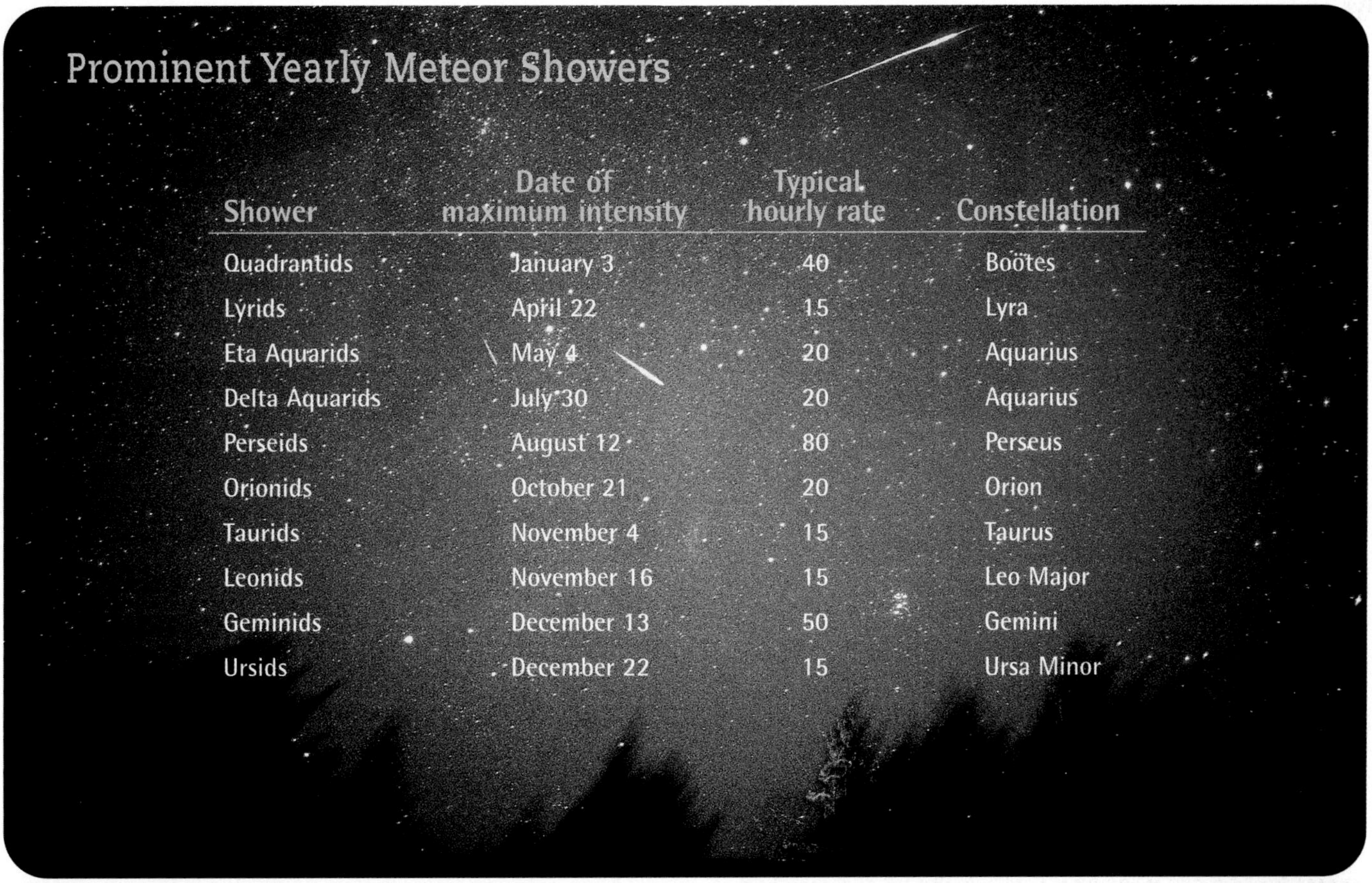

Prominent Yearly Meteor Showers

Shower	Date of maximum intensity	Typical hourly rate	Constellation
Quadrantids	January 3	40	Boötes
Lyrids	April 22	15	Lyra
Eta Aquarids	May 4	20	Aquarius
Delta Aquarids	July 30	20	Aquarius
Perseids	August 12	80	Perseus
Orionids	October 21	20	Orion
Taurids	November 4	15	Taurus
Leonids	November 16	15	Leo Major
Geminids	December 13	50	Gemini
Ursids	December 22	15	Ursa Minor

FIGURE 8-26 **Meteor Streaks Seen during a Meteor Shower** This table lists highly active, yearly meteor showers, which last for several days. The time exposure behind the table shows meteors streaking away from one place in the sky, in the constellation Perseus, during the Perseid meteor shower. (Akira Fujii, Chiro Astronomical Observatory, Koriyama, Japan)

R I V U X G

gious meteor showers as the Perseids, which take place in the summertime and are among the best light shows in astronomy. Except for the Lyrids, these showers are best seen after midnight.

 Meteors

8-8 Meteorites are space debris that land intact

Although most meteors completely vaporize in the atmosphere, some reach the ground before totally disintegrating. Such debris are called **meteorites,** and they come from a variety of sources: Many are believed to have been broken off from asteroids by collisions; some are debris that were never part of larger bodies; still others come from the Moon, Mars, and comets. We saw in Chapter 6, for example, how SNC meteorites offer clues to the chemistry of Mars.

Meteorites tell us the age of the solar system. Measuring the various radioactive elements in meteorites allows astronomers to determine how long ago they formed (see An Astronomer's Toolbox 4-2). The oldest known meteorites became solid bodies about 4.57 billion years ago.

People have been picking up debris from space for thousands of years. The descriptions of meteorites in historical Chinese, Indian, Islamic, Greek, and Roman literature show that early peoples placed special significance on these "rocks from heaven." They also have a practical "impact": Infalling space debris is increasing the Earth's mass by more than 300 tons per day on average.

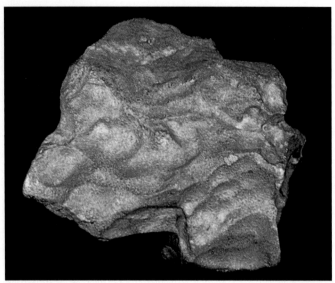

R I V U X G

FIGURE 8-27 A Stony Meteorite Most meteorites that fall to Earth are stones. Many freshly discovered specimens, like the one shown here, are coated with thin, dark crusts. This stony meteorite fell in Texas. (R. A. Oriti)

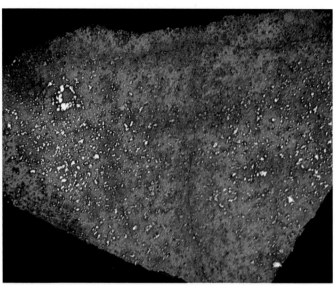

R I V U X G

FIGURE 8-28 A Cut and Polished Stone Some stony meteorites contain tiny specks of iron, which can be seen when the stones are cut and polished. This specimen was discovered in California. (R. A. Oriti)

Meteorites are classified as stones, stony-irons, and irons. Most **stony meteorites** look much like ordinary rocks, although some are covered with a dark *fusion crust,* created when the meteorite's outer layer melts during its fiery descent through the atmosphere (Figure 8-27). When a stony meteorite is cut in two and polished, tiny flecks of iron are sometimes found in the rock (Figure 8-28).

Meteorites with a high iron content can be located with a metal detector. They also look unusual and hence are more likely to be noticed. Consequently, the easily found iron and stony-iron meteorites dominate most museum collections. Nevertheless, stony meteorites account for about 95% of all meteoritic material that falls on the Earth.

Iron meteorites (Figure 8-29) may also contain from 10% to 20% nickel by weight. Iron is moderately abundant in the universe as well as being one of the most common rock-forming elements, so it is not surprising that iron is an important constituent of asteroids and meteoroids. Another element, iridium, is common in the iron-rich minerals of meteorites but rare in ordinary rocks because most iridium settled deep into the Earth eons ago. Measurements of iridium in the Earth's crust can thus tell us the rate at which meteoritic material has been deposited on the Earth over the ages.

In 1808, Count Alois von Widmanstätten, director of the Imperial Porcelain Works in Vienna, discovered a conclusive test for the most common type of iron meteorite. Most irons have a unique structure of long nickel-iron crystals called **Widmanstätten patterns,** which become visible

when the meteorites are cut, polished, and briefly dipped into a dilute solution of acid (Figure 8-30). Because nickel-iron crystals can grow to lengths of several centimeters only if the molten metal cools slowly over many millions of years,

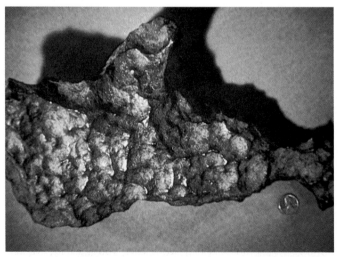

R I V U X G

FIGURE 8-29 An Iron Meteorite Irons are composed almost entirely of iron-nickel minerals. The surface of a typical iron is covered with thumbprintlike depressions created as the meteorite's outer layers vaporized during its high-speed descent through the atmosphere. This specimen was found in Australia. (R. A. Oriti)

R I V U X G

FIGURE 8-30 **Widmanstätten Patterns** When cut, polished, and etched with a weak acid solution, most iron meteorites exhibit interlocking crystals in designs called Widmanstätten patterns. This meteorite was found in Australia. (R. A. Oriti)

R I V U X G

FIGURE 8-31   **A Stony-Iron Meteorite** Stony-irons account for about 1% of all meteorites that fall to Earth. This specimen, a variety of stony-iron called a pallasite, was found in Chile. (Chip Clark)

Widmanstätten patterns are never found in counterfeit meteorites, or "meteorwrongs."

The final category of meteorites, the **stony-irons,** consists of roughly equal amounts of rock and iron. Figure 8-31, for example, shows the greenish mineral olivine suspended in a matrix of iron.

To understand why different types of meteorites exist, we consider their formation. Most meteorites were once pieces of asteroids. Heat from the rapid decay of radioactive isotopes melted newly formed asteroid interiors. Over the next few million years, differentiation occurred, just as in the young Earth. Iron sank toward the asteroid's center, while lighter rock floated up to the asteroid's surface. Iron meteorites are fragments of asteroid cores, and stones are samples of their crusts. Stony-irons are believed to come from the boundary regions between the iron cores and stony crusts.

A class of rare stony meteorites, **carbonaceous chondrites,** shows no evidence of ever having been melted as parts of asteroids. These rarities may therefore be primordial material from which our solar system was created. Carbonaceous chondrites contain complex carbon compounds and as much as 20% water bound into their minerals. The organic compounds would have been broken down and the water driven out if these meteorites had been significantly heated. Asteroid Mathilde, shown in Figure 8-8, has a very dark gray color and virtually the same spectrum as a carbonaceous chondrite meteorite, so it is likely composed of primordial material.

Amino acids, the building blocks of proteins upon which terrestrial life is based, are among the organic compounds occasionally found inside carbonaceous chondrites, although these may be contaminants acquired after the meteoroids entered the Earth's atmosphere. Nevertheless, some scientists suspect that carbonaceous chondrites may have played a role in the origin of life on Earth.

8-9 The Tunguska mystery and the Allende meteorite provide evidence of catastrophic collisions

At 7:14 A.M. local time on June 30, 1908, a spectacular explosion occurred over the Tunguska region of Siberia. The blast, comparable to a nuclear detonation of several megatons, knocked a man off his porch some 60 km away and was audible more than 1000 km away. Millions of tons of dust were injected into the atmosphere, darkening the air as far away as California.

Preoccupied with wars, along with political and economic upheaval, neither Russia nor its successor, the former Soviet Union, sent a scientific expedition to the site until 1927. At that time, Soviet researchers found that trees had been seared and felled radially outward in an area about 30 km in diameter (Figure 8-32). There was no clear evidence of a crater. In fact, the trees at "ground zero" were left standing upright, although they were completely stripped of branches and leaves. Because no significant meteorite samples were found, for many years scientists assumed that a small comet had struck the Earth.

R I V U X G

FIGURE 8-32 Aftermath of the Tunguska Event In 1908 a stony asteroid traveling at supersonic speed struck the Earth's atmosphere and exploded over the Tunguska region of Siberia. Trees were blown down for many kilometers in all directions from the impact site. (SOVFOTO)

Recently, however, several teams of astronomers have argued that a small comet, composed primarily of light elements and ice, breaks up too high in the atmosphere to cause significant damage on the ground. They argue that the Tunguska explosion was actually caused by a small asteroid or large meteoroid traveling at supersonic speed. The Tunguska event is consistent with an explosion of an asteroid about 80 m (260 ft) in diameter entering the Earth's atmosphere at 79,000 km/h (50,000 mph) and exploding in the air as a result of becoming exceedingly hot.

A chance to study debris immediately after impact came shortly after midnight on February 8, 1969, when a brilliant, blue-white light shot across the night sky around Chihuahua, Mexico. Hundreds of people witnessed the dazzling display. The light disappeared in a spectacular, noisy explosion that dropped thousands of rocks and pebbles over the terrified onlookers. Within hours, teams of scientists were on their way to collect specimens of carbonaceous chondrites, collectively named the *Allende meteorite,* after the locality in which they fell (Figure 8-33).

One of the most significant discoveries to come from the Allende meteorite was evidence of the detonation of a nearby supernova 4.6 billion years ago. Among nature's most violent and spectacular phenomena, a *supernova explosion* occurs when a massive star dies. A massive star blows apart in a cataclysm that hurls matter outward at tremendous speeds, as we will see in Chapter 12. During this detonation, violent collisions between atomic nuclei produce a host of radioactive elements, including a short-lived radioactive isotope of aluminum. Based on its decay products, scientists found unmistakable evidence that this isotope once lay within the

Allende meteorite. Some astronomers interpret this as evidence for a supernova in our vicinity at about the time the Sun was born. By compressing interstellar gas and dust, the supernova's shock wave may have helped stimulate the birth of our solar system.

8-10 An asteroid's impact with Earth apparently killed off the dinosaurs

In the late 1970s, the geologist Walter Alvarez and his father, physicist Luis Alvarez, discovered a different sort of shock wave closer to home. Working at a site of exposed marine limestone in the Apennine Mountains in Italy that had been on the Earth's surface 65 million years ago, the Alvarez team discovered an exceptionally high abundance of iridium in a dark-colored layer of clay between limestone strata (Figure 8-34).

Since this discovery was announced in 1979, a comparable layer of iridium-rich material has been uncovered at numerous sites around the world. In every case, geologic dating reveals that this apparently worldwide iridium-rich layer was deposited about 65 million years ago. Paleontologists were quick to realize the significance of this date, because it was 65 million years ago when all the dinosaurs rather suddenly became extinct. In fact, two-thirds of all the species on Earth disappeared within a brief span of time back then.

The Alvarez discovery suggests a startling explanation for the dramatic extinction of so much of the life that once inhabited our planet—an asteroid impact. An asteroid 10 km in diameter slamming into the Earth at high speed could have

R I V U X G

FIGURE 8-33 A Piece of the Allende Meteorite This carbonaceous chondrite fell near Chihuahua, Mexico, in February 1969. Note the meteorite's dark color, caused by a high abundance of carbon. Geologists believe that this meteorite is a specimen of primitive planetary material. The ruler is 15 cm long. (J. A. Wood)

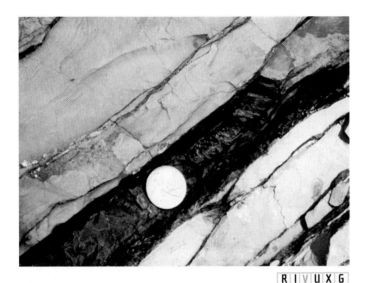

FIGURE 8-34 **Iridium-Rich Layer of Clay** This photograph of strata in the Apennine Mountains of Italy shows a dark-colored layer of iridium-rich clay sandwiched between white limestone (below) from the late Mesozoic era and grayish limestone (above) from the early Cenozoic era. The coin is the size of a U.S. quarter. (W. Alvarez)

thrown enough dust into the atmosphere to block out sunlight for several years. As the temperature dropped drastically and plants died for lack of sunshine, the dinosaurs would have perished, along with many other creatures in the food chain that were highly dependent on vegetation. The dust eventually settled, depositing an iridium-rich layer over the Earth. Tiny, rodentlike creatures capable of ferreting out seeds and nuts were among the animals that managed to survive this holocaust, setting the stage for the rise of mammals and, consequently, the evolution of humans.

In 1992, a team of geologists suggested that the hypothesized asteroid crashed into a site in Mexico. They based this conclusion on glassy debris and violently shocked grains of rock ejected from the multiringed, 195-km-diameter Chicxulub Crater buried under the Yucatán Peninsula in Mexico (Figure 8-35). From the known rate at which radioactive potassium decays, the scientists have pinpointed the date when the asteroid struck—64.98 million years ago. In 1998, geologists digging on the Pacific Ocean floor discovered a piece of meteoritic debris with precisely the same age, apparently a piece of the offending asteroid. Other geologists and paleontologists are not yet convinced that an asteroid impact

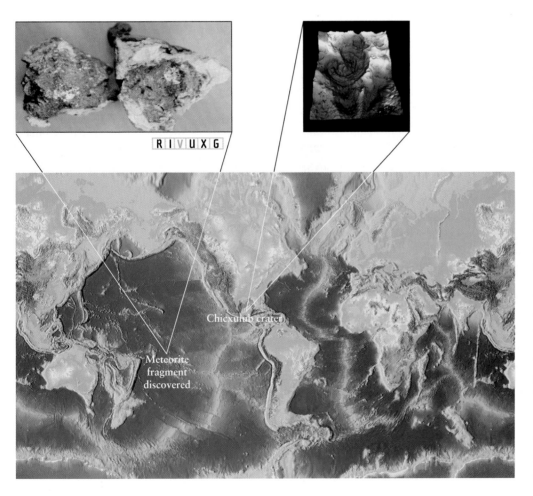

FIGURE 8-35 Confirming an Extinction-Level Impact Site **Right inset:** By measuring slight variations in the gravitational attraction of different materials under the Earth's surface, geologists create images of underground features. Concentric rings of the underground Chicxulub Crater, shown here, lie under a portion of the Yucatán Peninsula. This crater has been dated to 65 million years ago and is believed to be the site of the impact that led to the extinction of the dinosaurs. **Left inset:** A piece of 65-million-year-old meteorite discovered in the middle of the Pacific Ocean in 1998 and believed to be a fragment of the meteorite that struck the Yucatán Peninsula. The fragment, about a tenth of an inch long, was cut into two pieces, shown here. (Top left inset: Frank T. Kyte, UCLA; Top right inset: Virgil L. Sharpton, Lunar and Planetary Institute; Digital image by Peter W. Sloss, NOAA-NESDIS-NGDCD)

led to the extinction of the dinosaurs, but most agree that this hypothesis fits the available evidence better than any other explanation that has been offered so far.

The end of the dinosaurs' reign was not the only mass extinction caused by impacts from space. In 2001, evidence came to light from the Permian/Triassic boundary 250 million years ago that points to another devastating impact. This one wiped out 70 percent of the species of life living on land and 90 percent of those living in the oceans. Rocks from that time discovered in places from Japan to Hungary show evidence of coming from space in the form of fullerenes, soccer-ball-shaped molecules containing at least 60 carbon atoms. Trapped inside these fullerenes were gases that could only have been forced into them from stars. This signature of gas-filled fullerenes from space has also now been discovered in the layer of rock existing on the Earth's surface 65 million years ago.

Could such a catastrophic event happen again? So many asteroids cross the Earth's path that scientists agree that it is a matter of "when" rather than "if." The good news is that studies of craters show that larger asteroids strike the Earth significantly less often than do smaller ones. While asteroids large enough to create the Barringer Crater strike the Earth about once every 10,000 years, killer asteroids, like the one that struck the Yucatán, collide with Earth only once every 100 million years. The threat of a catastrophic impact by an asteroid or comet in our lifetimes, thankfully, is remote.

8-11 Frontiers yet to be discovered

The years of the *Pioneer* and *Voyager* spacecraft were the first golden era of solar system research. We are now in another such period, with spacecraft either in development or already going to planets, asteroids, comets, and even, as we will see shortly, out observing the weather in space. From the solar system debris astronomers hope to learn whether life on Earth was brought here from elsewhere by asteroid or meteoroid impacts. They also hope to answer such questions as the evolutionary history of the space debris, how much of Earth's water came here during the planet's formation and how much landed afterward from comet impacts, whether the asteroids have sufficiently valuable compositions to justify mining them, whether comets can be harvested to supply water and other materials for people colonizing the solar system, whether the Oort cloud really exists, and whether Pluto should be considered a Kuiper belt object.

 Further Reading on These Topics

WHAT DID YOU KNOW?

1 *Are the asteroids a planet that was somehow destroyed?* No. The gravitational pull from Jupiter prevented a planet from ever forming in the asteroid belt. Also, the total mass of the asteroids is much less than even the mass of tiny Pluto, the smallest planet.

2 *How far apart are the asteroids on average?* The distance between asteroids averages 10 million kilometers.

3 *Why do comets have tails?* Gas and dust that evaporate from the comet nucleus are pushed away from the Sun by sunlight and the solar wind. We see the tails by the sunlight they scatter in our direction.

4 *In what direction does a comet tail point?* Comets' gas tails point directly away from the Sun; their dust tails make arcs pointing away from the Sun.

5 *What is a shooting star?* A shooting star is a piece of space debris plunging through the Earth's atmosphere—a meteor. It is not a star.

KEY WORDS

amino acid, 241
Apollo asteroid, 229
asteroid (minor planet), 225
asteroid belt, 225
belt asteroid, 225
carbonaceous chondrite, 241
coma (of a comet), 232
comet, 231
dust tail (of a comet), 234
gas (ion) tail, 234

hydrogen envelope, 232
impact crater, 237
iron meteorite, 240
Kirkwood gaps, 227
Kuiper belt, 231
long-period comet, 236
meteor, 237
meteor shower, 238
meteorite, 239
meteoroid, 237

nucleus (of a comet), 232
Oort cloud, 232
radiation (photon) pressure, 234
short-period comet, 236
stable Lagrange points, 228
stony meteorite, 240
stony-iron meteorite, 241
tail (of a comet), 232
Trojan asteroid, 229
Widmanstätten patterns, 240

KEY IDEAS

Asteroids

• Tens of thousands of belt asteroids with diameters larger than a kilometer are known to orbit the Sun between the orbits of Mars and Jupiter. The gravitational attraction of Jupiter depletes certain orbits within the asteroid belt. The resulting gaps, called Kirkwood gaps, occur at simple fractions of Jupiter's orbital period.

• Jupiter's and the Sun's gravity combine to capture Trojan asteroids in two locations, called stable Lagrange points, along Jupiter's orbit.

• The Apollo asteroids move in highly elliptical orbits that cross the orbits of Mars and Earth. Many of these asteroids will eventually strike one of the inner planets.

Comets

• Comets are fragments of ice and rock that generally move in highly elliptical orbits about the Sun often at a great inclination to the plane of the ecliptic.

• As a comet approaches the Sun, its icy nucleus develops a luminous coma surrounded by a vast hydrogen envelope. A gas (or ion) tail and a dust tail extend from the comet, pushed away from the Sun by the solar wind and radiation pressure.

• Many comets orbit the Sun in the Kuiper belt, a doughnut-shaped region beyond Pluto. Billions of cometary nuclei are also believed to exist in the spherical Oort cloud located far beyond Pluto.

Meteoroids, Meteors, and Meteorites

• Boulders and smaller rocks in space are called meteoroids. When a meteoroid enters the Earth's atmosphere, it produces a fiery trail, and it is then called a meteor. If part of the object survives the fall, the fragment that reaches the Earth's surface is called a meteorite.

• Meteorites are grouped in three major classes according to their composition: iron, stony-iron, and stony meteorites. Rare stony meteorites called carbonaceous chondrites may be relatively unmodified material from the primitive solar nebula. These meteorites often contain organic hydrocarbon compounds, including amino acids.

• Fragments of rock from "burned-out" comets produce meteor showers.

• An analysis of the Allende meteorite suggests that a nearby supernova explosion may have been involved in the formation of the solar system some 4.6 billion years ago.

• An asteroid that struck the Earth 65 million years ago probably contributed to the extinction of the dinosaurs and many other species. Such devastating impacts occur on average every 100 million years.

REVIEW QUESTIONS

1   Why are asteroids, meteoroids, and comets of special interest to astronomers who want to understand the early history of the solar system?

2   Describe the asteroid belt.

 3   To test your understanding of the asteroid belt, do Interactive Exercise 8-1 on the Web or on the CD-ROM. You can print out your results, if required.

4   Why are there many small asteroids but only a few very large ones?

5   Does each comet always maintain its tails? Explain.

6   What are Kirkwood gaps and what causes them?

7   What are the Trojan asteroids, and where are they located?

8   Describe the three main classifications of meteorites. How do astronomers believe these different types of meteorites originated?

9   Suppose you found a rock that you suspect to be a meteorite. Describe some of the things you could do to see if it were a meteorite or a "meteorwrong."

10   Why do astronomers believe that many meteoroids come from asteroids, whereas the debris that creates meteor showers is related to comets?

11   Make a drawing that describes the structure of a comet.

 12   To test your understanding of comets, do Interactive Exercise 8-2 on the Web or on the CD-ROM. You can print out your results, if required.

13   Why is the phrase "dirty snowball" an appropriate characterization of a comet's nucleus?

14   What is the Kuiper belt, and how is it related to debris left over from the formation of the solar system?

15 Why do scientists think the Tunguska event was caused by a large meteoroid and not a comet?

16 Which tail in Figure 8-22 is the gas tail and which is the dust tail? Justify your answers.

ADVANCED QUESTIONS

The answers to computational problems, which are preceded by an asterisk (*), appear at the end of the book.

17 How did the regolith on Eros form?

18 Why are comets generally brighter after passing perihelion (closest approach to the Sun) than before reaching perihelion?

19 Can you think of another place in the solar system where a phenomenon similar to the Kirkwood gaps in the asteroid belt exists? Explain.

20 Where on Earth might you find large numbers of stony meteorites that have not been significantly changed by weathering?

*21 Assuming a constant rate of meteor infall, how much mass has the Earth gained in the past 4.6 billion years?

DISCUSSION QUESTIONS

22 Suppose it was discovered that the asteroid Hermes had been perturbed in such a way as to put it on a collision course with Earth. Describe what you would do to counter such a catastrophe using present technology.

23 From the abundance of craters on the Moon and Mercury, we know that numerous asteroids and meteoroids struck the inner planets during the early history of the solar system. Is it reasonable to suppose that numerous comets also pelted the planets 4.0 to 4.5 billion years ago? Speculate about the effects of such a cometary bombardment, especially with regard to the evolution of the primordial atmospheres of the terrestrial planets and oceans on Earth.

WHAT IF ...

24 An ocean on Earth were struck by a comet nucleus several kilometers across? What physical effects would occur to the Earth?

25 We passed through the tail of a comet? What would happen to the Earth and life on it?

26 As some astronomers have recently argued, passage of the solar system through an interstellar cloud of gas could perturb the Oort cloud, causing many comets to deviate slightly from their original orbits? What might be the consequences for Earth?

27 All the space debris (asteroids, meteoroids, and comets) had been cleared out of the solar system 3.8 billion years ago? What would have been different in the history of the Earth and in the history of life on the Earth?

WEB/CD-ROM QUESTIONS

28 Search the Web to find out why some scientists disagree with the idea that a tremendous impact led to the demise of the dinosaurs. (They do not dispute that the impact occurred, only what its consequences were.) What are their arguments? From what you learn, what is your opinion?

29 Several scientific research programs are dedicated to the search for near-Earth objects (NEOs), especially those that might someday strike our planet. Search the Web for information about at least one of these programs. How does the program search for NEOs? How many NEOs are known and how many has this program found? Will any of these NEOs pose a threat in your lifetime?

30 Search the Web to learn if there are any comets visible at present. List them. What constellations are they presently in? Are any visible to the naked eye? (Recall that the unaided human eye sees objects brighter than about sixth magnitude.)

OBSERVING PROJECTS

31 Make arrangements to view an asteroid. At opposition, some of the largest asteroids are bright enough to be seen through a modest telescope. Check the "Minor Planets" section of the current issue of the *Astronomical Almanac* to see if any bright asteroids are near opposition. If so, check the current issue as well as the most recent January issue of *Sky & Telescope* for a star chart showing the asteroid's path among the constellations. You will need such a chart to distinguish the asteroid from background stars. Observe the asteroid on at least two occasions separated by a few days. On each night, draw a star chart of the objects in your telescope's field of view. Has the position of one starlike object shifted between observing sessions? Does the position of the moving object agree with the path plotted on published star charts? Do you feel confident that you have in fact observed an asteroid?

32 Make arrangements to view a comet through a telescope. Because astronomers discover roughly a dozen comets each year, a comet is usually visible somewhere in the sky. Unfortunately, because most comets are quite dim, you will need access to a moderately large telescope. Consult the Web or recent issues of the *IAU Circular,* published by the International Astronomical Union's Central Bureau for Astronomical Telegrams, which contains predicted positions and the anticipated brightness of comets in the sky. Also, if there is an especially bright comet in the sky, the latest issue of *Sky & Telescope* might contain useful information. Is a comet visible and do you have a telescope at your disposal? If so, can you distinguish the comet from background stars? Can you see its coma? How many tails do you see?

33 Make arrangements to view a meteor shower. The date of maximum intensity in Figure 8-26 is the best time to observe a particular shower, although good displays can often be seen a day or two before and after the maximum. In order to see a fine meteor display, you need a clear, moonless sky. The Moon's presence above the horizon can significantly detract from the number of faint meteors you will be able to see.

34 Use *Starry Night Backyard*™ software to observe several comets. (a) Set the program to Atlas mode (*Go/Atlas*). Turn off the planets and Sun (*Sky/Planets/Sun*). Set the date to 3/22/97. Go to either +90° or −90° latitude (north or south pole). Choose Comet Hale-Bopp from the Planet list window (*Windows/Planet List/Comets*) by double-clicking on Hale-Bopp. Zoom in until you can see the comet and its tail. Predict in what direction the Sun is located relative to the comet and explain how you made your prediction. To verify that you are correct, zoom out to 90°, turn on the planets and Sun (*Sky/Planets/Sun*) to locate the Sun. If necessary use the hand cursor to move the sky in the direction of your prediction toward the Sun. Were you correct? (b) Using the Planet List, double-click on the name of another comet to center on it, but do not zoom in. Then move the sky with the hand cursor and locate the Sun. From your observations, predict the direction of the comet's tail on the sky, and explain how you made your prediction. Center again on the comet and zoom in until you can see its tail. Was your prediction correct?

9 THE SUN

IN THIS CHAPTER YOU WILL DISCOVER

- why the Sun is a typical star

- how today's technology has led to new understanding of solar phenomena, from sunspots to the powerful ejections of matter that sometimes enter our atmosphere

- that some features of the Sun generated by its varying magnetic field occur in cycles

- how the Sun generates the energy that makes it shine

- new insights into the nature of matter provided by solar neutrinos

The X-Ray Sun
Our momentary glances toward the Sun from day to day give the impression of a blindingly bright, uniform body. This sequence of X-ray images of the Sun taken by the *Yohkoh* spacecraft shows another, darker aspect. X rays are emitted near the solar surface and appear here in shades of yellow, orange, and red. The Sun's X-ray emissions are far less uniform than the visible light it emits. Sometimes its X rays are so intense as to be extremely danger-ous to astronauts. By studying the locations and intensities of X rays from the Sun, astronomers gain an understanding of the energy transfer from the Sun's interior through its atmosphere. This sequence shows how X-ray emissions from the Sun's outer layer change by a factor of about 100 over an 11-year cycle. (Stanford Lockheed Institute for Space Research and NASA)

RIVUXG

WHAT DO YOU THINK?

1   What percentage of the solar system's mass is in the Sun?

2   Does the Sun have a solid and liquid interior like the Earth?

3   What is the surface of the Sun like?

4   Does the Sun rotate?

5   What makes the Sun shine?

Without the Sun's energy, we on Earth could not exist. The Sun's ideal balance of heat, visible light, and ultraviolet radiation enabled life to form and flourish here. No wonder our star has been revered for millennia. No wonder, too, that astronomers have studied it intensively since Galileo first trained his telescope on the drama of sunspots.

We now complete our exploration of the solar system by studying the Sun. Today, the Sun holds the key to our understanding of stellar evolution because, for all its grandeur, the Sun is in fact a typical star. Figure 9-1 lists the Sun's properties. Comparing the mass of the Sun to the total mass of all the planets in Appendix Table A-2, you can see that

the Sun contains more than 99.85% of the total mass of the solar system.

THE SUN'S ATMOSPHERE

Although astronomers often speak of the solar "surface," the Sun is so hot that it has neither liquid nor solid matter anywhere inside it. Moving down through the Sun, one continually encounters ever denser and hotter gases.

9-1 The photosphere is the visible layer of the Sun

The Sun appears to have a surface only because most of its visible light comes from one specific gas layer (see Figure 9-1). This region, which is about 400 km thick, is appropriately called the **photosphere** ("sphere of light"). The density of the photosphere's gas is low by Earth standards, about 0.01% as thick as the air we breathe. The photosphere has a blackbody spectrum corresponding to an average temperature of 5800 K (review Figure 4-3).

The photosphere is the lowest of the three layers comprising the Sun's atmosphere. Because the upper two layers (discussed in the following two sections) are transparent to most wavelengths of visible light, we see through them down

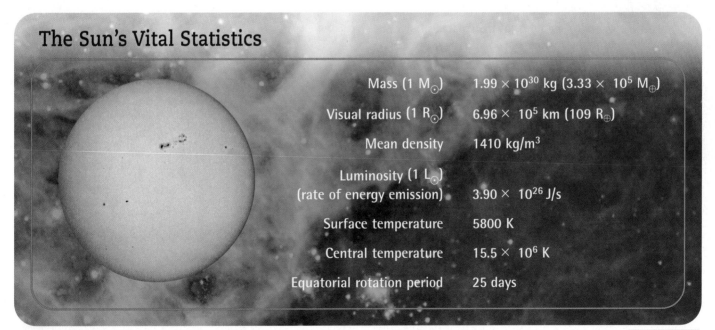

The Sun's Vital Statistics

Mass (1 $M_\odot$)	1.99×10^{30} kg (3.33×10^5 $M_\oplus$)
Visual radius (1 $R_\odot$)	6.96×10^5 km (109 $R_\oplus$)
Mean density	1410 kg/m^3
Luminosity (1 $L_\odot$) (rate of energy emission)	3.90×10^{26} J/s
Surface temperature	5800 K
Central temperature	15.5×10^6 K
Equatorial rotation period	25 days

R I V U X G

FIGURE 9-1 **Our Star, the Sun** The Sun emits most of its visible light from a thin layer of gas called the photosphere, shown here. Although the Sun has no solid or even liquid region, we see the bottom of the photosphere as its "surface." Astronomers think of the photosphere as the lowest level of the Sun's atmosphere. Astronomers always take great care when viewing the Sun by using extremely dark filters or by projecting the Sun's image onto a screen. (Celestron International)

WEB LINK 9.1

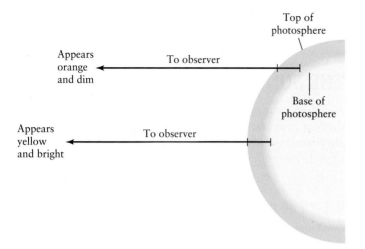

Top of
photosphere

Appears
orange
and dim

To observer

Base of
photosphere

Appears
yellow
and bright

To observer

FIGURE 9-2   **Limb Darkening** The Sun's edge or limb appears distinctly darker and more orange than does its center, as seen from Earth (see Figure 9-1). This occurs because we look through the same amount of solar atmosphere at all places. As a consequence, we see higher in the Sun's photosphere near its limb than when we look at its central regions. The higher photosphere is cooler and (since it is a blackbody) darker and more orange than the lower, hotter region of the photosphere.

to the photosphere. We cannot, however, see through the shimmering gases of the photosphere, so everything below the photosphere is called the Sun's interior.

As you can see in Figure 9-1, the photosphere appears darkest toward the edge, or **limb,** of the solar disk, a phenomenon called **limb darkening.** This occurs because we see light through roughly equal amounts of the Sun's atmosphere. But the Sun is spherical, so the light we see leaving the Sun at different places comes from different levels of the photosphere (Figure 9-2). Looking straight down at the center of the Sun's disk, we see farther into the Sun's atmosphere than when we look toward the limb. The photosphere is hottest (and brightest) at its base. Therefore, the center of the Sun's disk, where we see deepest into it, looks brighter than does its limb.

Under good observing conditions with a telescope and using special dark filters for protection, you can see a blotchy pattern called *granulation* on the photosphere (Figure 9-3). The lightly colored **granules** measure about 1000 km across and are surrounded by darkish boundaries. Time-lapse photography shows that granules form, disappear, and then reform in cycles lasting several minutes. At any single moment, several million granules cover the solar surface.

At the end of Chapter 4 we learned that radial motion of a light source affects its spectral lines through the Doppler effect (see Figures 4-16 and 4-17 and An Astronomer's Toolbox 4-3). By carefully measuring the wavelengths of spectral lines in various parts of individual solar granules, astronomers have determined that hot gases are convected upward

in the center of each granule. We saw in Chapter 5 that convection moves the continents on Earth and in Chapter 7 that convection also helps form the belts and zones on the giant planets. The Sun's gas radiates its energy out into space. We see this energy as visible light and other electromagnetic radiation. Upon radiating, the gas cools, spills over the edges of the granule, and plunges back down into the Sun along the boundaries between granules (see Figure 9-3 inset).

According to the Stefan-Boltzmann law (see Chapter 4), hotter regions emit more photons per square meter than do cooler regions. Note in Figure 9-3 that the centers of granules appear brighter than their edges. Measurements indicate that a granule's center is typically 100 K hotter than its edge, thus explaining the observed brightness difference.

9-2   The chromosphere is characterized by spikes of gas called spicules

Immediately above the photosphere is a dim layer of less dense stellar gas called the **chromosphere** ("sphere of color"). This unfortunate name suggests that it is the layer we normally see, but for centuries it was visible only when the photosphere was blocked during a total solar eclipse.

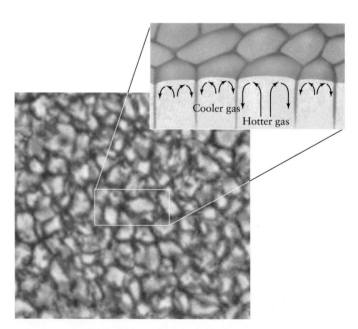

Cooler gas

Hotter gas

FIGURE 9-3   **Solar Granulation** High-resolution photographs of the Sun's surface reveal a blotchy pattern called granulation. Granules, which measure about 1000 km across, are convection cells in the Sun's photosphere. **Inset:** Gas rising upward produces bright granules. Cooler gas sinks downward along the darker, cooler boundaries between granules. This convective motion transports energy from the Sun's interior outward to the solar atmosphere. (MSFC/NASA; inset: Goran Scharmer, Lund Observatory)

During an eclipse, the chromosphere is visible as a pinkish strip some 2000 km thick around the edge of the dark Moon (Figure 9-4). Today, astronomers can also study the chromosphere through filters that pass light with specific wavelengths strongly emitted by it—but not by the photosphere—or through telescopes sensitive to nonvisible wavelengths that the chromosphere emits intensely.

High-resolution images of the chromosphere reveal numerous spikes, which are jets of gas called **spicules** (see Figure 9-5a). A typical spicule rises for several minutes at the rate of 72,000 km/h (45,000 mph) to a height of nearly 10,000 km (Figure 9-5b). Then it collapses and fades away. At any one time, roughly a third of a million spicules cover a few percent of the Sun's chromosphere.

Spicules are generally located on the boundaries of enormous regions of rising and falling chromospheric gas called **supergranules** (Figure 9-6). A typical supergranule has a diameter slightly larger than the Earth's and contains about 900 granules.

9-3 Temperatures increase higher in the Sun's atmosphere

It seems plausible that the temperature should fall as one rises through the Sun's atmosphere. After all, by moving upward, one moves farther from the sun's internal heat source. Indeed, starting in the photosphere at 5800 K, the tempera-

R I V U X G

FIGURE 9-4   **The Chromosphere** This photograph of the chromosphere was taken during an eclipse. It appears pinkish because the gas in the chromosphere emits only certain colors (wavelengths), among which the red one from hydrogen gas dominates. (NOAO)

ture drops to around 4000 K in the lower chromosphere. Surprisingly, however, the temperature then begins to rise much higher as one ascends, reaching about 10,000 K at the top of the chromosphere. The outermost region of the Sun's atmosphere, the **corona,** extends several million kilometers from the top of the chromosphere (see Figure 9-5). Between

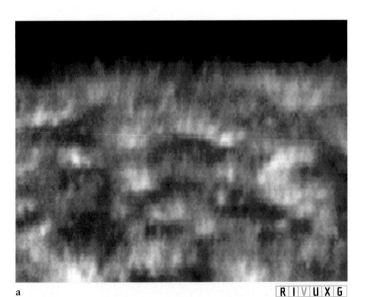

a   R I V U X G

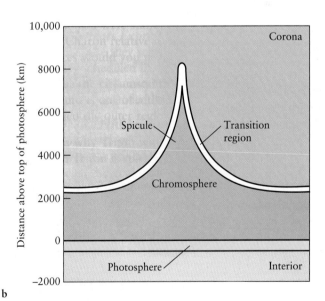

b

FIGURE 9-5   **Spicules** (a) Spicules are seen in this photograph of the Sun's chromosphere. The Sun appears rose-colored in this image because it was taken through an H_α filter that passes red light from hydrogen and effectively blocks most of the photosphere's light. (b) The spicules are jets of gas that surge upward into the Sun's outer atmosphere. This schematic diagram shows a spicule and its relationship to the solar atmosphere's layers. The photosphere is about 400 km thick. The chromosphere above it extends to an altitude of about 2000 km, with spicules jutting up to nearly 10,000 km above the photosphere. The outermost layer, the corona (discussed in Section 9-3), extends millions of kilometers above the photosphere. (a: NASA)

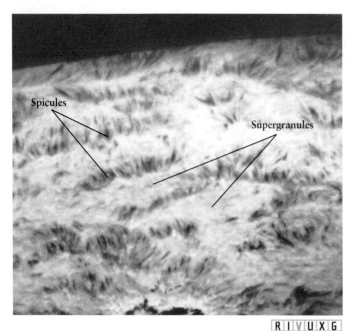

R I V U X G

FIGURE 9-6   **Supergranules** Surrounded by spicules, super-granules are regions of rising and falling gas in the chromosphere. Each supergranule spans hundreds of granules in the photosphere below. This image was taken through an H$_\alpha$ filter. (NASA)

the chromosphere and corona is a **transition zone** in which the temperature skyrockets to about 1 million K.

The unexpected increase in temperature was discovered around 1940 as a result of the high temperature's effect on the spectrum of the Sun's corona. The hotter a gas is, the more its electrons are stripped off the atoms in it. Astronomers discovered that the corona contains the emission lines of a number of highly ionized (therefore very hot) elements. For example, a prominent green line caused by the presence of Fe XIV, an iron atom stripped of 13 electrons, appears in the coronal spectrum. (Recall that Fe I is a neutral iron atom.) It is now known that coronal temperatures are typically in the range of 1 to 2 million kelvins, while some regions are even hotter.

Although the temperature is extremely high, the density of gas in the corona is very low, about 10 trillion times less dense than the air at sea level on Earth. The low density partly accounts for the dimness of the corona, which otherwise would outshine the photosphere. Astronomers have mounting evidence that the corona is heated by energy carried aloft and released there by the Sun's complex magnetic fields, discussed in Sections 9-5 and 9-6.

The total amount of visible light that we receive from the solar corona is comparable to the brightness of the full Moon—or only about one-millionth as bright as the photosphere. As with the chromosphere, the corona can be seen only when the photosphere is blocked out or through special filters or at nonvisible wavelengths (such as ultraviolet and X-ray), at which the corona is especially bright compared to the photosphere. Blocking the photosphere occurs naturally during a total eclipse or artificially with a specially designed telescope called a *coronagraph*. Figure 9-7a is a photograph of the corona taken during a total eclipse. See, also, Figure 1-25. Figure 9-7b and the photo that opens this chapter shows stunning X-ray images of the corona.

a   R I V U X G

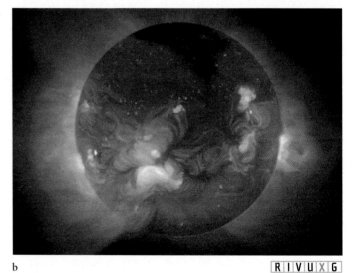

b   R I V U X G

FIGURE 9-7   **The Solar Corona** (a) This visible-light photograph was taken during the total solar eclipse of July 11, 1991. Numerous streamers are visible, extending millions of kilometers above the solar surface. (b) This X-ray image of the Sun's corona, taken by *Yohkoh* in 1999, provides hints of the complex activity taking place on and in the Sun. The million-degree gases in the corona emit the X rays visible here. (a: R. Christen and M. Christen, Astro-Physics Inc.; b: NASA)

Just as the Earth's gravity prevents most of our atmosphere from escaping into space, so, too, does the Sun's gravity keep most of its outer layers from leaving. However, some of the gas in the corona is moving fast enough—around a million kilometers per hour—to escape forever into space. As we saw in Chapter 5 in discussing Earth's magnetosphere and in Chapter 8 in discussing comets, such outflowing gas is called the **solar wind.** The *heliosphere* is a bubble in space containing the Sun and planets created by the solar wind. This outflow of gases from the Sun prevents most of the gases flowing in space from other stars from entering our solar system.

The Sun ejects around a million tons of matter each second as the solar wind. Even at this rate of emission, the mass loss due to the solar wind will amount to only a few tenths of a percent of the Sun's total mass throughout its lifetime. While electrons and hydrogen and helium nuclei comprise 99.9% of the solar wind—silicon, sulfur, calcium, chromium, nickel, neon, and argon ions have also been detected in it. The solar wind particles reach speeds up to 2.9×10^6 km/h (1.8×10^6 mph). The wind achieves these high speeds in part by being accelerated by the Sun's magnetic field, discussed in Section 9-5.

THE ACTIVE SUN

Granules, supergranules, spicules, and the solar wind occur continuously. These are features of the *quiet* Sun. But the Sun's atmosphere is periodically disrupted by magnetic fields that stir things up, creating the *active* Sun. The Sun's most obvious transient features are **sunspots,** regions of the photosphere that appear dark because they are cooler than the rest of the Sun's lower atmosphere. Sometimes sunspots occur in isolation (Figure 9-8a), but often they arise in clusters called *sunspot groups* (Figure 9-8b).

> **Insight into Science**   **Perception versus Reality** Our senses (and often our technology) are limited, and so we must be careful how we interpret what we perceive. For example, the sunspots in Figure 9-8 certainly look like black spots. However, they appear black only in contrast to the bright light around them. As we will discuss shortly, sunspots are actually red and orange.

9-4 Sunspots reveal the solar cycle and the Sun's rotation

Like other transient features of the active Sun, the average number and location of sunspots vary in fairly predictable cycles. As shown in Figure 9-9a, the average sunspot cycle lasts approximately 11 years. Within an 11-year cycle, most sunspots appear at a **sunspot maximum** (Figure 9-9b). Sunspot maxima occurred most recently in 1979, 1989, and 2000. During a **sunspot minimum,** the Sun is almost devoid of sunspots, as it was in 1976, 1986, and 1996 (Figure 9-9c). The next trough in the cycle is projected to occur in 2007.

A typical sunspot is 10,000 km across and lasts between a few hours and a few months. Each sunspot has two parts:

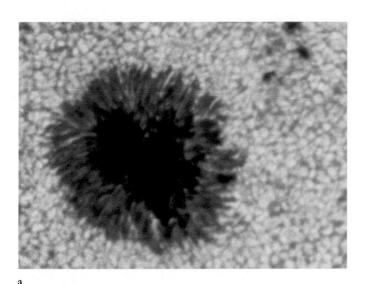

a

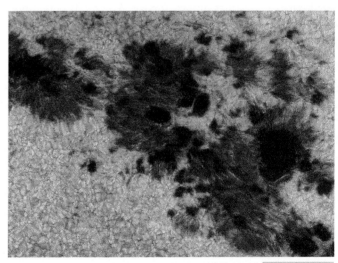

b   R I V U X G

FIGURE 9-8   **Sunspots** **(a)** This dark region on the Sun is a typical isolated sunspot. Granulation is visible in the surrounding, undisturbed photosphere. **(b)** This high-resolution photograph shows a sunspot group in which several sunspots overlap and others are nearby. (a: NOAO; b: National Solar Observatory)

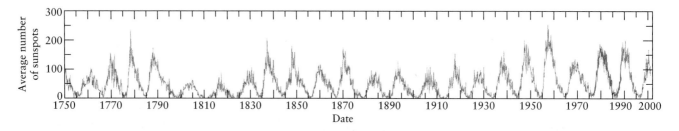

a

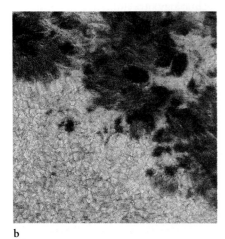

b

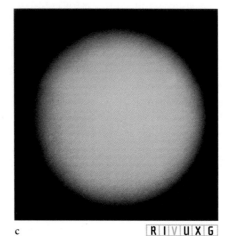

c

RIVUXG

WEB LINK 9.3 **FIGURE 9-9** The Sunspot Cycle **(a)** The number of sunspots on the Sun varies with a period of about 11 years. The most recent sunspot maximum occurred in 2000, and the most recent sunspot minimum occurred in 1996. **(b)** The active Sun has many sunspots and other features (this photo was taken in 1989). **(c)** The quiet Sun is devoid of such features (this photo was taken in 1989). (b & c: NOAO)

a dark, central region, called the *umbra,* and a brighter ring surrounding the umbra, called the *penumbra,* visible in Figures 9-8a and 9-8b. While these are the same names as the blocked areas of an eclipse (see Chapter 1), we will see in the next section that their causes are completely different—another good example of how words can have different meanings in different contexts.

Seen without the surrounding brilliant granules that outshine it, a sunspot's umbra appears red and its penumbra orange. From Wien's law (see Chapter 4), these colors indicate that the umbra is typically 4300 K and the penumbra 5000 K, both cooler than the normal photosphere.

On rare occasions a sunspot group is so large that it can be seen with the unaided eye. Chinese astronomers recorded such sightings 2000 years ago, and a huge sunspot group visible to the naked eye was seen in 2001. **Always use special dark filters or other means to protect your eyes when viewing the Sun. Looking directly at the Sun for more than a few moments can cause eye damage!** Of course, a telescope gives a much better view, so it was not until Galileo did so that anyone examined sunspots in detail.

By following sunspots as they moved across the solar disk (Figure 9-10), Galileo discovered that *the Sun rotates once in about four weeks.* Because a typical sunspot group lasts about two months, it can be followed for two solar rotations. Sunspot activity also lets us see the Sun's *differential rotation* (see Chapter 7): The equatorial regions rotate more rapidly than the polar regions. A sunspot near the solar equator takes 25 days to go once around the Sun, but a sunspot at 30° north or south of the equator takes about

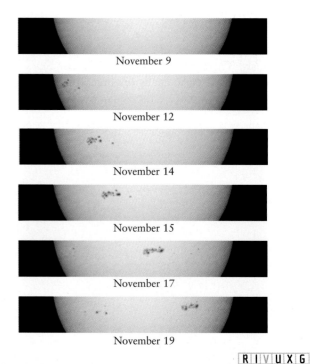

November 9

November 12

November 14

November 15

November 17

November 19

RIVUXG

FIGURE 9-10 The Sun's Rotation This series of photographs taken in 1999 shows the same sunspot group over nearly half a solar rotation. Note how the sunspot groups have changed over this time. By observing the same group of sunspots from one day to the next in this same manner, Galileo found that the Sun rotates once in about four weeks. Sunspot activity also reveals the Sun's differential rotation: The equatorial regions rotate faster than the polar regions. (Carnegie Observatories)

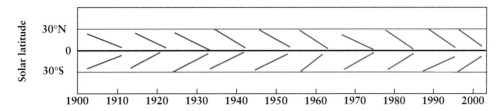

FIGURE 9-11 Average Latitude of Sunspots Throughout the Sunspot Cycle This graph shows that at the beginning of each sunspot cycle, most sunspots are at moderate latitudes, around 30° north or south. Sunspots arising later in each cycle typically form closer and closer to the Sun's equator.

27 days. The rotation period at 75° north or south of the equator is about 33 days, and near the poles it may be as long as 35 days.

Other stars rotate. In 2001, astronomers discovered that Altair, in the constellation Aquila, rotates at least once every 10.4 hours. This rotation rate is so great that the star bulges at its equator. Its equatorial diameter is 14 percent greater than its diameter measured from pole to pole. We will see later that there is evidence from pulsars that many stars rotate.

The average latitude at which new sunspots appear changes throughout the sunspot cycle. At the beginning of each cycle, the sunspots appear mostly at about 30° north and south latitudes. Ones that form later in the cycle typically occur closer to the equator. Figure 9-11 shows that the average latitude at which sunspots are located varies at the same 11-year rate as does the number of sunspots. Satellite measurements reveal that the Sun emits about 0.1% more energy at the peak of the sunspot cycle than at its minimum.

9-5 The Sun's magnetic fields create sunspots

In 1908, the American astronomer George Ellery Hale discovered that sunspots are directly linked to intense magnetic fields on the Sun. When Hale focused a spectroscope on sunlight coming from a sunspot (Figure 9-12a), he found that each spectral line in the normal solar spectrum is flanked by additional, closely spaced spectral lines not usually observed (Figure 9-12b). This "splitting" of a single spectral line into two or more lines is called the **Zeeman effect.** The Dutch physicist Pieter Zeeman, who first observed it in the laboratory in 1896, showed that an intense magnetic field splits the spectral lines of a light source inside the field. The more intense the magnetic field, the more the split lines are separated.

MORE TO KNOW 9-2 Zeeman Effect

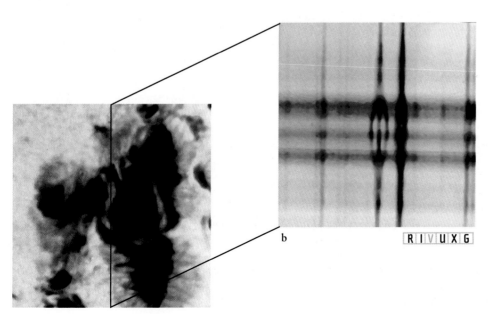

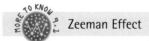

FIGURE 9-12 Zeeman Splitting by a Sunspot's Magnetic Field (a) The black line drawn across the sunspot indicates the location toward which the slit of the spectroscope was aimed. (b) In the resulting spectrogram, one line in the middle of the normal solar spectrum is split into three components. The separation between the three lines corresponds to a magnetic field roughly 5000 times stronger than the Earth's magnetic field. (a & b: NOAO)

Observationally, sunspots are areas where concentrated magnetic fields project through the hot gases of the photosphere. But how do the fields create the spots? The answer lies in the interaction between the magnetic fields and the photosphere's gases. Because of the photosphere's high temperature, many atoms in it are *ionized:* One or more of their electrons have been stripped off by high-energy photons there. As a result, the photosphere is a mixture of electrically charged ions and electrons called a **plasma.**

Plasmas are extremely good conductors of electricity and are repelled from regions of high magnetic field. Therefore, the magnetic field protruding through the photosphere prevents hot, ionized gases inside the Sun from rising to the surface as they normally do. Thus, such regions of the photosphere are left relatively devoid of hot gas and are therefore cooler and darker than the surrounding solar surface. We observe these darker, cooler regions as sunspots.

Recall from Chapter 5 that magnetic fields form complete loops, with each magnetic field having a north pole and a south pole. Likewise, when the Sun's magnetic field emerges through one sunspot or sunspot group, it forms a loop that reenters the Sun at another sunspot or sunspot group (Figure 9-13 insets). We associate the names "north pole" or "south pole" with each sunspot or sunspot group depending on whether the magnetic field there is leaving (a north pole) or entering (a south pole) the Sun. Sunspots and sunspot groups in each hemisphere are connected in pairs, one where the magnetic field leaves the Sun, the other where it returns.

During his observations, Hale found that on one hemisphere of the Sun, sunspots with a magnetic north pole always come into view before the corresponding sunspots with a magnetic south pole. At the same time on the other hemisphere, the order is reversed (see Figure 9-13 insets). Hale also found that this pattern reverses itself every 11 years. The hemisphere where north magnetic poles come first during one 11-year cycle has south magnetic poles coming first during the next. Astronomers therefore speak of the 22-year **solar cycle,** the time it takes the magnetic fields to return to the same orientation.

The Sun's magnetic field is created as a result of its rotation and the resulting motion of the ionized particles found throughout it. This was proposed in 1960 by another American astronomer, Horace Babcock, as the **magnetic dynamo** model, in an effort to explain the 22-year solar cycle. The Sun's magnetic field normally lies just below the surface in the highly conducting plasma located there, unlike the Earth's field, which passes through our planet's center.

As shown in Figure 9-13, the Sun's differential rotation causes the field to become increasingly stretched. Like expanding a rubber band, stretching the magnetic field this way causes it to store energy. This magnetic field becomes even more complicated as convection of the gases under the photosphere causes the fields to move vertically and horizontally, thereby developing tangles. Unlike a rubber band, however, magnetic field lines cannot break to release the energy stored in them. Rather, the fields must untangle themselves.

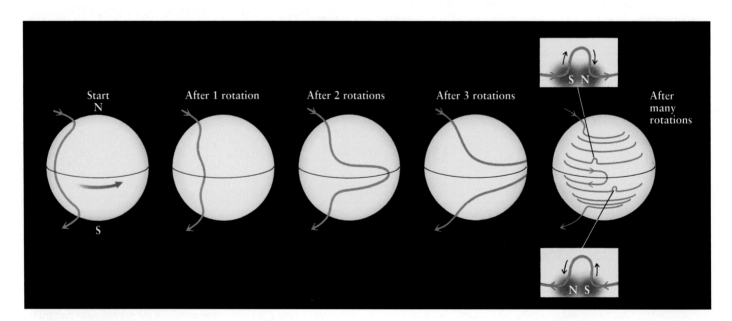

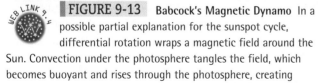

FIGURE 9-13   Babcock's Magnetic Dynamo In a possible partial explanation for the sunspot cycle, differential rotation wraps a magnetic field around the Sun. Convection under the photosphere tangles the field, which becomes buoyant and rises through the photosphere, creating sunspots and sunspot groups. **Insets:** In each group, the sunspot that appears first has the same polarity as the Sun's magnetic pole in that hemisphere (in this drawing, N in the upper hemisphere and S in the lower hemisphere).

This untangling process begins as the jumbled regions of magnetic field trap gases, which expand, becoming buoyant, and float up through the solar surface. Sunspots, with magnetic fields typically 1000 times stronger than the Earth's magnetic field, form where loops or tangles of solar magnetic field leave and reenter the Sun. This gas eventually leaks out, and the fields untangle, interact with other parts of the Sun's magnetic field, and gradually disappear along with their sunspots. This process of magnetic fields piercing the Sun's surface begins at high latitudes, and over the next 11 years it occurs closer and closer to the Sun's equator.

Because of how these fields interact and vanish, every 11 years the Sun's entire magnetic field is reversed—the Sun's north magnetic pole becomes its south magnetic pole and vice versa. After another 11-year cycle, the field is back to its original orientation. This is why the solar cycle is 22 years long. The most recent reversal of the Sun's magnetic field occurred in 2001.

Notable irregularities occur in the solar cycle. For example, the overall reversal of the Sun's magnetic field is often piecemeal and haphazard: One pole may reverse polarity long before the other, so that for weeks or months on end the Sun may exhibit two north poles but no south pole. The corresponding south poles remain under the surface during this time.

More intriguing still is the strong historical evidence that all traces of sunspots and the sunspot cycle have vanished for decades at a time. For example, in 1893 the British astronomer E. Walter Maunder used historical observations to conclude that virtually no sunspots occurred from 1645 through 1715. This period, called the *Maunder minimum,* coincides with a period of cold in Europe so extreme that it was called the Little Ice Age. At the same time, western North America was subject to severe drought.

Similar sunspot-free periods apparently occurred at irregular intervals in earlier times as well. Conversely, periods of increased sunspot activity in the eleventh and twelfth centuries coincided with periods of warmer-than-average temperatures. It remains to be seen, however, if scientists can find all the links between the number of sunspots and periods of extreme temperatures on Earth.

> **Insight into Science**   **Cause and Effect** While extremes of sunspot activity often occur at the same times as weather extremes on Earth, models of these phenomena must also take into account the possibility of other, terrestrial causes for these temperature changes. Remember, just because one event follows another does not mean that the first *causes* the second.

Theoretical calculations suggest that sunspots should last only for a matter of days before the magnetic fields collapse back into the Sun. Astronomers are just now beginning to understand how sunspots can persist for weeks or longer.

Observations indicate that the magnetic fields on the Sun's surface are replenished about once every two days by a mechanism in which the magnetic field lines are broken apart and reconnected (Figure 9-14). Current theoretical work in this area is providing many new insights into the behavior of the Sun's surface.

9-6 Solar magnetic fields also create other atmospheric phenomena

Figure 9-15 shows the active Sun's chromosphere and corona. The bright areas in this photograph are called **plages** (pronounced "plahzh," from the French word for "beaches"). Best seen in the light emitted by calcium or hydrogen atoms, they are hotter, and therefore brighter, than the surrounding chromosphere. Plages, which often appear just before nearby sunspots form, are believed to be created by magnetic fields under the photosphere crowding upward just before they emerge through the photosphere. In pushing upward, the fields compress the gases of the upper Sun, causing this gas to become hotter and therefore to glow more brightly.

The dark streaks in Figure 9-15 are features in the corona called **filaments.** These huge volumes of gas are lofted upward from the photosphere by the Sun's magnetic field. When viewed from the side rather than from above, filaments form gigantic loops or arches called **prominences**

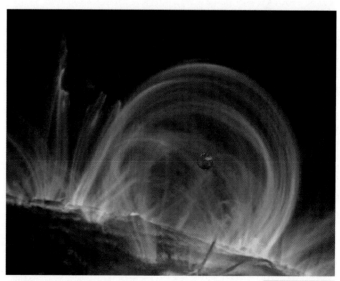

R I V U X G

 FIGURE 9-14   **Magnetic Fields of the Sun** The Sun's magnetic fields are revealed by the energy emitted from the gas they trap. This image shows large coronal loops (the Earth is superimposed for scale). Taken with the *TRACE* satellite, this ultraviolet image shows loops up to 160,000 km (100,000 mi) high, with gases moving along the magnetic field lines at speeds of 100 k/s (60 mi/s). (NASA)

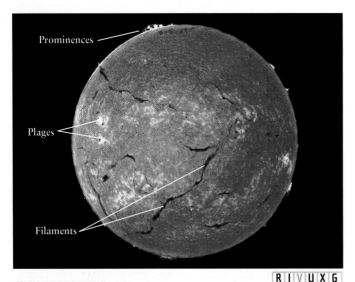

R I V U X G

FIGURE 9-15 Active Sun in H_α This photograph shows the chromosphere and corona during a solar maximum, when sunspots are abundant. The image was taken through a filter allowing only light from H_α emission to pass through. The hot upper layers of the Sun's atmosphere are strong emitters of H_α photons. (See Section 4-6 for details of the Balmer series, H_α – H_∞.) A few large sunspots are evident. Most notable are features that do not appear at the solar minimum, such as the snakelike features shown here called filaments, bright areas called plages, and prominences (filaments seen edge-on) observed at the solar limb. (NASA)

(Figures 9-15 and 9-16a). The temperature of gas in prominences can reach 50,000 K. These features are almost always associated with sunspots. Some prominences last only for a few hours, while others persist for months. The most energetic prominences escape the magnetic fields that confine them and surge out into space (Figure 9-16b). The Sun's corona typically has millions of prominences.

X-ray photographs also reveal numerous coronal bright and dark spots that are hotter or colder, respectively, than the surrounding corona (Figure 9-17). Temperatures in the bright regions occasionally reach 4 million K. Many of the bright coronal hot spots are located over sunspots. The darker, cooler **coronal holes** act as conduits for gases to flow out of the Sun. Therefore, when a coronal hole on the rotating Sun faces Earth, the solar wind in our direction increases dramatically.

Violent, eruptive events on the Sun, called **solar flares,** also send out vast quantities of high-energy particles, as well as X rays and ultraviolet radiation from the Sun (Figure 9-18). Flares are such powerful events that they leave the region of the surface of the Sun in their vicinity quaking for an hour or more. At the maximum of the sunspot cycle, there are about 1100 flares per year. Most flares last for less than an hour, but during that time, temperatures soar to 5 million K. By following the paths of particles emitted by flares as they are guided outward by the Sun's magnetic field, astronomers have recently begun mapping that field as it extends out beyond the Earth (Figure 9-19).

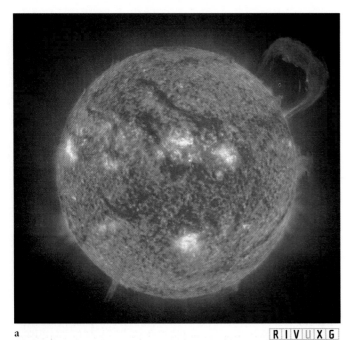

a   R I V U X G

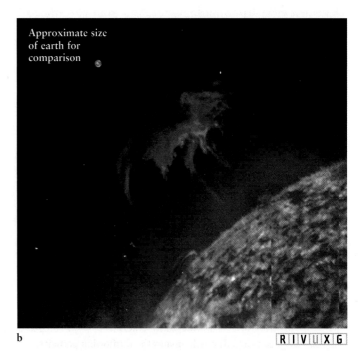

b   R I V U X G

FIGURE 9-16 Prominences (a) A huge prominence arches above the solar surface in this SOHO image taken in 2001. The radiation that exposed this picture is from singly ionized helium at a wavelength of 30.4 nm, corresponding to a temperature of about 50,000 K. (b) The gas in this prominence was so energetic that it broke free from the magnetic fields that shaped and confined it. This eruptive prominence occurred in 1999 and did not strike the Earth (shown for size). (a: ESA/NASA; b: Joseph B. Gurman, Solar Data Analysis Center, NASA)

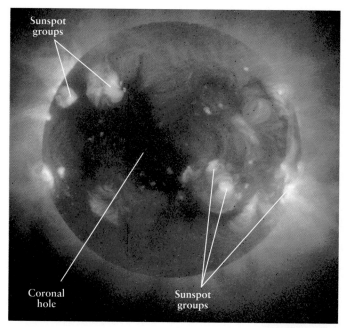

Sunspot groups

Coronal hole

Sunspot groups

R I V U X G

![WEB LINK 9.6]

FIGURE 9-17 **A Coronal Hole** This X-ray picture of the Sun's corona was taken by the *Yohkoh* satellite on February 23, 2000. A huge coronal hole dominates the central region of the corona. Gases from this hole caused spectacular aurora on Earth. The bright regions are emissions from sunspot groups. (NASA)

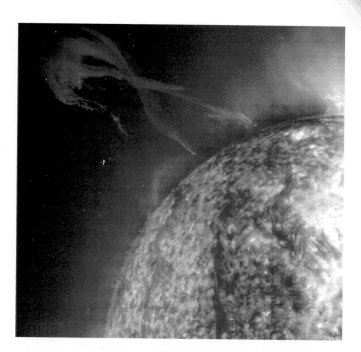

![WEB LINK 9.7]

FIGURE 9-18 **A Flare** Solar flares, which are associated with sunspot groups, produce energetic emission of particles from the Sun. This image, taken in 2000 by SOHO, shows a twisted flare in which the Sun's magnetic field lines are still threaded through the region of emerging particles. (NASA)

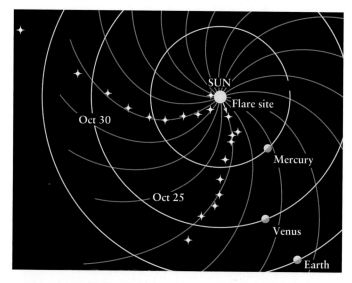

SUN
Flare site
Oct 30
Mercury
Oct 25
Venus
Earth

![WEB LINK 9.8]

FIGURE 9-19 **A Snapshot of the Sun's Global Magnetic Field** By following the paths of particles emitted by a solar flare, astronomers have begun mapping the solar magnetic field. The field guides the outflowing particles, which in turn emit radio waves that indicate the position of the field. These data were collected by the *Ulysses* spacecraft in 1994. *Ulysses* was the first spacecraft that explored interplanetary space from high above the plane of the ecliptic.

Insight into Science   **Accept Coincidences** As the word "serendipitous" suggests, purely chance occurrences imply no preplanning, natural, or supernatural influences. Literally as I was typing the last paragraph, NASA flashed news on the Internet that the Sun was seen to vibrate as a recent solar flare occurred (more about these vibrations in Section 9-8). All too often, people believe that simple coincidences have cause-and-effect relationships. Pseudosciences like astrology thrive on this sort of misconception.

Astronomers recently discovered that huge, balloon-shaped volumes of high-energy gas are ejected from the corona. These **coronal mass ejections** typically expel 2 trillion tons of matter at 400 km/s, and each one lasts for up to a few hours (Figure 9-20). Coronal mass ejections have enough energy to break through the Sun's magnetic fields that normally contain them or even to carry fields outward, altering the Sun's magnetic field. Flares have been observed to create shock waves that initiate some of the coronal mass ejections. The origins of others are still under investigation.

Some coronal mass ejections, solar flares, and prominences head toward Earth. It takes about 8 minutes for their

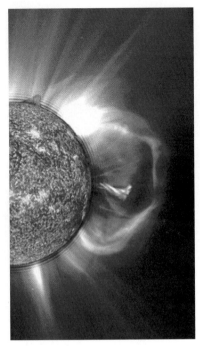

Coronal mass ejection

Two to four days later

FIGURE 9-20 A Coronal Mass Ejection **Left:** X-ray image of a coronal mass ejection from the Sun taken by SOHO. **Right:** Two to four days later, the highest energy gases from the ejection reach 1 AU. If they come our way, most particles are deflected by the Earth's magnetic field (in blue). However, as shown, some particles leak earthward causing auroras, disrupting radio communications and electrical power transmission, damaging satellites, and ejecting some of Earth's atmosphere into interplanetary space. (NASA)

R I V U X G

electromagnetic radiation to get here; their high-energy particles arrive a few days later. At times when these surges of particles are *not* coming to us, the normal solar wind particles are trapped by the Earth's magnetic fields in the Van Allen belts (Section 5-4). The additional particles from a coronal mass ejection or other solar event overwhelm the Van Allen belts, enabling matter in them to cascade earthward.

Several spacecraft now monitor the Sun and the space between it and the Earth. These include the Japanese *Yohkoh* spacecraft (launched 1991) that monitors X rays and gamma rays (the figure that opens this chapter is an X-ray image taken by *Yohkoh*), the *Solar and Heliospheric Observatory* (*SOHO*) launched in 1995 by the European Space Agency (ESA) and NASA, and the *Transition Region and Coronal Explorer* (*TRACE*) launched in 1998, among others. These spacecraft continuously monitor the Sun in gamma ray, X-ray, ultraviolet, and visible parts of the spectrum, and they provide both scientific and space weather information.

Space weather is especially important in this time of high technology, because strong surges of high-energy particles from the Sun can potentially damage satellites, disrupt radio communications, short out power grids, produce intense auroras in the Earth's atmosphere, and cause some of the Earth's atmosphere to gush into interplanetary space. The above satellites provide data, even from the side of the Sun facing away from the Earth, that allows astronomers to forecast several days in advance the arrival of dangerous levels of solar particles. This enables us to protect satellites and

other technology. We can determine the state of the Sun's far side by measuring ripples on the Earth-facing side of the Sun's surface and then use computer models of how activity on the far side affects these ripples.

The numbers of plages, prominences, flares, and coronal mass ejections vary with the same 11-year cycle as the number of sunspots. Coronal mass ejections, the major source of hazardous particles from the Sun, occur with varying frequency throughout the sunspot cycle, but they never completely cease.

THE SUN'S INTERIOR

During the nineteenth century, geologists and biologists found convincing evidence that the Earth must have existed in more or less its present form for at least hundreds of millions of years. This fact posed severe problems for astrophysicists, because at that time it seemed impossible to explain how the Sun could continue to shine for so long, radiating immense amounts of energy into space. If the Sun were shining by burning coal or hydrogen gas as we could burn these substances on Earth, the Sun would be ablaze for only 5000 years before consuming all its fuel.

Everyday experience tells us the Sun is the source of an enormous amount of energy. Formal observations reveal hot gas, intense magnetic fields, and a variety of continuous and transient features on the Sun's surface. But the energy does not come from the surface gas or the magnetic fields—they have no mechanisms to create it. The answer is deep within the Sun. To understand why the Sun shines, we must understand where its energy originates and how that energy is transported to its surface.

9-7 Thermonuclear reactions in the core of the Sun produce its energy

In 1905, Albert Einstein provided an important clue to the source of the Sun's energy with his special theory of relativity. One of the implications of this theory is that matter and energy are related by the simple equation:

$$E = mc^2$$

In other words, a mass (m) can be converted into an amount of energy (E) equivalent to mc^2, where c is the speed of light. Because c^2 ($c \times c$) is huge, 9×10^{16} km²/s², a small amount of matter can be converted into an awesome amount of energy.

Inspired by Einstein's work, astrophysicists began to wonder if the Sun's energy output might come from the conversion of matter into energy. But how? The answer came in the 1920s from the British astrophysicist Arthur Eddington. Eddington proposed that temperatures at the center of the Sun, its **core,** are much greater than had ever been imagined. Subsequent calculations revealed that the temperature in the Sun's core is about 15.5×10^6 K. At these high temperatures, the nuclei of the Sun's hydrogen fuse together to produce nuclei of the element helium. In this reaction a tiny amount of mass is lost: It is transformed into a very large amount of energy—the energy of the Sun.

The process of fusing nuclei at such extreme temperatures is called **thermonuclear fusion.** For more details on thermonuclear fusion, see An Astronomer's Toolbox 9-1. In particular, conversion of hydrogen into helium is called **hydrogen fusion,** and the same process provides the devastating energy of a hydrogen bomb. (We will encounter the thermonuclear fusion of helium and other elements in later chapters.) The energy generated by hydrogen fusion in the Sun's core eventually escapes through the photosphere into space. That energy makes the Sun shine.

You may have heard comments that matter is always conserved or that energy is always conserved. We know that both of these concepts are incorrect, because mass can be converted into energy and vice versa. What is true, however, is that the total amount of mass plus energy is conserved. So, the destruction of mass and the creation of energy by the Sun do not violate any laws of nature.

Hydrogen fusion is also called *hydrogen burning,* even though nothing is burned in the conventional sense. The ordinary burning of wood, coal, or any flammable substance is a chemical process involving only the electrons

orbiting the nuclei of the atoms. Thermonuclear fusion is a far more energetic process that involves violent collisions between the atomic nuclei themselves. An Astronomer's Toolbox 9-1 shows you just how energetic.

 Fusion

9-8 Solar models describe how energy escapes from the Sun's core

A scientific description of the Sun's interior, called a **solar model,** explains how the energy from nuclear fusion in the Sun's core gets to its photosphere. The model begins with the inward force of the Sun's gravity. This force raises the pressure and temperature in the Sun's core. However, because the Sun is not shrinking today, an outward force must counter the inward force of gravity everywhere inside it. That outward force is produced by the gamma-ray photons created during fusion.

These photons slam into nearby ions and electrons in the solar core. These particles absorb the energy of the photons and rebound at very high speeds. Because they are densely packed together, they do not travel far before striking other particles. In each collision, the particles exert forces on each other. They then rebound, emitting photons and colliding with other particles. The result of these frequent interactions is an outward force sufficient to counterbalance gravity. The balance between the inward force of gravity and the outward force from the motion of the hot gas is called **hydrostatic equilibrium.**

The photons emitted after particles collide in the Sun's core have slightly less energy than the photons that these particles had previously absorbed. The energy the photons lose provides the outward force keeping the Sun in equilibrium. Photons collide and new ones are emitted. Photon energies slowly diminish as the photons travel outward from the Sun's core to the photosphere, an odyssey that typically takes 170,000 years.

The outward transportation of energy by photons hitting particles, which then bounce off other particles and thereby reemit photons, is called *radiative transport,* because individual photons are responsible for carrying energy from collision to collision. Calculations show that radiative transport is the dominant means of outward energy flow in the **radiative zone,** extending from the core to 80% of the way out to the photosphere.

Near the photosphere, bulk motion of the hot gas, rather than the flying, energetic photons, carries most of the energy the remaining distance to the surface. As we discussed earlier, this circulation of gas blobs is called *convection.* We thus say that the Sun has a **convective zone.** Once hot gas convects up to the photosphere, it emits photons

AN ASTRONOMER'S TOOLBOX 9-1

Thermonuclear Fusion

What drives the thermonuclear fusion that powers the Sun? For nuclei to fuse, they must be brought together at incredibly high temperatures and pressures. That is exactly what occurs in the Sun's core, where the entire mass of the Sun compresses inward. The core's temperature is 15.5 million K, its pressure is about 3.4×10^{11} atm, and its density is 160 times greater than that of water.

Under less severe conditions, nuclei could not penetrate each other because the positive electric charge on each proton prevents nearby protons from coming together. Remember that like charges repel each other. But in the extreme heat and pressure of the Sun's center, the protons move so fast that they can stick, or fuse, together.

The nuclear transformations inside the Sun follow several routes, but each begins with the simplest atom, hydrogen (H). Most hydrogen nuclei consist of a single proton. The outcome of fusion is creation of the nucleus of the next simplest atom, helium (He), consisting of two protons and two neutrons. The fusion of hydrogen into helium takes several steps.

The accompanying diagram shows one common path for the reaction. This particular sequence, called the *proton-proton*, or *PP, chain,* is the most common fusion process occurring in the Sun.

Note that the proton-proton fusion releases positively charged electrons, e^+, called **positrons**. When these positrons encounter regular electrons in the Sun's core, both particles are annihilated and their mass is converted into energy in the form of gamma-ray photons. In addition, neutral, nearly massless particles called **neutrinos**, ν, are also emitted. The final step, in which the helium forms, returns two protons, which are then available to fuse again.

Because the PP chain produces neutrinos and the Sun's energy, we can summarize hydrogen fusion this way:

$$4H \rightarrow He + neutrinos + energy$$

Example: From our summary equation, we can easily calculate the energy released during a fusion reaction. We simply look at how much mass is converted into energy:

$$
\begin{aligned}
\text{Mass of 4 hydrogen atoms} &= 6.693 \times 10^{-27} \text{ kg} \\
- \text{ Mass of 1 helium atom} &= 6.645 \times 10^{-27} \text{ kg} \\
\hline
\text{Mass lost} &= 0.048 \times 10^{-27} \text{ kg}
\end{aligned}
$$

Thus, a small fraction (0.7%) of the mass of the hydrogen going into the nuclear reactions does not show up in the mass of the helium. This last mass is converted into energy, as predicted from Einstein's famous equation,

$$
\begin{aligned}
E &= mc^2 \\
&= (0.048 \times 10^{-27} \text{ kg}) \times (3 \times 10^8 \text{ m/s})^2 \\
&= 4.3 \times 10^{-12} \text{ J}
\end{aligned}
$$

Compare! This energy is released from the formation of a single helium atom. It would light a 10-watt lightbulb for almost one-half a trillionth of a second.

Now let's add up the energy emitted from the entire Sun. The Sun's mass, usually designated 1 $M_\odot$, is equal to 333,000 $M_\oplus$. Its total energy output per second, called the **solar luminosity** and denoted 1 $L_\odot$, is 3.9×10^{26} W. To produce this luminosity, the Sun converts 600 million metric tons of hydrogen into helium within its core each second. This prodigious rate is possible because the Sun contains a vast supply of hydrogen—enough to continue the present rate of energy output for another five billion years.

Try these questions: How many helium atoms would have to be created from hydrogen to release enough energy to light a 60 watt bulb for 12 hours? How many joules of energy would be released if a U.S. penny (2.5×10^{-3} kg) were converted entirely into energy?

(Answers appear at the end of the book.)

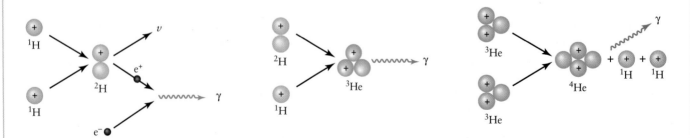

Steps to Fuse Hydrogen into Helium Note also that an electron, e^-, and a positron, or positively charged electron, e^+, annihilate each other and form a gamma ray, γ. Gamma rays, created in various fusion processes, are the source of the Sun's energy.

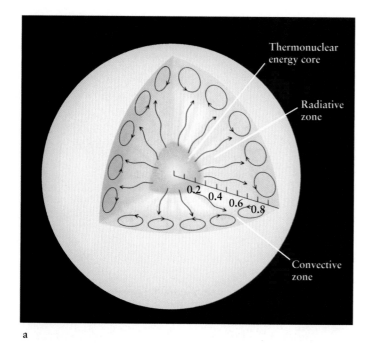

a

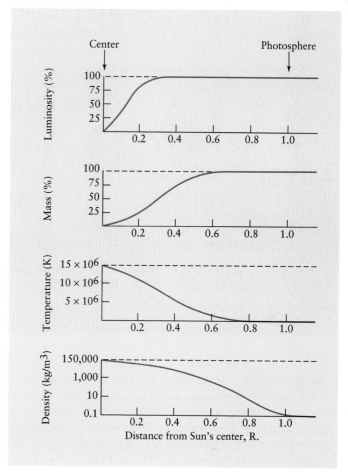

b

FIGURE 9-21   **The Solar Model (a)** Thermonuclear reactions occur in the Sun's core, which extends to a distance of 0.25 solar radius from the center. In this model, energy from the core radiates outward to a distance of 0.8 solar radius. Convection is responsible for energy transport in the Sun's outer layers. **(b)** The Sun's internal structure is displayed here with graphs that show how the luminosity, mass, temperature, and density vary with the distance from the Sun's center. A solar radius (the distance from the Sun's center to the photosphere) equals 696,000 km.

into space, cools, and settles back into the Sun. As noted earlier, this convective flow is the origin of solar granules, and the departing energy is the light that the Sun emits into space. Figure 9-21a sketches our model of the internal structure of the Sun.

Different gamma rays created in the Sun's core lose different amounts of energy as they and their successors travel upward through the Sun. Therefore, the photons emitted from the photosphere have a wide range of energies and, hence, wavelengths. The most intense emission is in the visible part of the electromagnetic spectrum. This process of energy loss by photons leaving the Sun's core is the origin of the blackbody nature of the photosphere's spectrum.

The solar model carefully charts the Sun's internal characteristics, such as its pressure, temperature, and density at various depths. Our model of the Sun's interior is expressed as a set of mathematical equations called the *equations of stellar structure*. These describe the conditions required inside any star to keep it stable, meaning that it neither expands nor collapses. Because the equations are so complex, astrophysicists today use computers to solve them. Figure 9-21b presents their results graphically. The four graphs show how the Sun's luminosity, mass, temperature, and density vary from the Sun's cen-

ter to its photosphere. For example, the upper graph gives the percentage of the Sun's luminosity created within that radius. The luminosity rises to 100% at about one-quarter of the way from the Sun's center to its photosphere. This tells us that all the Sun's energy is produced within a volume extending out to $1/4$ R$_\odot$, where R$_\odot$ is the radius of the Sun.

The mass curve rises to nearly 100% at about 0.6 R$_\odot$ from the Sun's center. Almost all of the Sun's mass is therefore confined to a volume extending only 60% of the distance from the Sun's center to its surface, which helps explain why the density of the gas in the photosphere is 10^4 times lower than the density of the air we breathe.

To learn more about the Sun's interior, astronomers record its vibrations—a study called **helioseismology**—just as geologists use earthquakes to study the Earth's interior structure. Although there are no true sunquakes, the Sun does vibrate at a variety of frequencies, somewhat like a ringing bell. These vibrations, first noticed in 1960, can be detected with sensitive Doppler shift measurements. They have shown that portions of the Sun's surface move up and down by about 10 km every 5 minutes (Figure 9-22a).

Slower vibrations with periods ranging from 20 minutes to nearly an hour were discovered in the 1970s. Oscillations

a

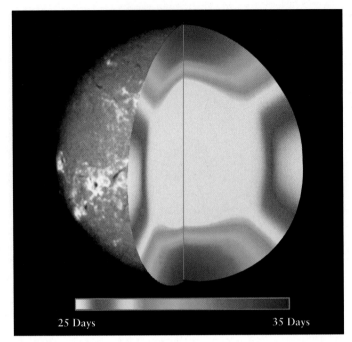

b

FIGURE 9-22 Helioseismology (a) This computer-generated image shows one of the millions of ways that the Sun vibrates because of sound waves resonating in its interior. The regions that are moving outward are colored blue; those moving inward are red. The cutaway shows how deep these oscillations are believed to extend. (b) This cutaway picture of the Sun shows how the rate of solar rotation varies with depth and latitude. Red and yellow denote faster-than-average motion; blue regions move more slowly than average. The pattern of differential surface rotation, which varies from 25 days at the equator to 35 days near the poles, persists at least 19,000 km down into the Sun's convective layer. Sunspots preferentially occur on the boundaries between different rotating regions. Earthlike jet streams and other wind patterns have also been discovered in the Sun's atmosphere.

lasting several days have been detected, as have pulses 16 months long. One important discovery from helioseismology is that the convective zone is twice as thick as prior solar models predicted. Another is that below the convective zone the Sun apparently rotates like a rigid body (Figure 9-22b).

Although the Sun is much more nearly spherical than the Earth, Doppler measurements reveal that it has "dimples" 1/2 kilometer deep, like a giant golf ball. Also, in 1998, astronomers first detected ripples in the Sun's atmosphere during solar flares. These rings of vibrating gas appear to be just like those you create when you throw a rock into a lake.

9-9 The mystery of the missing neutrinos inspired research into the fundamental nature of matter

As explained in An Astronomer's Toolbox 9-1, for every proton that changes into a neutron during thermonuclear fusion, a *neutrino* is released. Neutrinos have no electric charge, and they are extraordinarily difficult to detect because they rarely interact with ordinary matter. In fact, neutrinos interact so infrequently with matter that they can easily pass through the entire Earth as if it were not there. The Sun is also largely transparent to neutrinos, allowing these particles to stream outward, unimpeded, from its core.

According to our model of fusion in the Sun, nearly 10^{38} *solar neutrinos* are produced at the Sun's center each second. This output is so huge that, here on Earth, roughly 100 billion solar neutrinos pass through every square centimeter of your body every second!

On very rare occasions a solar neutrino strikes a neutron and converts it into a proton. If astronomers could detect even a few of these converted protons, it might be possible to build a "neutrino telescope" that could be used to "see" directly into the center of the Sun and the thermonuclear inferno there that is now hidden from ordinary view.

 Inspired by such possibilities, Raymond Davis of the Brookhaven National Laboratory designed and built a large neutrino detector. This device consists of a huge tank containing 100,000 gallons of perchloroethylene (C_2Cl_4, the fluid your local dry cleaner uses) buried deep in a mine in South Dakota. Because matter is virtually transparent to neutrinos, most of the solar neutrinos pass right through Davis's tank with no effect whatever. On rare occasions, however, a solar neutrino strikes the nucleus of one of the chlorine atoms in the cleaning fluid and converts one of its neutrons into a proton, creating a radioactive atom of argon. The rate at which argon is produced is therefore correlated with the number of solar neutrinos arriving at the Earth.

On average, solar neutrinos create one radioactive argon atom every three days in Davis's tank. To the consternation of astronomers, this rate corresponds to only one-third of the neutrinos predicted from the "standard" solar model presented in Figure 9-21b. This experiment, which began in the mid-1960s, has been repeated with extreme care by other researchers around the world during the past four decades using a variety of different detection techniques. There are three types of neutrinos, each associated with different nuclear reactions, but until recently all of the experiments were designed to detect just solar neutrinos (that is, those made when neutrons transform into protons). All these traditional neutrino detectors get comparable, low results.

Most astronomers believe that the solar neutrino experiments are sound, implying that there really are fewer solar neutrinos reaching the Earth than theory predicts. This means, of course, that the theory of neutrinos needs revision. Originally, neutrinos were postulated to be massless. In that case, there is no viable explanation for the lower-than-expected solar neutrino counts here on Earth. However, if neutrinos do have mass, then they have the property of being able to transform into other particles over time. Specifically, they can change into either of two other types of neutrinos, not normally created in the Sun. Furthermore, calculations reveal that many of the solar neutrinos would transform *before* reaching the Earth, which would explain the low solar neutrino detection rate.

While the first generation of neutrino detectors, described above, required analyzing the fluid for chemical changes, astronomers have recently devised a method of actually seeing flashes of light when neutrinos enter the detectors and interact with the liquid there. In that way, they immediately determine the presence of neutrinos and the direction from which they came. The first such experiment began in Japan (Figure 9-23), and it uses a water-based neutrino detector sensitive to the alternative types of neutrinos. It began discovering them in 1998, showing that neutrinos do, indeed, have mass. These observations were confirmed in 2001 by the Sudbury Neutrino Observatory in Ontario, Canada. At least a dozen other neutrino detectors have begun operations or are being built.

To understand how this new generation of neutrino detector works, note that water contains many protons, namely, the nuclei of hydrogen atoms. When one of the other

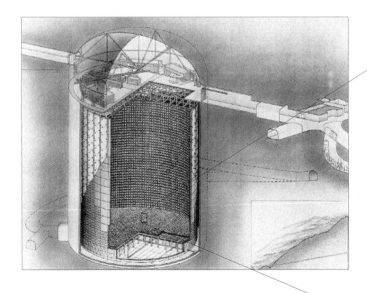

FIGURE 9-23   The Solar Neutrino Experiment Located in the Kamioka Mozumi mine in Japan, the Super-Kamiokande solar neutrino detector is centered around a tank containing 50,000 tons of ultrapure water. Occasionally, a neutrino entering the tank interacts with a proton, producing a positron, a positively charged electron. The positron moves so fast that it gives off a light flash called Cerenkov radiation. Some 13,000 light detectors (seen from below in the inset) detect this light. A serious accident in 2001 destroyed most of these light detectors, taking this detector off-line for now. (Inset: Super-Kamiokande Group, Institute for Cosmic Ray Research, University of Tokyo)

types of neutrino strikes a proton, it produces a *positron,* which is identical to an electron, except that it has a positive charge. The positron in turn emits a flash of light called **Cerenkov radiation.** As the Russian physicist Pavel A. Cerenkov (pronounced "Che-ren-kov") first observed, the flash occurs whenever a particle moves through water faster than light can. Such motion does not violate the tenet that the speed of light *in a vacuum* (3×10^5 km/s) is the ultimate speed limit in the universe. Light is slowed considerably as it passes through water, and high-energy particles can exceed this reduced speed without violating the laws of physics.

Thus, scientists detect neutrinos indirectly by observing Cerenkov radiation flashes with light-sensitive devices called *photomultipliers* mounted in the water (see Figure 9-23). This evidence that neutrinos have mass explains the earlier low rate of solar neutrino observations.

WHAT DID YOU KNOW?

|1| *What percentage of the solar system's mass is in the Sun?* The Sun contains about 99.85% of the solar system's mass.

|2| *Does the Sun have a solid and liquid interior, like the Earth?* No. The entire Sun is composed of hot gases.

|3| *What is the surface of the Sun like?* The Sun has no solid surface. Indeed, it has no solids anywhere. The level we see, the photosphere, is composed of hot, churning gases.

|4| *Does the Sun rotate?* The Sun's surface rotates differentially, varying between once every 35 days near its poles and once every 25 days at its equator.

|5| *What makes the Sun shine?* Thermonuclear fusion in the Sun's core is the source of the Sun's energy.

9-10 Frontiers yet to be discovered

Despite decades of observing and modeling it, the Sun holds innumerable secrets to be uncovered. That process is being aided by the armada of spacecraft specifically designed to observe the Sun in different parts of the spectrum. Among the concepts astronomers hope to soon understand is more about how the Sun's magnetic field is generated and how it changes throughout the solar cycle; how the Sun's atmosphere heats so dramatically in the transition zone; why the solar luminosity varies with time; just how the Sun's changing output affects the temperatures on the Earth; the details of how flares and coronal mass ejections occur; and why the Sun rotates differentially.

 Further Reading on These Topics

KEY WORDS

Cerenkov radiation, 266
chromosphere, 250
convective zone, 261
core (of the Sun), 261
corona, 251
coronal hole, 258
coronal mass ejection, 259
filament, 257
granules, 250
helioseismology, 263
hydrogen fusion, 261
hydrostatic equilibrium, 261

limb (of the Sun), 250
limb darkening, 250
magnetic dynamo, 256
neutrino, 262
photosphere, 249
plage, 257
plasma, 256
positron, 262
prominence, 257
radiative zone, 261
solar cycle, 256
solar flare, 258

solar luminosity (L$_\odot$), 262
solar model, 261
solar wind, 253
spicule, 251
sunspot, 253
sunspot maximum, 253
sunspot minimum, 253
supergranule, 251
thermonuclear fusion, 261
transition zone, 252
Zeeman effect, 255

KEY IDEAS

The Sun's Atmosphere

• The visible surface of the Sun is a layer at the bottom of the solar atmosphere called the photosphere. The gases in this layer shine as a blackbody. Convection of gas from below the surface produces features there called granules.

• Above the photosphere is a layer of hotter but less dense gas called the chromosphere. Jets of gas called spicules rise up into the chromosphere along the boundaries of supergranules.

• The outermost layer of thin gases in the solar atmosphere is called the corona, which extends outward to become the solar wind at great distances from the Sun. The gases of the corona are very hot but at low density.

The Active Sun

• Some surface features on the Sun vary periodically in an 11-year cycle, with the magnetic fields that cause these changes actually varying over a 22-year cycle.

• Sunspots are relatively cool regions produced by local concentrations of the Sun's magnetic field protruding through the photosphere. The average number of sunspots and their average latitude increases and decreases in an 11-year cycle.

• A prominence is gas lifted into the Sun's corona by magnetic fields. A solar flare is a brief, but violent, eruption of hot, ionized gases from a sunspot group. Coronal mass ejections send out large quantities of gas from the Sun. Coronal mass ejections and flares that head our way affect satellites, communication, electrical power, and cause auroras.

• The magnetic dynamo model suggests that many transient features of the solar cycle are caused by the effects of differential rotation and convection on the Sun's magnetic field.

The Sun's Interior

• The Sun's energy is produced by the thermonuclear process called hydrogen fusion, in which four hydrogen nuclei release energy when they fuse to produce a single helium nucleus.

• The energy released in a thermonuclear reaction comes from the conversion of matter into energy according to Einstein's equation $E = mc^2$.

• A stellar model is a theoretical description of a star's interior derived from calculations based on the laws of physics. The solar model suggests that hydrogen fusion occurs in a core that extends from the Sun's center to about 0.25 solar radius.

• Throughout most of the Sun's interior, energy moves outward from the core by radiative diffusion. In the Sun's outer layers, energy is transported to the Sun's surface by convection.

• Neutrinos generated and emitted by the Sun are detected at a lower rate than is predicted by our standard model of thermonuclear fusion. This occurs because neutrinos have mass; therefore, some of them change into other forms of neutrinos before they reach Earth. These alternative forms are now being detected.

REVIEW QUESTIONS

The answers to all computational problems, which are preceded by an asterisk (*), appear at the end of the book.

1 Describe the features of the Sun's atmosphere that are always present.

2 Describe the three main layers in the solar atmosphere and how you would best observe them.

3 Name and describe seven features of the active Sun. Which two are the same, seen from different angles? Which have a direct impact on the Earth? Explain.

4 Describe the three main layers of the Sun's interior.

*5 When will the next sunspot minimum and maximum occur after the maximum in 2000 and the minimum in 2007? Explain your reasoning.

6 Why is the solar cycle said to have a period of 22 years, even though the sunspot cycle is only 11 years long?

7 How do astronomers detect the presence of a magnetic field in hot gases such as those in the solar photosphere?

8 Describe the dangers in attempting to observe the Sun. How have astronomers learned to circumvent these hazards?

9 Give an everyday example of hydrostatic equilibrium.

10 Give some everyday examples of heat transfer by convection and radiative transport.

11 What do astronomers mean by "a model of the Sun"?

12 Why do thermonuclear reactions in the Sun take place only in its core?

13 What is hydrogen fusion? This process is sometimes called "hydrogen burning." How is hydrogen burning fundamentally unlike the burning of a log in a fireplace?

14 Describe the Sun's interior, including the main physical processes that occur at various depths within the Sun.

15 What is a neutrino, and why are astronomers so interested in detecting neutrinos from the Sun?

ADVANCED QUESTIONS

*16   Using the mass and size of the Sun, calculate the Sun's average density. Compare your answer to the average densities of the outer planets. *Hint:* The volume of a sphere of radius r is $(4/3)\pi r^3$.

*17   Assuming that the current rate of hydrogen fusion in the Sun remains constant, what fraction of the Sun's mass will be converted into helium over the next five billion years? How will this affect the chemical composition of the Sun?

*18   Calculate the wavelengths at which the photosphere, chromosphere, and corona emit the most radiation. Explain how the results of your calculations suggest the best way to observe these regions of the solar atmosphere. *Hint:* Use Wien's law and assume that the average temperatures of the photosphere, chromosphere, and corona are 5800 K, 50,000 K, and 1.5×10^6 K, respectively.

19   When we are near a sunspot maximum, the Hubble Space Telescope must be put in a higher orbit. Why? *Hint:* Think about how the increased solar energy affects the Earth's atmosphere.

DISCUSSION QUESTIONS

20   Discuss the extent to which cultures around the world have worshiped the Sun as a deity throughout history. Why do you think our star inspires such widespread veneration?

21   Discuss some of the difficulties of correlating solar activity with changes in the terrestrial climate.

22   Describe some advantages and disadvantages of observing the Sun (a) from space and (b) from Earth's south pole. What kinds of phenomena and issues do solar astronomers want to explore from both Earth-orbiting and Antarctic observatories?

WHAT IF ...

23   The Sun was not rotating? What about it would be different?

24   The typical solar wind was much stronger (say 100 times stronger) than it is now? What differences would there be in the solar system?

25   The typical solar wind was much weaker (say 100 times weaker) than it is now? What differences would there be in the solar system?

26   The Sun varied its output, as many stars do? What differences would there be in the solar system?

WEB/CD-ROM QUESTIONS

27   Search the Web for the latest information from the neutrino detectors at the Super-kamiokande Observatory and the Sudbury Neutrino Observatory. What happened to Super Kamiokande? What are the most recent results from Sudbury? What is the current thinking about the solar neutrino problem? What is the status of a new detector called Borexino?

28   Search the Web for information about features of the solar atmosphere called *sigmoids.* What are they? What causes them? How do they provide a way to predict coronal mass ejections?

 *29   **Determining the Lifetime of a Solar Granule** Access and view the video "Granules on the Sun's Surface" in Chapter 9 of the *Discovering the Universe* Web site or CD-ROM. You will use it to determine the approximate lifetime of a solar granule. Select an area on the Sun's image and slowly and rhythmically repeat "start, stop, start, stop" until you can consistently predict the appearance and disappearance of granules. While keeping your rhythm, move to a different area of the video and continue monitoring the appearance and disappearance of granules. When you are confident you have the timing right, move your eyes (or use a partner) to the clock shown in the video. Using your "start stop" cycle, determine the length of time between the appearance and disappearance of the granules and record your answer.

OBSERVING PROJECTS

30 Use a telescope to view the Sun by projecting the Sun's image onto a screen or sheet of white paper. **Do not look directly at the Sun! Looking at the Sun causes blindness.** Do you see any sunspots? If so, sketch their appearance. Can you distinguish between the umbras and penumbrae of the sunspots? Can you see limb darkening? Can you see granulation?

31 If you have access to an H_α filter attached to a telescope especially designed for viewing the Sun safely, use this instrument to examine the solar surface. How does the filtered appearance of the Sun differ from that in white light (see Observing Project 30)? What do sunspots look like in H_α? Can you see any prominences? Can you see any filaments? Are the filaments in the H_α image near any sunspots seen in white light?

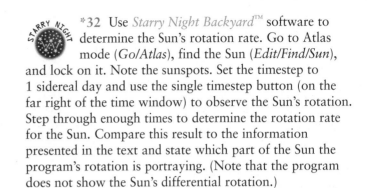

*32 Use *Starry Night Backyard*™ software to determine the Sun's rotation rate. Go to Atlas mode (*Go/Atlas*), find the Sun (*Edit/Find/Sun*), and lock on it. Note the sunspots. Set the timestep to 1 sidereal day and use the single timestep button (on the far right of the time window) to observe the Sun's rotation. Step through enough times to determine the rotation rate for the Sun. Compare this result to the information presented in the text and state which part of the Sun the program's rotation is portraying. (Note that the program does not show the Sun's differential rotation.)

III ESSENTIALS FOR UNDERSTANDING THE STARS

IN THIS SECTION OF THE BOOK YOU WILL DISCOVER

- that the distances to many nearby stars can be measured directly, while the distances to farther ones are determined indirectly (Essentials III)

- the observational properties of stars on which astronomers base their models of stellar evolution (Essentials III, Chapter 10)

- the nature of stars and just how special—and how ordinary—the Sun really is (Chapters 10 and 11)

- that stellar luminosities and spectra are clues to understanding stellar evolution (Chapters 10 and 11)

- that stars with different masses evolve differently (Chapters 11, 12, and 13)

- the properties of black holes (Chapter 13)

SCIENCE BY STARLIGHT

By analyzing starlight, an astronomer can determine stellar temperatures, chemistries, and masses; how far away the stars are; and how fast they are moving toward or away from us. This image shows some of the stars in Orion and adjoining constellations. These stars span virtually the entire range of star types. The brightest star in the night sky, Sirius (part of Canis Major), is toward the lower left. It is only 8.6 ly away, making it one of the stars closest to us. Most stars in this photograph are hundreds, or even thousands, of light-years away. The blue haze is starlight scattering off interstellar gas and dust. (Okiro Fujii, L'Astronomie).

R I ■ U X G

WHAT DO YOU THINK?

1️⃣   Do stars last forever?

2️⃣   How near is the closest star other than the Sun?

3️⃣   How luminous is the Sun compared with other stars?

Legend tells of a race of remarkable insects, the Ephemera, who inhabited a great forest. These noble creatures were blessed with great intelligence but were cursed with tragically short life spans. To the Ephemera, the forest seemed eternal and unchanging. Members of each generation lived out their brief lives without ever noticing any changes in their leafy world. Nevertheless, careful observations and reasoning led some Ephemera to postulate that the forest was not static. They began to suspect that small green shoots grew to become huge trees and that mature trees eventually died, toppled over, and littered the forest with rotting logs, enriching the soil for future trees. Although unable to witness the transformation personally, the Ephemera predicted the existence of life processes stretching over the mind-boggling periods of many years.

To us, the heavens, too, seem eternal and unchanging; the views that greet us every night are virtually indistinguishable from those seen by our ancestors. This permanence, however, is an illusion. We see the stars because they emit vast amounts of radiation, created at the expense of converting one element into another in their interiors. And the change is not limited to the stars' chemical composition. We will soon see that over the course of their existence, stars also change in size, luminosity (total energy emitted each second), temperature, and color. Astronomers call this process **stellar evolution.**

We do have one advantage over the Ephemera, who were so small that they could see only a few trees, leaves, shoots, and rocks. Astronomers are witnesses to literally billions of stars in various stages of evolution. Therefore, although we cannot see any one star go through more than a tiny fraction of the process, we see every stage of stellar evolution. By understanding the changes that stars undergo, we gain insight into our place in the cosmos. We begin our journey to the stars by learning how far across interstellar space we have to go to reach the nearest ones.

III-1 Distances to nearby stars are determined by stellar parallax

Nothing in the human experience prepares us for understanding the distances between the stars. On Earth, the greatest distance we encounter in our daily lives is at most a few thousand kilometers. From observing the Sun, Moon, and planets, we have some comprehension of hundreds of thousands or even millions of kilometers.

How far away do you think the nearest stars are? Ask ten people and you'll probably hear answers ranging from thousands to billions of kilometers or miles. In fact, the closest star other than the Sun, Proxima Centauri, in the constellation Centaurus, is about 40 trillion kilometers (25 trillion miles) away. It takes light about four years to get here from there. Most of the stars you see in the night sky are many times farther away than Proxima Centauri. (Centaurus, by the way, is not visible from most of the northern hemisphere.) How do we measure the distance to a nearby star?

Because Proxima Centauri is closer to us than other stars, its position among the background stars changes as the Earth orbits the Sun. This is precisely the same effect that we see when, for example, Mars appears to have retrograde motion as we pass between it and the Sun (see Figures 2-1 and 2-2) or that Tycho Brahe sought for the supernova of 1572 (see Figure 2-6). The motion of nearby stars among the background of more distant stars due to the Earth's motion around the Sun is called **stellar parallax.**

Parallax is an everyday phenomenon. We experience it when nearby objects appear to shift their positions against a distant background as we move (review Figure 2-5). We also experience it continuously when we are awake, because our eyes change angle when looking at objects at different distances. As you look at a tree 10 m away, your eyeballs cross only slightly. Looking at something closer, say this book, your eyes cross much more in order for both eyes to focus on the same word. The parallax angle formed between your eyes and the tree or the book lets your brain judge just how close they are to you. Working out distances of nearby stars is done similarly, but it requires painstaking angle measurements and a little geometry.

As the Earth moves from one side of its orbit around the Sun to the other, a nearby star's apparent position shifts among the more distant stars. Referring to Figure III-1a, the parallax angle, p, is half the angle by which it shifts, measured in arcseconds. The difference in parallax angles for stars at different distances can be seen by comparing Figures III-1a and b. An Astronomer's Toolbox III-1 explores some details of how distances are determined from parallax angles.

The first parallax measurement was made in 1838 by Friedrich Wilhelm Bessel, a German astronomer and mathematician. He found the parallax angle of the star 61 Cygni to be $1/3$ arcsec, and so its distance is about 3 pc. Because parallax angles smaller than about 0.01 arcsec are difficult to measure from Earth-based observatories, the stellar parallax method using Earth-based telescopes gives stellar distances only up to about 100 pc. The precision of stellar parallax measurements is limited by the angular resolution of the telescope, as discussed in Chapter 3.

Telescopes in space are unhampered by our atmosphere and therefore have higher resolutions than Earth-based telescopes. Parallax measurements made in space thus enable

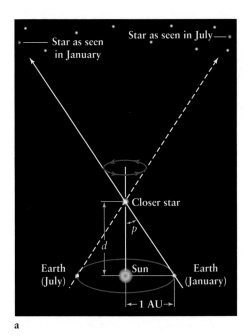

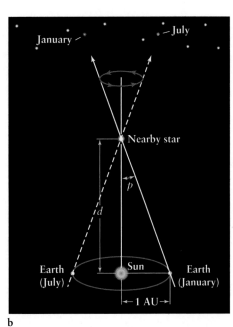

WEB LINK III.1

FIGURE III-1 Using **Parallax to Determine Distance** **(a)** As the Earth orbits the Sun, a nearby star appears to shift its position against the background of distant stars. The star's parallax angle (p) is equal to the angle between the Sun and Earth as seen from the star. The stars on the scale of this drawing are shown much closer than they are in reality. If drawn to the correct scale, the closest star, other than the Sun, would be about 5 km (3.2 mi) away. **(b)** The farther the star is from us, the smaller the parallax angle p. The distance to the star (in parsecs) is found by taking the inverse of the parallax angle p (in arcseconds), $d = 1/p$.

a

b

astronomers to determine the distances to stars well beyond the reach of ground-based observations.

WEB LINK III.2 In 1989, the European Space Agency (ESA) launched a satellite called *Hipparcos* (an acronym for *hi*gh *p*recision *par*allax *c*ollecting *s*atellite, named for Hipparchus, an astronomer in ancient Greece who created an early classification system for stars). Although the satellite failed to achieve its proper orbit, astronomers used it to measure the distances to over 2.5 million of the nearest stars up to 150 pc away. (This is also 500 ly; see An Astronomer's Toolbox I-2 to review astronomical distance units). The success of *Hipparcos* has led to plans for better satellites such as *FAME* (*F*ull-sky *A*strometric *M*apping *E*xplorer, scheduled for launch in 2004), *SIM* (*S*pace *I*nterferometry *M*ission, scheduled for launch in 2009), and *GAIA* (scheduled for launch in 2010) to collect parallax data from stars even farther away.

Despite the information gained from stellar parallax, astronomers need to know the distances to more remote stars for which parallax cannot yet be measured. Several of these methods of determining ever-greater distances will be introduced in Chapters 10, 11, 12, and 15. Having established the fact that different stars are at different distances from Earth, we now turn to considering the brightnesses that stars appear to have as seen from our planet. This will prepare us to calculate how much light stars actually emit.

III-2 Apparent magnitude measures the brightness of stars as seen from Earth

WEB LINK III.3 Greek astronomers, from Hipparchus in the second century B.C. to Ptolemy in the second century A.D., undertook the classification of stars strictly by evaluating how bright they appear to be, compared to each other. This made sense because the stars were all assumed to be at the same distance from us, and, therefore, differences in brightness due to different stellar distances from Earth were not expected. The brightnesses of stars without regard to their distances are called **apparent magnitudes,** denoted with a lowercase *m*.

The brightest stars were originally said to be of first magnitude, and their apparent magnitudes were designated $m = +1$. Those stars that appeared to be about half as bright as first-magnitude stars were said to be second-magnitude stars (designated $m = +2$), and so forth, down to sixth-magnitude stars, the dimmest ones still visible to the unaided eye. (Greek astronomers did not try to classify the Sun's dazzling brightness in this scheme.)

This magnitude scale was created before accurate measurements of the relative brightnesses of stars could be made, and it has since been refined. Specifically, careful measurements reveal that the original first-magnitude stars were about 100 times brighter than the original sixth-magnitude stars. Astronomers chose the brightness factor of exactly 100 to define the range of brightness between modern first- and sixth-magnitude stars. In other words, it takes 100 stars of apparent magnitude $m = +6$ to provide as much light as we receive from a single star of apparent magnitude $m = +1$.

To find out how much brighter each magnitude is from the next dimmer one, we note that there are five integer magnitudes between first and sixth magnitude. Going from $m = +6$ to $m = +5$ increases (multiplies) the brightness we see by the same factor as going from $m = +5$ to $m = +4$, and so on. The number we must multiply 5 times to get the range of brightness of 100 is $100^{1/5} \approx 2.512$ (or, put another way, $2.512 \times 2.512 \times 2.512 \times 2.512 \times 2.512 \approx 100$). In other

AN ASTRONOMER'S TOOLBOX III-1

Distances to Nearby Stars

Recall from An Astronomer's Toolbox I-2 that 1 parsec (1 pc) is the distance at which two objects 1 AU apart appear 1 arcsecond apart. This distance is 3.09×10^{13} km, or 206,265 AU. The name parsec (from *parallax second*) originated in the use of parallax to measure distance. Using parsecs, we can write down an especially simple equation for the distances to stars:

$$\text{distance to a star in parsecs} = \frac{1}{\text{parallax angle of that star in arcseconds}}$$

or

$$d = 1/p$$

where d is the distance to the star in parsecs and p is the parallax angle of that star in arcseconds.

The equation is only this simple in these units, which is one of the main reasons that many astronomers discuss cosmic distances in parsecs rather than light-years. We will continue to use light-years (ly) primarily throughout this book, however, as they are more intuitive. In these latter units, the same equation becomes approximately

$$\text{distance to a star in light years} = \frac{3.26}{\text{parallax angle of that star in arcseconds}}$$

or

$$d_{ly} = 3.26/p$$

where d_{ly} is the distance to a star in light-years.

Example: The nearest star, Proxima Centauri, has a parallax angle of 0.77 arcsec, and so its distance is 1/0.77, or approximately 1.3 pc. Equivalently, Proxima Centauri is 4.24 ly away. The parallax of Proxima Centauri is comparable to the angular diameter of a dime seen from a distance of 3 km. The parallax angles of the 25 nearest stars are listed in Appendix Table A-4.

Try these questions: What is the average distance from the Earth to the Sun in parsecs? What is the parallax angle of a star 100 ly away? How far away is a star with a parallax angle of 0.385″? Rigel is 773 ly away; how far away is it in parsecs? What is its parallax angle in arcseconds?

(Answers appear at the end of the book.)

---

words, each successively brighter magnitude is approximately 2.512 times brighter than the preceding magnitude. For example, an $m = +3$ star is approximately 2.512 times brighter than an $m = +4$ star. It takes 2.512 third-magnitude stars to provide as much light as we receive from a single second-magnitude star.

Rigorous measurements of stellar brightness also reveal that the brightest stars are actually brighter than apparent magnitude $m = +1.0$ (first magnitude). Therefore, astronomers resort to *negative* numbers for the magnitudes of the very *brightest* objects. Sirius, for example, the brightest star in the night sky, shown in the figure that opens this Essentials section, has an apparent magnitude of $m = -1.44$. With this convention we can describe other bright objects in the sky, such as the Sun, Moon, comets, and planets. At its brightest, Venus shines with an apparent magnitude of $m = -4.4$, the full Moon has an apparent magnitude of $m = -12.6$, and the Sun has an apparent magnitude of $m = -26.7$. Remember: *Stars with negative apparent magnitudes appear brighter than stars with positive apparent magnitudes—the more negative, the brighter.*

Insight into Science **Bigger Isn't Necessarily Brighter** Numbering as well as naming schemes may be counterintuitive in science. One would expect brighter stars to have larger, more positive numbers than dimmer stars, but the apparent magnitude scheme is just the opposite. Similarly, in Chapter 10 you will discover that on the standard plot of stars used by astronomers, the Hertzsprung-Russell diagram, the hottest stars fall on the *left* and the coolest stars on the *right*.

Astronomers also extended the magnitude scale to larger positive numbers to describe dimmer stars visible only through their telescopes. For example, the dimmest stars visible through a pair of binoculars have an apparent magnitude of about $m = +10$. Time-exposure photographs reveal even dimmer stars. Through telescopes such as the Keck telescopes and the Hubble Space Telescope, stars nearly as dim as magnitude $m = +30$ can be observed. Similarly, apparent magnitudes from entire groups of stars, such as distant galax-

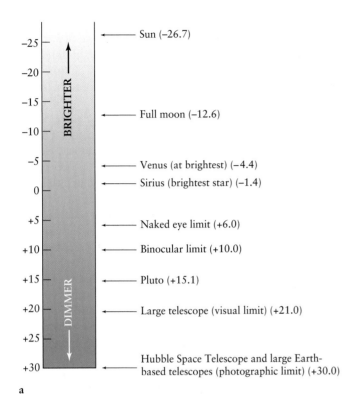

Sun (−26.7)

Full moon (−12.6)

Venus (at brightest) (−4.4)

Sirius (brightest star) (−1.4)

Naked eye limit (+6.0)

Binocular limit (+10.0)

Pluto (+15.1)

Large telescope (visual limit) (+21.0)

Hubble Space Telescope and large Earth-based telescopes (photographic limit) (+30.0)

a

R I V U X G

b

FIGURE III-2 **Apparent Magnitude Scale** **(a)** Astronomers denote the brightnesses of objects in the sky by their apparent magnitudes. Stars visible to the naked eye have magnitudes between $m = -1.44$ (Sirius) and about $m = +6$. CCD (charge-coupled device) photography through the Hubble Space Telescope or a large Earth-based telescope can reveal stars and other objects nearly as faint as magnitude $m = +30$. **(b)** Several stars in Orion are labeled with their names and apparent magnitudes. (Okiro Fujii, L'Astronomie)

ies, can be measured. Figure III-2a illustrates the modern apparent magnitude scale. Figure III-2b shows examples of apparent magnitudes for stars in the constellation of Orion.

The apparent magnitudes do not directly reveal fundamental stellar properties, because the brightnesses we measure are affected by the differing distances to the stars. All other things being equal, the closer of two identical stars appears brighter to us (has a smaller apparent magnitude) than the farther star. We now introduce another magnitude scale, which takes into account the different distances of stars.

III-3 Absolute magnitudes do not depend on distance

Suppose that we observe two identical stars, one twice as far away as the other. How much dimmer will the farther one appear to be as seen from Earth? As light moves outward from a source, it spreads out over increasingly larger areas of space and its brightness decreases. Thus, the farther away a source of light is, the dimmer it appears. The **inverse-**

square law provides the rule for just how quickly the brightnesses of objects change with distance.

Imagine a shell of light moving out from a star (Figure III-3). Start with a small square area of light on the shell when the shell has moved out a distance $d = 1$. The light in that square has a certain brightness. When the same shell has gone twice as far ($d = 2$), you can see that the light square has become 4 times larger. The light in each of the squares at $d = 2$ contains one-quarter of the photons that were in the single square at $d = 1$. Therefore, the small squares at $d = 2$ are one-quarter as bright as the same-sized square at $d = 1$. Similarly, when the shell has moved to $d = 3$, there are now nine small squares, each one-ninth as bright as the original square at $d = 1$. For example, the Sun emits a total of 3.83×10^{26} W (watts) of power. That energy spreads out as it travels through space. When it passes Mercury, sunlight provides 9140 W on every square meter of space. That same energy has spread out so much that by the time it passes us at 1 AU from the Sun, it provides only 1370 W per square meter.

The inverse-square law can be summarized mathematically as follows: Apparent brightness decreases inversely

FIGURE III-3   The Inverse-Square Law This drawing shows how the same amount of radiation from a light source must illuminate an ever-increasing area as the distance from the light source increases. Because the light spreads out as it moves away from the source, the apparent brightness of the source decreases. The decrease in brightness follows the inverse-square law, which means, for example, that tripling the distance decreases the brightness by a factor of 9.

with the square of the distance between the source and the observer. For example, consider two identical stars, one twice as far away as the other. The brightness of the one that is twice as far away decreases to $(1/2)^2 = 1/4$ the brightness of its closer twin by the time the starlight gets to us.

To determine the total energy emitted by stars, astronomers need to remove the effect of distance by calculating the brightnesses stars would have if they were all at the same distance from Earth, chosen to be 10 pc. The correction for distance is made by calculating the apparent magnitude each star would have if it were 10 pc from the Earth. The resulting number is a star's **absolute magnitude,** denoted by an uppercase M.

For example, if the Sun was moved to a distance of 10 pc from the Earth, it would have an apparent magnitude of $m = +4.8$. Therefore, the absolute magnitude of the Sun is $M = +4.8$. Absolute magnitude tells astronomers how much electromagnetic radiation stars actually emit and also how bright they are compared with each other. This information is used to evaluate models of stellar evolution, which we discuss in the next few chapters. Absolute magnitudes range from roughly $M = -10$ for the brightest stars to $M = +17$ for the dimmest.

Absolute magnitudes are usually determined by calculations based on apparent magnitudes. Of course, we cannot just move a star to 10 pc distance and then remeasure its ap-

parent magnitude. However, we can easily calculate the absolute magnitude of a nearby star. To do this, we first measure its apparent magnitude, and then we find its distance by measuring its parallax angle. An Astronomer's Toolbox III-2 gives the equation relating these two numbers and the star's absolute magnitude.

Unfortunately, absolute magnitudes use the same counterintuitive scale as apparent magnitudes. Fortunately, a more user-friendly scheme to measure stellar brightnesses exists. Recall from Chapter 4 that a star is very nearly a blackbody, and, thus, its brightness (and, hence, absolute magnitude) is directly related to the total amount of energy radiating from its surface each second, called its **luminosity.** The smaller or more negative a star's absolute magnitude, the greater its luminosity. For convenience, stellar luminosities are expressed in multiples of the Sun's luminosity, denoted $L_\odot$, which is roughly 3.83×10^{26} W. The intrinsically brightest stars ($M = -10$) have luminosities of $10^6 L_\odot$. In other words, each of these stars has the energy output of a million Suns. The dimmest stars ($M = +17$) have luminosities of $10^{-5} L_\odot$. We will provide both luminosity and absolute magnitude data in the following chapters.

The Sun is a wonderfully typical star. We will see in Chapter 12 that this is extremely fortunate for life on Earth: If the Sun was too much brighter (and hotter), it would not be able to shine long enough for life to have evolved here (see "What If ... The Earth Orbited a 1.5 $M_\odot$ Star?" on page 325). And if the Sun were too much dimmer (and cooler), the Earth at our present distance would not get nearly the energy needed for our life to flourish. To be warmed enough by a star with just half the Sun's mass for life to flourish, the Earth would have to be so close that the star's gravity would lock us into synchronous rotation, making one side of the planet too hot and the other too cold!

III-4   The brightnesses of stars help astronomers classify them and study their evolution

Armed with knowledge of stellar luminosities, astronomers early in the twentieth century searched for ways to use it to discover the nature of stars. In the following four chapters, we will learn how astronomers found a crucial relationship between stellar luminosities (or, equivalently, their absolute magnitudes) and their surface temperatures. Just as biologists graph the relationship between height and weight of each species of animal, stars lie in only certain regions of a graph of absolute magnitude versus surface temperature.

The information provided by the luminosity-surface temperature relationship gave astrophysicists insights into stellar evolution and led to development of the mathematical models of stars and stellar evolution that we explore in Chapter 10. These models are continually being tested and refined as they are compared to ever-improving observations.

AN ASTRONOMER'S TOOLBOX III-2

The Distance-Magnitude Relationship

The closer a star, the brighter it appears. The inverse-square law leads to a simple equation for absolute magnitude M. Suppose a star's apparent magnitude is m, and its distance from the Earth is d (measured in parsecs). Then

$$M = m - 5 \log (d/10)$$

where log stands for the base-10 logarithm. The distance-magnitude relation can be rewritten as

$$m - M = 5 \log d - 5$$

Example: Consider Proxima Centauri, the nearest star in the night sky. By measuring its parallax angle, we know this star is at a distance from the Earth of $d = 1.3$ pc. Its apparent magnitude is $m = +11.1$. Therefore, its absolute magnitude is

$$M = 11.1 - 5 \log (1.3/10) = 11.1 - (-4.4) = 15.5$$

Compare! The Sun is an average star with M = 4.8, so Proxima Centauri is an intrinsically dim star. If you know any two of d, m, and M, you can calculate the third variable. For example, if we know a star's absolute and apparent magnitude, the equation can be used to determine its distance.

Try these questions: A star is observed to have an apparent magnitude m = +0.268 and an absolute magnitude M = −0.01. How far from Earth is the star in parsecs and light-years? A star is observed to have an apparent magnitude $m = +1.17$ and is at a distance of 25.1 ly from Earth. What is its absolute magnitude? (Remember to convert to parsecs first.) A star is at a distance from Earth of 8.61 ly and has an absolute magnitude of $m = +1.45$. What is its apparent magnitude? You can check your results and identify the stars by referring to Appendix Tables A-4 and A-5.

(Answers appear at the end of the book.)

You will see in Chapters 11, 12, and 13 that stars with different masses evolve differently. Some, including the Sun, cease fusion after their cores are converted first from hydrogen into helium and then into carbon and oxygen. Other stars eventually convert their cores into neutrons; still others end up with their cores collapsing to become black holes.

III-5　Frontiers yet to be discovered

The legacy of *Hipparcos* is a better knowledge of the size of the universe. We will discover in the coming chapters that the more accurately we know the distance between stars and, indeed, between galaxies, the better we are able to understand the evolution of the universe. As noted above, parallax is the most accurate means of measuring these distances. This technique requires the fewest assumptions about how stars work and the nature of the material between us and the stars. Therefore, a major goal of space-based observational astronomy is to get more and better parallax measurements. To improve them, satellites will be sent far from the Earth to increase the baseline used to make these measurements (see Figure III-1) and hence the angular accuracy that is possible.

 Further Reading on These Topics

WHAT DID YOU THINK?

1 *Do stars last forever?* No, all stars go through evolutionary processes and eventually stop emitting radiation.

2 *How near is the closest star other than the Sun?* Proxima Centauri is about 40 trillion kilometers (25 trillion miles) away. It takes light about 4 years to reach the Earth from there.

3 *How luminous is the Sun compared with other stars?* The most luminous stars are about a million times brighter and the least luminous stars are about a hundred thousand times dimmer than the Sun.

KEY WORDS

KEY IDEAS

• Stars differ in size, luminosity, temperature, color, and chemical composition.

• Determining stellar distances from Earth is the first step to understanding the nature of the stars. Distances to the nearer stars can be determined by stellar parallax, the apparent shift of a star's location against the background stars while the Earth moves along its orbit around the Sun.

• The apparent magnitude of a star, denoted *m*, is a measure of how bright the star appears to Earth-based observers. The absolute magnitude of a star, denoted M, is a measure of the star's true brightness and is directly related to the star's energy output, or luminosity.

• The absolute magnitude of a star is the apparent magnitude it would have if viewed from a distance of 10 pc. Absolute magnitudes can be calculated from the star's apparent magnitude and distance.

• The luminosity of a star is the amount of energy emitted by it each second.

REVIEW QUESTIONS

The answers to all computational problems, which are preceded by an asterisk (*), appear at the end of the book.

1 Describe how the parallax method of finding a star's distance is similar to the binocular (two-eye) vision of animals.

2 What is stellar parallax?

3 How do astronomers use stellar parallax to measure the distances to stars?

4 Why do stellar parallax measurements work only with relatively nearby stars?

5 What is the difference between apparent magnitude and absolute magnitude?

6 Briefly describe how you would determine the absolute magnitude of a nearby star.

7 What does a star's luminosity measure?

8 Why is the magnitude scale "backward" from what common sense dictates?

9 Does the star Betelgeuse, whose apparent magnitude is *m* = 0.5, look brighter or dimmer than the star Pollux, whose apparent magnitude is *m* = 1.1?

*10 Consider two identical stars, with one star 5 times farther away than the other. How much brighter will the closer star appear than the more distant one?

ADVANCED QUESTIONS

11 What is the inverse-square law? Use it to explain why a headlight on a car can appear brighter than a star, even though the headlight emits far less light energy per second.

*12 Van Maanen's star, named after the Dutch astronomer who discovered it, has a parallax angle of 0.232 arcsec. How far away is the star?

13 Explain how sailors on a ship traveling parallel to a coastline at a known speed can use parallax angle measurements to determine the distance to the shore.

*14 Suppose that a dim star were located 2 million AU from the Sun. Find (a) the distance to the star in parsecs, and (b) the parallax angle of the star.

*15 How many times brighter is a star of apparent magnitude *m* = −1 than a star of apparent magnitude *m* = +7?

DISCUSSION QUESTION

16 Discuss the advantages and disadvantages of measuring stellar parallax from a space telescope in a large solar orbit, say, at the distance of Jupiter from the Sun.

WHAT IF ...

17 All the stars in our Milky Way Galaxy actually were the same distance from Earth, say, 10 pc? Describe what the night sky might look like. How about the daytime sky? *Hint:* If you are doing this quantitatively, assume for simplicity that all the stars have the same absolute magnitudes as the Sun and that there are roughly 200 billion stars in the galaxy.

18 All stellar parallax angles, p, were observed to be increasing? What would that imply about the motions of

stars? Is there another observational technique that could be used to confirm this motion?

19 A star's parallax was observed to oscillate—regularly increase in angle and then decrease in angle—over a period of 100 years? What would that imply about the star?

WEB/CD-ROM QUESTION

*20 **Distances to Stars Using Parallax.** Access the Active Integrated Media Module "Using Parallax to Determine Distance" in Essentials for Discovering the Stars of the *Discovering the Universe* Web

site or CD-ROM. Use this to determine the distance in parsecs and in light-years to each of the following stars: (a) Betelgeuse (parallax $p = 0.00763''$); (b) Vega ($p = 0.129''$); (c) Antares ($p = 0.00540''$); and (d) Sirius ($p = 0.379''$).

OBSERVING PROJECTS

21 Use your *Starry Night Backyard*™ software to locate the bright stars listed in Appendix Table A-5. Set the time in the program to the time you will be observing tonight. Display the outlines of the constellations (select constellations/*Boundaries,* constellations/*Labels,* and constellations/*Astronomical* in the Constellations menu). Locate the stars by selecting *Edit* and then *Find* and then typing the name of the object. Ignore the stars that the computer says are not visible tonight. Make a list of stars that are up tonight, including their apparent

magnitudes and the constellations they are in, and arrange to observe them. You may find it helpful to print out star charts showing them using the *Starry Night Backyard*™ program. By comparing stars of similar apparent magnitude, such as Antares and Spica, determine the difference in apparent magnitude for which you can tell stars have different apparent brightnesses. If you bring out a star chart, identify and observe other stars, estimating their apparent magnitudes based on the ones on your list.

An Astronomer's Almanac

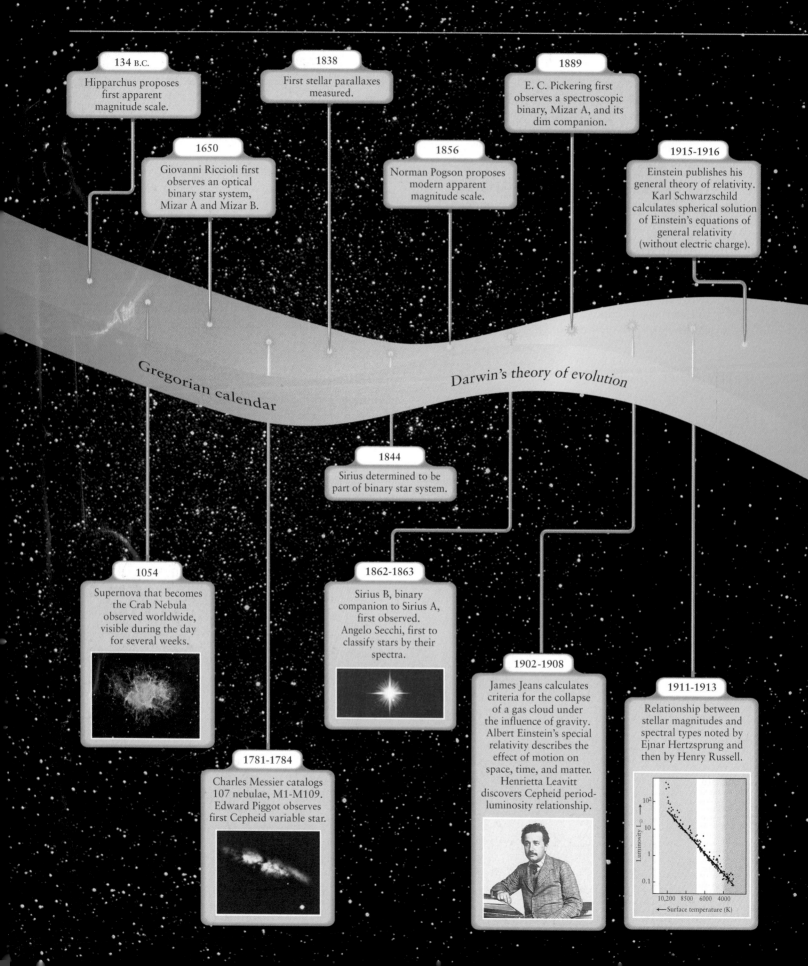

134 B.C.
Hipparchus proposes first apparent magnitude scale.

1650
Giovanni Riccioli first observes an optical binary star system, Mizar A and Mizar B.

1838
First stellar parallaxes measured.

1856
Norman Pogson proposes modern apparent magnitude scale.

1889
E. C. Pickering first observes a spectroscopic binary, Mizar A, and its dim companion.

1915-1916
Einstein publishes his general theory of relativity. Karl Schwarzschild calculates spherical solution of Einstein's equations of general relativity (without electric charge).

Gregorian calendar

Darwin's theory of evolution

1844
Sirius determined to be part of binary star system.

1054
Supernova that becomes the Crab Nebula observed worldwide, visible during the day for several weeks.

1862-1863
Sirius B, binary companion to Sirius A, first observed. Angelo Secchi, first to classify stars by their spectra.

1902-1908
James Jeans calculates criteria for the collapse of a gas cloud under the influence of gravity. Albert Einstein's special relativity describes the effect of motion on space, time, and matter. Henrietta Leavitt discovers Cepheid period-luminosity relationship.

1911-1913
Relationship between stellar magnitudes and spectral types noted by Ejnar Hertzsprung and then by Henry Russell.

1781-1784
Charles Messier catalogs 107 nebulae, M1-M109. Edward Piggot observes first Cepheid variable star.

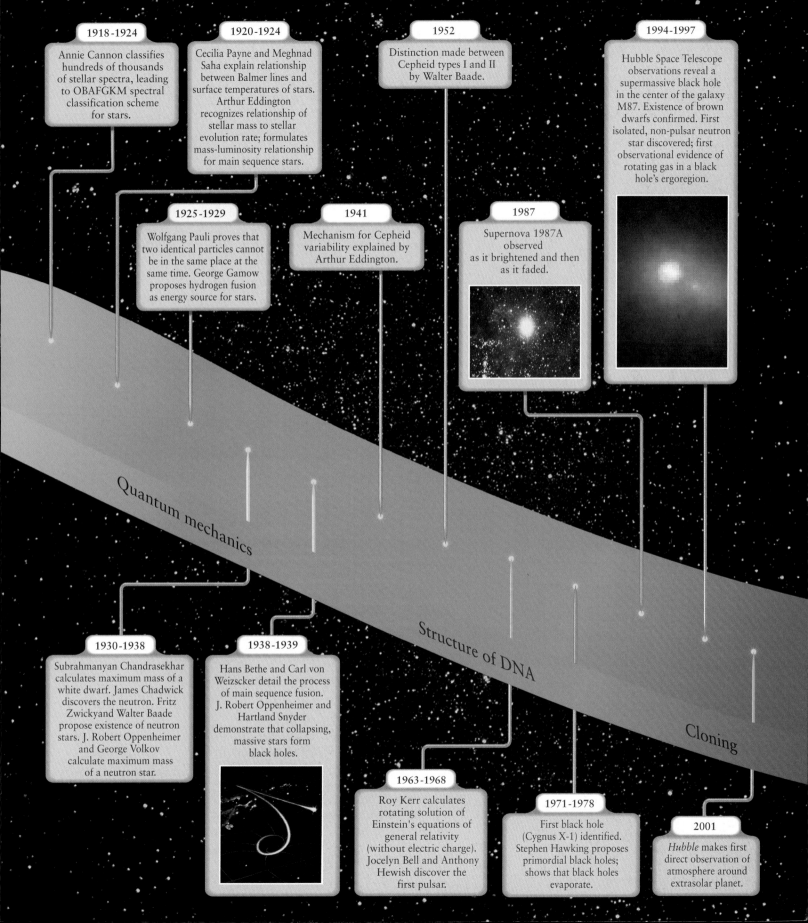

1918-1924
Annie Cannon classifies hundreds of thousands of stellar spectra, leading to OBAFGKM spectral classification scheme for stars.

1920-1924
Cecilia Payne and Meghnad Saha explain relationship between Balmer lines and surface temperatures of stars. Arthur Eddington recognizes relationship of stellar mass to stellar evolution rate; formulates mass-luminosity relationship for main sequence stars.

1952
Distinction made between Cepheid types I and II by Walter Baade.

1994-1997
Hubble Space Telescope observations reveal a supermassive black hole in the center of the galaxy M87. Existence of brown dwarfs confirmed. First isolated, non-pulsar neutron star discovered; first observational evidence of rotating gas in a black hole's ergoregion.

1925-1929
Wolfgang Pauli proves that two identical particles cannot be in the same place at the same time. George Gamow proposes hydrogen fusion as energy source for stars.

1941
Mechanism for Cepheid variability explained by Arthur Eddington.

1987
Supernova 1987A observed as it brightened and then as it faded.

Quantum mechanics

Structure of DNA

Cloning

1930-1938
Subrahmanyan Chandrasekhar calculates maximum mass of a white dwarf. James Chadwick discovers the neutron. Fritz Zwickyand Walter Baade propose existence of neutron stars. J. Robert Oppenheimer and George Volkov calculate maximum mass of a neutron star.

1938-1939
Hans Bethe and Carl von Weizscker detail the process of main sequence fusion. J. Robert Oppenheimer and Hartland Snyder demonstrate that collapsing, massive stars form black holes.

1963-1968
Roy Kerr calculates rotating solution of Einstein's equations of general relativity (without electric charge). Jocelyn Bell and Anthony Hewish discover the first pulsar.

1971-1978
First black hole (Cygnus X-1) identified. Stephen Hawking proposes primordial black holes; shows that black holes evaporate.

2001
Hubble makes first direct observation of atmosphere around extrasolar planet.

10 THE NATURE OF STARS

IN THIS CHAPTER YOU WILL DISCOVER

- some physical properties of stars

- how astronomers analyze starlight to determine a star's temperature and chemical composition

- how stellar luminosities and surface temperatures are related

- different classes of stars

- the variety and importance of binary star systems

- how astronomers calculate stellar masses

R I V U X G

Structure by Starlight

By analyzing the spectra of stars, astronomers determine details about them, such as their surface temperatures, chemical compositions, luminosities, and motion toward or away from us. The left image shows spectra of several stars in the Pleiades, a cluster of stars in the constellation Taurus. The Pleiades are shown, right. These colors reveal the stars' surface temperatures. Reddish stars are comparatively cool, with surface temperatures around 3000 K. Blue-white stars have much higher temperatures (15,000–30,000 K). (left: Jerry Lodriguss; right: © Dr. Jueng Alean, SPL, Photo Researchers, Inc.)

WHAT DO YOU THINK?

1 What colors are stars?

2 Are brighter stars hotter than dimmer stars?

3 What sizes are stars?

4 Are most stars isolated from other stars, as the Sun is?

Our understanding of the Sun has been aided immeasurably by the high-resolution images and electromagnetic spectra we can take of it, by the bits of solar matter that directly affect us via the solar wind, coronal mass ejections, and flares, and by the elusive solar neutrinos. This wealth of knowledge helps us build a sophisticated scientific model of why the Sun shines.

Now we are beginning to explore the rest of the stars, and our goal is to explain how they form, evolve, and finally cease their activity. There are roughly 200 billion stars in our Milky Way Galaxy alone. Many of them are hidden behind opaque interstellar clouds. Trying to collect data from all the stars and create models of each is clearly a hopeless and hopefully unnecessary task. Rather, we want to assemble observations of the relatively few stars we can see so that we can explain with just one or a few models how they all work. We begin organizing those data in this chapter. The insights we gain will then be applied in the next three chapters to testing stellar models.

THE TEMPERATURE OF STARS

Is there a way to group stars on a graph so that physically similar ones will be near each other? The patterns thus provided could lead to useful models. This search begins with a fact we might easily overlook: Stars are not all the same color.

10-1 A star's color reveals its surface temperature

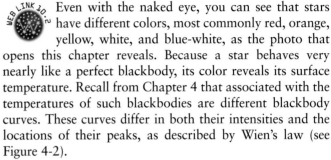

Even with the naked eye, you can see that stars have different colors, most commonly red, orange, yellow, white, and blue-white, as the photo that opens this chapter reveals. Because a star behaves very nearly like a perfect blackbody, its color reveals its surface temperature. Recall from Chapter 4 that associated with the temperatures of such blackbodies are different blackbody curves. These curves differ in both their intensities and the locations of their peaks, as described by Wien's law (see Figure 4-2).

Three typical blackbody curves are presented in Figure 10-1. The intensity of light from a relatively cool star peaks at long wavelengths, so the star looks red (Figure 10-1a). The intensity of light from a very hot star peaks at shorter wavelengths, making it look blue-white (Figure 10-1c). The maximum intensity of a star of intermediate temperature, such as the Sun, is found near the middle of the visible spectrum (Figure 10-1b).

To accurately determine the peaks of stars' blackbody spectra, and, hence, their surface temperatures, astronomers need to know which blackbody curve most accurately describes each star. **Photometry,** in which a telescope collects starlight that is then passed through one of a set of colored filters and recorded on a CCD (recall Figure 3-27), provides this information. At least three photometric images are taken of each star through filters passing different wavelengths of light. The intensity of a star's image is different at different wavelengths, and photometric data can thus be used to locate the peak of the star's blackbody radiation and, thereby, its surface temperature.

If a star's surface is very hot, for example, 10,000 K, its radiation is skewed toward the ultraviolet, which makes the star bright as detected through an ultraviolet-passing filter,

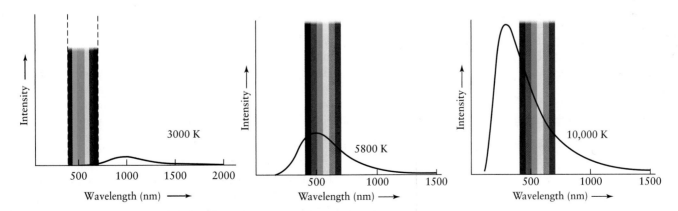

FIGURE 10-1   **Temperature and Color** This diagram shows the relationship between the color of a star and its surface temperature. The intensity of light emitted by three hypothetical stars is plotted against wavelength (compare with Figure 4-2). The range of visible wavelengths is indicated. Where the peak of a star's intensity curve lies relative to the visible light band determines the apparent color of its visible light.

GUIDED DISCOVERY   Star Names

Most of the stars never received fanciful names such as Betelgeuse or Aldebaran. Indeed, most are so dim that they have been observed only through telescopes in the last two centuries. To study the stars yourself, you must be able to keep track of them. Astronomers have created a system of labels for all of them.

Up to the 24 most prominent stars in each constellation are assigned Greek lowercase letters:

α	alpha	ι	iota	ρ	rho
β	beta	κ	kappa	σ	sigma
γ	gamma	λ	lambda	τ	tau
δ	delta	μ	mu	υ	upsilon
ε	epsilon	ν	nu	φ	phi
ζ	zeta	ξ	xi	χ	chi
η	eta	ο	omicron	ψ	psi
θ	theta	π	pi	ω	omega

A bright star's name is a Greek letter together with its constellation. We use the Latin possessive form of the constellations. For example:

Constellation	Possessive
Aries	Arietis
Taurus	Tauri
Gemini	Geminorum
Cancer	Cancri
Leo	Leonis
Virgo	Virginis

Constellation	Possessive
Libra	Librae
Scorpius	Scorpii
Ophiuchus	Ophiuchi
Sagittarius	Sagittarii
Capricornus	Capricorni
Aquarius	Aquarii
Pisces	Piscium

In most cases the brightest star in the constellation is α, the second brightest is β, the third is γ, and so on. For example, the brightest star in the constellation Leo is called α Leonis. This name is more informative than its common name, Regulus.

For the millions of stars extending beyond the 24 brightest, a variety of catalogs list the stars numerically. For example, HDE 226868 is a bright, blue star, the 226,868th star in the *Henry Draper Extended Catalogue of stars.*

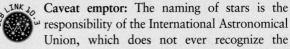 **Caveat emptor:** The naming of stars is the responsibility of the International Astronomical Union, which does not ever recognize the commercial sale of star names. Companies that offer to name a star for a price do not have any official standing or recognition in the astronomical community, nor do the names they promulgate. You can pay them to name a star for you, but the name is absolutely not official.

---

dimmer through a blue filter, and dimmer still through a yellow filter. Regulus, in Leo, is such a star. If a star is cool, say, 3000 K, its radiation peaks at long visible or even infrared wavelengths, making the star bright through a red filter, dimmer through a yellow filter, and dimmer still through a blue filter. Aldebaran and Betelgeuse are examples of such cool stars.

10-2 A star's spectrum also reveals its surface temperature

Another way to determine a star's surface temperature is from a study of its spectrum, a technique called **stellar spectroscopy** (see Chapter 4). Recall that spectral lines result from the absorption and scattering of starlight by gases in the star's atmosphere, in interstellar space, and in the Earth's atmosphere. Astronomers begin by taking the spectrum of a star and then identifying and discarding spectral lines due to interstellar gas and Earth's atmosphere. What is left are spectral lines created in the star's atmosphere.

Recall from Chapter 4 that the Sun's absorption lines were first observed by Joseph von Fraunhofer early in the nineteenth century. By the middle of that century, Italian astronomer Angelo Secchi had discovered spectral lines in the spectra of many stars. At first glance, stellar spectra seem to come in a bewildering variety, some of which are shown in Figure 10-2. Some stellar spectra show prominent absorption lines of hydrogen. Some exhibit many absorption lines of calcium and iron. Still others are dominated by broad absorption lines created by molecules such as titanium oxide.

Fortunately, this remarkable variety of spectra holds clues to each star's nature and, in particular, its surface temperature. Consider the abundant hydrogen gas found in nearly every star's atmosphere. Although this gas accounts for about three-quarters of the mass of a typical star, strong

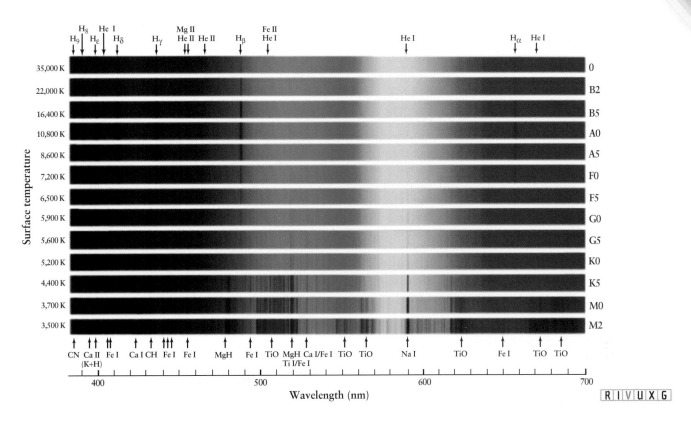

FIGURE 10-2   Principal Types of Stellar Spectra
This figure shows the spectra for stars with different surface
temperatures. The corresponding spectral types are indicated
on the right side of each spectrum. The hydrogen Balmer lines are
strongest in stars with surface temperatures of about 10,000 K (called
A-type stars). Cooler stars (G- and K-type stars) exhibit numerous atomic
lines caused by various elements, indicating temperatures from

4000 to 6000 K. The broad, dark bands in the spectrum of the coolest stars
(M-type stars) are caused by titanium oxide (TiO) molecules, which can
exist only if the temperature is below about 3500 K. Recall from section
4–5 that the roman numeral I after a chemical symbol means that the
absorption line is caused by a neutral atom; a numeral II means that the
absorption is caused by atoms that have each lost one electron. (R. Bell,
University of Maryland, and M. Briley, University of Wisconsin at Oshkosh)

(meaning dark) hydrogen absorption lines do not show up
in every star's spectrum. The strength of the absorption lines
depends on the star's temperature, and in the late 1920s,
Harvard astronomer Cecilia Payne and physicist Meghnad
Saha succeeded in explaining how a star's visible hydrogen
spectrum is affected by its surface temperature.

Niels Bohr's model of the hydrogen atom (recall Figure
4-13) explains why. As Bohr proposed in the early 1900s,
the visible hydrogen lines, called Balmer lines, are produced
when photons excite electrons in the second, $n = 2$, energy
level of hydrogen to a higher energy level. Because hydrogen
electrons are most strongly excited at a temperature of
10,000 K, a star with that surface temperature produces the
strongest Balmer lines.

Why do hotter or cooler stars produce weaker (less
dark) Balmer absorption lines? Suppose first that the star is
much hotter than 10,000 K. High-energy photons streaming
through the photosphere then completely strip away (ionize)

electrons from most of the hydrogen atoms there. Because
an ionized hydrogen atom cannot produce spectral lines, a
very hot star has very dim hydrogen Balmer lines, even
though it contains great quantities of hydrogen. Conversely,
if a star is much cooler than 10,000 K, most of the photons
escaping from it possess too little energy to boost many electrons from the $n = 2$ (first excited state) of the hydrogen
atoms, to higher energy states. Therefore, these stars also
produce dim Balmer lines. To produce strong Balmer lines,
a star must be hot enough to excite electrons out of the $n =$
2 state but not hot enough to ionize a significant fraction of
the atoms.

As you can see in Figure 10-2, at temperatures much
cooler or hotter than 10,000 K, the spectral lines of other elements dominate a star's spectrum. For example, the spectral
lines of neutral helium are pronounced at around 25,000 K,
because photons have enough energy to excite helium atoms
without tearing away the electrons. Conversely, the spectral

lines of neutral iron are especially strong at around 3500 K. By surveying the relative strengths of a variety of absorption lines, astronomers can determine a star's surface temperature to high precision.

10-3 Stars are classified by their spectra

 We have seen that astronomers can determine a star's surface temperature from either the peak of its blackbody or the strength of its various spectral lines. To make use of the diverse stellar spectra, astronomers since Secchi have grouped similar spectra into classes, or **spectral types**. According to one classification scheme popular in the late 1800s, a star was assigned a letter from A through P, depending on the strength of the Balmer hydrogen lines in the star's spectrum. It was assumed that the strength of these lines was uniquely related to the star's surface temperature, but as we have seen in the preceding section, this belief is unjustified.

In the early 1900s, Annie Cannon and her colleagues at Harvard Observatory set up the spectral classification scheme we use today. Many of the early A through P categories were dropped because the Balmer lines in a star's spectrum can be weak whether the star is very cool or very hot. The remaining Balmer-based spectral types were thus re-ordered by stellar surface temperature into the **OBAFGKM sequence**. This sequence is most easily learned with the aide of a mnemonic, such as: "Oh, Be A Fine Guy, Kiss Me!" or "Oh, Be A Fine Girl, Kiss Me!"

The hottest stars are the O type, with surface temperatures of more than 30,000 K; their spectra are dominated by He II and Si IV (triply ionized silicon). M stars are the coolest type, with surface temperatures around 3000 K. Table 10-1 includes representative examples of each spectral type, which you can compare to the spectra in Figure 10-2.

Exact Spectral Types

Astronomers have found it useful to subdivide the OBAFGKM temperature sequence further. Each spectral type is broken up into ten temperature subranges. These ten finer steps are indicated by adding an integer from 0 (hottest) through 9 (coolest). Thus, an A8 star is hotter than an A9 star, which is hotter than an F0 star, which is hotter than an F1 star, and so on. Test yourself on this: What class of star is just slightly cooler than a K9? (The caption for Figure 10-3 has the answer.) The Sun, whose spectrum is dominated by singly ionized metals (especially Fe II and Ca II), is a G2 star.

> **Insight into Science** **Tolerate Idiosyncrasies** In astronomy, the word "metal" applies to all elements other than hydrogen or helium. In chemistry, sodium and iron are metals, while carbon and oxygen are not.

TABLE 10-1 The Spectral Sequence

Spectral class	Color	Temperature (K)	Spectral lines	Examples
O	Blue-violet	30,000–50,000	Ionized atoms, especially helium	Naos (ζ Puppis), Mintaka (δ Orionis)
B	Blue-white	11,000–30,000	Neutral helium, some hydrogen	Spica (α Virginis), Rigel (β Orionis)
A	White	7500–11,000	Strong hydrogen, some ionized metals	Sirius (α Canis Majoris), Vega (α Lyrae)
F	Yellow-white	5900–7500	Hydrogen and ionized metals such as calcium and iron	Canopus (α Carinae), Procyon (α Canis Minoris)
G	Yellow	5200–5900	Both neutral and ionized metals, especially ionized calcium	Sun, Capella (α Aurigae)
K	Orange	3900-5200	Neutral metals	Arcturus (α Boötis), Aldebaran (α Tauri)
M	Red-orange	2500-3900	Strong titanium oxide and some neutral calcium	Antares (α Scorpii), Betelgeuse (α Orionis)

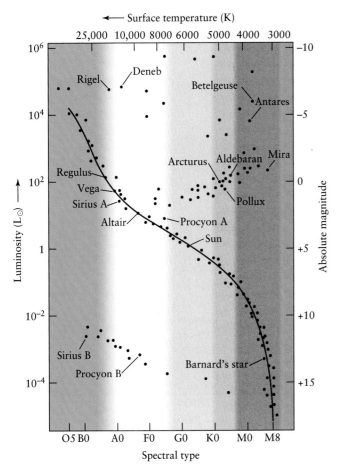

Surface temperature (K)

FIGURE 10-3   A Hertzsprung-Russell Diagram

On an H-R diagram, the luminosities of stars are plotted against their spectral types. Each dot on this graph represents a star whose luminosity and spectral type have been determined. Some well-known stars are identified. The data points are grouped in just a few regions of the diagram, revealing that luminosity and spectral type are correlated: Main-sequence stars fall along the red curve, giants are to the right, supergiants are on the top, and white dwarfs are below the main sequence. The absolute magnitudes and surface temperatures are listed at the right and top of the graph, respectively. These are sometimes used on H-R diagrams instead of luminosities and spectral types. *Answer to text question:* An M0 star is the next coolest after a K9.

TYPES OF STARS

Astronomers began observing stellar spectra in the early 1800s and classifying spectral types by mid-century. The first accurate measurements of stellar parallaxes were also being made at about the same time. Since then, observing techniques have vastly improved, and the spectral types and absolute magnitudes of millions of stars have been catalogued. Recall that the relationship between stellar parallaxes and absolute magnitudes is discussed in Essentials III.

10-4 The Hertzsprung-Russell diagram identifies distinct groups of stars

Around 1911, the Danish astronomer Ejnar Hertzsprung noticed that patterns emerge when the luminosities of stars (or their equivalent absolute magnitudes) are plotted against their surface temperatures or spectral types. Within two years, the American astronomer Henry Norris Russell independently discovered the same result. Graphs of stellar brightness (luminosity or absolute magnitude) against surface temperature (or, equivalently, spectral type) are now known as **Hertzsprung-Russell diagrams**, or **H-R diagrams**.

The H-R diagram is valuable because it shows that stars do not have random surface temperatures and brightnesses; the two factors are correlated. Figure 10-3 is a typical Hertzsprung-Russell diagram. Each dot represents a star whose luminosity and spectral type have been determined. To help you see the equivalence, the surface temperatures are plotted along the top of the figure and the absolute magnitudes along the right side.

Bright stars are near the top of the diagram; dim stars are near the bottom. Contrary to intuition, hot (O and B) stars are toward the left side and cool (M) stars are toward the right. Hertzsprung and Russell made this choice because of the standard sequence OBAFGKM.

Insight into Science From Patterns to Models Scientists look at patterns of behavior in related objects as valuable clues to underlying properties and their causes. Guided by this data, they then create theoretical models and make fresh predictions. For example, astronomers analyze the relationship between the luminosities and surface temperatures of stars as displayed on an H-R diagram to gain insight into the internal activities of stars.

The band of stars in Figure 10-3 stretching diagonally across the H-R diagram and on which a red curve is superimposed represents most of the stars we see in the nighttime sky. This band, called the **main sequence**, extends from the hot, bright, bluish stars in the upper left corner of the diagram down to the cool, dim, reddish stars in the lower right corner. Each star on this band is called a **main-sequence star**. Just over 91% of the stars surrounding the solar system fall into this category.

Observations reveal that the number of main-sequence stars decreases with increasing surface temperature. Therefore, along the main sequence, the cooler M, K, and G stars are the most common ones and the hot O stars are the rarest. The Sun (spectral type G2, absolute magnitude +4.8) is a main-sequence star.

To the right of the main sequence on the H-R diagram is a second major grouping of stars. These stars are bright but cool. From the Stefan-Boltzmann law (see An Astronomer's Toolbox 4-1), we know that a cool object radiates much less light from each unit of surface area than does a hot object. To be so bright, these cool stars must therefore be huge compared to main-sequence stars of the same temperature, so they are called **giant stars.** By contrast, main-sequence stars are often called **dwarf stars.** The dimmest, coolest main sequence stars are called *red dwarfs.*

Giants are typically 10 to 100 times the radius of the Sun and have surface temperatures between 3000 and 20,000 K. The cooler members of this class of stars (those with surface temperatures between 2000 and 4500 K) are often called **red giants** because they appear reddish in the nighttime sky. Aldebaran in the constellation Taurus and Arcturus in Boötes are examples of red giants that you can easily see with the naked eye.

 Stellar Sizes

A few rare stars are considerably bigger and brighter than typical giants. Located along the top of the H-R diagram, these superluminous stars are appropriately called **supergiants.** They can extend in radius up to about 1000 R$_\odot$. Betelgeuse in Orion and Antares in Scorpius are two examples that are visible in the nighttime sky. Together, giants and supergiants comprise less than 1% of the stars in our vicinity.

The remaining 8% of stars in our neighborhood of space fall in a final grouping toward the lower left and bottom of the Hertzsprung-Russell diagram. As their placement on the diagram shows, these stars are hot, dim, and tiny compared to the Sun. Called **white dwarfs,** we will see that they are actually remnants of stars. White dwarfs are roughly the same size as the Earth, and because of their great distances, they can be seen only with the aid of a telescope.

These results are summarized in Figure 10-4, where the dashed lines indicate the radii of stars. Notice that most main-sequence stars are roughly the same size as the Sun. It is important to keep in mind that while the H-R diagram is invaluable in helping astronomers *organize* stars by their physical properties (temperature and absolute magnitude), the diagrams do not *explain* the physical mechanisms that produce these characteristics. As we will see shortly, the locations of stars on the H-R diagram provided clues that astrophysicists used in developing and testing their theories of how stars shine.

10-5 Luminosity classes

We have seen that a star's surface temperature largely determines which lines are prominent in its spectrum. Therefore, classifying stars by spectral type is the same as categorizing

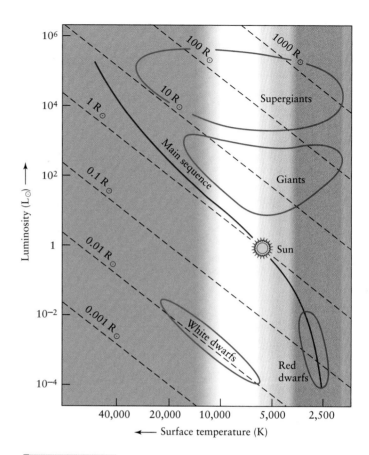

■ FIGURE 10-4   Determining the Sizes of Stars from an H-R Diagram   On this H-R diagram, done to the same scale as Figure 10-3, stellar luminosities are graphed against the surface temperatures of stars. The dashed diagonal lines indicate stellar radii. For a given stellar radius, as the surface temperature increases (corresponding to moving from right to left on the H-R diagram), the star glows more intensely and the luminosity increases (corresponding to moving upward on the diagram). While individual stars are not plotted, we show the regions of the diagram in which main-sequence, giant, supergiant, and white dwarf stars are found. Note that the Sun's size is intermediate in luminosity, surface temperature, and radius; it is very much a middle-of-the-road star.

them according to surface temperature. However, as we saw in Figure 10-3, stars of the same surface temperature can have different luminosities. As an example, a star with a surface temperature of 5800 K could be either a white dwarf, a main-sequence star, a giant, or a supergiant.

By studying the absorption lines in detail, astronomers can determine to which category a star belongs. This is possible primarily because absorption lines are also affected by the density and pressure of the gas in a star's atmosphere, both of which depend on whether the star is a white dwarf, main-sequence star, giant, or supergiant.

Based upon the differences in stellar spectra, W. W. Morgan and P. C. Keenan of the Yerkes Observatory developed a system of **luminosity classes** in the 1930s. Luminosity classes Ia and Ib include all the supergiants, giants of various luminosity are assigned classes II, III, and IV, and main-sequence stars are luminosity class V. Figure 10-5 summarizes these results. We will see in Chapter 11 that the different luminosity classes correspond to different stages of stellar evolution. White dwarfs are not given a luminosity class because, as we will see in Chapter 12, they are not creating energy by fusion like the Sun and other stars in the above five luminosity classes.

Plotting luminosity classes on the H-R diagram (see Figure 10-5) provides a useful subdivision of star types. In fact, astronomers commonly describe a star by both its spectral type and its luminosity class. The Sun, for example, is called a G2 V star. This notation supplies a great deal of information about the star, because its spectral type is correlated with its surface temperature. Thus, an astronomer knows immediately that a G2 V star is a main-sequence star with a luminosity of 1 $L_\odot$ and a surface temperature of around 5800 K. Similarly, knowing that Aldebaran is a K5 III star tells an astronomer that it is a red giant with a luminosity of around 370 $L_\odot$ and a surface temperature of about 4000 K (see Figure 10-3).

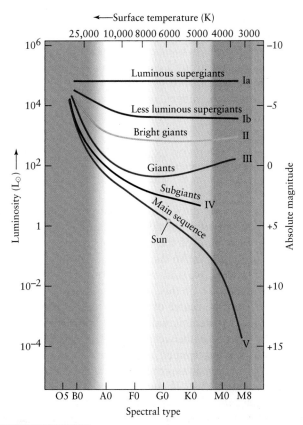

FIGURE 10-5 **Luminosity Classes** It is convenient to divide the H-R diagram into regions called luminosity classes. These subdivisions permit finer distinctions between giants and supergiants. Luminosity classes Ia and Ib encompass the supergiants. Luminosity classes II, III, and IV indicate giants of different brightness. Luminosity class V is the main-sequence stars. White dwarfs do not have their own luminosity class.

10-6 A star's spectral type and luminosity class provide a second distance-measuring technique

A star's spectral type and luminosity class give astronomers information necessary for determining the distances to stars millions of light years away, far beyond the maximum distance that can be measured using stellar parallax. The process works like this: First astronomers measure a distant star's apparent magnitude and spectral type. Then, from its spectrum they determine what luminosity class the star belongs to. This also tells them the star's absolute magnitude, since stars of a given spectral type and classification (such as main-sequence star or giant) each have a fairly well-defined absolute magnitude (see Figure 10-5). Using the distance-magnitude relationship (see An Astronomer's Toolbox III-2), the star's distance can then be calculated.

Consider, for example, the star Regulus in the constellation Leo. Its spectrum reveals Regulus to be a B7 V star (a hot, blue, main-sequence star). Placing it on the H-R diagram (see Figure 10-3), we can read off its luminosity as 140 $L_\odot$ and an absolute magnitude of –0.52. Given the star's apparent magnitude, we can use the distance-magnitude relationship to determine its distance from Earth (see An Astronomer's Toolbox III-2). Because both the spectral type and luminosity class are obtained spectroscopically, this method of determining distances is called **spectroscopic parallax.** The name is misleading, because no parallax angle is involved.

Spectroscopic parallax is limited in accuracy because of the spread of stars in each luminosity class—the stars in each class do not fall on a single, narrow line (hence the use of the words "fairly well-defined absolute magnitude" two paragraphs above). It is also limited in that spectra of distant stars become increasingly hard to determine. As a result, errors of 10% in distance are common using spectroscopic parallax. Accepting the larger errors, the power of this method is that it can be used for stars at much greater distances than those determined by stellar parallax. Indeed, it even provides distances to stars in other galaxies tens of millions of light-years away.

BINARY STARS AND STELLAR MASS

There is one more physical property of stars, their masses, that might provide insight into why stars only occur in limited places on the H-R diagram. The problem is that there is no way of determining stellar mass directly by examining isolated stars. The mass of a star must be determined by its gravitational effects on other bodies using Newton's law of gravity. When Newton derived the formula for Kepler's third law, he discovered that for any two objects in orbit, such as a planet around the Sun, a moon around a planet, or two stars around each other, the period of their orbits is related to the sum of their masses. For objects orbiting the Sun, we can

Star Wars -The Phantom Menace (20th Century Fox Lucas Film Ltd. May 1999 and *A New Hope* - BMG Classics/RCA 1997) *(The Kobal Collection)*

In two of the *Star Wars* movies series *Episode I–The Phantom Menace*, released in 1999, and *Episode IV–A New Hope*, released in 1977, the desert world of the planet, Tatooine, plays an important role. Tatooine is the home of Luke Skywalker and for many movie viewers where the adventure begins (although set some 30 years later than *Episode I*). Tatooine is also the primary setting for *Episode I–The Phantom Menace*.

In an early scene in *A New Hope*, the viewer sees a young Luke Skywalker sitting on a set under the planet's two stars (or Suns, if you like, although the word "Sun" describes just the star in our solar system). This establishes that the planet Tatooine is in a binary star system which is scientifically plausible, provided that the planet's orbit is appropriate. However, examine the scene from *The Phantom Menace* and explain what is scientifically wrong with the image. Assume that both stars are up in the sky. Also examine the shadow carefully and describe it. (Answers appear at the end of the book.)

ignore the mass of the smaller body, thereby getting Kepler's original third law in Chapter 2. For stars orbiting each other we use Newton's version, which includes the masses of both bodies.

Fortunately for astronomers, two-thirds of the stars near our solar system are members of star systems in which two stars orbit each other. This means that half the objects we see as single stars with our unaided eyes are actually pairs of stars in orbit around each other that are so distant they appear to us as one. Using telescopes to observe the periods of the orbits and the distances between stars, astronomers can determine the sum of the stellar masses of the pair.

10-7 Binary stars provide information about stellar masses

A pair of stars located at nearly the same position in the night sky is called a double star. Between 1782 and 1838, William Herschel and his son John catalogued thousands of them. Some *double stars* are not really near each other in space and do not orbit each other. These **optical doubles** (sometimes called *apparent binaries*) just happen to lie in the same direction as seen from Earth. For example, δ Herculis, visible with the naked eye, is an optical double with a dim star. Through a telescope these stars appear close together, but that is an optical illusion.

Other double stars are true **binary stars**—pairs in which two stars orbit each other. In the case of **visual binaries**, both stars can be seen, using a telescope if necessary (Figure 10-6). Astronomers can plot the orbit of one star about the other in a visual binary.

To see how astronomers determine the masses of stars, consider a visual binary. We rewrite Kepler's third law as a relation between the masses of the stars (in solar masses), their orbital period around each other in Earth years (this period is the same for both so it does not matter which star we assume orbits the other), and the average separation between the stars in AU (which equals the semimajor axis of the elliptical orbit):

$$\text{The sum of the masses} = \frac{\text{the cube of the semimajor axis}}{\text{the square of the orbital period}}$$

So, by observing the separation between a pair of stars and how long one of them takes to complete its orbit, we can calculate the sum of the masses of the two stars. The details are presented in An Astronomer's Toolbox 10-1.

In many cases, we can also determine the individual masses of the stars. To do this, we need to know the distances of the stars from the **center of mass** of the pair. Each of the stars in a binary system actually moves in an elliptical orbit around the same point between them. This point, called the center of mass, is determined by the masses of the stars and can be understood by analogy to a whirling wrench sliding on a table (see Figure 5-26). There is a point on the wrench that moves in a straight line. This is its center of mass. Just as the ends of the wrench orbit around its center of mass, the two stars in a binary system orbit around their center of mass under the influence of their mutual gravitational attraction (Figure 10-7).

The center of mass of a visual binary is located by plotting the separate orbits of the two stars, as in Figure 10-6, using the background stars as reference points. The center of

curve (Figure 10-11a), or total, creating a flat-bottomed trough (Figure 10-11b).

The light curve of an eclipsing binary can yield other useful information as well. For example, if an eclipsing binary is also a double-line spectroscopic binary, astronomers can calculate the masses, diameter, brightnesses, speeds, and stellar separation of each star from the light curves and the radial-velocity curves. These stars are rare, however, because the orbital planes of most spectroscopic binaries are tilted so that eclipses do not occur as seen from Earth.

Light curves can also reveal information about stellar atmospheres. Suppose that one star of a binary is a tiny white dwarf (stellar remnant the size of the Earth) and the other is a giant (late stage of stellar evolution in which the star has greatly expanded from its previous size). By observing exactly how the light from the white dwarf is gradually cut off as it begins to move behind the edge of the giant during an eclipse, astronomers can infer the pressure and density in the upper atmosphere of the giant. Such information is invaluable in testing models of stellar structure.

Many binary stars are separated by several AU or more. Other than orbiting each other, these stars behave as though they are isolated. That is, the models we develop in the coming chapters for the evolution of isolated stars apply to them as well. However, there are also **close binary** systems, with only a few stellar diameters separating the stars. Such stars are so close together that the gravity of each one dramatically affects the appearance and evolution of the other. If one member of a close binary is a giant, some of the gas of its outer layers is pulled onto its more compact companion—mass is transferred from one star to the other. We will explore more about such systems in Chapter 11.

This chapter has provided us with observational data that serve, in part, as the basis for the models of stellar activity and evolution presented in the next three chapters. The models make predictions, of course, that lead to new observations and refined models. We will take up some further observations thus driven as we proceed through the models.

10-11 Frontiers yet to be discovered

Classifying stars by luminosity and spectral type is an ongoing process providing an ever-growing body of data about the properties of stars. This information provides continual surprises. For example, there is a growing number of individual "stars" whose properties astronomers thought they understood but which turned out to be binary systems, each star having different properties from that of the single "star" they were originally believed to be.

Another significant piece of information astronomers need to better understand star formation is how many stars of each mass are formed. Observations strongly indicate that very high mass stars form quite infrequently, but is the same true of very low mass stars? Or do the numbers of stars formed increase with decreasing mass? Or perhaps there is a peak of star formation at some mass with decreasing numbers of stars with higher and lower mass. The number of stars formed with different masses is called the **initial mass function**. While there are several theoretical initial mass functions, in the end, observations will determine which is correct.

 Further Reading on These Topics

WHAT DID YOU KNOW?

1 *What colors are stars?* Stars are found in a wide range of colors, from red through violet as well as white.

2 *Are brighter stars hotter than dimmer stars?* Not necessarily. Many brighter stars, such as red giants, are cooler but larger, than hotter, dimmer stars, such as white dwarfs.

3 *What sizes are stars?* Stars range from more than 1000 times the Sun's diameter to less than 1/100 the Sun's diameter.

4 *Are most stars isolated from other stars, as the Sun is?* No. In the vicinity of the Sun, two-thirds of the stars are found in pairs or larger groups.

KEY WORDS

binary star, 290
center of mass, 290
close binary, 295
dwarf star, 288
eclipsing binary, 294
giant star, 288

Hertzsprung-Russell (H-R) diagram, 287
initial mass function, 295
light curve, 294
luminosity class, 289
main sequence, 287

main-sequence star, 287
mass-luminosity relation, 292
OBAFGKM sequence, 286
optical double, 290
photometry, 283
radial-velocity curve, 294

KEY IDEAS

Stars vary in their physical properties—color, temperature, mass, and size—a fact that helps astronomers understand stellar structure and evolution.

The Temperatures of Stars

• Stellar temperatures can be determined from stars' colors or stellar spectra.

• Stars are classified into spectral types (O, B, A, F, G, K, and M) based on their spectra. The spectral type of a star is directly related to its surface temperature.

Types of Stars

• The Hertzsprung-Russell (H-R) diagram is a graph on which luminosities of stars are plotted against their spectral types (or, equivalently, their absolute magnitudes are plotted against surface temperatures). The H-R diagram reveals the existence of four major groupings of stars: main-sequence stars, giants, supergiants, and white dwarfs.

• The mass-luminosity relation expresses a direct correlation between a main-sequence star's mass and the total energy it emits.

• Distances to stars can be determined using their spectral type and luminosity class.

Binary Stars and Stellar Masses

• Binary stars are surprisingly common. Those that can be resolved into two distinct star images by an Earth-based telescope are called visual binaries.

• The masses of the two stars in a binary system can be computed from measurements of the orbital period and orbital dimensions of the system.

• Some binaries can be detected and analyzed, even though the system may be so distant (or the two stars so close together) that the two star images cannot be resolved with Earth-based telescopes.

• A spectroscopic binary is a system detected from the periodic shift of its spectral lines. This shift is caused by the Doppler effect as the orbits of the stars carry them alternately toward and away from the Earth.

• An eclipsing binary is a system whose orbits are viewed nearly edge-on from the Earth, so that one star periodically eclipses the other. Detailed information about the stars in an eclipsing binary can be obtained by studying its light curve.

• Mass transfer occurs between binary stars that are close together.

REVIEW QUESTIONS

1   How and why is the spectrum of a star related to its surface temperature?

2   What is the primary chemical component of most stars?

3   A star of which spectral type has the strongest H absorption lines? At approximately what wavelength is this line normally found? *Hint:* The answer is represented graphically in the text.

4   Why does a G2 star have many more absorption lines than a B0 star?

5   Draw an H-R diagram and sketch the regions occupied by main-sequence stars, giants, supergiants, and white dwarfs. Briefly discuss the different ways you could have labeled the axes of your graph.

 6   To test your understanding of the H-R diagram, do Interactive Exercise 10-1. You can print out your answers if required.

7   How can observations of a visual binary lead to information about the masses of its stars?

8   What is a radial-velocity curve? What kinds of stellar systems exhibit such curves?

9   What is the difference between a single-line and a double-line spectroscopic binary?

10   What is meant by the light curve of an eclipsing binary? What sorts of information can be determined from such a light curve?

11   What is the mass-luminosity relation? To what kind of stars does it apply?

12   Refer to Figure 10-3: (a) What are the hottest and coolest named stars on the diagram? (b) What are the brightest and dimmest named stars on the diagram? (c) What are the hottest and coolest named main-sequence stars on the diagram? (d) What named stars are white dwarfs? giants? supergiants?

ADVANCED QUESTIONS

The answers to all computational problems, which are preceded by an asterisk (*), appear at the end of the book.

13 Sketch the radial-velocity curve of a binary whose stars are moving in nearly circular orbits that are (a) perpendicular and (b) parallel to our line of sight.

14 Sketch the light curve of an eclipsing binary whose stars are moving along highly elongated orbits with the major axes of the orbits (a) pointed toward the Earth and (b) perpendicular to our line of sight.

15 (a) What is the approximate mass of a main-sequence star that is 10,000 times as luminous as the Sun. (b) What is the approximate luminosity of a main-sequence star whose mass is one-tenth that of the Sun?

16 What is the approximate surface temperature of a main-sequence star with a luminosity 100 times as bright as the Sun.

DISCUSSION QUESTION

17 How does a star's rotation affect the appearance of its spectral lines? *Hint:* Assume we aren't looking down the star's rotation axis. Then, at every instant, half of the spinning star is approaching the Earth, while the other half is receding. Consider the resulting Doppler shift in the spectral lines.

WHAT IF ...

18 The Sun was part of a binary star system of two main-sequence stars and the Earth orbited around one of the stars, while the other star just changed the Earth's distance of orbit slightly? You might want to discuss such things as climate, tides, impacts from meteoroids and asteroids, and inhabitability.

19 The Sun was an M-type star, rather than a G-type star? Assuming that the Earth orbiting the M-type star had the same composition and orbit distance as it does today, what would be different here?

20 The Sun was a B-type star, rather than a G-type star? Assuming that the Earth orbiting the B-type star had the same composition and orbit distance as it does today, what would be different here? We will pick up this question again in Chapter 12 for further insights.

WEB/CD-ROM QUESTIONS

21 Search the Web for the periods of 10 binary star systems. Plot these on a graph of time (on the horizontal axis) versus number of systems. If possible, combine this information with similar data from your classmates. Do you see any patterns in the periods of these star systems?

22 To explore the range of periods of binary star systems locate on the Web binary star systems with periods of (a) less than 1 week, (b) between 1 day and 1 week, (c) between 1 and 2 years, (d) between 40 and 50 years, and (e) more than 400 years.

OBSERVING PROJECTS

23 Locate the stars Betelgeuse and Rigel in Orion. Observe them both by eye and through a small telescope. Are they the same color? To determine their color(s) it helps to compare them to their neighbors.

24 Locate one or more of the following double star systems: Regulus in Leo, Algieba in Leo, Irak (ε Boo) in Boötes, Ras Algethi in Hercules, Albireo in Cygnus, Vega in Lyra, Polaris in Ursa Minor, Rigel in Orion, Antares in Scorpius, Sirius in Canis Major, and Schedar in Cassiopeia. You can find their locations and print star charts using your *Starry Night Backyard*™ software. View them with and without a telescope. What are their colors?

11 THE LIVES OF STARS FROM BIRTH THROUGH MIDDLE AGE

IN THIS CHAPTER YOU WILL DISCOVER

- how stars form

- how astronomers use the physical properties of stars to learn about stellar life cycles

- the remarkable transformations of older stars into giants and supergiants

- that some dying stars eject material that creates new generations of stars, while others act as beacons that enable astronomers to pinpoint distant galaxies

- that the H-R diagram is your guide to the stellar life cycle

R I V U X G

Reflection and Emission Nebulae
Newly formed stars are found in and around interstellar clouds called nebulae. The two main bluish objects, NGC 6589 (top) and NGC 6590 (below), are reflection nebulae surrounding young, hot, main-sequence stars. Interstellar dust around these stars efficiently reflects their bluish light. Several smaller reflection nebulae are scattered around the large reddish patch of ionized hydrogen gas called IC 1283-4. Dust mixed with the gas dilutes the intense red emission of the hydrogen atoms with a soft blue haze. (Anglo-Australian Observatory)

WHAT DO YOU THINK?

1 How do stars form?

2 Are stars forming today?

3 Do stars with greater or lesser mass shine longer?

Stars emit huge amounts of radiation. As we saw for the Sun in Chapter 9, the price for radiating energy is change: Stars lose mass over time and their chemical makeup evolves. Major stages in the life of a star can last for millions or even billions of years.

The stars seem unchanging to us only because of the colossal time scales over which these cycles of change occur. You saw in Essentials III that by observing stars with different temperatures and brightnesses, astronomers have discovered groupings of stars with similar properties. We now begin to explore the scientific theories of **stellar evolution** that are the basis for our understanding of how stars form and how they "mature," "grow old," and "die." We will also expand our observational database to provide evidence for the validity of our theories.

> **Insight into Science** **Beware Poetic License** Anthropomorphism—assigning human attributes to nonhuman creatures or even to nonliving objects like stars—is descriptive but not scientific. In fact, scientists avoid this practice where possible, because it creates unjustified expectations. All too often, the names and descriptions we use for nonliving objects may appear to give them human qualities. For example, we use "birth" line to understand stellar "evolution"; we probe the "lives" and "deaths" of stars. This is poetic license of a sort: It piques our interest but muddles our science. As you study stellar evolution, always be mindful of that.

PROTOSTARS AND PRE–MAIN-SEQUENCE STARS

We saw in Essentials II that the solar system is believed to have formed from a collapsing, rotating cloud of gas and dust some 4.6 billion years ago. This theory requires that interstellar matter once existed. Taking this reasoning one step further, if such matter exists today, perhaps star formation continues.

11-1 Stars condense from clouds of gas and dust lying between existing stars

Radio telescopes have indeed revealed the existence of matter between the stars. We call it the **interstellar medium**, and it contains at least 10% of all the known mass in our Galaxy. From painstaking observations of the spectra of the interstellar medium, astronomers have determined that it is composed of atoms, molecules, and tiny pieces of dust.

Ninety percent of the particles comprising the interstellar medium are hydrogen atoms and molecules, 9% are helium atoms, and the remaining 1% are all the other naturally forming elements, often in the form of molecules or dust. Different atoms have different masses, and astronomers often need to know the percentages of the interstellar medium's mass in different elements. In that case, they present the same information in terms of masses: About 74% of the interstellar medium's mass is hydrogen, 25% is helium, and all the other naturally forming elements make up the remaining 1%.

Many types of molecules are found in interstellar space, including molecular hydrogen (H_2), carbon monoxide (CO), water (H_2O), ammonia (NH_3), and formaldehyde (H_2CO). These molecules can be identified by their unique spectral emissions, just as we can identify elements from atomic spectra.

The dust found in space has a variety of structures. Some are carbon-based particles, typically 0.005 micrometers across. Others, of about the same size, are similar to molecules found in engine exhaust and burnt hamburgers called polycyclic aromatic hydrocarbons (PAHs). PAHs are composed just of carbon and hydrogen atoms. Much larger space dust apparently has cores of carbon or silicon compounds surrounded with mantles of ice and other materials. These grow to more than .30 micrometers in diameter.

Working from the knowledge that new stars form from the gas and dust of the interstellar medium, astronomers map this matter to identify places to look for newly forming stars. Like clouds in our atmosphere scattering sunlight, gas and dust in space scatter visible light emitted by nearby stars—starlight illuminates nearby interstellar debris. But also like atmospheric clouds obscuring the sky above them, the light from nearby interstellar gas and dust obscures light from more distant regions, preventing us from seeing distant interstellar matter and distant stars. This problem of seeing distant regions of the interstellar medium in the visible part of the spectrum is compounded because most interstellar gas and dust is so cold that it emits very little visible light of its own.

Interstellar dust and its dramatic effects are clearly visible in the Pleiades star cluster (see the figure opening Chapter 10). Figure 11-1, the HR diagram of this cluster, reveals that it contains many more stars than just the

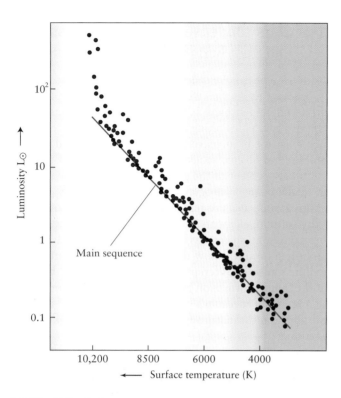

FIGURE 11-1 **Dating the Pleiades** The open cluster called the Pleiades (see Chapter 10 opener) can easily be seen with the naked eye in the constellation Taurus, the Bull. It lies about 375 ly (116 pc) from Earth. The stars are not shedding mass as are those in Figure 11-11. Their blue glow is a reflection nebula created as some of the stars' radiation scatters off preexisting dust grains in their vicinity. Each dot plotted on this H-R diagram represents a star in the Pleiades whose luminosity and surface temperature have been determined. Note that most of the cool, low-mass stars have arrived at the main sequence, indicating that hydrogen fusion has begun in their cores. The cluster has a diameter of about 5 ly, is about 100 million years old, and contains about 500 stars.

case, this is called **interstellar extinction.** Even when we can see a star through interstellar clouds, it appears redder than it actually is. As just discussed, this occurs because short-wavelength, mainly blue, starlight is scattered by dust grains in the cloud more than the longer-wavelength red light. Therefore, when we observe that star, we are seeing less blue light from it than we would if the cloud were not there (Figure 11-2). This **interstellar reddening** is different from reddening due to the Doppler shift (Chapter 4). The Doppler shift causes all wavelengths of electromagnetic radiation to lengthen equally, while interstellar reddening, due to the stronger scattering of shorter wavelengths, does not change the wavelengths of the starlight we receive—only their intensities.

However, many interstellar atoms and molecules emit radio or infrared photons, which are scattered relatively little by intervening gas and dust. Therefore, we search for the interstellar medium primarily in the radio and infrared parts of the electromagnetic spectrum. Because hydrogen is so common in the Galaxy and because the Sun is primarily hydrogen, we hypothesize that stars are composed mostly of hydrogen. However, star-forming molecular hydrogen is hard to detect in space. Therefore, radio astronomers often search instead for carbon monoxide, which emits lots of photons at a wavelength of 2.6 mm. Calculations based on the known abundances of elements reveal that there are about 10,000 hydrogen molecules (H_2) for every CO molecule in a typical interstellar cloud. Consequently, wherever

brightest ones in the Chapter 10 opening photograph. Easily visible to the unaided eye, the Pleiades is in the constellation Taurus. Note the distinctly bluish color of the nebulosity around these stars. This haze, called a **reflection nebula,** is caused by fine grains of interstellar dust that efficiently scatter and reflect blue light from surrounding stars. Indeed, reflection nebulae are blue for the same reason that Earth's sky is blue: Particles scatter short-wavelength light much more efficiently than longer-wavelength radiation. Blue light is therefore scattered back toward us much more intensely than is light of any other color.

Suppose there was a star behind a cloud of gas and dust. Like clouds in our sky blocking sunlight, sufficiently dense interstellar clouds or several thin clouds dim or even completely block the light from stars behind them. In either

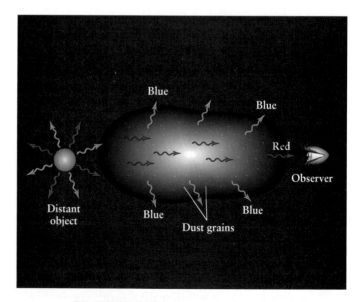

FIGURE 11-2 **Interstellar Reddening** Blue light from distant stars scatters off dust in interstellar clouds, leaving the light we see from the stars redder than it would otherwise appear.

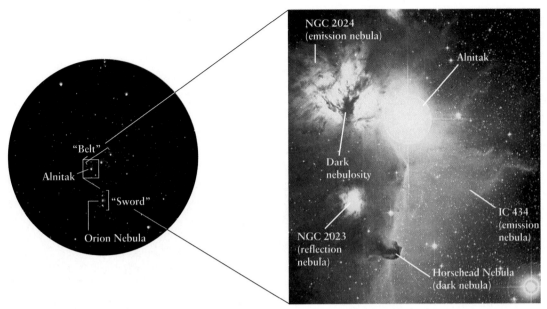

NGC 2024
(emission nebula)

Alnitak

Dark
nebulosity

IC 434
(emission
nebula)

NGC 2023
(reflection
nebula)

Horsehead Nebula
(dark nebula)

R I V U X G

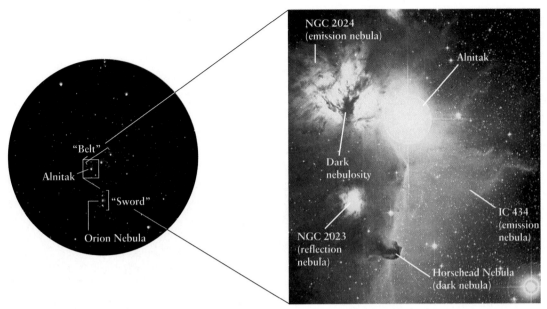

FIGURE 11-3   A Gas and Dust-Rich Region of Orion A variety of nebulae appear in the sky around Alnitak, also called ζ (zeta) Orionis, the easternmost star in the belt of Orion. To the left of Alnitak is a bright, red emission nebula called NGC 2024. The glowing gases in emission nebulae are excited by ultraviolet radiation from young, massive stars. Dust grains obscure part of NGC 2024, giving the appearance of black streaks, while the distinctively shaped dust cloud called the Horsehead Nebula blocks the light from the background nebula IC 434. The Horsehead is part of a larger complex of dark interstellar matter, seen in the lower left of this image. Above and to the left of the Horsehead Nebula is the reflection nebula NGC 2023, whose dust grains scatter blue light from stars between us and it more effectively than any other color. All this nebulosity lies about 1600 ly from Earth, while the star Alnitak is only 815 ly away from us. (Royal Observatory, Edinburgh; inset: R. C. Mitchell, Central Washington University)

astronomers detect strong emission of CO, they deduce that an enormous amount of hydrogen gas must also be present.

In mapping the locations of CO emission, astronomers came to realize that vast amounts of interstellar gas and dust are concentrated in **giant molecular clouds.** In some cases, these regions appear as dark areas silhouetted against a glowing background light, such as Orion's famous Horsehead Nebula (Figure 11-3). In other cases, the clouds appear as dark blobs that obscure the background stars (Figure 11-4). Some 6000 of these molecular clouds are known, with masses ranging from 10^5 to 2×10^6 $M_\odot$ and diameters ranging from 50 to 300 ly. The density inside one of these clouds ranges from 10^2 to 10^5 hydrogen molecules per cubic centimeter, several thousand times greater than the average density of the gas and dust dispersed throughout interstellar space, but some 10^{15} times less dense than the air you breathe.

11-2 Supernova explosions in cold, dark nebulae trigger the birth of stars

We will see in detail in Chapter 12 that a *supernova* is a violent detonation that ends the life cycle of a massive star. In

Barnard 86
(dark nebula)

NGC 6520
(star cluster)

R I V U X G

FIGURE 11-4   A Dark Nebula This dark nebula, Barnard 86, is located in Sagittarius. It is visible in this photograph simply because it blocks out light from the stars beyond it. The cluster of bluish stars to the left of the dark nebula is a star cluster called NGC 6520. (Anglo-Australian Observatory)

GUIDED DISCOVERY The Nebulae

Binoculars and the naked eye are enough to let you "get your hands dirty" exploring star dust. Distant nebulae—clusters of stars and glowing gases—are among the most impressive objects in the night sky.

You can observe the Great Nebula of Orion (M42) during winter in the northern hemisphere even with the naked eye. In all likelihood, you have seen it dozens of times without knowing it. To locate the Great Nebula, find the constellation Orion (using, for example, the star charts at the end of the book). Locate Orion's belt. Due south of the belt are three stars in a row making up Orion's sword. Examine the sword very carefully with your naked eye. Do any of the stars in it look at all odd? Now look at them through a pair of binoculars. Which one is different from the others? That one is the Great Nebula in Orion, not a star at all! How does what you see compare with Figure 11-13?

The North America Nebula and the Pelican Nebula in Cygnus are best spotted in the fall. Pick a dark, moonless night and use binoculars rather than a telescope. Higher magnification reveals too small a region of the sky for you to see the entirety of these vast, dim nebulae. To find them, first locate the bright star Deneb on the tail of Cygnus (using, for example, the star charts at the end of the book). The North America Nebula is located 3° east of Deneb, while the Pelican Nebula is located 2° southeast of it. These are both small angles, so sweep around the sky east of Deneb. If your binoculars are powerful enough, you should be able to see the outlines that give these nebulae their names.

The constellations of Orion and Monoceros encompass one of the most accessible regions of the sky for studying star formation and the interaction of young stars with the interstellar medium. Figure 11-5 shows a map of this region made with a radio telescope tuned to a wavelength of 2.6 mm, which is emitted by CO. Note the extensive areas covered by giant molecular clouds. Such comprehensive maps of CO emission help astronomers understand how the large-scale structure of the interstellar medium is related to the formation of stars.

RIVUXG

FIGURE 11-5 A Map of Carbon Monoxide Features in Orion This color-coded radio map of a large section of the sky shows the extent of giant molecular clouds in Orion and Monoceros. The intensity of carbon monoxide (CO) emission is displayed by colors in the order of the rainbow, from violet for the weakest to red for the strongest. Black indicates no detectable emission. The locations of four prominent star-forming nebulae are indicated on the star chart overlay. Note that the Orion and Horsehead nebulae are sites of intense CO emission, indicating that stars are forming in these regions. (R. Maddalena, M. Morris, J. Moscowitz, and P. Thaddeus)

a matter of seconds the core of the doomed star collapses, releasing vast quantities of particles and energy that blow the star apart. The star's outer layers are blasted into space at speeds of several thousand kilometers per second.

Astronomers find remains of many such dead stars scattered across the sky. These **nebulae** or **supernova remnants,** like the Cygnus Loop shown in Figure 11-6, have a distinctly arched appearance, as would be expected for a shell of gas expanding at supersonic speeds. As it passes through the surrounding interstellar medium, the supernova remnant excites the atoms and molecules there, causing the gases to glow. If the expanding shell of a supernova remnant rams into a giant molecular cloud, it can cause the cloud to contract, thus stimulating star birth. As we learned in Essentials II, there is evidence that such an event happened around the time the solar system formed.

A simple collision between two interstellar clouds can also create regions sufficiently dense and cool to collapse and form new stars. Likewise, radiation from an especially bright star or group of stars may also compress the sur-

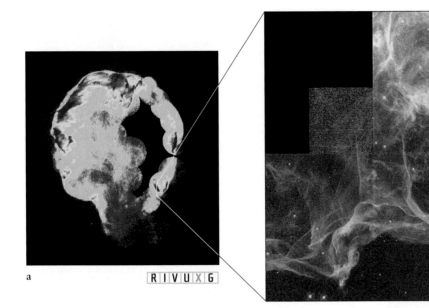

a   R I V U X G

FIGURE 11-6
A Supernova Remnant
(a) X-ray image of the Cygnus Loop, the remnant of a supernova that occurred nearly 20,000 years ago. The expanding spherical shell of gas now has a diameter of about 120 ly. The entire Cygnus Loop has an angular diameter in our sky 6 times wider than the Moon. **(b)** This visible-light Hubble Space Telescope image of part of the Cygnus Loop shows emission from different atoms: blue from oxygen, red from sulfur, and green from hydrogen. (a: Nancy Levenson/NASA; b: Jeff Hester, Arizona State University and NASA)

b   R I V U X G

rounding interstellar medium enough to give birth to stars (Figure 11-7).

Once a giant molecular cloud contracts and cools enough, gravitational attraction causes small regions of gas and dust in it to collapse until they become stars. This gas must be cool, because the higher the temperature, the faster its atoms and molecules move. As they move, they collide with other particles, driving them apart. In other words, the collisions between atoms and molecules create the pressure in the cloud. If the temperature is too high, the pressure overcomes any gravitational attraction between the particles, preventing them from drawing close enough together to form stars.

When a region of a cloud is sufficiently cold and dense, the gravitational attraction of the matter in this region overwhelms the pressure there and pulls the gas together to form a new star. This collapse is called a **Jeans instability,** after the British physicist James Jeans, who in 1902 calculated the conditions for it to occur.

Infrared observations show compact regions called **dense cores** inside many interstellar clouds. Their temperatures, around 10 K, are so low that the dense cores are destined to form stars. Often a giant molecular cloud has several hundred or even thousands of dense cores. In this case, hundreds or thousands of stars form together. Such stellar nurseries will become **open clusters** of stars like the Pleiades, shown in the figure opening Chapter 10.

At first, a collapsing dense core is just a cool, dusty region thousands of times larger than our solar system. The dense core actually collapses from the inside out. The inner region falls in rapidly, leaving the outer layers of the dense core to drift in at a more leisurely rate. This process of

increasing mass in the central region is called *accretion,* and the newly forming object at the center is called a **protostar.** Although fusion has not begun, a protostar glows from the heat generated by the compression of the gas it contains, under the influence of the gravitational attraction of this growing mass of hot gas.

R I V U X G

FIGURE 11-7   The Core of the Rosette Nebula The large, circular Rosette Nebula (NGC 2237) is near one end of a sprawling giant molecular cloud in the constellation Monoceros, the Unicorn. Radiation from young, hot stars has blown gas away from the center of this nebula. Some of this gas has become clumped in dark globules that appear silhouetted against the glowing background gases. New star formation is taking place within these globules. The entire Rosette Nebula has an angular diameter on the sky nearly 3 times that of the Moon, and it lies some 3000 ly from Earth. (Anglo-Australian Observatory)

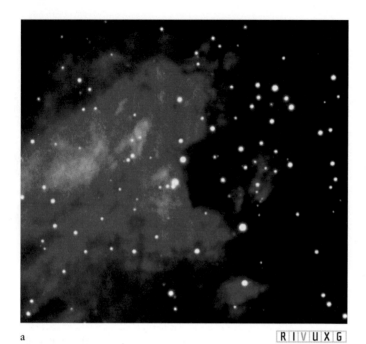

a   R I V U X G

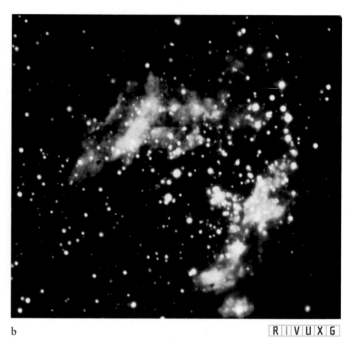

b   R I V U X G

FIGURE 11-8   Newborn Stars in the Swan Nebula   (a) This image at visible wavelengths shows an H II region called the Swan (or Omega) Nebula, so named because of its characteristic shape. It is 5500 ly from Earth in the constellation Sagittarius. A comparison of the visible and infrared views demonstrates that circumstellar and interstellar dust obscures many of the stars in this star-forming region, which contains around 800 $M_\odot$. (b) This infrared view, constructed from images taken at wavelengths of 1.2, 1.6, and 2.2 mm, reveals hundreds of stars that do not appear in (b). (a & b: NOAO)

Figure 11-8 shows views of an interstellar cloud in which stars are forming. The view at visible wavelengths (Figure 11-8a) is a familiar sight to many telescope observers, but it tells only part of the story. Visible light from the protostar never reaches us because it is absorbed by the surrounding shell of infalling dust. Only the infrared picture (Figure 11-8b) shows the hundreds of protostars.

If a dense core is not spinning, it collapses into a sphere, which ultimately becomes an isolated star. If it is spinning, it collapses into a disk, which may then condense into two or three stars. Or, if the disk has a low enough mass, it may become a single star with orbiting protoplanets. An example (one of many) is the disk of gas and dust observed around β Pictoris (see Figure II-13). This disk was the first sign of planetary development outside the solar system.

11-3   When a protostar ceases to accumulate mass, it becomes a pre–main-sequence star

Theories of star formation predict that after about 10^5 years of mass accretion, the protostar builds up to about the mass of the Sun. A protostar of 1 $M_\odot$ is about 5 times larger in diameter than the Sun. Its large size makes it brighter than the Sun, and it can be observed as an intense point of infrared radiation. Much matter is still slowly falling inward from the dense core's outer shell. However, the radiation and particles flowing off the protostar exert outward forces on this remaining gas and dust, preventing it from ever reaching the protostar. Now, mass accretion stops, and the protostar becomes a **pre–main-sequence star.**

A pre–main-sequence star contracts slowly, unlike the rapid collapse of a protostar. When the temperature at its core reaches 10^7 K, hydrogen fusion begins there. As we saw in Chapter 9, this thermonuclear process releases enormous amounts of energy. The outpouring of energy from hydrogen fusion creates pressure inside the pre–main-sequence star sufficient finally to halt its contraction.

The more massive a pre–main-sequence star is, the more rapidly it begins hydrogen fusion in its core. For example, calculations indicate that a 5-$M_\odot$ pre–main-sequence star starts fusing only 10^5 years after it first forms from a protostar, whereas a 1-$M_\odot$ pre–main-sequence star takes a few tens of millions of years to do the same. Stars more massive than about 7 $M_\odot$ start fusing so rapidly that they go directly from protostars to the main sequence.

In the final stages of pre–main-sequence evolution, the outer shell of gas and dust finally dissipates. For the first time, the star is directly revealed to the outside universe.

11-4 The evolutionary track of a pre–main-sequence star depends on its mass

Astrophysicists use computers and the equations of stellar structure (described in Chapter 9) to model the evolution of a pre–main-sequence star. By calculating changes in the energy that the contracting star emits, computer simulations can follow its changing position on a Hertzsprung-Russell diagram (Figure 11-9). Keep in mind that such an **evolutionary track** represents changes in a star's temperature and luminosity, not its motion in space.

Protostars transform into pre–main-sequence stars along a curve called the **birth line** (see blue line on Figure 11-9). A star's exact location on this curve depends primarily on its mass and, to a much smaller extent, on the amount of metal it contains. As a pre–main-sequence star of less than 2 $M_\odot$ contracts, its diminishing surface area causes its luminosity to drop significantly. On the H-R diagram, therefore, the track of the star drops below the birth line. Eventually, its surface temperature increases, and the star's track moves to the left in the diagram.

Pre–main-sequence stars more massive than 2 $M_\odot$ become hotter without much change in overall luminosity. The

FIGURE 11-10 A Brown Dwarf Located 18 ly (6 pc) from Earth in the constellation Lepus (the Hare), Gliese 229B (the smaller object in the image) was the first confirmed brown dwarf ever observed. With a surface temperature of about 1000 K, its spectrum is similar to that of Jupiter. Gliese 229B is in a binary star system. The overexposed image of part of its companion, Gliese 229A, is seen on the left. The two stars are separated by about 43 AU. Gliese 229B has from 20 to 50 times the mass of Jupiter, but the brown dwarf is compressed to the same size as our giant planet. The spike of light was produced when Gliese 229A overloaded part of the Hubble Space Telescope's electronics. (S. Kularni, California Institute of Technology; D. Golimowski, Johns Hopkins University; NASA)

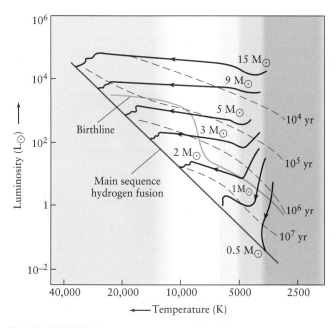

FIGURE 11-9 Pre–Main-Sequence Evolutionary Tracks The evolutionary tracks based on models of seven stars having different masses are shown in this H-R diagram. The dashed lines indicate the stage reached after the indicated number of years of evolution. The birth line, shown in blue, is the location where each protostar stops accreting matter and becomes a pre–main-sequence star. Note that all tracks terminate on the main sequence at points agreeing with the mass-luminosity relation.

evolutionary tracks of these pre–main-sequence stars thus traverse the H-R diagram horizontally, from right to left. A star more massive than about 7 $M_\odot$ has no pre–main-sequence phase at all. Its gravitational compression is so great that it begins to fuse hydrogen in its protostellar phase.

Calculations of the temperature required to start normal (see An Astronomer's Toolbox 9-1) hydrogen fusion in the core of a star indicate that pre–main-sequence stars less massive than about 0.08 $M_\odot$ do not have enough gravitational force compressing and heating their cores to initiate fusion. Instead, these small bodies contract to become planetlike orbs of hydrogen and helium, called **brown dwarfs** (Figure 11-10). See Guided Discovery: Extrasolar Planets and Brown Dwarfs for further discussion of these important, albeit nonstellar, bodies.

The upper limit to stellar mass is around 100 $M_\odot$. Some protostars initially have more than this mass. However, those protostars rapidly develop extremely high fusion rates in their cores, which lead to extremely high surface temperatures—

GUIDED DISCOVERY Extrasolar Planets and Brown Dwarfs

The lowest mass that an object can have and still maintain the fusion of normal hydrogen into helium as occurs in the Sun (see An Astronomer's Toolbox 9-1) is $0.08 \ M_\odot$ or about 75 times the mass of Jupiter. Astronomers have discovered hundreds of objects in our Galaxy with less than this mass. Like Jupiter, they are primarily composed of hydrogen and helium, with traces of other elements. Many of these objects are found in orbit around stars, while some are found as free-floating masses that apparently formed without ever orbiting a star. An intriguing question has arisen: What should they be called?

One school of thought is that all such low-mass objects orbiting stars should be called *extrasolar planets* and all the isolated ones should be called *brown dwarfs*. The second school of thought is that the distinction between extrasolar planets and brown dwarfs should be based on the fact that while normal hydrogen fusion does not occur in them, those bodies with more than 13 times Jupiter's mass do fuse deuterium (a rare form of hydrogen) into helium and those with more than 60 times Jupiter's mass also fuse lithium (three protons and four neutrons) into helium. Both of these types of fusion occur very briefly in cosmic terms because of the limited supplies of deuterium and lithium in any known object in space. Adherents of this second school of thought hold that all objects below 13 times Jupiter's mass should be considered as extrasolar planets regardless of their location, while all objects with more than this mass should be considered brown dwarfs.

An emerging third school of thought is that all objects above 13 times Jupiter's mass, no matter where they are located, are brown dwarfs. These astronomers suggest that objects orbiting stars with less than this mass be called extrasolar planets, while free-floating bodies with less than this mass be given a third name, perhaps *sub-brown dwarfs*. The debate rages in the literature and at professional meetings. We will hereafter use this third set of definitions.

Because they give off relatively little energy compared to stars, extrasolar planets and brown dwarfs are dim and therefore very challenging to observe. Those in orbit are detected by their gravitational or eclipsing effects on the stars they orbit. The first brown dwarf was discovered only in 1994. Named Gliese 229B, it is located in orbit around a star, Gliese 229A (Figure 11-10) in the constellation Lepus about 18 ly (6 pc) from Earth. Since then, over 100 more brown dwarfs have been found, along with over a dozen sub-brown dwarfs. Many of these are found in active star-forming regions, such as the Orion Nebula (Figure 11-13) and the rho Ophiuchi cloud (see figure). Astronomers have also found more than 70 extrasolar planets.

Brown dwarfs have the interesting feature that when they fuse deuterium or lithium (in the higher mass ones), the helium they create moves upward, out of the core where it is formed. This helium is replaced with fresh deuterium or lithium fuel to fuse. The upward motion of the helium and downward motion of deuterium and lithium-rich hydrogen are due to convection, and as a result of this motion, eventually all the deuterium and lithium are consumed. We say that brown dwarfs are *fully convective*. This is different behavior than we find in the Sun (see Chapter 9), which has a separate core, convective zone, and radiative zone that do not share atoms. Flares have been observed from brown dwarfs. By analogy to the Sun's flares caused by magnetic fields emerging from its surface, astronomers believe that some brown dwarfs rotate and have magnetic fields.

Based on the numbers and locations of the known brown dwarfs and sub-brown dwarfs, astronomers estimate that there may be as many of these bodies in our Milky Way Galaxy as there are stars. Even in these numbers, brown dwarfs do not contribute a substantial amount of mass or gravitational force in the Galaxy because they have such low individual masses. Nor are they now believed to contribute much mass to the total mass of the universe—a number that affects its fate.

R I V U X G

A Stellar Nursery Full of Brown Dwarfs Besides containing over 100 young stars, the rho Ophiuchi cloud, located 540 ly away in the constellation Ophiuchus, contains at least 30 brown dwarfs. By studying these objects, astronomers expect to learn more about early stellar evolution. This infrared image is color coded, with red indicating 7.7 micrometer radiation and blue indicating 14.5 micrometer radiation. (Infared Space Observatory, NASA)

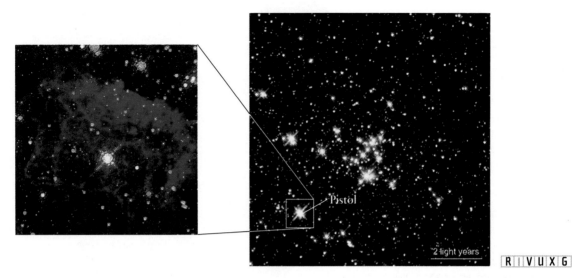

R I V U X G

FIGURE 11-11 **Mass Loss from a Supermassive Star** The Quintuplet Cluster is 25,000 ly from Earth. **Inset:** Within the cluster is the brightest known star, called the Pistol. Astronomers calculate that the Pistol formed nearly 3 million years ago and originally had 100–200 $M_\odot$. The structure of the gas cloud suggests the star ejected the gas we see in two episodes 6000 and 4000 years ago. The gas from any previous ejections is so thinly spread now that we cannot see it. The nebula shown in the inset is more than 4 ly (1.25 pc) across—it would stretch from the Sun nearly to the closest star, Proxima Centauri. The image of the Quintuplet Cluster was taken in the infrared. The name Pistol was given to the star based on early, low resolution radio images of its gas, which initially looked like an old-fashion pistol aimed to the left near the top of the inset. (D. Filger, NASA)

temperatures so great that their outer layers are super-heated and thereby expelled into interstellar space. This, in turn, lowers their masses and their temperatures. One of the brightest stars in our Galaxy, the Pistol Star (Figure 11-11), may have started its life with as much as 200 $M_\odot$. Observations reveal that every few thousand years it expels shells of gas, and it may have less than 10 $M_\odot$ left when the expulsion of matter stops. The entire process of mass loss from very massive stars takes only a few million years.

11-5   H II regions harbor young star clusters

We can detect the formation of a cluster of stars as a magnificent glow in the nebula. Figures 11-3 and 11-12 show these **emission nebulae.** Because these nebulae are predominantly ionized hydrogen, they are also called **H II regions.** To see why H II regions exist, remember that most massive pre–main-sequence stars, those of spectral types O and B, are exceptionally hot. Because their surface temperatures are typically 15,000 to 35,000 K, they emit vast quantities of ultraviolet radiation. This energetic radiation easily ionizes any surrounding gas. Photons from an O5 star can ionize hydrogen atoms up to 500 ly away.

But since H II denotes ionized hydrogen, how do we observe it? While some hydrogen atoms in the H II regions are being knocked apart by ultraviolet photons, some of the free protons and electrons manage to get back together. As these new hydrogen atoms assemble, their electrons return to their ground state (n = 1). This downward cascade through each atom's energy levels is what makes the nebula glow. Particularly prominent is the transition from n = 3 to n = 2, which produces H_α photons at 656 nm in the red portion of the visible spectrum (review the emission line spectrum in Figure 4-10). Thus, the nebula around a newborn star cluster often shines with a distinctive reddish hue (see Figure 11-12). Nebulae often look green when seen with the eye through a telescope because they contain oxygen gas, which has a green emission line at 501 nm. Because the eye is more sensitive to green light than red, the dimmer oxygen emission line appears brighter in our brains than does the H_α line, which shows up well on CCD images.

An H II region is a small, bright "hot spot" in a giant molecular cloud. The collection of hot, bright O and B stars that produces the ionizing ultraviolet radiation is called an **OB association.** The famous Orion Nebula (Figure 11-13) is an example. Four O and B stars at the heart of the Orion Nebula are responsible for the ionizing radiation that causes the surrounding gases to glow. The Orion Nebula is embedded in a giant molecular cloud whose mass is estimated at 500,000 $M_\odot$.

The OB association creating the H II region also affects the rest of the giant molecular cloud (see Figure 11-13 insets). Detailed models indicate that vigorous stellar winds, along with ionizing ultraviolet radiation from the O and B stars, carve out a cavity in the cloud. Where this outflow is supersonic, it creates a shock wave, like the sonic boom created by fast-flying aircraft. The shock wave forms along the outer edge of the expanding H II region, compressing hydrogen gas

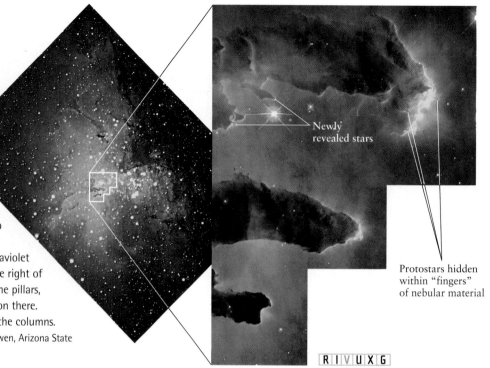

WEB LINK 11.9 ANIMATION 11.3

FIGURE 11-12

An H II Region This emission nebula, M16, called the Eagle Nebula because of its shape, surrounds a star cluster. Star formation is presently occurring in M16, which is located 7000 ly from Earth in the constellation of Serpens Cauda (Serpent's Tail). Several bright, hot O and B stars are responsible for the ionizing radiation that causes the gases to glow. **Inset:** Star formation is occurring inside these dark pillars of gas and dust. Intense ultraviolet radiation from existing massive stars off to the right of this image is evaporating the dense cores in the pillars, thereby prematurely terminating star formation there. Newly revealed stars are visible at the tips of the columns. (Anglo-Australian Observatory; J. Hester and P. Scowen, Arizona State University; NASA)

Newly revealed stars

Protostars hidden within "fingers" of nebular material

R I V U X G

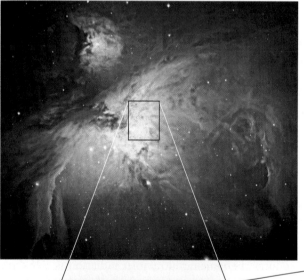

WEB LINK 11.10 ANIMATION 11.2

FIGURE 11-13 **The Orion Nebula** As you have observed, the middle "star" in Orion's sword is actually the Orion Nebula, a part of a huge system of interstellar gas and dust in which new stars are now forming. The Orion Nebula is an H II region visible to the naked eye. It is 1600 ly (490 pc) from Earth and has a diameter of roughly 16 ly (5 pc). This nebula's mass is about 300 $M_\odot$. **Inset (left)** This view at visible wavelengths shows the inner regions of the Orion Nebula. At the lower left are four massive stars, called the Trapezium, which cause the nebula to glow. **Inset (right)** This view of exactly the same region shows that infrared radiation penetrates interstellar dust that absorbs visible photons. Numerous infrared objects, many of which are probably stars in the early stages of formation, can be seen, along with shock waves caused by matter flowing out of protostars faster than the speed of sound waves in the nebula. Shock waves from the Trapezium may have helped trigger the formation of the protostars in this view. (Anglo-Australian Observatory; inset (left): C.R. O'Dell, S.K. Wong, and NASA; inset (right): R. Thompson, M. Rieke, G. Schneider, S. Stolovy, E. Erickson, D. Axon, and NASA)

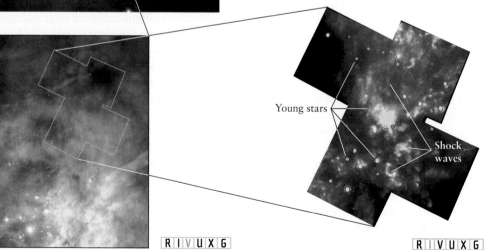

Young stars

Shock waves

R I V U X G

R I V U X G

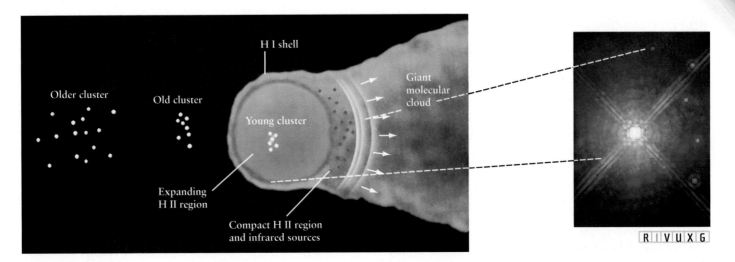

FIGURE 11-14 **The Evolution of an OB Association** High-speed particles and ultraviolet radiation from young O and B stars produce a shock wave that compresses gas farther into the molecular cloud, stimulating new star formation deeper into the cloud. Meanwhile, older stars are left behind. **Inset:** Stars forming around a massive star 2500 ly (770 pc) away in the constellation Monoceros's Cone Nebula. The six stars (small dots on the right side of the inset) arrayed around the bright, massive central star are believed to have formed as a result of the central star compressing surrounding gas with high-speed particles and radiation. The younger stars are just 0.04–0.08 ly from the central star. (Adapted from C. Lada, L. Blitz, and B. Elmegreen; inset: R. Thompson, M. Rieke, G. Schneider, and NASA)

as it passes and thereby stimulating a new round of star birth. As more O and B stars form, they power the expansion of the H II region still farther into the giant molecular cloud. Meanwhile, the older O and B stars left behind begin to disperse (Figure 11-14). In this way, an OB association "eats into" a giant molecular cloud, creating stars in its wake.

To test the theory of star birth just described, astronomers have looked for protostars adjacent to stars in an OB association. The inset in Figure 11-14 shows star formation around a single O star. Figure 11-13 shows star formation in the core of the Orion Nebula. Visible light observations easily reveal four O and B stars amid the glowing gas and dust, but infrared observations of the same region are showing us cocoons of warm dust still enveloping pre–main-sequence stars.

11-6 Plotting a star cluster on an H-R diagram reveals its age

The young open clusters we have studied so far offer astronomers a rich source of information about stars in their infancy. By measuring each star's apparent magnitude, color, and distance, an astronomer can deduce its luminosity and surface temperature. The data for all the stars in the cluster can then be plotted on an H-R diagram, as shown for the cluster NGC 2264 in Figure 11-15. (NGC stands for *New General Catalogue*, a listing of more than 7000 objects other than stars published in 1888. The original *General Catalogue* was compiled by William Herschel, his sister Caroline, and his son John during the first half of the nineteenth century.)

Stars with masses over the entire allowed range form in each cluster. Because all the stars in the cluster begin forming at the same time and because stars with different masses arrive on the main sequence at different times, astronomers can use the H-R diagram to determine the age of the cluster. Note that the hottest stars in NGC 2264, with surface temperatures around 20,000 K, lie on the main sequence. These hot stars are extremely bright and massive: Their radiation causes the surrounding gases to glow. Most of the stars cooler than about 10,000 K have not yet arrived at the main sequence. These less massive stars, which are in the final stages of pre–main-sequence contraction, are just now beginning to ignite thermonuclear reactions at their centers.

We will discover in Section 11-7 that stars with different masses evolve at different rates and thus also leave the main sequence at different times. The fact that none of the high-mass stars have left the main sequence yet, combined with the fact that many low-mass stars have not yet even reached it, can be applied to theories of stellar evolution to reveal that the open cluster in Figure 11-15 is roughly two million years old.

Spectroscopic observations of the cooler stars in NGC 2264 show that many are vigorously ejecting gas just before they reach the main sequence. Gas-ejecting stars in spectral classes G and cooler (that is, G, K, and M) are called **T Tauri stars,** after the first example discovered in the constellation of Taurus. Some astronomers propose that the onset of hydrogen fusion is preceded by vigorous chromospheric activity marked by enormous spicules and flares that propel the star's outermost layers back into space. In fact, an infant star going through its T Tauri stage can lose as much as 0.4 $M_\odot$ of

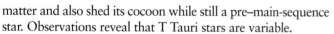

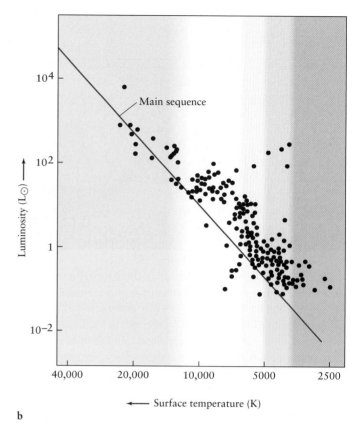

FIGURE 11-15 Plotting the Ages of Stars **(a)** This photograph shows an H II region and the young star cluster NGC 2264 in the constellation Monoceros. The nebulosity is located about 2600 ly from Earth and contains numerous stars that are about to begin hydrogen fusion in their cores. **(b)** Each dot plotted on this H-R diagram represents a star in this cluster whose luminosity and surface temperature have been measured. Note that most of the cool, low-mass stars have not yet arrived at the main sequence. Calculations of stellar evolution indicate that this star cluster started forming about two million years ago. (a: Anglo-Australian Observatory)

matter and also shed its cocoon while still a pre–main-sequence star. Observations reveal that T Tauri stars are variable.

In contrast to the H-R diagram for NGC 2264, nearly all the stars in the Pleiades (see Figure 11-1) have completed their pre–main-sequence stage. The cluster's age is about 100 million years, which is how long it takes for the least massive stars to finally begin hydrogen fusion in their cores.

Open clusters, such as the Pleiades and NGC 2264, possess barely enough mass to hold themselves together. A star moving faster than the average speed for the cluster occasionally escapes. Astronomers predict that in a few hundred million years, after their stars have separated from each other and mixed with the rest of the stars in the galaxy, most open clusters cease to exist.

MAIN-SEQUENCE AND GIANT STARS

As a pre–main-sequence star evolves, fusion begins in its core. Taking the Sun as our example of main-sequence activity, we saw in Chapter 9 that thermal pressure then pushes outward

from the core, balancing the inward force of gravity. We call that balance *hydrostatic equilibrium*. The pre–main-sequence star now ceases to collapse—and a star is born. It has reached the main sequence at last, as shown in Figure 11-9. *Main-sequence stars are those stars with nuclear reactions fusing hydrogen into helium in their cores at a constant rate.*

11-7 Stars spend most of their life cycles on the main sequence

The **zero-age main sequence** (**ZAMS**) is the location on the H-R diagram where the stellar model predicts that pre–main-sequence stars (fusing hydrogen in their cores) first become stable objects, neither shrinking nor expanding. This is the solid red line that appears on several of the H-R diagrams. As was the case with the birth line, the location of a ZAMS star on the main sequence depends primarily on its mass. Note that the evolutionary tracks in Figure 11-9 end at locations along the main sequence that agree with the mass-luminosity relation (recall Figure 10-8): The most massive main-sequence stars are the most luminous, while the least massive stars are the least luminous.

The equations of nuclear fusion predict that the more massive a star is, the faster it evolves and the less time it spends on the main sequence. This occurs because gravity presses down with greater force on a more massive star's core than on a less massive star's core. This tremendous pressure speeds up fusion so that some O and B stars consume all their core hydrogen in only a few million years, as shown in Table 11-1. Conversely, stars of very low mass take hundreds of billions of years to convert their cores from hydrogen into helium.

Whatever the rate of fusion, the equations reveal that the conversion of hydrogen into helium in a star's core takes a long time compared to any other stage of stellar evolution. (These stages are presented in the following sections.) That is why the vast majority of stars represented on an H-R diagram are located on the main sequence. Our theory of solar evolution predicts that the Sun's total lifetime on the main sequence will be about 10 billion years based on the time it will take to convert all the hydrogen in the core to helium.

11-8 Red dwarfs convert essentially their entire mass into helium

The lowest mass main-sequence stars, called **red dwarfs**, have masses between 0.08 and $0.4 M_\odot$. They have the lowest temperatures and pressures in their cores. Therefore, they fuse hydrogen slowest, and hence are the coolest, of all main-sequence stars. They are also distinct from all other main-sequence stars in that as red dwarfs create helium in their cores, this helium moves upward, while hydrogen from the outer layers move downward. This is carried out by the process of convection. As a result, these stars eventually convert their entire mass into helium. Convection occurs in the outer layers of the giant planets (see Section 7-1) and the Sun (see Section 9-8), but not throughout these entire bodies. Stars with more than $0.4 M_\odot$ *do not* convert their entire masses into helium because they do not convect the helium from their cores outward. The cores and the outer layers of these more massive stars remain essentially distinct from each other throughout their evolutions.

The lifetimes of red dwarfs on the main sequence range into the hundreds of billions of years and none of them have yet left the main sequence. They do not have enough pressure in their cores to fuse the helium they create into anything else and so they will stop fusing altogether. These helium bodies will then radiate the heat they generated while on the main sequence and thereby become cooler, moving down and to the right from the main sequence.

11-9 When core hydrogen fusion slows down, a main-sequence star with M > 0.4 M_☉ becomes a giant

When the hydrogen in the core of a main-sequence star more massive than a red dwarf is mostly converted into helium, fusion in the core slows down. Helium fusion does not occur at this point because the temperature of the core is not high enough to enable the helium to fuse into other elements. But the star continues to evolve. Let's see how the loss of the energy source can actually cause stars to *expand* in size to become *giants*.

Recall from our discussion of the Sun in Chapter 9 that while on the main sequence a star's outer layers are supported by thermal pressure, with energy supplied by fusion in the core. Without fusion, a star can no longer support the crushing weight of its outer layers, and so it begins compressing its helium core. We will return to the fate of the core shortly, but in the meantime, consider the hydrogen-rich gas just outside it. Under the influence of gravity, this gas is compressed and heated enough to begin fusing into helium. This is called **shell hydrogen fusion** because it occurs in a shell surrounding the core. Recall from An Astronomer's Toolbox 9-1 that the proton-proton chain is the major source of energy inside the Sun. This fusion process also occurs in shell hydrogen fusion.

TABLE 11-1 Main-Sequence Lifetimes

Mass ($M_\odot$)	Surface temperature (K)	Luminosity ($L_\odot$)	Time on main sequence (10^6 years)	Spectral class
25	35,000	80,000	3	O
15	30,000	10,000	15	B
3	11,000	60	500	A
1.5	7,000	5	3,000	F
1.0 (Sun)	6,000	1	10,000	G
0.75	5,000	0.5	15,000	K
0.50	4,000	0.03	200,000	M

Why then does the star expand to become a giant? Thermal pressure from photons created in shell fusion plays a part, but all the details are not yet understood. When our equations of stellar activity are used to create computer models of stars leaving the main sequence, the simulated stars swell up, just like real red giants. However, the equations are so complex that astrophysicists have been unable to isolate all the effects that go into creating red giants.

> ### Insight into Science   More than the Sum of Its Parts
> We have gotten to the point in science where some computer models that numerically solve complex sets of equations reproduce phenomena without explaining it. For example, computer models of the equations that describe how stars evolve show them expanding onto the giant phase. This gives us some confidence that the equations are meaningful representations of reality. However, the models are so complex that astrophysicists are unable to identify the specific physical effects taking place that cause the stars to expand.

Far from the fusing shell, the surface gases cool and soon the temperature of the star's bloated surface falls to between 3000 and 6000 K, depending on the star's total mass. In about 5 billion years, our Sun will swell to a red giant with a radius of about ½ AU, vaporizing Mercury and perhaps causing Venus to spiral into the Sun. The Earth will be scorched to a cinder.

Giant stars are so enormous that their bloated outer layers constantly leak gases into space. At times, this *mass loss* is significant (Figure 11-16), and it can be detected spectroscopically. The escaping gases exhibit narrow absorption lines, and the lines from gases coming toward us are slightly blueshifted, owing to the Doppler effect. This small shift corresponds to a speed of 10 km/s, typical of the expansion velocities with which gases leave the tenuous outer layers of giants. A typical rate of mass loss for a giant is roughly 10^{-7} $M_\odot$ per year. For comparison, in a main-sequence star such as the Sun, mass loss rates are only around 10^{-14} $M_\odot$ per year.

Although the surface of a giant is cooler than that of the main-sequence star from which it evolved, the giant is brighter: It can emit more photons each second because it has so much more surface area. As a full-fledged giant (Figure 11-17), our Sun will shine 2000 times brighter than it does today.

11-10 Helium fusion begins at the center of a giant

Consider now the helium core of a giant star. Helium is the product of hydrogen fusion. When a star first becomes a giant, its hydrogen-fusing shell surrounds a small, compact core of almost pure helium. In a moderately low-mass giant, the dense helium core is about twice the size of the Earth.

At first, no thermonuclear reactions occur in the helium-rich core of a giant because the temperature there is too low to fuse helium nuclei. The hydrogen-fusing shell creates more helium, which becomes part of the core. The ever more massive helium core continues to contract and heat up, while the hydrogen-fusing shell surrounding it consumes more of the star's hydrogen.

When the central temperature reaches about 100 million K, **core helium fusion** begins at the giant's center. The aging star thus has a central energy source for the first time since it left the main sequence. The *triple-alpha process* provides the dominant path for helium in the core to fuse into carbon. First, two helium nuclei combine to create beryllium. Then, within 10^{-8} seconds, a third helium nucleus is added to the short-lived beryllium to create carbon. This process also releases energy and can be summarized as follows:

$$^4\text{He} + {}^4\text{He} + {}^4\text{He} \rightarrow {}^{12}\text{C} + \gamma$$

R I V U X G

FIGURE 11-16   A Mass-Loss Star This unusual star in the constellation Norma (the Carpenter's Square) is the brightest of a system of three stars orbiting each other. This extremely hot, massive star, named HD 148937, is continuously shedding its outer layers. Other vigorous outbursts in the past gave rise to the two symmetric shells of material, called NGC 6164 and NGC 6165, on either side of the star. These shells absorb ultraviolet radiation from the star, causing them to glow with the characteristic red color of excited hydrogen gas. (Anglo-Australian Observatory)

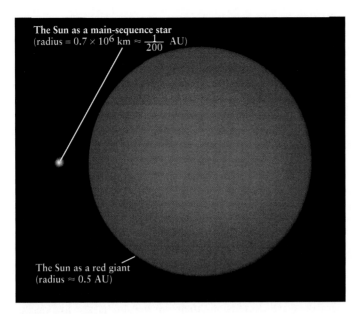

The Sun as a main-sequence star
(radius = 0.7×10^6 km $\approx \frac{1}{200}$ AU)

The Sun as a red giant
(radius ≈ 0.5 AU)

FIGURE 11-17   The Sun Today and as a Giant In about five billion years, when the Sun expands to become a giant, its diameter will increase a hundredfold from what it is now, while its core becomes more compact. Today, the Sun's energy is produced in a hydrogen-fusing core whose diameter is about 200,000 km. When the Sun becomes a giant, it will draw its energy from a hydrogen-fusing shell surrounding a compact helium-rich core. The helium core will have a diameter of only 30,000 km. The Sun's diameter will be about 100 times bigger, and it will be about 2000 times more luminous as a giant than it is today.

where γ denotes energy emitted as photons. But the fusion in giants does not stop there. Some of the carbon created can then fuse with another helium nucleus to produce oxygen:

$$^{12}C + {}^4He \rightarrow {}^{16}O + \gamma$$

Again, energy is released as gamma rays.

The energy released in core helium fusion reestablishes hydrostatic equilibrium—but only briefly. A mature giant fuses helium in its core for about 20% of the time that it spent fusing hydrogen as a main-sequence star. While fusion is again occurring in the core, further gravitational contraction of the giant star ceases. Starting in the distant future, the Sun will consume the helium it is now creating in its core for about two billion years.

The theories of stellar evolution reveal that how helium fusion begins at a giant's center depends on the mass of the star. In high-mass stars, helium fusion begins gradually as temperatures in the star's core approach 100 million K. In low-mass stars, helium fusion begins explosively and suddenly in an event called the **helium flash.** Depending on whose calculations of stellar evolution you choose to believe, the upper limit to stars that undergo the helium flash

is between 2 $M_\odot$ and 4 $M_\odot$. For convenience, we will adopt the limit of 2 $M_\odot$ in what follows. The helium flash is the result of unusual conditions that develop in the core of a low-mass star after it leaves the main sequence. To appreciate these conditions, we must first understand how an ordinary gas behaves, then explore how the densely packed electrons at the star's center alter this behavior.

When an ordinary gas is compressed, it heats up; when it expands, it cools down. The helium in the core of a giant star with mass greater than about 2 $M_\odot$ behaves the same way. If energy production overheats its core, the core expands, cooling the gases and slowing the rate of thermonuclear reactions. Conversely, if too little energy is being created to support the star's overlying layers, they move inward, compressing the core. The resulting increase in temperature speeds up the thermonuclear reactions and thus increases the energy output, which stops the contraction. Either way, the star has a "safety valve" to keep it from collapsing or exploding.

A giant with less than 2 $M_\odot$ lacks that safety valve because densely packed electrons in its core change its behavior. The lower a giant's mass, the more it has to compress to drive temperatures high enough to ignite helium fusion. In fact, at the extreme pressures deep inside a giant with less than 2 $M_\odot$, the atoms are completely ionized, separating into nuclei and electrons. Furthermore, the nuclei are squeezed into a regular pattern, a crystal-like solid. The electrons, distributed between the nuclei, are so closely crowded together that another law of physics called the **Pauli exclusion principle** becomes important.

According to the Pauli exclusion principle, first formulated in 1925 by the Austrian physicist Wolfgang Pauli, two identical particles cannot exist in the same place at the same time. As the electrons are pressed closer and closer together by the star's gravitational force, the exclusion principle prevents them from freezing in place. Instead, many of them must vibrate faster and faster so that they do not become "identical," here meaning moving with the same speeds, to adjacent electrons.

The faster the electrons vibrate, the more energy they have, and the more they can provide an outward pressure to slow the contraction of the core. Astronomers say that electrons in this state are *degenerate* and that the helium-rich core of a low-mass giant is supported by **electron degeneracy pressure,** which provides a greater outward force than did the normal pressure in the star before it became degenerate. Therefore, electron degeneracy pressure prevents the core from collapsing further. It sits there and heats up under the gravitational influence of the star's mass.

The equations describing degenerate matter such as the electrons in this star predict that *degeneracy pressure, unlike the pressure of an ordinary gas, does not change with temperature.* This means that as the core's temperature grows, the pressure there does not increase. Without the "safety valve" of increasing pressure, the star's core cannot expand and cool.

Eventually, the core temperature reaches about 10^8 K, the point where helium begins to fuse. This fusion creates energy, so the core's temperature goes up, but the degenerate electrons still do not increase the pressure. Therefore, the core does not immediately expand, and thus for several hours the core temperature and fusion rate rise dramatically. This is the helium flash.

In the helium flash, temperatures become so high (around 3.5×10^8 K) that the helium again becomes an ordinary gas. Suddenly, the usual safety valve operates once again. In a matter of hours, the star's core expands and cools, decreasing the fusion rate. After the helium flash subsides, the star's energy output declines, so its outer layers again contract. A low-mass, helium-fusing giant is left smaller, dimmer, and hotter than it was before it began fusing helium.

11-11 As stars evolve, their positions on the H-R diagram shift

It is enlightening to plot the theoretical evolutionary tracks of post–main-sequence stars on the Hertzsprung-Russell dia-

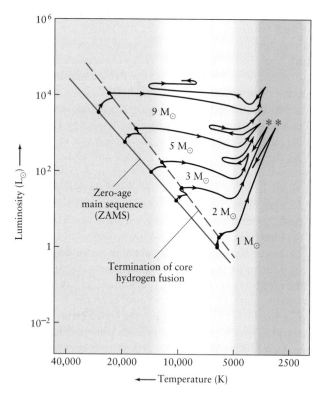

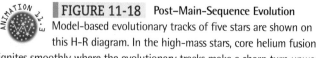

FIGURE 11-18 Post–Main-Sequence Evolution Model-based evolutionary tracks of five stars are shown on this H-R diagram. In the high-mass stars, core helium fusion ignites smoothly where the evolutionary tracks make a sharp turn upward into the giant region of the diagram. The red asterisks on the 1-$M_\odot$ and 2-$M_\odot$ curves indicate where the helium flash in these stars occurs.

gram (Figure 11-18). Each star arrives on the main sequence as a zero-age main-sequence star. As it ages, its core is being converted into helium. This and related internal changes cause its surface brightness and temperature to change, so its track slowly inches away from its ZAMS location. The dashed line in Figure 11-18 shows the locations of the stars when all their core hydrogen has been consumed. As we saw in Section 11-7, it takes a few million years for the most massive main-sequence stars to reach this point, while a 1-$M_\odot$ star takes 10 billion years. Calculations indicate that lower-mass stars take tens or even hundreds of billions of years to get there. Because, as we will see, the universe is roughly 15 billion years old, no main-sequence stars with masses less than about 0.75 $M_\odot$ have yet moved into the giant stage.

After core hydrogen fusion ceases, the points representing high-mass stars move rapidly from left to right across the H-R diagram. This is when each star's core begins to contract and its outer layers expand. Although the stars' surface temperatures are decreasing, their surface areas are increasing. Therefore, their overall luminosities remain roughly constant.

Just before core helium fusion begins, the evolutionary tracks of high-mass stars turn upward into the giant region of the H-R diagram. After the core helium fusion begins, however, the evolutionary tracks back away from these peak luminosities. The tracks then wander back and forth in the giant region while the stars readjust to their new energy sources.

We saw that after the helium flash, low-mass stars become dimmer but hotter. On the H-R diagram in Figure 11-18, note that the two low-mass stars move down and to the left.

11-12 Globular clusters are bound groups of old stars

Post–helium-flash stars are often found in old star clusters, called **globular clusters,** so named because of their spherical shapes. A typical globular cluster, like the one shown in Figure 11-19, may contain up to a million stars in a volume 300 ly across. Like open clusters, the stars in globular clusters all form at about the same time. Unlike open clusters of young stars, globular clusters are gravitationally bound groups of stars that do not disperse.

Astronomers know that globular clusters are old because they contain no high-mass main-sequence stars. If you measure the luminosity and surface temperature of many stars in a globular cluster and plot the data on a color-magnitude diagram, as shown in Figure 11-20, you will find that the upper half of the main sequence is missing. All the high-mass main-sequence stars evolved long ago into giants, leaving behind only lower-mass, slowly evolving stars still undergoing core hydrogen fusion.

The H-R diagram of a globular cluster typically shows a horizontal grouping of stars to the left of the center portion

FIGURE 11-19 A Globular Cluster A globular cluster is a spherical cluster that typically contains a few hundred thousand stars. This cluster, called M13, is located in the constellation Hercules, roughly 23,000 ly from Earth. The stars in M13 are all crowded within a diameter of only 150 ly, making the average density of stars there about 100 times greater than in the solar neighborhood. (NASA)

of the diagram (see Figure 11-20). These stars, called **horizontal branch stars,** are post–helium-flash stars, and they have luminosities of about 50 $L_\odot$. Eventually, these stars will move back toward the giant region as core helium fusion and shell hydrogen fusion devour their fuel.

As noted earlier, an H-R diagram of a cluster can also be used to determine its age, assuming that all of the stars in the cluster formed at the same time. In the H-R diagram for a young open cluster (review Figure 11-15b), almost all the stars are on the main sequence. As a cluster gets older, like the Pleiades, stars begin to leave the main sequence (see Figure 11-1). The high-mass, high-luminosity stars are the first to become giants.

Over time, the main sequence in a cluster gets shorter and shorter. The top of the surviving portion of the main sequence is called the *turnoff point* (see Figure 11-20). Stars at the turnoff point are just beginning to exhaust the hydrogen in their cores. The theory of stellar evolution enables us to predict how long stars with different masses last on the main sequence before moving toward the giant stage. Therefore, the mass of the stars currently at the turnoff point is used to determine the age of the cluster (recall Table 11-1).

Consider the plot for the globular cluster M55 shown in Figure 11-20 (so named because it was fifty-fifth in the *Messier Catalogue* of astronomical objects). In M55, 0.8-$M_\odot$ stars are just leaving the main sequence, so, according to Table 11-1, the cluster's age is approximately 14 billion years.

Using data collected from the *Hipparcos* satellite, Figure 11-21a is a composite diagram of isolated and cluster stars, reflecting the overall structure of the H-R diagram. Using data from stellar evolution theory as presented in Table 11-1, the ages of star clusters can be estimated from the turnoff points, such as those depicted in Figure 11-21b.

The youngest star clusters in the Milky Way (those with their main sequences still intact) are found in open clusters. These clusters exist in the plane of the Galaxy, that is, along the band of light sweeping majestically across the night sky. Stars in these young clusters are said to be *metal-rich,* because their spectra contain many prominent spectral lines of heavy elements. (Recall from Chapter 10 that all elements other than hydrogen and helium are considered metals by astronomers.) This material originally came from stars that exploded long ago, enriching the interstellar gases with the heavy elements formed in their cores. The young clusters are therefore formed from the debris of older

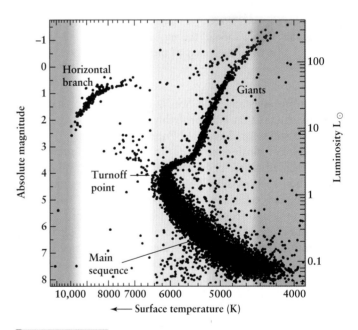

FIGURE 11-20 An H-R Diagram of a Globular Cluster Each dot on this graph represents the apparent magnitude and surface temperature of a star in the globular cluster called M55. Note that the upper half of the main sequence is missing. The horizontal branch stars are low-mass stars that recently experienced the helium flash in their cores and now exhibit core helium fusion and shell hydrogen fusion.

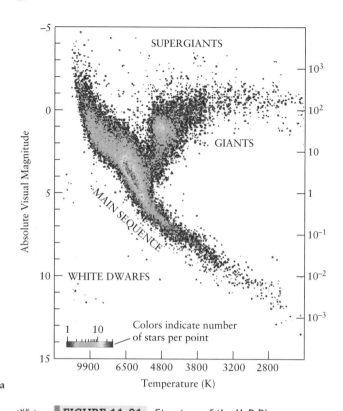

a

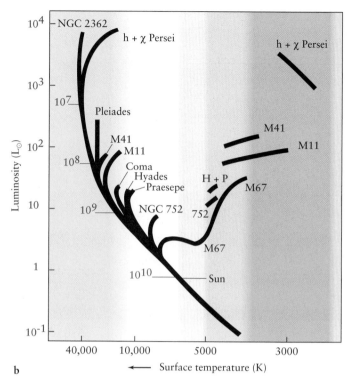

b

FIGURE 11-21   Structure of the H-R Diagram
(a) Data taken by the *Hipparcos* satellite place
41,453 stars more precisely on the H-R diagram than any previous observations. This figure shows the overall structure of the H-R diagram. The thickness of the main sequence is due in large part to stars of different ages turning off the main sequence at different

places, as shown in (b). **(b)** The black bands indicate where data from various star clusters fall on the H-R diagram. The ages of turnoff points (in years) are listed in red alongside the main sequence. The age of a cluster can be estimated from the location of the turnoff point, where the cluster's most massive stars are just now leaving the main sequence.

generations of stars. The Sun is an example of a young, metal-rich star that was probably formed in an open cluster. Such stars are also called **Population I stars.**

Globular clusters contain the oldest stars and are generally located above or below the plane of our Galaxy. Because their spectra show only weak lines of heavy elements, these ancient stars are said to be *metal-poor.* They were created long ago from gases that had not yet been substantially enriched with heavy elements. They are also called **Population II stars.** Figure 11-22 compares the spectra from a Population II star and the Sun.

VARIABLE STARS

After core helium fusion begins, the evolutionary tracks of mature stars move across the middle of the H-R diagram. In Figure 11-18, we saw the evolutionary tracks of post–main-sequence stars. During these excursions across the H-R diagram, a star can become unstable and pulsate. In fact, the region on the H-R diagram between the main sequence and the giant branch is called the **instability strip** (Figure 11-23). When a star's evolutionary track carries it through this region, the star slowly pulsates. As it does so, its brightness

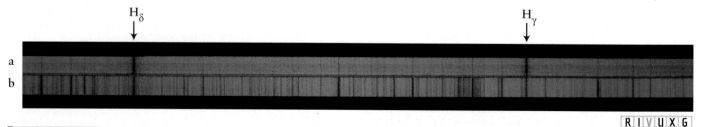

R I V U X G

FIGURE 11-22   Spectra of a Metal-Poor and a Metal-Rich Star
These spectra compare **(a)** a metal-poor (Population II) and **(b)** a metal-rich (Population I) star (the Sun) of the same surface temperature. Numerous spectral lines prominent in the solar spectrum are caused by

elements heavier than hydrogen and helium. Note that corresponding lines in the metal-poor star's spectrum are weak or absent. Both spectra cover a wavelength range that includes two strong hydrogen absorption lines labeled H_γ (434 nm) and H_δ (410 nm). (a & b: Lick Observatory)

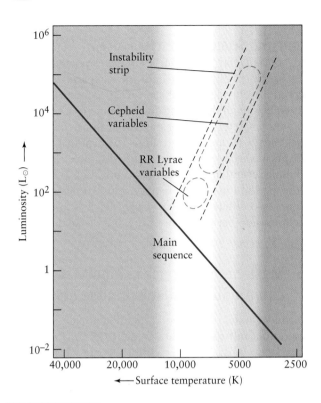

FIGURE 11-23   **The Instability Strip** The instability strip occupies a region between the main sequence and the giant branch on the H-R diagram. A star passing through this region along its evolutionary track becomes unstable and pulsates.

varies periodically. These so-called **variable stars** can be easily identified by their changes in brightness amid a field of stars of constant luminosity.

Low-mass, post–helium-flash stars pass through the lower end of the instability strip as they move in the horizontal branch along their evolutionary tracks. These stars become **RR Lyrae variables,** named after the prototype in the constellation of Lyra (the Lyre). RR Lyrae variables all have periods shorter than one day, and all have roughly the same average brightness as stars on the horizontal branch. High-mass stars pass back and forth through the upper end of the instability strip on the H-R diagram. These stars become **Cepheid variables,** often simply called Cepheids.

11-13   A Cepheid pulsates because it is alternately expanding and contracting

A Cepheid variable is characterized by the way in which its light output varies—rapid brightening followed by gradual dimming. A Cepheid variable brightens and fades because the star's outer layers cyclically expand and contract. Lines in the spectrum of δ (delta) Cephei shift back and forth with the same 5.4-day period that

characterizes its variations in magnitude. According to the Doppler effect, these shifts mean that the star's surface is alternately approaching and receding from us.

When a Cepheid variable pulsates, the star's surface oscillates up and down like a spring. Consequently, the star's gases alternately heat up and cool down; the surface temperature changes from about 6300 K to about 5000 K and back. Thus, the characteristic light curve (luminosity versus time curve, as in Figure 10-11) of a Cepheid variable results from changes in both size and surface temperature.

Just as a bouncing ball eventually comes to rest, a pulsating star would soon stop pulsating without some mechanism to keep its oscillations going. In 1941 the British astronomer Arthur Eddington explained that a Cepheid variable feeds energy into its pulsations. According to this theory, a star is more opaque, or "light tight," when compressed than when expanded. When the star compresses, trapped heat pushes the star's surface outward. When the star expands, the heat escapes, and the star's surface, which is no longer supported, falls inward.

11-14   Cepheids enable astronomers to estimate vast distances

Cepheids are important to astronomers because there is a direct relationship between a Cepheid's period of pulsation and its average luminosity. This relationship is called, appropriately enough, the **period-luminosity relation.** Dim Cepheid variables pulsate rapidly with periods of one to two days and have average brightnesses of a few hundred Suns. The most luminous Cepheids have the longest periods of all Cepheids, with variations occurring over 100 days and average brightnesses equaling 10,000 $L_\odot$. Because the changes in brightness of Cepheids can be seen even in distant galaxies where many other techniques for measuring distance fail, the period-luminosity relation plays an important role in determining the overall size and structure of the universe, as we will see in Chapter 15.

The details of a Cepheid's pulsation depend on the abundance of heavy elements in its atmosphere. The average luminosity of metal-rich Cepheids is roughly 4 times greater than the average luminosity of metal-poor Cepheids having the same period. Thus, there are two classes: **Type I Cepheids** (also called δ Cephei stars), which are the brighter, metal-rich stars, and **Type II Cepheids** (also called W Virginis stars), which are the dimmer, metal-poor stars. The period-luminosity relation for both types of variables is shown in Figure 11-24.

In rare cases, stellar pulsations can be quite substantial. In some instances, the expansion velocity is so high that the star's outer layers are ejected completely. There are several other implications of stars shedding matter. We end this chapter by relating the effect of mass loss in a close binary system, and in the next chapter we explore the mass shedding that occurs at the ends of stellar life cycles.

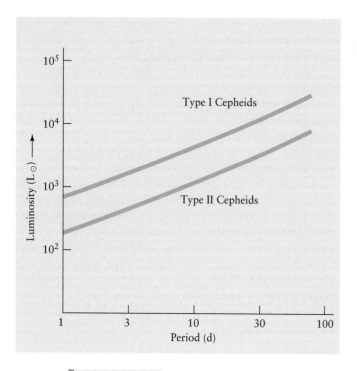

 FIGURE 11-24 **The Period–Luminosity Relation** The period of a Cepheid variable is directly related to its average luminosity: The more luminous the Cepheid, the longer its period and the slower its pulsations. Type I Cepheids (δ Cephei stars) are metal-rich Population I stars. They are brighter than the Type II Cepheids (W Virginis stars), which are metal-poor Population II stars.

Insight into Science   **Firm Foundations?** To understand concepts far from our everyday size and time scales, scientists often must base their explanations of observations on one or perhaps several complex theories even before all the testing and refinement of the theories are complete. An error anywhere along the chain of ideas leads to incorrect results. For example, using the Cepheid variable stars to measure distances requires combining several intermediate concepts, such as knowing the period-luminosity relationship, knowing that there are two different types of Cepheids, and knowing the relationship between luminosity and distance.

11-15 Mass transfer in close binary systems can produce unusual double stars

In the mid-1800s, the French mathematician Édouard Roche pointed out that the atmospheres of two stars in a binary system must remain within a pair of teardrop-shaped regions surrounding the stars. Otherwise, the gas escapes from the star of its origin, either transferring to the other star, where the teardrops make contact, or escaping completely from the binary in the opposite direction. Cut through, these **Roche lobes** take on a figure-eight shape (dashed lines in Figure 11-25). The more massive star is always located inside the larger Roche lobe.

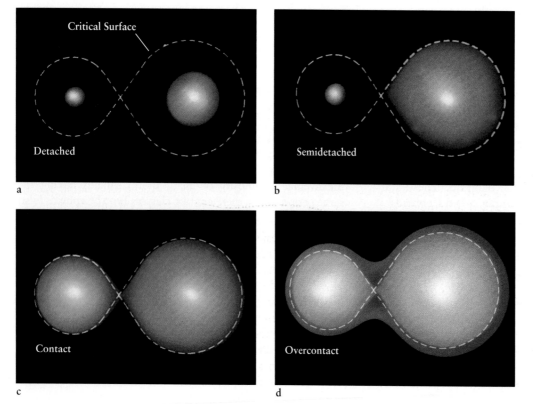

FIGURE 11-25 **Detached, Semidetached, Contact, and Overcontact Binaries** These figures show the various types of binary star systems. (a) In a detached binary, neither star fills its Roche lobe. (b) If one star fills its Roche lobe, the binary is semidetached. Mass transfer is often observed in semidetached binaries. (c) In a contact binary, both stars fill their Roche lobes. (d) The two stars in an overcontact binary both overfill their Roche lobes. The two stars actually share the same outer atmosphere.

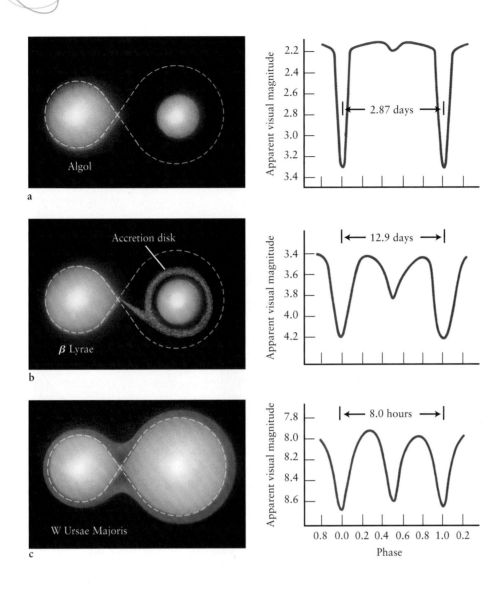

FIGURE 11-26 **Three Close Binaries** Sketches of and light curves for three eclipsing binaries are shown. The phase denotes the fraction of the orbital period from one primary minimum to the next. **(a)** Algol, also known as β Persei, is a semidetached binary. The deep eclipse occurs when the giant star (on the left) blocks the light from the smaller, but more luminous, main-sequence star. **(b)** β Lyrae is a semidetached binary in which mass transfer has produced an accretion disk surrounding the detached star. This disk is so thick and opaque that it renders the secondary star almost invisible. **(c)** W Ursae Majoris is an overcontact binary. Both stars therefore share their outer atmospheres. The short, 8-hour period of this binary indicates that the stars are very close to each other.

The stars in many binaries are so far apart that even during their giant stages the stars' surfaces remain well inside their Roche lobes. Other than orbiting one another, each star in such systems lives out its life cycle as if it were single and isolated, and the system is referred to as a **detached binary** (Figure 11-25a). If two stars are relatively close together, however, one star may fill or overflow its Roche lobe as it expands into the giant phase. This system is called a **semidetached binary,** and gases then flow across the point where the two Roche lobes touch and fall onto the companion star (Figure 11-25b). When both stars completely fill their Roche lobes, the system is called a **contact binary,** because the two stars actually touch and exchange gas (Figure 11-25c). It is quite unlikely, however, that both stars will exactly fill their Roche lobes at the same time. It is more likely that they overflow their lobes, giving rise to a common atmospheric envelope. Such a system is called an **overcontact binary** (Figure 11-25d).

Semidetached and contact binaries are easiest to detect if they are also eclipsing binaries (recall Figure 10-11). Their light curves have a distinctly rounded appearance caused by these tidally distorted egg-shaped stars. The eclipsing binary called β Persei, or Algol (from an Arabic term for "demon"), is a semidetached binary that can easily be seen with the naked eye in the constellation Perseus. From β Persei's light curve (Figure 11-26a), astronomers have determined that the binary contains a star that fills its Roche lobe. Sometime in the past, as it expanded and became a giant, this star dumped a significant amount of gas onto its companion.

Mass transfer is still occurring in a semidetached eclipsing binary called β Lyrae in the constellation Lyra. Like β Persei, β Lyrae contains a giant that fills its Roche lobe (Figure 11-26b). For many years astronomers were puzzled by the fact that the detached companion star in β Lyrae is severely underluminous, contributing virtually no light at all to the visible radiation coming from the system. Furthermore, the spectrum of β Lyrae contains unusual features, some of which are consistent with gas flowing between the stars and around the system as a whole.

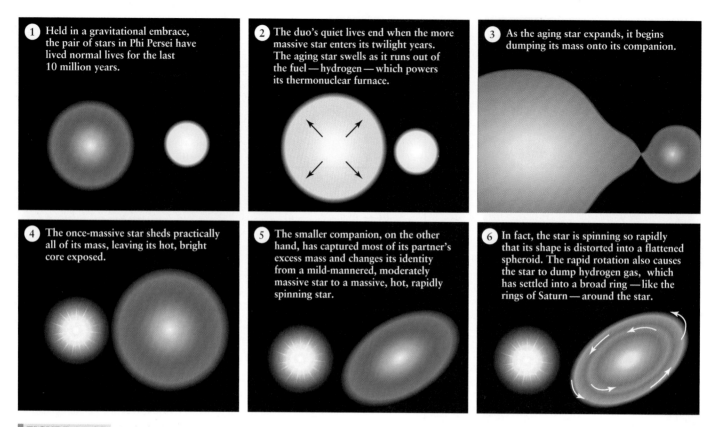

1 Held in a gravitational embrace, the pair of stars in Phi Persei have lived normal lives for the last 10 million years.

2 The duo's quiet lives end when the more massive star enters its twilight years. The aging star swells as it runs out of the fuel — hydrogen — which powers its thermonuclear furnace.

3 As the aging star expands, it begins dumping its mass onto its companion.

4 The once-massive star sheds practically all of its mass, leaving its hot, bright core exposed.

5 The smaller companion, on the other hand, has captured most of its partner's excess mass and changes its identity from a mild-mannered, moderately massive star to a massive, hot, rapidly spinning star.

6 In fact, the star is spinning so rapidly that its shape is distorted into a flattened spheroid. The rapid rotation also causes the star to dump hydrogen gas, which has settled into a broad ring — like the rings of Saturn — around the star.

FIGURE 11-27 **Mass Exchange between Close Binary Stars** This sequence of drawings shows how close binary stars can initially be isolated but, as they age, grow and exchange mass. Such mass exchange leads to different fates than if the same stars had evolved in isolation.

The β Lyrae system was explained in 1963, when Su-Shu Huang proposed that the underluminous star in β Lyrae is enveloped in a huge **accretion disk** of gas captured from its bloated companion. The disk is so large and thick that it completely shrouds the secondary star, making it impossible to observe at visible wavelengths. The primary star is overflowing its Roche lobe, with gases streaming onto the disk at the rate of 1 $M_\odot$ per hundred thousand years.

The fate of a semidetached system like β Persei or β Lyrae depends primarily on how fast its stars evolve. If the detached star expands to fill its Roche lobe while the companion star fills its own Roche lobe, then the result is an overcontact binary. An example is W Ursae Majoris, in which two stars share the same photosphere (Figure 11-26c).

Theories of the evolution of binary systems that transfer mass reveal that some remarkable transformations are possible. This is seen in the case of φ (phi) Persei. As shown in Figure 11-27, the more massive star in a binary can actually be transformed into the lower-mass star and vice versa! When stars in such systems gain mass, their increased gravitational force increases their rate of fusion, which increases the rate at which they evolve. Conversely, the stars that lose mass begin evolving more slowly than they were before their mass loss. In Chapters 12 and 13 we will see that mass transfer produces some of the most extraordinary objects in the sky.

11-16 Frontiers yet to be discovered

Stars, whether individually, in pairs, or in much larger groups, are our primary source of information about the universe. Details of each step of stellar evolution remain to be explored: the pre–main-sequence evolution of stars is being revealed in sources like the Eagle Nebula (Figure 11-12) as the pre–main-sequence stars there are being revealed to our telescopes (recall that they are usually surrounded by dust). These observations provide data with which to compare the theories of early star formation. Over the past two decades, observations have revealed that more stars lose mass while on the main sequence than previously expected. Consequently, we have a lot to learn about why such stars shed mass.

As noted, we still need to understand the details of why stars swell into the red giant phase. Also, the mechanisms that cause stars to be variable are still under investigation.

Further Reading on These Topics

WHAT DID YOU KNOW?

|1   *How do stars form?* Stars form from the mutual gravitational attraction between gas and dust inside giant molecular clouds.

|2   *Are stars forming today?* Yes. Astronomers have seen stars that have just arrived on the main sequence, as well as infrared images of gas and dust clouds in the process of forming stars.

|3   *Do stars with greater or lesser mass shine longer?* Lower-mass stars last longer because the lower gravitational force inside them causes fusion to take place at slower rates compared to the fusion inside higher-mass stars.

KEY WORDS

accretion disk, 320
birth line, 305
brown dwarf, 305
Cepheid variable, 317
contact binary, 319
core helium fusion, 312
dense core, 303
detached binary, 319
electron degeneracy pressure, 313
emission nebula, 307
evolutionary track, 305
giant molecular cloud, 301
globular cluster, 314
H II region, 307
helium flash, 313

horizontal branch star, 314
instability strip, 316
interstellar extinction, 300
interstellar medium, 299
interstellar reddening, 300
Jeans instability, 303
nebulae (*singular* nebula), 302
OB association, 307
open cluster, 303
overcontact binary, 319
Pauli exclusion principle, 313
period–luminosity relation, 317
Population I star, 316
Population II star, 316
pre–main-sequence star, 304

protostar, 303
red dwarf, 311
reflection nebula, 300
Roche lobe, 318
RR Lyrae variable, 317
semidetached binary, 319
shell hydrogen fusion, 311
stellar evolution, 299
supernova remnant, 302
T Tauri stars, 309
Type I Cepheid, 317
Type II Cepheid, 317
variable stars, 317
zero-age main sequence (ZAMS), 310

KEY IDEAS

Protostars and Pre–Main-Sequence Stars

• Enormous, cold clouds of gas and dust, called giant molecular clouds, are scattered about the disk of the Galaxy.

• Star formation begins when gravitational attraction causes clumps of gas and dust called protostars to coalesce within a giant molecular cloud. As a protostar contracts, its matter begins to glow. When the contraction slows down, the protostar becomes a pre–main-sequence star. When the pre–main-sequence star's core temperature becomes high enough to begin hydrogen fusion and stop contracting, it becomes a main-sequence star.

• The most massive pre–main-sequence stars take the shortest time to become main-sequence stars (O and B stars). They emit strong ultraviolet radiation that ionizes hydrogen in the surrounding molecular cloud, creating reddish emission nebulae called H II regions.

• In the final stages of pre–main-sequence contraction, when hydrogen fusion is about to begin in the core, the pre–main-sequence star may undergo vigorous

chromospheric activity that ejects large amounts of matter into space. Such gas-ejecting stars are called T Tauri stars.

• Ultraviolet radiation and stellar winds from the OB association at the core of an H II region create shock waves that compress the gas cloud, triggering the formation of more protostars. Supernova explosions also compress gas clouds and trigger star formation.

• A collection of a few hundred or a few thousand newborn stars is called an open cluster. Stars escape from open clusters, most of which eventually dissipate.

Main-Sequence and Giant Stars

• The Sun has been a main-sequence star for 4.6 billion years and should remain so for about another 5 billion years. Less massive stars than the Sun evolve more slowly and have longer main-sequence lifetimes. More massive stars than the Sun evolve more rapidly and have shorter main-sequence lifetimes.

• Main-sequence stars with less than $0.4 M_\odot$ convert all of their mass into helium and then stop fusing. Their lifetimes are hundreds of billions of years and so none of these stars have yet left the main sequence.

• Core hydrogen fusion ceases when hydrogen is exhausted in the core of a main-sequence star with $M > 0.4 M_\odot$, leaving a core of nearly pure helium surrounded by a shell where hydrogen fusion continues. Shell hydrogen fusion adds more helium to the star's core, which contracts and becomes hotter. The outer atmosphere expands considerably, and the star becomes a giant.

• When the central temperature of a giant reaches about 100 million K, the thermonuclear process of helium fusion begins. This process converts helium to carbon, then to oxygen. In a massive giant, helium fusion begins gradually. In a less massive giant, it begins suddenly in a process called the helium flash.

• The age of a stellar cluster can be estimated by plotting its stars on an H-R diagram. The upper portion of the main sequence disappear first, because more massive main-sequence stars become giants before low mass stars do. Relatively young stars are metal-rich; ancient stars are metal-poor.

• Giants undergo extensive mass loss, sometimes producing shells of ejected material that surround the entire star.

Variable Stars

• When a star's evolutionary track carries it through a region called the instability strip in the H-R diagram, the star becomes unstable and begins to pulsate.

• RR Lyrae variables are low-mass, pulsating variables with short periods. Cepheid variables are high-mass, pulsating variables exhibiting a regular relationship between the period of pulsation and luminosity.

• Mass can be transferred from one star to another in close binary systems. When this occurs the evolution of the two stars changes.

REVIEW QUESTIONS

1   What is a giant molecular cloud, and what role do these clouds play in star formation?

2   Why are low temperatures necessary for dense cores to form and contract into protostars?

3   What is an H II region?

4   Why don't thermonuclear reactions occur on the surface of a main-sequence star?

5   What is an evolutionary track, and how can such tracks help us interpret the H-R diagram?

6   Why are most stars we see in the sky main-sequence stars?

7   Draw the pre–main-sequence evolutionary track of the Sun on an H-R diagram. Briefly describe what was occurring throughout the solar system at various stages along this track. (You may find reviewing Essentials II useful.)

8   On what grounds are astronomers able to say that the Sun has about 5 billion years remaining in its main-sequence stage?

9   What will happen inside the Sun 5 billion years from now when it begins to evolve into a giant?

10   How is the evolution of a main sequence star with less than $0.4\ M_\odot$ fundamentally different from that of a main-sequence star with more than $0.4 M_\odot$?

11   Draw the post–main-sequence evolutionary track of the Sun on an H-R diagram up to the point when the Sun becomes a helium-burning giant. Briefly describe what might occur throughout the solar system as the Sun undergoes this transition.

12   What does it mean when an astronomer says that a star "moves" from one place to another on an H-R diagram?

13   What is the helium flash and what causes it?

14   Explain how and why the turnoff point on the H-R diagram of a cluster is related to the cluster's age.

15   Why do astronomers believe that globular clusters are made of old stars?

16   What are Cepheid variables and how are they related to the instability strip?

17   Which are older stars: Type I or Type II Cepheids? Justify your answer.

18   What are RR Lyrae variables and how are they related to the instability strip?

19   What is a Roche lobe and what is its significance in close binary systems?

20   What are the differences between detached, semidetached, contact, and overcontact binaries?

ADVANCED QUESTIONS

The answers to all computational problems, which are preceded by an asterisk (*), appear at the end of the book.

21   How is a degenerate gas different from an ordinary gas?

22   If you took a spectrum of a reflection nebula, would you see absorption lines, emission lines, or no lines? Explain your answer.

23   Why is it useful to plot the apparent magnitudes of stars in a single cluster on an H-R diagram rather than their absolute magnitudes?

24   What observations would you make of a star to determine whether its primary source of energy was fusing hydrogen or helium?

25   What might happen to the massive outer planets when the Sun becomes a giant?

*26   How many 1.5-$M_\odot$ main-sequence stars would it take to equal the luminosity of one 15-$M_\odot$ star?

*27   How many times longer does a 1.5-$M_\odot$ star fuse hydrogen in its core than does a 15-$M_\odot$ star?

28   Why does a shock wave from a supernova produce relatively few high-mass O and B stars compared to the lower-mass A, F, G, K, and M stars?

29   How would you distinguish a newly formed protostar from a giant, given that they occupy the same location on the H-R diagram?

30   What observational consequences would we find in H-R diagrams for star clusters if the universe had a finite age? Could we use these consequences to establish constraints on the possible age of the universe? Explain.

DISCUSSION QUESTIONS

31   Discuss the possibility of life-forms and biological processes occurring in giant molecular clouds. In what ways might conditions favor or hinder biological evolution?

32   Is there any evidence that the Earth has ever passed through a star-forming region in space?

WHAT IF ...

33   The solar system passed through a giant molecular cloud? How would this encounter affect the Earth and life on it?

34   The Sun had already entered its giant phase? What would happen to the Earth and to life on it?

35   The Earth were orbiting a 0.5-$M_\odot$ star at a distance of 1 AU? What would be different for the Earth and life

on it? What effects would moving the Earth closer to the lower-mass Sun have?

36   The Sun were a variable star? How would this change life on Earth and, assuming we could live in orbit around such a star, how might it change our perspective of the cosmos?

WEB/CD-ROM QUESTIONS

 37   To test your understanding of where stars are formed, do Interactive Exercise 11-1 on the Web or CD-ROM. You can print out your results, if required.

38   To test your understanding of variable stars, do Interactive Exercise 11-2 on the Web or the CD-ROM. You can print out your results, if required.

 39   To test your understanding of close binary star systems, do Interactive Exercise 11-3 on the Web or the CD-ROM. You can print out your results, if required.

OBSERVING PROJECTS

 40 Use a telescope to observe at least two of the following interstellar gas clouds: M42 (Orion), M43, M20 (Trifid), M8 (Lagoon), M17 (Omega). You can easily locate them with the aid of your *Starry Night Backyard*™ program, star charts published in such magazines as *Astronomy* and *Sky and Telescope*, or the list of coordinates below. In each case, can you identify the stars responsible for the ionizing radiation that causes the nebula to glow? Draw a picture of what you see through the telescope and compare it with a photograph of the object. Which portions of the nebula are not visible through your telescope? Why are they not visible?

Nebula	Right ascension	Declination
M42 (Orion)	5^h 35.4^m	–5° 27'
M43	5^h 35.6^m	–5° 16'
M20 (Trifid)	18^h 02.6^m	–23° 02'
M8 (Lagoon)	18^h 03.8^m	–24° 23'
M17 (Omega)	18^h 20.8^m	–16° 11'

41 Several star clusters can be seen quite well with a good pair of binoculars. You can easily locate many of them with the aid of your *Starry Night Backyard*™ program, star charts published during the summer months in such magazines as *Astronomy* and *Sky & Telescope*, or the list of coordinates below. Observe as many of these clusters as you can. Look at them through a telescope, if available. Note the overall distribution of stars in each cluster. Can you see any of these clusters with the naked eye? What difference do you note between binocular and telescopic images of individual clusters?

Star cluster	Right ascension	Declination
M45 (Pleiades)	3^h 47.0^m	+24° 07'
Hyades	4^h 27.0^m	+16° 00'
Praesepe	8^h 40.1^m	+19° 59'
Coma	12^h 25.0^m	+26° 00'
M11	18^h 51.1^m	–06° 16'

 42 Use the *Starry Night Backyard*™ program to examine a star-forming region. First put the program into Atlas mode (*Go/Atlas*). Center the field on M20 (*Edit/Find/Trifid*). Zoom out to the maximum field of view. Set the timestep to 1 lunar month and step through the months of the year. (a) In which month is M20 highest in the sky at noon? Explain how you determined this. (b) In which month is M20 highest in the sky at midnight, so that it is best placed for observing with a telescope. Explain how you determined this.

THE EARTH ORBITED A 1.5M$_\odot$ SUN?

The Earth is at a perfect distance from a wonderful star. The Sun provides just enough heat so that liquid water, necessary to sustain life, can exist here. Would our planet still be suitable for the evolution of life if the Sun were 1.5 M$_\odot$ rather than 1.0 M$_\odot$? To begin with, we need to know how far from the new Sun, let's call it Sol II, to place the Earth.

A Very Sunny Day Sol II's surface temperature would be 8400 K and it would appear blue-white in our sky. Sol II would give off 7 times as much energy per second as our present Sun, due to the combination of a higher temperature and a 20% larger radius than the Sun. The effect of Sol II's increased energy emission would require that Earth be located much farther away from Sol II than our present distance from the Sun. To understand why, consider the increase in infrared (heat) output from Sol II. That extra heat would have the initial effect of raising the average global temperature on Earth at our present distance by about 10 K (about 20°F). This does not seem like a lot, but the impact of that slight increase would actually boost the atmospheric temperature much higher.

The extra heat from Sol II would cause more ocean water to evaporate into the atmosphere. Because water is a greenhouse gas (that is, it traps infrared radiation), the air temperature would rise, causing even more water to evaporate from the oceans, which, in turn, would cause the air to heat even more. This vicious cycle, called a runaway greenhouse effect, would make the Earth's surface so hot and dry that it would be uninhabitable.

By moving the Earth about 2.6 times farther away from Sol II, the temperature would become suitable for life. At that distance, the year would be 1249 days long. While moving away from the heat is one thing, moving away from the ultraviolet radiation emitted by Sol II is something else altogether.

Ultraviolet Excess Just by increasing the Sun's mass by 50%, the ultraviolet radiation emitted would be several thousand times stronger. This is because the energy output of stars with different surface temperatures varies with wavelength. While the output of visible light would change slightly, the output of ultraviolet radiation would be vastly greater. Therefore, even though the Earth's surface temperature would be suitable farther from Sol II, the flood of ultraviolet radiation would be so strong that the ozone layer would be overwhelmed, and the level of ultraviolet radiation at the Earth's surface would be much higher than it is today. This would be so even with the greater concentration of ozone created by the increased ultraviolet from Sol II.

Life would have to evolve greater protection from ultraviolet radiation than it has today. And if this were not a great enough challenge, suppose intelligent life-forms were evolving on the Earth orbiting Sol II 4.6 billion years after the solar system formed. They would discover that their star had evolved so rapidly that it was just about to expand into the giant phase!

(John Elk III/Bruce Coleman)

12 THE DEATHS OF STARS

IN THIS CHAPTER YOU WILL DISCOVER

- what happens to stars after helium fusion in their cores ceases

- how heavy elements are created

- the end of stellar evolution, when stars eject matter back into space

- that lower-mass stars expel their outer layers relatively gently

- that higher-mass stars explode violently

- how astronomers have observed neutron stars

Eta Carinae
This spectacular image is Eta Carinae, believed to be a binary star system, with each star having more than 70 $M_\odot$. This Hubble Space Telescope image details the remnants of an ejection that was first seen in 1841 and that temporarily made Eta Carinae the second brightest object in the night sky. Despite losing more than 1 $M_\odot$ in this event, the stars survived. Because of their high masses, these stars are expected to self-destruct as supernovae within the next few hundred million years. Eta Carinae is 2450 pc (8000 ly) from Earth. Among its other mysteries is the mechanism by which it has created the two lobes of expanding gas. We will encounter other hourglass-shaped distributions later in this chapter. (J. Morse, University of Colorado and NASA)

R I V U X G

WHAT DO YOU THINK?

1 Will the Sun someday cease to exist? If so, how?

2 What is a nova?

3 What are the origins of the carbon, silicon, oxygen, iron, uranium, and other heavy elements on Earth?

4 What are cosmic rays?

5 What is a pulsar?

Change is the underlying theme of stellar evolution. We have seen how gas and dust come together to form stars; how hydrogen fuses into helium; how post–main-sequence stars expand, contract, and expand again; and how helium transforms into carbon and oxygen. Does this spiral of changing elements and changing stellar properties continue indefinitely, or are there limits to what can be created inside stars? In this chapter and Chapter 13, we learn that there are indeed limits to stellar fusion processes, limits that are determined by stellar masses. Furthermore, we investigate the often spectacular finales to stellar evolution.

LOW-MASS STARS AND PLANETARY NEBULAE

In Chapter 11, we discovered what happens when shell hydrogen fusion first begins: The outpouring of energy causes a star to expand and become a giant. In this process, low-mass stars (under 8 $M_\odot$) move over to, and ascend, the giant branch on the H-R diagram for the first time (Figure 12-1a). This period is accompanied by the expulsion of mass into space in the form of stellar winds, which reduce the masses of these stars. Then comes helium fusion, with stars of less than 2 $M_\odot$ undergoing a core helium flash in the giant stage (Figure 12-1a). After fusion begins, these stars shrink and move onto the *horizontal branch* (Figure 12-1b). Eventually, the core is converted into carbon and oxygen. Helium fusion there ceases and these stars undergo another stage that closely parallels the end of core hydrogen fusion.

12-1 Low-mass stars expand into the supergiant phase before expanding into planetary nebulae

Calculations indicate that carbon and oxygen require a temperature of at least 600 million K to fuse. Because the core of a low-mass giant on the horizontal branch is only about 200 million K, fusion of these elements in the core does not occur. Photon production therefore drops off, and the inner regions of the star again contract, compressing and heating the shell of helium-rich gas just outside the core. As a result, **shell helium fusion** begins outside the core; this shell is itself surrounded by a hydrogen-fusing shell. All this takes place within a volume roughly the size of the Earth.

Once shell helium fusion commences, the new outpouring of energy pushes the outer envelope of the star out again.

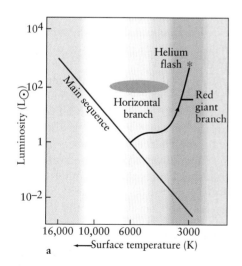
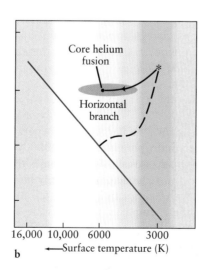
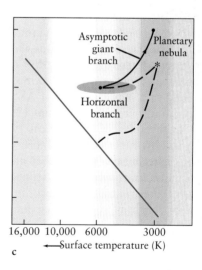

FIGURE 12-1 Post–Main-Sequence Evolution of Low-Mass Stars (a) The H-R track as a star makes the transition from the main sequence to the giant phase. The asterisk shows the helium flash occurring in a low-mass star. (b) After the helium flash, the star converts its helium core into carbon and oxygen. While doing so, its core reexpands, decreasing shell fusion. As a result, the star's outer layers recontract. (c) After the helium core is completely transformed into carbon and oxygen, the core recollapses, and the outer layers reexpand, powered up the asymptotic giant branch by shell helium fusion.

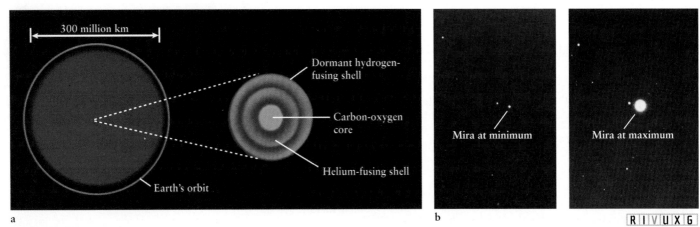

a

b   R I V U X G

■ FIGURE 12-2   The Structure of an Old Low-Mass Star
(a) Near the end of its life, a low-mass star like the Sun becomes a supergiant, with a diameter almost as large as the diameter of the Earth's orbit. The star's core, the dormant hydrogen-fusing shell, and an active helium-fusing shell are contained within a volume roughly the size of the Earth. (b) The variable star Mira, also called O (omicron) Ceti, changes brightness significantly with a roughly 332-day period. These variations occur because Mira pulsates and thus varies in both luminosity and surface temperature. (b: Lowell Observatory)

A low-mass star thus leaves the horizontal branch and ascends the giant branch for a second time. Powered by the fusion from two shells, it becomes brighter than ever before. Such **asymptotic giant branch stars,** or **AGB stars,** have diameters as big as the orbit of Mars and shine with the brightness of 10^4 $L_\odot$. These stars are now low-temperature, red supergiants (Figure 12-1c).

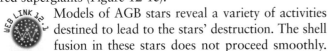

Models of AGB stars reveal a variety of activities destined to lead to the stars' destruction. The shell fusion in these stars does not proceed smoothly. Sometimes conditions of temperature and pressure are right for only one or the other shell to operate, so they switch on and off with increasing ferocity. Therefore, as these stars proceed up the asymptotic giant branch, they irregularly change brightness with periods ranging from about 100 to 700 days. These are called *Mira* or *long-period variables.* As the Sun moves along the asymptotic giant branch, it will expand to a radius of about 1 AU, enveloping the Earth in its tenuous outer layers (Figure 12-2).

Giant stars of all spectral types have strong stellar winds. Low-mass stars expel their outer layers more slowly than higher-mass stars, but they all reduce their masses significantly from what they were on the main sequence. During their initial ascent up the giant branch, stars can lose as much as 30% of their masses. On the asymptotic giant branch they lose even more, often surrounding themselves with thickening cocoons of gas and dust. A star of the Sun's mass loses about 10^{-5} $M_\odot$ per year at this stage, eventually dumping half of its mass back into space.

This mass loss limits the amount of gravity available to compress the star's core and regions of shell fusion. A fine balance is struck: The core of our low-mass AGB star is compressed until it again becomes degenerate, meaning that the electrons in the core provide a growing repulsive force that stops its contraction. However, even for the most massive of these low-mass stars (now much less than their original 8 $M_\odot$ due to their stellar winds), the core temperature cannot reach the 600 million K necessary to fuse carbon or oxygen. Therefore, no further core fusion occurs.

The final act in the drama of a low-mass star begins with a "thermal runaway" (meaning a rapid rise in temperature) in the helium shell, like the helium flash in its core earlier in its life. This occurs because the triple alpha process (see Section 11-10) is extremely sensitive to temperature, and as the temperature goes up slightly, the fusion rate skyrockets. The increase in energy output from the thin helium-fusing shell, called a **helium shell flash,** expands the star, thereby briefly decreasing the temperature in the shell and slowing the rate of fusion.

A star undergoes several helium shell flashes as its helium shell thickens. During each flash, the helium shell's energy output jumps a thousandfold. These brief outbursts are separated by relatively quiet intervals lasting about 100,000 years, during which the helium shell gradually becomes thicker. The energy from the increased fusion during the helium shell flashes causes the star's outer layers to expand and, therefore, cool further. Eventually, the outer gases are sufficiently cool so that electrons and ions there can recombine. This is exactly the opposite of ionization (discussed in Section 4-5). Whereas ionization requires an electron to absorb a photon, recombination forces an electron to emit a photon.

The pressure provided by the recombination photons, combined with the pressure provided by the photons from the helium flashes, generates enough energy to eject more and more of the star's outer layers into space. As the ejected material expands and cools, some of it condenses to form dust grains that are propelled outward by radiation pressure from the star's hot, burned-out core. The expanding dust and gases are called a **planetary nebula** (see Figure 12-1c). As the outer layers of the star are shed, an increasingly hot interior is revealed, and the star moves to the left across the H-R diagram (Figure 12-3). Including the mass lost prior to the planetary nebula phase, low-mass stars lose as much as 80% of their masses by shedding their outer layers.

> **Insight into Science** Names Don't Change As Understanding Improves Planetary nebulae have *nothing* to do with planets. This term was coined by the astronomer William Herschel in the eighteenth century. When viewed through the small telescopes of the day, these glowing objects, often green in appearance, looked like planets, such as Uranus.

Planetary nebulae come in a breathtaking variety of shapes (Figures 12-4 and 12-5). The Hourglass Nebula (Figure 12-5c) appears initially to have shed mass in a doughnut shape around itself. In the star's final death throes, this gas and dust forced the final outflow to go in two directions perpendicular to the plane of the doughnut, creating what is called a *bipolar planetary nebula*.

Our theories about why the gas and dust in planetary nebulae emerge in such a variety of patterns are incomplete. Evidence that many of the nonspherical planetary nebulae are created in binary star systems is growing. Astronomers

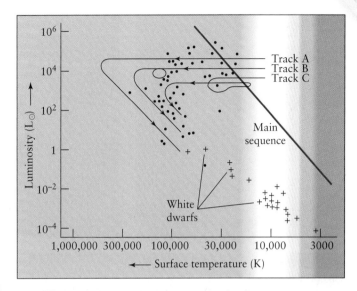

Evolutionary track	Mass ($M_\odot$)		
	Red supergiant	Ejected nebula	White dwarf
A	3.0	1.8	1.2
B	1.5	0.7	0.8
C	0.8	0.2	0.6

FIGURE 12-3 Evolution from Supergiants to White Dwarfs The evolutionary tracks of three low-mass, red supergiants are shown as they eject planetary nebulae. The table gives their masses as red supergiants, the amount of mass they lose as planetary nebulae, and their remaining (white dwarf) masses. The dots on this graph represent the central stars of planetary nebulae whose surface temperatures and luminosities have been determined. The crosses are white dwarfs for which similar data exist.

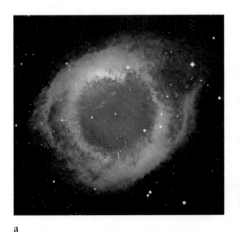

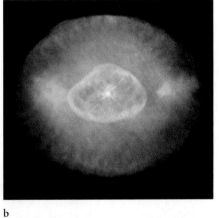

a   b   c

R I V U X G

FIGURE 12-4 Some Shapes of Planetary Nebulae The outer shells of dying low-mass stars are ejected in a wonderful variety of patterns. **(a)** NGC 7293, the Helix Nebula, is located in the constellation Aquarius. The star that ejected these gases is seen at the center of the glowing shell. This nebula, located about 215 pc (700 ly) from Earth, has an angular diameter equal to about half that of the full Moon. **(b)** NGC 6826 shows jets of gas (in red) whose origin is as yet unknown. **(c)** Mz 3 (Menzel 3), in the constellation Norma (the carpenter's level) is 900 pc (3000 ly) from Earth. The dying star, creating these bubbles of gas, may be part of a binary system. (a: Anglo-Australian Observatory; b: Howard Bons, STScI; Robin Ciardullo, Pennsylvania State University; NASA; c: AURA / STScI / NASA)

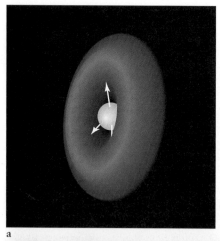

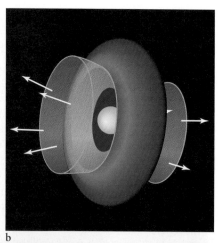

a   b

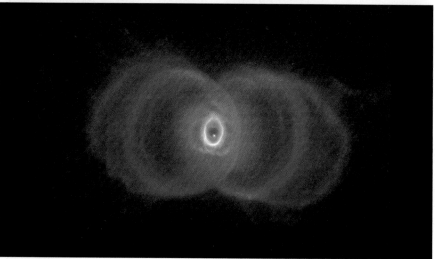

c

R I V U X G

FIGURE 12-5   Formation of a Bipolar Planetary Nebula Bipolar planetary nebulae may form in two steps. Astronomers hypothesize that **(a)** first, a doughnut-shaped cloud of gas and dust is emitted, **(b,)** followed by outflow that is channeled by the original gas to squirt out perpendicular to the plane of the doughnut. **(c)** The Hourglass Nebula appears to be a textbook example of such a system. The bright ring is believed to be the doughnut-shaped region of gas, lit by energy from the planetary nebula. The Hourglass is located about 2500 pc (8000 ly) from Earth. (c: R. Sahai and J. Trauer, JPL, the WFPC-2 Science Team, and NASA)

believe that the companion star, along with stellar rotation and stellar magnetic fields, propels the outgoing mass into the patterns we see.

Planetary nebulae are quite common in our Galaxy. More than 1000 have been identified, and astronomers estimate that 20,000 to 100,000 exist in the Milky Way. Indeed, as we shall see shortly, such will be the fate of our Sun.

Spectroscopic observations of planetary nebulae show bright emission lines of hydrogen, carbon, neon, magnesium, oxygen, and nitrogen. From the Doppler shifts of these lines, we conclude that the expanding shell of gas is moving outward from the dying low-mass star at speeds of 10 to 30 km/s. A typical planetary nebula has a diameter of roughly 1 ly, which means that it began expanding about 10,000 years ago.

By astronomical standards, a planetary nebula is short-lived. After about 50,000 years, the nebula spreads over distances so far from the cooling central star that its nebulosity simply fades from view. The gases then mingle and mix with the surrounding interstellar medium, enriching it with mod-

erately heavy elements, such as carbon. Astronomers estimate that planetary nebulae return a total of 5 $M_\odot$ to the disk of our Galaxy each year. This amounts to about 15% of all matter expelled by all types of stars each year. This material goes into the formation of new, metal-rich Population I stars and associated planets, and so planetary nebulae play an important role in the chemical and physical evolution of the Galaxy.

12-2 The burned-out core of a low-mass star becomes a white dwarf

Our models predict that main-sequence stars less massive than about $8M_\odot$ never develop the central pressures or temperatures necessary to ignite thermonuclear reactions of the carbon or oxygen in their cores. Instead, as we have seen, the process of mass ejection strips away the stars' outer layers and exposes the carbon-oxygen cores, which simply cool off. Such burned-out hulks are called **white dwarfs.** The matter in these stars is degenerate (see section 11-10). This is to be the fate of the Sun's core.

Theoretical models predict that degenerate matter, such as the electrons in a white dwarf, can withstand only a finite amount of pressure, above which that matter cannot exert enough outward force to prevent the core from collapsing further. This limits the mass of a white dwarf to less than 1.4 $M_\odot$. This mass is called the **Chandrasekhar limit,** after the astrophysicist Subrahmanyan Chandrasekhar, who received a Nobel Prize for his pioneering theoretical studies of white dwarfs in the 1940s.

WEB LINK 12.4 Theoretical evolutionary tracks of three burned-out stellar cores are shown in Figure 12-3. These particular white dwarfs evolved from main-sequence stars of between 0.8 and 3.0 $M_\odot$. During the ejection phase, the appearance of these stars changes rapidly. These objects appear to race along their evolutionary tracks on the H-R diagram, sometimes executing loops corresponding to thermal pulses (tracks B and C in Figure 12-3). Finally, as the ejected planetary nebulae fade and the stellar cores cool, the evolutionary tracks of the dying stars take a sharp turn downward toward the white dwarf region of the diagram.

As noted above, the temperature inside a white dwarf is too low to ignite the carbon or oxygen. Electron degeneracy pressure (discussed in Section 11-10) prevents the crushing weight of the carbon and oxygen from contracting the remnant core further, thereby creating a stable white dwarf roughly the size of the Earth. The density of matter in one of these stellar "corpses" is typically 10^9 kg/m^3. In other words, a teaspoonful of white dwarf matter brought to Earth would weigh 5 tons.

As billions of years pass, an isolated white dwarf radiates its stored energy into space, cooling and decreasing in luminosity. It moves down and to the right on the H-R diagram. The Sun will become a cold, dark, dense sphere of degenerate gases rich in oxygen and carbon, about the size of the Earth. We can actually follow the theory a little further. It predicts that after billions of years of radiating away their heat, the interior temperatures of white dwarfs will decrease to about 4000 K, and the carbon and oxygen in them will solidify, transforming them into giant crystals.

Many white dwarfs are found in our solar neighborhood, but all are too faint to be seen with the naked eye. The first white dwarf to be discovered is a companion to the bright star Sirius. The binary nature of Sirius was first deduced in 1844 by the German astronomer Friedrich Bessel, who noticed that the star was moving back and forth, as if orbited by an unseen object. This companion, called Sirius B, was first glimpsed in 1862 (Figure 12-6). Recent satellite observations at ultraviolet wavelengths—where white dwarfs emit most of their light—demonstrate that the surface temperature of Sirius B is about 30,000 K. Recalling our discussion of binary star systems in Section 11-15, the fact that Sirius B is already a white dwarf while its companion is not means that Sirius B must have originally been more massive than Sirius A. Sirius A is the more massive of the two today. Because most stars occur in binary star systems, it is important to take into account any mass transfer that occurs when determining the evolution of these stars.

12-3 White dwarfs in close binary systems can create powerful explosions

Occasionally, a star in the sky suddenly becomes between ten thousand and a million times brighter. This phenomenon is called a **nova.** Novae are fairly common, with about 20 estimated to occur in our Galaxy each year. We do not see all of them because of the gas and dust that obscures many parts of the Galaxy. Astronomers typically observe two or three novae in our Galaxy each year. Their abrupt rise in brightness is followed by a gradual decline that may stretch over several months or more (Figures 12-7 and 12-8).

Painstaking observations strongly suggest that novae occur in close binary systems containing a white dwarf. The ordinary companion star presumably fills its Roche lobe, so it gradually deposits fresh hydrogen onto the white dwarf. This new mass becomes a dense layer covering the hot surface of the white dwarf. As more gas is deposited and compressed, the temperature in the hydrogen layer continues to increase. Finally, at about 10^7 K, hydrogen fusion ignites throughout the layer, blowing it into interstellar space. This explosion is the nova. After a nova, fusion ceases on the white dwarf. The companion

R I V U X G

FIGURE 12-6   **Sirius and Its White Dwarf Companion** Sirius, the brightest-appearing star in the night sky, is actually a double star. The secondary star is a white dwarf, seen here at the five o'clock position in the glare of Sirius. The spikes and rays around Sirius are created by optical effects within the telescope. (R. B. Minton)

12

a   b   R I V U X G

star, however, may retain enough mass to supply a new layer of surface hydrogen, enabling some novae to reoccur.

Recent visible observations by the Hubble Space Telescope and X-ray observations by *Chandra* reveal that novae are much more complex than astronomers had thought. Rather than smooth shells of expanding gas, they are composed of thousands of clumps of gas. These observations are stimulating astrophysicists to reexamine the nova mechanism described above to try and explain these newly observed effects.

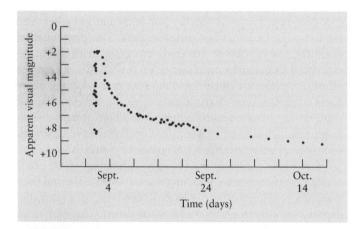

FIGURE 12-8   The Light Curve of a Nova This graph shows the history of Nova Cygni 1975, a nova that was observed to blaze forth in the constellation of Cygnus in September 1975. The rapid rise in magnitude followed by a gradual decline is characteristic of many novae, while some oscillate in intensity as they become dimmer.

HIGH-MASS STARS AND SUPERNOVAE

In high-mass main-sequence stars (greater than 8 $M_\odot$) sufficient gravitational force compresses the core to heat it enough to enable the carbon and oxygen there to fuse. This leads to the creation of a host of other elements that are eventually ejected into interstellar space.

12-4 A series of fusion reactions in high-mass stars leads to luminous supergiants

When helium fusion ends in the core of a massive star, gravitational compression again collapses the core, driving the star's central temperature beyond 600 million K. Now, carbon fusion begins, producing such elements as neon and magnesium. When a star compresses itself enough to drive its central temperature to 1.2 billion K, neon fusion occurs. When the central temperature of the star then reaches 1.5 billion K, oxygen fusion begins. As the star consumes increasingly heavier nuclei, it produces sulfur and isotopes of silicon and phosphorus, among other elements.

As the core temperature increases, its fusion rate soars. Also, each succeeding core stage has fewer and fewer atoms than the previous one. For example, because each helium atom is made up of four hydrogen atoms, there are only one quarter as many helium atoms in a star's core than there were hydrogen atoms. Combining these two effects reveals that each cycle of core thermonuclear fusion occurs more rapidly than the last. For example, detailed calculations for a

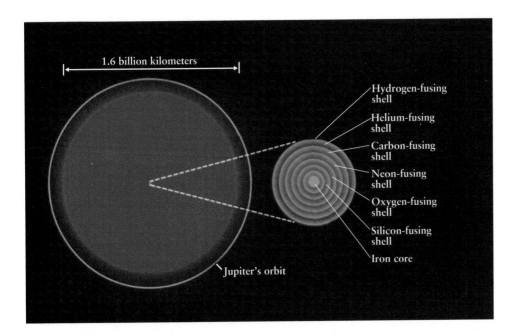

1.6 billion kilometers

Hydrogen-fusing shell
Helium-fusing shell
Carbon-fusing shell
Neon-fusing shell
Oxygen-fusing shell
Silicon-fusing shell
Iron core

Jupiter's orbit

FIGURE 12-9   The Structure of an Old High-Mass Star Near the end of its life, a high-mass star becomes a supergiant with a diameter almost as wide as the orbit of Jupiter. The star's energy comes from six concentric fusing shells, all contained within a volume roughly the same size as the Earth.

star that was initially 25 $M_\odot$ on the main sequence demonstrate that carbon fusion occurs for 600 years, neon fusion for 1 year, and oxygen fusion for only 6 months. After half a year of core oxygen fusion, gravitational compression forces the central temperature up to 2.7 billion K and silicon fusion begins. This thermonuclear process proceeds so furiously that the entire core supply of silicon in a 25-$M_\odot$ star is used up in 1 day.

Each stage of fusion adds a new shell of matter outside the core (Figure 12-9), creating a structure resembling the layers of an enormous onion. Together, the tremendous numbers of fusion-generated photons created in all these shells, as well as in the core, push the outer layers of the star outward. It expands to become a luminous supergiant almost as wide across as Jupiter's orbit around the Sun.

Luminous supergiant stars, which are brighter than 10^5 $L_\odot$, emit winds throughout most of their existence, with mass-loss rates exceeding those of giants (see Chapter 11). Figure 11-11 shows a supergiant star losing mass. Betelgeuse (see the figure that opens Essentials III), 470 ly away in the constellation of Orion, is another good example of a supergiant experiencing mass loss. Spectroscopic observations show that Betelgeuse is losing mass at the rate of 1.7×10^{-7} $M_\odot$ per year and is surrounded by ejected gas, its *circumstellar shell*. This huge shell is expanding at 10 km/s, and escaping gases have been detected at distances of 10,000 AU from the star. The expanding circumstellar shell has an overall diameter of 1/3 ly.

Silicon fusion in high-mass stars involves many hundreds of nuclear reactions, but its major final product is iron. Eventually, the core is converted entirely into iron, which is surrounded by layers of shell fusion that consume the star's remaining reserves of fuel (see Figure 12-9). Whereas the star's enormously bloated atmosphere is nearly as big as the orbit of Jupiter, its entire energy-producing region is again contained in a volume the size of the Earth. The buildup of an inert, iron-rich core signals the impending violent death of a massive star.

12-5 High-mass stars violently blow apart in supernova explosions

Unlike lighter elements, iron cannot fuel further thermonuclear reactions. The protons and neutrons inside iron nuclei are already so tightly bound together that no further energy can be extracted by fusing still more nuclei together. The sequence of fusion stages in the cores of high-mass stars therefore ends.

Because iron atoms do not fuse and emit energy, the electrons in the core must now support the star's outer layers by the strength of electron degeneracy pressure alone. Soon, however, the continued deposition of fresh iron from the silicon-fusing shell causes the core's mass to exceed the Chandrasekhar limit. Electron degeneracy suddenly fails to support the star's enormous weight, and the core collapses.

Any isolated main-sequence star of greater than 8 $M_\odot$ develops an iron core as a luminous supergiant. When the core collapses, a rapid series of cataclysms is triggered that tears the star apart in a few seconds. Let us see how this happens in the death of a 25-$M_\odot$ star according to computer simulations (Table 12-1).

In a 25-$M_\odot$ star, electron degeneracy pressure fails when the density is sufficiently high inside the iron core. The core, only some 3000 km in diameter, then collapses immediately.

TABLE 12-1   Evolutionary Stages of a 25M$_\odot$ Star

Stage	Central temperature (K)	Central density (kg/m^3)	Duration of stage
Hydrogen fusion	4×10^7	5×10^3	7×10^6 yr
Helium fusion	2×10^8	7×10^5	5×10^5 yr
Carbon fusion	6×10^8	2×10^8	600 yr
Neon fusion	1.2×10^9	4×10^9	1 yr
Oxygen fusion	1.5×10^9	1×10^{10}	6 mo
Silicon fusion	2.7×10^9	3×10^{10}	1 d
Core collapse	5.4×10^9	3×10^{12}	0.2 s
Core bounce	2.3×10^{10}	4×10^{17}	milliseconds
Supernova explosion	about 10^9	varies	10 s

In roughly $1/10$ second, the central temperature exceeds 5 billion K. Gamma-ray photons associated with this intense heat have so much energy that they begin to break apart the iron nuclei, a process called **photodisintegration**. Although millions of years passed from the time the star arrived on the main sequence with a hydrogen- and helium-filled core until its core became iron, it takes less than a second to convert the core back into elemental protons, neutrons, and electrons.

Within another $1/10$ second, as the density continues to climb, the electrons in the core are forced to combine with protons there to produce neutrons, and the process releases a flood of neutrinos. As we saw in Chapter 9, neutrinos have no electric charge, very little mass, and most pass through the Earth or Sun without interaction. Nevertheless, the matter deep inside a collapsing high-mass star is so fantastically dense that the newly created neutrinos there cannot immediately escape from the star's core.

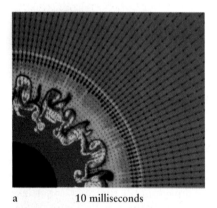

a   10 milliseconds

b   20 milliseconds

FIGURE 12-10   Supernovae Proceed Irregularly Images **(a)** and **(b)** are computer simulations showing how chaotic the supernova is deep inside the star as it begins to explode. This helps account for the globs of iron and other heavy elements emitted from deep inside, as well as the lopsided distribution of all elements in the supernova remnant, as shown in **(c)**, **(d)**, and **(e)**. The latter three are X-ray images of a supernova remnant taken by *Chandra* at different wavelengths. (a and b: Courtesy of Adam Burrows, University of Arizona and Bruce Fryxell, NASA/GSFC; c, d, and e: U. Hwang et al., NASA/GSFC)

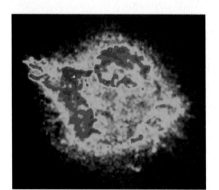

c   Silicon

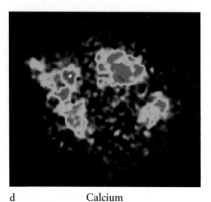

d   Calcium

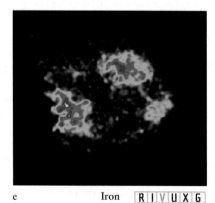

e   Iron   R I V U X G

According to the calculations, about ¼ second after the collapse begins, the density of the entire core reaches 4×10^{17} kg/m³, which is *nuclear density,* the density at which neutrons and protons are normally packed together inside atomic nuclei. (Compare this with the density of water, 10^3 kg/m³.) At nuclear density, matter is virtually incompressible. Thus, when the neutron-rich material of the core reaches nuclear density, it suddenly stiffens. The collapse of the core halts so abruptly that it rebounds and begins rushing back out in a process called *core bounce.*

The model predicts that during this critical stage, the star's unsupported layers of shell-fusing matter are plunging inward at up to 15% of the speed of light. The outward-flowing neutrinos and rebounding core slam into this matter, causing it to reverse course. In just a fraction of a second, a tremendous volume of matter begins to move back up toward the star's surface. This matter accelerates rapidly as it encounters less and less resistance, and soon it forms an outgoing shock wave. After a few hours, this shock wave reaches the star's surface, lifting the star's outer layers away from the core in a mighty blast. The star becomes a **supernova.**

As the supernova expands, the star's luminosity suddenly increases by a factor of 10^8. The energy emitted by a supernova is truly staggering—as much as all 200 billion stars in the Milky Way Galaxy combined! In other words, for a few days following the explosion, a supernova may shine as brightly as an entire galaxy. Over its lifetime, our model 25-$M_\odot$ star ejects more than 20 $M_\odot$ of its mass back into space via stellar winds, gas ejected during unstable periods, and during its supernova phase. Less massive stars return proportionately less mass to the interstellar medium.

As the layers of a massive dying star are blasted into space, they are compressed so much by the neutrinos and the shock wave that fusion actually occurs during the explosion, creating a broad assortment of elements. Indeed, the products of shell fusion and this final burst of fusion during supernovae are where much of the elements on the Earth and other terrestrial planets come from, including many of the atoms in our bodies. Computer simulations indicate that the onion skins of different elements do not just move out of a star going supernova in lockstep (Figures 12-10a, b). Therefore, the ejected gases are not expected to be uniform shells of matter. Indeed, observations reveal that the material emerging in a supernova comes out quite irregularly. Figures 12-10c, d, and e shows the distributions of some of the elements in the supernova Cassiopeia A (Cas A). Other observations show that some of the dense elements, like iron, ejected in supernovae are actually created deep inside their innermost silicon and oxygen layers and then shot out like bullets or geysers. Other elements are created throughout the expanding star. However, calculations and observations reveal that supernovae do not create *enough* of the heaviest elements, like gold, silver, and platinum. We will explore their possible origin in Section 12-11.

> **Insight into Science**   **The Process of Science** The best scientific theories and models also provide explanations about matters that they were not initially designed to study. In modeling stellar evolution and supernovae, the processes that create and eject metals (that is, elements heavier than hydrogen and helium) from stars also explain how the elements necessary for life are both created and made available.

12-6 Remnants of supernova explosions can be detected for millennia afterward

Astronomers find the debris of supernova explosions scattered across the sky. A beautiful example is the Cygnus Loop (see Figure 11-6). The doomed star's outer layers were blasted into space 20,000 years ago so violently that they are still traveling at supersonic speeds. As this expanding shell of gas plows through the interstellar medium, it collides with atoms and molecules, making the gases glow.

 Many supernova remnants cover sizable fractions of the sky. The largest is the Gum Nebula, with an angular diameter of 60° (Figure 12-11). This nebula

R I V U X G

FIGURE 12-11   **The Gum Nebula** The Gum Nebula is the largest known supernova remnant, spanning 60° across the sky, centered roughly on the southern constellation of Vela. The nearest portions of this expanding nebula are only 300 ly from the Earth. The supernova explosion occurred about 11,000 years ago, and its remnant now has a diameter of about 2300 ly. Only the central regions of the nebula are shown here. (Royal Observatory, Edinburgh)

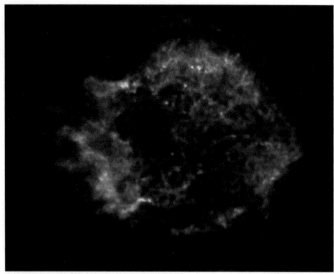

a

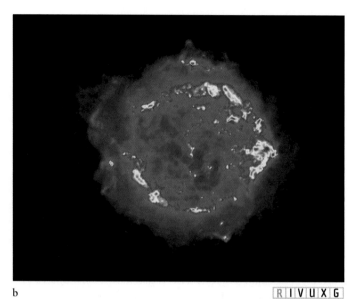

b   R I V U X G

FIGURE 12-12 Cassiopeia A Supernova remnants such as Cassiopeia A are typically strong sources of X rays and radio waves. **(a)** An X-ray picture of Cassiopeia A taken from the *Chandra* satellite, a European Space Agency mission. **(b)** A corresponding radio image produced by the Very Large Array (VLA). Radiation from the supernova that produced this nebula first reached Earth 300 years ago. The explosion occurred about 10,000 ly from here. (a: Rutgers/J. Hughes/CXC/SAO/NASA; b: Very Large Array)

looks so big because it is so close: Its near side is only about 300 ly from Earth. Studies of the Gum Nebula's expansion rate suggest that the supernova exploded about 11,000 years ago. At maximum brilliance, the exploding star probably was as bright as the Moon at first quarter.

Many supernova remnants can be detected only at nonvisible wavelengths, ranging from X rays through radio waves. For example, Figure 12-12 shows images of the supernova remnant Cassiopeia A as seen in X-ray (Figure 12-12a) and radio (Figure 12-12b). Visible-light photographs of this part of the sky reveal only a few small, faint wisps. Thus, radio searches for supernova remnants are more fruitful than visual searches. Only two dozen supernova remnants have been found in photographs, but more than 100 remnants have been discovered by radio astronomers.

From the expansion rate of the nebulosity in Cassiopeia A, astronomers calculate that the light from the supernova explosion first reached Earth about 300 years ago. Although telescopes were in wide use by the late 1600s, no record of the event is known. In fact, the last supernova of a star near its maximum brightness in our Galaxy was seen by Johannes Kepler in 1604. In 1572, Tycho Brahe also recorded the sudden appearance of an exceptionally bright star in the sky. To find previous accounts of supernova explosions, we must delve into astronomical records that are almost a thousand years old.

Besides the gas and dust enriched with heavy elements, supernovae may leave a heritage long after their clouds have dissipated. Astronomers have detected particles flying through space at speeds exceeding 90% the speed of light. These high-speed particles are the **cosmic rays.** *Cosmic rays are not rays at all!* They are primarily hydrogen nuclei, along with nuclei of more massive elements like helium and carbon, and a few percent of electrons and positrons (positively charged electrons).

 Origins of Cosmic Rays

The origins of cosmic rays are just beginning to become clear. At first many astronomers thought that these particles were emitted by supernovae. This made sense because these explosions are the best-known source of the energy necessary to give these particles their tremendous speeds. However, observations by the *Advanced Composition Explorer* satellite reveal that the isotopes of nickel coming to Earth in the form of cosmic rays were not the isotopes emitted by supernovae. Instead, evidence mounts that most cosmic rays are created when supernovae slam into gas already in space, causing some of this preexisting matter to accelerate and become cosmic rays.

Most cosmic rays coming in our direction collide with gas in the atmosphere some 15 km above the Earth's surface. This is fortunate because each cosmic ray packs an enormous amount of energy that potentially could harm living tissue. The collision in the air divides a cosmic ray's energy among several gas particles, which are shoved earthward. These, in turn, often hit other gas atoms, creating a cascade of lower-energy particles that eventually reach the Earth's surface as a **cosmic ray shower.** These cosmic rays created from atoms in our atmosphere are called **secondary cosmic rays.**

12-7 Supernova 1987A offered a detailed look at a massive star's death

Supernovae are often seen in remote galaxies with the aid of a telescope. In 1885 a supernova in the Andromeda Galaxy was just barely visible to the unaided eye. On February 23, 1987, a supernova was observed in the Large Magellanic Cloud, a galaxy near our own Milky Way. The supernova, designated SN 1987A, was the first to be observed that year (Figure 12-13), and it was so bright that it could easily be seen with the naked eye. What made SN 1987A such an exciting discovery was that it gave astronomers a rare opportunity to study the death of a nearby massive star using modern equipment and thereby provide data with which to test the theory of supernovae described above.

At first, SN 1987A reached only a tenth of the luminosity predicted for an exploding massive star. For the next 85 days, it gradually brightened before slowly dimming, as is characteristic of an ordinary supernova. Fortunately, the doomed star had been observed before it became a supernova. The Large Magellanic Cloud is about 160,000 ly from Earth—near enough to us that many of its stars have been individually observed and catalogued. The doomed star had been identified as a B3 I supergiant. The theory of the evo-

R I V U X G

FIGURE 12-13 **Supernova 1987A** A supernova was discovered in a nearby galaxy called the Large Magellanic Cloud (LMC) in 1987. This photograph shows a portion of the LMC that includes the supernova and a huge H II region called the Tarantula Nebula. At its maximum brightness, observers at southern latitudes saw the supernova without a telescope. (European Southern Observatory)

lution of such stars provided an explanation for SN 1987A's unusually slow brightening.

When this star was on the main sequence, its mass was about 20 $M_\odot$, although by the time it exploded it had shed many solar masses. The evolutionary track for an aging 20-$M_\odot$ star wanders back and forth across the top of the H-R diagram, so the star alternates between being a hot (blue) supergiant and cool (red) supergiant. The star's size changes significantly as its surface temperature changes. A blue supergiant is only 10 times larger in diameter than the Sun, but a red supergiant of the same luminosity is 1000 times larger. Because the doomed star was a relatively small blue supergiant when it exploded, it reached only a tenth of the brightness that it would have attained had it been a red supergiant at that time.

Ordinary telescopic observations of a supernova explosion can only show us the expanding outer layers of the dying star. Even X-ray or radio observations fail to see through the hot gases being blasted into space. Thus, using ordinary techniques, we cannot observe the extraordinary events occurring in and around the doomed star's core. However, most of the energy of a supernova explosion is carried away in a characteristic form: Recall that neutrinos are produced in great profusion as the star's core collapses. By detecting these neutrinos and measuring their properties, astronomers can learn many details about the star's collapsing core, especially about core bounce. Several neutrino detectors were operating when the neutrinos from SN 1987A reached Earth.

Nearly a day before SN 1987A was first observed in the sky, teams of scientists at neutrino detectors in Japan and the United States reported finding Cerenkov flashes (see Section 9-9) from a burst of neutrinos. The Kamiokande II detector (see Figure 9-23) in Japan detected 12 neutrinos at about the same time that 8 were found by the IMB (Irvine-Michigan-Brookhaven) detector in a salt mine under Lake Erie. Neutrinos preceded the visible supernova outburst because they escaped from the dying star before the shock wave from the collapsing core reached the star's surface. They were detected in the Earth's northern hemisphere, where the supernova is always below the horizon, after having passed through the Earth. The discovery of these neutrinos provided strong support for the theory that supernovae are caused by the collapse of the star's core, its subsequent bounce, and the flow of neutrinos that help cause the supernova, as described in Section 12-5.

About three and a half years after SN 1987A's detonation was first seen from Earth, astronomers using the Hubble Space Telescope obtained a picture showing several rings of glowing gas around the exploded star (Figure 12-14). This gas had been emitted by the star 20,000 years before the supernova, when a hydrogen-rich stellar envelope was ejected by stellar winds from the doomed star, as discussed earlier in this chapter. This gas expanded

a   R I V U X G

FIGURE 12-14 Shells of Gas Around SN 1987A (a) Intense radiation from the supernova explosion caused three rings of gas surrounding SN 1987A to glow in this Hubble Space Telescope image. This gas was ejected from the star 20,000 years before it detonated. All three rings lie in parallel planes. The inner ring is about 1.3 ly across. The white spots are unrelated stars. (b) When the progenitor star of SN 1987A was still a red supergiant, a slowly moving wind from the star filled the surrounding space with a thin gas. When the

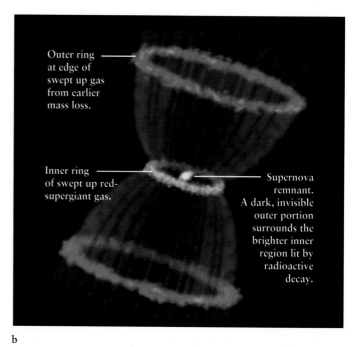

b

star contracted into a blue supergiant, it produced a faster-moving stellar wind. The interaction between the fast and slow winds somehow caused gases to pile up along an hourglass-shaped shell surrounding the star. The burst of ultraviolet radiation from the supernova ionized the gas in the rings, causing the rings to glow. The supernova itself, at the center of the hourglass, glows because of energy released from radioactive decay. (Robert P. Kirshner, Harvard-Smithsonian Center for Astronomy)

in an hourglass shape, because it was blocked from expanding around the star's equator either by a preexisting ring of gas there or by the orbit of an as-yet-unseen companion star.

While emitting these gases, the doomed star was a red supergiant. In its final blue giant phase, the same star emitted higher-speed gas that compressed the hourglass gas into three rings—a narrow one around the equator and wider ones at the top and bottom of the hourglass. This relic gas is now being illuminated by photons from the supernova. In 1998, astronomers began to observe gas ejected from the supernova striking the preexisting circumstellar gas. This collision has caused the ring of gas to brighten and it should eventually illuminate more of the earlier gas that had been ejected.

SN 1987A provided astronomers with invaluable information, in part because the doomed star had been studied before it exploded and its distance from Earth was known. The supernova was also located in an unobscured part of the sky, and neutrino detectors happened to be operating at the time of the outburst. Astronomers will be monitoring the progress of this supernova for years to come.

12-8 Accreting white dwarfs in close binary systems can also explode as supernovae

 Supernovae such as SN 1987A, which are the death throes of massive stars, are known as **Type II supernovae**, but not every supernova originated as a high-mass star. Observations reveal that others, known as **Type Ia supernovae**, begin as white dwarfs.

A supernova's spectrum indicates its type: Hydrogen lines are prominent in Type II supernovae but absent in Type Ia. Both types begin with a sudden rise in brightness (Figure 12-15). A Type Ia supernova typically reaches an absolute magnitude of −19 at peak brightness, while a Type II supernova usually peaks at −17. Type Ia supernovae then decline gradually for more than a year, whereas Type II supernovae alternate between periods of steep and gradual declines in brightness. Type II light curves therefore have a steplike appearance.

A Type Ia supernova begins with a high-mass carbon-oxygen–rich white dwarf in a close, semi-detached binary system. As we saw in Chapters 10 and 11, mass transfer can occur in a close binary if one star overflows its Roche lobe (recall Figures 11-25 and 11-27). To trigger a Type Ia supernova, a swollen giant companion star dumps gas onto a white dwarf. When the white dwarf's mass gets close to the Chandrasekhar limit, the increased core pressure created by the additional mass enables carbon fusion to begin. In a catastrophic runaway process reminiscent of the helium flash, the rate of carbon fusion skyrockets, and the star blows up.

A Type Ia supernova is powered by nuclear energy. What we see is simply the fallout from a gigantic thermonuclear explosion, which produces a wide array of radioactive isotopes. Especially abundant is an unstable isotope of nickel that decays into a radioactive isotope of cobalt. Most of the electromagnetic display of a Type Ia supernova, including the smooth decline of its light curve, results directly from the radioactive decay of this nickel and cobalt. The spectrum lacks hydrogen because the star that explodes is mostly carbon and oxygen to begin with.

Astronomers have seen more than 600 supernovae in distant galaxies. These observations suggest that in a typical galaxy like the Milky Way, Type Ia supernovae occur approximately once every 36 years, while Type II supernovae occur about once every 44 years. Thus, there should be about

five supernovae exploding in our Milky Way Galaxy each century. Where are they?

Vigorous stellar evolution in our Galaxy occurs primarily in the Galaxy's disk, where the giant molecular clouds are located. Our Galaxy's disk is therefore the place where massive stars are born and supernovae explode. However, this region of our Galaxy is so filled with interstellar gas and dust that we simply cannot see very far into it. Supernovae probably do erupt every few decades in remote parts of our Galaxy, but their detonations are hidden by interstellar debris.

NEUTRON STARS AND PULSARS

When main-sequence stars of between approximately 8 $M_\odot$ and 25 $M_\odot$ explode as supernovae, their remnant cores are highly compressed clumps of neutrons called **neutron stars.** Isolated neutron stars are stable stellar remnants with masses between 1.4 $M_\odot$ and about 3 $M_\odot$.

12-9 The cores of many Type II supernovae become neutron stars

The neutron was discovered during laboratory experiments in 1932. Within a year, two astronomers had predicted the existence of neutron stars. Inspired by the realization that white dwarfs are supported by electron degeneracy pressure, Fritz Zwicky and Walter Baade proposed that a highly compact ball of neutrons could similarly support a stellar corpse. They noted that the Pauli exclusion principle prevents neutrons with identical properties from packing too closely together. When neutrons are pressed together, many of them have to move rapidly (so as not to be identical to their neighbors). This motion provides a pressure, called **neutron degeneracy pressure**, that equations predict should be even greater than electron degeneracy pressure (see Chapter 11) and thus allows stellar remnants with masses beyond the Chandrasekhar limit to be stable. Zwicky and Baade wrote, "We advance the view that supernovae represent the transition from ordinary stars into neutron stars, which in their final stages consist of extremely closely packed neutrons."

Most scientists ignored Zwicky and Baade's theory for years. After all, a neutron star would have to be a rather unlikely object. To transform protons and electrons into neutrons, the density in the star would have to be equal to nuclear density, about 10^{17} kg/m^3. Thus, a teaspoon full of neutron star matter brought back to Earth would weigh 1 billion tons. Furthermore, an object compacted to nuclear density would be very small. A 2-$M_\odot$ neutron star would have a diameter of only 20 km, making it only as large as Saturn's smallest known moons. The surface gravity on one of these neutron stars would be so strong that the escape

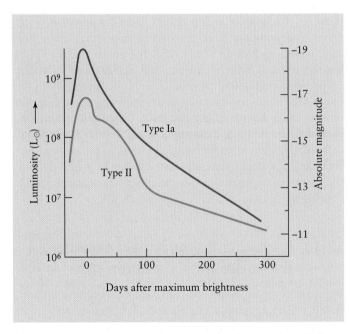

FIGURE 12-15   Supernova Light Curves A Type Ia supernova, which gradually declines in brightness, is caused by an exploding white dwarf in a close binary system. A Type II supernova is caused by the explosive death of a massive star and usually has alternating intervals of steep and gradual declines in brightness.

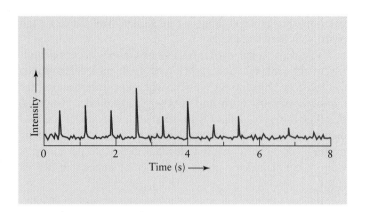

FIGURE 12-16 A Recording of a Pulsar This chart recording shows the intensity of radio emissions from one of the first pulsars to be discovered, PSR 0329+54. Note that some of the pulses are weak and others are strong. Nevertheless, the spacing between pulses is so regular (0.714 seconds) that it is more precise than most clocks on Earth.

velocity would equal one-half the speed of light. All these conditions seemed so outrageous that few astronomers paid any serious attention to the subject of neutron stars—until 1968.

As a young graduate student at Cambridge University, Jocelyn Bell spent many months helping to construct an array of radio antennas covering $4^{1}/_{2}$ acres in the English countryside. By the fall of 1967 the instrument was completed, and Bell and her colleagues began detecting radio emissions from various celestial sources. In November, while scrutinizing data from the new telescope, Bell noticed that the antennas had detected regular pulses from one particular location in the sky. Careful repetition of the observations demonstrated that the radio pulses were arriving with a regular period of 1.3373011 seconds.

The regularity of this pulsating radio source was so striking that the Cambridge team suspected that they might be detecting signals from an advanced alien civilization, which accounts for the name first assigned to it: LGM1. (LGM was short for Little Green Men.) This possibility was soon discarded as several more of these pulsating radio sources, which soon came to be known as **pulsars**, were discovered across the sky. In all cases, the periods were extremely regular, ranging between 0.2 second and 1.5 seconds (Figure 12-16).

When the discovery of pulsars was officially announced in early 1968, astronomers around the world began proposing all sorts of explanations. Many of these theories were bizarre, and arguments raged for months. We have already ruled out alien civilizations. The broad distribution of pulsars would require incredibly widespread civilizations, inconsistent with what we have found in our

searches for alien intelligence (Chapter 18). Astronomers therefore looked for some recurring event in a star's life. We know that a star sheds matter as it swells to a giant. Perhaps some stars alternately expand and contract, emitting energy as they pulsate. However, this explanation also fails, because any star expanding and contracting as fast as a pulsar pulses would explode. Other astrophysicists suggested another kind of periodic event in a star's life: It might be spinning, giving off pulses with each revolution. But how?

Whatever the explanation, most astronomers doubted that pulsars were associated with dead stars. It was generally assumed that all dying stars somehow manage to eject enough matter so that their corpses do not exceed the Chandrasekhar limit. There seemed to be a sufficient number of white dwarfs in the sky to account for all the stars that have died since our Galaxy was formed. However, by late 1968, all controversy was laid to rest with the discovery of a pulsar in the middle of the Crab Nebula.

In A.D. 1054, American Indian, Chinese, and possibly other peoples saw and recorded the appearance of a "guest star" in the constellation of Taurus. When we turn a telescope toward this location, we find the Crab Nebula, shown in Figure 12-17a. It looks like the residue of an explosion and is, in fact, a supernova remnant. At the center of the Crab Nebula is the Crab pulsar (Figure 12.17b).

If the Crab pulsar is spinning, it is spinning too fast to be a white dwarf. Its period is 0.033 seconds, which means that it rotates 30 times each second. At that speed, something as wide as a white dwarf would immediately fly apart. Astronomers have discovered numerous other pulsars, including PSR 1937+21 (PSR for pulsar and 1937+21 for the object's right ascension and declination), a stellar corpse that pulses 640 times each second. Staying with the belief that they are spinning, astronomers concluded that pulsars must be incredibly compact. Calculations revealed that if neutron stars exist, they have a sufficiently small diameter to remain intact rotating as fast as pulsars pulsate. Rejecting the belief that all stars end up as white dwarfs, the theory of neutron stars, as described above, was quickly developed.

12-10 A rotating magnetic field explains the pulses from a neutron star

We now consider how neutron stars can become pulsars. Based on observations of the Sun, Betelgeuse, and other nearby stars, astronomers theorize that many, perhaps most, stars on the main sequence have some angular momentum (rotation). The Sun, for example, takes nearly a full month to rotate once about its axis. But the rotation rate of collapsing stars increases, just as pirouetting ice skaters speed up when they pull in their arms. (Recall from Section 2-5

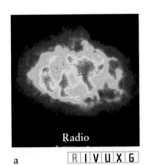
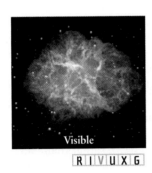

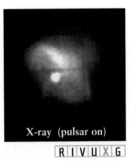

| Radio | Visible | Ultraviolet | X-ray (pulsar on) | Gamma ray |

R I V U X G   R I V U X G   R I V U X G   R I V U X G   R I V U X G

a

R I V U X G

b

FIGURE 12-17 The Crab Nebula and Pulsar **(a)** This beautiful nebula, named for the crablike appearance of its filamentary structure in visible light, is the remnant of a supernova seen in A.D. 1054. The distance to the nebula is about 6000 ly, and its present angular size (4 by 6 arcmin) corresponds to linear dimensions of about 7 by 10 ly. Observations at different wavelengths give astronomers information about the nebula's chemistry, motion, history, and interactions with preexisting gas and dust. **(b)** The Crab pulsar in its "on" (left) and "off" (right) states. Both its radio pulses and visible flashes have identical periods of 0.033 second. (a: NASA; b: Harvard-Smithsonian Center for Astrophysics)

that the total amount of angular momentum in an isolated system always remains constant.) Even an ordinary star rotating once a month would spin faster than once a second if it collapsed to the size of a neutron star.

In addition to rapid rotation, most neutron stars are believed to have intense magnetic fields. In an average star like our Sun, the magnetic field is spread out over millions upon millions of square kilometers just under the star's surface (see Chapter 9). However, if a star of solar dimensions collapses down to a neutron star, its magnetic field becomes very concentrated. The strength of the magnetic field increases a millionfold.

The axis of rotation of a typical neutron star is not the same as the axis connecting its north and south magnetic poles (Figure 12-18), just as the magnetic and rotation axes of the planets and Sun are different axes that are inclined to each other. As the neutron star rotates, its powerful magnetic field therefore rapidly changes direction. Like a giant electric generator, the star creates intense electric fields, which act on protons and electrons near its surface. The powerful electric fields channel these charged particles, causing them to flow out from the neutron star's polar regions, as sketched in Figure 12-18. As the particles stream along the field, they accelerate and emit energy. The result is two very thin beams of radiation pouring out of the neutron star's north and south magnetic polar regions—a pulsar.

This explanation for pulsars is often called the **lighthouse model.** A rotating, magnetized neutron star is somewhat like a lighthouse beacon. The magnetic field causes nearby gases to

accelerate and radiate. Calculations indicate that this radiation is beamed and as the star rotates, it sweeps around the sky. If the Earth happens to be located in the right direction, a brief flash can be observed each time a beam whips past our line of sight. As a result of its neutron star's rotation, the Crab Nebula is flashing on and off 30 times each second (see Figure 12-17b).

Belief in the lighthouse model was strengthened in 1999 when the *Chandra* X-ray observatory imaged bright X-ray emission at the heart of the Crab Nebula (X-ray image in Figure 12-17a) consistent with the predictions of this model. Detailed observations from this satellite are providing astronomers the opportunity to test and refine the lighthouse model.

The Crab pulsar is one of the youngest known pulsars, its creation having been observed some 950 years ago. Also visibly flashing is the Vela pulsar at the core of the Gum Nebula (see Figure 12-11). The Vela pulsar, with a period of 0.089 second, is the slowest pulsar ever detected at visual wavelengths. Because they use some of their energy of rotation in the pulses they emit, *isolated pulsars slow down as they get older.* Only the very youngest pulsars are energetic enough to emit visible flashes along with their radio pulses, and the Vela pulsar was created about 11,000 years ago. Astronomers are observing SN 1987A with the hope of seeing pulses that would indicate a rapidly spinning neutron star there.

Astronomers have developed a theoretical model for the surface and internal structure of a neutron star. Its interior has a radius of about 10 km and is composed of a fluid of neutrons with the strange property that they flow without any friction. They are *superfluid* neutrons. Superfluids are not

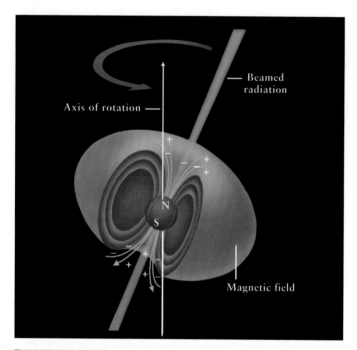

FIGURE 12-18 A Rotating, Magnetized Neutron Star It is reasonable to conclude that neutron stars rotate rapidly and possess powerful magnetic fields. Charged particles are accelerated near a star's magnetic poles and produce two oppositely directed beams of radiation. As the star rotates, the beams sweep around the sky. If the Earth happens to lie in the path of a beam, we see a pulsar.

hypothetical—they have been created in the laboratory. The surface of the neutron star is a solid crust of neutrons about ⅓ km thick. The gravitational force of the neutron star is so great at its surface that climbing a bump just a millimeter high there would take more energy than it takes to climb Mt. Everest.

The Earth's crust has earthquakes. Calculations show that rotating neutron stars should have equivalent events. When a rotating neutron star (pulsar) undergoes such an event, its rotation rate suddenly changes to conserve its angular momentum. This creates a **glitch** in the rate at which the pulsar is slowing down. Such events have been observed in a number of pulsars (Figure 12-19).

Insight into Science   **Test and Retest** New discoveries almost invariably allow for a variety of theories to explain them. Theories that are inconsistent with observations or make incorrect predictions are modified or discarded, while theories that explain what is observed are tested and retested until they break down. Such testing shows the range of situations over which each theory can be used reliably. Consider how the theories proposed above to explain pulsars were analyzed, tested, and, for the most part, rejected.

While most pulsars appear to be isolated bodies, more than 50 have been discovered with companions. Some, like the one labeled PSR 1916+13, have a massive companion, which are probably another neutron star. In the case of PSR 1916+13, the two neutron stars orbit each other with a period of about 8 hours. Because they are so close together and so incredibly dense, these two bodies follow orbits predicted by Einstein's general theory of relativity (coming up in Chapter 13), rather than Newton's laws of motion (Chapter 2). Close binary pulsars are spiraling toward each other and will eventually collide. Indeed, their behavior is a strong confirmation that general relativity is the more accurate model of the behavior of matter. Another *binary pulsar*, labeled PSR B1257+2, is even more unusual because its Doppler shift reveals the presence of at least three planets in orbit around it.

12-11 Colliding neutron stars may provide some of the heavy elements in the universe

We saw in Sections 12-4 and 12-5 how a variety of elements up to iron are created inside massive stars and then ejected into space during supernovas, while even heavier elements are created and ejected during the supernovas themselves. However, models of *nucleosynthesis* (the formation of elements) indicate that supernovas do not create enough of the

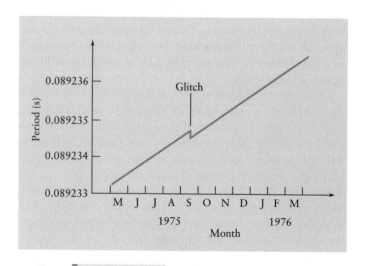

FIGURE 12-19 A Glitch Interrupts a Pulsar's Spindown Rate Isolated pulsars slow down as they lose energy by radiating it. This "spin down" is not always smooth. When the spinning, solid, neutron surface readjusts itself, the angular momentum of the pulsar suddenly jumps, changing the pulsar's rotation period. This event is called a glitch.

elements heavier than iron to account for the amounts of these elements found in the universe.

Computer simulations of what happens during collisions of two neutron stars predict that these events should lead to the observed amounts of the heaviest elements. In these impacts, some of the neutrons are splashed into space, where many of them decay back into protons and form various elements. The interactions that follow create a variety of heavy elements in amounts consistent with observations. The observational question remains, however, whether there are enough close binary neutron star systems that have collisions to pepper the Galaxy with heavy elements.

12-12 Pulsating X-ray sources are neutron stars in close binary systems

Neutron stars in binary systems may also hold the key to another regular pulse from the sky—pulses from X-ray sources. During the 1960s astronomers obtained tantalizing X-ray views of the sky during short rocket and balloon flights that briefly lifted X-ray detectors above the Earth's atmosphere. Several strong X-ray sources were discovered, and each was named after the constellation in which it was located. For example, Scorpius X-1 is the first X-ray source found in the constellation of Scorpius.

Astronomers were so intrigued by these preliminary discoveries that NASA built and launched *Explorer 42*, an X-ray-detecting satellite that could make observations 24 hours a day. The satellite was launched from Kenya (to place it in an orbit above the Earth's equator) in 1970, on the seventh anniversary of Kenyan independence. To commemorate the occasion, *Explorer 42* was renamed *Uhuru*, which means "freedom" in Swahili.

Uhuru gave us our first comprehensive look at the X-ray sky. As the satellite slowly rotated, its X-ray detectors swept across the heavens. Each time an X-ray source came into view, signals were transmitted to receiving stations on the ground. Before its battery and transmitter failed in early 1973, *Uhuru* had succeeded in locating 339 X-ray sources.

The discovery of pulsars was still fresh in everyone's mind when the *Uhuru* team discovered X-ray pulses coming from Centaurus X-3 in early 1971. Figure 12-20 shows data from one sweep of *Uhuru*'s detectors across Centaurus X-3. The pulses have a regular period of 4.84 seconds. A few months later, similar pulses were discovered coming from a source called Hercules X-1, which has a period of 1.24 seconds. Because the periods of these two X-ray sources are so short, astronomers began to suspect that they had found rapidly rotating neutron stars.

It soon became clear, however, that systems such as Centaurus X-3 and Hercules X-1 are not ordinary pulsars like the Crab or Vela pulsars. Every 2,087 days, Centaurus X-3 completely turns off for almost 12 hours. This fact suggests that Centaurus X-3 is an eclipsing binary and that the X-ray source takes nearly 12 hours to pass behind its companion star.

The case for the binary nature of Hercules X-1 is even more compelling. It has an off state corresponding to a 6-hour eclipse every 1.7 days, and careful timing of the X-ray pulses shows a periodic Doppler shift every 1.7 days. This information provides direct evidence of orbital motion about a companion star. When the X-ray source is approaching us, its pulses are separated by slightly less than 1.24 seconds. When the source is receding from us, slightly more than 1.24 seconds elapse between the pulses.

Careful visual searches around the location of Hercules X-1 soon revealed a dim star named HZ Herculis. The apparent magnitude of this star varies between +13 and +15, with a period of 1.7 days. Because this period is exactly the same as the orbital period of the X-ray source, astronomers conclude that HZ Herculis is the companion star around which Hercules X-1 orbits.

Astronomers now realize that systems such as Centaurus X-3 and Hercules X-1 are examples of binary stars in which one is a neutron star. Because all these binaries have very short orbital periods, the distance between the ordinary star

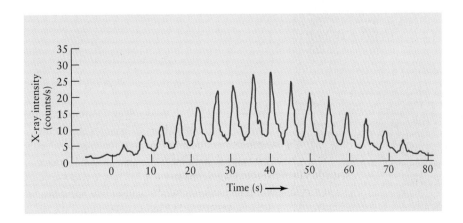

FIGURE 12-20   X-Ray Pulses from Centaurus X-3 This graph shows the intensity of X rays detected by *Uhuru* as Centaurus X-3 moved across the satellite's field of view. The successive pulses are separated by 4.84 seconds. The variation in the height of the pulses was a result of the changing orientation of *Uhuru*'s X-ray detectors toward the source as the satellite rotated.

and the neutron star must be small. This proximity enables the neutron star to capture gas escaping from the companion star.

To explain the pulsations of X-ray sources such as Centaurus X-3 or Hercules X-1, astronomers assume that the ordinary star either fills or nearly fills its Roche lobe. Either way, matter escapes from the star. If the star fills its lobe, as in the case of Hercules X-1, mass loss results from direct overflow through the Roche lobe. If the star's surface lies just inside its lobe, as with Centaurus X-3 (Figure 12-20), a stellar wind carries off the mass. A typical rate of mass loss for the ordinary star in such a pair is approximately 10^{-9} M$_\odot$ per year.

The neutron star in a pulsating X-ray source, like an ordinary pulsar, rotates rapidly and has a powerful magnetic field inclined to the axis of rotation (recall Figure 12-18). As the gas from the companion star falls toward the neutron star, the magnetic field funnels the incoming matter down onto its magnetic polar regions. The neutron star's gravity is so strong that the gas is traveling at nearly half the speed of light by the time it crashes onto the star's surface. This violent impact creates hot spots at both poles, with temperatures of about 10^8 K, that emit abundant X rays with a luminosity nearly 100,000 times brighter than the Sun. As the neutron star rotates, two beams of X rays from the polar caps sweep around the sky. If the Earth happens to be in the path of one of the beams, we can observe this pulsating X-ray source. The pulse period is thus equal to the neutron star's rotation period. For example, the neutron star in Hercules X-1 is spinning at the rate of once every 1.24 seconds.

12-13 Other neutron stars in binary systems emit powerful jets of gas

In a binary system, gas captured by a neutron star's gravity goes into orbit around the neutron star, as shown in Figure 12-21. Recall from Chapter 10 that the result is a revolving disk of material called an *accretion disk*. Accretion disks have been detected in many close binary systems where mass transfer is occurring (recall Figure 11-25b).

In 1995, NASA launched the *Rossi X-Ray Timing Explorer*. This spacecraft is able to detect X-ray events in space that vary as often as a thousand times a second. In 1998, it discovered a binary pulsar with an X-ray source that pulses 401 times a second. The neutron star and its companion orbit each other once every 2 hours, and the companion, losing mass to the neutron star, has decreased to about 0.1 M$_\odot$. Calculations indicate that the companion may completely evaporate within 100 million years.

Isolated pulsars are expected to rotate at rates like that of the Crab pulsar (30 times a second). Binary pulsars can spin much more rapidly. The accretion of mass by the rapidly spinning neutron star discovered by *Rossi* accounts for its very high rotation rate. To conserve angular momentum, the neutron star speeds up as it acquires the swirling gases from its companion.

With pulsating X-ray sources like Hercules X-1 and the one detected by *Rossi*, the rate at which gas falls onto the neutron star from the inner edge of the accretion disk is low enough to allow the resulting X rays to escape. If the com-

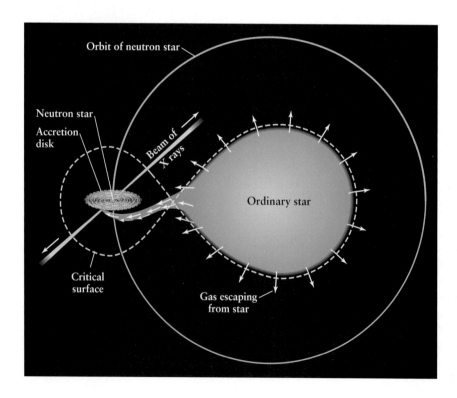

FIGURE 12-21 **A Model of a Pulsating X-Ray Source** Gas transfers from an ordinary star to the neutron star. The infalling gas is funneled down onto the neutron star's magnetic poles, where it strikes the star with enough energy to create two X-ray–emitting hot spots. As the neutron star spins, beams of X rays from the hot spots sweep around the sky.

panion star is dumping vast amounts of material onto the neutron star, however, the resulting energy cannot escape easily. In the plane of the accretion disk, the newly arrived gases are constantly spiraling in toward the neutron star. X-rays striking the incoming gases create tremendous pressures in them. Some of this gas is thereby ejected perpendicular to the accretion disk. It squirts out as two powerful jets of high-velocity gas.

12-14 Neutron stars in binary systems can also emit powerful isolated bursts of X rays

Neutron stars can acquire additional mass from a companion star and flare up. Beginning in late 1975, astronomers analyzing data from X-ray satellites detected sudden, powerful bursts of radiation. The record of a typical burst is shown in Figure 12-22. The source, called an **X-ray burster,** emits X rays at a constant low level; then, suddenly and without warning, there is an abrupt increase followed by a gradual decline. A typical burst lasts for 20 seconds. Several dozen X-ray bursters have been located, most lying in the plane of the Milky Way Galaxy. Their locations on the celestial sphere indicate that they are actually in the Milky Way.

X-ray bursters, like novae, are believed to arise from mass transfer in binary star systems. With a burster, however, the stellar corpse is a neutron star rather than a white dwarf. Gases escaping from the ordinary companion star fall onto the neutron star. The energy released as this gas crashes down onto the neutron star's surface produces the low-level X rays that are continuously emitted by the burster.

Most of the gas falling onto the neutron star is hydrogen, which becomes compressed against the hot surface of the star by the star's powerful surface gravity. In fact, temperatures and pressures in this accreting layer are so high that the arriving hydrogen is promptly converted by fusion into helium. Constant hydrogen fusion soon produces a layer of helium that covers the entire neutron star.

When the helium layer is about 1 m thick, helium fusion ignites explosively, and we observe a sudden burst of X rays. In other words, whereas explosive hydrogen fusion on a white dwarf produces a nova, explosive helium fusion on a neutron star produces an X-ray burster. In both cases, the fusion is explosive, because the fuel is so strongly compressed against the star's surface that it is degenerate, like the star itself. As we saw with the helium flash inside giants, ignition of a degenerate thermonuclear fuel always involves a sudden thermal runaway, because the usual safety valve between temperature and pressure is not operating.

In death as well as in life, the mass of a star determines its fate. Just as there is an upper limit to the mass of a white dwarf (the Chandrasekhar limit), there is also an upper limit to the mass of a neutron star, called the *Oppenheimer–*

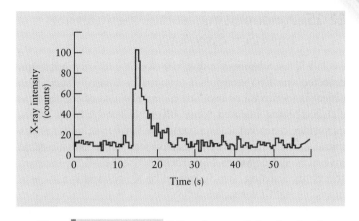

FIGURE 12-22 X Rays from an X-Ray Burster A burster emits X rays with a constant low intensity interspersed with occasional powerful bursts. This burst was recorded in September 1975 by an X-ray telescope that was pointed toward the globular cluster NGC 6624. About one-third of all known X-ray bursters are located in globular clusters.

Volkov limit. Above this limit, neutron degeneracy pressure cannot support the overbearing weight of the star's matter pressing in from all sides. The Chandrasekhar limit for a white dwarf is 1.4 $M_\odot$, and the Oppenheimer–Volkov limit for a neutron star is about 3 $M_\odot$. What happens if a dying massive star fails to eject enough matter to get below the Oppenheimer–Volkov limit?

In 2002, astronomers observed X-rays from two very dense objects, 3C58 and RXJ1856.5-3754, that are too small and too cool to fit the accepted model of neutron stars. These may be the first examples of stellar remnants with slightly more mass than the Oppenheimer-Volkov limit. According to calculations first done in the 1980s, neutrons can become so compressed that they dissolve into quarks. Neutron and protons are each comprised of 3 quarks, which normally cannot exist as separate objects. But at the densities that exist in the high mass neutron stars, they may be able to, thereby creating *quark stars.* The upper limit for the mass of quark stars is unknown, but it won't be much greater than the Oppenheimer-Volkov limit.

The gravity associated with a neutron star or quark star is so strong that the escape velocity is roughly half the speed of light. With a stellar corpse slightly greater than 3$M_\odot$, there is so much matter crushed into such a small volume that the neutron star or quark star collapses and the escape velocity exceeds the speed of light. Because nothing can travel faster than light, nothing—not even light—can leave the dead star. The star therefore disappears. The discovery of neutron stars thus inspired astrophysicists to consider seriously one of the most bizarre and fantastic objects ever predicted by modern science—the black hole, which we examine in Chapter 13. Figure 12-23 summarizes the evolutionary history of stars.

12-15 A nonpulsing, isolated neutron star has been discovered

In 1997, astronomers using the *ROSAT* X-ray telescope detected the first nonpulsar, nonbinary neutron star. This object is identified as a neutron star because of its physical characteristics. Located some 400 pc (1300 ly) from Earth, its surface temperature is a sizzling 6.5×10^5 K, and it is less than 28 km across. Only a neutron star in the astronomical inventory is known to have these properties. This neutron star emits a strong, steady stream of X rays and is also seen in visible light telescopes.

12-16 Frontiers yet to be discovered

The incredible telescopes now in existence or scheduled to go on-line soon are revolutionizing our understanding of the final stages of star formation. In coming years, astronomers hope to pin down the details of how planetary nebulae, novae, and supernovae work. Among specific questions to be answered are: What are the physical elements that cause planetary nebulae? Why is there such a variety of shapes of planetary nebulae? What causes the hourglass shape of so many stellar remnants? Why are novae composed of many blobs of gas? Do quark stars really exist, and if so, what are their properties?

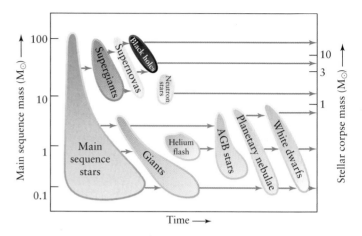

FIGURE 12-23 **A Summary of Stellar Evolution** The evolution of isolated stars depends on their masses. The higher the mass, the shorter the lifetimes. The scale on the left indicates the mass of a star when it is on the main sequence. The scale on the right gives the mass of the resulting stellar corpse. Stars less massive than about 8 $M_\odot$ can eject enough mass to become white dwarfs. High-mass stars can produce Type II supernovae and become neutron stars or black holes. The horizontal (time) axis is not to scale, but the relative lifetimes are accurate.

 Further Reading on These Topics

WHAT DID YOU KNOW?

1 *Will the Sun someday cease to exist? If so, how?* The Sun will shed matter as a planetary nebula in about six billion years and then cease nuclear fusion. Its remnant white dwarf will dim over the succeeding billions of years.

2 *What is a nova?* A nova is a relatively gentle explosion of hydrogen gas on the surface of a white dwarf in a binary star system.

3 *What are the origins of the carbon, silicon, oxygen, iron, uranium, and other heavy elements on Earth?*

These elements are created during stellar evolution, by supernovae, and by colliding neutron stars.

4 *What are cosmic rays?* Cosmic rays are high-speed particles (mostly hydrogen and other atomic nuclei) in space. Many of them are believed to have been created as a result of supernovae.

5 *What is a pulsar?* A pulsar is a rotating neutron star in which the magnetic field's axis does not coincide with the rotation axis. The beam of radiation it emits sweeps across our region of space.

KEY WORDS

asymptotic giant branch (AGB) star, 328
Chandrasekhar limit, 331
cosmic rays, 336
cosmic ray shower, 337
glitch, 342
helium shell flash, 328

lighthouse model, 342
neutron degeneracy pressure, 339
neutron star, 339
nova (*plural* novae), 331
photodisintegration, 334
planetary nebula, 329
pulsar, 341

secondary cosmic rays, 337
shell helium fusion, 327
supernova, 335
Type Ia supernova, 338
Type II supernova, 338
white dwarf, 330
X-ray burster, 345

KEY IDEAS

• Stars with different masses fuse different elements.

• Stars lose mass via stellar winds throughout their giant and supergiant phases.

Low-Mass Stars and Planetary Nebulae

• A low-mass (below 8 $M_\odot$) main-sequence star becomes a giant when shell hydrogen fusion begins. It becomes a horizontal branch star when core helium fusion begins. It enters the asymptotic giant branch and becomes a supergiant when shell helium fusion starts.

• Stellar winds during the thermal pulse phase can also eject mass from the star's outer layers.

• The burned-out core of a low-mass star becomes a dense carbon-oxygen body, called a white dwarf, with about the same diameter as the Earth. The maximum mass of a white dwarf (the Chandrasekhar limit) is 1.4 $M_\odot$.

• Explosive hydrogen fusion may occur in the surface layer of a white dwarf in a close binary system, producing the sudden increase in luminosity that we call a nova.

• An accreting white dwarf in a close binary system can also become a supernova when carbon fusion ignites explosively throughout such a degenerate star. Such a detonation is called a Type Ia supernova.

High-Mass Stars and Supernovae

• After exhausting its central supply of hydrogen and helium, the core of a high-mass (above 8 $M_\odot$) star undergoes a sequence of other thermonuclear reactions.

These include carbon fusion, neon fusion, oxygen fusion, and silicon fusion. This last fusion eventually produces an iron core.

• A high-mass star dies in a supernova explosion that ejects most of the star's matter into space at very high speeds. This Type II supernova is triggered by the gravitational collapse of the doomed star's core.

• Neutrinos were detected from Supernova 1987A, which was visible to the naked eye.

Neutron Stars and Pulsars

• A high-mass star's dead core becomes a neutron star or even a black hole. A neutron star is a very dense stellar corpse consisting of closely packed neutrons in a sphere roughly 20 km in diameter. The maximum mass of a neutron star, called the Oppenheimer-Volkov limit, is about 3 $M_\odot$.

• A pulsar is a rapidly rotating neutron star with a powerful magnetic field that makes it a source of periodic radio and other electromagnetic pulses. Energy pours out of the polar regions of the neutron star in intense beams that sweep across the sky.

• Some X-ray sources exhibit regular pulses. These objects are thought to be neutron stars in close binary systems with ordinary stars.

• Explosive helium fusion may occur in the surface layer of a companion neutron star, producing the sudden increase in X-ray radiation called an X-ray burster.

REVIEW QUESTIONS

1   What is the difference between a giant and a supergiant?

2   Why is knowing the temperature in a star's core so important in determining which nuclear reactions can occur there?

3   What determines the temperature in the core of a star?

4   What is a planetary nebula and how does it form?

5   What is a white dwarf?

6   What is the Chandrasekhar limit?

7   What is a neutron star?

8   Compare a white dwarf and a neutron star. Which of these stellar corpses is more common? Why?

 9   To test your understanding of the stages of stellar evolution, do Interactive Exercise 12-1. You can print out your results, if required.

10   What is the Oppenheimer-Volkov limit?

11   On an H-R diagram, sketch the evolutionary track that the Sun will follow between the time it leaves the main sequence and when it becomes a white dwarf. Approximately how much mass will the Sun have when it becomes a white dwarf? Where will the rest of the mass have gone?

12   Why are all the white dwarfs known to astronomers relatively close to the Sun?

13   Why have radio searches for supernova remnants been more fruitful than searches at visible wavelengths?

14 To test your understand of how stars "die," do Interactive Exercise 12-2. You can print out your results, if required.

 15 To test your understanding of neutron stars, do Interactive Exercise 12-3. You can print out your results, if required.

16 Why do astronomers believe that pulsars are rapidly rotating neutron stars?

 17 To test your understanding of rotating neutron stars, pulsars, and novae, do Interactive Exercise 12-4. You can print out your results, if required.

18 What is the difference between Type Ia and Type II supernovae?

19 Compare a nova with a Type Ia supernova. What do they have in common? How are they different?

20 Compare a nova and an X-ray burster. What do they have in common? How are they different?

21 Describe what radio pulsars, X-ray pulsars, and X-ray bursters have in common. How are they different manifestations of the same type of astronomical object?

ADVANCED QUESTIONS

The answers to all computational problems, which are preceded by an asterisk (*), appear at the end of the book.

22 What prevents thermonuclear reactions from occurring at the center of a white dwarf? If no thermonuclear reactions take place in its core, why doesn't the star collapse?

23 Suppose you wanted to determine the age of a planetary nebula. What observations would you make, and how would you use the resulting data?

24 What explains why the rate of expansion of the gas shell in a planetary nebula is often not uniform in all directions?

25 What kinds of stars would you monitor if you wished to observe a supernova explosion from its very beginning? Look up the tabulated lists of the brightest and nearest stars in Appendix Tables A-4 and A-5. Which, if any, of these stars are possible supernova candidates? Explain.

26 To determine the period of a pulsar accurately, astronomers must take the Earth's orbital motion about the Sun into account. Explain why.

*27 The distance to the Crab Nebula is about 2000 pc. When did it actually explode?

 28 To test your overall understanding of stellar evolution, do Interactive Exercise 12-5. You can print out your results, if required.

DISCUSSION QUESTIONS

29 Suppose that you discover a small glowing disk of light while searching the sky with a telescope. How would you decide just from your observations if this object was a planetary nebula? Could your object be something else? Explain.

30 Immediately after the first pulsar was discovered, one explanation offered was that the pulses were signals from an extraterrestrial civilization. Why did astronomers discard this idea?

WHAT IF ...

31 The Earth passed through an old supernova remnant? What would happen to the Earth and life on it?

32 The Sun were a B-type star, rather than a G-type star? Assuming that the Earth orbiting the B-type star had the same composition and orbital distance that it has today, what would be different here? If you answered this question in Chapter 10, you might want to see how the material in this chapter enhanced your previous answer.

33 The Sun were an M-type star, rather than a G-type star? Assuming that the Earth orbiting the M-type star had the same composition and orbital distance that it has today, what would be different here?

34 The Sun were expanding into a giant star today? How might we cope with this change?

WEB/CD-ROM QUESTIONS

35   It has been claimed that the Dogon tribe in western Africa has known for thousands of years that Sirius is a binary star. Search the Web for information about these claims. What is their basis? Why are scientists skeptical, and how do they refute these claims?

36   Search the Web for recent information about SN 1987A. Sketch the shape of the supernova remnant. Has a pulsar yet been detected in the center of this supernova remnant? If so, how fast is it spinning? Has the supernova debris thinned out enough to give a clear view of the neutron star?

37   Search the Web for information about SN 1994I, a supernova that occurred in the galaxy M51 (NGC 5194). Why was this supernova unusual? Was it bright enough to have been seen by amateur astronomers?

 38   **Types of Supernovae** Access and view the animations "In the Heart of a Type II Supernova" and "A Type Ia Supernova" in Chapter 12 of *Discovering the Universe* Web site or CD-ROM. Describe how these two types are fundamentally different in their origin.

OBSERVING PROJECTS

39   Planetary nebulae are found all across the sky. Some of the brightest are listed in Chapter 12 of the *Discovering the Universe* Web site and CD-ROM. With the help of star maps or your *Starry Night Backyard*™ software, locate and observe as many planetary nebulae as possible on clear, moonless nights, using the largest telescope at your disposal. Some of the more notable ones include: Little Dumbbell (M76), NGC 1535, Eskimo, Ghost of Jupiter, Owl (M97), Ring (M57), Blinking Planetary, Dumbbell, Saturn Nebula, and NGC 7662. Note and compare the various shapes of the different nebulae. In which cases can you see the central star?

40   Two supernova remnants can be seen through modest telescopes, one in the winter sky and the other in the summer sky. Both are quite faint, however, so you should schedule your observations for a moonless night. The winter sky contains the Crab Nebula, which is discussed in detail in this chapter. It is located near the star marking the eastern horn of Taurus (the Bull) and can be found using *Starry Night Backyard*™

software. Its coordinates are R.A. = 5^h 34.5^m and Dec. = $+22°$ $00'$. Whereas the entire Crab Nebula easily fits into the field of view of an eyepiece, the Veil (or Cirrus) Nebula in the summer sky is so vast that you can see only a small fraction of it at a time. The easiest way to find the Veil Nebula is to aim the telescope at the star 52 Cygni (R.A. = 20^h 45.7^m and Dec. = $+30°43'$, which lies on one of the brightest portions of the nebula. This star is also accessible in your *Starry Night Backyard*™ program. If you then move the telescope slightly north or south until 52 Cygni is just out of the field of view, you should see giant wisps of glowing gas.

41   Use a telescope to observe the triple star system 40 Eridani, whose coordinates are R.A. = 4^h 15.3^m and Dec. = $-7°$ $39'$. The primary, a 4.4 magnitude yellowish star like the Sun, has a 9.6 magnitude white dwarf companion, the most easily seen white dwarf in the sky. On a clear, dark night with a moderately large telescope, you should also see that the white dwarf has an 11th-magnitude companion, which completes this interesting trio.

A SUPERNOVA EXPLODED NEAR EARTH?

The Flash What would happen here if a supernova occurred only 50 ly away? (Considering the titanic forces supernovae release, it should come as no surprise that the high-energy electromagnetic radiation from such an explosion detonating much closer than this distance would immediately kill virtually all life on Earth.) Without warning, the doomed star would grow tremendously luminous, 50 times brighter than the Moon and only 8000 times less bright than the Sun. It would be brighter than the light from all the other stars in the night sky combined.

Neutrinos would foreshadow by a few hours the visible flash and pending flood of X rays and gamma rays from the supernova. These lethal radiations would be the first causes of death on Earth. Within days individuals from virtually all species of plants and animals would begin dying of radiation poisoning. Entire radiation-sensitive species might be annihilated immediately. The radiation would also cause many cancers and other internal diseases over succeeding years, leading to many more plant and animal deaths. At the same time, ultraviolet radiation reaching the Earth's surface would cause an astronomical jump in the rates of skin cancer and cataracts.

The first blast of ultraviolet radiation from the supernova would destroy the Earth's ozone layer within a matter of days, transforming the ozone primarily into atomic oxygen. With the removal of this protective barrier, ultraviolet radiation from both the supernova and the Sun would saturate the Earth's surface. After the intensity of ultraviolet radiation from the supernova diminished, sunlight would begin repairing the ozone layer, eventually returning it to normal levels.

The brightness and emission of all the electromagnetic radiation from the supernova would peak after a month. It would then fade, but the supernova remnant would be visible for millennia as an expanding cloud of gas and dust in the night sky. Fortunately for life that survived the onslaught of high-energy radiation from the supernova, the remnant cloud would primarily emit harmless visible light.

The Aftermath The most energetic particles from the supernova and its environs, the so-called cosmic rays, would also affect the Earth's surface. Moving at 90% of the speed of light, cosmic rays would arrive here only five to ten years after the photons. Cosmic ray energies are much higher than those of any particles normally existing on Earth. For example, a typical cosmic ray has enough energy to light a small lightbulb for one second; it takes over 600,000 trillion electrons flowing through wiring in your house to do the same thing.

Some cosmic rays would slam into nitrogen or oxygen molecules in the air, shattering the air molecules and creating the cosmic ray showers. The highest energy cosmic rays from the supernova would reach the Earth intact, as do the highest energy cosmic rays from other sources today. These would break up atoms of the objects they strike on the Earth's surface, enhancing the earlier biological damage caused by the supernova's electromagnetic radiation.

The bulk of the matter ejected into space by the supernova would travel at speeds of 16,000 km/sec (60 million km/hr), nearly 20 times slower than the emitted photons. Thus, the bulk of the supernova remnant would take at least a thousand years longer to reach us than did its initial radiation. By the time the bulk of the blast wave reaches the solar system, it would have become so thin and diffuse that it would probably do little damage to the Earth's atmosphere. However, we could expect this material to deposit a thin but exotic mix of elements into the upper atmosphere. All this material would eventually fall to Earth and alter the chemistry of both the ocean and the soil.

The Outcome The damage done to life by both the electromagnetic radiation and cosmic ray particles would disrupt the global food chain. In the oceans large quantities of plankton and other microscopic organisms at the foundation of the chain would die off. As a result, many of the larger aquatic species that feed on these smaller organisms would starve. Similarly, on the surface, most plant life would wither and many herbivorous animals would starve as a result. This would, of course, lead to the death and dislocation of animals throughout the food chain. Surviving plants and animals would undergo genetic mutations due to the supernova's radiation. Most of these genetic changes would lead to immediate death of the altered plant or animal, but a few changes would be beneficial and enable the mutated organisms to thrive.

How would the human race fare after the supernova? Our long-term survival would depend, in part, on whether surviving humans could coexist with surviving plants and animals, and with each other.

13 BLACK HOLES

IN THIS CHAPTER YOU WILL DISCOVER

- that Einstein's general theory of relativity predicts regions of space and time that are severely distorted by the extremely dense mass they contain

- how black holes arise

- the surprisingly simple theoretical properties of black holes

- the apparent bizarre fate of a black hole

The Behavior of Light near a Black Hole
Black holes distort spacetime nearby. In this artist's rendition, two light beams fired in slightly different directions from the upper left travel on different paths. They both move in straight lines, but the meaning of "straight" near a black hole defies our intuition. The upper beam, aimed slightly away from the black hole, is virtually unaffected by it. The lower beam, inclined toward the black hole, travels through space so distorted that it twists the light into a spiral, which is swept down into the black hole. Don't let this painting (or similar images in science fiction) mislead you into believing that black holes have a "top" or a "bottom." Theory predicts that they can be entered from any direction.

WHAT DO YOU THINK?

1   Are black holes empty holes in space?

2   What exists at the surface of a black hole?

3   What power or force enables black holes to draw things into them?

4   Do black holes last forever?

The extreme gravitational force acting on a neutron star is hard to imagine. But for a neutron star smaller than 3 $M_\odot$ that force is counterbalanced by an equal, outward force from neutron degeneracy pressure. Prevented from collapsing further, a neutron star is incredibly compact, barely 20 km across. But not even neutron degeneracy pressure can stand up to masses greater than 3 $M_\odot$ and their inward gravitational force.

Once a massive star overcomes neutron degeneracy pressure (or the pressure keeping the recently discovered quark stars stable), what prevents it from collapsing until nothing—not even light—can manage to escape its gravitational attraction? The answer according to contemporary theories of physics is "Nothing"! It becomes a **black hole**. We must turn to Einstein's powerful theories of relativity to understand these exotic objects.

THE EVIDENCE FOR BLACK HOLES

Black holes inspire awe, fear, and uncertainty. Many people harbor the mistaken belief that black holes are destined to "swallow up" all the matter in the universe. Happily, the equations describing them reveal that black holes are more benign than that, but they truly are strange.

13-1   Special relativity changes our conception of space and time

In 1905, Albert Einstein began a revolution in physics with his **special theory of relativity,** a description of how motion affects our measurements of distance, time, and mass. He was guided by two innovative ideas. While its implications proved revolutionary, the first notion seems simple: *Your description of physical reality is the same regardless of the (constant) velocity at which you move.* In other words, as long as you are moving in a straight line at a constant speed, you experience the same laws of physics as anyone else moving at any other constant speed and in any other direction. To illustrate, suppose you were inside a closed boxcar moving in a straight line at 100 km/h. Any scientific measurements you make in the moving boxcar would yield the same results as the measurements taken from the same experiment while the boxcar was at rest (or moving at any other constant speed).

The second idea seems more bizarre: *Regardless of your speed or direction, you always measure the speed of light to be the same.* Suppose that you are in a car moving toward a distant street lamp. Even if you are moving at 95% of the speed of light, you will still measure the speed of the oncoming light photons to be the same if you were standing by the roadside.

Einstein's special theory of relativity expresses these two assumptions mathematically, and the results of these assumptions have been confirmed in innumerable experiments. Here are three fundamental results of special relativity. The first is that the length of an object decreases as its speed increases. In other words, the faster an object moves past you, the shorter its dimension is in the direction of its motion. If a boxcar moves past you, it is actually shorter, from your perspective on the ground, than when it is at rest (Figure 13-1). However, if you measure the length of the moving boxcar while you are inside it and moving with it, your measurement of its length will be the same as measured on the ground when it was at rest. The word *relativity* emphasizes the importance of the relative speed between the observer and the measured object.

Einstein's second result is just as strange: Clocks passing by you run more slowly than do clocks at rest. This is called *time dilation.* Indeed, the faster a clock passes by, the slower it appears to tick from your perspective. For example, pilots and cabin attendants actually age more slowly than they would if they didn't fly (although because of the relatively low speeds of aircraft travel compared to the speed of light, this difference is imperceptibly small). People flying in aircraft do not feel time moving more slowly, however, because their biological activities slow down at the same rate as the clocks around them. Only an observer moving more slowly sees that the clock and the activities of the high-speed travelers have slowed. These connections between motion and clocks mean that space and time cannot be considered as two separate concepts. Relativity requires us to combine them in a single entity, thus creating the concept of **spacetime.**

Finally, Einstein showed that the mass of an object increases as it moves faster. A body approaching the speed of light becomes infinitely massive. To push mass faster than the speed of light would therefore require an infinite amount of energy. The speed of light is the universal speed limit.

13-2   General relativity predicts black holes

First published in 1915, Einstein's **general theory of relativity** extends his special theory of relativity to include the effects of gravity and accelerating systems. It describes how matter creates gravitational force by distorting the very fabric, or geometry, of space and time.

At rest

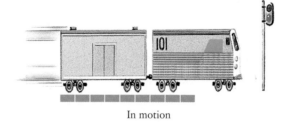

In motion

FIGURE 13-1 **Movement and Space** According to the special theory of relativity, the faster an object moves, the shorter it becomes in its direction of motion as observed by someone not moving with

the object. It becomes infinitesimally small as its speed approaches the speed of light. The dimensions perpendicular to the object's motion are unchanged.

He found that matter affects the nature of space and the rate at which time passes.

Newton's laws of motion and his universal law of gravitation are accurate only for objects with relatively small masses, slow velocities, and low densities. We use Newton's laws to determine the force necessary to put the Space Shuttle in orbit around the Earth, for example. But Newton's laws are inadequate to accurately predict Mercury's orbit around the Sun or the paths of objects near a collapsing neutron star. In such cases, general relativity correctly predicts how objects move.

In general relativity, unlike in Newtonian physics, the path of light is affected by nearby mass (Figure 13-2). As seen by observers far away, massive objects actually curve space and slow down the flow of time. The more massive or dense the object, the more it curves nearby space and the more it slows down time in its vicinity. Just by being near matter, clocks run slower than they do in empty space. Furthermore,

photons leaving the vicinity of a star lose energy in climbing out of the star's gravitational field. They show this change by shifting to longer wavelengths, an effect we see in the spectra of some white dwarfs, whose light appears redder than it would if this effect did not occur. This shift of wavelengths leaving the vicinity of a massive object is called **gravitational redshift.**

Einstein's description of gravity is radically different from that of Newton. According to Newtonian mechanics, space is perfectly flat and extends infinitely far in all directions. Similarly, Newtonian clocks monotonously tick at an unchanging rate, never speeding up or slowing down. In this rigid, unalterable framework of space and time, gravity is described as a "force" that acts at a distance. The planets literally pull on each other across empty space with a strength described by Newton's universal law of gravitation.

The general theory of relativity does not treat gravity as a force at all. Instead, gravity causes space to become curved and time to slow down. A planet or spacecraft passing near the Sun is deflected from a straight-line path because the space the object passes through is curved by the presence of the Sun.

General relativity has been confirmed again and again, as seen in these observations:

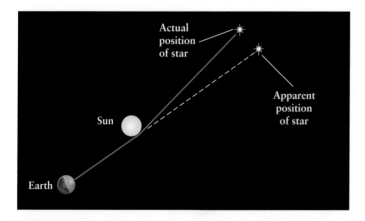

FIGURE 13-2 **Warped Space and the Curved Path of Light** That the warping of space by matter deflects light was one of the first predictions of general relativity to be confirmed, in 1919. This confirmation came when stars near the Sun were observed during an eclipse. Starlight was deflected around the Sun by up to 1.75 arcsec, a small but measurable amount.

- Light is measurably deflected by the Sun's gravitational curving of space (see Figure 13-2).

- The perihelion position of Mercury as seen from the Sun shifts or precesses by 43 arcsec per year more than is predicted by Newtonian gravitational theory (Figure 13-3).

- Extremely accurate clocks run more slowly when being flown in airplanes than identical clocks remaining on the ground.

- The spectra of some stars are observed to have gravitational redshifts.

These observations are consistent with the general theory of relativity but not with the laws of Newtonian physics.

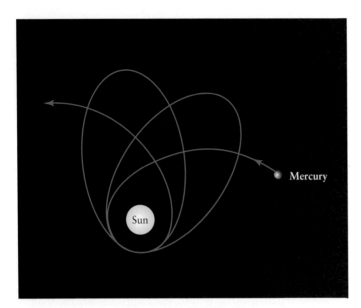

FIGURE 13-3   Mercury's Orbit Explained by General Relativity
In this greatly exaggerated figure, the location of Mercury's perihelion (its position closest to the Sun) changes with each orbit because of the gravitational influences of the other planets as well as effects predicted by Einstein's general theory of relativity. The amount of this change is inconsistent with the predictions made by Newton's law of gravity alone.

hole, matter compresses to infinite density, a state called a **singularity.** However, general relativity and our other theories of physics are known to be invalid in such a situation. A more comprehensive theory of nature must be developed to explain the state of matter in a black hole's singularity. Efforts are under way to develop such a theory. The best known nascent theory is called *superstrings,* but its mathematics are daunting and much remains to be done. All we can say with confidence is that the matter in black holes becomes incredibly dense and compact.

13-3 Several binary star systems contain black holes

Black holes are more than fine points of relativity theory: They are real, their presence has been observed by their effects on the orbits of other stars and on gas and dust near them, and more of them are being located all the time. To find evidence for black holes, we look first to binary star systems.

As we saw in Chapter 10, many stars occur in binary star systems. The technique for detecting black holes formed from collapsing stars is based on the interaction between the

Let us now return to stellar evolution and consider what happens when main-sequence stars with more than 25 $M_\odot$ explode as supernovae. As with main-sequence stars between $8_\odot$ and 25 $M_\odot$, extremely high-mass stars create neutron (and possibly quark) stars. These remnant stars, however, have more than 3 $M_\odot$. This is problematic because above this mass, the gravitational force of the neutron or quark star overcomes the repulsion created by the interactions between its particles. This causes such neutron stars to collapse. There is no force in nature that can stop their infall.

It is necessary to use the equations of general relativity to understand the fate of collapsing neutron or quark stars. We just saw that all matter warps the space around itself (see Figure 13-2). When matter gets sufficiently dense, it actually causes nearby space to curve so much that it closes in on itself (Figure 13-4). Photons flying outward at an angle from such a collapsing star arc back inward. Photons flying straight outward lose all their energy, and they cease to exist.

If light, the fastest moving of all known things, cannot escape from the vicinity of such dense matter, then nothing can! Such regions out of which no matter or any form of electromagnetic radiation can escape are called black holes. As we will see next, their discovery provided dramatic confirmation of Einstein's theories.

Plummeting in on itself, a collapsing neutron or quark star becomes so dense that it ceases to consist of neutrons or quarks. General relativity predicts that in creating a black

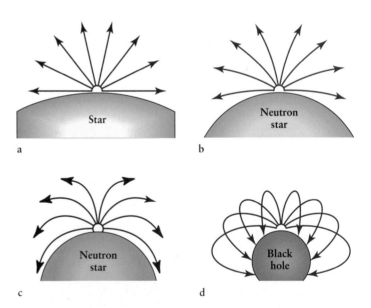

FIGURE 13-4   Trapping of Light by a Black Hole (a) Light departing from a main-sequence, giant, or supergiant star is affected very little by the star's gravitational force. (b) However, as a neutron or quark star collapses, its gravitational force increasingly curves the path of departing light until (c) some of the photons actually return to the star's surface. (d) When the star becomes a black hole, all light leaving its surface remains trapped. Most photons curve back in, except those flying straight upward, which become infinitely redshifted, thereby disappearing.

black hole and a binary companion. When one star in a close binary becomes a black hole, its gravitational attraction pulls off some of its companion's atmosphere. However, such black holes have diameters of only a few kilometers, so there is not room for all this gas to fall straight in. Rather, the infalling gas swirls into the black hole like water going down a bathtub drain. The gas waiting to fall in forms an accretion disk (see Chapter 12). Calculations reveal that this gas is compressed and heats so much that it gives off X rays (Figure 13-5). Thus, if a visible star has a sufficiently tiny, sufficiently massive X-ray–emitting companion, we have located a black hole.

Shortly after the *Uhuru* X-ray satellite was launched in the early 1970s, astronomers found a promising black-hole candidate—an X-ray source called Cygnus X-1. This source is highly variable and irregular. Its strong X-ray emission flickers on time scales as short as a hundredth of a second. If different parts of an X-ray source grew bright at different times, its emission would be a continuous stream. In order for Cygnus X-1 to flicker, the entire star must brighten and dim as a unit. Therefore, light must have time to travel across Cygnus X-1 between pulses. Because light travels 3000 km in

a hundredth of a second, Cygnus X-1 must be smaller in diameter than the Earth.

Cygnus X-1 occasionally emits radio radiation, and in 1971 radio astronomers succeeded in associating these emissions with the visible star HDE 226868 (Figure 13-6). Spectroscopic observations revealed that HDE 226868 is a B0 supergiant with a surface temperature of about 31,000 K. Because such stars do not emit significant X rays, HDE 226868 alone cannot be the Cygnus X-1 X-ray source.

Further spectroscopic observations soon showed that the lines in the spectrum of HDE 226868 shift back and forth with a period of 5.6 days. This behavior is characteristic of a single-line spectroscopic binary (see Section 10-9), HDE 226868's companion is too dim to produce its own set of spectral lines. The clear implication is that HDE 226868 and Cygnus X-1 are the two components of a binary star system.

The B0 supergiant HDE 226868's mass is estimated at about 30 $M_\odot$, like other B0 supergiants. As a result, Cygnus X-1 must have at least 7 $M_\odot$; otherwise, it would not exert enough gravitational pull to make the B0 star wobble by the amount deduced from the periodic Doppler shift of its spectral lines. Cygnus X-1 cannot be a white dwarf or a neutron

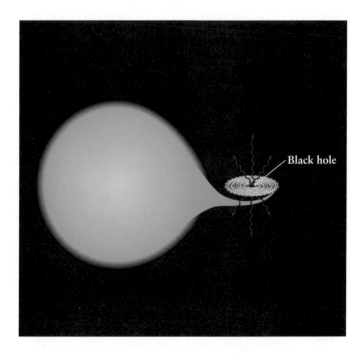

FIGURE 13-5   X Rays Generated by Accretion of Matter near a Black Hole Stellar-remnant black holes, such as Cygnus X-1, LMC X-3, V404 Cygni, and probably A0620-00, are detected in close binary star systems. This drawing of such a system shows how gas from the companion star transfers to the black hole, thereby creating an accretion disk. As the gas spirals inward, friction and compression heat it so much that the gas emits X rays, which astronomers can detect.

R I V U X G

FIGURE 13-6   HDE 226868 This star is the optical companion of the X-ray source Cygnus X-1. This binary system is located about 8000 ly from Earth and contains a 7-$M_\odot$ black hole in orbit with HDE 226868, a B0 blue supergiant star. The photograph was taken with the 200-in. telescope at Palomar Observatory on Palomar Mountain, north of San Diego. The slightly dimmer star above is an optical double that is not part of the binary system. (J. Kristian, Carnegie Observatories)

star, because its mass is too large for either of these objects. The only remaining possibility is that it must be a fully collapsed star—a black hole.

In the early 1980s, a similar binary system was identified in the nearby galaxy called the Large Magellanic Cloud. The X-ray source, called LMC X-3, exhibits rapid fluctuations, just like those of Cygnus X-1. LMC X-3 orbits a B3 main-sequence star every 1.7 days. From its orbital data, astronomers conclude that the mass of LMC X-3 is about 10 $M_\odot$, which would make it a black hole.

Another black-hole candidate is a spectroscopic binary in the constellation Monoceros that contains the flickering X-ray source A0620-00. The visible companion of A0620-00 is an orange dwarf star of spectral type K that orbits the X-ray source every 7.75 hours. From orbital data, astronomers conclude that the mass of A0620-00 must be greater than 5 $M_\odot$, more probably about 9 $M_\odot$. To date, in the Milky Way galaxy alone, detailed observations and calculations such as those just described have identified at least nine black holes in binary star systems. There is also growing evidence that black holes can collide with each other or with different types of objects, such as neutron stars.

13-4 Black holes range in mass up to billions of solar masses

Based on the equations of general relativity, the fate of massive neutron stars led to the idea of black holes as early as 1939. Since then, calculations have supported at least two other sources of black hole formation: Early in the life of the universe, black holes could have formed from the condensing of vast amounts of gas during the process of galaxy formation thereby creating **supermassive black holes.** Also, they could have formed during the explosive beginning of the universe as tiny amounts of matter were compressed sufficiently to form **primordial black holes.**

Supermassive black holes, with millions or even billions of solar masses, have been observed at the centers of many galaxies. In May 1994, the Hubble Space Telescope obtained compelling evidence for a black hole at the very center of the galaxy M87. In the nucleus of M87 is a tiny, bright source of light. Spectra showed that nearby gas and stars are orbiting it extremely rapidly. They can be held in place only if the bright object contains some three *billion* solar masses (Figure 13-7). Given that its size is only slightly larger than the solar system, the source can only be a black hole.

Since 1994, numerous black holes in the centers of galaxies have been identified by their gravitational effect on surrounding gas and stars. While spectroscopy reveals that the dust in Centaurus A orbits in an accretion disk seen edge-on, a distinct, hubcap-shaped accretion disk around the supermassive black hole in NGC 7052 was seen by the

Hubble Space Telescope in 1998 (Figure 13-8). As we will see in the next chapter, a black hole containing several million solar masses has even been found at the center of our Milky Way, only 25,000 ly from the Earth. Indeed, in a survey done by the Hubble Space Telescope of 27 nearby galaxies and in follow-up studies, evidence for central, supermassive black holes was found in most of these galaxies!

As we will explore in more detail in Chapter 17, galaxies were created from condensing gas and stars in the early universe. Our technology is now providing us with observations of this epoch of the universe, and the formation process of supermassive black holes is now being studied. These black holes apparently formed in part from the gas that was condensing to form galaxies and in part from the collisions of black holes formed shortly after the beginning of time. Indeed, there is observational evidence that the for-

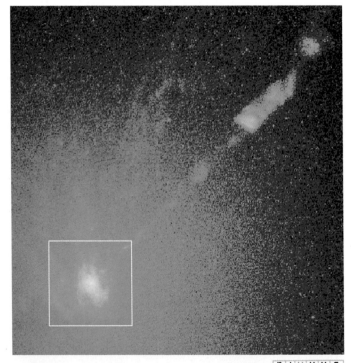

R I V U X G

FIGURE 13-7   **Supermassive Black Hole** Notice the bright region in the center of this galaxy M87, where stars and gas are held in tight orbits by a black hole. M87's bright nucleus (center of the region in the white box) is only about the size of the solar system and pulls on the nearby stars with so much force that astronomers calculate that it is a 3-billion-$M_\odot$ black hole. (Holland Ford, STScI/Johns Hopkins University; Richard Harms, Applied Research Corp.; Zlatan Tsvetanov, Arthur Davidsen, and Gerard Kriss at Johns Hopkins University; Ralph Bohlin and George Hartig at STScI; Linda Dressel and Ajay K. Kochhar at Applied Research Corp. in Landover, Md; and Bruce Margon, University of Washington, Seattle)

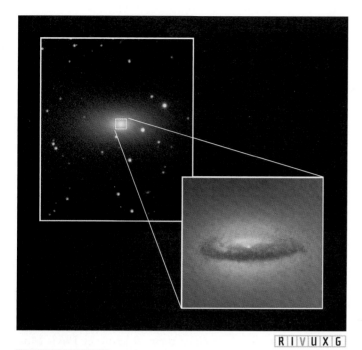

R I V U X G

FIGURE 13-8   Accretion Disk around a Supermassive Black Hole Swirling around a 300-million-M$_\odot$ black hole in the center of the galaxy NGC 7052, this disk of gas and dust is 3700 ly across. The gas is cascading into the black hole, which will consume it over the next few billion years. NGC 7052 is 191 million ly from Earth in the constellation of Vulpecula. (R. P. van der Marel, STScI; F. C. van den Bosch, University of Washington; NASA)

mation of supermassive black holes is still ongoing. We will explore this further in Chapter 15.

 Astronomers were taken by surprise at the end of the twentieth century when black holes with masses between 10^2 and 10^4 solar masses began to be discovered. For example, at the center of the galaxy M82, astronomers found a 500 M$_\odot$ black hole (Figure 13-9), while at the center of the galaxy NGC 4395 lies a 50,000 M$_\odot$ black hole. Whether these are remnant black holes from the first generation of massive black holes described above or derive from a different source remains to be seen.

 As indicated earlier, even more exotic black holes may have formed along with the universe itself. British astrophysicist Stephen Hawking has proposed that the Big Bang explosion from which most astronomers believe the universe emerged may have been chaotic and powerful enough to have compressed tiny knots of matter into primordial black holes. Their masses may have ranged from a few grams to greater than the mass of the Earth. Astronomers have not yet observed evidence of primordial black holes.

INSIDE A BLACK HOLE

The formation of a black hole is complicated, but its nature is surprisingly simple. It has a boundary shaped like a sphere, it either rotates or it does not, and it slowly disappears.

R I V U X G   R I V U X G

FIGURE 13-9   A "Mid-Mass" Black Hole M82 is an unusual galaxy in the constellation Ursa Major. The inset shows an image of the central region of M82 from the *Chandra* X-ray Observatory. The bright, compact X-ray source shown varies in its light output over a period of months. The properties of this source suggest that it is a black hole of roughly 500 M$_\odot$ (Subaru Telescope, National Astronomical Observatory of Japan; inset: NASA/SAO/CXC)

13-5 Matter in a black hole becomes much simpler than elsewhere in the universe

A black hole is separated from the rest of the universe by a boundary called the **event horizon.** Within this shell, even light cannot escape the black hole's enormous gravitational attraction. The event horizon is not like the surface of a solid body. No matter exists at this location except for the instant it takes all infalling mass to cross the event horizon and enter the black hole.

We cannot look inside a black hole because no electromagnetic radiation escapes from it. Our understanding of its structure comes from the equations of general relativity. According to Einstein's theory, the event horizon is a sphere. The distance from the center of the black hole to its boundary is called the **Schwarzschild radius** (abbreviated R_{Sch}), after the German physicist Karl Schwarzschild, who first determined its properties. An Astronomer's Toolbox 13-1 shows how to calculate this distance, which depends only on the black hole's mass. The more massive the black hole, the larger its event horizon.

MORE TO KNOW 13-1   Gravitational Waves

WEB LINK 13-7 According to the equations of general relativity, when a stellar remnant collapses to a black hole, it loses its magnetic field. The field's energy radiates away in the form of **gravitational waves,** which are ripples in the very fabric of spacetime. These waves are also created when neutron stars or black holes collide or even when they are in close orbits around each other. The *binary pulsar* mentioned in Section 12-10 is a pair of neutron stars orbiting each other and emitting gravitational waves. The energy in these waves comes from the orbital energy of the neutron stars, so they are spiraling toward each other.

The ripples in spacetime created by gravitational waves are incredibly tiny—each meter-wide volume of space changes by less than 10^{-20} m as the waves pass by. Astronomers around the world have begun building *gravitational wave detectors* to measure these small changes. Once they become functional in the middle of this decade, they will provide a new, unique way of observing the cosmos. Although gravitational radiation has not yet been directly detected, its effects on the orbits of the neutron stars in the binary pulsar are in complete agreement with the predictions of general relativity for such a system. This agreement earned the Nobel Prize in 1993 for Joseph Taylor and Russell Hulse, who discovered the binary pulsar in 1974 and measured the changes in its neutron stars' orbits.

Besides losing its magnetic field, matter within a black hole loses almost all traces of its composition and origin. It retains only three properties that it had before entering the black hole: its *mass,* its *angular momentum,* and its *electrical charge.* All other attributes of matter vanish inside a black hole, and familiar concepts, such as proton, neutron, electron, atom, and molecule, no longer apply. In addition, because few large bodies appear to have a net charge, it is also doubtful that black holes do. We therefore predict that there are only two different types of black holes: those that rotate and those that do not.

If the mass creating a black hole is not rotating, the black hole it forms does not rotate either. We call nonrotating black holes **Schwarzschild black holes** (Figure 13-10). General relativity predicts that all the mass in such a black hole collapses to a point of infinite density at its center, the singularity mentioned earlier. The rest of the volume inside the Schwarzschild radius of a Schwarzschild black hole is empty space. Indeed, there is nothing at the "surface" of the black hole, either.

When the matter creating a black hole possesses angular momentum, that matter collapses to a ring-shaped singularity located inside the black hole between its center and the event horizon (Figure 13-11). Such rotating black holes are called **Kerr black holes** in honor of the New Zealand mathematician Roy Kerr, who first calculated their structure in 1963. Most Kerr black holes should be spinning thousands of times every second, even faster than the pulsars we studied in Chapter 12.

Unlike Schwarzschild black holes, the equations indicate that Kerr black holes possess a doughnut-shaped region directly *outside* their event horizons in which objects cannot

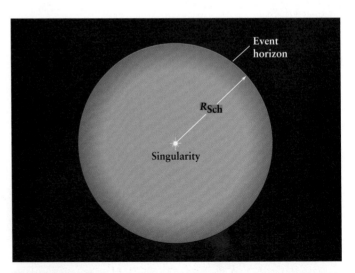

FIGURE 13-10   **Structure of a Schwarzschild (Nonrotating) Black Hole** A nonrotating black hole has a simple structure. In fact, it has only two notable parts: its singularity and its boundary. Its mass, called a singularity because it is so dense, collects at its center. The spherical boundary between the black hole and the outside universe is called the event horizon. The distance from the center to the event horizon is the Schwarzschild radius, R_{Sch}. There is no solid, liquid, or gas surface at the event horizon. In fact, except for its location at the boundary of the black hole, an event horizon lacks any features at all.

AN ASTRONOMER'S TOOLBOX 13-1

The Sizes of Black Holes

In 1918, Karl Schwarzschild, a German physicist, discovered the first solution of Einstein's equations. His solution describes the nature of nonrotating black holes. According to Einstein's general theory of relativity, the Schwarzschild radius R_{Sch} of any black hole can be found from its mass M:

$$R_{Sch} = \frac{2GM}{c^2}$$

where R_{Sch} is measured in meters; M is the black hole's mass in kilograms; c is the speed of light, 3×10^8 m/s²; and G is the gravitational constant, 6.67×10^{-11} m³/(kg • s²). This equation can be approximated by:

$$R_{Sch} \approx 3\, M'$$

where $M_{\odot}$ is the black hole's mass in solar masses and R_{Sch} is in kilometers.

Example: A 5-$M_{\odot}$ main-sequence star has a radius of 3×10^6 km, while the above equation reveals that a black hole with the same mass has a 15-km Schwarzschild radius.

Try these questions: A primordial black hole with the mass of Mount Everest would have a Schwarzschild radius of just 1.5×10^{-15} m! What is Mount Everest's mass? What is the Schwarzschild radius of a galactic black hole containing 3 billion solar masses? A black hole with a Schwarzschild radius of the Earth has how much mass?

(Answers appear at the end of the book)

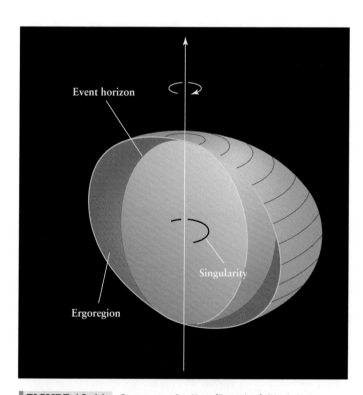

FIGURE 13-11 Structure of a Kerr (Rotating) Black Hole
Rotating black holes are only slightly more complex than nonrotating ones. The singularity of a Kerr black hole is located in an infinitely thin ring around the center of the hole. It appears as an arc in this cutaway drawing. The event horizon is again a spherical surface. There is also a doughnut-shaped region, called the ergoregion, just outside the event horizon, in which nothing can remain at rest. Space in the ergoregion is being curved or pulled around by the rotating black hole.

remain at rest without falling into the black hole. Called **ergoregions,** they are regions of spacetime that the rotating black hole drags around, like so much batter in a blender. In 1997, matter orbiting a black hole was observed to behave in a fashion consistent with the existence of an ergoregion surrounding the hole. If it is moving fast enough, an object entering the ergoregion can fly out of it again; if it stops in the ergoregion, however, it must fall into the black hole.

13-6 Falling into a black hole is an infinite voyage

Imagine being in a spacecraft orbiting only 1000 Schwarzschild radii (15,000 km) from an isolated 5-$M_{\odot}$ black hole (see An Astronomer's Toolbox 13-1 for the equation of the Schwarzschild radius). You are held in orbit by the black hole's gravitational force. Even at that short distance, the only effect the black hole has on you is its gravitational attraction. It is only when you get very, very close to the event horizon that bizarre things begin to happen. To investigate these changes, you send a cube-shaped probe toward the black hole, with the same side of the cube always facing "downward" toward the black hole. The probe emits a blue glow so that you can follow its progress. What happens to the cube as it approaches the black hole?

From the time you eject it until it reaches about 100 Schwarzschild radii (1500 km), you see the probe descend as if it were falling toward a planet or moon (Figure 13-12a). At 100 Schwarzschild radii, however, the probe begins to respond to a severe tidal effect from the black hole. The face

Probe far from
black hole

Probe close to black hole

Black
hole

Event
horizon

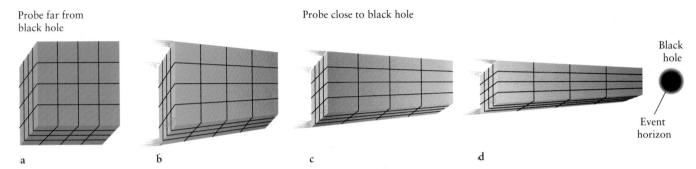

a   b   c   d

FIGURE 13-12 Effect of a Black Hole's Tidal Force on Infalling Matter (a) A cube-shaped probe intact at 1500 km from a 5-$M_\odot$ black hole. (b, c, d) Near the Schwarzschild radius, the probe is pulled long and thin by the difference in the gravitational forces felt by its different sides. This tidal effect is a greatly magnified version of the Moon's gravitational force on the Earth. The probe changes color as its photons undergo extreme gravitational redshift.

of the probe closest to the event horizon receives more gravitational pull from the black hole than its farther-away parts, and it begins to stretch apart.

By the time the probe comes within a few Schwarzschild radii of the event horizon, the tidal forces on it are so great that it violently elongates. The part of the probe closest to the black hole accelerates downward and away from the rest of the probe. Furthermore, the sides of the probe are drawn together: They are falling in straight lines toward a common center. The net gravitational effect of moving close to the event horizon is for the probe to be pulled long and thin. From a practical perspective, this means that the probe would be violently torn apart, because it is not composed of perfectly elastic material.

As the probe nears the black hole, the blue photons leaving it must give up more and more energy to escape the increasing gravitational force. However, unlike a projectile fired upward, photons cannot slow down. Rather, they lose energy by increasing their wavelengths (see An Astronomer's Toolbox 3-1). This is another example of the gravitational redshift predicted by general relativity. The closer the probe gets to the event horizon, the more its light is redshifted—first to green, then yellow, then orange, then red, then infrared, and finally radio waves (see Figures 13-12b, c, d).

Stranger still is the black hole's effect on time. General relativity predicts that when the probe approaches within a few Schwarzschild radii of the black hole, its infall rate will slow down as seen from far away. Also, signals from the probe show that its clocks are running more slowly than they did when it left your spacecraft. Time dilation is so great near the event horizon that the probe will appear to hover above it and its clocks will stop. That is what we

MOVIE MISCONCEPTIONS

The Black Hole (Walt Disney & Company, 1979)
(Laguna Design Science Photo Library/Photo Researchers)

In 1979, Disney ventured into new, big budget science fiction territory with the film release of *The Black Hole*. This film was considered to have some of the most impressive and beautiful special effects to grace theater screens at the time of its time. Even the opening credits, which depict a computer-animated representation of a black hole set to John Barry's whirling musical score, are paramount.

Ever since the release of *The Black Hole* (and possibly before), black holes have been shown in movies and television shows as funnels into which objects can fall if they follow a path into the mouth of the funnel. In what way is this Hollywood construct wrong? Also, in all the movies, some things do *not* happen to the objects and people entering the black hole that should happen. What should these infalling objects and people actually experience as they approach the event horizon?
(Answers appear at the end of the book.)

observe from far away. Someone in the probe observes something else altogether: They see the probe actually cross the event horizon and continue falling toward the black hole's singularity. Pulled apart by tidal effects, it disintegrates as it falls inward. Contrary to the science fiction concept of traveling great distances quickly by passing through a black hole, calculations indicate that objects entering them could not survive passage through, even if there were a way to come out somewhere else.

Could a black hole be connected to another part of spacetime or even some other universe? General relativity predicts such connections, called **wormholes,** for Kerr black holes, but astrophysicists are skeptical. Their conviction is called **cosmic censorship:** Nothing can leave a local region of space containing a singularity (that is, a black hole).

13-7 Black holes evaporate

In exploring the fate of black holes, astrophysicists find that, once again, these objects confound common sense. With a black hole cloaked behind the event horizon and its mass collapsed into a singularity, it must seem that there is no way of getting mass from the black hole back out into the universe again. No way, that is, until one recalls that mass and energy are two sides of the same coin. What if there was a way of converting the mass into a form of energy that *could* get out of the black hole, such as gravitational energy? The key to the conversion is Einstein's equation $E = mc^2$, and the conversion does occur, according to an idea first proposed by Stephen Hawking.

Black holes convert their mass into energy by a process called **virtual particle** production. Quantum mechanics (see Chapter 4) allows pairs of particles to form spontaneously. Each pair always consists of a particle and its antiparticle, such as a proton and an antiproton, an electron and a positron, or a pair of photons (because a photon is its own antiparticle). Normally, virtual particles form, come back together, and annihilate each other without a trace, all within the incredibly short time of about 10^{-21} seconds. For example, an electron and its antiparticle, a positron, might form and destroy each other in this process. This is occurring everywhere in the universe, not just near black holes. Astronomers visualize the entire volume of the universe, even "empty space," as a seething foam of virtual particles spontaneously appearing and disappearing.

Virtual particles have actually been observed. This is done in high-energy particle accelerators called synchrotrons and linear accelerators. These machines accelerate electrons, protons, or other particles to nearly the speed of light. Occasionally, one of these high-speed particles encounters a pair of virtual particles *before* the two virtual particles annihilate each other. The collision forces the virtual particles apart, preventing them from annihilating each other and making the two particles real and observable.

The virtual particles that form *just outside* the event horizon of a black hole feel an exceptionally powerful gravitational field. If one particle is created slightly farther from the hole than its companion, the two virtual particles feel a tidal force from the black hole's tremendous gravitational pull (recall the blue probe discussed earlier). When the gravitational tidal force is strong enough, it can pull the two virtual particles apart, making them real (Figure 13-13).

To conserve momentum, at least one of the newly formed real particles always falls into the black hole. But the other particle will sometimes have enough energy to escape from the vicinity of the black hole. The particle flies free into space, and the black hole has effectively emitted energy equal to $E = mc^2$, where m is the mass of the freed particle or $E = \frac{hc}{\lambda}$, where λ is the wavelength of the photon. Where the virtual particles become real is a temporary void in space outside the event horizon. The black hole fills the void with energy that comes from its mass. Some of the black hole's mass is converted into gravitational energy. This energy is then transmitted outside the event horizon to replace the energy taken away by the escaped particle! This is called the **Hawking process,** named after its proposer.

The time it takes a black hole to completely evaporate by the Hawking process increases with the mass of the hole. More massive black holes have lower tidal forces at their event horizons than do less massive and, therefore, smaller-diameter, black holes. Therefore, the rate at which virtual particles are converted into real particles around bigger black holes is actually slower than it is around smaller ones. The faster the real particles are produced, the faster they wear down the mass of the black hole; thus, lower-mass black

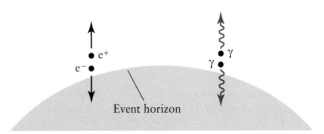

FIGURE 13-13 Evaporation of a Black Hole Throughout the universe, pairs of virtual particles spontaneously appear and disappear so quickly that they do not violate any laws of nature. The tidal force just outside the event horizon of a black hole is strong enough to tear apart two virtual particles that appear there before they destroy each other. The gravitational energy that goes into separating them makes them real and, therefore, permanent. Some of these newly created particles fall into the black hole, while some escape into interstellar space. Because the gravitational energy used to create the particles came from the black hole, the hole loses mass and shrinks, eventually evaporating completely. Here we see just a few particles in the making: an electron (e^-) and positron (e^+), and a pair of photons (γ).

holes evaporate more rapidly. Indeed, calculations indicate that in its final moments of evaporation, a black hole should create particles so quickly that it appears to explode.

A stellar black hole of 5 $M_\odot$ would take more than 10^{62} years to evaporate (much longer than the present age of the universe), while a 10^{10}-kg primordial black hole (equivalent to the mass of Mount Everest) would take only about 15 billion years. Depending on the precise age of the universe, a black hole of this size and age should be in the final throes of evaporating. Astronomers are trying to identify events in space corresponding to the violent, particle-generating deaths of primordial black holes.

13-8 Frontiers yet to be discovered

Black holes take us to the limits of science. They represent a state of matter that our science is not yet capable of explaining. This state of affairs makes black holes an especially intriguing field of study. Understanding the nature of black hole singularities is a major goal of science that has yet to be fulfilled. Numerous other frontiers exist in this realm. By directly observing gravitational radiation, we will have a new tool for exploring violent activity in the universe. We still need to understand the details of black hole formation, both from stars and by other means. We have yet to understand in detail why mid-mass and high-mass black holes have the masses they have.

If, as is now expected, black holes merge, the details of these events and their effects on the rest of the universe have still to be understood. It also remains to be shown whether or not the cosmic censorship theorem is correct. If not, the implications of the existence of wormholes need to be examined in more depth. Among the more intriguing issues surrounding black holes is whether primordial black holes exist. While the evaporation process has not yet been seen, we should soon have the technology to discover it, if it is occurring.

    Further Reading on These Topics

WHAT DID YOU KNOW?

1 *Are black holes empty holes in space?* No, black holes contain highly compressed matter—they are not empty.

2 *What exists at the surface of a black hole?* The surface of a black hole, called the event horizon, is empty space. No stationary matter exists there.

3 *What power or force enables black holes to draw things into them?* The only force that pulls things in is the gravitational attraction of the matter in the black hole.

4 *Do black holes last forever?* No, black holes evaporate.

KEY WORDS

black hole, 352
cosmic censorship, 360
ergoregion, 360
event horizon, 358
general theory of relativity, 352
gravitational redshift, 353

gravitational waves, 358
Hawking process, 362
Kerr black hole, 360
primordial black hole, 357
Schwarzschild black hole, 360
Schwarzschild radius, 358

singularity, 354
spacetime, 352
special theory of relativity, 352
supermassive black hole, 357
virtual particle, 361
wormhole, 360

KEY IDEAS

• If a stellar corpse is more massive than about 3 $M_\odot$, gravitational compression overcomes neutron degeneracy or the equivalent quark pressure and forces it to collapse further and become a black hole.

• A black hole is an object so dense that the escape velocity from it exceeds the speed of light.

The Evidence for Black Holes

• According to general relativity, mass causes space to curve and time to slow down. These effects are significant only near large masses or compact objects.

• Observations indicate that some binary star systems harbor black holes. In such systems, gases captured by the black hole from the companion star heat up and emit detectable X rays.

• Supermassive black holes originated in the cores of some galaxies. Very low mass black holes may have formed at the beginning of the universe.

Inside a Black Hole

• The event horizon of a black hole is a spherical boundary where the escape velocity equals the speed of light. No matter or electromagnetic radiation can escape from inside the event horizon. The distance from the center of the black hole to the event horizon is called the Schwarzschild radius.

• The matter inside a black hole collapses to a singularity. The singularity for nonrotating matter is a point at the center of the black hole. For rotating matter, the singularity is a ring inside the event horizon.

• Matter inside a black hole has only three physical properties: mass, angular momentum, and electrical charge.

• Nonrotating black holes are called Schwarzschild black holes. Rotating black holes are called Kerr black holes. The event horizon of a Kerr black hole is surrounded by an ergoregion in which all matter must constantly move to avoid being pulled into the black hole.

• Matter approaching a black hole's singularity is torn by extreme tidal forces generated by the black hole, light from the matter is redshifted, and time appears to slow.

• Black holes can evaporate by the Hawking process, in which virtual particles near the black hole become real. This transition decreases the mass of a black hole until eventually it disappears.

REVIEW QUESTIONS

1   Under what conditions do all outward pressures on a collapsing star fail to stop its inward motion?

2   In what way is a black hole blacker than black ink or a black piece of paper?

3   If the Sun suddenly became a black hole, how would the Earth's orbit be affected?

4   What is cosmic censorship?

5   What are the differences between rotating and nonrotating black holes?

6   Why are all the observed black hole candidates that are stellar remnants members of close binary systems?

7   If light cannot escape from a black hole, how can we detect X rays from such objects?

ADVANCED QUESTIONS

The answers to all computational problems, which are preceded by an asterisk (*), appear at the end of the book.

*8   What is the Schwarzschild radius of a black hole, measured in kilometers, containing 3 $M_\odot$? 30 $M_\odot$?

9   If more massive stars evolve and die before less massive ones, why do some black-hole candidates have lower masses than their normal stellar companions?

10   Under what circumstances might a neutron star in a binary star system become a black hole?

11   Which type of black hole, nonrotating or rotating, do science fiction writers (implicitly) use in sending spaceships from one place to another through the hole? Why would the other type not be suitable?

WHAT IF ...

12   A black hole of 5 $M_\odot$ passed by the Earth at, say, Pluto's average distance from the Sun? What would happen to our planet's orbit and to life here?

13   A primordial black hole with the mass of our Moon approached, passed through, and exited the Earth? What might happen to our planet and to life here?

14   A primordial black hole exploded nearby? What would astronomers observe?

WEB/CD-ROM QUESTIONS

15   To test your understanding of black hole structure, do Interactive Exercise 13-1 on the Web or CD-ROM. You can print out your answers if required.

16   Search the Web for information about the stellar-mass black hole candidate named V4641 Sgr. In what ways does it resemble other black hole candidates such as Cygnus X-1 and V404 Cygni? In what ways is it different and more dramatic? How do astronomers currently explain why V4641 Sgr is different?

IV ESSENTIALS FOR UNDERSTANDING THE UNIVERSE

IN THE CHAPTERS AHEAD YOU WILL DISCOVER

the Milky Way Galaxy—billions of stars along with gas and dust bound together by mutual gravitational attraction (Chapter 14)

how astronomers determined that myriad galaxies lie beyond the Milky Way (Essentials IV and Chapter 15)

how the period-luminosity law for Cepheid variables and the light curves of supernovae enable astronomers to calculate distances to other galaxies (Essentials IV and Chapter 15)

other distant objects, the large-scale structure of the universe, and its ultimate fate (Chapters 16 and 17)

a way to estimate whether advanced life may have formed on planets orbiting distant stars (Chapter 18)

THE REALM OF THE GALAXIES

Beyond the solar system, beyond the Milky Way, lie myriad galaxies. Just as stars orbit in their respective galaxies, so, too, many galaxies orbit each other. This image shows hundreds of galaxies grouped together some 325 million light-years from Earth. This computer-enhanced photograph shows only a tiny sliver of the universe, yet it provides a valuable glimpse at the variety and staggering numbers of galaxies that exist. This region of the universe is part of the so-called Great Wall of galaxies stretching across more than one-third of the sky. (© 1994 Smithsonian Institution/Corbis "Over the Great Wall," digital image opener by Margaret Geller and enhanced by Roger Resselmeyer.)

R I V U X G

WHAT DO YOU THINK?

1 Are all the stars in the universe located in the Milky Way?

2 How many galaxies exist?

You have discovered stars singly, in binary systems, and in the relatively small groups called open and globular star clusters. In binary systems and in globular clusters, the stars are *gravitationally bound:* They are held together by their mutual attraction. Now we will see that the same force creates far larger groupings of stars. This final part of the book explores how matter is distributed throughout the universe, what the universe is, where it came from, and where it is going. We start with the group of stars most familiar to us, the Milky Way, and the tools that astronomers used to establish its place in the vast universe.

IV-1 The Milky Way is only one among billions of galaxies

For those of us fortunate enough to live away from bright outdoor lights, the Milky Way appears as a filmy white band overlaid with the glow of individual stars. As children we are all told—correctly, it turns out—that our solar system is part of the **Milky Way Galaxy,** an enormous assemblage of stars. But just what else is a part of it? Perhaps, we might imagine, all the stars in the universe belong to the Milky Way. If so, then "the galaxy" and "the universe" would mean the same thing. Indeed, that is precisely what most astronomers believed until the early decades of the twentieth century.

Today, we know that the universe contains myriad galaxies, of which the Milky Way is just one. Each **galaxy** is a grouping of millions, billions, or even trillions of stars, all gravitationally bound together. These stars are sometimes accompanied by huge quantities of interstellar gas and dust. In fact, as we saw in Chapter 10, new stars are forming in giant molecular clouds every day.

As early as 1755, the German philosopher Immanuel Kant suggested that vast collections of stars lie far beyond the confines of the Milky Way. Less than a century later, an Irish astronomer observed the structure of some of those "island universes" proposed by Kant. William Parsons was the third Earl of Rosse in Ireland. He was rich, he liked machines, and he was fascinated with astronomy. Accordingly, he set about building gigantic telescopes. In February 1845, his pièce de résistance was finished. This telescope's massive mirror measured 1.8 m (6 ft) in diameter and was mounted at one end of an 18-m (60-ft) tube controlled by cables, straps, pulleys, and cranes

(Figure IV-1a). For many years, this triumph of nineteenth-century engineering enjoyed distinction as the largest telescope in the world.

With this new telescope, Lord Rosse examined many of the glowing interstellar clouds discovered and catalogued by William Herschel. Recall from your study of stars that Herschel, his sister Caroline, and his son John, among others, discovered and recorded details of many astronomical objects, including double star systems and, as is relevant here, fuzzy-looking objects they called **nebulae** (*singular* **nebula**). Lord Rosse observed that some of these nebulae have a distinct spiral structure. Perhaps the best example is M51, also called the Whirlpool or NGC 5194.

Lacking photographic equipment, Lord Rosse made drawings of what he observed. Figure IV-1b shows his drawing of M51. Views like this inspired Lord Rosse to echo Kant's proposal of "island universes." Figure IV-1c shows a modern photograph of M51. It is interesting to note the differences between the perception and interpretation of astronomical objects on the one hand and the camera's recording of them on the other.

Most astronomers of Rosse's day did not agree with the notion of island universes outside our Galaxy. Most astronomers thought that the Milky Way contained all the stars in the universe—the Milky Way *was* the universe. A considerable number of the "nebulae" listed in the *New General Catalogue* were, in fact, interstellar clouds and star clusters scattered throughout the Milky Way. For those astronomers who believed that the Milky Way contains all the matter in the universe, it naturally followed that all the intriguing nebulae were in the Milky Way, too.

The astronomical community became increasingly divided over the nature of spiral nebulae. Finally, in April 1920, a formal discussion, now known as the **Shapley–Curtis debate,** was held at the National Academy of Sciences in Washington, D.C. On one side was Harlow Shapley, an astronomer renowned for his determination of the size of the Milky Way Galaxy (see Section 14-1). Shapley believed the spiral nebulae to be relatively small, nearby objects scattered around our Galaxy like the globular clusters he had studied. Opposing Shapley was Heber D. Curtis of the Lick Observatory near San Jose, California. Curtis championed the island universe theory, arguing that each of these spiral nebulae is a separate rotating system of stars much like our own Galaxy.

The Shapley–Curtis debate generated much heat but little light. While focusing scientific attention on the size of the universe, nothing was decided, because no one had any firm evidence to demonstrate exactly how far away the spiral nebulae are. Astronomers desperately needed to devise a way to measure the distances to them. Such a measurement was the first great achievement of a young teacher and former basketball player who moved from Kentucky to Chicago to study astronomy. His name was Edwin Hubble.

a

b

c

FIGURE IV-1 High-Tech Telescope of the Mid-Nineteenth Century (a) Built in 1845, this structure housed a 1.8-m-diameter telescope, the largest of its day. The improved resolution it provided over other telescopes is like the improvement that the Hubble Space Telescope provided over Earthbound optical instruments when it was launched. The telescope, as shown here, was restored to its original state during 1996–1998. (b) Using his telescope, Lord Rosse made this sketch of the spiral structure of M51. (c) A modern photograph of M51 (NGC 5194). This spiral galaxy in the constellation of Canes Venatici is known as the Whirlpool Galaxy because of its distinctive appearance. Its distance from Earth is about 20 million light-years. The blob on the right, at the end of one of its spiral arms, is a companion galaxy. (a: Birr Castle Demesne; b: Lund Humphries; c: NOAO)

IV-2 Studies of Cepheid variable stars led to the discovery of the distances to other galaxies

After completing his studies, Edwin Hubble joined the staff of the Mount Wilson Observatory in Pasadena, California, and in 1923 he took an historic photograph of M31, then called the Andromeda Nebula. It was one of the spiral nebulae around which controversy raged. Hubble carefully examined his photographic plate and discovered what he first thought to be a nova. Referring to previous plates of that region, he soon realized that the object was actually a Cepheid variable star. As we saw in Chapter 11, these pulsating stars vary in brightness periodically. Further scrutiny over the next several months revealed many other Cepheids. Figure IV-2 shows a Cepheid in the galaxy M100 at different stages of brightness.

Only a decade before, in 1912, the American astronomer Henrietta Leavitt had published an important study of Cepheid variables. Leavitt studied numerous Cepheids in the Small Magellanic Cloud, then believed to be a nebula; now known to be a galaxy orbiting the Milky Way. On a clear night, this galaxy is visible to the naked eye in the southern hemisphere as a thin haze near the Large Magellanic Cloud galaxy. As we learned in Chapter 11, there are two kinds of Cepheid variables: the metal-rich Type I Cepheids, which Leavitt studied, and the slightly dimmer, metal-poor Type II Cepheids. The latter were not discovered until the 1940s, when U.S. cities were blacked out during World War II and the sky was, therefore, especially dark.

Leavitt's study led her to the period-luminosity law for Type I Cepheids. As we saw in Figure 11-24, a direct relationship exists between a Cepheid's luminosity (or absolute magnitude) and its period of oscillation. Suppose you find a Cepheid variable, measure its period, and use a graph such as

FIGURE IV-2 M100 One of the most reliable ways to determine the distance to moderately remote galaxies is to locate Cepheid variable stars in them, as discussed in the text. At 17 Mpc (56 Mly), the galaxy M100 in the constellation Coma Berenices is among the most remote objects whose distances have been determined using Cepheids. **Insets:** The Cepheid in this view, one of 20 located to date in M100, is shown at different stages in its brightness cycle, which recurs over several weeks. (Dr. Wendy L. Freedman, Observatories of the Carnegie Institution of Washington; NASA)

R I V U X G

Insight into Science **Room for Debate** Lacking definitive data, competent scientists can develop and believe strikingly different explanations for the same observations. At the time of the Shapley–Curtis debate, the observations allowed both points of view. It was only with the advent of the distance measurement technique used by Hubble that the spiral nebulae were definitely shown to be outside the Galaxy.

IV-3 Discovering the universe helps us understand our place in it

Hubble's results, which he presented at the end of 1924, settled the Shapley–Curtis debate once and for all. The universe was recognized to be far larger and populated with far bigger objects than most astronomers had imagined. Hubble had discovered the realm of the galaxies.

In Chapter 14, we apply Hubble's method to find Earth's place in the Milky Way Galaxy. Recall that the Sun's proximity to us makes it our best-understood star. It might seem that the nearness of the stars and clouds in the Milky Way would make it the best understood galaxy. However, we will see that the clouds of gas and dust surrounding the solar system make it very challenging for astronomers to survey the Galaxy completely.

Moving out beyond the Milky Way, observations of distant galaxies by the Hubble Space Telescope lead astrono-

---

Figure IV-3 to determine the star's average luminosity. The star's brightness can be expressed as an absolute magnitude. Meanwhile, you observe the star's apparent magnitude. Because you now know both the apparent and the absolute magnitudes, you can calculate the star's distance (see An Astronomer's Toolbox III-2). By observing the period, Hubble was able to determine the absolute magnitude of the Cepheids in M31 and to calculate the distances to them and, therefore, the distance to the nebula containing them.

Straightforward calculations using modern data on the distance of Type I Cepheids demonstrate that M31 is some 2.2 million light-years *beyond* the Milky Way. This proves that M31 is not an open or globular cluster (what were traditionally called nebulae) in our Galaxy but rather an enormous separate stellar system—a separate galaxy. M31, now called the *Andromeda Galaxy,* is the most distant object in the universe that can be seen with the naked eye. Similar calculations are done for all galaxies in which Cepheids can be observed. In a similar way, observations of Type Ia supernovae provide distance measurements to even more distant galaxies (An Astronomer's Toolbox IV-1).

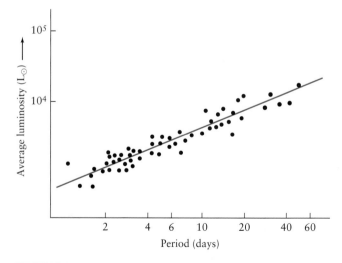

FIGURE IV-3 **The Period–Luminosity Relation** This graph shows the relationship between the periods and average luminosities of classical (Type I) Cepheid variables and the closely related RR Lyrae stars (discussed in Chapters 11 and 14). Each dot represents a Cepheid or RR Lyrae whose brightness and period have been measured.

AN ASTRONOMER'S TOOLBOX IV-1

Cepheids and Supernovae as Indicators of Distance

Because their periods are directly linked to their luminosities, Cepheid variables are one of the most reliable tools astronomers have for determining the distances to galaxies. To this day, astronomers use this link—much as Hubble did back in the 1920s—to measure intergalactic distances. More recently, they have begun to use Type Ia supernovae, which are far more luminous and thus can be seen much farther away, to determine the distances to very remote galaxies.

Example: In 1992, a team of astronomers used Cepheid variables in a galaxy called IC 4182 to deduce that galaxy's distance from the Earth. The team of astronomers used the Hubble Space Telescope on 20 separate occasions to record images of the stars in IC 4182. By comparing these images, the astronomers could pick out which stars vary in brightness. In this way, they discovered 27 Cepheids in IC 4182. Using their observations, the astronomers estimated the brightnesses of the Cepheid variables and plotted their light curves. The Hubble Space Telescope is particularly well suited for studies of this kind, because its extraordinary angular resolution makes it possible to pick out individual stars at great distances. One such Cepheid has a period of 42.0 days and an average apparent magnitude (m) of +22.0. (See Section III-2 for an explanation of the apparent magnitude scale.) By comparison, the dimmest star you can see with the naked eye has $m = +6$; this Cepheid in IC 4182 appears less than one one-millionth as bright. The star's spectrum shows that it is a metal-rich Type I Cepheid variable.

According to the period-luminosity relation shown in Figure IV-3, such a Type I Cepheid with a period of 42.0 days has an average luminosity of 33,000 $L_\odot$. This can be expressed by saying that this Cepheid has an average absolute magnitude (M) of −6.5. (This compares to $M = +4.8$ for the Sun). Hence, the difference between the Cepheid's apparent and absolute magnitudes, called its **distance modulus,** is

$$m - M = (+22.0) - (-6.5) = 22.0 + 6.5 = 28.5$$

From An Astronomer's Toolbox III-2, we see that the **distance modulus** of a star is related to its distance in parsecs (d) by

$$m - M = 5 \log d - 5$$

This can be rewritten as

$$d = 10^{(m - M + 5)/5} \text{ parsecs}$$

Inserting the value for the distance modulus in this equation, we get the distance to the Cepheid variable and, hence, the distance to the galaxy of which the star is part:

$$d = 10^{(28.5 + 5)/5} \text{ parsecs} = 10^{6.7} \text{ parsecs} = 5\ 4 \times 10^6 \text{ parsecs}$$

The galaxy is 5 Mpc (1 Mpc = 1 megaparsec = 10^6 parsecs), or 16 million (1.6×10^7) light-years, from Earth.

Astronomers are interested in IC 4182 because a Type Ia supernova was observed there in 1937. All Type Ia supernovae are exploding white dwarfs that reach nearly the same maximum brightness at the peak of their outburst (see Section 12-8). Once astronomers know the peak absolute magnitude of Type Ia supernovae, they can use these supernovae as distance indicators. Because the distance to IC 4182 is known from its Cepheids, the 1937 observations of the supernova in that galaxy allow us to calibrate Type Ia supernovae as distance indicators. At maximum brightness, the 1937 supernova reached an apparent magnitude of $m = +8.6$. Since the distance modulus of the galaxy ($m - M$) is 28.5, we see that when a Type Ia supernova is at maximum brightness, its absolute magnitude is

$$M = m - (m - M) = 8.6 - 28.5 = -19.9$$

Whenever astronomers find a Type Ia supernova in a remote galaxy, they can combine this absolute magnitude with the observed maximum apparent magnitude to get the galaxy's distance modulus, from which the galaxy's distance can be easily calculated (just as we did above for the Cepheids in IC 4182). This technique has been used to determine the distances to galaxies hundreds of millions of parsecs away.

Try these questions: At what distance in parsecs is a star with distance modulus 20? Epsilon (ε) Indi has an apparent magnitude of +4.7 and distance of 3.6 pc. What is its absolute magnitude? What would the Sun's apparent magnitude and distance modulus be as seen from 100 pc away? Its absolute magnitude is +4.83.

(Answers appear at the end of the book.)

mers to estimate that some 50 billion galaxies exist throughout the visible universe. In Chapter 15 we learn about their different types. By looking at neighboring galaxies not concealed by clouds, we learn about the activity in their centers. We also discover that galaxies are themselves often found in groups, and the large-scale structure of the universe does not stop there. Most groups of galaxies are themselves parts of larger groups, distributed as if on the surfaces of enormous, (nearly) empty bubbles in space. These groupings and their movements away from Earth offer us clues to the structure and origin of the universe.

Quasars and active galaxies, exceptionally powerful sources of radiation, are examined in Chapter 16. We learn of their structure and the causes of their emissions. In Chapter 17 we consider the beginning, evolution, and fate of the whole universe. Because this subject involves the most distant observations, it entails the greatest uncertainty of all the topics we have covered. Finally, in Chapter 18 we take a brief look at the possible existence of life elsewhere in the universe.

IV-4 Frontiers yet to be discovered

While astronomers now know that there are billions of galaxies in the visible universe, we do not yet have a complete map of the cosmos. Just as mapping the human genome seemed a distant dream only a few decades ago, mapping the universe seems so today. However, we are already making strides in that direction with the Sloan sky survey, and from the Hubble Space Telescope, *Chandra* X-ray observatory, and other deep field observations. However, a full catalog of all galaxies and other matter remains to be made. Furthermore, we have yet to determine why the universe has the foamy structure that is observed.

WHAT DID YOU THINK?

1 *Are all the stars in the universe located in the Milky Way?* While all the stars we can see without the aid of telescopes are in the Milky Way, most stars in the universe are in other galaxies.

2 *How many galaxies exist?* There are an estimated 50 billion galaxies.

KEY WORDS

distance modulus, 369
galaxy, 366

Milky Way Galaxy, 366
nebula (*plural* nebulae), 366

Shapley–Curtis debate, 366

KEY IDEAS

• A century ago, astronomers were divided on whether all stars and nebulae are part of the Milky Way Galaxy.

• The Shapley–Curtis debate was the first major discussion among astronomers of whether the Milky Way contains all the stars in the universe.

• Cepheid variable stars are important in determining the distance to other galaxies.

• Edwin Hubble first determined that there are other galaxies far outside the Milky Way.

REVIEW QUESTIONS

1 What was the Shapley–Curtis debate all about? Was a winner declared at the end of the debate? Whose ideas turned out to be correct?

2 How did Edwin Hubble prove that M31 is not a nebula in our Milky Way Galaxy?

WHAT IF . . .

3 The billions of what we now call galaxies, with their myriad stars, were actually as close as the stars in the Milky Way? What would be different about the night sky? Can you think of any other things that would be different under such conditions?

4 The Milky Way was, in fact, the entire universe?

WEB/CD-ROM QUESTION

5 Search the Web for information on the Shapley–Curtis debate. Who proposed the debate in the first place? Briefly outline the relevant scientific beliefs held by the two men. What points did they agree upon and what did they disagree on? Why was the debate considered inconclusive? Support your argument with examples, if possible.

OBSERVING PROJECTS

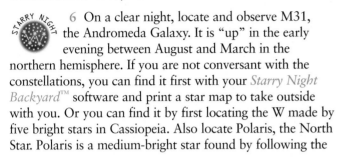

6 On a clear night, locate and observe M31, the Andromeda Galaxy. It is "up" in the early evening between August and March in the northern hemisphere. If you are not conversant with the constellations, you can find it first with your *Starry Night Backyard*™ software and print a star map to take outside with you. Or you can find it by first locating the W made by five bright stars in Cassiopeia. Also locate Polaris, the North Star. Polaris is a medium-bright star found by following the line from the two stars in the Big Dipper farthest from the handle upward until you reach a medium-bright star. Draw a mental line from Polaris to the second star on the right side of Cassiopeia, if you see it as the letter W, or the second star on the left of Cassiopeia, if you see it as the letter M. Keep moving south along that line one-half as far again, and you should see a fuzzy "star," which is M31. Describe what you see. Examine M31 through binoculars and again through the largest telescope available. Describe what you see.

1755

Milky Way proposed by Immanuel Kant to be only one among many galaxies.

1914-1922

Redshift observed in spectra of galaxies taken by Vesto Slipher.

1927-1929

Bertil Lindblad calculates the solar system's motion through the Milky Way. BL Lac discovered.

1943

First Seyfert galaxy detected by Carl Seyfert.

1785

William Herschel calculates the location of solar system within the Milky Way, but incorrectly.

1917-1918

Harlow Shapley observes that Milky Way is surrounded by a spherical distribution of globular clusters not centered on Earth; determines, essentially correctly, the location of the solar system in the Milky Way and the location of Galactic center.

Photography, Radio

Television, Radar

Transistors

964

Andromeda galaxy discovered by Abd al-Rahman al-Sufi.

1922

Alexander Friedmann discovers expanding universe solution to Einstein's equations of general relativity.

1930-1936

Robert Trumpler establishes presence of interstellar dust. Karl Jansky detects intense radio waves from direction of Galactic center. Grote Reber builds first radio telescope.

1912-1915

Henrietta Leavitt discovers the period–luminosity relationship for Cepheid Type I stars. Einstein's general theory of relativity proposes a dynamic universe.

1924-1929

Edwin Hubble uses Cepheid variable stars to prove that nebulae are galaxies external to the Milky Way; proposes "tuning fork" classification scheme for galaxies based on their shapes; determines the relationship between the distance to superclusters of galaxies and their speed away from the Milky Way.

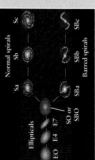

1948

Ralph Alpher, Hans Bethe, and George Gamow predict the cosmic microwave background radiation.

1845

Lord Rosse observes and sketches spiral-shaped nebulae later proved to be external galaxies.

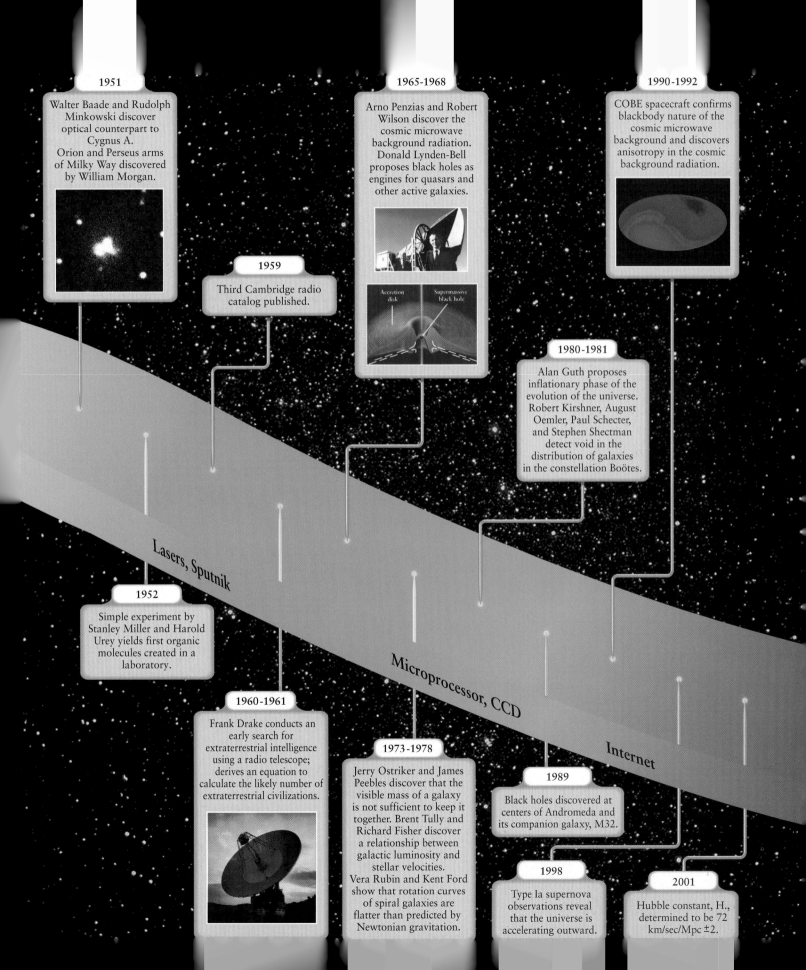

1951

Walter Baade and Rudolph Minkowski discover optical counterpart to Cygnus A.
Orion and Perseus arms of Milky Way discovered by William Morgan.

1959

Third Cambridge radio catalog published.

1965-1968

Arno Penzias and Robert Wilson discover the cosmic microwave background radiation.
Donald Lynden-Bell proposes black holes as engines for quasars and other active galaxies.

Accretion disk

Supermassive black hole

1990-1992

COBE spacecraft confirms blackbody nature of the cosmic microwave background and discovers anisotropy in the cosmic background radiation.

1980-1981

Alan Guth proposes inflationary phase of the evolution of the universe. Robert Kirshner, August Oemler, Paul Schecter, and Stephen Shectman detect void in the distribution of galaxies in the constellation Boötes.

Lasers, Sputnik

1952

Simple experiment by Stanley Miller and Harold Urey yields first organic molecules created in a laboratory.

Microprocessor, CCD

1960-1961

Frank Drake conducts an early search for extraterrestrial intelligence using a radio telescope; derives an equation to calculate the likely number of extraterrestrial civilizations.

1973-1978

Jerry Ostriker and James Peebles discover that the visible mass of a galaxy is not sufficient to keep it together. Brent Tully and Richard Fisher discover a relationship between galactic luminosity and stellar velocities.
Vera Rubin and Kent Ford show that rotation curves of spiral galaxies are flatter than predicted by Newtonian gravitation.

Internet

1989

Black holes discovered at centers of Andromeda and its companion galaxy, M32.

1998

Type Ia supernova observations reveal that the universe is accelerating outward.

2001

Hubble constant, H., determined to be 72 km/sec/Mpc ±2.

14 THE MILKY WAY GALAXY

IN THIS CHAPTER YOU WILL DISCOVER

- the properties of our Milky Way Galaxy

- Earth's location in the Milky Way

- how interstellar gas and dust enable star formation to continue in our Galaxy

- that observations reveal the presence of significant mass in the Milky Way that astronomers have yet to identify

- that a black hole exists at the center of our Galaxy

Our Galaxy

This wide-angle photograph spans half the Milky Way as seen from the equatorial latitudes. The Northern Cross is at the left, and the Southern Cross is at the right. The center of the Galaxy is in the constellation Sagittarius, in the middle of this photograph. The dark lines and blotches are caused by hundreds of interstellar clouds of gas and dust that obscure the light from background stars, rather than by a lack of stars. (Dirk Hoppe).

R I V U X G

WHAT DO YOU THINK?

|1   Where in the Milky Way is the solar system located?

|2   How many stars does the Milky Way Galaxy contain?

|3   Is the Sun moving through the Milky Way and, if so, how fast?

Like the curtain in a theater, our nighttime view of the Milky Way raises expectations. What breathtaking natural displays lie beyond the translucent glow of stars and gas? The Milky Way stretches all the way around the sky in a continuous band that is almost perpendicular to the plane of the ecliptic. Galileo, the first person to look at the Milky Way through a telescope, discovered that it contains countless dim stars. How far do they extend and what exists in the space between them? Today, we know that most of the stars, gas, and dust in the Galaxy lie in an enormous, rotating, disk-shaped assemblage, very much like the spiral nebulae that Lord Rosse first sketched more than 150 years ago. To discover this distribution of matter, astronomers first had to find our place in the Milky Way.

THE STRUCTURE OF OUR GALAXY

Because the band of the Milky Way completely encircles us, astronomers long ago suspected that the Sun and all the stars are part of it. In the 1780s, William Herschel took the first steps toward mapping its structure. He attempted to deduce the Sun's location in the Galaxy by counting the number of stars in 683 regions of the sky. He reasoned that the greatest density of stars should be seen toward the Galaxy's center and a lesser density seen toward the edge. However, Herschel found roughly the same density of stars all along the Milky Way. He therefore concluded that we are at the center of the Galaxy.

Herschel was wrong: The Earth has no privileged place in the Milky Way. The Sun is 26,000 ly (8000 pc) from the Galaxy's center, the **galactic nucleus.** Herschel's physical understanding of the cosmos was incomplete, so he misinterpreted his observations and thus came to an incorrect conclusion.

14-1 Interstellar dust hides the true extent of the Milky Way

While studying star clusters in the 1930s, R. J. Trumpler discovered the reason for Herschel's mistake. Herschel did not know about interstellar gas and dust, which affected his counts of the stars. Trumpler noticed that remote star clus-

ters appear dimmer than would be expected just from their distance alone. Something must be blocking starlight on its way toward the Earth. He concluded that interstellar space is not a perfect vacuum. Instead, it contains dust that absorbs light from distant stars. Great patches of this dust are clearly visible in wide-angle photographs, such as the figure that opens this chapter. Like the stars, this dust is concentrated in the plane of the Galaxy.

This interstellar dust completely obscures from view visible light emanating from the center of our Galaxy. Any visual photons from there are absorbed or scattered before they reach us. Therefore, Herschel was seeing only nearby stars, and he measured apparent magnitudes that were dimmer than they would have been had there been no interstellar dust. Without adjusting for the effects of the dust, he concluded the stars were farther away than they really are. He also had no idea of either the true size of the Galaxy or the vast number of stars located in the general direction of the galactic center and invisible to us.

> **Insight into Science** **A Little Knowledge** Incomplete information often leads to incorrect interpretation of data and, therefore, to incorrect conclusions. Herschel's lack of knowledge about the matter in interstellar space prevented him from correctly interpreting the distribution of stars surrounding the Earth and, thus, led to an inaccurate position for the Sun within the Galaxy.

Because interstellar dust is concentrated in the plane of the Galaxy, the absorption of starlight is strongest in those parts of the sky covered by the Milky Way. Above or below the plane of the Galaxy, our view is relatively unobscured. Knowledge of our true position in the Galaxy eventually came from observations of globular clusters (see Section 11-12) in these unobscured portions of the sky.

As we saw in Essentials IV, Henrietta Leavitt published her studies of Type I Cepheids in 1912. Ever since, these variable stars have been important tools in determining distances to the stars. Shortly after Leavitt's discovery, Harlow Shapley, who later debated the existence of other galaxies with Heber Curtis, turned his attention to the family of pulsating stars known as RR Lyrae variables, which are quite similar to Cepheid variables (see Figure IV-3). RR Lyrae variables are commonly found in globular clusters (Figure 14-1). We saw in Chapter 11 that RR Lyrae variable stars also have a distinctive period-luminosity relationship. Therefore, they, too, can be used to determine distances.

Shapley used the period-luminosity relationship to determine the distances to the then-known 93 globular clusters in the sky. (About 150 are known today.) From their directions and distances, he mapped out the distribution of these

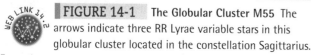

FIGURE 14-1   The Globular Cluster M55 The arrows indicate three RR Lyrae variable stars in this globular cluster located in the constellation Sagittarius. From the average apparent magnitude (as seen in this photograph) and the average absolute magnitudes of these stars (known to be roughly 100 $L_\odot$), astronomers have deduced that the distance to this cluster is 20,000 ly. (Harvard Observatory)

FIGURE 14-2   A View Toward the Galactic Center More than a million stars fill this view, which covers a relatively clear window just 4° south of the galactic nucleus in Sagittarius. Note the two prominent globular clusters. Although most regions of the sky toward Sagittarius are thick with dust, there is very little obscuring matter in this tiny section of the sky. (Harvard Observatory)

clusters in three-dimensional space. By 1917, Shapley had discovered that the globular clusters are located in a spherical distribution centered not on the Earth, but about a point in the Milky Way toward the constellation Sagittarius. Figure 14-2 shows two globular clusters in that part of the sky. Shapley then made a bold conjecture: *The globular clusters orbit the center of the Milky Way located in Sagittarius.* His pioneering research has since been observationally verified. The Earth is not at the center of the Galaxy.

14-2 Nonoptical observations help map the galactic disk

Efforts to observe into the dust-filled plane of the Milky Way are made mainly using radio wave, infrared, X-ray, and gamma ray telescopes. These wavelengths are scattered much less by the interstellar gas and dust located throughout the Galaxy's disk than are visible or ultraviolet wavelengths. (In the same way, the *Magellan* spacecraft used radio waves to pierce the veil of clouds around Venus, enabling astronomers to create a detailed map of that planet's surface; see Figures 6-14 through 6-17.) A notable exception is the region photographed at visible wavelengths in Figure 14-2.

Observations of the distant parts of the Galaxy were first made using radio telescopes. Radio waves penetrate the Earth's atmosphere, so radio observations can be made from anywhere you can build a radio telescope. Infrared observations must be made at high altitudes or in space, and X-ray and gamma observations must all be made from space.

Detecting the radio emission directly from interstellar hydrogen, by far the most abundant element in the universe, is a primary means of mapping the Galaxy. Unfortunately, the major transitions of electrons in the hydrogen atom (see Figure 4-15) produce photons at ultraviolet and visible wave-

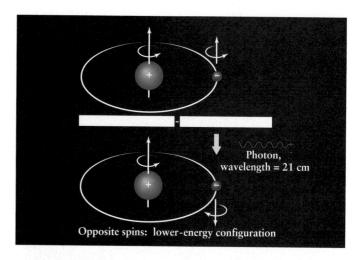

FIGURE 14-3   Electron Spin and the Hydrogen Atom Due to their spin, electrons and protons are both tiny magnets. When an electron and the proton it orbits are spinning in the same direction, their energy is higher than when they are spinning in opposite directions. When the electron flips from the higher-energy to the lower-energy configuration, the atom loses a tiny amount of energy that is radiated as a radio photon with a wavelength of 21 cm.

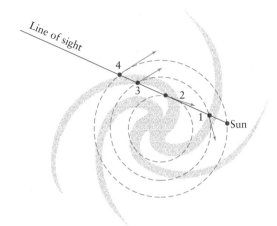

FIGURE 14-4 **A Technique for Mapping the Galaxy**
Hydrogen clouds at different locations along our line of sight are moving around the center of the Galaxy at different speeds. Radio waves from the various gas clouds therefore exhibit slightly different Doppler shifts, permitting astronomers to sort out the gas clouds and map the Galaxy.

lengths that do not penetrate the interstellar medium. How, then, can radio telescopes detect all this hydrogen? The answer lies in atomic physics.

In addition to mass and charge, particles such as protons and electrons possess a tiny amount of angular momen-

tum commonly called **spin**. An electron or a proton is something like a tiny spinning sphere. According to the laws of quantum mechanics, the electron and proton in a hydrogen atom can only spin in either parallel or opposite directions (Figure 14-3); they can have no other spin orientations. If the electron in a hydrogen atom flips from one orientation to the other, the atom must gain or lose a tiny amount of energy. In particular, when flipping from parallel to opposite spins, the atom emits a low-energy radio photon whose wavelength is 21 cm. In 1951, a team of astronomers succeeded in detecting the faint hiss of 21-cm radio static from spin flips of interstellar hydrogen.

The detection of **21-cm radio radiation** was a major breakthrough in mapping the **disk** of the Galaxy. To see why, suppose that you aim your radio telescope across the Galaxy as sketched in Figure 14-4. Your radio receiver picks up 21-cm emission from hydrogen clouds at points 1, 2, 3, and 4. However, the radio waves from these various clouds are Doppler shifted (see Section 4-7) by slightly different amounts, because they are moving at different speeds as they orbit around the center of the Galaxy. These various Doppler shifts smear the 21-cm radiation over a range of wavelengths. Because these radio waves from gas clouds in different parts of the Galaxy arrive at our radio telescopes with slightly different wavelengths, it is possible to identify which radio signals come from which gas clouds and thus to produce an initial map of the Galaxy, such as that shown in Figure 14-5.

Our radio map reveals numerous arched lanes of neutral hydrogen gas. If this were the overall structure of the Galaxy,

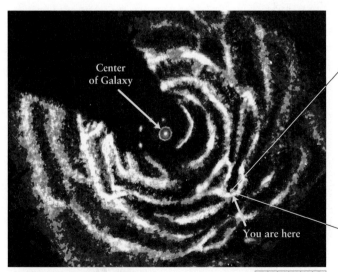

R I V U X G

FIGURE 14-5 **A Map of the Galaxy** This map, based on radio telescope surveys of 21-cm radiation, shows the distribution of hydrogen gas in a face-on view of the Galaxy. This view just hints at spiral structure. The galactic nucleus is marked with a dot surrounded by a circle. Details in the

large, blank, wedge-shaped region toward the upper left of the map are unknown, because gas in this part of the sky is moving perpendicular to our line of sight and thus does not exhibit a detectable Doppler shift. **Inset:** Our solar system lies just inside the Orion arm of the Milky Way Galaxy. (Courtesy of G. Westerhout; inset: National Geographic)

then the Milky Way would appear unlike any other observed galaxy (see Chapter 15 for typical images of other galaxies). Indeed, the other disk-shaped galaxies we observe have spiral arms. We need more information from space to improve our understanding of the Galaxy's disk features. Note that photographs of disk galaxies (Figure 14-6) show spiral arms outlined by "spiral tracers"—bright, population I stars and emission nebulae. As we saw in Chapter 11, these features indicate active star formation. If the Milky Way is spiral, then another useful way to further chart its structure and show that it has **spiral arms** is to map the locations of star-forming complexes marked by H II regions, giant molecular clouds, and massive, hot, young stars in OB associations.

Dust absorption limits the range of visual observations in the plane of the Galaxy to less than 10,000 ly from the Earth. Astronomers use visible observations of nearby, bright OB associations and associated H II regions to plot the spiral arms near the Sun. Radio observations of CO and hydrogen have been used to chart more remote star-forming regions of the Galaxy. Taken together, all these observations indicate that our Galaxy has about 200 billion stars located in and between at least four major spiral arms and several short arm segments (Figure 14-7a). Recent observations also suggest that a bar of stars and gas may cross the nuclear

bulge. If so, the Milky Way resembles the galaxy in Figure 14-7b more than it does Figure 14-7a. We will explore the origin of the spiral arms in Chapter 15, when we present more evidence about the cause of spiral structure from observations of other galaxies.

The observable disk of our Galaxy is about 100,000 ly in diameter and about 2000 ly thick (Figure 14-8). The Sun is located near a relatively short arm segment called the Orion arm, which includes the Orion Nebula (see Figure 11-13) and neighboring sites of vigorous star formation in that constellation. Two major spiral arms border either side of the Sun's position. On the side toward the galactic center is the Sagittarius arm, which stargazers in the northern hemisphere see in the summer when they look at the portion of the Milky Way stretching across Scorpius and Sagittarius (see the figure that opens this chapter). Directed away from the galactic center is the Perseus arm, which is visible in the northern hemisphere in the winter. The remaining two major spiral arms are usually referred to as the Centaurus arm and the Cygnus arm, neither of which can be seen at visible wavelengths because of obscuring dust in the interstellar medium.

 Observations reveal that the Galaxy's arms spiral out from a flattened sphere of stars, called the **nuclear bulge,** that is about 20,000 ly in diameter.

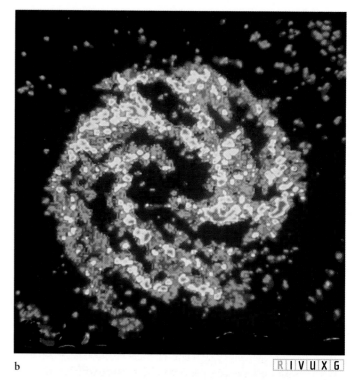

a   R I V U X G

b   R I V U X G

FIGURE 14-6   Two Views of a Spiral Galaxy The galaxy M83 is in the southern constellation of Centaurus, about 12 million light-years from Earth. **(a)** At visible wavelengths, spiral arms are clearly illuminated by young stars and glowing H II regions. **(b)** A radio view at 21-cm wavelength shows the emission from neutral hydrogen gas. Note that the spiral arms are more clearly demarcated by visible stars and H II regions than by 21-cm radio emission. (a & b: Anglo-Australian Observatory, VLA, NRAO)

a

b

FIGURE 14-7 **Our Galaxy Seen Face-on** **(a)** Our Galaxy has four major spiral arms and several shorter arm segments. The Sun is located near the short Orion arm, between two major spiral arms. The Galaxy's diameter is about 100,000 ly, and the Sun is about 26,000 ly from the galactic center. **(b)** If the Milky Way has a bar of stars crossing its nuclear bulge, as some recent observations suggest, then it probably looks more like this. (a & b: Illustration by Dennis Davidson, courtesy of AMNH/Hayden Planetarium)

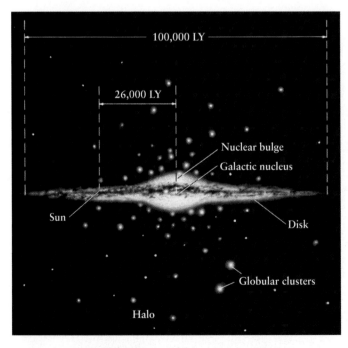

FIGURE 14-8 **Edge-on View of Our Galaxy** As seen from the side, the three major components of our Galaxy are a thin disk, a nuclear bulge, and a halo. The disk contains gas and dust along with Population I (young, metal-rich) stars. The halo is composed almost exclusively of Population II (old, metal-poor) stars.

This feature is seen in Figure 14-9, a wide-angle infrared image of the Galaxy taken by the *COBE* spacecraft. The nuclear bulge is centered on the galactic nucleus 26,000 ly away from us.

14-3 The galactic nucleus is an active, crowded place

If you lived on a planet near the very center of the Galaxy, called the nucleus, you would see a million stars as bright as Betelgeuse. The total intensity of starlight from all those nearby stars would be equivalent to 200 of our full Moons. Night would never really fall. Stranger still, the galaxy around you would be filled with intense activity.

Figure 14-10 shows three infrared views looking toward the nucleus of the Galaxy. Figure 14-10a is a wide-angle view covering a 50° segment of the Milky Way through Sagittarius and Scorpius. The prominent band across this image is a thin layer of dust in the plane of the Galaxy. The numerous knots and blobs along the dust layer are interstellar clouds heated by young O and B stars. Figure 14-10b is an IRAS view of the galactic center. Numerous streamers of dust (in blue) surround it. The strongest infrared emission (in white) comes from **Sagittarius A**, which is a grouping of several powerful sources of radio waves. One of these sources, called Sagittarius A* (pronounced A-star), is believed to be the galactic

R I V U X G

FIGURE 14-9 Infrared **View of the Milky Way** Taken by the *COBE* satellite in 1997, this infrared image shows the disk and nuclear bulge of our Galaxy. Most of the sources scattered above and below the disk are nearby stars. Stars appear white, while interstellar dust appears red. Note that the dust that obscures light from more distant stars in the image opening this chapter is quite bright in this infrared image. (NRL/RSD/NASA)

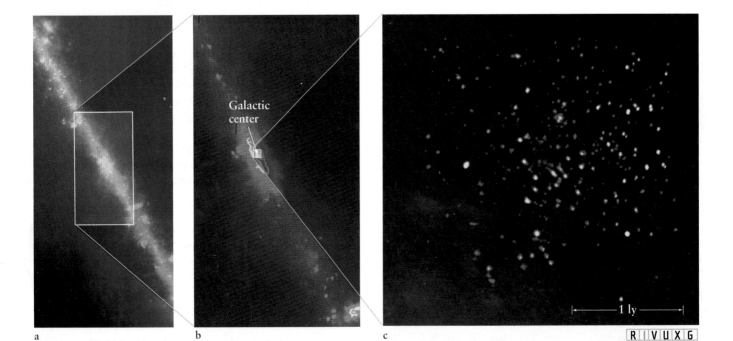

Galactic center

1 ly

a   b   c   R I V U X G

FIGURE 14-10 The Galactic Center (a) This wide-angle view at infrared wavelengths shows a 50° segment of the Milky Way centered on the nucleus of the Galaxy. Black represents the dimmest regions of infrared emission, with blue the next strongest, followed by yellow and red; white represents the strongest emission. The prominent band across this photograph is a layer of dust in the plane of the Galaxy. Numerous knots and blobs along the plane of the Galaxy are interstellar clouds of gas and dust heated by nearby stars. (b) This close-up infrared view of the galactic center covers the area outlined by the white rectangle in (a). (c) This infrared image shows about 300 of the brightest stars less than 1 ly from Sagittarius A*, which is at the center of the picture. The distribution of stars and their observed motions around the galactic center imply a very high density (about a million solar masses per cubic light-year) of less luminous stars. (a & b: NASA; c: A. Eckart, R. Genzel, R. Hofman, B.J. Sams, and C.E. Tocconi-Garman, ESO)

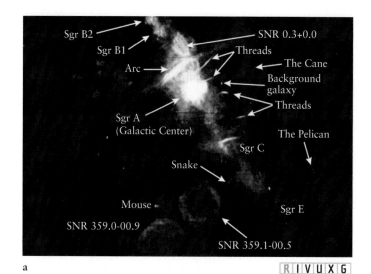

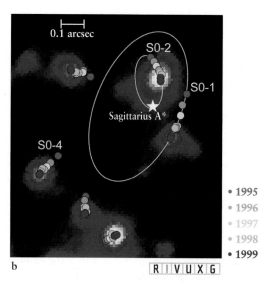

FIGURE 14-11 Two Views of the Galactic Nucleus **(a)** A radio image taken at the VLA of the galactic nucleus and environs. This image covers an area of the sky 8 times wider than the Moon. SNR means supernova remnant. The numbers following each SNR are its right ascension and declination. The Sgr features are radio-bright objects. **(b)** The colored dots superimposed on this infrared image show the motion of six stars around the unseen massive object at the position of the radio sources Sagittarius A*, part of Sag A in (a). The orbits were measured over a 5-year period. Sag A* is actually at one focus of each star's orbit. It appears otherwise only because the orbits are not oriented face-on to us. This plot indicates that the stars are held in orbit by a 2.5×10^6 $M_\odot$ black hole. (a: The Kobe Project/DIRBE/NASA; b: Andrea Ghez, Department of Astronomy, UCLA)

nucleus. Figure 14-10c shows stars within 1 ly of Sagittarius A*, with resolution of 0.02 ly.

Radio observations give a different picture of the center of our Galaxy. In 1960, Doppler shift measurements of 21-cm radiation revealed two enormous arms of hydrogen. One arm, which is located between Earth and the galactic nucleus, is approaching us at a speed of 53 km/s. The other arm, on the far side of the galactic nucleus, is receding from us at a rate of 135 km/s. The total amount of hydrogen in these expanding arms is at least several million solar masses. Given their speed and how far they have traveled, astronomers conclude that something quite extraordinary must have happened about 10 million years ago to expel such an enormous amount of gas from the central region of the Galaxy.

In addition to 21-cm radiation from neutral hydrogen gas, astronomers have also detected radio noise coming from the galactic center. This radio emission, which is produced by high-speed electrons spiraling around a magnetic field, is called **synchrotron radiation.** Despite its small size, Sagittarius A is one of the brightest sources of synchrotron radiation in the entire sky.

Some of the most detailed radio images of the galactic nucleus come from the Very Large Array (VLA). Figure 14-11a is a wide-angle view of Sagittarius A and surrounding features, including arcs of gas, at least three supernova remnants, and localized radio sources. Huge filaments, labeled "Arc," lie perpendicular to the plane of the Galaxy

and stretch 200 ly northward of the galactic disk, then abruptly arch southward toward Sagittarius A. The orderly arrangement of these filaments suggests that a magnetic field may be controlling the distribution and flow of ionized gas, just as magnetic fields on the Sun funnel such gas to create solar prominences.

If the center of our Galaxy isn't active and bizarre enough, recent gamma-ray observations reveal positrons being ejected from that region. (Recall that positrons have positive electrical charges but are otherwise identical to electrons.) The source of these positrons is still unknown. Positrons can be detected because when a positron and an electron collide, they annihilate each other and release their energy as gamma rays with well-defined wavelengths. These special gamma rays have been observed emanating from the galactic center.

Stars and gas have been observed to orbit around Sagittarius A, shown in Figure 14-11b. Something must be holding this high-speed gas in such tight orbit about the galactic nucleus. Using Kepler's third law, astronomers calculate that 2.5×10^6 $M_\odot$ is needed to prevent this gas from flying off into interstellar space. The observed broadening of spectral lines suggests that an object with the mass of two and a half million Suns in a volume only the size of the solar system is concentrated at Sagittarius A*. Astronomers believe that the object is a supermassive black hole. In 2001, the *Chandra* X-ray telescope observed a burst of x-rays from this black hole indicating that it is indeed no longer than 1 AU across, which is consistent with theoretical calculations.

The x-rays were generated by gas heating up as it fell into the black hole.

As we saw in Chapter 13, extraordinary activity is also occurring in the nuclei of many other galaxies, implying the presence of supermassive black holes at their centers as well. Astronomers are actively studying these regions in an effort to understand the complex, intriguing activities that are happening there.

14-4 Our Galaxy's disk is surrounded by a spherical halo of stars and other matter

The spherical distribution of globular clusters that Harlow Shapley mapped highlights a spherical distribution of stars called the **halo** of the Galaxy. Looking out of the plane of the Milky Way's disk and between globular clusters we see apparently unobstructed views of distant galaxies such as the Andromeda Galaxy observed by Lord Rosse (Section IV-1). Appearances can be deceiving. About 99 percent of the halo's stars are isolated *halo field stars* spread all through the halo, while the localized concentrations of stars in globular clusters account for only 1 percent of the halo's stars.

The various components of the Milky Way Galaxy overlap and interpenetrate each other. For example, the disk slices through the nuclear bulge, while globular clusters and halo field stars periodically pass through the plane of the disk.

In 1994, astronomers discovered that another galaxy is orbiting in the halo and is slowly being consumed by the Milky Way. Named the Sagittarius Dwarf (Figure 14-12), it periodically passes through the disk of the Milky Way and loses stars due to the disrupting influence of our Galaxy's strong gravitational attraction. Within the next 100 million years, that galaxy is expected to cease to exist, its stars becoming part of the Milky Way. There are other galaxy-galaxy collisions awaiting the Milky Way. We will explore these in Chapter 15, after presenting information about our neighboring galaxies.

14-5 The Galaxy is rotating

Just as the orbital motion of the planets keeps them from falling into the Sun, the motion of the stars and interstellar clouds around the galactic center keeps these bodies apart. If the stars and clouds in our Galaxy were not in orbit, their mutual gravitational forces would have caused them to fall into one massive lump billions of years ago. In that case, our Galaxy would not have spiral arms and we would not be here. Just as detecting the positions of the stars and clouds has been difficult, so, too, has been measuring the orbital motion of the stars and gas.

Radio observations of 21-cm radiation from hydrogen gas provide important clues about our Galaxy's rotation. By

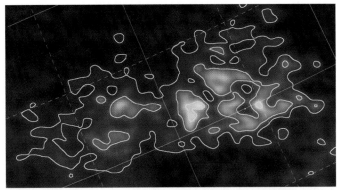

RIVUXG

FIGURE 14-12 **The Nearest Galaxy** The Sagittarius Dwarf is a dwarf elliptical galaxy that lies some 80,000 ly from the Milky Way. The curved lines in this image outline areas of equal brightness and emphasize the structure of the galaxy. Discovered in 1994, the Sagittarius Dwarf is the closest known galaxy to the Milky Way. It is so close that within the next 100 million years or so, the Milky Way will absorb this galaxy. (M. J. Irwin, Royal Greenwich Observatory, and A. B. Whiting and G. K. T. Hau, Institute of Astronomy, Cambridge University)

measuring Doppler shifts, astronomers can determine the speed of objects toward or away from us across the Galaxy. These observations clearly indicate that our Galaxy does not rotate like a rigid body such as Earth or Mars, but rather exhibits *differential rotation,* meaning that stars at different distances from the galactic center orbit the Galaxy at different rates.

The Sun itself must be moving around the center of the Galaxy. Otherwise, the solar system would be pulled by gravity into the heart of the Milky Way. Observations show that the Sun has a nearly circular orbit around the center of the Galaxy. Because of the stars' differential rotation in the Galaxy, the Sun is like a car on a circular freeway with the fast lane on one side and the slow lane on the other. As sketched in Figure 14-13, stars in the fast lane are passing the Sun and thus appear from our vantage point to be moving in one direction, while stars in the slow lane are being overtaken by the Sun and therefore appear to be moving in the opposite direction. This is like the retrograde motion of the planets discussed in Chapter 2.

Unfortunately, like the 21-cm observations, studying the motion of nearby stars and gas reveals only how fast they are moving relative to the Sun. To get a complete picture of the Galaxy's rotation, we must find out how fast the Sun itself is orbiting the center of the Galaxy. The Swedish astronomer Bertil Lindblad proposed a method of computing this speed. He noted that not all the stars in the sky move in the orderly pattern shown in Figure 14-13. Different globular clusters in the halo of our Galaxy orbit in different planes and so they do not participate in the organized rotation of the objects in the Galaxy's disk. The combined velocities of the globular clusters around the center of the Galaxy must average to zero or else they would drift, en masse, relative to the rest of the Galaxy. Using the motions of globular clusters as a reference, astronomers have calculated that the Sun orbits the galactic center at a speed of 230 km/s, or about 828,000 km (half a million miles) per hour.

Given the Sun's speed and its distance from the galactic center, astronomers can calculate the Sun's orbital period. Traveling at 828,000 km per hour, our Sun takes about 230 million years to complete one trip around the Galaxy. This result demonstrates the vastness of the Milky Way. When the solar system last passed our present location in the Galaxy, early dinosaurs of the Triassic period ruled the Earth.

By combining the true speed of the Sun with the relative speed of the stars around us, as measured by radio astronomers, we can determine the actual orbital speeds of the stars. This computation gives us the **rotation curve** of the Galaxy, a graph showing the orbital speeds of stars and interstellar clouds at various distances from the center of the Milky Way (Figure 14-14).

Knowing the Sun's velocity around the Galaxy from the rotation curve, we can use Kepler's third law to estimate the mass of the Galaxy. Putting in the numbers, we obtain a mass of about $1.1 \times 10^{11} \ M_\odot$ located inside the Sun's orbit around the Galaxy.

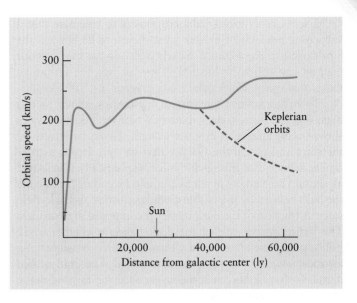

FIGURE 14-14 **The Galaxy's Rotation Curve** The blue curve shows the orbital speeds of stars and gas in the Galaxy out to a distance of 60,000 ly from the galactic center. The dashed red curve labeled *Keplerian orbits* indicates how this orbital speed should decline beyond the confines of most of the Galaxy's visible mass. Because the data (blue curve) do not show any such decline, there apparently is an abundance of invisible dark matter that extends to great distances from the galactic center. This additional mass gives the outer stars higher speeds than they would have otherwise.

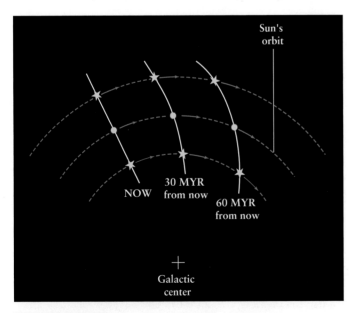

FIGURE 14-13 **Differential Rotation of the Galaxy** Stars at different distances from the galactic center have different angular speeds. It takes stars farther from the center longer to go around the Galaxy than it does stars closer to the center. As a result, stars closer to the Galaxy's center than the Sun are overtaking the solar system, while stars farther from the center are lagging behind us.

MYSTERIES AT THE GALACTIC FRINGES

Looking at the Milky Way extending all around the Earth, we know that stars and other matter exist beyond the Sun's orbit. In recent years, astronomers have been astonished to discover just how much matter lies beyond the orbit of the solar system. The nature of most of it has yet to be discovered.

14-6 Most of the matter in the Galaxy has not yet been identified

According to Kepler's third law, the farther a star or cloud is from the center of the Milky Way, the slower it should be moving, just as the orbital speeds of the planets decrease with increasing distance from the Sun (see the dashed red line in Figure 14-14). Observations reveal, however, that galactic orbital speeds continue to *climb* well beyond the visible edge of the galactic disk. This means there must be more gravitational force from the Galaxy acting on the distant stars and clouds than we can see or have taken into account.

These observations led astronomers to calculate that a surprising amount of matter must therefore lie beyond the Sun's orbit in the Galaxy. Nearly 90% of the mass of our Galaxy has yet to be located. If we include the unknown mass, the Milky Way's total mass exceeds 1×10^{12} M$_{\odot}$.

To make matters even more mysterious, much of this mass is in the form of **dark matter:** Whatever it is, it does not show up on images at any wavelength. Given its effects on the stars and gas in the Galaxy that we can detect, astronomers deduce that this dark matter is spherically distributed all around the Galaxy in a massive halo extending far beyond the halo defined by the visible globular clusters and halo field stars. A fraction of the dark matter is composed of neutrinos. The nature of the rest of the dark matter—whether black holes, gas, Jupiterlike bodies, brown dwarfs, or something far more exotic—has yet to be determined. The unidentified matter in our Galaxy is sometimes called the **missing mass.** This is a poor name because the matter is not missing; we just do not yet know its location or nature.

Some searches for the dark matter rely on the fact from general relativity that matter causes passing light to change direction (see Figure 13-2). The Galaxy's dark matter may contain relatively massive objects, like brown dwarfs or black holes. If so, this dark matter will cause the light from more distant stars to change direction as the dark objects pass between those stars and the Earth (Figure 14-15a). Indeed, as a dark object moves directly between a more distant star and the Earth, the star's light is focused toward us, causing the star to brighten. Called **microlensing,** this brightening of distant stars for several weeks has been observed (Figure 14-15b). Microlensing observations indicate that only a small fraction of the Galaxy's dark matter is in the form of these compact, massive objects.

14-7 Frontiers yet to be discovered

Despite our being part of it, our knowledge of the Milky Way is limited by the difficulty in making observations through its gas and dust. We have yet to locate all the globular clusters. Even more challenging will be locating and categorizing all the stars and clouds of gas and dust in the Galaxy. Are there any more small galaxies that the Milky Way is consuming? We also have to verify whether or not there is a bar-shaped distribution of stars in the nuclear bulge and to determine fully and accurately the structure and extent of the spiral arms.

Our observations of the Galaxy's nucleus hint at remarkable activity there that has yet to be discovered. What, for example, is creating the positrons, whose resulting gamma rays we observe? In its farthest reaches, what else is there in the Galaxy's halo creating all the gravitational force we detect, but whose source is not yet observed. In other words, what is the dark matter?

Further Reading on These Topics

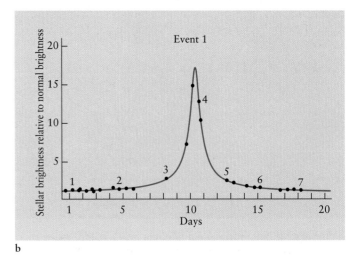

a

b

FIGURE 14-15   Microlensing by Dark Matter in the Galactic Halo (a) Gravitational fields cause light to change direction. A brown dwarf or black hole in the Galaxy's halo passing between the Earth and a more distant star will focus the starlight in our direction, making distant objects appear brighter than they do normally. (b) The light curve of the gravitational microlensing of light from a star in the Galaxy's nuclear bulge by an intervening object. Observational evidence indicates that some of the intervening objects are white dwarfs.

WHAT DID YOU KNOW?

1 *Where in the Milky Way is the solar system located?* The solar system is near the Orion spiral arm, about 26,000 ly from the center of the Galaxy.

2 *How many stars does the Milky Way Galaxy contain?* The Milky Way has about 200 billion stars.

3 *Is the Sun moving through the Milky Way and, if so, how fast?* The Sun orbits the center of the Milky Way Galaxy at a speed of 828,000 km per hour.

KEY WORDS

dark matter (missing mass), 384
disk (of a galaxy), 377
galactic nucleus, 375
halo (of a galaxy), 382

microlensing, 384
nuclear bulge, 378
rotation curve (of a galaxy), 383
Sagittarius A, 379

spin (of an electron or proton), 377
spiral arm, 378
synchrotron radiation, 381
21-cm radio radiation, 377

KEY IDEAS

The Structure of Our Galaxy

• Our Galaxy has a disk about 100,000 ly in diameter and about 2000 ly thick, with a high concentration of interstellar dust and gas. It contains around 200 billion stars.

• Interstellar dust obscures our view into the plane of the galactic disk at visual wavelengths. Despite the intervening interstellar dust, hydrogen clouds can be detected by the 21-cm radio waves emitted by changes in the relative spins of electrons and protons.

• The galactic nucleus has been studied at infrared and radio wavelengths, which pass readily through intervening interstellar dust and H II regions that illuminate the spiral arms. These observations have revealed many details of the galactic nucleus, but much remains unexplained.

• A supermassive black hole of about 2.5×10^6 $M_\odot$ exists in the galactic nucleus.

• The center, or nucleus, of the Milky Way is surrounded by a flattened sphere of stars called the nuclear bulge. The entire Galaxy is surrounded by a halo of matter that includes a spherical distribution of globular clusters, as well as large amounts of dark matter.

• Young OB associations, H II regions, and molecular clouds in the galactic disk outline huge spiral arms where stars are forming.

• The Sun is located about 26,000 ly from the galactic nucleus, between two major spiral arms. The Sun moves in its orbit at a speed of about 828,000 km per hour and takes about 230 million years to complete one orbit about the center of the Galaxy.

• From studies of the rotation of the Galaxy, astronomers estimate that its total mass is about 1×10^{12} $M_\odot$. Much of this mass is still undetectable.

REVIEW QUESTIONS

The answers to computational problems, which are preceded by an asterisk (*), appear at the back of the book.

1 As seen in the northern hemisphere, why is the Milky Way far more prominent in July than in December? Your *Starry Night Backyard*™ software may be useful in answering this question.

2 What observations led Harlow Shapley to conclude that we are not at the center of the Galaxy?

3 Explain why globular clusters spend more time in the galactic halo than in the plane of the disk, even though their eccentric orbits take them across the disk of the Galaxy.

4 How do hydrogen atoms generate 21-cm radiation? What do astronomers learn about our Galaxy from observations of that radiation?

5 Why do astronomers believe that vast quantities of dark matter surround our Galaxy?

6 What is synchrotron radiation?

7   Why are there no massive O and B stars in globular clusters?

8   How would you estimate the total number of stars in the Galaxy?

9   What evidence indicates that a supermassive black hole is located at the center of our Galaxy?

*10   Approximately how many times has the solar system orbited the center of the Galaxy since the Sun and planets were formed?

ADVANCED QUESTIONS

11   Why don't astronomers detect 21-cm radiation from the hydrogen in giant molecular clouds?

12   Describe the rotation curve you would get if the Galaxy rotated like a rigid body.

13   Compare the apparent distribution of open clusters, which contain young stars, with the distribution of globular clusters relative to the Milky Way. Why are open clusters also referred to as galactic clusters?

*14   The visible disk of the Galaxy is about 100,000 ly in diameter and 2000 ly thick. If about five supernovae explode in the Galaxy each century, how often on the average would you expect to see a supernova within 1000 ly of the Sun?

15   Give reasons for the rapid rise in the Galaxy's rotation curve (see Figure 14-14) at distances close to the galactic center.

WHAT IF ...

16   The solar system was located in a globular cluster 26,000 ly above the plane of the Galaxy's disk? Make a drawing of what the Milky Way might look like from that location.

17   The solar system were located at the edge of the galactic disk? How would the Milky Way appear to us?

18   The solar system were located at the center of the Milky Way galaxy? What would we see and how would things be different on Earth?

19   A globular cluster were crossing the disk of the Milky Way in our vicinity right now? What would we see and how might things be different here?

WEB/CD-ROM QUESTIONS

 20   To test your understanding of our Galaxy's rotation, do Interactive Exercise 14-1 on the Web or CD-ROM. You can print out your results, if required.

 21   To test your understanding of the Milky Way's structure, do Interactive Exercise 14-2 on the Web or CD-ROM. You can print out your results, if required.

22   Clouds of positively charged electrons, called positrons (see An Astronomer's Toolbox 9-1), have been discovered near the center of the Galaxy. Search the Web for information about this discovery. When was it made? How were they discovered? What is thought to be the origin of these particles? What happens when ordinary electrons and positrons collide?

OBSERVING PROJECTS

23   To determine the angle between the equatorial coordinate system and the Milky Way, turn on your *Starry Night Backyard*™ software and set to Atlas mode (Go/Atlas). Turn on the equatorial coordinates (Guides/Celestial Grid). Make sure that the Milky Way is visible (Sky/Sky settings/Milky Way Colour). You can change its color and brightness, if necessary. Next, drag the screen around until you see the Milky Way crossing the celestial equator. Put the celestial equator in the middle of the screen and either print the screen or directly measure the angle between the celestial equator and roughly the middle of the Milky Way with a protractor (some estimation is necessary here). What is the right ascension of the crossing between the Milky Way and the celestial equator? Now rotate the celestial sphere by 12h around the celestial equator. Is the Milky Way again crossing the celestial equator? Measure the angle between the two. How does it compare with the previous measurement you made? Explain.

15 GALAXIES

IN THIS CHAPTER YOU WILL DISCOVER

- how Edwin Hubble categorized galaxies by their shapes

- why some galaxies have spiral arms

- that galaxies are found in clusters and are surrounded by halos of dark matter

- that clusters of galaxies are clustered in superclusters

- how some galaxies devour others in dramatic collisions

- the apparent large-scale structure of the universe and its observed rate of expansion

WEB LINK 15.1

A Spiral Galaxy
Andromeda, a beautiful spiral galaxy, is the only galaxy visible to the naked eye from the Earth's northern hemisphere. Without a telescope, it appears to be a fuzzy blob in the constellation of Andromeda. Located only 0.9 Mpc (2.9 Mly) from us, Andromeda is gravitationally bound to the Milky Way, and it covers an area in the sky roughly 5 times as large as the full Moon. (Bill and Sally Fletcher/Tom Stack and Associates)

R I V U X G

WHAT DO YOU THINK?

1   Do all galaxies have spiral arms?

2   Are most of the stars in a spiral galaxy located in its arms?

3   Are galaxies isolated objects?

4   Are all other galaxies moving away from the Milky Way?

Off in a galactic suburb, far from the center of the action, the Sun is just one of about 200 billion stars in the Milky Way. And our Galaxy is but one of an estimated 50 billion galaxies in the visible universe. By studying galaxies, we enter the realm of truly cosmic systems. While nearby stars are separated from us by light-years (trillions of miles), neighboring galaxies are hundreds of thousands of times farther away, and the most distant galaxies are billions of light-years from Earth.

Recall from Chapter 10 that astronomers classify stars using the Hertzsprung-Russell diagram to help them understand stellar properties. For precisely the same reason, astronomers classify galaxies by various physical properties. We will see that it is truly remarkable how few overall shapes galaxies have. This classification scheme is where we begin our study of the large-scale structure of the universe.

TYPES OF GALAXIES

1   Unlike the Milky Way, most galaxies do not have spiral arms. However, despite the incredible number of galaxies, there is surprising consistency in their overall shapes. Edwin Hubble began cataloging their appearance in the 1920s, after his measurements of Cepheid variables proved that they lie far outside the Milky Way. The **Hubble classification** of galaxies—spirals, barred spirals, ellipticals, irregulars, and their subclasses—is still used today.

15-1 The winding of a spiral galaxy's arms is correlated to the size of its nuclear bulge

Spiral galaxies are characterized by a nuclear bulge and by arched lanes of stars and glowing interstellar clouds, which appear as spiral arms. The closest spiral galaxy to the Milky Way is the Andromeda Galaxy, M31, pictured at the opening of this chapter. Andromeda is visible to the naked eye as a fuzzy blob in the constellation of Andromeda.

Hubble noted that how tightly the spiral arms are wound varies among different spiral galaxies. Further, the size of nuclear bulges varies. Hubble also observed that these two variations are correlated: The tighter the spiral, the larger the nuclear bulge.

As shown in Figure 15-1, spirals with tightly wound spiral arms (and fat nuclear bulges) are called *Sa* (for spiral type *a*) *galaxies*. Those with moderately wound spiral arms (and a moderate nuclear bulge) are *Sb galaxies*. Loosely wound spirals (with tiny nuclear bulges) are *Sc galaxies*. A typical spiral galaxy contains an estimated 100 billion stars and measures nearly 10^5 ly in diameter.

Because not all spiral galaxies are oriented face-on to the Earth, their spiral arms are not always evident. However, we can still classify many of them just from observing the size of their central bulges. For example, M104 (Figure 15-2a) has

| Sa | NGC 1357 | Sb | M81 | Sc | NGC 4321 | R I V U X G |

FIGURE 15-1   Spiral Galaxies (Nearly Face-on Views)
Edwin Hubble classified spiral galaxies according to the tightness of the spiral arms and the size of the nuclear bulge. Sa galaxies have the largest nuclear bulges and the most tightly wound spiral arms, while Sc galaxies have the smallest nuclear bulges and the least tightly wound arms. The images are different colors because they were taken through filters passing different colors. (left: STScI Digital Sky Survey; middle: © Robert Gendler; right: Anglo-Australian Observatory)

a   M104   b   NGC891   c   NGC4631   R I V U X G

FIGURE 15-2   Spiral Galaxies Tilted with Respect to the Milky Way   **(a)** Because of its large nuclear bulge, this galaxy is classified as an Sa. If we could see it face-on, the spiral arms would be tightly wound around a voluminous bulge. **(b)** Note the smaller nuclear bulge in this Sb galaxy. **(c)** At visible wavelengths, interstellar dust obscures the relatively insignificant nuclear bulge of this Sc galaxy. (a: NOAO; b: Anglo-Australian Observatory; c: Dr. Rudy Schild, Smithsonian Astrophysical Observatory)

a huge central bulge. It must therefore be an Sa galaxy with tightly wound arms. An Sb galaxy (Figure 15-2b) has a smaller central bulge. The tiny central bulge of a Sc galaxy (Figure 15-2c) is hardly noticeable at all in this galaxy, which is steeply tilted toward our view.

Besides the degree of winding, the overall appearance of individual spiral arms varies from galaxy to galaxy. In some galaxies, called **flocculent spirals** (from the word meaning "fleecy"), the spiral arms are broad, fuzzy, chaotic, and poorly defined (Figure 15-3a). Other galaxies, called **grand-design spirals**, exhibit beautiful arching arms outlined by brilliant H II regions and OB associations. In these latter galaxies, the spiral arms are thin, delicate, graceful, and well defined (Figure 15-3b).

a   M33   b   M74   R I V U X G

FIGURE 15-3   Variety in Spiral Arms   The differences in spiral galaxies suggest that at least two mechanisms create spiral arms. **(a)** This galaxy has the fuzzy, poorly defined spiral arms hypothesized to be created by self-propagating star formation. **(b)** This galaxy has the thin, well-defined spiral arms hypothesized to be created by a spiral density wave. (a: Tony and Daphne Hallas; b: P. Seiden, D. Elmegreen, B. Elmegreen, and A. Mobarak, IBM)

15-2 Self-propagating star formation and spiral density waves produce spiral arms

For many years the very existence of spiral arms confounded astronomers. They cannot come about simply from the orbital motions of stars. As shown in Figure 14-14, most stars and interstellar gas and dust orbit a galaxy's center at roughly the same linear (straight-line) velocity. Consequently, stars farther from the galactic center take longer to complete their orbits than stars closer in, because more distant stars have farther to travel. As a result, the spiral arms should eventually "wind up," from the inside out, wrapping themselves tightly around the nucleus, much more tightly than is seen in any galaxy (Figure 15-4). Indeed, after a few galactic rotations, the spiral structure should disappear altogether. Why, then, do we observe spiral arms in our Galaxy and many others? This *winding dilemma* suggests that we must look for other physical processes, rather than just the motions of stars, to generate spiral arms.

Imagine first a dense interstellar cloud somewhere in the disk of a young galaxy. Stars are just being created in the cloud, and the galaxy does not yet have spiral arms. As hot, massive stars form, their radiation and stellar winds compress the nearby interstellar gas, triggering the formation of additional stars. Moreover, massive stars quickly explode as supernovae and produce shock waves, which further compress surrounding gases. Shock waves are very powerful compressions of the gas, like the sonic booms created by supersonic aircraft. As new stars thus trigger the birth of still other stars, the star-forming region grows, a process called *self-propagating star formation.*

The continuing birth of stars also helps account for spiral arms. The galaxy's differential rotation drags the inner edges of the young region ahead of the outer edges. As the newly formed stars spread out, they form a spiral arm highlighted by bright O and B stars and glowing nebulae. The high-mass O and B stars explode and then dim before the spiral arms wind up, while newer arms continually develop.

The existence of both flocculent and grand design spiral structures suggests that more than one mechanism gives rise to the creation of a spiral galaxy. If self-propagating star formation were the only process leading to spiral arms, bits and pieces of spiral arms would appear only to disappear as the bright, massive stars in them die off. This process creates chaotic spiral arms, such as those observed in flocculent galaxies (see Figure 15-3a), but not the smooth spirals of grand-design galaxies. So, self-propagating star formation cannot be the whole story.

Our search for a second mechanism begins in a pond. If you throw a rock into the water, you create ripples. As you know from experience, the ripples move outward in concentric rings from where the rock struck. But suppose the pond water is rotating when you throw in a rock. *Now* how do the ripples move? Could they be spiral-shaped? According to the astronomer Bertil Lindblad, working in the 1920s, spiral ripples can travel around the disks of some galaxies.

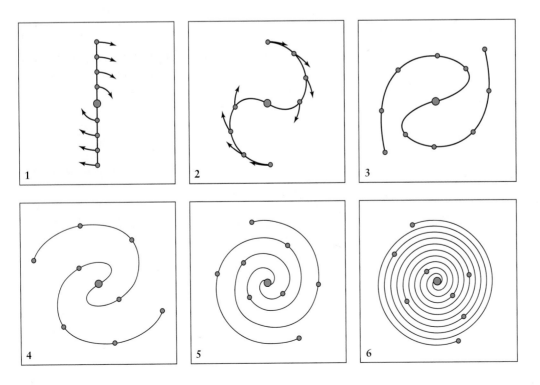

FIGURE 15-4 **The Winding Dilemma** The rotation curve of the disk stars in our Galaxy indicates that most of them have the same linear (straight-line) speed. The ones farther from the center take longer to go around because they have a greater distance to travel at the same speed as stars closer to the center of the Galaxy, which orbit in smaller circles. All the dots in these drawings have circular orbits at the same linear speed. Think of them as a few stars in an initially straight line of stars **(1)**. As time goes on, the outer stars are left behind, creating a spiral shape that becomes more and more tightly wound **(2)–(6)**. Such tightening is not observed in our Galaxy or in other galaxies.

MOVIE MISCONCEPTIONS

Mission to Mars (Touchstone Pictures, 2000)
(Jack Newton)

The movie *Mission to Mars*, set in 2020, presents the heroic Mars recovery mission launched after the first manned mission to the Red Planet meets with disaster. The mission, led by characters Jim McConnell and Woodrow "Woody" Blake, to investigate the tragedy and bring back any survivors is fraught with dangers that confront the crew members as they journey through space.

In the last scene, the movie shows a spacecraft heading from Mars toward a beautiful spiral galaxy spanning most of the screen. We are not told whether this is supposed to be the Milky Way or some external galaxy. For the following questions, ignore the fact that the spacecraft was allegedly created by beings from Mars. (a) Assuming that it is supposed to represent the Milky Way, what is wrong with the image of the Galaxy? (b) Assuming that it is an external galaxy, what is wrong with the image of the galaxy?
(Answers appear at the end of the book.)

Like water in a pond, interstellar gas and dust sustain these ripples, called **spiral density waves.**

Russian scientists in the 1970s tested Lindblad's model. Lacking the computer power to create simulations of galaxies, they threw pebbles into pie pans of water rotating on phonograph turntables! The resulting water wave patterns were indeed spirals. Spiral density waves never wind up. Instead, they orbit in a rigid spiral pattern at about half the speed of the interstellar medium through which they move.

But can we explain why density waves plus rotation leads to spiral galaxies? In the mid-1960s, two American astronomers looked carefully at how galactic spiral density waves traveling through the disk of a galaxy lead to spiral arms. C. C. Lin and Frank Shu argued that density waves cause interstellar gas and dust to pile up temporarily. This occurs because, unlike water waves, which move up and down as they travel along, spiral density waves are compressional waves, just like sound. As a result, they cause interstellar gas and dust to become denser as this material passes through the waves. Lin and Shu demonstrated that a spiral arm is a galactic traffic jam. Imagine workers painting a line down a busy freeway. The cars normally cruise at 65 mph, but the crew of painters causes a bottleneck. The cars slow down temporarily to avoid hitting other cars and the slowly moving paint truck. As seen from the air, there is a noticeable congestion of cars around the painters (Figure 15-5). An individual car spends only a few moments in the moving traffic jam before resuming its usual speed, but the traffic jam itself lasts all day long.

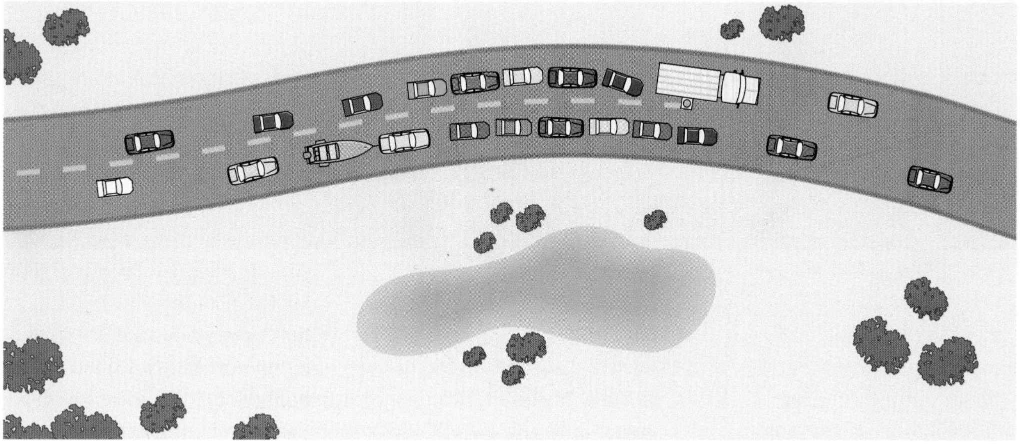

FIGURE 15-5

Compression Wave in Traffic Flow When normal traffic flow is slowed down, cars bunch together. In a grand-design galaxy, a density wave moves through the stars and gas. The wave is merely a region of slightly denser matter, which, in turn, creates more gravitational force. This force compresses the gas and enhances star formation, which highlights the spiral density wave.

The vehicles in the traffic jam represent the interstellar gas and dust entering a spiral arm (paint truck). This interstellar medium sweeps through the more slowly moving spiral density waves. The waves compress some of this interstellar gas and dust, creating new clouds, which then collapse to form new, metal-rich stars. We see spiral arms because they contain numerous, bright O and B stars and copious quantities of dust and gas that these stars illuminate. The sprawling dust lanes in Figure 15-3b attest to the recent passage of that material through a shock wave. While all this is going on, the density wave maintains its shape, and, overall, the spiral arms in these galaxies are better defined than the ones discussed above.

Stars as dim as the Sun contribute virtually nothing to the brightness of spiral arms, which explains why the winding dilemma does not occur in the spiral density wave model: The massive stars highlighting the wave explode before they finish passing through it. The remaining, longer-lived, lower-mass stars, like our Sun, fill the space between the spiral arms without emitting much light. Indeed, as pronounced as spiral arms may appear, only about 5% more stars are found in them than are found between them.

It takes an enormous amount of energy to compress interstellar gas and dust. After a billion years or so, even spiral density waves would begin to fade away. Some driving mechanism must keep them going, like throwing more rocks in the pond. The most likely explanation for reinvigorating spiral structure is the passage of a nearby companion galaxy. As this latter galaxy periodically passes close by, its gravitational attraction pulls on the gas, stars, and dust of the spiral galaxy to generate new density waves. Indeed, grand-design galaxies are usually found in the presence of a companion galaxy.

Although most spiral galaxies have two arms, a sizable minority have more. The Milky Way, for example, has at least four spiral arms. Astronomers have not yet established why the numbers of arms vary.

Since new stars are formed from interstellar gas and dust, it is plausible that galaxies with different amounts of this material will have different star formation rates and therefore different overall structures. Infrared and radio observations reveal that Sa galaxies typically contain about 4% gas and dust, while Sb galaxies contain 8% and Sc galaxies contain 25%. Astronomers hope that computer simulations of galaxies will reveal the impact of this gas and dust on galactic structure.

15-3 Bars of stars run through the nuclear bulges of barred spiral galaxies

Most astronomers believe that the Milky Way is a spiral galaxy. As we noted in Chapter 14, however, some evidence suggests that it is a **barred spiral,** a spiral galaxy with a bar of stars crossing through the nuclear bulge. The arms in barred spirals extend from the ends of the bar rather than from the nuclear bulge itself. Computer models suggest that bars may arise in galaxies with less dark matter than in unbarred spirals.

Hubble found that the winding of the spiral arms in barred spirals again correlates with the size of the nuclear bulge (Figure 15-6). An *SBa* (for spiral, barred, type *a*) galaxy has a large central bulge (and tightly wound spiral arms). Likewise, a barred spiral with moderately wound spiral arms (and a moderate central bulge) is an *SBb* galaxy, and an *SBc* galaxy has loosely wound spiral arms (and a

SBa   NGC 4650   SBb   M83   SBc   NGC 1365   R I V U X G

FIGURE 15-6 Barred Spiral Galaxies As with spiral galaxies, Edwin Hubble classified barred spirals according to the tightness of their spiral arms (which correlates with the sizes of their nuclear bulges). SBa galaxies have the most tightly wound spirals and largest nuclear bulges, SBb have moderately tight spirals and medium-sized nuclear bulges, while SBc galaxies have the least tightly wound spirals and the smallest nuclear bulges. (left: AURA/NOAO/NSF; middle: Anglo-Australian Observatory, VLA, NRAO; right: Anglo-Australian Observatory)

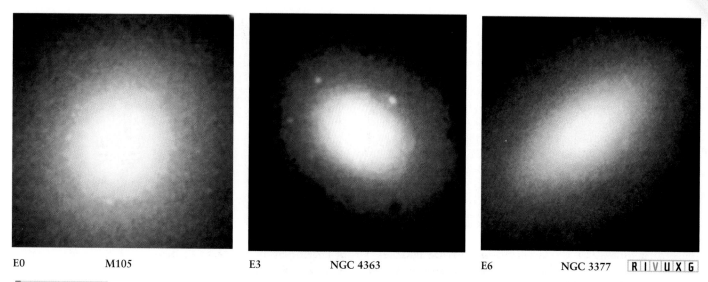

| E0 | M105 | E3 | NGC 4363 | E6 | NGC 3377 | R I V U X G |

FIGURE 15-7 **Elliptical Galaxies** Hubble classified elliptical galaxies according to how round or elongated they appear. An E0 galaxy is round; a very elongated elliptical galaxy is an E7. Three examples are shown here. (left, middle, and right: J. D. Wray, McDonald Observatory)

tiny central bulge). Observations indicate that ordinary spiral galaxies outnumber barred spirals by about 2 to 1.

The motions of the stars in many nearby galaxies have been measured from Doppler shifts of their spectra. In all spiral and barred spiral galaxies discovered to date except one, the arms trail around behind as the galaxies rotate. They are thus called **trailing-arm spirals**. The exception is the galaxy labeled NGC 4622. This galaxy has the points of its spirals leading the motion of the arms. It is a *leading-arm spiral.*

15-4 Elliptical galaxies display a wide variety of sizes and masses

Elliptical galaxies, named for their distinctive shapes, have no spiral arms. Hubble subdivided elliptical galaxies according to how round or oval they look. The roundest elliptical galaxies are called *E0 galaxies,* while the most elongated are *E7 galaxies.* Elliptical galaxies of intermediate elongation are numbered E1 to E6 (Figure 15-7).

The Hubble scheme classifies galaxies solely by their appearance from our Earth-bound view. But whenever we observe anything in the sky, we are seeing a two-dimensional view of a three-dimensional object. In the case of elliptical galaxies, we observe length and width, but we have no way of knowing anything about the third dimension of depth. What looks like an E0 galaxy (basically circular) might actually be egg-shaped when viewed from another angle. Conversely, an elongated E7 galaxy might look circular when viewed face-on. Observational evidence suggests that many elliptical galaxies are indeed shaped differently in virtually every direction.

Elliptical galaxies look far less dramatic than their spiral and barred spiral cousins because they contain relatively little interstellar gas and dust. Because stars form in interstellar clouds, few stars should be forming in ellipticals. Observations of spectra confirm the hypothesis that these galaxies contain primarily Population II, low-mass, long-lived stars.

Elliptical galaxies range tremendously in size and mass—from the biggest to the smallest galaxies in the universe. Figure 15-8 shows two **giant elliptical galaxies** that form part of a cluster of galaxies in the constellation Virgo. These enormous giant elliptical galaxies are each about 2 million light-years in diameter, 20 times the diameter of the Milky Way Galaxy.

Giant ellipticals, containing some 10 trillion solar masses, are rare compared to other types of galaxies, while **dwarf elliptical galaxies** are extremely common. Dwarf ellipticals are only a fraction of the size of an average elliptical galaxy and contain so few stars—only a few million—that we see these galaxies as nearly transparent. Because you can actually see straight through the center of a dwarf elliptical galaxy and out the other side (Figure 15-9), distant ones are hard to detect. Hence, many more dwarf ellipticals undoubtedly exist than have been identified.

15-5 Hubble represented galaxies with different shapes in a tuning fork diagram

Edwin Hubble connected the three regularly shaped types of galaxies—spirals, barred spirals, and ellipticals—in a diagram shaped like a tuning fork (Figure 15-10). According to

R I V U X G

FIGURE 15-8   **Giant Elliptical Galaxies** The Virgo cluster is a rich, sprawling collection of more than 2000 galaxies about 50 million light-years from Earth. Only the center of this huge cluster appears in this photograph. The two largest galaxies in the cluster are the giant elliptical galaxies M84, on the right, and M86, on the left. (Royal Observatory, Edinburgh)

R I V U X G

FIGURE 15-9   **A Dwarf Elliptical Galaxy** This nearby E4 galaxy, called Leo I, is about 600,000 light-years from Earth. It is only about 3000 ly in diameter and so sparsely populated with stars that you can see right through its center. It is a satellite of the Milky Way. (Anglo-Australian Observatory)

his scheme, S0 or SB0 galaxies, called **lenticulars** (lens-shaped), are an intermediate type between ellipticals and the two kinds of spirals. Although they look somewhat like ellipticals, lenticular galaxies have both a central bulge and a disk, like spiral galaxies, but they lack spiral arms.

Hubble found some galaxies that cannot be classified as spirals, barred spirals, or ellipticals. He called these **irregular galaxies.** They are generally rich in interstellar gas, dust, and both young and old stars. Irregular galaxies with only hints of organized structure and numerous OB associations are

denoted Irr I. The nearby Large Magellanic Cloud (LMC) (Figure 15-11a) and the Small Magellanic Cloud (SMC), which orbit the Milky Way, are examples of Irr I. Both can be seen with the naked eye from southern latitudes.

Occasionally, irregulars are observed that appear highly distorted and completely asymmetrical, as though created by collisions between galaxies or by violent activity in their nuclei. These are denoted Irr II, as shown by NGC 4485 and NGC 4490 (Figure 15-11b). Irregular galaxies are typically smaller and less massive than spirals, containing between

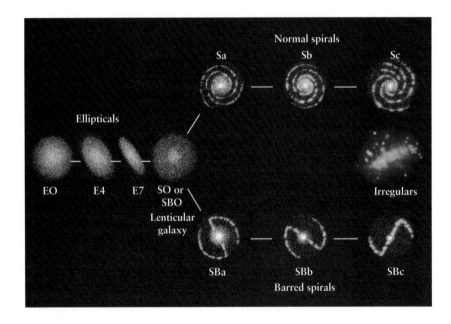

FIGURE 15-10   **Hubble's Tuning Fork Diagram** Hubble summarized his classification scheme for galaxies with this tuning fork diagram. Elliptical galaxies are classified by how oval they appear, while spirals and barred spirals are classified by the sizes of their central bulges and the correlated winding of their spiral arms. An S0 or SB0 galaxy, also called a lenticular galaxy, is an intermediate type between ellipticals and spirals.

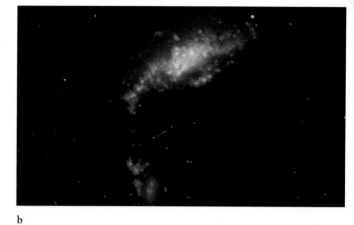

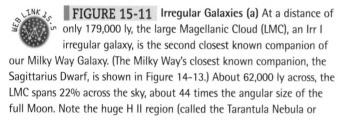

a

b

FIGURE 15-11 Irregular Galaxies (a) At a distance of only 179,000 ly, the large Magellanic Cloud (LMC), an Irr I irregular galaxy, is the second closest known companion of our Milky Way Galaxy. (The Milky Way's closest known companion, the Sagittarius Dwarf, is shown in Figure 14-13.) About 62,000 ly across, the LMC spans 22% across the sky, about 44 times the angular size of the full Moon. Note the huge H II region (called the Tarantula Nebula or 30 Doradus) toward the left side of this image. Its diameter of 800 ly and mass of 5 million Suns makes it the largest known H II region. (b) The small irregular (Irr II) galaxy NGC 4485 (bottom galaxy) interacts with the highly distorted Sc galaxy NGC 4490, also called the Cocoon galaxy. This pair is located in the constellation Canes Venatici. (a: Anglo-Australian Observatory; b: Hoher Observatory)

10^8 and about 3×10^{10} M$_\odot$. For want of any better scheme, the irregular galaxies are sometimes placed between the ends of the tuning fork tines of the Hubble diagram.

The idea that one type of galaxy may change into another has been in and out of favor with astronomers for decades. Indeed, this was part of Hubble's motivation in relating the different types of galaxies as he did. As we will see later in this chapter and in Chapter 17, extensive observations by the Hubble Space Telescope and by ground-based telescopes now indicate that interactions between galaxies sometimes do lead to changes in their structures. However, there is no evidence that most galaxies change overall shape. The properties of the different types of galaxies are summarized in Table 15-1.

CLUSTERS AND SUPERCLUSTERS

As you consider the vastness of the universe, it is hard to imagine that structures as huge and apparently isolated as galaxies orbit each other in groups. But they do.

TABLE 15-1   Some Properties of Galaxies

	Spiral (S) and barred spiral (SB) galaxies	Elliptical galaxies (E)	Irregular galaxies (Irr)
Mass (M$_\odot$)	10^9 to 4×10^{11}	10^5 to 10^{13}	10^8 to 3×10^{10}
Luminosity (L$_\odot$)	10^8 to 2×10^{10}	3×10^5 to 10^{11}	10^7 to 10^9
Diameter (ly)	$16 \times 10^3 - 8 \times 10^5$	$3 \times 10^3 - 6.5 \times 10^5$	$3 \times 10^3 - 3 \times 10^4$
Stellar populations	disk: young Population I central bulge and halo: Population II and old Population I	Population II and old Population I	mostly Population I
Percentage of observed galaxies	77%	*20%	3%

*This percentage does not include dwarf elliptical galaxies that are as yet too dim and distant to detect. Hence, the actual percentage of galaxies that are ellipticals may be higher than shown here.

15-6 Galaxies occur in clumps called clusters, which occur in clumps called superclusters

Galaxies are not scattered randomly throughout the universe but rather group together in **clusters**. The Fornax cluster (Figure 15-12), so called because it is located in the constellation of Fornax, the Furnace, is typical. Observations of galactic motion indicate that members of a cluster of galaxies are gravitationally bound together: The galaxies in a cluster orbit each other and occasionally even collide.

Clusters of galaxies are themselves grouped together in huge associations called **superclusters**. A typical supercluster contains dozens of individual clusters spread over a volume 150 million light-years across. Figure 15-13 shows the distribution of clusters in our vicinity of the universe. The nearer ones, out to the Virgo cluster, are members of our *Local Supercluster*. The others in this figure are members of other superclusters. Observations indicate that most superclusters are not gravitationally bound units—most clusters in each supercluster are drifting away from most of the other clusters in that same supercluster. Furthermore, the superclusters are all moving away from each other.

How superclusters are arranged in space became clearer beginning in the early 1980s, when astronomers first discovered enormous **voids** between superclusters where exceptionally few galaxies are observed. These voids are roughly

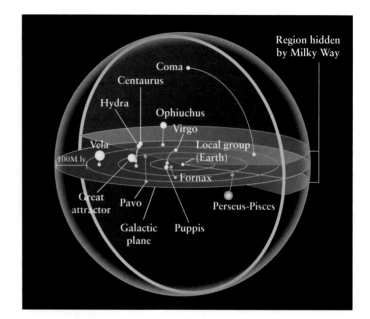

FIGURE 15-13 **Clusters of Galaxies in Our Neighborhood** This is a drawing of a sphere of space 800 Mly across centered on the Earth in the Local Cluster. The spherical dots represent the locations of the nearby clusters of galaxies, while the flat circles represent the projection of the cluster locations onto the plane of the Milky Way. To better see the 3-dimensionality of this figure, yellow arcs are drawn from each cluster down to the green projection of the Milky Way's plane extended out through the universe. Note that clusters of galaxies are unevenly distributed here, as elsewhere in the universe.

spherical and measure between 100 million and 400 million light-years in diameter. They are not completely empty, however. Observations reveal hydrogen clouds in some of them, while others may be subdivided by strings of dim galaxies. Recent surveys show that most galaxies are distributed on the surfaces between these voids (Figure 15-14). Two notable nearby clumps of galaxies are labeled in Figure 15-14. The Milky Way, and the rest of the local supercluster, are heading toward the Great Attractor.

To better understand the distribution of galaxies throughout the universe, several surveys are underway to map the three-dimensional locations of many galaxies. The Sloan Digital Sky Survey is observing 100 million galaxies, stars, and other celestial objects and measuring redshifts of at least the nearest 1 million galaxies. This survey, made in the northern hemisphere, will cover about one quarter of the sky and give us a three-dimensional map through 100 times the volume of space as we have mapped today. The Two Degree Field (2dF) Survey is observing regions of both the northern and southern hemispheres.

The distribution of clusters of galaxies throughout the universe is sometimes said to be "sudsy" because it resem-

FIGURE 15-12 **A Cluster of Galaxies** This group of galaxies, called the Fornax cluster, is about 60 million light-years from Earth. Both elliptical and spiral galaxies are easily identified. The barred spiral galaxy at the lower left is NGC 1365, the largest and most impressive member of the cluster. For a closer view of NGC 1365, see Figure 15-6. (Anglo-Australian Observatory, photos by David Malin, © 1984 Royal Observatory, Edinburgh)

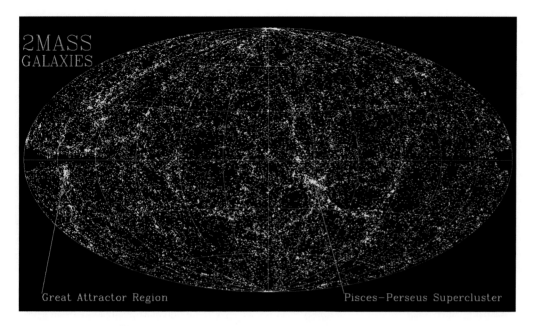

Great Attractor Region   Pisces–Perseus Supercluster

FIGURE 15-14 **Structure in the Nearby Universe** This map of the entire sky (called 2MASS for 2-Micron All Sky Survey) shows the locations of over 30,000 of the nearest galaxies. They are all located within 500 million light years of the Milky Way. The locations of two concentrations of galaxies, the Great Attractor and the Pisces-Perseus supercluster, are labeled. This map used infrared telescopes to locate these galaxies, including many whose visible light is blocked by the Milky Way. Note that there are regions almost devoid of galaxies, surrounded by thin regions full of them. (Stephen Schneider, University of Massachusetts)

bles a collection of giant soap bubbles. Galaxies surround voids in the same way that soap suds are thin films surrounding bubbles of air. Astronomers believe that this sudsy pattern contains important clues about conditions in the early universe.

15-7 Clusters of galaxies may appear densely or sparsely populated and regular or irregular in shape

A cluster of galaxies is said to be either a **poor cluster** or a **rich cluster,** depending on how many galaxies it contains. For example, the Milky Way Galaxy, the Andromeda Galaxy, and the Large and Small Magellanic Clouds belong to a poor cluster called the **Local Group.** The Local Group contains some 40 galaxies, over a third of which are dwarf ellipticals. Figure 15-15 shows a map covering most of the Local Group.

New galaxies in the Local Group continue to be found. This is happening primarily because astronomers are developing techniques to observe objects that lie in the same plane as, but beyond, the Milky Way. Besides the 1994 discovery of the Sagittarius Dwarf Galaxy (see Figure 14-12), the dwarf galaxy Antlia was first observed in 1997. It is about 3 million light-years from the Milky Way and contains about one million stars (Figure 15-16).

Astronomers also categorize clusters of galaxies according to their shapes. A **regular cluster** is distinctly spherical, with a marked concentration of galaxies at its center. Numerous gravitational interactions over the ages have spread the galaxies into a distinctive spherical distribution. In contrast, galaxies in an **irregular cluster** are more randomly scattered about a sprawling region of the sky.

The nearest example of a rich, regular cluster is the Coma cluster, located about 300 million light-years from us in the constellation of Coma Berenices (Figure 15-17; see also Figure 15-13). Despite its great distance, more than 1000 bright galaxies within it are easily visible from Earth. It

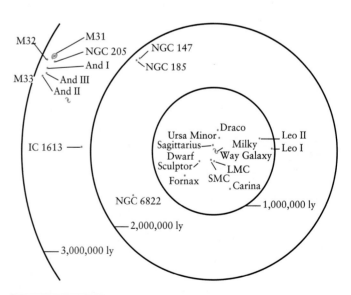

FIGURE 15-15 **The Local Group** Our Galaxy belongs to a poor, irregular cluster consisting of about 40 galaxies called the Local Group. This map shows the distribution of about ³/₄ of the galaxies. The Andromeda Galaxy (M31) is the largest and most massive galaxy in the Local Group. The second largest is the Milky Way itself. M31 and the Milky Way are each surrounded by a dozen satellite galaxies. The recently discovered Sagittarius Dwarf Galaxy is the Milky Way's nearest known neighbor.

R I V U X G

FIGURE 15-16 **The Most Recently Discovered Member of the Local Group** The galaxy Antlia was first detected in 1997. It lies about 3 million ly away, outside the region depicted in Figure 15-15. This galaxy contains only about a million stars. (M. J. Irwin, Royal Greenwich Observatory, and A. B. Whiting and G. K. T. Hau, Institute of Astronomy, Cambridge University)

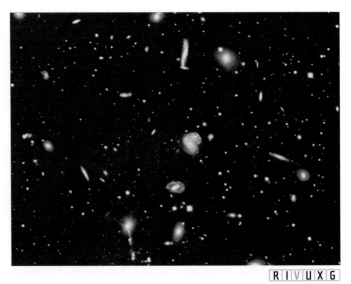

R I V U X G

FIGURE 15-18 **The Hercules Cluster** This irregular cluster, which is about 700 million light-years from Earth, contains a high proportion of spiral galaxies, often associated in pairs and small groups. (NOAO)

NGC 4881

R I V U X G

FIGURE 15-17 **The Coma Cluster** This rich, regular cluster containing thousands of galaxies is about 300 million light-years from Earth. Regular clusters are composed mostly of elliptical and lenticular galaxies and are common sources of X rays. The brightest member of the Coma cluster shown here is the elliptical galaxy NGC 4881. (Hubble Space Telescope WFPC Team, Caltech, and NASA)

is likely that the Coma cluster contains many thousands of dwarf ellipticals soon to be detected—maybe as many as 10,000 galaxies overall.

Rich, regular clusters like the Coma cluster contain mostly elliptical and lenticular galaxies. Only 15% of the Coma cluster's galaxies are spirals and irregulars. Irregular clusters, such as the Virgo cluster and the Hercules cluster (Figure 15-18), display a more even mixture of galaxy types. Two-thirds of the 200 brightest galaxies in the Hercules cluster are spirals, nearly a fifth are ellipticals, and the rest are irregular.

15-8 Galaxies in a cluster can collide and combine

Occasionally, two galaxies in a cluster or two galaxies from different, nearby clusters pass through one another (Figure 15-19). There is so much space between stars that the probability of two stars crashing into each other is extremely small. On the other hand, the galaxies' huge clouds of interstellar gas and dust are so large that they do collide, slamming into each other and producing strong shock waves. The colliding interstellar clouds are stopped in their tracks.

Collisions can merge galaxies together—or hurl their stars far into space. A violent collision can strip both galaxies of their interstellar gas and dust while the stars in each

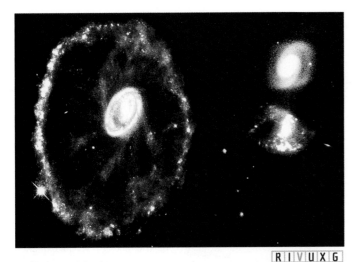

R I V U X G

FIGURE 15-19   The Cartwheel Galaxy This ring-shaped assemblage 500 million light-years from Earth is the likely result of one galaxy, probably the blue-white one on the right, having passed through the middle of the larger one. Astronomers suspect that the passage created a circular density wave in the Cartwheel that stimulated a burst of star formation, creating many bright blue and white stars. (Kirk Borne, STScI; and NASA)

R I V U X G

FIGURE 15-20   A Starburst Galaxy A ring of vigorous star formation 2400 ly wide highlights this stunning ultraviolet, visible light, and infrared composite image of the barred spiral galaxy NGC 1512 taken by *Hubble*. Located 30 Mly away in the constellation Horologium, this galaxy is 70,000 ly across. Such rings of star formation are common in starburst galaxies. (D. Maoz, ESA and NASA)

one keep right on going. The violence of the collision heats the gas stripped from these galaxies to extremely high temperatures. This process may be a major source of the hot **intergalactic gas** often observed in rich, regular clusters.

In a less violent collision between two galaxies or in a near miss, the compressed interstellar gas may cool sufficiently over time to allow new stars to form. Such collisions can thus stimulate prolific star formation, which may account for the **starburst galaxies** that blaze with the light of numerous newborn stars. These galaxies are characterized by bright centers surrounded by clouds of warm interstellar dust, indicating a recent, vigorous episode of star birth (Figure 15-20). There is observational evidence that globular clusters are formed during collisions from interstellar gas and dust. Unlike the globular clusters in our Galaxy, some galaxies contain very young globular clusters. Whether the globular clusters in our Galaxy formed as a result of an early collision of the Milky Way has yet to be determined. The warm dust is so abundant that starburst galaxies are among the most luminous objects in the universe at infrared wavelengths.

The Irr II, starburst galaxy M82 is one member of a nearby cluster of about a dozen galaxies that includes the beautiful spiral galaxy M81 and a fainter companion called NGC 3077 (Figure 15-21a). A radio survey of that region of the sky revealed enormous streams of hydrogen gas connecting these and other, smaller galaxies in that cluster (Figure 15-21b). The loops and twists in these streamers suggest that the three galaxies have had several close encounters over the ages, most recently about 600 million years ago.

The effects of the gravitational interactions between colliding galaxies can also hurl thousands of stars out into intergalactic space along huge, arching streams as seen in Figure 15-22a.

In rich clusters, astronomers observe near misses between galaxies (Figure 15-22b). If galaxies are surrounded by extended halos of dim stars, these near misses could strip the galaxies of these outlying stars. In this way, a loosely dispersed sea of dim stars might come to populate the space between galaxies in a cluster. One of the projects assigned to the Hubble Space Telescope is to search for these dim stars in extended halos and in intergalactic space.

Similarly, after galaxies collide, some interior stars are flung far and wide, scattering material into intergalactic space. Such isolated stars have been observed by the Hubble Space Telescope. However, other stars slow down, often causing the remnants of the galaxies to merge. Several dramatic examples of **galactic mergers** have been discovered (Figure 15-23). Astronomers also speak of **galactic cannibalism,** which occurs when a large galaxy captures and "devours" a smaller one. Cannibalism differs from mergers in that the dining galaxy is significantly bigger than its dinner, whereas merging galaxies are about the same size.

Many astronomers suspect that giant ellipticals are the product of galactic mergers and cannibalism, including spiral-spiral collisions. As we have seen, giant galaxies typically occupy the centers of rich clusters. Once formed, giant ellipticals can continue to grow because smaller

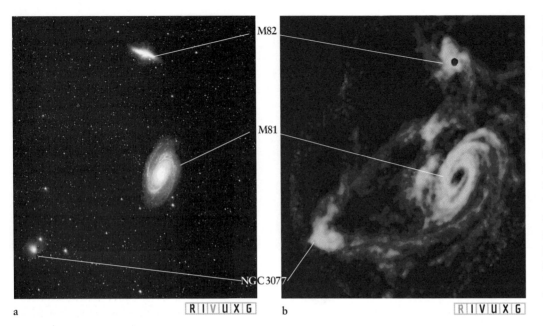

a

b

FIGURE 15-21 The M81 Group The Irr II, starburst galaxy M82 is in a nearby cluster of about a dozen galaxies, including the spectacular spiral M81. Several of the galaxies in this cluster are connected by streamers of hydrogen gas. **(a)** This photograph shows the three brightest galaxies at visual wavelengths. **(b)** This radio image, created from data taken by the Very Large Array, shows the streamers of hydrogen gas that connect the bright galaxies and also several dim ones, seen as regions of bright orange here. (a: Palomar Sky Survey; b: M. S. Yun, VLA, and Harvard)

galaxies are often located around them (see Figure 15-8). As they pass through the extended halo of a giant elliptical, these smaller galaxies slow down and are eventually consumed by the larger galaxy.

Figure 15-24, a computer simulation, shows a large, disk-shaped galaxy devouring a small satellite galaxy. The large galaxy consists of 90% stars (in blue) and 10% gas (in white) by mass. It is surrounded by a halo of dark matter having a mass about 3.3 times that of the disk. The satellite galaxy, which has a tenth of the mass of the large galaxy, contains only stars (in orange). Initially, the satellite is in circular orbit about the large galaxy. Note that spiral arms appear in the large galaxy as the collision proceeds. Two billion years elapse as the satellite spirals in toward the core of the large galaxy. Although much material is stripped from the satellite, most of its stars plunge into the nucleus of the large galaxy.

 Computer Simulation

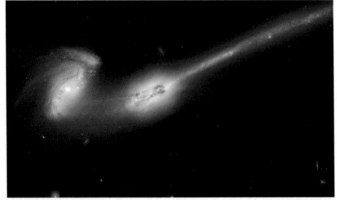

FIGURE 15-22 Interacting and Colliding Galaxies **(a)** Pairs of colliding galaxies often exhibit long "antennae" of stars ejected by the collision, as shown at left. This particular system is known as NGC 4676 or "the Mice" (because of their tails of stars and gas). It is 300 million ly from Earth in the constellation Coma Berenices. The collision has stimulated a firestorm of new star formation, as can be seen in the bright blue regions. Mass can also be seen flowing between the two

galaxies, which will eventually merge. **(b)** These two galaxies, NGC 2207 (right) and IC 2163, are orbiting and tidally distorting each other. Their most recent close encounter occurred 40 Myr ago when the two were perpendicular to each other and about one galactic diameter apart. Computer simulations indicate that they should eventually coalesce. (a: STS/NASA; b: NASA)

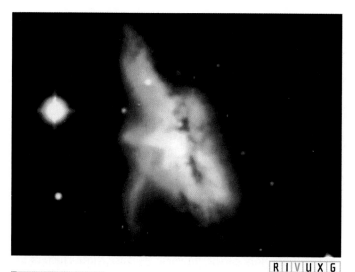

R I V U X G

FIGURE 15-23 Merging Galaxies This contorted object in the constellation of Ophiuchus is the result of two spiral galaxies in the process of merging. The widespread blue area reveals that the collision between the two galaxies has triggered an immense burst of star formation. (W. C. Keel, University of Alabama)

15-9 Galactic halos may account for some dark matter in the universe

What prevents the rapidly moving galaxies in clusters from wandering away from each other? There must be sufficient matter to provide the necessary gravitational force to bind galaxies together in clusters and in some cases clusters into superclusters. However, *no cluster of galaxies contains enough visible matter to stay bound together.* A lot of nonluminous (dark) matter must be scattered about each cluster of galaxies, or else the galaxies would have long ago wandered apart in random directions, and the clusters would no longer exist. Recall that a similar problem arose in Chapter 14 regarding dark matter in individual galaxies. Analyses of the total visible mass demonstrate that the total mass needed to bind a typical rich cluster is 10 times greater than the mass of material that shows up on visible-light images.

Extensive evidence supports the idea of extended halos surrounding galaxies. Many galaxies have rotation curves similar to that of our Milky Way (recall Figure 14-14). These rotation curves remain remarkably flat out to surprisingly great distances from the galaxies' centers. For example, Figure 15-25 shows the rotation curves of four spiral galaxies. In all cases, the orbital speed is fairly constant, even out where stars and nebulae are too dim and widely scattered for reliable measurements. According to Kepler's third law, we should see a decline in orbital speed toward the outer portions of a galaxy. The fact that in most cases we do not implies that we have still not detected the true edge of these and many similar galaxies. Astronomers conclude that a considerable amount of dark matter must extend well beyond the visible portion of a galaxy's disk.

Astronomers using X-ray telescopes have discovered the nature of some of the remaining dark matter. Satellite observations have revealed X rays pouring from the space between galaxies in rich clusters. This radiation is emitted by substantial amounts of hot intergalactic gas at temperatures between

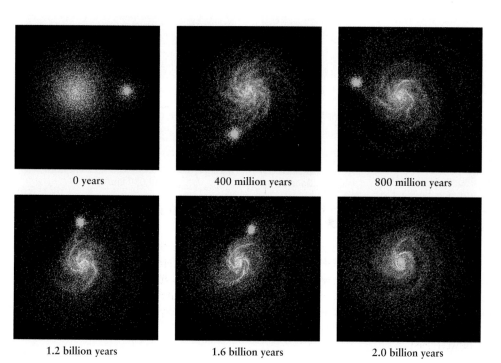

0 years   400 million years   800 million years

1.2 billion years   1.6 billion years   2.0 billion years

ANIMATION 15.2

FIGURE 15-24 Simulated Galactic Cannibalism This computer simulation shows a small galaxy (stars in yellow) being devoured by a larger, disk-shaped galaxy (stars in blue, gas in white). Note how spiral arms are generated in the disk galaxy by its interaction with the satellite galaxy. (Lars Hernquist, Institute for Advanced Study with simulations performed at the Pittsburgh Supercomputing Center)

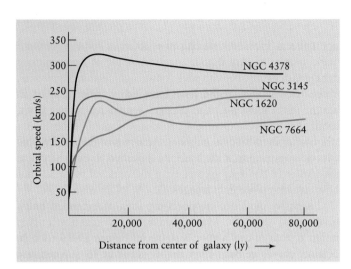

FIGURE 15-25 The Rotation Curves of Four Spiral Galaxies This graph shows how the orbital speed of material in the disks of four spiral galaxies varies with the distance from the center of each galaxy. If most of each galaxy's mass was concentrated near the center of the galaxy, these curves would fall off at large distances. But these and many other galaxies have flat rotation curves that do not fall off. This indicates the presence of extended halos of dark matter. See Figure 14-14 to compare these to the Milky Way's rotation curve.

10 million and 100 million K. The mass of this hot intergalactic medium is typically as great as the combined mass of all the visible galaxies in a rich cluster.

SUPERCLUSTERS IN MOTION

Whenever an astronomer finds an object in the sky, one of the first tasks is to determine its composition. As we saw in Chapters 3 and 4, that means attaching a spectrograph to a telescope and recording the object's spectrum. As long ago as 1914, V. M. Slipher, working at the Lowell Observatory in Arizona, took spectra of "spiral nebulae." He was surprised to discover that the spectral lines of 11 of the 15 spiral nebulae he studied were substantially redshifted. These redshifts indicate that they are moving away from us at significant speeds. This marked dominance of redshifts was presented by Heber D. Curtis in the Shapley–Curtis debate as evidence that the "spiral nebulae" could not be part of our Milky Way Galaxy. It also left scientists with a major puzzle: Is the entire universe expanding?

15-10 The redshifts of superclusters indicate that the universe is expanding

 During the 1920s, Edwin Hubble and Milton Humason recorded the spectra of many galaxies

with the 100-in. telescope on Mount Wilson, confirming that most galaxies are rapidly receding from the Milky Way. Using the Doppler effect (recall Figure 4-16), Hubble calculated the speed at which each galaxy is moving away from us.

Using techniques such as the brightness of Cepheid variables (Section IV-2), Hubble also estimated the distances to a number of these galaxies. He found a direct correlation between the distance to a galaxy and the size of its redshift: *Galaxies in distant clusters and superclusters are moving away from us more rapidly than galaxies in nearby clusters and superclusters.* Figure 15-26 shows this correlation for four elliptical galaxies. This recessional motion pervades the universe and is now called the **Hubble flow.** The separation of clusters in the same supercluster is slower than the general Hubble flow of the universe because these clusters are relatively close to each other, so their gravitational attractions slow each other down. Keep in mind that the Hubble

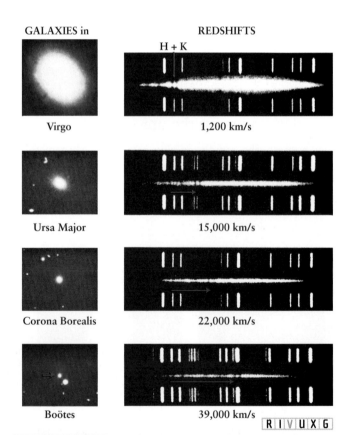

FIGURE 15-26 Four Galaxies and Their Spectra The photographs of these four elliptical galaxies were all taken at the same magnification. They are labeled according to the constellation in which each galaxy is located. The spectrum of each galaxy is the hazy band between the comparison spectra at the top and bottom of each plate. In all five cases, the so-called H and K lines of calcium are seen. The recessional velocity (calculated from the Doppler shifts of the H and K lines) appears below each spectrum. Note that the fainter—and thus more distant—a galaxy is, the greater is its redshift. (Carnegie Observatories)

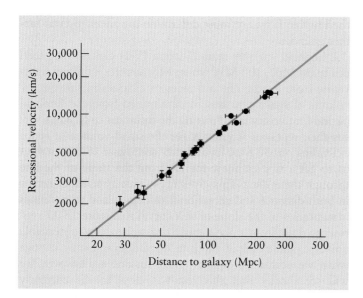

FIGURE 15-27 **The Hubble Law** The distances and recessional velocities of 18 relatively nearby spiral galaxies are plotted on this graph. The short horizontal and vertical lines around each point, called error bars, represent uncertainties in the distances and velocities, respectively. The straight line is the "best fit" for the data. This linear relationship between distance and speed is called the Hubble law.

flow does not occur for galaxies in any given cluster, since all those galaxies are gravitationally bound to each other. Specifically, the Hubble flow applies to the separation of superclusters from each other.

When Hubble plotted the redshift data on a graph of distance versus speed, he found that the points lie nearly along a straight line (Figure 15-27). The relationship between the distances to galaxies and their recessional motion is most easily stated as a formula called the **Hubble law:**

$$\text{Recessional velocity} = H_0 \times \text{distance}$$

This is simply the equation of a straight line, with the slope of the line denoted by the constant H_0, commonly called the **Hubble constant.** An Astronomer's Toolbox 15-1 shows how the Hubble law works.

The relationship between the distances to galaxies and their redshifts is one of the most important astronomical discoveries of the twentieth century. It tells us that we are living in an expanding universe, and the Hubble law reveals the speed of that expansion. This discovery raised the question of whether the universe will expand forever or whether the gravitational attraction of all its parts is enough to cause it someday to stop and recollapse. This is like trying to escape forever from the Earth. A normal jump might

AN ASTRONOMER'S TOOLBOX 15-1

The Hubble Law

The Hubble law describes our expanding universe. It is a simple formula:

$$v = H_0 \times d$$

where v is the recessional velocity, d is the distance, and H_0 is the Hubble constant. It is the formula for the straight line displayed in Figure 15-27, and the Hubble constant is the slope of this line.

As Figure 15-27 shows, we usually measure a galaxy's velocity in kilometers per second and its distance in megaparsecs. You will see in section 15-11 that astronomers have determined Hubble's constant $H_0 \approx 72$ km/s/Mpc.

Example: A galaxy 1 Mpc from us is moving away at a rate of

$$v = (72 \text{ km/s/Mpc}) \times (1 \text{ Mpc}) = 72 \text{ km/s}$$

A galaxy 2 Mpc away is therefore receding at 144 km/s, and so on. A galaxy located 100 million parsecs from Earth rushes away from us with a speed of 7200 km/s! Uncertainty in the distances of remote galaxies leads to uncertainty in the value of H_0.

Compare! While a supercluster of galaxies 10 Mpc away recedes at 720 km/s, the solar system orbits the center of our Galaxy at a comparable 230 km/s.

Try these questions: At what speed is a galaxy observed to be 8 billion light-years away moving from us? How far away from us is a galaxy observed to be moving away at 216 km/s? If H_0 were 50 km/s/Mpc, would a galaxy 4 billion light-years away be moving away from us faster or slower than the same galaxy with $H_0 = 72$ km/s/Mpc?

(Answers appear at the end of the book.)

raise you up a foot or so, but if you had a cannon to shoot you upward fast enough, you could escape the Earth's gravitational attraction completely and enter interplanetary space. To determine whether the superclusters are expanding away from each other fast enough to escape their mutual gravitation forever, astronomers have worked hard to determine an exact value of H_0. We explore that now and then pick up the question of the fate of the universe in Chapter 17.

> **Insight into Science** **Power in Numbers** Much of the power of science comes from expressing systematic data in mathematical form. By themselves, the spectra of galaxies give a qualitative picture of their motion, and the pattern of recessional velocities suggests a deeper insight into nature. When the data are converted into the equation called the Hubble law, it becomes clear that the universe is expanding, and we can begin to look for the reason behind this expansion.

15-11 Different techniques determine the expansion of the universe at different distances from Earth

To determine the Hubble constant, astronomers must measure the redshifts and distances to many galaxies. Although redshift measurements from spectra can be quite precise, it is very difficult to measure the distances to remote galaxies accurately. We cannot use the method of stellar parallax at the distances of galaxies (recall from Essentials III that the distance to the stars in our Galaxy within about 150 pc (500 ly) can be determined very precisely this way).

Distances to other galaxies are determined from observations of the apparent magnitudes of objects in them or of the entire galaxy. These observations are compared to known absolute magnitudes for similar objects. Having the apparent magnitude and the absolute magnitude enables astronomers to calculate distances, as discussed in An Astronomer's Toolbox III-2. Astronomers use the term **standard candle** to denote any object whose absolute magnitude is known.

Spectroscopic parallax (Section 10-6) provides distances to stars up to 10 kpc (33 kly, where kly denotes 1000 light-years). This distance is still within the Milky Way. To determine distances to other galaxies, we need sources in them that become very bright compared to the luminosity of normal stars, and whose absolute magnitudes are well known. Using the brightnesses of RR Lyrae variable stars (Section 14-1), astronomers can measure distances to the Magellanic clouds or about 100 kpc (330 kly). To measure

the Hubble flow, we must determine distances to even farther galaxies.

Cepheid variable stars (Section IV-2) can now be seen out to 30 Mpc (100 Mly, where Mly denotes a million light-years) from Earth. The distances to galaxies in this nearby volume of space can thus be determined from the Cepheid period-luminosity law. Plotting the distances to galaxies versus their recessional velocity for the small volume of space extending out 30 Mpc from Earth (see Figure 15-27) enables us to get a first estimate for H_0 from the slope of the line through these data. This gives $H_0 \approx 75$ km/sec/Mpc. Errors in both distance and recessional velocities lead to potential inaccuracies in the slope of this curve. As Figure 15-27 represents data from only a small fraction of the observable universe, the slope of the line on this graph may change when we add more data at larger distances. However, we will see shortly that this result is close to the currently accepted value of H_0.

Beyond 30 Mpc, even the brightest Cepheid variables, which have absolute magnitudes of about −6, are not visible with current technology. In the 1970s, the astronomers Brent Tully and Richard Fisher developed another method for determining distances. They discovered that the width of the hydrogen 21-cm emission line of a spiral galaxy (discussed in Chapter 14) is related to the galaxy's absolute magnitude: *The broader the line, the brighter the galaxy.* This correlation is called the **Tully-Fisher relation.** By measuring the width of the 21-cm line, we can determine a galaxy's absolute magnitude. Combining this with observations of the galaxy's apparent magnitude allows us to calculate its distance from us, just as we did with spectroscopic parallax (see Section 10-6).

To understand the Tully-Fisher relation, we begin by noting that more massive galaxies rotate faster than less massive ones. Tully and Fisher then made the plausible assumption that brighter galaxies are more massive than dimmer ones. The Tully-Fisher relation works like this: As a galaxy rotates, some of its stars move toward us and the light from these stars is blueshifted. Other stars are moving away and the light from the receding stars is redshifted. Taken together, the motions of the stars in a rotating galaxy spread out the 21-cm emission line. (The wider the emission line, then, the faster the galaxy rotates, so the more massive it must be.)

Because line widths can be measured quite accurately, astronomers can use the Tully-Fisher relation to determine the luminosities (absolute magnitudes) of spiral galaxies and thus their distances. Measurements of 21-cm-line widths for distant, receding galaxies yield a Hubble constant of 71 km/s/Mpc.

A key method for determining the distance to the most remote galaxies is to measure the brightness of supernova explosions within them. The brightest supernovae reach an absolute magnitude of −19 at the peak of their outbursts

(Figure 15-28). Using this method, astronomers in 2001 measured a supernova nearly 3 billion parsecs (10 billion light-years) away.

Figure 15-29 summarizes the distances for which all the measuring techniques described above are valid. Including data from all the methods yields a distance-velocity graph as shown in Figure 15-30. Using this diagram, astronomers have calculated that the Hubble constant is

$$H_0 = 72 \pm 8 \text{ km/s/Mpc}$$

The plus or minus 8 km/sec/Mpc indicates the possible range for H_0, when taking into account all the errors that may still remain in the observations and calculations. A significant source of error is the effect of intergalactic gas causing the standard candles to appear dimmer than they would if the gas were not present. Astronomers determine the amount of interstellar gas and take its effects into account, something that is often hard to do.

15-12 Astronomers are looking back to a time when galaxies were first forming

The Hubble Space Telescope has been looking deeper and deeper into the universe. In 1998, working in concert with the Keck I telescope, it observed galaxies 8 Bly (where Bly denotes a billion light-years) away, many of them in the process of merging (Figure 15-31a). The same year it saw galaxies nearly 12 Bly away (Figure 15-31b). In 2002, galaxies up to 13.6 Bly away were observed. We are seeing these objects as they were only a

R I V U X G

FIGURE 15-28 Two Supernovae in NGC 664 In 1997 the rare occurrence of two supernovae in the same galaxy was observed in the spiral galaxy NGC 664, located about 300 Mly (90 Mpc) from Earth. Supernovae observed in remote galaxies are important standard candles used by astronomers to determine the distances to these faraway objects. The two supernovae overlap each other as shown. The upper, yellow–orange supernova was observed to occur two months before the hotter, blue one, which was observed to occur less than two weeks before this image was made and had not yet achieved maximum brightness. (Perry Berlind and Peter Garnavich, Harvard Smithsonian Center for Astrophysics)

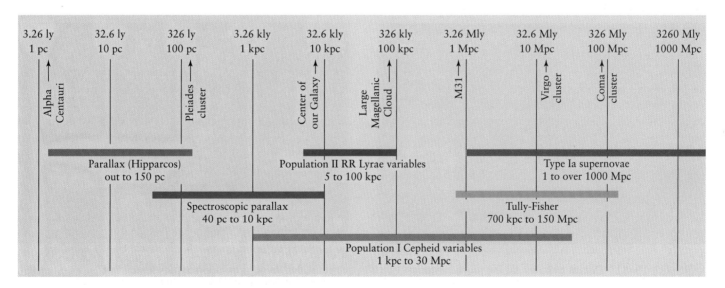

FIGURE 15-29 Techniques for Measuring Cosmological Distances Astronomers use different methods to determine different distances in the universe. All the methods shown here are discussed in the text.

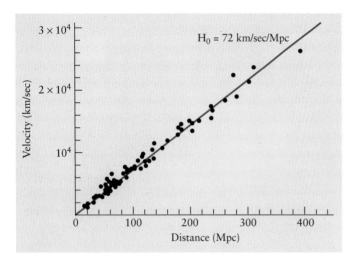

FIGURE 15-30 **Velocity-Distance Diagram** The points on this diagram represent observations of distance and velocity for a variety of galaxies. Hubble's constant is the slope of the red line, which represents the line that best fits the data.

billion years or so after the universe came into existence. While many of the early galaxies are not as well-formed as nearby galaxies, the fact that there are galaxy-sized collections of stars so far back in time is an important clue to the processes of star and galaxy formation that occurred in the early universe.

Nature provides astronomers with assistance in visualizing distant galaxies through the gravitational warping of space, as described by Einstein's general theory of relativity (Section 13-2). We saw in Chapter 14 how the same effect occurs on small scales here in the Milky Way. While the local distortion is called microlensing, when entire galaxies or clusters of galaxies focus light from more distant galaxies, the effect is called **gravitational lensing.** Figure 15-32 shows the gravitationally lensed images of distant galaxies. Indeed, the most distant galaxies, mentioned in the preceding paragraph, were detected as a result of gravitational lensing.

The amount of gravitational lensing by a galaxy or cluster of galaxies provides astronomers with more than just enlarged images of distant objects. It tells how much mass the galaxy or cluster that does the lensing contains. The greater the mass of the lensing galaxy or cluster, the greater the effect on the distant object. The visible matter in the galaxies could only account for about 10% of the gravitational lensing that occurs. In other words, the amount of lensing confirms that at least 90% of the matter in the universe is still unaccounted for.

 The observational data now being accumulated from the period of time when the universe was less than a few billion years old are helping astronomers develop and test theories of the evolution of the early universe. We will explore this point further in Chapter 17.

15-13 Frontiers yet to be discovered

The farther out in the universe we explore, the more there is to discover. In the realm of the galaxies, we have yet to understand fully how the different types of galaxies form. Are nuclear bulges formed as a result of collisions between galaxies? Are globular clusters the remnants of previous galaxies that have been consumed by bigger galaxies? How

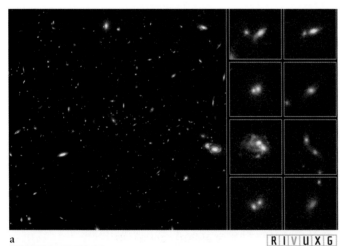

a R I V U X G

FIGURE 15-31 **Distant Galaxies** **(a)** The young cluster of galaxies MS1054-03, shown on the left, contains many orbiting pairs of galaxies, as well as remnants of recent galaxy collisions. Several of these systems are shown at the right. This cluster is located 8 billion light-years away from Earth. **(b)** This image of more than 300 spiral,

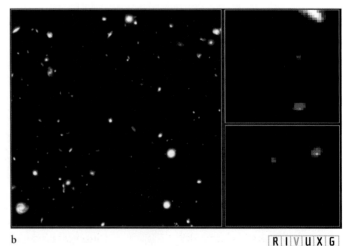

b R I V U X G

elliptical, and irregular galaxies contains several that are an estimated 12 billion light-years from Earth. Two of the most distant galaxies are shown in the images on the right, colored in red at the centers of the pictures. (a & b: P. Van Dokkum, Uner of Granengen, ESA and NASA)

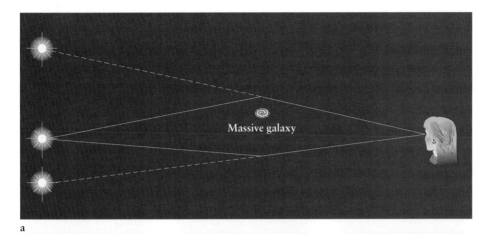

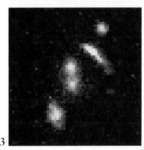

FIGURE 15-32 Gravitational Lensing of Extremely Distant Galaxies **(a)** Schematic of how a gravitational lens works. Light from the distant object changes direction due to the gravitational attraction of the intervening galaxy. It appears in two places to the observer on the right. **(b)** Three examples of gravitational lensing: **(1)** The bluer arc has been lensed by the redder elliptical galaxy. **(2)** A pair of bluish images of the same object lensed symmetrically by the brighter, redder galaxy between them. **(3)** The lensed object appears as a blue arc under the gravitational influence of the group of four galaxies. (b: Kavan Ratnatunga, Carnegie Mellon University, ESA and NASA)

long does it take interacting galaxies to combine? Do supermassive black holes form before, simultaneously with, or after galaxies come into existence? Moving into the realm of the global structure of the universe, what is the overall distribution of clusters and superclusters of galaxies in the universe? Are most of them really found on soap-bubble–like surfaces and, if so, what is that structure telling us about the underlying distribution of dark matter?

WHAT DID YOU KNOW?

1 *Do all galaxies have spiral arms?* No. Galaxies may be either spiral, barred spiral, elliptical, or irregular. Only spirals and barred spirals have arms.

2 *Are most of the stars in a spiral galaxy located in its arms?* No. The spiral arms contain only 5% more stars than the regions between the arms.

3 *Are galaxies isolated objects?* No. Galaxies are grouped in clusters, and clusters are grouped in superclusters.

4 *Are all other galaxies moving away from the Milky Way?* All galaxies except those in our local cluster are receding from us.

KEY WORDS

barred spiral galaxy, 392
cluster (of galaxies), 396
dwarf elliptical galaxy, 393
elliptical galaxy, 393
flocculent spiral galaxy, 389
galactic cannibalism, 399
galactic merger, 399
giant elliptical galaxy, 393
grand-design spiral galaxy, 389
gravitational lensing, 406

Hubble classification, 388
Hubble constant, 403
Hubble flow, 402
Hubble law, 403
intergalactic gas, 399
irregular cluster (of galaxies), 397
irregular galaxy, 394
lenticular galaxy, 394
Local Group, 397
poor cluster (of galaxies), 397

regular cluster (of galaxies), 397
rich cluster (of galaxies), 397
spiral density wave, 391
spiral galaxy, 388
standard candle, 403
starburst galaxy, 399
supercluster (of galaxies), 396
trailing-arm spiral galaxy, 393
Tully-Fisher relation, 403
void, 396

KEY IDEAS

Types of Galaxies

• The Hubble classification system groups galaxies into four major types: spiral, barred spiral, elliptical, and irregular.

• The arms of spiral and barred spiral galaxies are sites of active star formation.

• According to the theory of self-propagating star formation, spiral arms of flocculent galaxies are caused by the births and deaths of stars over extended regions of a galaxy. Differential rotation of a galaxy stretches the star-forming regions into elongated arches of stars and nebulae that we see as spiral arms.

• According to the spiral density wave theory, spiral arms of grand-design galaxies are caused by density waves. The gravitational field of a spiral density wave compresses the interstellar clouds that pass through it, thereby triggering the formation of stars, including OB associations, which highlight the arms.

• Elliptical galaxies contain much less interstellar gas and dust than do spiral galaxies; little star formation is occurring in elliptical galaxies.

Clusters and Superclusters

• Galaxies group into clusters rather than being randomly scattered through the universe.

• A rich cluster contains hundreds or even thousands of galaxies; a poor cluster may contain only a few dozen. A regular cluster has a nearly spherical shape with a central concentration of galaxies; in an irregular cluster, the distribution of galaxies is asymmetrical.

• Our Galaxy is a member of a poor, irregular cluster called the Local Group.

• Rich, regular clusters contain mostly elliptical and lenticular galaxies; irregular clusters contain more spiral and irregular galaxies. Giant elliptical galaxies are often found near the centers of rich clusters.

• No cluster of galaxies has an observable mass large enough to account for the observed motions of its galaxies; a large amount of unobserved mass must be present between the galaxies.

• Hot intergalactic gases emit X rays in rich clusters.

• When two galaxies collide, their stars initially pass each other but their interstellar gas and dust collide violently, either stripping the gas and dust from the galaxies or triggering prolific star formation. The gravitational effects of a galactic collision can cast stars out of their galaxies into intergalactic space.

• Galactic mergers occur; a large galaxy in a rich cluster may grow steadily through galactic cannibalism, perhaps producing a giant elliptical galaxy.

Superclusters in Motion

• A simple linear relationship exists between the distance from the Earth to galaxies in other superclusters and the redshifts of those galaxies (a measure of the speed at which they are receding from us). This relationship is the Hubble law, recessional velocity $= H_0 \times$ distance, where H_0 is the Hubble constant.

• Astronomers use standard candles—Cepheid variables, the brightest supergiants, globular clusters, H II regions, supernovae in a galaxy—and the Tully-Fisher relation to estimate intergalactic distances. Because of difficulties in measuring the distances to remote galaxies, the value of the Hubble constant, H_0, is not known with complete certainty.

REVIEW QUESTIONS

1 What is the Hubble classification scheme? Which category includes the biggest galaxies? Into which category do the smallest galaxies fall? Which type of galaxy is the most common?

2 In which types of galaxies are new stars most commonly forming? Describe the observational evidence that supports your answer.

3 What is the difference between a flocculent spiral galaxy and a grand-design spiral galaxy?

4 Briefly describe how the theory of self-propagating star formation accounts for the existence of spiral arms in some spiral galaxies.

5 Briefly describe how the spiral density wave theory accounts for the existence of spiral arms in some spiral galaxies.

6 How is it possible that galaxies in our Local Group still remain to be discovered? In what part of the sky are these galaxies located? What sorts of observations might reveal these galaxies?

7 Can any galaxies besides our own be seen with the naked eye? If so, which?

8 What is the difference between a rich cluster of galaxies and a poor one? What is the difference between a regular cluster of galaxies and an irregular one?

9 How can a collision between galaxies produce a starburst galaxy?

10 Why do astronomers believe that considerable quantities of dark matter must exist in clusters of galaxies?

11 Explain why the dark matter in galaxy clusters cannot be neutral hydrogen.

12 What is the Hubble law?

13 Some galaxies in the Local Group exhibit blueshifted spectral lines. Why aren't these blueshifts violations of the Hubble law?

14 What is a standard candle? Why are standard candles important to astronomers trying to measure the Hubble constant?

15 What kinds of stars would you expect to find populating space between galaxies in a cluster?

ADVANCED QUESTIONS

The answers to all computational problems, which are preceded by an asterisk (*), appear at the end of the book.

*16 Suppose a spectrum of a distant galaxy showed that its redshift corresponds to a speed of 22,000 km/s. How far away is the galaxy?

*17 A cluster of galaxies in the southern constellation of Pavo (the Peacock) is located 100 Mpc from Earth. How

fast, on average, are galaxies in this cluster receding from us? Why do different galaxies in the cluster show different velocities?

18 How would you determine what fraction of a galaxy's redshift is caused by the galaxy's orbital motion about the center of mass of its cluster?

DISCUSSION QUESTIONS

19 Discuss the advantages and disadvantages of using the various standard candles to determine extragalactic distances.

20 Discuss whether the various Hubble types of galaxies actually represent some sort of evolutionary sequence.

21 Discuss the sorts of phenomena that can occur when galaxies collide. Do you think that such collisions can change the Hubble type of a galaxy? Explain.

22 From what you know about stellar evolution, the interstellar medium, and the spiral density wave theory, explain the appearance and structure of the spiral arms of grand-design spiral galaxies.

WHAT IF ...

23 The solar system was located in an active star-forming region in a spiral arm, rather than on the edge of the Orion arm? How would the solar system be different?

24 The Milky Way collided with the Andromeda Galaxy? What might we experience here on Earth? (This merger is actually expected to occur in the distant future.)

25 No distant galaxies showed any redshift or blueshift? Discuss what that would imply about the universe.

26 All distant galaxies showed a blueshift, rather than a redshift. Discuss what that would imply about the universe.

WEB/CD-ROM QUESTIONS

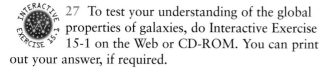 27 To test your understanding of the global properties of galaxies, do Interactive Exercise 15-1 on the Web or CD-ROM. You can print out your answer, if required.

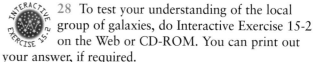 28 To test your understanding of the local group of galaxies, do Interactive Exercise 15-2 on the Web or CD-ROM. You can print out your answer, if required.

29 To test your understanding of the various types of galaxies, do Interactive Exercise 15-3 on the Web or CD-ROM. You can print out your answer, if required.

 30 To test your understanding of star formation in the spiral density wave model, do Interactive Exercise 15-4 on the Web or CD-ROM. You can print out your results, if required.

OBSERVING PROJECTS

 31 Using a telescope with an aperture of at least 30 cm (12 in.) in a dark location, observe as many of the spiral galaxies listed in Table A-8, in the back of the book, as you can. You can locate them and print out their locations on a star chart using your *Starry Night Backyard*™ software. Many of these galaxies are members of the Virgo cluster, which can best be seen from the northern hemisphere during the spring. Because all galaxies are quite faint, be sure to schedule your observations for a moonless night. The best view is obtained when a galaxy is high in the sky. While at the eyepiece, make a sketch of what you see. Can you distinguish any spiral structure? After completing your observations, compare your sketches with photographs found in such popular books as *Galaxies* by Timothy Ferris (Sierra Club Books, 1980) and *The Color Atlas of the Galaxies* by James Wray (Cambridge University Press, 1988).

32 Using a telescope with an aperture of at least 30 cm (12 in.), observe as many of the interacting galaxies listed in Table A-8, in the back of the book, as you can. You can locate them and print out their locations on a star chart using your *Starry Night Backyard*™ software. As in the previous Observing Project, be sure to schedule your observations for a moonless night when the galaxies you wish to observe will be near the meridian. While at the eyepiece, make a sketch of what you see. Can you distinguish hints of interplay among the galaxies? After completing your observations, compare your sketches with photographs found in popular books, such as *Galaxies* by Timothy Ferris (Sierra Club Books, 1980).

WHAT IF . . .

HE SOLAR SYSTEM WERE LOCATED CLOSER TO THE CENTER OF THE GALAXY?

It is a clear, moonless night. Stars by the thousands twinkle serenely against the ebony darkness of space. Overhead the soft white span of the Milky Way catches your attention, and you try to see individual stars in its glowing haze.

Most of the 6000 stars visible throughout the year are within 300 ly of Earth. The rest of the Galaxy's 200 billion or so stars are too dim or too obscured by interstellar gas and dust to be easily observed. What if the solar system were one-third of its present distance from the center of our Galaxy? At that distance, we would still be in the realm of the spiral arms, extremely close to the nuclear bulge, and none of the stars we see now would be visible.

Population Explosion Out in the galactic suburbs, where we live today, there is about one star per 300 cubic light-years. The displaced Earth would be surrounded by nearly 5 times as many stars, or one per 60 cubic light-years. And the solar system would pass much more frequently through the dust-rich spiral arms of the Galaxy. Scattering starlight, this interstellar matter would glow as diaphanous wisps throughout our nighttime sky. Furthermore, whenever the solar system was actually in a spiral arm, several nearby stars would be of the high-mass, high-luminosity variety. The combined light from these stars and the shimmering clouds would be so great that for millions of years at a time night would never fall.

Evolution Interruption Recall from Chapter 12 that high-mass stars evolve much more rapidly than average-mass stars like the Sun. If we were closer to the center of the Milky Way and frequently passing through the spiral arms of the Galaxy, massive stars would explode near us much more frequently than they do today. Those nearby supernovae, occurring within 50 ly of Earth, would deposit lethal radiation, damage life, and cause mass extinctions. As a result, the direction of evolution would change more frequently.

At Earth's current location, the closest star, Proxima Centauri, is more than 4 ly away. Would Earth be in danger of colliding with a star if the solar system was closer to the center of the Galaxy? Several stars would certainly be much closer than Proxima Centauri is now. However, stars are so small compared to the vastness of a galaxy that the likelihood of a collision would still approach nil.

Much more likely would be the passage of a star so close to the Sun that the Earth's orbit would be disturbed. If the Earth's orbit became more elliptical, say, the change of seasons would be noticeably affected. With the seasonal effect of the Earth's tilt compounded by greater changes in distance between the Earth and Sun, one hemisphere of the Earth would suffer much more extreme temperatures than it does today, while the other would see less variation. This would affect the evolution of life, of course, and the distribution of life-forms on the planet.

Close Encounters Earth-evolved life at our new location in the Galaxy would be more likely to encounter sentient beings on planets orbiting nearby stars. Today, after nearly a century of broadcasting radio and television signals, a nearly 200-ly-diameter sphere of space centered on the Earth is filled with such signals. There are about 6300 stars in that sphere. At our new location there would be 31,500 stars in the same volume and the probability of many more stars with life-supporting planets orbiting them. Perhaps one of the reasons that life on Earth has been able to evolve so long unaffected by other intelligent life is that the solar system exists near the fringe of the Galaxy.

16 QUASARS, ACTIVE GALAXIES, AND GAMMA-RAY BURSTERS

IN THIS CHAPTER YOU WILL DISCOVER

- distant, luminous quasars

- the unusual spectra and small volumes of quasars

- bright and unusual objects, called active galaxies

- supermassive black holes that probably serve as central engines for quasars and active galaxies

- how observations from the central regions of quasars and active galaxies have helped astronomers devise theories about their energy sources

- gamma-ray bursters

2704 BATSE Gamma-Ray Bursts

+90

-180

+180

-90

The Most Powerful Known Bursts
Gamma-ray bursters have been observed everywhere in the sky, indicating that, unlike X-ray busters, they do not originate in the disk of the Milky Way Galaxy. This map of the entire sky "unfolded" onto the page shows 2704 bursts detected by the Burst and Transient Source Experiment (BATSE) aboard the Compton Gamma-Ray Observatory. The colors indicate the brightness of the bursts; they are coded brightest in red, dimmest in violet. Gray dots indicate incomplete information about the burst strength. (NASA)

WHAT DO YOU THINK?

1 What do quasars look like?

2 What powers a quasar?

The Sun releases far more energy than anything on Earth, and its brightness is reassuring. The energy emitted into space by O- and B-type stars is truly spectacular, and the momentary energy output of supernovae boggles the mind. But all of these are nothing compared to the power of quasars, active galaxies, and gamma-ray bursters.

As we will explore in this chapter, quasars are the resulting emission from supermassive black holes in the centers of some galaxies that are accreting (pulling in) large amounts of gas. Quasars emit more energy each second than the Sun does in 200 years. And they continue to do so for millions, even billions, of years. Some truly remarkable activity must occur deep within to make them so luminous.

QUASARS

Our knowledge of quasars began in an amateur astronomer's backyard. Grote Reber built the first radio telescope in 1936 in his backyard in Illinois, opening the realm of nonvisual astronomy. By 1944, Reber had detected strong radio emissions from sources in the constellations of Sagittarius, Cassiopeia, and Cygnus. Two of these sources, Sagittarius A (Sgr A) and Cassiopeia A

(Cas A), are in our Galaxy. The first is the galactic nucleus (see Chapter 14) and the second is a supernova remnant (see Chapter 12). However, Reber's third source, called Cygnus A (Cyg A), proved hard to categorize (Figure 16-1). The mystery only deepened in 1951, when Walter Baade and Rudolph Minkowski, using the 200-in. optical telescope on Mount Palomar, discovered a strange-looking galaxy at the same position (Figure 16-1 inset).

The galaxy associated with Cyg A is very dim. Nevertheless, Baade and Minkowski managed to photograph its spectrum. They detected a redshift corresponding to a speed of 14,000 km/s. According to the Hubble law, this speed indicates that Cyg A lies 635 million light-years (194 Mpc) from Earth!

Because it is one of the brightest radio sources in the sky, the enormous distance to Cyg A astounded astronomers. Although barely visible through the giant optical telescope at Palomar, Cyg A's radio waves can be picked up by amateur astronomers with backyard equipment. Its energy output must therefore be colossal. In fact, Cyg A shines with a radio luminosity 10^7 times as bright as that of an ordinary galaxy, such as M31 in Andromeda, yet it is far more distant. The object creating the Cyg A radio emissions has to be something extraordinary.

16-1 Quasars look like stars but have huge redshifts

WEB LINK 16-1 Cygnus A is not the only powerful radio source in the far-distant sky. Starting in the 1950s, radio astronomers busily made long lists of radio sources.

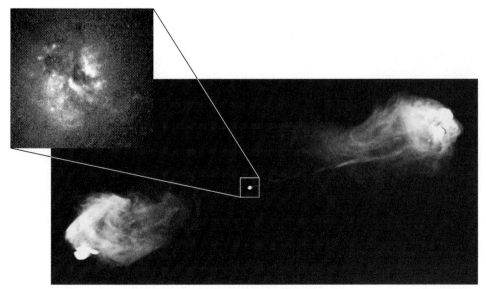

FIGURE 16-1 Cygnus A (3C 405) This radio image was produced from observations made at the Very Large Array. Most of the radio emission from Cygnus A comes from the radio lobes located on either side of the peculiar galaxy seen in the inset. The two radio lobes each extend about 160,000 ly from the optical galaxy, and each contains a brilliant, condensed region of radio emission. **Inset:** At the heart of this system of gas lies a strange-looking galaxy, which has a redshift corresponding to a recessional speed of 5% of the speed of light. According to the Hubble law, this corresponds to a distance of about 635 million light-years from Earth. Because Cygnus A is one of the brightest radio sources in the sky, this remote galaxy's energy output must be enormous. (R. A. Perley, J. W. Dreher, J. J. Cowan, NRAO; inset: William C. Keel)

R I V U X G

One of the most famous lists, the *Third Cambridge Catalogue,* was published in 1959. (The first two catalogs produced by this British team were filled with inaccuracies.) Even today, astronomers often refer to its 471 radio sources by their "3C numbers." Cyg A, for example, is designated 3C 405, because it is the 405th source on the Cambridge list. Because of the extraordinary luminosity of Cyg A, astronomers were eager to learn whether any other sources in the 3C catalog had similar properties.

One interesting case was 3C 48. In 1960, Allan Sandage used the Palomar telescope to discover a "star" at the location of this radio source (Figure 16-2). Recall that stars are blackbodies whose peak intensity typically is visible light and whose radio emission is much less intense. Because ordinary stars are not strong sources of radio emission, 3C 48 had to be something unusual. Indeed, its spectrum showed a series of emission lines that, initially, no one could identify. Although 3C 48 was clearly an oddball, many astronomers thought it was just another strange star in our Galaxy.

Another such "star," called 3C 273, was discovered in 1962. Like 3C 48, this object and the luminous "jet" of bright gas found protruding from one side of it emit a series of bright spectral emission lines that no one could then identify (Figure 16-3). These spectral lines are brighter than the background radiation at other wavelengths, called the continuum (recall Kirchhoff's laws in section 4-4).

R I V U X G

FIGURE 16-3 **The Quasar 3C 273** This greatly enlarged radio view shows the starlike object associated with the radio source 3C 273. Note the luminous jet on one side. By 1963, astronomers determined that the redshift of this quasar is so great that, according to the Hubble law, it is nearly 2 billion light-years from Earth. (Chandra X-ray Observatory, Harvard-Smithsonian Center for Astrophysics)

R I V U X G

FIGURE 16-2 **The Quasar 3C 48** For several years astronomers erroneously believed that this object (arrow) is simply a peculiar, nearby star that happens to emit radio waves. Actually, the redshift of this starlike object is so great that, according to the Hubble law, it must be roughly 4 billion light-years away. (Alex G. Smith, Rosemary Hill Observatory, University of Florida)

A breakthrough finally came in 1963, when Maarten Schmidt at the California Institute of Technology found that four of the brightest spectral lines of 3C 273 are positioned relative to one another, just as are four spectral lines of hydrogen. However, these emission lines from 3C 273 are found at much longer wavelengths than the usual wavelengths of hydrogen lines. The intense emission lines mean that something unusual is heating the gas. Schmidt reasoned that they are hydrogen lines, but subjected to a substantial redshift.

Spectra for stars in our Galaxy exhibit comparatively small Doppler shifts, because these stars cannot move extremely fast relative to the Sun without soon escaping from the Galaxy. Schmidt thus concluded that 3C 273 is not a nearby star after all. Pursuing this conclusion, he promptly found that its redshift corresponds to a speed of almost 15% the speed of light. According to the Hubble law, this huge redshift implies the incredible distance to 3C 273 of roughly 2 billion light-years (Bly).

Figure 16-4 shows the spectrum of 3C 273. Remember from Chapter 4 that a spectrograph records energy intensity at different wavelengths. The emission lines appear as peaks, and absorption lines appear as valleys. The emission lines

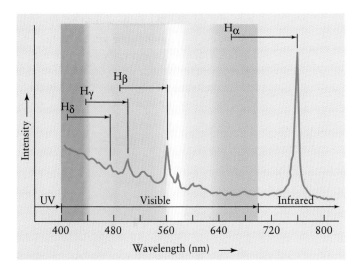

FIGURE 16-4   The Spectrum of 3C 273 Four bright emission lines caused by hydrogen dominate the spectrum of 3C 273. The arrows indicate how far these spectral lines are redshifted from their usual wavelengths.

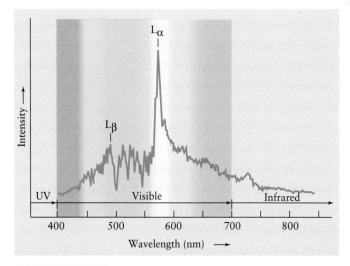

FIGURE 16-5   The Spectrum of a High-Redshift Quasar The light from this quasar, known as PKS 2000-330, is so highly redshifted that spectral emission lines normally in the far-ultraviolet (L_α and L_β) can be seen at visible wavelengths. Note the many deep absorption lines on the short-wavelength side of L_α. These lines, collectively called the "Lyman-alpha forest," are believed to be created by remote clouds of gas along our line of sight to the quasar. Hydrogen in these clouds absorbs photons from the quasar at wavelengths less redshifted than the quasar's L_α emission line.

are caused by excited gas atoms emitting radiation at specific wavelengths.

Inspired by Schmidt's success, astronomers looked again at the spectral lines of 3C 48. The redshift corresponds to a velocity of nearly one-third the speed of light. Therefore, 3C 48 must be nearly twice as far away as 3C 273, or about 4 Bly from Earth, assuming that the Hubble constant, H_0, is 72 km/s/Mpc.

Because of their starlike appearances and strong emissions, 3C 48 and 3C 273 were dubbed **quasi-stellar radio sources,** a term soon shortened to **quasars.** Quasars, also called **quasi-stellar objects (QSOs),** look like stars but have energy outputs that are incredibly higher. Furthermore, quasars need not specifically be radio sources; in fact, only 10% are strong radio emitters. Others have strong emissions in all different parts of the electromagnetic spectrum, with most being strongest in the infrared. Nevertheless, the name quasar has stuck, and more than 10,000 have been discovered since the pioneering days of the early 1960s.

All quasars have redshifts greater than about 0.06 or 6% the speed of light. The upper limit corresponds to speeds away from Earth of greater than 90% the speed of light. Figure 16-5 shows the spectrum of a quasar whose redshift corresponds to 92% of the speed of light. From the Hubble law, it follows that the distances to these high-redshift quasars are typically in the range of 10 billion to over 13 billion light-years. When we look at these quasars, we are seeing objects as they appeared when the universe was very young. In 2002, using a cluster of galaxies as a gravitational lens to brighten its light, astronomers discovered a quasar at a distance corresponding to when the universe was only 800 million years old.

16-2 A quasar emits a huge amount of energy from a small volume

Galaxies are big and bright. A typical large galaxy, like our own Milky Way, contains hundreds of billions of stars and shines with the luminosity of 10 billion Suns. The most gigantic and most luminous galaxies, the giant ellipticals, are only 10 times brighter. But beyond 8 billion light-years from Earth, even the brightest galaxies are too faint to be easily detected. Most ordinary galaxies are too dim to be detected at half that distance. If quasars can be seen more than 13 billion light-years away, they must be far more luminous than galaxies. Indeed, a typical quasar is 100 times brighter than our Milky Way. Stranger still, quasars fluctuate in brightness.

In the mid-1960s, several astronomers discovered evidence of these fluctuations. Going back over old images, they discovered that some newly identified quasars had actually been photographed in the past but they were not considered as anything special because they looked like stars. One photograph of 3C 273, for example, dates back to 1887. By carefully examining these old images, as well as recent ones, astronomers could see that *quasars occasionally flare up.* As Figure 16-6 shows for another quasar, 3C 279, energy from prominent outbursts reached Earth around 1937 and 1943. During these outbursts, the luminosity of 3C 279 increased

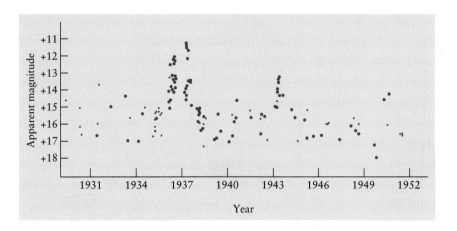

by a factor of at least 25. Because of the enormous distance to this quasar, during those peak periods it must have been emitting at least 10,000 times as much energy as the entire Milky Way.

Because quasars fluctuate in brightness, astronomers have been able to place strict upper limits on their size. As we saw in Chapter 13, an object cannot vary in brightness faster than the time it takes light to travel across that object. For example, an object that is 1 ly in diameter cannot vary in brightness with a period of less than 1 year.

Many quasars vary in brightness over only a few months, weeks, or days. In fact, X-ray observations reveal large variations in as little as 3 hours! This rapid flickering means that the source of the quasar's energy must be quite small by galactic standards. The energy-emitting region of a typical quasar—the "powerhouse" that blazes with the luminosity of 100 galaxies—is less than 1 light-day in diameter. Something must be producing the luminosity of 100 galaxies from a volume with approximately the same diameter as our solar system.

ACTIVE GALAXIES

16-3 Active galaxies bridge the energy gap between ordinary galaxies and quasars

Astronomers have discovered objects called **active galaxies,** with luminosities between those of ordinary galaxies and remote quasars. Active galaxies are exotic. Some have unusually bright, starlike (that is a pinpoint as seen from Earth) nuclei; others have strong emission lines in their spectra; still others are highly variable. Some have jets and beams of radiation emanating from their cores like the emissions observed for 3C 273, and most of these objects are more luminous than ordinary galaxies. Some, called **peculiar galaxies** (denoted pec), appear to be blowing themselves apart. Any of the Hubble classes of galaxies (see Chapter 15) can be peculiar.

 Carl Seyfert at the Mount Wilson Observatory discovered the first active galaxies in 1943 while surveying spiral galaxies. Now called **Seyfert galaxies,** these spirals reveal exceptionally bright, starlike nuclei and strong emission lines in their spectra. For example, the rich spectrum of NGC 4151 has many prominent emission lines. Some of these lines are produced by iron atoms with a dozen or more electrons stripped away, indicating that NGC 4151 contains some extremely hot gas. Seyfert galaxies also vary in brightness. For example, at times the magnitude of NGC 4151 changes over the course of a few days.

Another example of a Seyfert galaxy is NGC 1566, shown in Figure 16-7. At infrared wavelengths this galaxy shines with the brilliance of 10^{11} Suns. This extraordinary luminosity has been observed to vary by as much as $7 \times 10^9 L_\odot$ over only a few weeks. In other words, the

R I V U X G

FIGURE 16-7 The Seyfert Galaxy NGC 1566 This Sc galaxy is a Seyfert galaxy some 50 Mly (16 Mpc) from Earth in the southern constellation Dorado (the Goldfish). The nucleus of this galaxy is a strong source of radiation whose spectrum shows emission lines of highly ionized atoms. (Anglo-Australian Observatory)

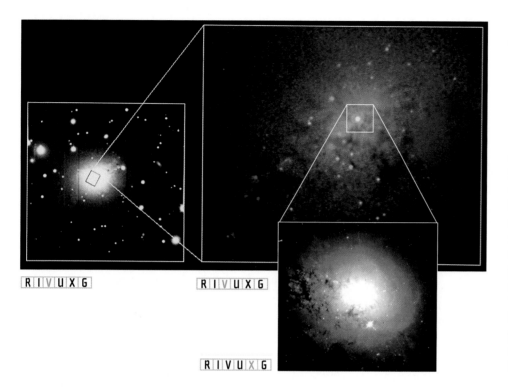

FIGURE 16-8 The Active Galaxy NGC 1275 (3C 84) This ground-based image (far left) is of the Seyfert galaxy NGC 1275, the largest and brightest member of the Perseus cluster of galaxies located about 250 million light-years from Earth. Spectroscopic studies indicate that NGC 1275 is actually two galaxies in collision. **Inset (near left):** This Hubble Space Telescope image at visible wavelengths shows about 50 small, bright, blue objects surrounding NGC 1275. They are believed to be young globular clusters formed as a result of the collision. The small white spot is the nucleus of the galaxy. **Inset (bottom):** NGC 1275 is also a strong source of X rays and radio waves. This X-ray image from the *ROSAT* X-ray Observatory shows that most of the galaxy's X-ray emission comes from its nucleus. (J. Holtzman, NASA; inset bottom: Swinburne University of Technology)

infrared power output of the nucleus of NGC 1566 rises and falls by an amount nearly equal to the total luminosity of our entire Galaxy.

Many more Seyfert galaxies have been discovered in recent years. Approximately 10% of the most luminous galaxies in the sky are Seyfert galaxies. Some of the brightest Seyfert galaxies shine as brightly as faint quasars, which leads most astronomers to suspect that the nuclei of Seyfert galaxies are, in fact, low-luminosity quasars.

Some Seyfert galaxies show vestiges of violent explosions in their nuclei. Filaments of gas tens of thousands of light-years long protrude from the nucleus of NGC 1275 in all directions (Figure 16-8). Spectroscopic studies indicate that this gas is being blasted away from the galaxy's nucleus at 3000 km/s. In 1977, Vera Rubin and her colleagues reported observations demonstrating that NGC 1275 actually consists of two colliding galaxies, a spiral and an elliptical. As we saw in Chapter 15, a collision or close encounter between two galaxies can eject matter into intergalactic space. More and more of this intergalactic medium is being discovered. Table 16-1 summarizes the relative power emitted by the Sun, the Milky Way Galaxy, quasars, Seyfert galaxies, and radio galaxies.

Other active galaxies are the **BL Lacertae objects.** The name comes from their prototype, BL Lacertae (also called BL Lac) in the constellation of Lacerta (the Lizard). BL Lac (Figure 16-9) was discovered in 1929, when it was mistaken for a variable star, largely because its brightness varies by a factor of 15 in only a few

months. A BL Lac's most intriguing characteristic is a *totally featureless spectrum,* exhibiting neither absorption nor emission lines. A BL Lac object is thought to be an elliptical galaxy with a bright quasarlike center, much as a Seyfert galaxy is a spiral galaxy with a quasarlike center. BL Lac objects change intensities—they are variable. Those that have periods of a day or less are also called **blazars.**

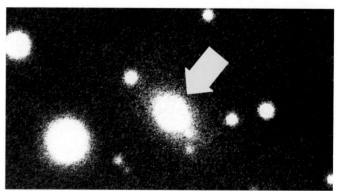

FIGURE 16-9 BL Lacertae This photograph shows fuzz around BL Lacertae (arrow). The redshift of this fuzz indicates that BL Lacertae is about 900 Mly (280 Mpc) from the Earth. BL Lac objects appear to be giant elliptical galaxies with bright quasar-like nuclei, much as Seyfert galaxies are spiral galaxies with quasar-like nuclei. BL Lac objects contain much less gas and dust than Seyfert galaxies. (T. D. Kinman, Kitt Peak Observatory)

TABLE 16-1 Galaxy and Quasar Luminosities

Object	Luminosity (watts)
Sun	4×10^{26}
Milky Way Galaxy	10^{37}
Seyfert galaxies	$10^{36} - 10^{38}$
Radio galaxies	$10^{36} - 10^{38}$
Quasars	$10^{38} - 10^{42}$

16-4 Active galaxies lie at the center of double radio sources

Much of the ejected material from active galaxies emits little visible radiation. Therefore, searches for high-speed ejecta are frequently carried out at nonvisible wavelengths. For example, Figure 16-10 is a combined optical and radio image of the galaxy NGC 5128 in the southern constellation of Centaurus. An unusually broad dust lane stretches across the galaxy. NGC 5128 is therefore classified as an "E0 (pec)." The gas ejected from this galaxy creates the two **radio lobes** shown on this figure.

Seyfert galaxies resemble dim, radio-quiet quasars. Certain elliptical galaxies, called **radio galaxies** because of their strong radio emission, are like dim, radio-loud quasars. Radio galaxy NGC 5128, also called Centaurus A, is one of the brightest sources of radio waves in the sky. In part, the brightness of Centaurus A at radio wavelengths comes from its proximity to Earth, only 13 million light-years away. It was one of the first radio sources discovered when radio telescopes were erected in Australia.

Detailed observations of Centaurus A reveal both radio and X-ray (Figure 16-10 inset) jets originating in the galaxy's nucleus. Particles and energy stream out of the galaxy's nucleus toward the radio lobes at nearly the speed of light. Such galaxies, with two radio lobes, are now called **double radio sources.**

By 1970, radio astronomers had discovered dozens of other double radio sources. An active galaxy, usually resembling a giant elliptical, is often found between the two radio lobes. The visible galaxy associated with Cyg A is also located between two radio lobes (see Figure 16-1). At radio wavelengths, double radio sources are among the brightest objects in the universe.

All double radio sources seem to have a central "engine" that ejects charged particles and magnetic fields outward along two oppositely directed jets. After traveling many thousands or even millions of light-years, this material slows down, allowing ejected electrons and magnetic fields to produce the radio radiation that we detect. This type of radio emission, called synchrotron radiation (see Section 14-3),

occurs whenever energetic electrons move in a spiral within a magnetic field. The radio waves that come from the lobes of a double radio source have all the characteristics of synchrotron radiation.

The idea that a double radio source emits jets of particles is supported by the existence of **head-tail sources,** so named because each such source appears to have a region of concentrated radio emission (the head) with two tails of gas streaming from it in opposite directions and then sweeping back. NGC 1265, an active elliptical galaxy in the Perseus cluster of galaxies, shows these properties. NGC 1265 is known to be moving at a high speed (2500 km/s) relative to the Perseus cluster as a whole. In a radio map (Figure 16-11), its radio emission has a distinctly windswept appearance.

Particles ejected in the two jets from NGC 1265 are deflected by the galaxy's passage through the sparse intergalactic medium. The head-tail source leaves behind a trail of

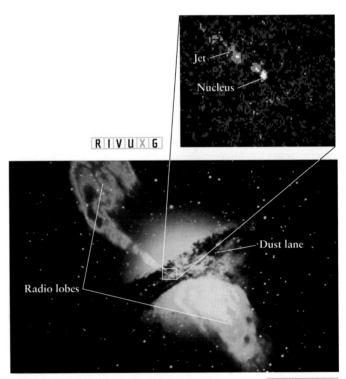

R I V U X G

R I V U X G

FIGURE 16-10 The Peculiar Galaxy NGC 5128 (Centaurus A) This extraordinary radio galaxy is located in the constellation of Centaurus, roughly 13 million light-years from Earth. At visible wavelengths a dust lane crosses the face of the galaxy. Superimposed on this visible image, a false-color radio image shows that vast quantities of radio radiation pour from matter ejected from the galaxy perpendicular to the dust lane. **Inset:** This X-ray image from the Einstein Observatory shows that NGC 5128 has a bright X-ray nucleus. An X-ray jet protrudes from the nucleus along a direction perpendicular to the galaxy's dust lane. (Jack Burns, University of Missouri; inset: Harvard Smithsonian Center for Astrophysics)

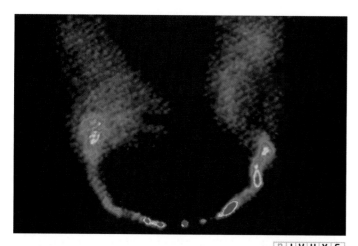

R I V U X G

FIGURE 16-11 **The Head-Tail Source NGC 1265** This active elliptical galaxy is moving at a high speed through the intergalactic medium. Because of this motion, the two tail jets trail the galaxy at its head, giving this radio source a distinctly windswept appearance. (NRAO)

particles, like the trail of smoke pouring from a rapidly moving steam train. But what is an active galaxy's locomotive, its "central engine"? What gives rise to its enormous jets?

SUPERMASSIVE CENTRAL ENGINES

As long ago as 1968, the British astronomer Donald Lynden-Bell suggested how quasars and active galaxies could produce such enormous amounts of energy from such small volumes.

Lynden-Bell argued that the gravitational field of a supermassive black hole (see Chapter 13) could be the central engine.

16-5 Supermassive black holes exist at the centers of most galaxies

As we saw in Chapter 13, black holes with a wide range of masses are being discovered throughout the universe. Telescopes are providing increasingly precise observations in support of the theoretical models of how matter behaves around black holes. Supermassive black holes have been detected in a rapidly growing number of galaxies, including M31, M32, M87, M104, and the Milky Way.

The Andromeda Galaxy (M31) is the largest, most massive galaxy in the Local Group (see the figure that opens Chapter 15). In the mid-1980s, several astronomers made careful spectroscopic observations of the core of M31. Using measurements of Doppler shifts, they determined that stars within 50 ly of the galactic core are orbiting the nucleus at exceptionally high speeds, which suggests that a massive object is located at the galaxy's center. Without the gravity of such an object to keep the stars in their high-speed orbits, they would have escaped from the core region long ago. From such observations, astronomers estimate the mass of the central object in M31 to be about 50 million solar masses. That much matter confined to such a small volume strongly suggests the existence of a supermassive black hole.

Located near M31 is a small elliptical galaxy called M32 (Figure 16-12). High-resolution spectroscopy indicates

R I V U X G

R I V U X G

FIGURE 16-12 **The Elliptical Galaxy M32** This small galaxy is a satellite of M31, a portion of which is seen at the left of this wide-angle photograph. Both galaxies are roughly 2.2 million light-years from Earth. **Inset:** This high-resolution image from the Hubble Space Telescope shows the center of M32. Note the concentration of stars at the nucleus of the galaxy. The area covered in this view is only 175 ly across. (Palomar Observatory; inset: NASA, ESA)

R I V U X G

FIGURE 16-13 The Sombrero Galaxy (M104) This spiral galaxy in Virgo is nearly edge-on to our Earth-based view. Spectroscopic observations indicate that a billion-solar-mass black hole is located at the galaxy's center. (ESO)

that stars close to the center of M32 are also orbiting this galaxy's nucleus at unusually high speeds, which could be explained by the presence of a supermassive black hole there. Furthermore, a picture taken by the Hubble Space Telescope (Figure 16-12 inset) shows that the concentration of stars at the core of M32 is truly remarkable. The density of stars there is more than 100 million times greater than the density of stars in the Sun's neighborhood. The concentration of stars and their high speeds strongly support the belief that a supermassive black hole exists at the center of M32.

Astronomers have uncovered evidence of supermassive black holes in other, more remote galaxies (Figure 16-13). John Kormendy used a 3.6-m telescope on Mauna Kea to examine the core of M104 spectroscopically. Once again, high-speed gas orbiting the galaxy's nucleus was found. These observations suggest that the center of this galaxy is dominated by a supermassive black hole containing a billion solar masses. Similar spectroscopic observations of the edge-on S0 galaxy NGC 3151 also reveal evidence for a billion-solar-mass black hole at its core.

Observations with the Hubble Space Telescope show evidence for supermassive black holes in an ever-growing number of galaxies. One early candidate galaxy for study by Hubble was the giant elliptical galaxy M87. Located some 50 million light-years from Earth, M87 is an active galaxy that has long been recognized as unusual. In 1918, Heber Curtis at the Lick Observatory reported that the center of M87 has "a curious straight ray ... apparently connected with the nucleus by a thin line of matter." Figure 16-14a shows M87 and several neighboring galaxies; M87's nucleus and jet are buried in that galaxy's glare. Figure 16-14b is a radio image showing the gas emission from the radio galaxy M87, which is also a powerful source of X rays.

The Hubble Space Telescope image of M87 in Figure 16-14c shows an exceptionally bright, starlike nucleus and a surrounding disk of gas with trailing spiral arms. To produce this fiery glow, stars must be packed so tightly at the center of M87 that their density is at least 300 times greater than that normally found at the centers of giant ellipticals. The motions of the stars in this central clustering support the hypothesis that a black hole with a mass of nearly 3 bil-

a   R I V U X G

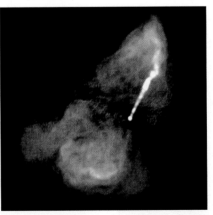

b   R I V U X G

R I V U X G   c

WEB LINK 16.6

FIGURE 16-14 The Giant Elliptical Galaxy M87 M87 is located near the center of the sprawling, rich Virgo cluster, which is about 50 million light-years from Earth. **(a)** This long exposure shows the extent of M87 and some smaller, neighboring galaxies. The numerous fuzzy spots that surround M87 are globular clusters; each contains about a million stars. **(b)** This radio image shows the intergalactic gas emitted by M87. These gas

clouds dwarf the giant galaxy, the bright spot near the center of the picture. **(c)** This Hubble Space Telescope image of M87 shows the gas disk in its nucleus. M87's extraordinarily bright nucleus and the gas jets result from a 3-billion-solar-mass black hole, whose gravity causes huge amounts of gas and an enormous number of stars to crowd around it. (a: © 1987 Anglo-Australian Observatory; b: NASA; c: STScI)

lion Suns resides at the center of M87. Some astronomers have asserted the belief that black holes lie at the hearts of most galaxies.

Black Hole Masses

16-6 Jets of matter ejected from around a black hole may explain quasars and active galaxies

If supermassive black holes lie at the centers of galaxies, the gravitational energy associated with them could well give rise to quasars and active galactic nuclei. Here is one scenario many astronomers find convincing: Consider a black hole at the center of a young galaxy filled with gas and dust. Because the centers of galaxies are congested places, any black hole there must capture a massive accretion disk of this interstellar debris. According to Kepler's third law, the material in the inner regions of such a disk orbit the most rapidly. The inner matter therefore constantly rubs against the more slowly moving gases in the outer regions, heating them up. As energized gases spiral toward the black hole, they release their energy violently, giving the hole its brilliant luminosity.

The crowding of infalling matter would also explain the jets of gas we observe streaming out from active galaxies. In our scenario, some of the infalling material is swallowed up by the black hole, but not all. Rather, the pressure in the accretion disk becomes so great that before much of the gas crosses the event horizon, it is compressed enough to be violently ejected, perpendicular to the accretion disk, where the hot gas experiences the least resistance. The result is two oppositely directed beams, depicted in Figure 16-15.

In 1995, the Hubble Space Telescope took a picture of the giant elliptical galaxy NGC 4261 (Figure 16-16) that astronomers interpret as showing an accretion disk about 800 ly in diameter orbiting a supermassive black hole. Indeed, the speed of the gas and dust in the disk indicates that it is held in orbit by a 1.2 billion-solar-mass object. The inset in Figure 16-16 is a Hubble Space Telescope image of the galaxy's nucleus. The dark, horizontal oval in the picture is our oblique view of the accretion disk, because ground-based radio and optical observations show a double-lobed structure (Figure 16-16) that is bisected by the oval.

What confines the ejected matter to narrow jets? Theory predicts that the ejecta is initially forced into such columns by magnetic fields that spiral out from the region of the black hole (Figure 16-17). These magnetic fields are generated by the gases orbiting around the black hole. The gas still falling toward the black hole then prevents escaping matter from spreading. To see why, consider water squirting out of the nozzle of an ordinary garden hose. The stream of water broadens and the spray fans out through a wide angle in the air. If the nozzle is placed in a swimming pool, however, the stream of water does not fan out as much. Similarly, as the two jets of hot gas leave the vicinity of the black hole, they must blast their way through the gas that is still crowding inward. Passage through this material causes the jets to become extremely narrow, concentrated beams. This "rocket nozzle" may explain not only double radio sources but also the jets and beams we see protruding from active galaxies and some quasars.

The main difference between double radio sources, quasars, and BL Lac objects is apparently just the angle at which the central engine is viewed. As Figure 16-18 shows, an observer sees a double radio source when the accretion disk is viewed nearly edge-on, because the jets are nearly in the plane of the sky. At a steeper angle, the observer sees a quasar. If one of the jets is aimed almost directly at Earth, a BL Lac object is seen.

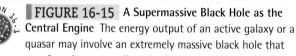

FIGURE 16-15 A Supermassive Black Hole as the Central Engine The energy output of an active galaxy or a quasar may involve an extremely massive black hole that captures matter from its surroundings. In the scenario depicted here, the inflow of material through an accretion disk is redirected to produce two powerful jets of particles traveling at nearly the speed of light.

Insight into Science   **Occam's Razor Revisited**   It is possible to create *separate* theories for quasars, BL Lac objects, and radio galaxies. Scientists prefer a comprehensive single theory that explains all three, because it incorporates common properties where only unconnected phenomena existed before. This search for a simple, more powerful explanation is the central engine of progress in science.

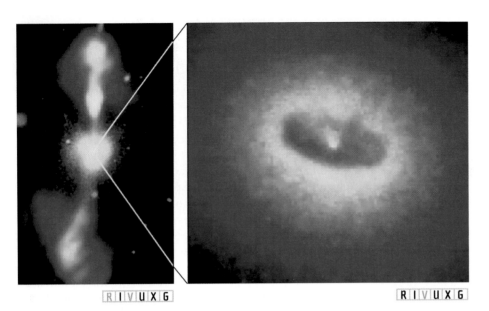

R I V U X G   R I V U X G

WEB LINK 16.7   ANIMATION 16.2

FIGURE 16-16

The Core of an Active Galaxy The giant elliptical galaxy NGC 4261 is a double radio source located in the Virgo cluster, about 100 million light-years from Earth. **(Far left)** An optical photograph of the galaxy (white) is combined with a radio image (orange and yellow) to show both the visible galaxy, which does not emit much radio energy, and its jets, which do. **Inset:** This Hubble Space Telescope image of the nucleus of NGC 4261 shows a disk of gas and dust about 800 ly (250 pc) in diameter. (NASA; inset: ESA)

The light emitted by distant quasars is subject to Einstein's general theory of relativity. Like the light from stars behind the Sun or behind the dark matter in our Galaxy, the light from distant objects such as quasars is therefore deflected as it travels past other galaxies toward us. This distortion of light, or gravitational lensing, can lead to our receiving several images of the distant quasar or galaxy, as discussed in Section 15-12. Figure 16-19a shows such images. Figure 16-19b shows four images of a quasar in a configuration called an **Einstein cross.** If the background object is exactly behind an intervening galaxy, the light should actually be focused as a ring, called an **Einstein ring,** rather than as several separate images. Conversely, gravitational lensing can reveal the existence of relatively nearby clusters of galax-

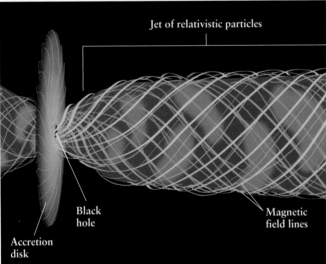

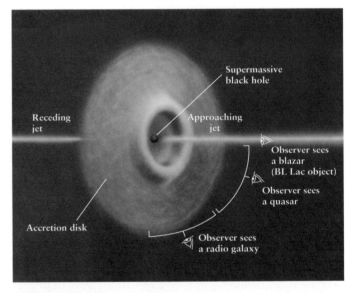

ANIMATION 16.3

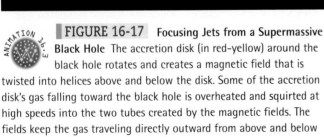

FIGURE 16-17 Focusing Jets from a Supermassive Black Hole The accretion disk (in red-yellow) around the black hole rotates and creates a magnetic field that is twisted into helices above and below the disk. Some of the accretion disk's gas falling toward the black hole is overheated and squirted at high speeds into the two tubes created by the magnetic fields. The fields keep the gas traveling directly outward from above and below the disk, thus creating the two jets.

FIGURE 16-18 The Orientation of the Central Engine and Its Jets Double radio sources, quasars, and BL Lacertae objects may be the same type of object viewed from different directions. If one of the jets is aimed almost directly at the Earth, we see a BL Lac object. If the jet is somewhat tilted to our line of sight, we see a quasar. If the jets are nearly perpendicular to our line of sight, we see a double radio source. The central region of the system is shown in Figure 16-17.

a

R I V U X G

FIGURE 16-19 Gravitational Lensing of Galaxies and Quasars **(a)** The five distorted, blue loops in this Hubble Space Telescope image are all the same spiral galaxy lensed by the cluster of yellow, intervening elliptical and spiral galaxies. The cluster is about 5 Bly away in the constellation Pisces. The distorted galaxy is about twice that distance. **(b)** This image from the Hubble Space Telescope shows the gravitational lensing of a quasar in the constellation of Pegasus. The quasar, about 8 billion

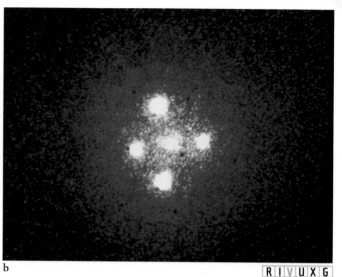

b

R I V U X G

light-years from Earth, is seen as four separate images surrounding a galaxy that is only 400 million light-years away. This pattern is called an Einstein cross. The diffuse image at the center of this Einstein cross is the core of the intervening galaxy. The physical effect that creates these multiple images is a large-scale version of microlensing, depicted in Figure 14-15. (a: W.N. Colley and E. Turner (Princeton University) J.A. Tyson (Bell Labs, Lucent Technologies) and NASA; b: NASA/ESO)

ies that are ordinarily too dim to see. In 2001, astronomers discovered such a cluster by observing more distant background galaxies being lensed by it.

GAMMA-RAY BURSTERS

Supermassive black holes are not the only source of extremely powerful emissions from the remote reaches of the universe. Astronomers have also discovered pulses of gamma-ray energy from equally distant, yet smaller, sources. These short-lived sources have astonishing energy output.

16-7 Gamma-ray bursters are the most powerful explosions in the known universe

Because atomic blasts emit pulses of gamma rays, the U.S. military put the gamma-ray-detecting Vela satellites in orbit around the Earth to monitor illegal nuclear explosions in the 1960s. In 1973, astronomers were told that these satellites were detecting bursts of gamma rays from objects in space. This unexpected news stunned the astronomical community, in part because there was no known mechanism to emit such gamma rays and in part because the energies involved in these blasts are staggering.

Gamma-ray bursters emit energy for between a few milliseconds and about 1000 seconds. Unlike X-ray bursters (see Section 12-14), each gamma-ray burster emits only one burst. By plotting the length of time that the observed gamma-ray bursters radiate, astronomers have discovered that there appear to be at least two types of gamma-ray bursters. One group typically lasts for a few tenths of a second, while those in the other group typically last for about 40 seconds.

More than 3000 gamma-ray bursters have been observed, and more are being discovered at a rate of about one per day. If gamma-ray bursters were located in the plane of the Milky Way's disk, we would expect them to be found in localized regions of the sky. However, as this chapter's opening photo shows, gamma-ray bursters are located all over the celestial sphere. This indicates that either these bursters are in a spherical distribution in the immediate neighborhood of the solar system or that they are scattered throughout the far reaches of the universe.

To help determine their locations, astronomers have searched for optical and infrared objects located in the same place as the bursters. Since 1997, optical and infrared "afterglows" of many gamma-ray bursters have been observed. These often persist for days, which allows astronomers to study their spectra in detail. Observations of the spectra of visible light that has accompanied several bursts (Figure 16-20) reveal that the light has passed through intergalactic gas

22 seconds

48 seconds

73 seconds

R I V U X G

FIGURE 16-20 Gamma-Ray Bursts In 1999, astronomers photographed the visible-light counterpart of a 100-second gamma-ray burst some 9 billion light-years away in the constellation Boötes. The times indicated are after the burst began. (Carl Akerlof, University of Michigan, Los Alamos National Laboratory, Lawrence Livermore National Laboratory)

clouds. Combining this with observations of the bursters' spectra and redshifts reveals that bursters typically originate more than half the distance to the edge of the observable universe. They are billions of light-years away. As of 2002, the most distant gamma-ray burster was more than 12 Bly away. Taking their distance into account, a typical gamma-ray burster emits as much energy in 100 seconds as the Sun will emit over its entire 10-billion-year lifetime.

Deep sky observations at the locations of some gamma-ray bursters have begun to show galaxies in the same places. As Figure 16-21 shows, the bursters do not appear in the centers of their host galaxies—they are probably not associated with the intermediate-mass or supermassive black holes in the galactic cores. This suggests that the gamma-ray bursters are instead associated with some stellar cataclysm.

The origins of gamma-ray bursters are still being debated. Two theories in the forefront are gamma-ray emission from a black hole swallowing a neutron star or from pairs of colliding neutron stars. Calculations confirm that the collision of such dense objects should indeed lead to a rapid, intense burst of gamma rays. A third possibility under investigation is the emission from especially energetic supernovae called *hypernovae*. In this scenario, a supermassive star in a dense, star-forming region evolves very rapidly and explodes, sending out a high speed fireball that slams into the surrounding gas. Given the observations, discussed above, that there appear to be two types of gamma-ray bursters, it is likely that more than one mechanism exists to create them.

16-8 Frontiers yet to be discovered

WEB LINK 16.10 Quasars, active galaxies, and gamma-ray bursters are rich areas for observational and theoretical research. Why do different quasars emit most strongly in different parts of the spectrum? Are quasars an early phase of most galaxies, including the Milky Way, as many astronomers are coming to believe? What causes quasar variability? Do we have the proper explanations that blazars, quasars, and radio galaxies are caused by supermassive black holes seen from different orientations? Are there actually two (or more) sources of gamma-ray bursters, and, if so, what are they? How does so much energy get transformed so quickly into electromagnetic radiation in a gamma ray burster? New discoveries in this realm are being made nearly every week.

FIGURE 16-21 The Host Galaxy of a Gamma-Ray Burster The Hubble Space Telescope recorded this visible-light image in 1999, 16 days after a gamma-ray burster was observed at this location. The image shows a faint galaxy that is presumed to be the home of the gamma-ray burster. The galaxy has a bright blue color, indicating the presence of many recently formed stars. (Andrew Fruchter, STScI; and NASA)

Host galaxy

"Afterglow" of gamma-ray burster

1 arcsec

R I V U X G

WHAT DID YOU KNOW?

1 *What do quasars look like?* They look like stars, but they spew out much more energy than any star.

2 *What powers a quasar?* A quasar is believed to be powered by a supermassive black hole with millions or even billions of solar masses at the center of a galaxy.

KEY WORDS

active galaxy, 416

blazar, 417

BL Lacertae (BL Lac) object, 417

double radio source, 418

Einstein cross, 422

Einstein ring, 422

gamma-ray burster, 423

head-tail source, 418

peculiar galaxy (pec), 416

quasar (quasi-stellar radio source), 415

quasi-stellar object (QSO), 415

radio galaxy, 418

radio lobe, 418

Seyfert galaxy, 416

KEY IDEAS

• The development of radio astronomy in the late 1940s opened the realm of nonvisual astronomy.

Quasars

• A quasar, or quasi-stellar radio source, is an object that looks like a star but has a huge redshift. This redshift corresponds to an extreme distance from the Earth, according to the Hubble law.

• To be seen from Earth, a quasar must be very luminous, typically about 100 times brighter than an ordinary galaxy. Relatively rapid fluctuations in the brightnesses of some quasars indicate that they cannot be much larger than the diameter of our solar system.

Active Galaxies

• An active galaxy is an extremely luminous galaxy that has one or more unusual features: an unusually bright, starlike nucleus; strong emission lines in its spectrum; rapid variations in luminosity; or jets or beams of radiation emanating from its core.

• An active spiral galaxy with a bright, starlike nucleus and strong emission lines in its spectrum is categorized as a Seyfert galaxy.

• BL Lacertae (BL Lac) objects (some of which are called blazars) have bright nuclei whose cores show relatively rapid variations in luminosity.

• Most double radio sources contain an active galaxy located between two characteristic radio lobes. A head-tail radio source shows evidence of jets of high-speed particles emerging from an active galaxy.

• Most quasars are probably very distant active galaxies.

Supermassive Central Engines

• Spectroscopic observations indicate huge concentrations of matter at the centers of certain galaxies.

• The strong energy emission from quasars, active galaxies, and double radio sources is probably produced as matter falls toward a supermassive black hole at the center.

• Some matter spiraling in toward a supermassive black hole would be squeezed into two oppositely directed beams that carry particles and energy into intergalactic space.

Gamma-Ray Bursters

• Gamma-ray bursters are events believed to be caused by the collisions of dense objects, such as neutron stars or black holes. They are observed to be billions of light-years away from Earth.

• Typical gamma-ray bursters occur for a few tens of seconds and emit more energy than the Sun will radiate over its entire 10-billion-year lifetime.

REVIEW QUESTIONS

1 Suppose you suspected a certain object in the sky to be a quasar. What sort of observations would you perform to confirm your hypothesis?

2 Explain why astronomers do not use any of the standard candles described in Chapter 15 to determine the distances to quasars.

3 Explain how the rate of variability of a source of light can be used to place an upper limit on the size of the source.

4 What is an active galaxy? List the different kinds of active galaxies. How do they differ from one another?

5 Why do astronomers believe that the energy-producing region of a quasar is very small?

6 How is synchrotron radiation produced?

7 What evidence indicates that quasars are extremely distant active galaxies?

8 What is a double radio source?

9 What is a supermassive black hole? What observational evidence suggests that supermassive

black holes might be located at the centers of many galaxies?

10 Why do many astronomers believe that the engine at the center of a quasar is a supermassive black hole surrounded by an accretion disk?

11 How does the orientation of the jets emanating from the center of a galaxy relative to our line of sight relate to the type of active galaxy that we observe?

ADVANCED QUESTIONS

12 In the 1960s, some astronomers suggested that quasars might be compact objects ejected at high speeds from the centers of nearby ordinary galaxies. Why does the absence of blueshifted quasars disprove this hypothesis?

13 When quasars were first discovered, many astronomers were optimistic that these extremely luminous

objects could be used to probe distant regions of the universe. For example, it was hoped that quasars would provide high-redshift data from which the Hubble constant could be accurately determined. Why have these hopes not been realized?

DISCUSSION QUESTIONS

14 Explore the belief that quasars, double radio sources, and giant elliptical galaxies represent an evolutionary sequence.

15 Some quasars show several sets of absorption lines whose redshifts are less than the redshift of the quasars'

emission lines. For example, the quasar PKS 0237-23 has five sets of absorption lines, all with redshifts somewhat less than the redshift of the quasar's emission lines. Propose an explanation for these sets of absorption lines.

WHAT IF ...

16 Earth passed through a jet emitted by a radio galaxy? What would we see and what might happen to the Earth?

17 A jet of gas suddenly appeared all across the night sky? What might that indicate has recently happened?

18 Sirius suddenly became ring-shaped for a few hours and then returned to normal? What might cause such an event?

19 A gamma-ray burst occurred between here and Alpha Centauri? What might have caused such an event and what might it mean for Earth?

WEB/CD-ROM QUESTION

20 To test your understanding of active galaxies, do Interactive Exercise 16-1 on the Web or CD-ROM. You can print out your answer if required.

17 COSMOLOGY

IN THIS CHAPTER YOU WILL DISCOVER

- cosmology, which seeks to explain how the universe began and whether it will end

- the best theory we have for the origin of the universe—the Big Bang

- how astronomers trace the emergence of matter and the formation of galaxies

- how astronomers explain the overall structure of the universe

- our understanding of the fate of the universe

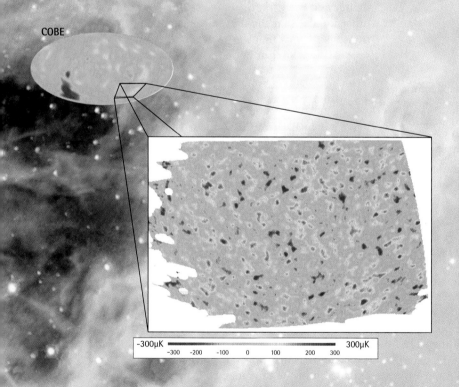

COBE

-300µK ▬▬▬▬▬▬▬▬▬▬ 300µK
-300 -200 -100   0   100   200   300

Structure of the Early Universe
This microwave map of the entire sky, produced from data taken by NASA's Cosmic Background Explorer (COBE), shows temperature variations in the cosmic microwave background. Yellow regions are about 0.00003 K warmer than the average temperature of 2.73 K; blue regions are about 0.00003 K cooler than the average. These tiny temperature fluctuations are related to the large-scale structure of the universe today. The radiation detected to make this map is from a time 500,000 years after the Big Bang. Inset: This map was made by the BOOMERANG telescope, which has about 30 times the resolving power of COBE. This image shows much finer detail in the temperature variation of the cosmic microwave background. (NSF/NASA)

WHAT DO YOU THINK?

|1| What is the universe?

|2| Did the universe have a beginning?

|3| Will the universe last forever?

People have speculated about the origins of the universe since before the dawn of recorded time, but until modern science and technology were developed, these theories were untestable and, therefore, matters of faith. Using ever-improving observations, astronomers have developed scientific theories of the evolution of the universe over the past two centuries. Only one theory of the universe's origin proposed so far—the Big Bang—has survived the scrutiny of the scientific process. With the aid of this theory, we can now describe scientifically how the universe has evolved from about 10^{-43}s after it began up to today. Observations made at the very close of the twentieth century have also illuminated the fate of the universe. Contrary to what virtually all astronomers believed even a few years ago, we now have compelling evidence that the universe is expanding outward faster and faster. Consequently, the universe will last forever, even as the objects in it evolve and fade away.

THE BIG BANG

|1| The **universe** consists of all matter, energy, and spacetime that we can ever detect. **Cosmology** is the study of the origin, structure, and evolution of the universe. As a science, modern cosmology got off the ground in 1915, when Einstein published his general theory of relativity. To his surprise and dismay, the equations predicted that the universe is not static: It should either be expanding or contracting.

17-1 The universe is expanding

The prediction of a changing universe flew in the face of the belief in an infinite, static universe, a concept promoted by Isaac Newton more than two centuries earlier. Newton believed that each star is fixed in place and held there under the influence of a uniform gravitational pull from every part of the cosmos. If the stars were not uniformly distributed, he argued, one region would have more mass than another. The denser region's gravity would then attract other stars, causing them to clump together further. Because he did not observe this clumping, Newton concluded that the mass of the universe must be distributed uniformly over an infinite space.

Newtonian static cosmology and prevailing theology made Einstein doubt the implications of his own theory, so he missed the opportunity to propose that we live in a changing universe. Instead, he adjusted his elegant equations to yield a static cosmos. He did this by adding a repulsive (outward-pushing) term called the **cosmological constant** to his equations so that gravity's attractive force would be counterbalanced and the universe would be static.

After observations revealed an expanding universe, Einstein said that adding the cosmological constant was the biggest blunder of his career. He may have been wrong about that. Observations since 1997 indicate that the universe is not just expanding, but actually *accelerating* outward. This means that there must be an outward pressure that counteracts the effects of normal gravitation. One of the two current theories that can explain this is the presence of a cosmological constant in nature. We discuss this further in Section 17-14.

Edwin Hubble is credited with discovering that we live in an **expanding universe.**
He was the first person to observe that the clusters of galaxies in our local supercluster are redshifted as seen from Earth. Since redshift implies motion away from us, this implies that the clusters are all moving away from us and, presumably, from one another (see section 15-10). Subsequent observations have extended this result to clusters in other superclusters. They are all moving away from the Local Group. Since superclusters are the largest groupings of matter and energy in the universe, their separation from each other implies that the universe is expanding. These observations also reveal that most, if not all, clusters of galaxies in superclusters are not bound to each other.

17-2 The expansion of the universe creates a Dopplerlike redshift

The separation of clusters and superclusters could be interpreted as due to the Doppler effect. Recall (see Section 4-7) that the normal Doppler shift is caused by an object moving toward or away from us through fixed spacetime. However, using Einstein's general theory of relativity, one finds that spacetime, the fabric of the universe, is not fixed, but is actually expanding. The universe can change shape and size. This expansion carries the superclusters away from each other. Gravitationally bound objects like planets and stars, and gravitationally bound systems such as galaxies and clusters of galaxies, are not individually expanding due to this effect, however. The expansion of space just acts to increase the separation between large ensembles of galaxies.

The redshift that Hubble observed, caused by the expansion of the universe, is properly called the **cosmological redshift.** Because space is expanding, photons we observe from galaxies in other superclusters are all redshifted. To understand why, consider a wave drawn on a

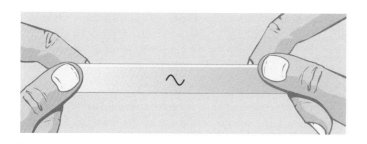

a

FIGURE 17-1   Cosmological Redshift (a) Just as the wave drawn on this rubber band is stretched along with the rubber band,

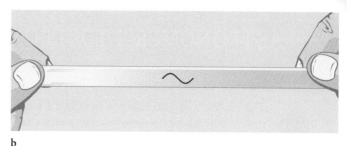

b

(b) so too are the wavelengths of photons stretched as the universe expands.

rubber band (Figure 17-1). The wave gets longer as the rubber band stretches. Imagine a photon coming toward us from a distant galaxy. As the photon travels through space, space is expanding, and this stretches the photon's wavelength. When the photon reaches our eyes, we see a drawn-out wavelength—the photon is redshifted. The longer the photon's journey, the more its wavelength is stretched by the expansion of the universe. Therefore, astronomers observe larger redshifts in photons from distant galaxies than in photons from nearby galaxies.

As we saw in Chapter 15, Hubble discovered the linear relationship between the distances to galaxies in other superclusters and the redshifts of those galaxies' spectral lines. Specifically, a galaxy twice as far from Earth as another galaxy has twice the redshift of the closer one. This implies that remote galaxies are moving away from us at speeds proportional to their distances from us (review Figure 15-27). Except for the most distant galaxies and quasars, the normal Doppler shift and the cosmological redshift predict the same relationship between redshift and motion. That is why Hubble was justified in using the Doppler equation to calculate the recession of galaxies.

17-3 The Hubble constant is related to the age of the universe

Hubble's law gives us a way to estimate the age of the universe. Imagine watching a movie of two superclusters receding from each other. If we then run the film backward, time runs in reverse, and we observe the superclusters approaching each other. The time it would take for them to collide is the time since the Big Bang. *Assuming the universe has always expanded at a constant rate,* we can use a simple equation to estimate the age of the universe:

$$\text{Time since the Big Bang} = \frac{\text{separation distance}}{\text{recessional velocity}}$$

Recall that Hubble found the relationship

$$\text{Recessional velocity} = H_0 \times \text{separation distance}$$

which we can rewrite as

$$H_0 = \frac{\text{recessional velocity}}{\text{separation distance}}$$

Comparing the first and last equations (see An Astronomer's Toolbox 17-1), we see that Hubble's constant is the inverse of the time since the universe began. Using a Hubble constant of 72 km/s/Mpc, we find:

$$1/H_0 = 1/72 \text{ km/s/Mpc} = 13.6 \text{ billion years}$$

The universe has *not* been expanding at a constant rate, and we do not yet know the complete history of its motion. Nevertheless, the age of the universe given by $1/H_0$ is probably within 2 billion years of the correct answer. For convenience, we hereafter assume the universe is 14 billion years old.

17-4 Remnants of the initial expansion, the Big Bang, have been detected

In 1927, the Belgian astrophysicist and Catholic priest Georges Lemaître proposed that, as a consequence of the equations of Einstein's general theory of relativity, the universe is expanding. Using Hubble's subsequent observations of an expanding universe, Lemaître logically concluded that the superclusters must have expanded from a much smaller volume. He proposed that the universe began as an extraordinarily dense *primordial atom* of energy. This idea led to the concept of the **Big Bang,** when all of spacetime, matter, and energy were created. Indeed, Lemaître is considered the "father of the Big Bang."

As noted earlier, scientists do not have a definitive explanation for the Big Bang. However, a growing body of observations supports the belief that this event occurred. Because the universe is expanding, the volume of space that exists is also increasing. Ironically, Sir Fred Hoyle, who aptly named this explosion the Big Bang in 1949, was one

AN ASTRONOMER'S TOOLBOX 17-1

H_0 and the Age of the Universe

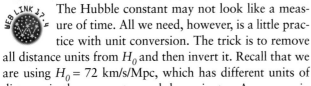

The Hubble constant may not look like a measure of time. All we need, however, is a little practice with unit conversion. The trick is to remove all distance units from H_0 and then invert it. Recall that we are using $H_0 = 72$ km/s/Mpc, which has different units of distance in the numerator and denominator. As we saw in Chapter 15, this mix of units makes it easy to determine how fast a galaxy is receding once its distance is known.

Example: A galaxy 10 Mpc away is receding at

$$72 \text{ km/s/Mpc} \times 10 \text{ Mpc} = 720 \text{ km/s}$$

We know that

$$1 \text{ pc} = 3.09 \times 10^{13} \text{ km}$$

and so, 1 Mpc = 3.09×10^{19} km. Because there are 3.156×10^7 seconds in a year, we get the conversion factors we need:

$$H_0 = \frac{1}{72 \text{ km/s/Mpc}} \times (3.09 \times 10^{19} \text{ km/Mpc})$$

$$\times \frac{1}{3.156 \times 10^7 \text{ s/yr}} = 13.6 \times 10^9 \text{ yr}$$

If the universe had always been expanding at the same speed, then $1/H_0$ would be the age of the universe. However, for the first several billion years of its existence, the expansion of the universe was slowing down, and since then it has been speeding up.

Compare! Decades of observations by many astronomers have confirmed that all distant galaxies are moving away from the Earth. Not only that, *the Hubble law relating velocities and distances is the same in all directions.* The Hubble constant, H_0, is the same no matter where you aim your telescope. The fact that the recession rates are the same in all directions is a condition called **isotropy**. The universe is isotropic: Its expansion is the same in every direction.

The general expansion away from us suggests that the Earth is at the center of the universe. After all, where else could we be if all the distant galaxies are moving away from us? In fact, the answer is that we could be practically anywhere in the universe! To understand why, see Guided Discovery: The Expanding Universe.

---

of the very few astrophysicists who never believed that the Big Bang occurred.

Hoyle was justified in questioning the Big Bang theory back then because Hubble's observations that the universe is expanding are insufficient evidence for accepting the Big Bang theory. In 1948, Hoyle, Herman Bondi, and Thomas Gold proposed a plausible alternative called the *steady state theory.* In this theory the universe has existed forever, is continuously expanding, and, as it expands, new matter is created that takes the place of the receding matter. The steady state theory proposes that throughout all time and space, the newly created matter precisely compensates for the matter moving away due to the expansion. As a result, the average density of the universe remains constant. This new matter eventually forms new galaxies. Where this new matter comes from is unknown, but the laws of physics do not necessarily exclude its creation. So, why do virtually all astronomers now accept the Big Bang theory rather than the steady state? The answer lies in later observations that are consistent with the former, but not with the latter.

Shortly after World War II, the astrophysicists Ralph Alpher and George Gamow proposed that seconds after the Big Bang, the universe was an incredibly hot blackbody—far

in excess of 10^{12} K. It must therefore have been filled with high-energy, electromagnetic radiation. As the universe expanded, the cosmological redshift stretched the wavelengths of the radiation created during the Big Bang so that most of the photons left over from this event are now radio waves. Recall from Section 3-3 that photon energies decrease with increasing wavelength. Because the photons are being redshifted, the energy packed into each cubic meter of the universe is decreasing, which is another way to say that the universe is cooling off.

Calculations indicated that if the universe began as the Big Bang, the remnants of that energy should still fill all space today, giving the universe a background temperature only a few kelvin above absolute zero. This radiation is called the **cosmic microwave background** because the peak of its blackbody spectrum should lie in the microwave part of the radio spectrum. In the early 1960s, Robert Dicke, P. J. E. Peebles, David Wilkinson, and their colleagues at Princeton University began designing a microwave radio telescope to detect it.

Meanwhile, just a few miles from Princeton, two physicists had already detected the cosmic background radiation. Arno Penzias and Robert Wilson of Bell Laboratories

GUIDED DISCOVERY   The Expanding Universe

To understand the universe, scientists build models, mathematical representations of the world about us. To develop such pictures, it often helps to begin with a simple analogy. Because the expansion of the universe is hard to visualize, imagine for a moment the batter for a chocolate chip cake floating in an oven in the Space Shuttle. As the cake bakes, it expands and the chocolate chips move farther apart. They do not get larger, nor do they move through the batter. Each chocolate chip remains at rest in its own little bit of cake and is carried along as the cake spreads out.

Now, think of each chocolate chip as a supercluster of galaxies and the batter between the chocolate chips as the rest of spacetime. As the universe expands, the distance between widely separated superclusters of galaxies grows larger and larger. The expansion of the universe is the expansion of spacetime.

Suppose you were on one particular chocolate chip, say the chocolate chip labeled 3 at the left in the figure. You would see chocolate chips 2 and 4 moving away from you with equal velocities. Every 10 minutes chocolate chips 2 and 4 get 1 inch farther away as the cake expands, as shown on the right. In that same time interval, chocolate chips 1 and 5 move twice as far from you as chocolate chips 2 and 4. To go twice as far away in the same time, chocolate chips 1 and 5 must be moving twice as fast as the closer chocolate chips 2 and 4. This is exactly what Hubble's law of cosmological expansion (recessional velocity = $H_0 \times$ distance) says: *Double the distance and you double the velocity.*

To see why we need not occupy the center of the universe, put yourself on chocolate chip 4. You see chocolate chips 3 and 5 moving away at the same rate—in fact, they recede at the same rate as you saw chocolate chips 2 and 4 move away when you were on chocolate chip 3. From chocolate chip 4, chocolate chips 2 and 6 are moving away twice as fast as chocolate chips 3 and 5. From any chocolate chip, you will see all the other chocolate chips moving away. Similarly, no matter where we are in the universe, we see all the other superclusters receding from us.

The Expanding Chocolate Chip Cake Analogy The expanding universe can be compared to a chocolate chip cake baking and expanding in the Space Shuttle. Just as all the chocolate chips move apart as the cake rises, all the superclusters of galaxies recede from each other as the universe expands.

were working on a new horn antenna designed to relay telephone calls to Earth-orbiting communications satellites (Figure 17-2). Penzias and Wilson were deeply puzzled. No matter where in the sky they pointed their antenna, they detected a faint background noise. Even the careful removal of static noise-generating pigeon droppings from the antenna failed to eliminate the noise completely. Thanks to a colleague, they soon learned of the then-theoretical cosmic microwave background and the work of Dicke, Peebles, and Wilkinson in trying to locate it. Communicating with the Princeton astronomers, Penzias and Wilson presented their finding and thus were able to claim the first detection.

Precise measurements of the cosmic microwave background were first made by the *Cosmic Background Explorer* (COBE) satellite, which operated between 1989 and 1994 (Figure 17-3a). Data from COBE's spectrometer shown in Figure 17-4 demonstrate that this

FIGURE 17-2   **The Bell Labs, Horn Antenna** This Bell Laboratories horn antenna at Holmdel, New Jersey, was used by Arno Penzias (on the right) and Robert Wilson in 1965 to detect the cosmic microwave background. (Lucent Technologies, Bell Laboratories)

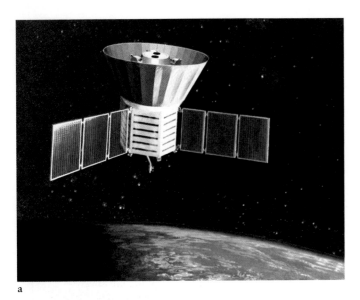

a

b

FIGURE 17-3 In Search of Primordial Photons **(a)** The *Cosmic Background Explorer* (*COBE*) satellite, which operated from 1989 to 1994, measured the spectrum and angular distribution of the cosmic microwave background over a wavelength range of 1 mm to 1 cm. *COBE* (pronounced "co-bee") found local temperature variations across the sky but no overall deviation from a perfect blackbody spectrum. **(b)** The balloon-carried telescope BOOMERANG orbited above Antarctica for 10 days collecting data used to resolve the cosmic microwave background with 10 times higher resolution than that of *COBE*. Results from these telescopes are shown in the figure that opens this chapter. (a: J. Mather, NASA; b: The BOOMERANG Group, University of California, Santa Barbara)

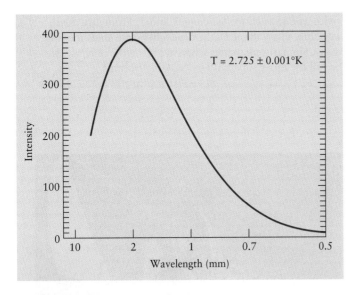

FIGURE 17-4 The Spectrum of the Cosmic Microwave Background The intensity of radiation from the cosmic microwave background of the universe as measured by the Far Infrared Absolute Spectrometer (FIRAS) on the *COBE* satellite. The data fall along a blackbody curve of 2.735 K. The peak of the curve occurs at about a wavelength of 1.9 mm, in accordance with Wien's law for this temperature. The precision of the measurements was so great that the error bars, indicating the possible errors for the 34 data points taken by FIRAS, are too small to be seen on this figure.

ancient radiation indeed had the spectrum of a blackbody, with a temperature of approximately 2.73 K. Cosmic background radiation, also commonly called the *3-degree background radiation*, is required by the Big Bang theory but is neither required nor predicted by the steady-state theory, lor any other competing cosmological theory. Detection of the cosmic microwave background is one reason why the Big Bang is accepted by astronomers as the origin of the universe.

17-5 The universe has two symmetries— isotropy and homogeneity

Observations show that the cosmic microwave background is almost perfectly **isotropic:** Its intensity is nearly the same in every direction in the sky. Figure 17-5 shows a map of the microwave sky. It is very slightly warmer than average in the direction of the constellation Leo and slightly cooler in the opposite direction, toward Aquarius. The tiny temperature variation depicted in Figure 17-5 results from the Earth's overall motion through the cosmos. We are moving through the background radiation because of our rotation and orbit around the Sun and the solar system's motion around the Milky Way. More important, the Milky Way, and even the Local Group of galaxies, are themselves in motion (Figure 17-6).

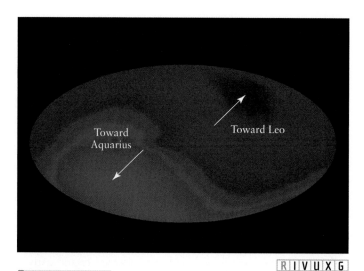

R I V U X G

FIGURE 17-5   **The Microwave Sky** This map of the microwave sky was produced from data taken by instruments on board *COBE*. The galactic center is in the middle of the map, and the plane of the Milky Way runs horizontally across the map. Color indicates temperature—red is warm, blue is cold. The temperature variation across the sky is caused by the Earth's motion through the microwave background. The variation is quite small, only 0.0033 K above the average radiation temperature of 2.726 K (which we have rounded to 2.73 K in the text). (NASA)

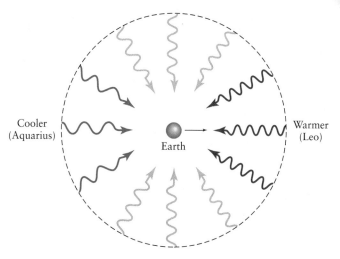

FIGURE 17-6   **Our Motion Through the Microwave Background** Because of the Doppler effect, we detect shorter wavelengths in the microwave background and a higher temperature of radiation in that part of the sky toward which we are moving. This part of the sky is the area shown in red in Figure 17-5. In the opposite part of the sky, shown in blue in Figure 17-5, the microwave radiation has longer wavelengths and a cooler temperature.

The observed temperature differences shown in Figure 17-5 mean that the solar system is at present moving toward Leo at a speed of 390 km/s. Taking into account the known velocity of the Sun around the center of our Galaxy, astronomers calculate that the entire Milky Way Galaxy is moving relative to the cosmic microwave background at 600 km/s—some 1.3 million miles per hour—in the general direction of the Centaurus cluster. The gravitational attraction of four nearby clusters of galaxies, most significantly the *Great Attractor*, and an enormous supercluster called the *Shapley concentration* are believed to be pulling us in that direction. Subtracting the motion of the Earth, Sun, Milky Way, and the Local Cluster with respect to the microwave background, the average intensity of radiation is found to be the same in all directions to about 1 part in 100,000. While the existence of the cosmic background radiation provides compelling support for the Big Bang theory, its near-isotropy was, until recently, a major problem.

 Isotropy is not limited to the blackbody radiation observed throughout the universe. It is also found on sufficiently large scales when exploring the numbers of galaxies in different directions. Astronomers have counted the numbers of galaxies at different distances from us and in a variety of directions. The number of galaxies stays roughly constant, allowing, as we saw in Chapter 15, for the distribution of galaxies on the boundaries of bubbles throughout the universe. Because of this *uniformity with distance*, we say that the universe is also **homogeneous.** Isotropy and homogeneity must be explained by any viable theory of cosmology such as the Big Bang.

The need to explain isotropy and homogeneity, among other things, has led to numerous refinements in the Big Bang theory. As a result, this theory now provides a scenario for the evolution of the universe from a tiny fraction of a second after it formed and onward.

A BRIEF HISTORY OF SPACETIME, MATTER, ENERGY, AND EVERYTHING

The composition of the universe during its first few minutes of existence was profoundly different from what it is today. To understand why, we need to expand briefly on the nature of the four known physical forces: gravity, electromagnetism, and the strong and weak nuclear forces.

17-6 All physical forces in nature were unified at first

Only the effects of electromagnetism and gravity can extend over infinite distances. The electromagnetic force holds electrons in orbit about the nuclei in atoms. But over large volumes of space, the net effects of electromagnetism cancel out because a negative electric charge exists for every positive charge and a south magnetic pole exists for every north magnetic pole.

While gravity is the weakest of all four forces, it is the only force whose effect is infinite and attractive for all ordinary matter. That is why gravity determines the evolutionary behavior of stars, the orbits of planets, the existence of galaxies, and many other large-scale phenomena. Indeed, the attractive force of gravity dominates the universe at astronomical distances. (Despite all this, we will learn shortly that for certain forms of energy, gravity can be repulsive.)

In contrast to gravity and electromagnetism, the strong and the weak nuclear forces are both extremely short range, because their influences extend only over distances less than about 10^{-15} m. The **strong nuclear force** holds protons and neutrons together inside the nuclei of atoms. Without this force, nuclei would disintegrate because of the electromagnetic repulsion of the positively charged protons. Thus, the strong nuclear force overpowers the electromagnetic force inside nuclei.

The **weak nuclear force** is at work in certain kinds of radioactive decay, such as the transformation of a neutron into a proton. Protons and neutrons are composed of more basic particles called **quarks**. A proton is composed of two "up" quarks and one "down" quark, whereas a neutron is made of two "down" quarks and one "up" quark. The weak nuclear force is at work whenever a quark changes from one variety to another. For example, in one kind of radioactive decay, a neutron transforms into a proton, and one of the neutron's "down" quarks changes into an "up" quark.

To examine details of the physical forces, scientists use particle accelerators that hurl high-speed electrons and protons at targets. In such experiments, physicists find that as the speeds of the particles approach the speed of light, the different forces begin to behave the same way. In fact, during experiments at the CERN accelerator in Europe in the 1980s, particles were slammed together with such violence that the electromagnetic force and the weak nuclear force were indistinguishable. Thus was confirmed the *electroweak* force, first proposed by Steven Weinberg, Sheldon Glashow, and Abdus Salam, and for which they received the 1979 Nobel Prize in physics.

The equations describing the electromagnetic, weak, and strong forces indicate that they all have the same strength when particles have energies or, equivalently, temperatures much greater than will ever be possible to create in the laboratory on Earth. But recall from the discussion above that when the universe was young, it was very hot—hot enough for this unifying of the forces to occur. Detailed calculations reveal that in the first 10^{-35} of existence, the temperature of the universe was more than high enough for the electromagnetic as well as the weak and strong nuclear forces to have the same strength.

The relationship between gravitation and the other three forces is still uncertain. Scientists have not yet been able to include its effects, although it is generally expected that the next generation of theories that explain elementary particles will include it. Efforts to create such a theory are under way under the general name of **superstring** theories. Assuming gravity will be joined to the other forces, we include its expected effects here. Figure 17-7 shows that during the beginning moments of the universe (from 0 to 10^{-43} s, called the **Planck era**), the four forces are believed to have been unified—only one physical force existed in nature. However, this remains to be proven because the equations we use to describe nature break down when applied to this era.

Because physicists have no hope of building accelerators powerful enough to test the unified theory of the fundamental forces, they have joined with astrophysicists and astronomers to make the entire universe a large laboratory. If there is a way to see what happened during the earliest moments of the universe, they could use those observations to test the theory. Amazingly, such observations can be, and are being, made.

17-7 Equations explain the evolution of the universe even before matter and energy as we know it existed.

The earliest time at which our current equations can explain the behavior of the universe is about 10^{-43} seconds, called the **Planck time** in honor of Max Planck, who helped derive some of the fundamental quantum physics that was used to

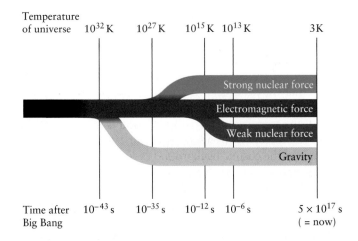

FIGURE 17-7   **Unification of the Four Forces** The strength of the four physical forces depends on the speed or energy with which particles interact. As shown in this figure, the higher the temperature of the universe, the more the forces resemble each other. Also included here are the ages of the universe at which the various forces were equal.

calculate this time. The Planck time denotes the end of the Planck era. It was at the Planck time that gravity is believed to have become a separate force, leaving electromagnetism and the weak and strong nuclear forces still united as one, called the **GUT** (for **grand unified theory**) force. The Planck time corresponded to when the universe had a temperature of about 10^{32} K and the currently observable universe was much smaller than the size of an atom today. There were no separate particles and electromagnetic radiation then, just gravitation and the GUT force, so matter and energy as we know them did not exist.

The matter in the early universe was chaotically distributed. Some places had higher temperatures than other places. *If* the universe had continued expanding only at the rate calculated for the very beginning, today we should be able to see different parts of the universe with significantly different temperatures. As we saw in Figure 17-5, the temperatures of every region of the universe are the same to within a tiny fraction of 1° K. This is called the **isotropy problem** or **horizon problem.** Realizing how uniform the universe's temperature is, astronomers have hypothesized that very early in its existence, the universe underwent a short period of dramatically higher expansion, called the **inflationary epoch,** that smoothed out the universe we can see.

For the first 10^{-35} seconds of its existence, calculations indicate that the universe expanded at the initial rate and that its temperature dropped to 10^{27} K. At this point, the energy in each volume of the universe had declined to the point at which the strong nuclear force could no longer remain unified with the electromagnetic and weak nuclear forces. At this instant the strong nuclear force emerged, distinct from the other two forces, which remained together as the electroweak force.

Many astronomers hypothesize that before the strong force decoupled from the electroweak force, the universe was in an unstable state called a *false vacuum*. It was unstable in the same sense as you would be if you woke up one day to find yourself on a slippery, one-foot-wide shelf on the face of a cliff thousands of feet above the cliff bottom. The slightest move and you fall. Analogous to the transition you make by falling off the shelf, the young universe is believed to suddenly have made a transition (Figure 17-8) into a stable *true vacuum* state. The universe's transition to the true vacuum state caused it to momentarily accelerate.

As the universe began moving toward the true vacuum state, it was already expanding. The transition caused the expansion rate to increase dramatically. From about 10^{-35} until about 10^{-33} seconds after the Big Bang, the universe ballooned outward about 10^{30} times larger. The currently observable universe expanded to about the size of a soccer ball. This huge rate of expansion of the universe has come to be called **inflation.** The end of the inflationary epoch heralded the first formation of elementary particles: quarks, electrons, photons, and their antiparticles.

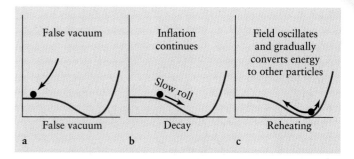

FIGURE 17-8 The Cause of Inflation (a) The universe formed in an unstable energy state that (b) quickly began to transition to a stable configuration (c).

You might think that the energy in space would be very uniform after inflation spread it out so much and so rapidly. However, quantum theory played a vital role during the inflationary epoch. Normally, quantum fluctuations occur on the size scales of pairs of particles, as we have seen in Section 13-7 with the creation and annihilation of virtual particles throughout all space. However, quantum fluctuations on microscopic scales at the beginning of the inflationary period were stretched by inflation to scales that spanned clusters of galaxies and beyond! Indeed, the theory predicts that these tiny fluctuations would ultimately be the seeds for the formation of the large-scale structures of the universe that exist today. Since inflation is part of a scientific theory, its occurrence leads to predictable observations. As we will see shortly, telescopes have indeed seen its predicted distributions of matter.

When inflation ended, the universe resumed the relatively leisurely expansion initiated by the Big Bang. Arriving at the time 10^{-12} s, when the temperature of the universe had dropped to 10^{16} K, the electromagnetism force separated from the weak nuclear force. From that moment on, all four forces interacted with particles essentially as they do today. At that time, the universe's matter was primarily quarks, antiquarks, neutrinos, antineutrinos, electrons, and positrons.

17-8 During the first second, most of the matter and antimatter in the universe annihilated each other

The next significant event is calculated to have occurred at 10^{-6} s, when the universe's temperature was 10^{13} K. Before this moment, the temperature was so high that it was energetically favorable for quarks to exist as isolated particles, rather than to combine in twos and threes to create the particles that exist in the universe today. We know this occurred because seas of quarks were recreated under equivalent conditions in the laboratory in 2001 by smashing dense atoms together. After 10^{-6} s, appropriately called **confinement,** the

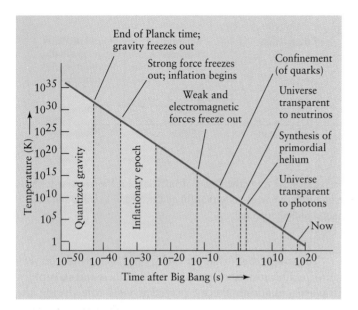

FIGURE 17-9 **The Early History of the Universe** Current theory holds that, as the universe cooled, the four forces "froze out" of their initial unified state. The inflationary epoch lasted from 10^{-35} s to 10^{-24} s after the Big Bang. Neutrons and protons froze out one-millionth of a second after the Big Bang. The universe became transparent to light (that is, photons decoupled from matter) when the universe was 500,000 years old.

universe was sufficiently cool so that quarks could finally stick together to form individual protons, neutrons, and their antiparticles. Figure 17-9 summarizes the connections between particle physics and cosmology.

Equations show that early in the first second the universe was so hot that photons were gamma rays possessing incredibly high energies. These energies were high enough to

create matter and antimatter, according to Einstein's equation E = mc^2. As we saw in Chapter 9, to make a particle of mass m, you need an amount of energy E at least as great as mc^2, where c is the speed of light.

The creation of matter from energy is routinely observed in laboratory experiments involving high-energy gamma rays. A highly energetic photon colliding with an atomic nucleus can create a pair of particles, as sketched in Figure 17-10. This process, called **pair production**, always creates one ordinary particle and its so-called antiparticle. For example, if one of the particles is an electron, the other is an antielectron or positron (recall from Chapter 13 that similar pairs of particles are also believed to be produced near a black hole).

A positron has the same mass as an electron but the opposite charge. When an electron and a positron meet, they annihilate each other and their energy is converted into photons. Theory predicts that at times earlier than 1 second in the life of the universe, pair production occurred continually. In other words, shortly after the Big Bang, all space was chock full of protons, neutrons, electrons, and their antiparticles, all immersed in a phenomenally hot bath of high-energy photons.

As the universe expanded, the temperature declined until the photons could no longer create pairs of particles and antiparticles. After that first second, pair production ceased. Thereafter, particles and antiparticles often collided, annihilating each other and converting their masses back into high-energy photons. But nothing could replenish the dwindling supply of particles and antiparticles.

If particles and antiparticles are always created or destroyed in pairs, why aren't they found in equal numbers in the universe today? For every proton created, there should be an antiproton; for every electron there should be an antielectron. Furthermore, if all particles had annihilated their antiparticles in the early universe, no matter would be

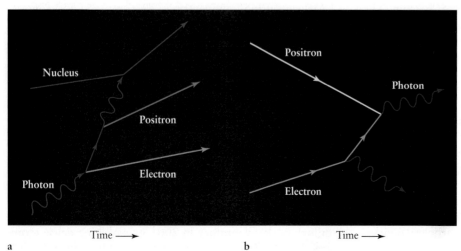

a   b

FIGURE 17-10 **Pair Production and Annihilation** (a) A particle and an antiparticle can be created when a high-energy photon collides with a nucleus. (b) Conversely, a particle and an antiparticle can annihilate each other and emit energy in the form of gamma rays.

left at all. And yet, here we are and astronomers observe very few antiparticles in the universe.

Physicists therefore theorize that the symmetry or equality between the number of particles and the number of antiparticles was broken—a *symmetry breaking*. This means that the number of particles was actually slightly greater than the number of antiparticles. For every billion antiprotons, perhaps a billion plus one protons formed; for every billion antielectrons, a billion plus one electrons formed. As far as theorists can tell, virtually all the antiparticles created in the first second of time annihilated normal particles shortly thereafter. The remaining particles in the universe were the slight excess of normal matter created as a result of this symmetry breaking.

17-9 The universe changed from being controlled by radiation to being controlled by matter

As the universe expanded and cooled further, the remaining protons and neutrons began colliding and fusing together. For the most part, they were quickly separated again by the gamma rays in which they were bathed. However, during the first 3 minutes, enough fusions occurred that were not subsequently split apart to create most of the helium and at least 10% of the trace element lithium that exists today. Hydrogen, helium, and lithium are the three lowest-mass elements. After a few minutes, the universe was too cool to allow fusion to create more massive elements, such as carbon and oxygen. All elements other than the three lowest-mass ones formed later as a result of stellar evolution, as discussed in Chapters 9, 11, and 12.

> **Insight into Science**   **Share Information** While the evolution of the universe is a truly large-scale event, the activity in the first 3 minutes was nevertheless intimately connected to the creation and evolution of elementary particles, the smallest objects in the universe. Physicists who study elementary particles and astrophysicists who study cosmology now regularly hold meetings to compare notes and connect their fields.

Calculations indicate that for tens of thousands of years, photons dominated the behavior of the universe in two important ways. The first effect derives from a prediction of general relativity, that the energy in photons contributes to the gravitational attraction in the universe. This contradicts intuition, because photons are massless and mass is what we normally associate with causing gravitational attraction. We now know that gravitation *can* have other sources.

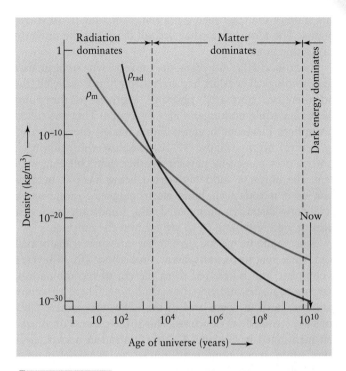

FIGURE 17-11   **The Evolution of Density** For approximately 30,000 years after the Big Bang, the gravitational effects from photons (ρ_{rad}, shown in red) exceeded the effects of the matter density (ρ_m, shown in blue). This early period is said to have been radiation-dominated. Later, however, continued expansion of the universe caused ρ_{rad} to become less than ρ_m, at which time the universe became matter-dominated.

When the universe was young, the energy density, and hence the gravitational effect, of photons was greater than the gravitational effect of matter. This was called the **radiation-dominated universe**. The major effect of the radiation-dominated universe was that photons prevented matter from collapsing together and forming stars and galaxies. As the universe expanded, the gravitational effects of both photons and matter decreased. Calculations show that the effect of photons fell faster than that of matter (Figure 17-11), until eventually, after about 30,000 years, matter came to dominate the gravitational behavior of the universe, a situation that has persisted ever since. Thus developed the **matter-dominated universe**. It was only after the universe became matter-dominated that gas could respond to gravitational interactions and begin clumping where the force of gravity was sufficiently strong.

The second way that photons dominated the early universe occurred for about the first 500,000 years. During this time, photons prevented ions and electrons from connecting to form neutral atoms. During that epoch, the universe was completely filled with a shimmering expanse of high-energy photons colliding vigorously with protons and electrons.

This state of matter, called a *plasma,* is opaque to all wavelengths. The term **primordial fireball** describes the universe during this time.

By around 500,000 years, the universe was so large that the cosmological redshift had stretched photon wavelengths and therefore decreased photons' energies to the point where they could no longer ionize hydrogen. This was called **decoupling.** Electrons began orbiting nuclei, thereby creating neutral atoms (Figure 17-12). Because the primordial photons now passed by particles, rather than colliding with them, the universe suddenly ceased being hazy. It became clear, as it is today, enabling electromagnetic radiation to travel long distances unimpeded and, important from our point of view, enabling us to see far into the universe.

This dramatic period, when the universe transformed from being opaque to transparent, took about 100,000 years to complete and is referred to as the **era of recombination,** referring to nuclei and electrons combining. There are many instances, both in space and in the laboratory, where electrons are stripped off of atoms. This is the process of ionization mentioned in Chapter 4. After ionization occurs, electrons often recombine with nuclei to recreate neutral atoms. However, the gases created early in the life of the universe were never neutral, so the "re" in recombination is misleading. Because the universe was opaque in its first 500,000 years, we cannot see any further into the past than the era of recombination with telescopes that detect electromagnetic radiation of any wavelength.

We can use Wien's law to calculate the temperature of the cosmic background radiation at this time. The 3000 K temperature at decoupling corresponds to a peak microwave wavelength of 0.001 mm. The temperature history of the universe is graphed in Figure 17-9.

Ever since the primordial photons formed, the expansion of the universe has been increasing their wavelengths, so today we see these same photons as the cosmic microwave background. The universe is about 1000 times larger than it was at decoupling. Although the universe is dominated by matter today, the number of photons left over from the Big Bang is still immense. From the physics of blackbody radiation, astronomers calculate that there are now 400 million cosmic background microwave photons in every cubic meter of space. In contrast, if all the visible matter in the universe were uniformly spread throughout space, there would be roughly one hydrogen atom in every 3 cubic meters of space. In other words, photons outnumber atoms by more than a billion to one. The microwave background contains the most ancient photons we expect ever to be able to observe. This microwave background is a ghostly relic of the former dazzling splendor of the universe.

17-10 Inflation explains why the universe is isotropic and homogeneous

We can now explore the issues raised earlier in the chapter of how the Big Bang cosmology explains the universe's isotropy and homogeneity. Recall that there is no reason to expect that the entire universe created at the Big Bang was everywhere the same. It plausibly had different temperatures, pressures, and densities in different places. Because light takes time to travel to Earth, the farther we look into space, the further back in time we are seeing. We can only see a finite distance into it, namely back to the time of decoupling, about 14 billion light-years away.

Since the universe has a finite age of about 14 billion years, we cannot see anything beyond 14 billion light-years away. At that distance, we see objects as they were just being formed. Therefore, all that we can see from the Earth is contained in an enormous sphere (Figure 17-13). The boundary of this sphere is called the **cosmic light horizon.** This sphere, with a radius of 14 billion light-years (assuming a 14-billion-year-old universe), is our entire observable universe.

Light from all the luminous objects in this volume has had time to reach the Earth, but light from objects beyond

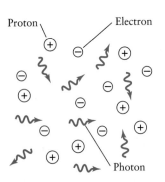

a **Before recombination**

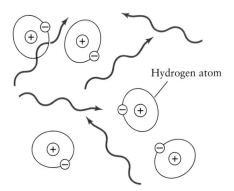

b **After recombination**

FIGURE 17-12 The Era of Recombination
(a) Before recombination, the energies of photons in the cosmic background were high enough to prevent protons and electrons from forming hydrogen atoms. The universe was opaque and the temperature of the matter was the same as the temperature of the radiation. (b) As soon as the energy of the background radiation became too low to ionize hydrogen, neutral atoms came into existence. The universe became transparent, the remaining photons traveled without affecting the gas, and the temperature of the matter and radiation were no longer the same.

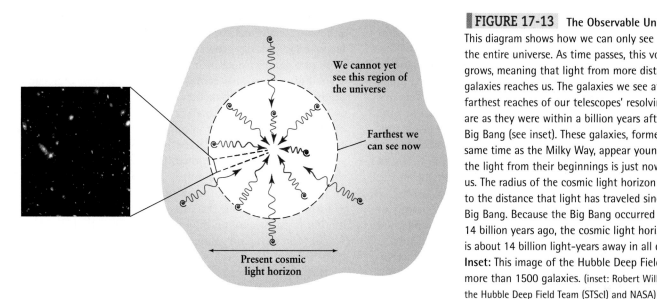

FIGURE 17-13 **The Observable Universe**
This diagram shows how we can only see a part of the entire universe. As time passes, this volume grows, meaning that light from more distant galaxies reaches us. The galaxies we see at the farthest reaches of our telescopes' resolving power are as they were within a billion years after the Big Bang (see inset). These galaxies, formed at the same time as the Milky Way, appear young because the light from their beginnings is just now reaching us. The radius of the cosmic light horizon is equal to the distance that light has traveled since the Big Bang. Because the Big Bang occurred about 14 billion years ago, the cosmic light horizon today is about 14 billion light-years away in all directions. Inset: This image of the Hubble Deep Field shows more than 1500 galaxies. (inset: Robert Williams and the Hubble Deep Field Team (STScI) and NASA)

the cosmic light horizon has not yet arrived here, so we are unaware of objects in these regions. If the universe had not undergone a period of inflation, we should be able to see different regions of the universe with different densities and temperatures in different directions. To understand why, consider looking in two opposite directions out into space. The objects we see on these two sides are roughly 28 billion light-years apart, and, therefore, light in the 14-billion-year-old universe traveling from one side has not yet reached the other side. Therefore, without inflation, there is no way that the two sides could know what temperatures the other is at. By the same reasoning, without inflation, the two sides could not thermalize, meaning come to the same temperature. And yet we see all parts of the universe with virtually the same temperature.

But because inflation did occur, the universe has expanded so much, so swiftly, that our cosmic light horizon only encompasses what was initially a tiny volume of the whole universe, a volume so small that it was initially isotropic and homogeneous.

17-11 Galaxies formed from huge clouds of primordial gas

The Big Bang theory explains the existence of spacetime, matter, and energy. It accounts for the hydrogen, most of the helium, and some of the lithium that exist today. But how do we explain the existence of superclusters of galaxies, clusters of galaxies, and individual galaxies? As noted above, these must have formed after the beginning of the matter-dominated era of the universe.

To explain large-scale structures in the early matter-dominated universe, recall the quantum fluctuations that were stretched by inflation. These clumps throughout the inflated universe were initially extremely small differences in density with a wide range of sizes in the microwave background. We know this because the *COBE* satellite observed the cosmic microwave background and discovered these variations in density back when the universe was only 500,000 years old. *COBE* data show that they were only about 1 part in 100,000 denser than the average density of matter in the early universe (see the figure that opens this chapter). But these small variations on various size scales are all that is needed to explain the present large-scale structure of the universe.

When the universe was radiation-dominated, the particles in it were moving too fast to clump together and form structure. When the universe became matter-dominated and particles had cooled down sufficiently, the extra gravitational attraction in regions where the concentration of matter was slightly greater than average caused the particles to draw together and ultimately form galaxies, clusters, and superclusters. In the first stage of that process, the contracting matter collapsed inward, heated up, and rebounded outward due to the increased pressure between particles. Gravity drew them back inward and the process repeated itself over many cycles. This produced oscillation in the gas of primordial particles analogous to sound waves in our air.

The oscillations should still be visible today as regions of slightly higher and lower density than average, but with smaller sizes than those discovered by *COBE*. Since 1999, the telescopes Maxima, BOOMERANG (Figure 17-3b), MAT/TOCO, and Dasi (pronounced Daisy), each with 10 times better resolution than *COBE*, discovered such variations in the cosmic microwave background (see the figure that opens this chapter). As we will see, these observations have an important bearing on the overall shape of the universe.

Virtually all the matter in the early universe consisted of clouds of hydrogen and helium. Between the time of decoupling and the first burst of star formation was a period of little starlight, called the **dark ages**. Note, however, that the dark ages were bright due to light from energy created in the Big Bang! There is growing observational evidence that star formation began within a few hundred million years of the Big Bang (Figure 17-14a) and that by 600 million years after the Big Bang, galaxies were beginning to coalesce (Figure 17-14b). We also have observational evidence from *Chandra* that galaxies were in clusters within 3 billion years of the Big Bang (Figure 17-14c).

During the first billion years, it appears that clumps of gas containing millions or billions of solar masses collapsed to create supermassive black holes. These would have attracted other matter into orbit around them and thereby served as the seeds for the growth of some galaxies. By observing remote galaxies, astronomers have discovered that most galaxies initially emitted energy we associate with quasars and other active galaxies. This implies that most young galaxies have supermassive black holes at their centers. Recent observations reveal that massive and supermassive black holes typically grow until they have about 0.2% of the mass of a galaxy's nuclear bulge.

Observations also reveal that galaxies were bluer and brighter in the past than they are today. These changes in color and brightness suggest a high abundance of young, bright, hot, massive stars in newly formed galaxies (Figure 17-15a). As galaxies age, these blue O and B stars become supergiants and eventually die off. Therefore, galaxies grow somewhat redder and dimmer. This is especially true of those elliptical galaxies that formed early on. They appear to form nearly all their stars in one vigorous burst of activity that lasts for about a billion years (Figure 17-15b).

In contrast, spiral galaxies have been forming stars for at least the past 10 billion years, although at a gradually decreasing rate. There is still plenty of interstellar hydrogen in the disks of spiral galaxies like our Milky Way to fuel star formation today. That is why O and B stars still highlight their spiral arms. Figure 17-15b compares the rates at which spiral

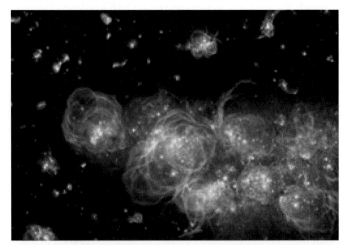

a

b

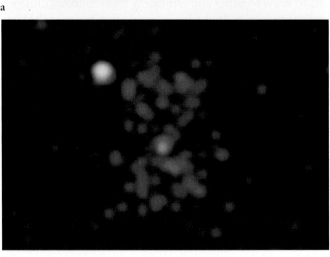

c

FIGURE 17-14 Galaxies Forming by Combining Smaller Units (a) This painting indicates how astronomers visualize the burst of star formation that occurred within a few hundred million years after the Big Bang. The arcs and irregular circles represent interstellar gas that is illuminated by supernovae. (b) Using the Hubble and Keck telescopes, astronomers discovered two groups of stars (arrows in inset) 13.4 Bly away that are believed to be protogalaxies, from which bigger galaxies grew. These protogalaxies were discovered because they were enlarged by the gravitational lensing of an intervening cluster of galaxies. (c) The *Chandra* X-ray telescope imaged gravitationally bound gas around the distant galaxy 3C 294. The X-ray emission from this gas is the signature of an extremely massive cluster of galaxies, in this case at a distance of about 11.2 Bly from us. (a: K. Lanzetta (SUNY), NASA; b: Richard Ellis (Caltech) and Jean-Paul Kneib (Observatorie Midi-Pyrenees, France) NASA, ESA; c: NASA)

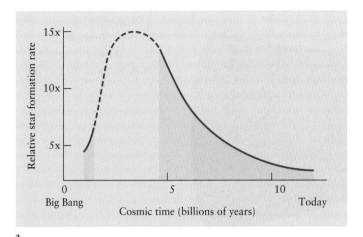

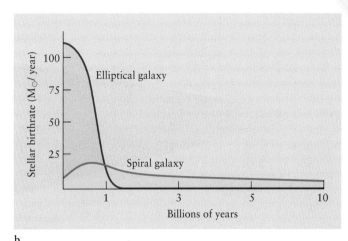

a

b

FIGURE 17-15 Stellar Birth Rates (a) Based on observations of star formation at very early times in the universe and also at times from 5 billion years after the universe formed and onward, this figure shows our present knowledge of the overall star-formation history of the universe. Astronomers hope to fill the data gap between 2 billion years and 5 billion years shortly. They expect it to show a peak of star formation activity. (b) Most of the stars in an elliptical galaxy are created in a brief burst of star formation when the galaxy is very young. In spiral galaxies, stars form at a more leisurely pace that extends over billions of years.

and elliptical galaxies form stars. Evidence suggests that, overall, star formation peaked at about 15 times its present rate when the universe was about one-third its present age.

17-12 Star formation activity determines a galaxy's initial structure

Imagine a developing galaxy, called a *protogalaxy*, forming from a cloud of gas. Theory proposes that the rate of star formation determines whether this protogalaxy becomes a spiral or an elliptical. If stars form slowly enough, then the gas surrounding them has plenty of time to settle by collision with other infalling gas into a flattened disk, just like the early solar system. Star formation continues because the protogalactic disk contains an ample supply of hydrogen, and a spiral or lenticular galaxy is created. If, however, the initial stellar birthrate is high in the protogalaxy, the theory predicts that virtually all pregalactic gas is used up in the creation of stars before a disk can form. In this case, an elliptical galaxy is created. Figure 17-16 depicts these contrasting scenarios. Figure 17-17 summarizes the major evolutionary activities in the universe through the present day.

Not all galaxies have maintained their initial structure over the life of the universe. As we discussed in Chapter 15, some galaxies collide, and these collisions can change a galaxy's structure from spiral, for example, to elliptical. Indeed, astronomers using the Hubble Deep Field images have observed elliptical galaxies during the first few billion years of the universe's existence that are far less uniform in color than are closer ellipticals. They observed blue stars in the young ellipticals consistent with the merger of spirals

a

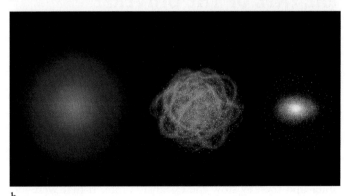

b

FIGURE 17-16 The Creation of Spiral and Elliptical Galaxies A galaxy begins as a huge cloud of primordial gas that collapses gravitationally. (a) If the rate of star birth is low, then much of the gas collapses to form a disk, and a spiral galaxy is created. (b) If the rate of star birth is high, then the gas is converted into stars before a disk can form, resulting in an elliptical galaxy.

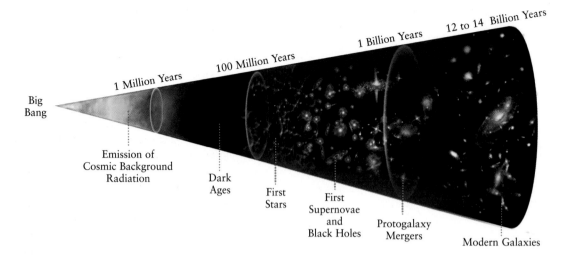

FIGURE 17-17

A Cosmic Timeline This figure shows our current thinking about the evolution of star and galaxy formation in the universe. (Don Dixon)

Big Bang

1 Million Years

100 Million Years

1 Billion Years

12 to 14 Billion Years

Emission of Cosmic Background Radiation

Dark Ages

First Stars

First Supernovae and Black Holes

Protogalaxy Mergers

Modern Galaxies

and a resulting burst of star formation. Our understanding of galactic formation and evolution is far from complete. For instance, why wasn't gas in the early universe eventually transformed into systems of stars a million times bigger or smaller than galaxies?

Even more troublesome is the puzzle of dark matter (see Sections 14-6 and 15-9). The observable stars, gas, and dust in a galaxy account for only about 10% to 20% of its mass. Other than massive neutrinos, we have very little idea of what the remaining 80% to 90% looks like, how it is distributed in space, or what it is made of. With this level of ignorance, any discussion of galactic evolution must be incomplete. Add in the great distances involved, and it is easy to see why observing the early universe is among the greatest challenges facing astronomers.

THE FATE OF THE UNIVERSE

We now turn to the future. Will the universe last forever? Or will it someday stop expanding and collapse?

17-13 The average density of matter is one factor that determines the future of the universe

As different superclusters move apart in the expanding universe, their mutual gravitational attractions act on each other and thereby slow the expansion rate. If that were all that mattered in determining the fate of the universe, one would simply have to add up all the matter in existence and see if its gravitational force is enough to eventually stop the expansion.

This is analogous to what engineers do in calculating, for example, how high a cannonball shot upward from the surface of the Earth will travel. The Earth's gravitational force slows the ball's ascent. If the cannonball's speed upward is less than the escape velocity from the Earth's gravity (about 11 km/s straight up), it will fall back to Earth. If the cannonball's speed equals the escape velocity, it will just barely escape falling back to Earth and come to rest an infinite distance away. And if the ball's speed exceeds the escape velocity, it will leave the Earth and continue outward forever, despite the relentless pull of the Earth's gravity.

It would seem at first glance that if the universe is moving outward too slowly to overcome the mutual gravitational attraction of its parts, it will stop expanding and someday collapse. If it is moving outward at its escape velocity, it will expand until coming to a stop an infinite time in the future. Or, if it is moving outward fast enough, it will continue to expand forever.

However, the laws of physics pertaining to the evolution of the universe, as spelled out in the general theory of relativity and in recent observations of distant objects, reveal that reality is more complex. There are two additional effects we now know must be considered: the effect of the matter on the shape of the universe and the presence of the repulsive gravitational force called **dark energy** (not to be confused with dark matter). We begin by considering the effect that the universe's mass has beyond simple gravitational attraction.

During the 1920s, Alexandre Friedmann in Russia, Georges Lemaître in Belgium, Willem de Sitter in the Netherlands, and, of course, Einstein himself applied the general theory of relativity to the expanding universe. General relativity predicts that the presence of matter curves the fabric of spacetime as we saw in Chapters 13, 14, and 16 in the form of gravitational lensing and the distortion of spacetime around black holes. Similarly, the presence of energy also curves spacetime—recall that matter and energy are related by $E = mc^2$. It is important to note that while the presence of matter and energy curves spacetime (the four-dimensional combination of space plus time), the overall curvature of just the spatial dimensions can be zero. Such a space is called *flat*.

There are three possibilities for the overall shape of the universe, determined by the amount of mass and energy it contains and how fast it is expanding. Imagine shining two powerful laser beams out into space. Suppose that we can align these two beams so that they are perfectly parallel as they leave the Earth. Suppose, further, that nothing gets in the way of these two beams. We follow the beams across the spacetime whose shape we wish to determine. The light beams will begin to diverge due to the expanding universe carrying them apart. This will happen regardless of any properties of the spacetime. It is not what we are interested in, so we compensate for it (that is, we ignore it) in what follows. The three possibilities for the paths and, hence, the shape of the universe, are:

1.   Two beams of light remain parallel even after traversing billions of light-years. In this case, space is not curved. That is, space is flat and it extends without limit. This is the structure that meets our commonsense beliefs.

2.   Two beams of light, starting out parallel, gradually get closer and closer together as they move across the universe, eventually intersecting at some enormous distance from Earth. In this case, we say that space is *positively curved*. Recall that lines of longitude on the Earth's surface are parallel at the equator but intersect at the poles. Thus, if the universe has this effect on light, the shape of the universe is like the surface of a sphere: Space is spherical and the universe has positive curvature. This is what would happen if the universe had such a high density of mass that the mass literally curved space back in on itself. In the absence of any outward-pushing force, such as would be supplied by the cosmological constant (see section 17-1) or other form of dark energy, such a universe does not have enough energy to keep expanding forever. It would someday recollapse.

It is easiest to understand such a universe by visualizing a two-dimensional version of the universe as being the surface of a balloon on whose surface you must remain. In such a universe, you could, in principle, keep going in a straight line (which is a curve on the balloon's surface) and eventually end up back where you started. Like the surface of a balloon, a three-dimensional universe with positive curvature has no outside or center and is said to be a **closed universe.**

3.   Two initially parallel beams of light gradually diverge farther and farther apart as they move across the universe. This is what would happen if the energy of the Big Bang was sufficiently great to assure that the universe is going to expand forever. In this case, the universe has negative curvature. A horse's saddle is a good example of a negatively curved or *hyperbolic* surface. Parallel lines drawn on a saddle always diverge. Thus, in a negatively curved universe,

we would describe space as hyperbolic. A hyperbolic universe extends without limit and is called an **open universe.**

Figure 17-18 illustrates these cases with two-dimensional analogies: a sphere, a plane, and a saddle. Of course, real space is three-dimensional. Both the flat and the hyperbolic universes are open and will therefore grow larger forever.

Most astronomers have a distinct preference for the flat universe. The reason for this expectation that the universe is flat stems from the observations that the universe is homogeneous and isotropic. The only mechanism we have at present to explain these properties of the matter distribution in space is inflation, as discussed above. However, the equations predict that if inflation occurred, the universe must be very nearly flat. This arises because the period of inflation stretched the volume of the universe and the matter in it so much that it became virtually flat.

Telescopic observations strongly support the belief that the universe is flat. This conclusion comes from examining the sizes of the regions of slightly higher and lower temperature in the cosmic microwave background (see the figure that opens this chapter) and comparing them to the sizes predicted for a flat universe. The observations are consistent with the theoretical variations. Furthermore, if the universe

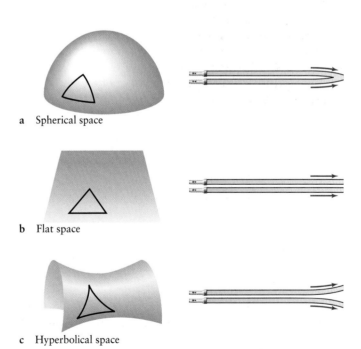

a   Spherical space

b   Flat space

c   Hyperbolical space

FIGURE 17-18 **The Geometry of the Universe** The shape of space (represented here as two-dimensional for ease of visualization) is determined by the matter and energy contained in the universe. The curvature is either **(a)** positive, **(b)** zero, or **(c)** negative, depending on whether the average matter and energy density throughout space is greater than, equal to, or less than a critical value.

had positive curvature, light from the hot regions of the early universe would be curved and thereby focused, creating bigger, brighter images than are observed. Figure 17-19 summarizes the effects of space curvature on observations of the cosmic microwave background.

17-14 Dark energy is causing the universe to accelerate outward

Until the past few years, there was a major inconsistency between the distribution of observed matter and energy in the universe and the flatness of space. Just as there is not enough visible matter to account for galaxies and clusters of galaxies remaining bound systems, there is not enough observed matter or energy to account for a flat universe. Visible matter accounts for only 0.5% of the required mass. The cosmic microwave background photons only add 0.005% of the required gravitational effects needed for flatness. And calculations of the mass necessary to keep galaxies and clusters bound, combined with gravitational lensing by dark matter of distant galaxies and quasars (Chapters 14 and 16), reveal that the dark matter known to exist accounts for only about 30% of the required mass. Therefore, the universe has only 30% of the required mass and energy necessary to make it flat. Allowing for the errors that still exist in all these observations, the possible range of mass and energy from all matter and photons in the universe is still only between 20% and 40% of that required for flatness. Yet flat it certainly appears to be.

By the mid-1990s, combined evidence of the microwave background and large-scale structure had forced astronomers to the conclusion that the remaining energy required to make the universe flat must be a form of dark energy with the remarkable feature that it causes the universe to accelerate outward. Corroborating evidence for the existence of such

dark energy was found in the recent observations of Type Ia supernovae in extremely distant galaxies (Figure 17-20).

The light curves (Section 12-8) for Type Ia supernovae (explosions of white dwarfs in binary star systems) are very well known. Studies show that these supernovae in nearby galaxies behave similarly to those observed in the Milky Way, regardless of their environment. Because they are so bright, such supernovae are excellent standard candles for objects billions of light-years away. They can be used to calculate the distances to remote galaxies. Assuming that supernovae in distant galaxies also behave similarly to those in our Galaxy, observations made in 1998 revealed that the supernovae in distant galaxies appear dimmer than they would if the universe had been continually decelerating. In other words, the universe has begun expanding faster than it had been, moving the distant galaxies farther away than they would have been and making the supernovae in them appear dimmer than in a continually slowing universe. The universe is *accelerating* outward. It will expand forever.

There must be some kind of repulsive force acting to increase the rate at which superclusters separate today. It is hypothesized that this is due to some kind of dark energy that has a repulsive gravitational effect, as introduced above. There are two theoretical explanations for dark energy. Let's consider first what would happen if the cosmological constant that Einstein introduced and then rejected actually did exist. It would add energy that would come from the vacuum of space.

Recall from Chapter 13 that even in a vacuum, particles are constantly appearing and disappearing. They add energy to the cosmos—even empty space has energy. This form of energy has the property of contributing a repulsive gravitational force that competes with the attractive gravitational force from matter and radiation, which slows down the expansion. Which gravitational force wins depends on whether there is more vacuum

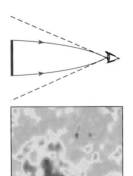

a If universe is closed, hot spots appear larger than actual size

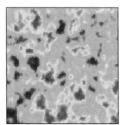

b If universe is flat, hot spots appear actual size

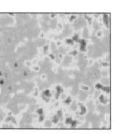

c If universe is open, hot spots appear smaller than actual size

FIGURE 17-19 The Cosmic Microwave Background and the Curvature of Space Temperature variations in the early universe appear as "hot spots" in the cosmic microwave background. The apparent sizes of these spots depend on the curvature of space. **(a)** In a closed universe with positive curvature, light rays from opposite sides of a hot spot bend toward each other. Hence, the hot spot appears larger than it actually is, as shown by the dashed lines. **(b)** The light rays do not bend in a flat universe. **(c)** In an open universe, light rays bend apart. The dashed lines show that a hot spot would appear smaller than its actual size. (The BOOMERANG Group, University of California, Santa Barbara)

a

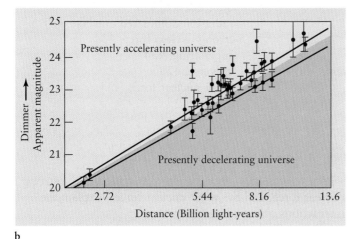

b

FIGURE 17-20   **The Most Distant Supernova (a)** These Hubble
Space Telescope images show the galaxy in which the supernova
SN 1997ff occurred. This supernova, more than 10 Bly away, was dimmer
than expected, indicating that the distance to it is greater than the
distance it would have if the universe had been continually slowing
down since the Big Bang. This supports the notion that an outward
(cosmological) force is acting over vast distances in the universe. The
arrow on the first inset shows the galaxy in which the supernova was
discovered. The bright spot on the second inset shows the supernova by
subtracting the constant light emitted by all the other nearby objects.
(b) The distances and brightnesses of many very distant supernovae are
plotted on this diagram. The location of the most distant supernovae in
the region labeled "presently accelerating universe" strongly indicates
that the universe has been accelerating outward for the past 6 billion
years. (a: Adam Riess, Space Telecope Science Institute, NASA)

energy or more matter and radiation. As the universe
expands, the density of matter and energy decreases, while
the density of vacuum energy remains constant. Eventually,
the repulsive gravitational force would come to dominate.

The unlikely part of the theory that the cosmological
constant exists is that the repulsive force it creates has just
the right strength to have allowed the universe to slow down
for several billion years and then to slowly cause it to accelerate outward, as seen today. To explain this apparent coincidence would require fine tuning the ratio of the vacuum
energy to the matter and radiation to an unacceptably high
degree of precision. In other words, scientists do not yet
have a good reason why the repulsion created by a cosmological constant is not a hundred times greater, in which case
structure throughout the universe such as stars and galaxies
would not have formed yet, or a billion times less, in which
case the repulsive effects of the cosmological constant would
be negligible today.

The alternative explanation for dark energy is
called **quintessence.** Scientists are exploring a variety of mathematical descriptions of quintessence.
It differs from the vacuum energy created by the cosmological constant in that the energy of quintessence is not constant, but changes with the expansion of the universe and
can also be changed by the flow of energy and matter
through the universe. Because it can change, quintessence
need not have been fine-tuned to a strength to make the universe flat and to create just the slight acceleration we are seeing. Instead, it could have started out with some arbitrary
strength and then been adjusted by physical properties of
the universe to match the growth of the universe.

As the universe grew, the energy in quintessence would
eventually come to dominate the gravitational energy from
matter. If this took place a few billion years ago, quintessence would have begun exerting a negative gravitational
force and so the universe would have begun accelerating.
This model does not require the fine tuning that the presence
of a cosmological constant requires.

The unexpected discovery that the universe is accelerating outward completely alters astronomers' perspectives
about the fate of the cosmos. At the same time, however, it
neatly solves the mystery of why the universe is flat. Dark
energy provides the remaining 70% of the energy needed to
account for the universe being flat.

17-15 Frontiers yet to be discovered

Some of the unanswered questions about the cosmos follow
from the discoveries presented in this chapter. What caused
the Big Bang? What happened during the Planck time before
10^{-43} seconds? Is our universe the only one, or are there others, perhaps with profoundly different properties than ours,
that were created at the same time but that are inaccessible
to us? What caused the universe to come into existence in a

false vacuum? What are the details of the formation process of galaxies? Did black holes grow with the evolution of the galaxies, as is currently believed? What are the origins of the dark energy? Which, if either, of the current candidates for dark energy, the cosmological constant or quintessence, is correct? One of the things that makes this time in human existence so fascinating is that it is likely we will have answers to most of these questions within your lifetime.

 Further Reading on These Topics

WHAT DID YOU KNOW?

1 *What is the universe?* It is all the matter, energy, and spacetime that will ever be detectable from the Earth or that will ever affect us.

2 *Did the universe have a beginning?* Yes, it probably occurred between 13 billion and 15 billion years ago in the Big Bang.

3 *Will the universe last forever?* Current observations support the belief that the universe will last forever.

KEY WORDS

Big Bang, 429
closed universe, 443
confinement, 436
cosmic microwave background, 430
cosmic light horizon, 438
cosmological constant, 428
cosmological redshift, 428
cosmology, 428
dark ages, 440
dark energy, 442
decoupling, 438

era of recombination, 438
expanding universe, 428
grand unified theory (GUT), 435
homogeneous, 433
horizon problem, 435
inflation, 435
inflationary epoch, 435
isotropy, 430
isotropy problem, 435
matter-dominated universe, 437
open universe, 443

pair production, 436
Planck era, 434
Planck time, 434
primordial fireball, 438
quark, 434
quintessence, 445
radiation-dominated universe, 437
strong nuclear force, 434
superstring, 434
universe, 428
weak nuclear force, 434

KEY IDEAS

The Big Bang

• Most astronomers believe that the universe began as an exceedingly dense cosmic singularity that expanded explosively in an event called the Big Bang. The Hubble law describes the ongoing expansion of the universe and the rate at which superclusters of galaxies move apart.

• The observable universe extends about 14 billion light-years in every direction from the Earth to what is called the cosmic light horizon. We cannot see any objects that may exist beyond the cosmic light horizon because light from these objects has not had enough time to reach us.

• According to the theory of inflation, matter originally near our location in the universe spread throughout a volume of the universe so large that we cannot yet observe much of it. The observable universe today is thus a growing volume of space containing matter and radiation that was in close contact with our matter and radiation during the first instant after the Big Bang. This explains the isotropic appearance of the universe.

A Brief Physical History of Matter

• Four basic forces—gravity, electromagnetism, the strong nuclear force, and the weak nuclear force—explain the interactions observed in the universe.

• According to current theory, all four forces were identical just after the Big Bang. At the end of the Planck time (about 10^{-43} s after the Big Bang), gravity "froze out" to become a separate force. A short time later, the strong nuclear force became a distinct force. A final "freeze-out" separated the electromagnetic force from the weak nuclear force.

• Before the Planck time, the universe was so dense that known laws of physics do not properly describe the behavior of spacetime, matter, and energy.

• In its first 500,000 years, the universe was radiation-dominated, during which time photons prevented matter from forming clumps. Then it was matter-dominated, during which time superclusters and smaller clumpings of matter formed. Today it is dark-energy-dominated. Dark energy apparently supplies a repulsive gravitational force that causes superclusters to accelerate away from each other.

• During the first 500,000 years of the universe, astronomers believe that matter and energy formed an opaque plasma called the primordial fireball. Cosmic microwave background radiation is the greatly redshifted remnant of the universe as it existed about 500,000 years after the Big Bang.

• By 500,000 years after the Big Bang, spacetime expansion caused the temperature of the universe to fall below 3000 K, enabling protons and electrons to combine to form hydrogen atoms. This event is called the era of recombination. The universe became transparent during the era of recombination, meaning that the microwave background radiation contains the oldest photons in the universe.

• Clusters of galaxies and individual galaxies formed from pieces of enormous hydrogen and helium clouds, each of which became a separate supercluster of galaxies.

• All the superclusters and most of the clusters of galaxies are moving away from each other.

• During the matter-dominated era, structure formed in the universe. As the universe goes farther into the dark-energy-dominated era, the large-scale structure of superclusters of galaxies will fade away.

The Fate of the Universe

• The average density of matter and dark energy in the universe determines the curvature of space and the ultimate fate of the universe.

• Observations are that the universe is flat and that the cosmic microwave background is almost perfectly isotropic, probably resulting from a brief period of very rapid expansion (the inflationary epoch) in the very early universe.

• It appears that the universe is accelerating outward and that it will expand forever.

REVIEW QUESTIONS

1 What does it mean when astronomers say that we live in an expanding universe?

2 Explain the difference between a Doppler redshift and a cosmological redshift.

3 In what ways are the fate of the universe, the shape of the universe, and the average density of the universe related?

4 Assuming that the universe will expand forever, what will eventually become of the microwave background radiation?

5 What does it mean to say that the universe is dark energy dominated? When was the universe radiation-dominated? When was it matter-dominated? How did radiation domination show itself?

6 Explain the difference between an electron and an antielectron.

7 Where do astronomers believe most of the photons in the cosmic microwave background come from?

8 Describe an example of each of the four basic physical forces in the universe.

9 What is the observational evidence for (a) the Big Bang, (b) the inflationary epoch, and (c) the confinement of quarks?

ADVANCED QUESTIONS

10 Explain why the detection of cosmic microwave background radiation was a major blow to the steady state theory.

11 The separation of nuclei by energetic photons in the early universe was caused by the same mechanism as what other transformation covered earlier in this book?

DISCUSSION QUESTIONS

12 Discuss the theological implications of the idea that we cannot use science to tell us what existed before the Big Bang.

13 Explain why gravitational attraction dominated the behavior of the early universe and why the dark energy will determine the fate of the universe.

WHAT IF ...

14 The universe were destined to collapse, and that collapse were under way? What would be different in space and on Earth under those conditions?

15 Our solar system formed very early in the life of our Galaxy, when the universe was just two billion years old? What would be different in space and on Earth?

16 Our solar system formed much later in the life of the universe? How would observations of stars and galaxies be different than they are now?

17 Our solar system formed with the first generation of stars? What would be different about the solar system? Would the Earth exist as an inhabitable world? Why or why not?

WEB/CD-ROM QUESTIONS

18 Search the Web for information about the Microwave Anisotropy Probe mission. What is the purpose of this mission? Has the spacecraft been launched? What information is this mission expected to provide that the BOOMERANG and MAXIMA experiments could not?

19 Temperatures in the Early Universe. Access the Active Integrated Media Module "Black Body Curves" in the *Discovering the Universe* Web site or CD-ROM. (a) Use the module to determine by trial and error the temperature at which a blackbody spectrum has its peak at a wavelength of 1 μm. (b) At the time when the temperature of the cosmic background radiation was equal to the value you found in (a), was the universe matter-dominated or radiation-dominated? Explain your answer. *Hint:* Figure 17-9 might help.

18 THE SEARTH FOR EXTRATERRESTRIAL LIFE

IN THIS CHAPTER YOU WILL DISCOVER

- how scientists search for life that may exist elsewhere in the universe

- how scientists estimate how many planets orbiting other stars could support complex life

- the requirements scientists believe are necessary for life to exist

- the instruments used to search for life beyond our own solar system—and the results of those searches

- how we are trying to communicate with extraterrestrial life

The Prevalence of Life
This painting, entitled *DNA Embraces the Planets*, artistically expresses the suspicion of many scientists that carbon-based life may be a common phenomenon in the universe. Other scientists argue, however, that we may be unique and no intelligent alien civilizations exist. In either case, humanity has the clear mandate to preserve and protect the abundance of life-forms with which we share our planet. (J. Lomberg)

WHAT DO YOU THINK?

1   How do astronomers search for extraterrestrial intelligence?

2   Have astronomers located any extraterrestrial civilizations?

The heavens inspire us to contemplation. We want to look beyond ourselves, to the formation of the Earth, the nature of the stars, and even the creation of the universe. Perhaps no subject is as compelling, however, as life elsewhere in the universe. Are we alone? What are the chances that we might someday make contact with an alien civilization?

18-1 The existence of life depends on chemical and physical properties of matter

There are no firm answers yet to these questions. To build a physical foundation for understanding the issues about life existing off Earth, let's begin by considering some of the fundamental requirements for the existence of life. First is the presence of liquid water. This medium enables many atoms and molecules to bond and separate in large numbers. Such activity was essential for the creation and evolution of the complex molecules and systems of molecules that exist in living organisms. Biologists believe that life on Earth began developing in primordial bodies of water.

Second is the existence of atoms that can bond suitably with a variety of different elements to create life. There is only one element that can serve as the foundation for life as we know it: carbon. Carbon (C) is one of only five elements, along with boron (B), nitrogen (N), silicon (Si), and phosphorus (P), capable of bonding with three or more other atoms. Three is the minimum number of bonds necessary to allow for the complexity and diversity of biological molecules. Carbon is unique among elements that can bond to at least three other atoms, because its bonds are flexible yet strong. Silicon-silicon bonds come apart under the slightest disturbance, in comparison, while silicon-oxygen bonds are so strong they create rocks. The other three elements that might serve as the backbone of life—boron, nitrogen, and phosphorus—have similar bonding problems. Despite the enormous range of conditions under which terrestrial life exists, all of it is based on the unique properties of carbon.

Carbon atoms form chemical bonds that can combine to create especially long, complex molecules. These molecules can be further linked in elaborate chains, lattices, and fibers. It is for these reasons that carbon-based compounds, called **organic molecules,** are the stuff of life. Indeed, biologists have already identified more than six million carbon compounds.

The variety and stability of living organisms depend on the complex, self-regulating, chemical reactions of organic molecules. Carbon is also among the most abundant elements in the universe. The unique versatility and abundance of carbon imply that extraterrestrial biology would also be based on organic chemistry.

Third is the presence of the other elements used to create life. Other constituents of organic molecules, such as hydrogen, oxygen, nitrogen and sulfur, are also plentiful in our Galaxy. In interstellar clouds, carbon atoms have combined with other elements to produce an impressive variety of organic compounds. Since the 1960s, radio astronomers have detected telltale microwave emission lines from interstellar clouds that help to identify dozens of these carbon-based compounds. Examples include polycyclic aromatic hydrocarbons (PAHs—molecules composed just of carbon and hydrogen atoms), as well as ethyl alcohol (CH_3CH_2OH), formaldehyde (H_2CO), methyl cyanoacetylene (CH_3C_3N), and acetaldehyde (CH_3CHO). Since stars and planets condense from interstellar gas and dust enriched in these elements, they are available elsewhere.

The chemicals needed to form life may not just condense on planets as they form. Some astronomers have proposed that after planets come into existence, organic material from space may rain down on these worlds, thereby expediting the evolution of life. This belief is supported by the fact that spectra taken of comets in our solar system reveal that they contain an assortment of organic compounds. Such data was collected by, among others, the spacecraft *Giotto* as it passed close to the nucleus of Halley's comet. In 1997, NASA launched the *Stardust* mission to rendezvous with Comet Wild 2 in January 2004, collect dust and other debris from the comet, and return to Earth in 2006. Scientists hope to study this material for organic compounds uncontaminated by exposure to the Earth's atmosphere and surface.

Further evidence of extraterrestrial organic molecules comes from meteorites called carbonaceous chondrites (Figure 18-1). Many of the carbonaceous chondrites that have fallen to Earth contain a variety of organic substances. As noted in Chapter 8, carbonaceous chondrites are ancient meteorites dating from the formation of the solar system. If they contained organic molecules while in space, it seems reasonable to conclude that even from their earliest days, the planets have been continually bombarded with organic compounds.

Interstellar space is not the only source of organic material. In a classic experiment performed in 1952, the American chemists Stanley Miller and Harold Urey demonstrated that simple chemicals can combine to form

a

FIGURE 18-1   **A Carbonaceous Chondrite**   **(a)** Carbonaceous chondrites are ancient meteorites that date back to the formation of the solar system. This sample is a piece of the Allende meteorite, a large carbonaceous chondrite that fell in Mexico in 1969.

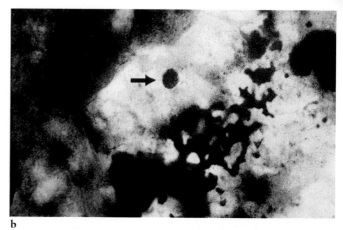

b

(b) Chemical analysis of newly fallen carbonaceous chondrites discloses that they are rich in organic compounds (arrow). (a: From the collection of Ronald A. Oriti; b: Harvard Smithsonian Center for Astrophysics)

organic compounds. Hoping to capture conditions on the primitive Earth and in its atmosphere, Miller and Urey combined hydrogen, ammonia, methane, and water vapor in a closed container. They then subjected their mixture to an electric arc to simulate lightning bolts. At the end of a week, the inside of the container had become coated with a reddish-brown substance rich in compounds essential to life.

Since that time, geochemists have come to realize that Earth's early atmosphere was more likely to have been a mixture of carbon dioxide, nitrogen, and hydrogen, along with water vapor "outgassed" from volcanoes and deposited by comet impacts. Modern versions of the Miller-Urey experiment have therefore used these common gases (Figure 18-2), and, once again, organic compounds are produced. Could these compounds be forming elsewhere in the universe today?

Keep in mind that scientists have not yet created life in a test tube. Biologists have yet to figure out, among other things, how simple organic molecules gathered into cells and developed systems for self-replication.

The fourth issue surrounding the existence of life elsewhere relates to physical properties of planets and life. If a planet is too close to a star, the temperature and ultraviolet radiation levels will be too high for life to evolve. If the planet is too far away, it will be too cold to have liquid water on its surface. If a planet's environment is hostile, life may either never get started or quickly become extinct. If the star is too massive, it will explode before life on the planets orbiting it has a chance to evolve. If the star is too small, the planet would have to be very close to it to be warm enough for life, but then tidal forces from the star would lock the planet in synchronous rotation, making most of its surface uninhabitable.

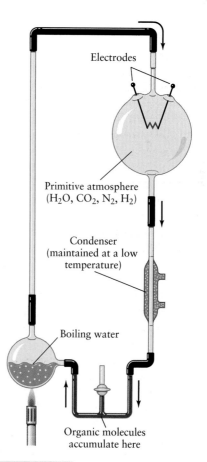

Electrodes

Primitive atmosphere
(H_2O, CO_2, N_2, H_2)

Condenser
(maintained at a low
temperature)

Boiling water

Organic molecules
accumulate here

FIGURE 18-2   **The Miller–Urey Experiment Updated** Modern versions of this classic experiment prove that numerous organic compounds important to life can be synthesized from gases that were present in Earth's primordial atmosphere. This experiment supports the hypothesis that life on Earth arose as a result of ordinary chemical reactions.

WEB LINK 18.2

18-2 Evidence is mounting that life exists elsewhere in our solar system

Scientists have strong evidence that everything does not have to be optimal in order for life to form. Geologists and biologists have discovered life on Earth in some incredibly challenging environments, such as on the ocean bottom, in a lake far under the Antarctic ice pack, deep inside the planet's crust, and even in hot geothermal vents. It therefore seems reasonable to believe that life could have originated off Earth under similarly challenging conditions. There are at least two places here in our solar system that scientists are considering as possible habitats for life past or present. They are Mars and Jupiter's moon Europa.

Astronomers have discovered (Section 6-12) that Mars once had surface water and that it may still have underground bodies of water. We saw in Section 7-5 that Europa is also likely to have liquid water under its frozen surface. These two bodies, then, are candidates for having evolved simple life-forms. Indeed, we saw in Chapter 6 what may be fossilized bacterial life from Mars (see Figure 6-27).

Confirmation that life evolved on either of these worlds will demonstrate that the formation of complex molecules and structures are possible off Earth. However, these two locations are unsuitable for the evolution of sufficiently complex life, like ourselves, that is self-aware and able to create advanced civilizations. Are there other advanced civilizations in the cosmos? Despite all the alleged UFO sightings and alleged personal encounters, no one has produced evidence convincing to scientists that an intelligent extraterrestrial life-form has *ever* visited Earth. Yet the existence of life elsewhere in the universe is so important to the human psyche that high-tech searches for extraterrestrial life are now under way.

18-3 Searches under way for advanced civilizations try to detect their radio signals

We have not yet discovered Earthlike planets orbiting Sun-like stars. Furthermore, we saw in Essentials II that many stars have Jupiter-mass planets orbiting very close to them. The presence of massive planets near stars would make it very hard for those stars to support habitable planets. The gravitational tugs from the massive planets would put Earthlike planets in highly elliptical orbits. However, not all Sunlike stars have close-orbiting, massive planets.

There are billions of Sunlike stars in our Galaxy to be studied, and most astronomers and many other people expect that we will eventually discover inhabitable planets around some of them. This belief helps sustain the search for extraterrestrial intelligence, or **SETI.** Unlike the fossil evidence for life on Mars, the effort to locate life outside the solar system is based on searching for high-tech evidence of advanced life.

How might we ascertain whether extraterrestrial civilizations exist, given the tremendous distances that separate us from other stars and the huge number of stars with the potential to support life-bearing planets? Looking for individual, inhabitable, extrasolar planets is technically demanding and time consuming. As we saw in Section II-6, it requires careful observation of data available only in the past 20 years. Furthermore, the discovery of a planet does not imply that it harbors life. SETI astronomers therefore need a way to bypass the daunting search for habitable planets, and starting in the 1950s and 1960s, they identified a promising approach to searching directly for extraterrestrial intelligence. They look for radio transmissions from distant civilizations.

As we have seen, radio waves can travel immense distances without being significantly altered by the interstellar medium. Because they penetrate gas and dust, radio waves are a logical choice for interstellar communication. Even if alien civilizations are not trying to communicate with us, they probably use radio waves in their own technology. But at what frequencies should we look to maximize the odds of detecting alien signals?

We could try random frequencies in the hopes of hearing something—anything. But the number of directions to look and frequencies to check are both staggering. We need to find some way to improve the odds. It turns out that some radio wavelengths travel farther without being absorbed by interstellar gas and dust, or without competing with noise from other sources, than do other wavelengths. The SETI pioneer Bernard Oliver pointed out that a range of relatively clear frequencies exists in the neighborhood of the microwave emission lines of hydrogen and the hydroxyl radical (OH) (Figure 18-3). This region of the microwave part of the radio spectrum is called the **water hole,** a humorous reference to the H and OH lines being so close together. If extraterrestrial beings *are* purposefully sending messages into space, it seems reasonable that they would choose to transmit on these frequencies.

Alternatively, aliens might reason that other civilizations are already listening at certain scientifically important frequencies. For example, astronomers who study the distribution of hydrogen around the Galaxy have their radio telescopes tuned to 21-cm radiation (Chapter 14). Alien civilizations might therefore try to send signals toward us at this wavelength.

Even if only a few alien civilizations are scattered across the Galaxy, we have the technology to detect radio transmissions from them. One of the first searches occurred in 1960, when astronomer Frank Drake carried out Project Ozma (named after the land in L. Frank Baum's Oz series of

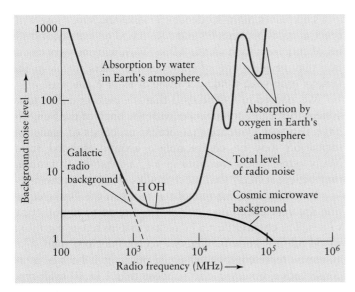

FIGURE 18-3   **The Water Hole** The so-called water hole is a range of radio frequencies from about 1000 to 10,000 megahertz (MHz) that happens to have relatively little cosmic noise. Some scientists suggest that this noise-free region would be well suited for interstellar communication.

FIGURE 18-4   **A Deep-Space Tracking Antenna** This dish-shaped antenna, located in the Mojave Desert in California, was originally built by NASA to track interplanetary spacecraft. It was recently used in an all-sky survey, along with an antenna at the Arecibo Observatory in Puerto Rico, to search for extraterrestrial intelligence. In 1996, an antenna in Canberra, Australia, joined the network. (JPL)

books). In it he used a radio telescope at the National Radio Astronomy Observatory in West Virginia to "listen" to two Sunlike stars, τ (tau) Ceti and ε (epsilon) Eridani, but without success.

SETI searches use radio telescopes along with sophisticated electronic equipment and powerful computers to conduct both a targeted search and an all-sky survey. While NASA has created astrobiology centers to study issues regarding life in the solar system and Galaxy, most of the SETI programs in operation today are privately funded. Several are actively listening and analyzing enormous volumes of data. Indeed, there is so much data to analyze that one SETI project, called SETI@home, has enlisted the help of home computer users. These SETI researchers provide actual data from radio telescopes (Figure 18-4) to be analyzed, along with a data-analysis program. While the computer's screensaver is on, the program runs, the data are analyzed, and the results are sent back to the researchers to be combined with data from other home-computer users. The processed data is replaced with fresh data to be analyzed. Another SETI program uses an array of small, inexpensive TV satellite antennas to look for signals from space.

Figure 18-5 shows the region of the Galaxy from which radio signals similar to the kind of transmissions we make could be detected. Occasionally, a SETI project does detect powerful or unusual signals from space. However, because none of these signals is ever repeated, SETI researchers do not believe they have yet discovered any extraterrestrial civ-

ilization. Despite more than 40 unsuccessful searches to date using radio telescopes around the world, the search for extraterrestrial intelligence continues.

The detection of a message from an alien civilization would be one of the greatest events in human history. Such a message could dramatically change the course of civilization through the sharing of scientific information or an awakening of social or humanistic enlightenment. In only a few years, our industry and social structure could advance

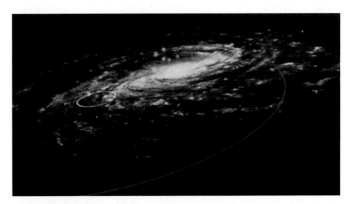

FIGURE 18-5   **The Region of the Search** The yellow circle, centered on the solar system, shows the region of the Milky Way in which SETI searches can reasonably expect to detect radio emissions from alien civilizations. This assumes that the signals are similar in nature to the kinds of radio emissions we generate here on Earth. (STS, NASA)

centuries into the future. The effects of such change would touch every person on Earth.

18-4 The Drake equation predicts how many civilizations are in the Milky Way

Just how many planets are likely to harbor complex life throughout our Galaxy? The first person to tackle this question quantitatively was Frank Drake. Drake proposed that the number of technologically advanced civilizations in the Galaxy (designated by the letter N) could be estimated by what is now called the **Drake equation:**

$$N = R^* f_p n_e f_l f_i f_c L$$

in which

R^* = rate at which solar-type stars form in the Galaxy

f_p = fraction of stars that have planets

n_e = number of planets per star system suitable for life

f_l = fraction of those habitable planets on which life actually arises

f_i = fraction of those life-forms that evolve into intelligent species

f_c = fraction of those species that develop adequate technology and then choose to send messages out into space

L = lifetime of that technologically advanced civilization

The Drake equation expresses the number of extraterrestrial civilizations quantitatively as a product of terms, some of which can be estimated from what we know about stars and stellar evolution. For instance, thanks to new evidence for extrasolar planets, astronomers can now hope to determine the first two terms, R^* and f_p, by observation. We should probably exclude stars larger than about 1.5 $M_\odot$, because they have main-sequence lifetimes shorter than the time it took for intelligent life to develop here on Earth—some 3.8 to 4.0 billion years. If that is typical of the time needed to evolve higher life-forms, then a massive star becomes a giant or even explodes as a supernova before self-aware creatures can evolve on any of its planets.

Although low-mass stars have much longer lifetimes, they, too, seem unsuited for life because they are so cool. Only planets very near a low-mass star would be sufficiently warm for water to be a liquid. But, as mentioned earlier, a planet that close would become tidally coupled to the star, developing synchronous rotation. One side would have continuous daylight, while the other would be in perpetual, frigid darkness.

This leaves main-sequence stars like the Sun, those with spectral types between F5 and M0. Based on statistical studies of star formation in the Milky Way, astronomers calculate that roughly one of these Sunlike stars forms in the Galaxy each year, yielding a value of $R^* \approx 1$ per year.

We learned in Essentials II that the planets in our solar system formed in conjunction with the birth of the Sun. We have also seen evidence that similar processes of planetary formation may be commonplace around isolated stars. Many astronomers therefore give f_p a value of 1, meaning they believe it likely that most Sunlike stars have planets.

Unfortunately, the rest of the terms in the Drake equation are very uncertain. Let's work with some hypothetical values. The chances that a planetary system has an Earthlike world are not known. Were we to consider our own solar system as representative, we could put n_e at 1. Let's be more conservative, however, and suppose that 1 in 10 solar-type stars is orbited by a habitable planet, making $n_e = 0.1$.

From what we know about the evolution of life on Earth, we might assume that, given appropriate conditions, the development of life is a certainty, which would make $f_l = 1$. This is, of course, an area of intense interest to biologists. For the sake of argument, we might also assume that evolution naturally leads to the development of intelligence (a conjecture that is hotly debated) and also make $f_i = 1$. It is anyone's guess as to whether these intelligent extraterrestrial beings would attempt communication with other civilizations in the Galaxy, but if we assume they all would, f_c would also be put at 1.

The last variable, L, the longevity of technological civilization, is the most uncertain of all—it cannot be tested! Looking at our own example, we see a planet whose atmosphere and oceans are increasingly polluted, potentially destroying the food chain. When we add in how close we have come to destroying ourselves with weapons of mass destruction, it may be that we humans are among the lucky few technological civilizations to squeak through its first years. In other words, L may be as short as 100 years. Putting all these numbers together, we arrive at

$$N = 1 \times 1 \times 0.1 \times 1 \times 1 \times 1 \times 100 = 10$$

Therefore, out of the hundreds of billions of stars in the Galaxy, there may be only ten civilizations technologically advanced enough to communicate with us.

 Of course, this is just an estimate. A wide range of values has been proposed for the terms in the Drake equation, and these various numbers produce vastly different estimates of N. Some scientists argue that there is exactly one advanced civilization in the Galaxy and that we are it. Others speculate that there may be tens of millions of planets inhabited by intelligent creatures. Although we don't know yet, science enables us to hone in on the number.

18-5 Humans have been sending signals into space for more than a century

We have been doing more than just listening passively for voices from the cosmos. Astronomers have intentionally broadcast well-focused signals through radio telescopes toward likely star systems. For more than a century, we humans have been transmitting other messages about ourselves, too. It all began with the first radio broadcast. Space within 100 ly of the solar system is now filled with signals from radio and television shows. If other advanced civilizations fall within that sphere, they could very well be listening to radio programs or watching television shows from decades past, trying to make sense of our species.

Launched in 1977, the two *Voyager* spacecraft are now making their way into interstellar space, carrying a more pedestrian message. Each has a plaque describing who and where we are, as well as phonograph records (have you ever seen one?) with human voices (Figure 18-6). Because space is so vast and the spacecraft are so small, the likelihood of their being discovered is truly remote. And yet millions, perhaps billions, of years from now, one of the spacecraft just might reach another race of intelligent creatures. From its path and the information on board, those creatures might be able to determine where the *Voyager* spacecraft came from. If their travel budgets are sufficiently large, they might even decide to come and visit us. At the least, they could send us radio messages. Will our descendants be here to receive them?

18-6 Frontiers yet to be discovered

We still have much to learn about the formation and evolution of life both here and elsewhere in the cosmos. We need to confirm whether life existed on Mars and Europa, and whether it still exists on either of those worlds. Astronomers are still working to refine their ideas of how organic compounds formed in space, and biologists still need to understand crucial steps taken by life as it began and evolved here on Earth. A paramount frontier yet to be discovered is detecting life elsewhere in the universe.

Further Reading on These Topics

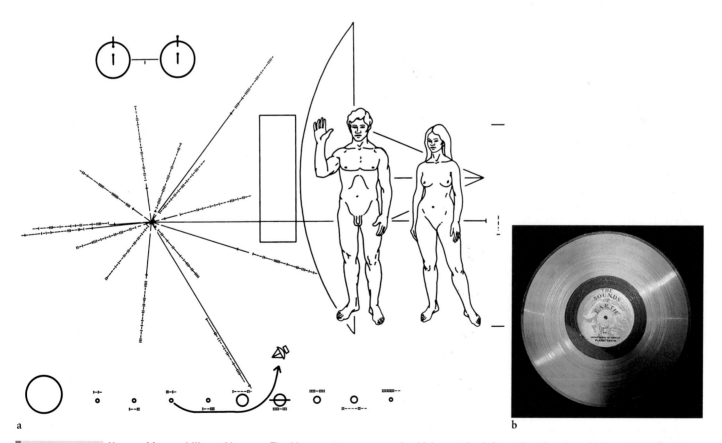

a   b

▌FIGURE 18-6   **Human Memorabilia on Voyager**   The *Voyager 1* and *Voyager 2* spacecraft, now in interstellar space, each carries a plaque **(a)** with images of a man and a woman, as well as a phonograph record **(b).** There are also instructions for playing the record, which contains information about our biology, our technology, and our knowledge base. Each record also contains the sounds of children's voices. It is remotely possible that another race might someday discover the spacecraft. (a & b: NASA)

WHAT DID YOU KNOW?

1 *How do astronomers search for extraterrestrial intelligence?* They search for radio signals from other advanced civilizations.

2 *Have astronomers located any extraterrestrial civilizations?* No extraterrestrial civilizations have yet been discovered.

KEY WORDS

Drake equation, 454
organic molecule, 450

SETI, 452
water hole, 452

KEY IDEAS

• The chemical building blocks of life exist throughout the Milky Way Galaxy.

• Organic molecules have been discovered in interstellar clouds, in some meteorites, and in comets.

• The Drake equation is used to estimate the number of technologically advanced civilizations in the Galaxy whose radio transmissions we might discover.

• Astronomers are using radio telescopes to search for signals from other self-aware life in the Galaxy. This effort

is called the search for extraterrestrial intelligence, or SETI. SETI is primarily done at frequencies where radio waves pass most easily through the interstellar medium. So far, these searches have not detected any life off Earth.

• Everyday radio and television transmissions from Earth, along with intentional broadcasts into space, may be detected by other life-forms.

DISCUSSION QUESTIONS

1 What information would you have put on the record sent out on the *Voyager* spacecraft?

2 What arguments could you make against sending messages via radio or spacecraft into interstellar space?

3 Try using the Drake equation with values that you find reasonable. How many civilizations do you estimate there are in our Galaxy?

4 Discuss the possible biological effects of our probes visiting other life-sustaining worlds.

5 What social effects might probes from other worlds have on us?

6 List some of the pros and cons in the argument that alien spacecraft have visited the Earth.

7 Why would most of a planet locked in synchronous rotation with its star be uninhabitable?

WHAT IF ...

8 Our planet were the moon of a much larger, Jupiterlike planet orbiting the Sun at 1 AU? What might the Earth and life inhabiting it be like?

9 Our Moon were Earth-sized and also had life on it?

10 We discover living organisms in Europa's subsurface oceans? What precautions should we take in the search for such life, and why?

11 We develop life in a Miller-Urey–like experiment? Explore some of the implications this event would have.

WEB/CD-ROM QUESTIONS

 12 To test your understanding of the Miller-Urey experiment, do Interactive Exercise 18-1 on the Web or CD-ROM. You can print out your answer if required.

 13 To test your understanding of the Drake equation, do Interactive Exercise 18-2 on the Web or CD-ROM. You can print out your answer if required.

WHAT IF . . .

LIFE HAD BEGUN ON AN OLDER EARTH?

Our view of the universe is hardly more than a snapshot in time. The universe evolves, and Earth's surface, the eclipses, and the stars in the sky appear the way they do only because we experience them at this unique time. What would the universe look like if evolutionary processes on Earth had taken twice as long to produce creatures who were aware of their own existence? What if humans had evolved 9 billion years after the solar system formed, rather than after only 4.5 billion years?

A Different Earth As you have learned, our planet's crust is composed of tectonic plates floating on Earth's mantle. The continents are still drifting, and 4.5 billion years from now their distribution will be completely different. No pattern of land that we recognize today will remain intact. The Mediterranean Sea will be gone, and the Atlantic Ocean may be as wide as the Pacific Ocean is today, while the Pacific would probably be much narrower. The islands of the Pacific, from Japan south to Australia, may be a single connected unit. New islands will grow over hot spots in the mantle, just as the Hawaiian Islands do today. The Andes and the Himalayas will have tried to rise higher, only to come tumbling down due to weathering.

Another major change is the length of a day on an older Earth. It will be almost twice as long as our day, and the Moon will be about 1.5 times farther away, appearing correspondingly smaller in the sky and taking 1.8 times as long to orbit the planet. Both the longer day and more distant Moon will result from the gravitational interplay between ocean tides and the Moon. As tidal waters rub against the solid Earth, this friction slows the Earth's rotation and currently causes the day to grow 0.002 seconds longer each century. The energy lost as Earth spins down must be retained in the Earth-Moon system, so the Moon must gain energy in its orbit around Earth. This increase in energy currently forces the Moon to move farther away from Earth at a rate of 3.8 centimeters per year.

Because the Moon will appear smaller in the sky, "humans" on the older Earth will never see a total eclipse of the Sun. The best they will see is an annular eclipse, with a broad ring of the Sun showing around the edges of the Moon. Total lunar eclipses will still occur because the older Earth's shadow will remain wider than the Moon.

Changes in the Night Sky Currently, Polaris appears almost exactly over Earth's north rotation pole. But citizens of the older Earth will not see a pole star for two reasons. First, Earth's rotation axis precesses, or wobbles, by 23.5° over a period of 25,800 years, making the pole wander around the sky. It is just by chance that in recent centuries there has been a bright star to mark the north pole. It is unlikely that such a bright star would mark the pole when humans evolve on the older Earth. Second, many bright stars that could be near the pole in the future may no longer be visible, having used up their nuclear fuels in the next 4.5 billion years and exploded.

Perhaps new bright stars will form that will take turns as future polar guide stars as Earth's rotation axis precesses, much as Polaris does now and Thuban in Draco did in the past. Asterisms (patterns of stars), such as the Big Dipper or Orion, will all be different, too, because many stars that exist in our sky now will have exploded within the next 4.5 billion years, while the others will all move relative to each other.

Older Earth's astronomers will, like Edwin Hubble, discover that distant galaxies appear to move away from us because of the expansion of the universe. However, the Hubble constant is not at all constant. When those astronomers measure the speed of distant galaxies moving away from us, they will get a very different number for the Hubble constant.

A Brighter Sun The Sun in the time of the older Earth will still have nearly the same surface temperature and yellowish hue that it has now, but it will appear about twice as bright because it will be about 60% bigger. The extra 4.5 billion years will have begun to take their toll on the Sun's nuclear fuel supply.

The Sun has fused hydrogen into helium throughout its present lifetime of 4.5 billion years, using up less than half of the available hydrogen in its core. In another 4.5 billion years, a total of 80% to 90% of the available hydrogen in the core will have been converted into helium. While the Sun's giant phase would still be a billion years off, the humans of the older Earth would face changes to the planet and the solar system as the Sun's output increased. These changes, such as a runaway greenhouse effect in which much of Earth's surface water becomes vaporized, could potentially force those humans to devise a way to quickly leave the planet and colonize elsewhere in the Galaxy.

APPENDIX

Tables of Data

Table A-1
The Planets: Orbital Data

Planet	Semimajor axis (AU)	Semimajor axis (10⁶ km)	Sidereal period (yr)	Sidereal period (d)	Synodic period (d)	Mean orbital speed (km/s)	Orbital eccentricity	Inclination of orbit to ecliptic (°)
Mercury	0.3871	57.9	0.2408	87.97	115.88	47.9	0.206	7.00
Venus	0.7233	108.2	0.6152	224.70	583.92	35.0	0.007	3.39
Earth	1.0000	149.6	1.0000	365.26	——	29.8	0.017	0.00
Mars	1.5237	227.9	1.8809	686.98	779.94	24.1	0.093	1.85
(Ceres)	2.7656	413.7	4.603	1,681.3	466.6	17.9	0.097	10.61
Jupiter	5.2028	778.3	11.862	4,332.7	398.9	13.1	0.048	1.31
Saturn	9.5388	1427.0	29.458	10,760	378.1	9.6	0.056	2.49
Uranus	19.1914	2871.0	84.01	30,685	369.7	6.8	0.046	0.77
Neptune	30.0611	4497.1	164.79	60,082	367.5	5.4	0.010	1.77
Pluto	39.5294	5913.5	248.5	90,767	366.7	4.7	0.248	17.15

Table A-2
The Planets: Physical Data

Planet	Equatorial diameter (km)	Equatorial diameter (Earth = 1)	Mass (kg)	Mass (Earth = 1)	Average density (kg/m³)	Rotation period* (days)	Inclination of equator to orbit (°)	Surface gravity (Earth = 1)	Albedo	Escape speed (km/s)
Mercury	4,879	0.383	3.302×10^{23}	0.055	5430	58.646	2 (?)	0.39	0.106	4.3
Venus	12,104	0.949	4.869×10^{24}	0.815	5240	243.01^R	177.3	0.91	0.65	10.4
Earth	12,756	1.000	5.974×10^{24}	1.000	5515	1.000	23.45	1.000	0.37	11.2
Mars	6,794	0.533	6.419×10^{23}	0.107	3940	1.026	25.19	0.38	0.15	5.0
Jupiter	142,984	11.209	1.899×10^{27}	317.83	1330	0.414	3.12	2.5	0.52	59.5
Saturn	120,536	9.449	5.685×10^{26}	95.16	700	0.444	26.73	1.1	0.47	35.5
Uranus	51,118	4.007	8.662×10^{25}	14.50	1300	0.718^R	97.86	0.90	0.50	21.3
Neptune	49,528	3.883	1.028×10^{26}	17.204	1640	0.671	29.56	1.1	0.5	23.5
Pluto	2,300	0.18	1.3×10^{22}	0.002	2030	6.387^R	118	0.07	0.5	1.3

* For Jupiter, Saturn, Uranus, and Neptune, the internal rotation period is given. A superscript R means that the rotation is retrograde (opposite the planet's orbital motion).

Table A-3
Satellites of the Planets

Planet	Satellite	Discoverers	Average distance from center of planet (km)	Orbital (sidereal) period* (days)	Orbital eccentricity	Diameter of satellite (km)	Mass (kg)
EARTH	Moon	—	384,400	27.322	0.0549	3476	7.348×10^{22}
MARS	Phobos	Hall (1877)	9,380	0.319	0.01	$28 \times 23 \times 20$	1.1×10^{16}
	Deimos	Hall (1877)	23,460	1.263	0.00	$16 \times 12 \times 10$	1.8×10^{15}
JUPITER	Metis	Synott (1979)	127,960	0.2948	0.00	40	1×10^{17}
	Adrastea	Jewitt et al. (1979)	128,980	0.2983	0 (?)	$24 \times 16 \times 20$	1.9×10^{16}
	Amalthea	Barnard (1892)	181,300	0.4981	0.00	$270 \times 200 \times 155$	7.2×10^{18}
	Thebe	Synott (1979)	221,900	0.6745	0.01	100	8×10^{17}
	Io	Galileo (1610)	421,600	1.769	0.00	3630	8.94×10^{22}
	Europa	Galileo (1610)	670,900	3.551	0.01	3138	4.92×10^{22}
	Ganymede	Galileo (1610)	1,070,000	7.155	0.00	5262	1.48×10^{23}
	Callisto	Galileo (1610)	1,883,000	16.689	0.01	4800	1.08×10^{23}
	S/2000 J1	Sheppard et al. (2002)	7,435,000	131	0.20	5	(?)
	Leda	Kowal (1974)	11,094,000	238.72	0.15	16	6×10^{15}
	Himalia	Perrine (1904)	11,480,000	250.57	0.16	180	1×10^{19}
	Lysithea	Nicholson (1938)	11,720,000	259.22	0.11	40	8×10^{16}
	Elara	Perrine (1905)	11,737,000	259.65	0.21	80	8×10^{17}
	S/2000 J11	Sheppard et al. (2002)	12,654,000	290	0.22	4	(?)
	S/2000 J10	Sheppard et al. (2002)	20,375,000	591^R	0.16	4	(?)
	S/2000 J3	Sheppard et al. (2002)	20,733,000	605^R	0.27	5	(?)
	S/2000 J5	Sheppard et al. (2002)	21,019,000	618^R	0.20	4	(?)
	S/2000 J7	Sheppard et al. (2002)	21,162,000	626^R	0.15	7	(?)
	Ananke	Nicholson (1951)	21,200,000	631^R	0.17	30	4×10^{16}
	S/2000 J9	Sheppard et al. (2002)	21,734,000	652^R	0.25	5	(?)
	S/2000 J4	Sheppard et al. (2002)	21,948,000	661^R	0.35	3	(?)
	Carme	Nicholson (1938)	22,600,000	692^R	0.21	44	1×10^{17}
	S/2000 J6	Sheppard et al. (2002)	22,806,000	704^R	0.28	4	(?)
	Pasiphae	Melotte (1908)	23,500,000	735^R	0.38	70	2×10^{17}
	S/2000 J8	Sheppard et al. (2002)	23,521,000	733^R	0.53	5	(?)
	Sinope	Nicholson (1914)	23,700,000	758^R	0.28	40	8×10^{16}
	S/2000 J2	Sheppard et al. (2002)	24,164,000	766^R	0.32	5	(?)
SATURN	Pan	Showlater (1990)	133,570	0.573	0.00	20	(?)
	Atlas	Terrile (1980)	137,640	0.602	0 (?)	$40 \times 30 \times 30$	(?)
	Prometheus	Collins et al. (1980)	139,350	0.613	0.00	$140 \times 80 \times 100$	3×10^{17}
	Pandora	Collins et al. (1980)	141,700	0.629	0.00	$110 \times 70 \times 100$	2×10^{17}
	Epimetheus	Walker (1966)	151,422	0.694	0.01	$140 \times 100 \times 100$	6×10^{17}
	Janus	Dolfuss (1966)	151,472	0.695	0.01	$220 \times 160 \times 200$	2×10^{18}
	Mimas	Herschel (1789)	185,520	0.942	0.02	390	3.8×10^{19}
	Enceladus	Herschel (1789)	238,020	1.370	0.00	500	7.3×10^{19}
	Tethys	Cassini (1684)	294,660	1.888	0.00	1050	6.2×10^{20}
	Calypso	Smith et al. (1980)	294,660	1.888	0 (?)	$30 \times 20 \times 25$	(?)

Table A-3 (continued)
Satellites of the Planets

Planet	Satellite	Discoverers	Average distance from center of planet (km)	Orbital (sidereal) period* (days)	Orbital eccentricity	Diameter of satellite (km)	Mass (kg)
	Telesto	Smith et al. (1980)	294,660	1.888	0 (?)	24	(?)
	Dione	Cassini (1684)	377,400	2.737	0.00	1120	1.1×10^{21}
	Helene	Laques et al. (1980)	377,400	2.737	0.01	$40 \times 30 \times 30$	(?)
	Rhea	Cassini (1672)	527,040	4.518	0.00	1530	2.3×10^{21}
	Titan	Huygens (1655)	1,221,850	15.945	0.03	5150	1.4×10^{23}
	Hyperion	Bond (1848)	1,481,000	21.277	0.10	$410 \times 260 \times 220$	2×10^{19}
	Iapetus	Cassini (1671)	3,561,300	79.331	0.03	1440	1.6×10^{21}
	Phoebe	Pickering (1898)	12,952,000	550.48^{R}	0.16	220	4×10^{18}
URANUS	Cordelia	Voyager 2 (1986)	49,750	0.335	0 (?)	40	(?)
	Ophelia	Voyager 2 (1986)	53,760	0.376	0 (?)	30	(?)
	Bianca	Voyager 2 (1986)	59,160	0.435	0 (?)	40	(?)
	Cressida	Voyager 2 (1986)	61,770	0.464	0 (?)	70	(?)
	Desdemona	Voyager 2 (1986)	62,660	0.474	0 (?)	60	(?)
	Juliet	Voyager 2 (1986)	64,360	0.493	0 (?)	80	(?)
	Portia	Voyager 2 (1986)	66,100	0.513	0 (?)	110	(?)
	Rosalind	Voyager 2 (1986)	69,930	0.558	0 (?)	60	(?)
	Belinda	Voyager 2 (1986)	75,260	0.624	0 (?)	70	(?)
	Puck	Voyager 2 (1986)	86,010	0.762	0 (?)	150	(?)
	Miranda	Kuiper (1948)	129,780	1.414	0.00	470	6.6×10^{19}
	Ariel	Lassell (1851)	191,240	2.520	0.00	1160	1.4×10^{21}
	Umbriel	Lassell (1851)	265,790	4.144	0.00	1170	1.2×10^{21}
	Titania	Herschel (1787)	435,840	8.706	0.00	1580	3.5×10^{21}
	Oberon	Herschel (1787)	582,600	13.463	0.00	1520	3.0×10^{21}
	Caliban	Gladman, et al. (1997)	5,753,400	415^{R}	0.16	80	(?)
	Sycorax	Gladman, et al. (1997)	8,000,000	415^{R}	0.52	160	(?)
NEPTUNE	Naiad	Voyager 2 (1989)	48,230	0.296	0 (?)	60	(?)
	Thalassa	Voyager 2 (1989)	50,070	0.312	0 (?)	80	(?)
	Despina	Voyager 2 (1989)	52,530	0.333	0 (?)	150	(?)
	Galatea	Voyager 2 (1989)	61,950	0.429	0 (?)	180	(?)
	Larissa	Voyager 2 (1989)	73,550	0.554	0 (?)	190	(?)
	Proteus	Voyager 2 (1989)	117,640	1.121	0 (?)	415	(?)
	Triton	Lassell (1846)	354,800	5.877^{R}	0.00	2700	2.15×10^{22}
	Nereid	Kuiper (1949)	5,513,400	359.2	0.75	340	(?)
PLUTO	Charon	Christy (1978)	19,640	6.387	0.00	1190	1.9×10^{21}

A superscript R means that the satellite orbits in a retrograde direction (opposite to the planet's rotation).

Table A-4
The Nearest Stars

Name*	Parallax (arcsec)	Distance (light-years)	Spectral type	Radial velocity** (km/s)	Proper motion (arcsec/year)	Apparent visual magnitude	Absolute visual magnitude	Luminosity (Sun = 1)
Sun			G2 V			−26.7	+4.83	1.00
Proxima Centauri	0.772	4.22	M5.5 V	−22	3.853	+11.01	+15.45	8.2×10^{-4}
Alpha Centauri A	0.742	4.40	G2 V	−25	3.710	−0.01	+4.34	1.77
Alpha Centauri B	0.742	4.40	K0 V	−21	3.724	+1.35	+5.70	0.55
Barnard's Star	0.549	5.94	M4 V	−111	10.358	+9.54	+13.24	3.6×10^{-3}
Wolf 359	0.418	7.80	M6 V	+13	4.689	+13.45	+16.6	3.5×10^{-4}
Lalande 21185	0.392	8.32	M2 V	−84	4.802	+7.49	+10.46	0.023
L 726-8 A	0.381	8.56	M5.5 V	+29	3.366	+12.41	+15.3	9.4×10^{-4}
L 726-8 B	0.381	8.56	M6 V	+32	3.366	+13.2	+16.1	5.6×10^{-4}
Sirius A	0.379	8.61	A1 V	−9	1.339	−1.44	+1.45	26.1
Sirius B	0.379	8.61	white dwarf	−9	1.339	+8.44	+11.3	2.4×10^{-3}
Ross 154	0.336	9.71	M3.5 V	−12	0.666	+10.37	+13.00	4.1×10^{-3}
Ross 248	0.316	10.32	M5.5 V	−78	1.626	+12.29	+14.8	1.5×10^{-3}
Epsilon Eridani	0.311	10.49	K2 V	+17	0.977	+3.72	+6.18	0.40
Lacaille 9352	0.304	10.73	M1.5 V	+10	6.896	+7.35	+9.76	0.051
Ross 128	0.300	10.87	M4 V	−31	1.361	+11.12	+13.50	2.9×10^{-3}
L 789-6	0.294	11.09	M5 V	−60	3.259	+12.33	+14.7	1.3×10^{-3}
61 Cygni A	0.287	11.36	K5 V	−65	5.281	+5.20	+7.49	0.16
61 Cygni B	0.285	11.44	K7 V	−64	5.172	+6.05	+8.33	0.095
Procyon A	0.286	11.40	F5 IV–V	−4	1.259	+0.40	+2.68	7.73
Procyon B	0.286	11.40	white dwarf	−4	1.259	+10.7	+13.0	5.5×10^{-4}
BD +59° 1915 A	0.281	11.61	M3 V	−1	2.238	+8.94	+11.18	0.020
BD +59° 1915 B	0.281	11.61	M3.5 V	+1	2.313	+9.70	+11.97	0.010
Groombridge 34 A	0.280	11.65	M1.5 V	+12	2.918	+8.09	+10.33	0.030
Groombridge 34 B	0.280	11.65	M3.5 V	+11	2.918	+11.07	+13.3	3.1×10^{-3}
Epsilon Indi	0.276	11.82	K5 V	−40	4.704	+4.69	+6.89	0.27
GJ 1111	0.276	11.82	M6.5 V	−5	1.288	+14.79	+17.0	2.7×10^{-4}
Tau Ceti	0.274	11.90	G8 V	−17	1.922	+3.49	+5.68	0.62
GJ 1061	0.270	12.08	M5.5 V	−20	0.836	+13.03	+15.2	1.0×10^{-3}
L 725-32	0.269	12.12	M4.5 V	+28	1.372	+12.10	+14.25	1.7×10^{-3}
BD +05° 1668	0.263	12.40	M3.5 V	+18	3.738	+9.84	+11.94	0.011
Kapteyn's star	0.255	12.79	M0 V	+246	8.670	+8.86	+10.89	0.013
Lacaille 8760	0.253	12.89	M0 V	+28	3.455	+6.69	+8.71	0.094
Krüger 60 A	0.250	13.05	M3 V	−33	0.990	+9.85	+11.9	0.010
Krüger 60 B	0.250	13.05	M4 V	−32	0.990	+11.3	+13.3	3.4×10^{-3}

* Stars that are components of binary systems are labeled A and B.

** A positive radial velocity means the star is receding; a negative radial velocity means the star is approaching.

Compiled from the Hipparcos General Catalogue and from data reported by the Research Consortium on Nearby Stars.
The table lists all known stars within 4.00 parsecs (13.05 light-years).

Table A-5
The Visually Brightest Stars

Name	Designation	Distance (light-years)	Spectral type	Radial velocity* (km/s)	Proper motion (arcsec/year)	Apparent visual magnitude	Apparent visual brightness** (Sirius = 1)	Absolute visual magnitude	Luminosity (Sun = 1)
Sirius A	α CMa A	8.61	A1 V	−9	1.339	−1.44	1.000	+1.45	26.1
Canopus	α Car	313	F0 I	+21	0.031	−0.62	0.470	−5.53	1.4×10^4
Arcturus	α Boo	36.7	K2 III	−5	2.279	−0.05	0.278	−0.31	190
Rigel Kentaurus	α Cen A	4.4	G2 V	−25	3.71	−0.01	0.268	+4.34	1.77
Vega	α Lyr	25.3	A0 V	−14	0.035	+0.03	0.258	+0.58	61.9
Capella	α Aur	42.2	G8 III	+30	0.434	+0.08	0.247	−0.48	180
Rigel	β Ori A	773	B8 Ia	+21	0.002	+0.18	0.225	−6.69	7.0×10^5
Procyon	α CMi A	11.4	F5 IV–V	−4	1.259	+0.4	0.184	+2.68	7.73
Achernar	α Eri	144	B3 IV	+19	0.097	+0.45	0.175	−2.77	5250
Betelgeuse	α Ori	427	M2 Iab	+21	0.029	+0.45	0.175	−5.14	4.1×10^4
Hadar	β Cen	525	B1 II	−12	0.042	+0.61	0.151	−5.42	8.6×10^4
Altair	α Aql	16.8	A7 IV–V	−26	0.661	+0.76	0.132	+2.2	11.8
Aldebaran	α Tau A	65.1	K5 III	+54	0.199	+0.87	0.119	−0.63	370
Spica	α Vir	262	B1 V	+1	0.053	+0.98	0.108	−3.55	2.5×10^4
Antares	α Sco A	604	M1 Ib	−3	0.025	+1.06	0.100	−5.28	3.7×10^4
Pollux	β Gem	33.7	K0 III	+3	0.627	+1.16	0.091	+1.09	46.6
Fomalhaut	α PsA	25.1	A3 V	+7	0.368	+1.17	0.090	+1.74	18.9
Deneb	α Cyg	3230	A2 Ia	−5	0.002	+1.25	0.084	−8.73	3.2×10^5
Mimosa	β Cru	353	B0.5 III	+20	0.05	+1.25	0.084	−3.92	3.4×10^4
Regulus	α Leo A	77.5	B7 V	+4	0.249	+1.36	0.076	−0.52	331

Data in this table was compiled from the Hipparcos General Catalogue.

* *A positive radial velocity means the star is receding; a negative radial velocity means the star is approaching.*

** *This is the ratio of the star's apparent brightness to that of Sirius, the brightest star in the night sky.*

Note: *Acrux, or α Cru (the brightest star in Crux, the Southern Cross), appears to the naked eye as a star of apparent magnitude +0.87, the same as Aldebaran, but it does not appear in this table because Acrux is actually a binary star system. The blue-white component stars of this binary system have apparent magnitudes of +1.4 and +1.9, and thus they are dimmer than any of the stars listed here.*

Table A-6
Some Important Astronomical Quantities

Astronomical unit:	$1\ \text{AU} = 1.496 \times 10^{11}\ \text{m}$
Light-year:	$1\ \text{ly} = 9.461 \times 10^{15}\ \text{m} = 63{,}240\ \text{AU}$
Parsec:	$1\ \text{pc} = 3.086 \times 10^{16}\ \text{m} = 3.262\ \text{ly}$
Solar mass:	$1\ M_\odot = 1.989 \times 10^{30}\ \text{kg}$
Solar radius:	$1\ R_\odot = 6.960 \times 10^{8}\ \text{m}$
Solar luminosity:	$1\ L_\odot = 3.827 \times 10^{26}\ \text{W}$
Earth's mass:	$1\ M_\oplus = 5.974 \times 10^{24}\ \text{kg}$
Earth's equatorial radius:	$1\ R_\oplus = 6.378 \times 10^{6}\ \text{m}$
Moon's mass:	$1\ M_{\text{Moon}} = 7.348 \times 10^{22}\ \text{kg}$
Moon's equatorial radius:	$1\ R_{\text{Moon}} = 1.738 \times 10^{6}\ \text{m}$

Table A-7
Some Important Physical Constants

Speed of light:	$c = 2.998 \times 10^{8}\ \text{m/s}$
Gravitational constant:	$G = 6.668 \times 10^{-11}\ \text{N m}^2\ \text{kg}^{-2}$
Planck constant:	$h = 6.625 \times 10^{-34}\ \text{J s}$
	$= 4.136 \times 10^{-15}\ \text{eV s}$
Boltzmann constant:	$k = 1.380 \times 10^{-23}\ \text{J K}^{-1}$
	$= 8.617 \times 10^{-5}\ \text{eV K}^{-1}$
Stefan–Boltzmann constant:	$\sigma = 5.669 \times 10^{-8}\ \text{W m}^{-2}\ \text{K}^{-4}$
Mass of electron:	$m_e = 9.108 \times 10^{-31}\ \text{kg}$
Mass of ^{1}H atom:	$m_H = 1.673 \times 10^{-27}\ \text{kg}$

Table A-8

Spiral Galaxies

Spiral galaxy	R.A.	Decl.	Hubble type
M31 (NGC 224)	$0^h\ 42.7^m$	$+41°\ 16'$	Sb
M58 (NGC 4579)	12 37.7	+11 49	Sb
M61 (NGC 4303)	12 21.9	+ 4 28	Sc
M63 (NGC 5055)	13 15.8	+42 02	Sb
M64 (NGC 4826)	12 56.7	+21 41	Sb
M74 (NGC 628)	1 36.7	+15 47	Sc
M83 (NGC 5236)	13 37.0	−29 52	Sc
M88 (NGC 4501)	12 32.0	+14 25	Sb
M90 (NGC 4569)	12 36.8	+13 10	Sb
M94 (NGC 4736)	12 50.9	+41 07	Sb
M98 (NGC 4192)	12 13.8	+14 54	Sb
M99 (NGC 4254)	12 18.8	+14 25	Sc
M100 (NGC 4321)	12 22.9	+15 49	Sc
M101 (NGC 5457)	14 03.2	+54 21	Sc
M104 (NGC 4594)	12 40.0	−11 37	Sa
M108 (NGC 3556)	11 11.5	+55 40	Sc

Interacting Galaxies

Interacting galaxies	R.A.	Decl.
M51 (NGC 5194)	$13^h\ 29.9^m$	$+47°12'$
NGC 5195	13 30.0	+47 16
M65 (NGC 3623)	11 18.9	+13 05
M66 (NGC 3627)	11 20.2	+12 59
M81 (NGC 3031)	9 55.6	+69 04
M82 (NGC 3034)	9 55.8	+69 41
M95 (NGC 3351)	10 44.0	+11 42
M96 (NGC 3368)	10 46.8	+11 49
M105 (NGC 3379)	10 47.8	+12 35

AN ASTRONOMER'S TOOLBOX A-1

Temperature Scales

Three temperature scales are in common use. Throughout most of the world, temperatures are expressed in degrees Celsius (°C), named in honor of the Swedish astronomer Anders Celsius, who proposed it in 1742. The **Celsius temperature scale** (also known as the "centigrade scale") is based on the behavior of water, which freezes at 0°C and boils at 100°C at sea level on Earth.

Scientists usually prefer the **Kelvin scale**, named after the British physicist Lord Kelvin (William Thomson), who made many important contributions to our knowledge about heat and temperature. On the Kelvin temperature scale, water freezes at 273 K and boils at 373 K. Note that we do not use the degree symbol with the Kelvin temperature scale.

Because water must be heated by 100 K or 100°C to go from its freezing point to its boiling point, you can see that the size of a Kelvin is the same as the size of a degree Celsius. When considering temperature *changes,* measurements in Kelvins and in degrees Celsius lead to the same number.

A temperature expressed in Kelvins is always equal to the temperature in degrees Celsius plus 273. Scientists prefer the Kelvin scale because it is closely related to the physical meaning of temperature. All substances are made of atoms, which are very tiny (a typical atom has a diameter of about 10^{-10} m) and constantly in motion. The temperature of a substance is directly related to the average speed of its atoms. If something is hot, its atoms are moving at high speeds. If a substance is cold, its atoms are moving much more slowly.

The coldest possible temperature is the temperature at which atoms move as slowly as possible (they can never quite stop completely). This minimum possible temperature, called absolute zero, is the starting point for the Kelvin scale. Absolute zero is 0 K, or −273°C. Because it is impossible for anything to be colder than 0 K, there are no negative temperatures on the Kelvin scale.

In the United States, many people still use the now-archaic Fahrenheit scale, which expresses temperatures in degrees Fahrenheit (°F). When the German physicist Gabriel Fahrenheit introduced this scale in the early 1700s, he intended 0°F to represent the coldest temperature then achievable (with a mixture of ice and saltwater) and 100°F to represent the temperature of a healthy human body. On the Fahrenheit scale, water freezes at 32°F and boils at 212°F. Because there are 180 degrees Fahrenheit between the freezing and boiling points of water, a degree Fahrenheit is only 100/180 (= 5/9) the size of the other scales.

The following equation converts from degrees Fahrenheit to degrees Celsius:

$$T_C = \frac{5}{9}\ (T_F - 32)$$

To convert from Celsius to Fahrenheit, a simple rearrangement of terms gives the relationship

$$T_F = \frac{9}{5}\ T_C + 32$$

where T_F is the temperature in degrees Fahrenheit and T_C is the temperature in degrees Celsius.

Example: Consider a typical room temperature of 68°F. Using the first equation, we can convert this measurement to the Celsius scale as follows:

$$T_C = \frac{5}{9}\ (68 - 32) \approx 20°C$$

To arrive at the Kelvin scale, we simply add 273 degrees to the value in degrees Celsius. Thus, 68°F ≈ 20°C = 293 K

Compare! The accompanying figure displays the relationships among these three temperature scales.

Try these questions! The Sun's surface temperature is about 5800 K. What is its temperature in Celsius and Fahrenheit? The temperature of empty space is about 3 K. What is its temperature in Celsius and Fahrenheit?

(Answers appear at the end of the book.)

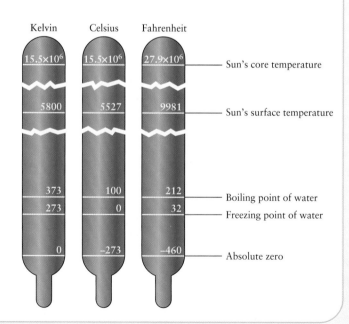

GLOSSARY

A ring The outermost ring of Saturn visible from Earth; it is located just beyond Cassini's division.

absolute magnitude The apparent magnitude that a star would have if it were 10 parsecs from Earth.

absorption line spectrum Dark lines superimposed on a continuous spectrum.

acceleration A change in the direction or magnitude of a velocity.

accretion The gradual accumulation of matter by an astronomical body, usually caused by gravity.

accretion disk An orbiting disk of matter spiraling in toward a star or black hole.

achromatic lens A compound lens designed to minimize the effect of chromatic aberration.

active galaxy A very luminous galaxy, often containing an active galactic nucleus.

active optics A system that adjusts a reflecting telescope in response to changes in temperature and shape of the mount; it helps optimize an image.

adaptive optics Primary telescope mirrors that are continuously and automatically adjusted to compensate for the distortion of starlight due to the motion of the Earth's atmosphere.

AGB star *See* **asymptotic giant branch (AGB) star.**

albedo The fraction of sunlight that a planet, asteroid, or satellite scattered directly back into space.

amino acids A class of chemical compounds that are the building blocks of proteins.

angle The opening between two straight lines that meet at a point.

angular diameter (angular size) The arc angle across an object.

angular momentum A measure of how much energy an object has stored in its rotation and/or revolution.

angular resolution The angular size of the smallest detail of an astronomical object that can be distinguished with a telescope.

annular eclipse An eclipse of the Sun in which the Moon is too distant to cover the Sun completely so that a ring of sunlight is seen around the Moon at mideclipse.

anorthosite A light-colored rock found throughout the lunar highlands and in some very old mountains on Earth.

aphelion The point in its orbit where a planet or other solar system body is farthest from the Sun.

Apollo asteroid An asteroid that is sometimes closer to the Sun than the Earth is.

apparent magnitude A measure of the brightness of light from a star or other object as seen from Earth.

arc angle The measurement of the angle between two objects or two parts of the same object.

arcminute (arcmin) One-sixtieth of a degree of arc.

arcsecond (arcsec) One-sixtieth of a minute of arc.

asteroid Any of the rocky objects larger than a few hundred meters in diameter (and not classified as a planet or moon) that orbits the Sun.

asteroid belt (minor planet) A $1^1/2$-astronomical-unit-wide region between the orbits of Mars and Jupiter in which most of the asteroids are found.

astronomical unit (AU) The average distance between the Earth and the Sun: 1.5×10^8 kilometers $\approx$ 93 million miles.

astronomy The branch of science dealing with objects and phenomena that lie beyond the Earth's atmosphere.

astrophysics That part of astronomy dealing with the physics of astronomical objects and phenomena.

asymptotic giant branch (AGB) star A red giant star that has completed core helium fusion and has re-expanded for a second time.

atom The smallest particle of an element that has the properties characterizing that element.

atomic number The number of protons in the nucleus of an atom.

aurora (*plural* **aurorae**) Light radiated by atoms and ions in the Earth's upper atmosphere; seen most commonly in the polar regions.

autumnal equinox The intersection of the ecliptic and the celestial equator where the Sun crosses the equator moving from north to south.

average density The mass of an object divided by its volume.

B ring The brightest of the three rings of Saturn visible from Earth; it lies just inside the Cassini division.

barred spiral galaxy A spiral galaxy in which the spiral arms begin from the ends of a bar running through the nuclear bulge.

belt asteroid An asteroid whose orbit lies in the asteroid belt.

belts (of Jupiter) Dark, reddish bands in Jupiter's cloud cover.

Big Bang An explosion that took place roughly 15 billion years ago, creating all space, matter, and energy in which the universe emerged.

binary star Two stars revolving about each other; a double star.

birth line A line on the Hertzsprung-Russell diagram corresponding to where stars with different masses transform from protostars to pre–main-sequence stars.

BL Lacertae (BL Lac) object A type of active galaxy; a blazar.

black hole An object whose gravity is so strong that the escape velocity from it exceeds the speed of light.

blackbody A hypothetical perfect radiator that absorbs and re-emits all radiation falling upon it.

blackbody curve The curve obtained when the intensity of radiation from a blackbody at a particular temperature is plotted against wavelength.

blackbody radiation Electromagnetic radiation emitted by a blackbody.

blazar A BL Lacertae object.

blueshift A shift of all spectral features toward shorter wavelengths; the Doppler shift of light from an approaching source.

Bohr atom A model of the atom, described by Niels Bohr, in which electrons revolve about the nucleus in various circular orbits.

brown dwarf Any of the planetlike bodies with less than $0.08 \ M_\odot$ and more than about $13 \ M_{Jupiter}$; such bodies do not have enough mass to sustain fusion in their cores.

C ring The faint, inner portion of Saturn's main ring system.

caldera The crater at the summit of a volcano.

Callisto One of the four Galilean satellites of Jupiter.

capture theory The idea that the Moon was created at a different location in the solar system and subsequently captured by Earth's gravity.

carbon fusion The thermonuclear fusion of carbon nuclei to produce nitrogen.

carbonaceous chondrites A class of extremely ancient, carbon-rich meteorites.

Cassegrain focus An optical arrangement in a reflecting telescope in which light rays are reflected by a secondary mirror through a hole in the primary mirror.

Cassini division A prominent gap between Saturn's A and B rings discovered in 1675 by J. D. Cassini.

celestial equator A great circle on the celestial sphere 90° from the celestial poles.

celestial poles Points about which the celestial sphere appears to rotate.

celestial sphere A hypothetical sphere of very large radius centered on the observer; the apparent sphere of the night sky.

Celsius temperature scale *See* **temperature (Celsius).**

center of mass The point around which a rigid system is perfectly balanced in a gravitational field; also, the point in space around which mutually orbiting bodies have elliptical orbits.

Cepheid variable star One of two types of yellow, supergiant, pulsating stars.

Cerenkov radiation Radiation produced by particles traveling through a substance faster than light can.

Ceres The largest known asteroid and the first to be discovered.

Chandrasekhar limit The maximum mass of a white dwarf, about $1.4 \ M_\odot$.

charge-coupled device (CCD) A type of solid-state silicon wafer designed to detect photons.

chromatic aberration An optical property whereby different colors of light passing through a lens are focused at different distances from it.

chromosphere The layer in the solar atmosphere between the photosphere and the corona.

circumpolar stars All the stars that never set at a given latitude; all the stars between Polaris and the northern horizon.

close binary A binary star whose members are separated by a few stellar diameters.

closed universe A universe that contains enough matter to cause it to recollapse. It is finite in extent and has no "outside."

cluster (of galaxies) A collection of a few hundred to a few thousand galaxies bound by gravity.

cocreation theory The theory that the Moon formed simultaneously with the Earth and in orbit around it.

collision-ejection theory The theory that the Moon was created by the impact of a planet-sized object with the Earth; presently considered the most plausible theory of the Moon's formation.

color-magnitude diagram A plot of the surface temperatures (colors) versus the absolute magnitudes of stars.

coma (of a comet) The nearly spherical, diffuse gas surrounding the nucleus of a comet near the Sun.

comet A small body of ice and dust in orbit about the Sun. While passing near the Sun, a comet's vaporized ices give rise to a coma, tails, and a hydrogen envelope.

conduction (thermal) The transfer of heat by passing energy directly from atom to atom.

configuration (of a planet) A particular geometric arrangement of the Earth, a planet, and the Sun.

confinement The moment shortly after the Big Bang when quarks bound together to form particles like protons and neutrons.

conic section The curve of intersection between a circular cone and a plane. This curve can be a circle, ellipse, parabola, or hyperbola.

conjunction The alignment of two bodies in the solar system so that they appear in the same part of the sky as seen from Earth.

conservation of angular momentum The law of physics stating that the total amount of angular momentum in an isolated system remains constant.

constellation Any of the 88 contiguous regions that cover the entire celestial sphere, including all the objects in each region; also, a configuration of stars often named after an object, a person, or an animal.

contact binary A close binary system in which both stars fill or overflow their Roche lobes.

continental drift The gradual movement of the continents over the surface of the Earth due to plate tectonics.

continuous spectrum A spectrum of light over a range of wavelengths without any spectral lines.

continuum *See* **continuous spectrum.**

convection The transfer of energy by moving currents of fluid or gas containing that energy.

convective zone A layer in a star where energy is transported outward by means of convection; also known as the *convective envelope* or *convection zone.*

core The central portion of any astronomical object.

core helium fusion The fusion of helium to form carbon and oxygen at the center of a star.

core hydrogen fusion The fusion of hydrogen to form helium at the center of a star.

corona The Sun's outer atmosphere.

coronagraph A specially designed telescope with a baffle that blocks out the solar disk so that the corona can be photographed.

coronal hole A dark region of the Sun's inner corona as seen at X-ray wavelengths.

coronal mass ejection Large volumes of high energy gas released from the Sun's corona.

cosmic censorship The belief that the only connection between a black hole and the universe is the black hole's event horizon.

cosmic light horizon A sphere, centered on the Earth, whose radius equals the distance traveled by light since the Big Bang.

cosmic microwave background Photons from every part of the sky with a blackbody spectrum at 2.73 K; the cooled-off radiation from the primordial fireball that originally filled all space.

cosmic ray High speed particles traveling through space.

cosmic ray shower Groups of particles from Earth's atmosphere propelled Earthward by the impact of a cosmic ray.

cosmological constant A number sometimes inserted in the equations of general relativity that represents a pressure that opposes gravity throughout the universe.

cosmological redshift An increase in wavelength from distant galaxies and quasars caused by the expansion of the universe.

cosmology The study of the formation, organization, and evolution of the universe.

coudé focus A reflecting telescope in which a series of mirrors direct light to a remote focus away from the moving parts of the telescope.

crater A circular depression on a celestial body caused by the impact of a meteoroid, asteroid, or comet or by a volcano.

crescent Moon A lunar phase during which the Moon appears less than half full.

crust The solid surface layer of some astronomical bodies, including the terrestrial planets, the moons, the asteroids, and some stellar remnants.

dark ages The age of the universe between the time of decoupling and the first burst of star formation.

dark energy A repulsive gravitational effect that is causing the universe to accelerate outward.

dark matter (missing mass) The as-yet-undetected matter in the universe that is underluminous and probably quite different from ordinary matter.

dark nebula A cloud of interstellar gas and dust that obscures the light of more distant stars.

declination The coordinate on the celestial sphere exactly analogous to latitude on Earth; measured north and south of the celestial equator.

decoupling The epoch in the early universe when electrons and ions first combined to create stable atoms; the time when electromagnetic radiation ceased to dominate over matter.

deferent A fixed circle in the Earth-centered universe along which a smaller circle (an epicycle) moves carrying a planet, the Sun, or the Moon.

degree (°) A unit of angular measure or a temperature measure.

dense core Any of the regions of interstellar gas clouds that are slightly denser than normal and destined to collapse to form one or a few stars.

density The ratio of the mass of an object to its volume.

density wave A spiral-shaped compression of the gas and dust in a spiral galaxy.

density-wave theory An explanation of spiral arms in galaxies elaborated by C. C. Lin and F. Shu. (*See* **spiral density wave.**)

detached binary A binary system in which the surfaces of both stars are inside their Roche lobes.

differential rotation The rotation of a nonrigid object in which parts at different latitudes or different radial distances move at different speeds.

differentiation *See* **planetary differentiation.**

diffraction grating An optical device consisting of closely spaced lines ruled on a piece of glass that is used like a prism to disperse light into a spectrum.

direct motion The gradual, eastward apparent motion of a planet against the background stars as seen from Earth.

disk (of a galaxy) A flattened assemblage of stars, gas, and dust in a spiral galaxy like the Milky Way.

distance modulus The difference between the apparent and absolute magnitudes of an object.

diurnal motion Cyclic motion with a one-day period.

Doppler effect (or Doppler shift) The change in wavelength of radiation due to relative motion between the source and the observer along the line of sight.

double-line spectroscopic binary A spectroscopic binary whose spectrum exhibits spectral lines of both stars.

double radio source An extragalactic radio source characterized by two large lobes of radio emission, often located on either side of an active galaxy.

Drake equation A mathematical equation used to estimate the number of extraterrestrial civilizations that may exist in our Galaxy.

dust devil Whirlwind found in dry or desert areas on both Earth and Mars.

dust tail A comet tail caused by dust particles escaping from the comet's nucleus.

dwarf elliptical galaxy A small elliptical galaxy with far fewer stars than a typical galaxy.

dwarf star Any star smaller than a giant, such as a main-sequence star or a white dwarf.

dynamo theory The generation of a magnetic field by circulating electric charges.

eccentricity *See* **orbital eccentricity.**

eclipse The blocking of part or all of the light from the Moon by the Earth (lunar eclipse) or from the Sun by the Moon (solar eclipse).

eclipse path The track of the tip of the Moon's shadow along the Earth's surface during a total or annular solar eclipse.

eclipsing binary A double star system in which stars periodically pass in front of each other as seen from Earth.

ecliptic The annual path of the Sun on the celestial sphere; the plane of the Earth's orbit around the Sun.

Einstein cross The appearance of four images of the same galaxy or quasar due to gravitational lensing by an intervening galaxy.

Einstein ring The circular or arc-shaped image of a distant galaxy or quasar created by gravitational lensing by an intervening galaxy.

ejecta blanket The ring of material surrounding a crater that was ejected during the crater-forming impact.

electromagnetic radiation Radiation consisting of oscillating electric and magnetic fields, namely gamma rays, X rays, visible light, ultraviolet and infrared radiation, and radio waves.

electromagnetic spectrum The entire array or family of electromagnetic radiation.

electron A negatively charged subatomic particle usually found in orbit about the nucleus of an atom.

electron degeneracy pressure A powerful pressure produced by repulsion of closely packed (degenerate) electrons.

element A substance that cannot be decomposed by chemical means into simpler substances.

ellipse A closed curve obtained by cutting completely through a circular cone with a plane; the shape of planetary orbits.

elliptical galaxy A galaxy with an elliptical shape, little interstellar matter, and no spiral arms.

elongation The angle between a planet and the Sun as seen from Earth.

emission line spectrum A spectrum that contains only bright emission lines.

emission nebula A glowing gaseous nebula whose light comes from fluorescence caused by a nearby star.

Encke division A thin gap in Saturn's A ring, possibly first seen by J. F. Encke in 1838.

energy The ability to do work.

energy flux The amount of energy emitted from each square meter of an object's surface per second.

energy level (in an atom) A particular amount of energy possessed by an electron in orbit around a nucleus.

epicycle In the Earth-centered universe, a moving circle about which planets revolve.

equations of stellar structure A set of relationships that describe the interactions of matter, energy, and gravity inside a star.

equinox Either of the two days of the year when the Sun crosses the celestial equator and is therefore directly over the Earth's equator; *see also* **autumnal equinox** *and* **vernal equinox.**

era of recombination The time, roughly 500,000 years after the Big Bang, when the universe became transparent.

ergoregion The region of space immediately outside the event horizon of a rotating black hole where it is impossible to remain at rest.

Europa One of the Galilean satellites of Jupiter.

event horizon The location around a black hole where the escape velocity equals the speed of light; the boundary of a black hole.

evolutionary track On the Hertzsprung-Russell diagram, the path followed by a point representing an evolving star.

expanding universe The motion of the superclusters of galaxies away from each other.

eyepiece lens A magnifying lens used to view the image produced at the focus of a telescope.

F ring A thin ring just beyond the outer edge of Saturn's main ring system.

Fahrenheit scale *See* **temperature (Fahrenheit).**

filament A dark curve seen above the Sun's photosphere that is the top view of a solar prominence.

first quarter Moon A phase of the waxing Moon when Earth-based observers see half of the Moon's illuminated hemisphere.

fission theory The theory that the Moon formed from matter flung off the Earth because the planet was rotating extremely fast.

flare *See* **solar flare.**

flocculent spiral galaxy A spiral galaxy whose spiral arms are broad, fuzzy, and poorly demarcated.

focal length The distance from a lens or concave mirror to where converging light rays meet.

focal plane The plane at the focal length of a lens or concave mirror on which an extended object is focused.

focal point *See* **focus.**

focus (of a lens or concave mirror) The place at the focal length where light rays from a point object (that is, one that is too distant or tiny to resolve) are converged by a lens or concave mirror.

focus (*plural* **foci**) (of an ellipse) The two points inside an ellipse, the sum of whose distances from any point on the ellipse is constant.

force That which can change the momentum of an object.

full Moon A phase of the Moon during which its full daylight hemisphere can be seen from Earth.

galactic cannibalism A collision between two galaxies of unequal mass and size in which the smaller galaxy is absorbed by the larger galaxy.

galactic merger A collision and subsequent merger of two roughly equal-sized galaxies.

galactic nucleus The center of a galaxy; the center of the Milky Way Galaxy.

galaxy A large assemblage of stars, gas, and dust bound together by their mutual gravitational attraction.

Galilean satellite (Galilean moon) Any one of the four large moons of Jupiter that is visible from Earth through a small telescope.

gamma ray The most energetic form of electromagnetic radiation.

gamma-ray burster An object that emits a short burst of gamma rays; they are believed to be outside our Galaxy.

Ganymede One of the Galilean satellites of Jupiter.

gas (ion) tail The relatively straight tail of a comet produced by the solar wind acting on ions in a comet's coma.

general theory of relativity A description of spacetime formulated by Einstein explaining how gravity affects the geometry of space and the flow of time.

geocentric cosmology The belief that the Earth is at the center of the universe.

giant elliptical galaxy A very large, extremely massive elliptical galaxy, usually located near the center of a rich cluster of galaxies.

giant molecular cloud A large interstellar cloud of cool gas and dust in a galaxy.

giant star A star whose diameter is roughly 10 to 100 times that of the Sun.

gibbous Moon A phase of the Moon in which more than half, but not all, of the Moon's daylight hemisphere is visible from Earth.

glitch A sudden speedup in the period of a pulsar.

globular cluster A large spherical cluster of gravitationally bound stars usually found in the outlying regions of a galaxy.

grand-design spiral galaxy A spiral galaxy whose spiral arms are thin, graceful, and well defined.

grand unified theory (GUT) A theory that describes and explains the four physical forces.

granulation The rice-grain–like structure of the solar photosphere due to convection of solar gases.

granules Lightly colored convection features about 1000 kilometers in diameter seen constantly in the solar photosphere.

gravitation (gravity) The tendency of all matter to attract all other matter.

gravitational lensing The distortion of the appearance of an object by a source of gravity between it and the observer.

gravitational redshift The redshift of photons leaving the gravitational field of any massive object, such as a star or black hole.

gravitational waves (gravitational radiation) Ripples in the overall geometry of space produced by nonspherical moving objects.

gravity *See* **gravitation.**

Great Dark Spot A large, dark, oval-shaped storm that used to be in Neptune's southern hemisphere.

Great Red Spot A large, red-orange, oval-shaped storm in Jupiter's southern hemisphere.

Great Wall A huge arc of galaxies between two voids in the cosmos.

greatest elongation The largest possible angle between the Sun and an inferior planet.

greenhouse effect The trapping of infrared radiation near a planet's surface by the planet's atmosphere.

ground state The lowest energy level of an atom.

H II region A region of ionized hydrogen in interstellar space.

halo (of a galaxy) A spherical distribution of globular clusters, isolated stars, and possibly dark matter that surrounds a galaxy.

Hawking process The formation of real particles from virtual ones just outside a black hole's event horizon; the means by which black holes evaporate.

head-tail source A radio galaxy whose radio emission is deflected from the galaxy.

heliocentric cosmology A theory of the formation and evolution of the solar system with the Sun at the center.

helioseismology The study of vibrations of the solar surface.

helium flash The explosive ignition of helium fusion in the core of a low-mass, giant star.

helium fusion The thermonuclear fusion of helium to produce carbon and oxygen.

helium shell flash The explosive ignition of helium fusion in a thin shell surrounding the core of a low-mass star.

Hertzsprung-Russell (H-R) diagram A plot of the absolute magnitude or luminosity of stars versus their surface temperatures or spectral classes.

highlands Heavily cratered, mountainous regions of the lunar surface.

horizon problem (isotropy problem) The difficulty in explaining why seemingly disconnected regions of the universe have the same temperature.

horizontal branch stars A group of post–helium-flash stars near the main sequence on the Hertzsprung-Russell diagram of a typical globular cluster.

hot-spot volcanism The creation of volcanoes on a planet's surface caused by a reservoir of hot magma in the planet's mantle under a thin part of the crust.

Hubble classification A system of classifying galaxies according to their appearance into one of four broad categories: spirals, barred spirals, ellipticals, and irregulars.

Hubble constant (H_0) The constant of proportionality in the relation between the recessional velocities of remote galaxies and their distances; the correct value will determine the age of the universe.

Hubble flow The recession of the galaxies caused by the expansion of the universe.

Hubble law The relationship that states that the redshifts of remote galaxies are directly proportional to their distances from Earth.

hydrocarbon A molecule based on hydrogen and carbon.

hydrogen envelope An extremely large, tenuous sphere of hydrogen gas surrounding the head of a comet.

hydrogen fusion (hydrogen burning) The thermonuclear fusion of hydrogen to produce helium.

hydrostatic equilibrium A balance between the weight of a layer in a star and the pressure that supports it.

hyperbola An open curve obtained by cutting a cone with a plane.

impact breccia A rock consisting of various fragments cemented together by the impact of a meteoroid.

impact crater A crater on the surface of a planet or moon produced by the impact of an asteroid, meteoroid, or comet.

inferior conjunction The configuration when Mercury or Venus is directly between the Sun and the Earth.

inflation A sudden expansion of space.

inflationary epoch A brief period shortly after the Big Bang during which the scale of the universe increased very rapidly.

infrared radiation Electromagnetic radiation of a wavelength longer than visible light but shorter than radio waves.

initial mass function The numbers of stars on the main sequence at all different masses.

instability strip A region on the Hertzsprung-Russell diagram occupied by pulsating stars.

interferometry A method of increasing resolving power by combining electromagnetic radiation obtained by two or more telescopes.

intergalactic gas Gas located between the galaxies within a cluster of galaxies.

interstellar dust Microscopic solid grains of various compounds in interstellar space.

interstellar extinction The dimming of starlight as it passes through the interstellar medium.

interstellar gas Sparse gas in interstellar space.

interstellar medium Interstellar gas and dust.

interstellar reddening The reddening of starlight passing through the interstellar medium resulting from the scattering of short wavelength light more than long wavelength light.

inverse-square law The gravitational attraction between two objects and the apparent brightness of a light source are both inversely proportional to the square of its distance.

Io One of the Galilean satellites of Jupiter.

ion An atom that has become electrically charged due to the loss or addition of one or more electrons.

ionization The process by which an atom loses or gains electrons.

iron meteorite A meteorite composed of iron with a small admixture of nickel; also called an *iron*.

irregular cluster (of galaxies) An unevenly distributed group of galaxies bound together by their mutual gravitational attraction.

irregular galaxy An asymmetrical galaxy having neither spiral arms nor an elliptical shape.

isotope Any of several forms of the same chemical element whose nuclei all have the same number of protons but different numbers of neutrons.

isotropy The fact that the average number of galaxies at different distances from Earth is the same in all directions; also, the fact that the temperature of the cosmic microwave background is essentially the same in all directions.

isotropy problem (horizon problem) The difficulty in explaining why seemingly disconnected regions of the universe have the same temperature.

Jeans instability The condition under which gravitational forces overcome thermal forces to cause part of an interstellar cloud to collapse and form stars and planets.

kelvin *See* **temperature (Kelvin).**

Kepler's laws Three statements, formulated by Johannes Kepler, that describe the motions of the planets.

Kerr black hole Any rotating, uncharged black hole.

kiloparsec (kpc) One thousand parsecs; about 3260 light-years.

kinetic energy The energy an object has as a result of its motion.

Kirchhoff's laws Three statements formulated by Gustav Kirchhoff describing spectra and spectral analysis.

Kirkwood gaps Gaps in the spacing of asteroid orbits discovered by Daniel Kirkwood.

Kuiper belt A doughnut-shaped ring of space around the Sun beyond Pluto that contains many frozen comet bodies, some of which are occasionally deflected toward the inner solar system.

Large Magellanic Cloud (LMC) An irregular galaxy, companion to the Milky Way.

last quarter Moon A phase of the waning Moon when Earth-based observers see half of the Moon's illuminated hemisphere.

law of equal areas Kepler's second law.

law of inertia The physical law that an object will stay at rest or move at a constant speed in a fixed direction unless acted upon by an outside force.

lenticular galaxy A disk-shaped galaxy without spiral arms.

light Electromagnetic radiation, which travels in packets called photons.

light curve A graph that displays variations in the brightness of a star or other astronomical object over time.

light-gathering power A measure of how much light a telescope intercepts and brings to a focus.

light-year (ly) The distance that light travels in a vacuum in one year.

lighthouse model The explanation that a pulsar pulses by rotating and funneling energy outward via magnetic fields that are not aligned with the rotation axis.

limb (of the Sun) The apparent edge of the Sun as seen in the sky.

limb darkening The phenomenon whereby the Sun is darker near its limb than near the center of its disk.

line of nodes The line along which the plane of the Moon's orbit intersects the plane of the ecliptic.

liquid metallic hydrogen A metallike form of hydrogen that is produced under extreme pressure.

Local Group The cluster of galaxies of which our own Galaxy is a member.

long-period comet A comet that takes tens of thousands of years or more to orbit the Sun once.

luminosity The rate at which electromagnetic radiation is emitted from a star or other object.

luminosity class The classification of a star of a given spectral type according to its luminosity; the classes are supergiant, bright giant, giant, subgiant, and main sequence.

lunar Referring to the Moon.

lunar eclipse An eclipse during which the Earth blocks light that would have struck the Moon.

lunar month *See* **synodic month**.

lunar phases The names given to the apparent shapes of the Moon as seen from Earth.

Lyman series A series of spectral lines of hydrogen produced by electron transitions to and from the lowest energy state of the hydrogen atom.

magnetic dynamo A theory that explains phenomena of the solar cycle as a result of periodic winding and unwinding of the Sun's magnetic field in the solar atmosphere.

magnetic field A region of space near a magnetized body within which magnetic forces can be detected.

magnetosphere The region around a planet occupied by its magnetic field.

magnification (magnifying power) The number of times larger in angular diameter an object appears through a telescope than when it is seen by the naked eye.

magnitude *See* **absolute magnitude; apparent magnitude**.

magnitude scale The system of denoting magnitudes.

main sequence A grouping of stars on the Hertzsprung-Russell diagram extending diagonally across the graph from the hottest, brightest stars to the dimmest, coolest stars.

main-sequence star A star, fusing hydrogen to helium in its core, whose surface temperature and luminosity place it on the main sequence on the Hertzsprung-Russell diagram.

mantle (of a planet) That portion of a terrestrial planet located between its crust and core.

mare (*plural* **maria**) Latin for "sea," a large, relatively crater-free plain on the Moon.

mare basalt Dark, solidified lava that covers the lunar maria.

mascons Regions of high density matter near the surface of the Moon.

mass A measure of the total amount of material in an object.

mass-luminosity relation The linear relationship between the masses and luminosities of main-sequence stars.

matter-dominated universe A universe in which the radiation field that fills all space is unable to prevent the existence of neutral atoms.

megaparsec (Mpc) One million parsecs.

mesosphere The layer in the Earth's atmosphere above the stratosphere.

metal-poor star *See* **Population II star**.

metal-rich star *See* **Population I star**.

meteor The streak of light seen when any space debris vaporizes in the Earth's atmosphere; a "shooting star."

meteor shower Frequent meteors that seem to originate from a common point in the sky.

meteorite A fragment of space debris that has survived passage through the Earth's atmosphere.

meteoroid A small rock in interplanetary space.

microlensing The gravitational focusing of light from a distant star by a closer object to give a brighter image of the star.

Milky Way Galaxy The galaxy in which our solar system resides.

minor planet *See* **asteroid**.

missing mass *See* **dark matter**.

model A hypothesis that has withstood observational or experimental tests.

molecule A bound combination of two or more atoms.

momentum A measure of the inertia of an object; an object's mass multiplied by its velocity.

neap tide The least change from high to low tide during a day; it occurs during the first and third quarter phases of the Moon.

nebula (*plural* **nebulae**) A cloud of interstellar gas and dust.

neon fusion The thermonuclear fusion of neon nuclei.

neutrino A subatomic particle, with no electric charge and little mass, that is important in many nuclear reactions and in supernovae.

neutron A nuclear particle with no electric charge and with a mass nearly equal to that of the proton.

neutron degeneracy pressure A powerful pressure produced by degenerate neutrons.

neutron star A very compact, dense stellar remnant composed almost entirely of neutrons.

New General Catalogue (NGC) A catalog of star clusters, nebulae, and galaxies, first published in 1888.

new Moon The phase of the Moon when it is nearest the Sun in the sky.

Newtonian laws of motion A branch of physics based on Newton's laws of mechanics and gravitation.

Newtonian reflector An optical arrangement in a reflecting telescope in which a small, flat mirror reflects converging light rays to a focus on one side of the telescope tube.

nonthermal radiation *See* **synchrotron radiation**.

north celestial pole The location on the celestial sphere directly above the Earth's northern rotation pole.

northern lights (aurora borealis) Light radiated by atoms and ions in the Earth's upper atmosphere due to high-energy particles from the Sun and seen mostly in the northern polar regions.

northern vastness (northern lowlands) Relatively young, crater-free terrain in the northern hemisphere of Mars.

nova (*plural* **novae**) A star that experiences a sudden outburst of radiant energy, temporarily increasing its luminosity by a factor of between 10^4 and 10^6.

nuclear Referring to the nucleus of an atom.

nuclear bulge A distribution of stars in the shape of a flattened sphere that surrounds the nucleus of a spiral galaxy like the Milky Way.

nuclear density The density of matter in the nucleus of an atom; about 10^{17} kilograms per cubic meter (kg/m^3).

nucleus (of an atom) The massive part of an atom, composed of protons and neutrons, about which electrons revolve.

nucleus (of a comet) A collection of ices and dust that constitute the solid part of a comet.

nucleus (of a galaxy) *See* **galactic nucleus**.

OB association An unbound group of very young, massive stars predominantly of spectral types O and B.

OBAFGKM sequence The sequence of stellar spectral classifications from hottest to coolest stars.

objective lens The principal lens of a refracting telescope.

observable universe All space that is nearer to us than the distance traveled by light since the time of the Big Bang.

Occam's razor The principle of choosing the simplest scientific theory that correctly explains any phenomenon.

occultation The eclipsing of an astronomical object other than the Moon or Sun by another astronomical body.

Oort cloud A spherical region of the solar system beyond the Kuiper belt where most comets are believed to spend most of their time.

open cluster A loose association of young stars in the disk of the galaxy; a galactic cluster.

open universe A universe with a hyperbolic shape; lacks the mass necessary to someday stop expanding and recollapse. It will expand forever.

opposition The configuration of a planet when it is at an elongation of 180° and thus appears opposite the Sun in the sky.

optical double A pair of stars that appear to be near each other but are unbound and at very different distances from Earth.

optics The branch of physics dealing with the behavior and properties of light.

orbit The path of an object that is moving about a second object.

orbital eccentricity A measure between 0 and 1 indicating how close to circular a planet's orbit is (the eccentricity of a circular orbit is 0).

organic molecule A carbon-based compound.

overcontact binary A close binary system in which the two stars share a common atmosphere.

oxygen fusion The thermonuclear fusion of oxygen nuclei.

ozone layer The lower stratosphere, where most of the ozone in the air exists.

pair production The creation of a particle and an antiparticle from energetic photons.

Pallas The second asteroid to be discovered.

parabola An open curve formed by cutting a circular cone at an angle parallel to the sides of the cone.

parallax The apparent displacement of an object relative to more distant objects caused by viewing it from different locations.

parsec (pc) A unit of distance equal to 3.26 light-years.

partial eclipse A lunar or solar eclipse in which the eclipsed object does not appear completely covered.

Pauli exclusion principle A principle of quantum mechanics that says that two identical particles cannot simultaneously have the same position and momentum.

peculiar galaxy Any Hubble class of galaxy that appears to be blowing apart.

penumbra The portion of a shadow in which only part of the light source is covered by the shadow-making body.

penumbral eclipse A lunar eclipse in which the Moon passes only through the Earth's penumbra.

perihelion The point in its orbit where a planet is nearest the Sun.

period The interval of time between successive repetitions of a periodic phenomenon.

period-luminosity relation A relationship between the period and average luminosity of a pulsating star.

periodic table A listing of the chemical elements according to their properties; invented by D. Mendeleev.

phase (of the Moon) The appearance of the Moon at different points in its orbit of the Earth.

photodisintegration The breakup of nuclei in the core of a massive star due to the effects of energetic gamma rays.

photometry The measurement of light intensities.

photon A discrete unit of electromagnetic energy.

photon pressure The force per unit area exerted by photons on stellar or interstellar gas.

photosphere The region in the solar atmosphere from which most of the visible light escapes into space.

physics Basic principles that govern the behavior of physical reality.

pixel A contraction of the term "picture element"; usually refers to one square of a grid into which the light-sensitive component of a charge-coupled device is divided.

plage A bright spot on the Sun believed to be associated with an emerging magnetic field.

Planck era Time from the Big Bang until the Planck time (10^{-43} sec).

Planck time The brief interval of time, about 10^{-43} second, immediately after the Big Bang, when all four forces (gravity, electromagnetism, weak, strong) had the same strength.

Planck's law A relationship between the energy carried by a photon and its wavelength.

planetary differentiation The process early in the life of each planet whereby denser elements sank inward and lighter ones rose.

planetary nebula A luminous shell of gas ejected from an old, low-mass star.

planetesimal Primordial asteroidlike object from which the planets accreted.

plasma A hot, ionized gas.

plate techtonics The motions of large segments (plates) of the Earth's surface over the underlying mantle.

polymer A long molecule composed of many smaller molecules.

poor cluster (of galaxies) A cluster of galaxies with only a few members.

Population I star A star, such as the Sun, whose spectrum exhibits spectral lines of many elements heavier than helium; a metal-rich star.

Population II star A star whose spectrum exhibits comparatively few spectral lines of elements heavier than helium; a metal-poor star.

positron An electron with a positive rather than negative electric charge; an antielectron.

potential energy The energy stored in an object as a result of its location in space.

powers of ten A convenient method of writing large and small numbers that uses a number between 1 and 10 multiplied by a power of 10.

pre–main-sequence star The stage of star formation just before the main sequence; it involves slow contraction of the young star.

precession (of the Earth) A slow, conical motion of the Earth's axis of rotation caused by the gravitational pull of the Moon and Sun on the Earth's equatorial bulge.

precession of the equinoxes The slow westward motion of the equinoxes along the ecliptic because of the Earth's precession.

primary mirror The large, concave, light-gathering mirror in a reflecting telescope, analogous to the objective lens on a refracting telescope.

prime focus The point in a reflecting telescope where the primary mirror focuses light.

primordial black hole A relatively low-mass black hole hypothetically formed at the beginning of the universe.

primordial fireball The extremely hot gas that filled the universe immediately following the Big Bang.

prism A wedge-shaped piece of glass used to disperse white light into a spectrum.

prograde orbit An orbit of a moon or satellite around a planet that is in the same direction as the planet's rotation.

prominence Flamelike protrusion seen near the limb of the Sun and extending into the solar corona.

proper motion The change in the location of a star on the celestial sphere.

proton A heavy, positively charged nuclear particle.

protoplanet The embryonic stage of a planet when it is growing because of collisions with planetesimals.

protoplanetary disk (proplyd) A disk of material encircling a protostar or a newborn star.

protostar The earliest stage of a star's life before fusion commences and when gas is rapidly falling onto it.

protosun The Sun prior to the time when hydrogen fusion began in its core.

pulsar A pulsating source associated with a rapidly rotating neutron star with an off-axis magnetic field.

quantum mechanics The branch of physics dealing with the structure and behavior of atoms and their interactions with each other and with light.

quark A particle that is a building block of the heavy nuclear particles such as protons and neutrons.

quarter Moon A phase of the Moon when it is located 90% from the Sun in the sky.

quasar (quasi-stellar radio source) A starlike object with a very large redshift.

quasi-stellar object (QSO) A quasar.

quintessence One of the explanations of the dark energy causing the universe to accelerate outward.

radial velocity That portion of an object's velocity parallel to the line of sight.

radial-velocity curve A plot showing the variation of radial velocity with time for a binary star or variable star.

radiation Electromagnetic energy; photons.

radiation (photon) pressure The transfer of momentum carried by radiation to an object on which the radiation falls.

radiation-dominated universe The time at the beginning of the universe when the electromagnetic radiation prevented ions and electrons from combining to make neutral atoms.

radiative zone A region inside a star where energy is transported outward by the movement of photons through a gas from a hot location to a cooler one.

radio astronomy The branch of astronomy dealing with observations at radio wavelengths.

radio galaxy A galaxy that emits an unusually large amount of radio waves.

radio lobes Vast regions of radio emission on opposite sides of a radio galaxy.

radio telescope A telescope designed to detect radio waves.

radio wave Long-wavelength electromagnetic radiation.

radioactivity The process whereby certain atomic nuclei naturally decompose by spontaneously emitting particles.

red giant A large, cool star of high luminosity.

red supergiant An extremely large, cool star of high luminosity; a star in the upper right corner of the Hertzsprung-Russell diagram.

redshift The shifting to longer wavelengths of the light from remote galaxies and quasars; the Doppler shift of light from any receding source.

reflecting telescope (reflector) A telescope in which the principal light-gathering component is a concave mirror.

reflection The rebounding of light rays off a smooth surface.

reflection nebula A comparatively dense cloud of gas and dust in interstellar space that is illuminated by a star between it and the Earth.

refracting telescope (refractor) A telescope in which the principal light-gathering component is a lens.

refraction The bending of light rays when they pass from one transparent medium to another.

regolith The powdery, lifeless material on the surface of a moon or planet.

regular cluster (of galaxies) An evenly distributed group of galaxies bound together by mutual gravitational attraction.

resolution The degree to which fine details in an optical image can be distinguished.

resolving power A measure of the ability of an optical system to distinguish fine details in the image it produces.

resonance The large response of an object to a small periodic gravitational tug from another object.

retrograde motion The occasional backward (that is, westward) apparent motion of a planet against the background stars as seen from Earth. Retrograde motion is an optical illusion.

retrograde orbit The orbit of a moon or satellite around a planet that is in the direction opposite to the planet's rotation.

retrograde rotation The rotation of a planet opposite to its direction of revolution around the Sun. Only Pluto, Uranus, and Venus have retrograde rotation.

revolution The orbit of one body about another.

rich cluster (of galaxies) A cluster of galaxies with many members.

right ascension The celestial coordinate analogous to longitude on Earth and measured around the celestial equator from the vernal equinox.

rille A winding crack or depression in the lunar surface.

ringlet Any one of numerous, closely spaced, thin bands of particles in Saturn's ring system.

Roche limit The shortest distance from a planet or other object at which a second object can be held together by its own gravitational forces.

Roche lobe The teardrop-shaped regions around each star in a binary star system inside of which gas is gravitationally bound to that star.

rotation The spinning of a body about an axis passing through it.

rotation curve (of a galaxy) A graph showing how the orbital speed of material in a galaxy depends on the distance from the galaxy's center.

RR Lyrae variable star A type of pulsating star with a period less than one day.

Sa, Sb, Sc Categories of spiral galaxies determined by the sizes of their nuclear bulges or how tightly wound their spiral arms are; an Sa galaxy is the most tightly wound.

Sagittarius A The strong radio source associated with the nucleus of the Milky Way galaxy.

satellite A body that revolves about a larger one.

SBa, SBb, SBc Categories of barred spiral galaxies determined by how tightly wound their spiral arms are; SBa galaxies are the most tightly wound.

scarp A cliff on Mercury believed to have formed when the planet cooled and shrank.

Schmidt corrector plate A specially shaped lens used with spherical mirrors that corrects for spherical aberration and provides an especially wide field of view.

Schwarzschild black hole Any nonrotating, uncharged black hole.

Schwarzschild radius The distance from the center to the event horizon in any black hole.

scientific method The method of doing science based on observation, experimentation, and the formulation of hypotheses (theories) that can be tested.

scientific notation The style of writing large and small numbers using powers of ten.

scientific theory An idea about the natural world that is subject to testing and refinement.

seafloor spreading The process whereby magma upwelling along rifts in the ocean floor causes adjacent segments of the Earth's crust to separate.

secondary cosmic rays Particles from the Earth's atmosphere given high speeds Earthward by cosmic rays from space.

secondary mirror A relatively small mirror used in reflecting telescopes to guide the light out the side or bottom of the telescope.

seeing disk The size that a star appears to have on a photographic or charge-coupled-device image as a result of the changing refraction of the starlight passing through the Earth's atmosphere.

seismic waves Vibrations traveling through or around an astronomical body usually associated with earthquakelike phenomena.

seismograph A device used to record and measure seismic waves, such as those produced by earthquakes.

seismology The study of earthquakes and related phenomena.

self-propagating star formation The process whereby the deaths of stars in one part of a galaxy stimulates star formation in a neighboring region of that galaxy.

semidetached binary A close binary system in which one star fills or is overflowing its Roche lobe.

semimajor axis (of an ellipse) Half of the longest dimension of an ellipse.

SETI The search for extraterrestrial intelligence.

Seyfert galaxy A spiral galaxy with a bright nucleus whose spectrum exhibits emission lines.

Shapley-Curtis debate An inconclusive debate between Harlow Shapley and Heber Curtis in 1920 about whether certain nebulae were beyond the Milky Way.

shell helium fusion Helium fusion that occurs in a thin shell surrounding the core of a star.

shell hydrogen fusion Hydrogen fusion that occurs in a thin shell surrounding the core of a star.

shepherd satellite (moon) A small satellite whose gravitational tug is responsible for maintaining a sharply defined ring of matter around a planet such as Saturn or Uranus.

shock wave An abrupt, localized region of compressed gas caused by an object traveling through the gas at a speed greater than the speed of sound.

shooting star *See* **meteor.**

short-period comet A comet that orbits the Sun in the vicinity of the planets, thereby reappearing with tails every 200 years or less.

sidereal month The period of the Moon's revolution about the Earth measured with respect to the Moon's location among the stars.

sidereal period The orbital period of one object about another measured with respect to the stars.

single-line spectroscopic binary A spectroscopic binary whose periodically varying spectrum exhibits the spectral lines of only one of its two stars.

singularity A place of infinite curvature of spacetime in a black hole.

Small Magellanic Cloud (SMC) An irregular galaxy that is a companion to the Milky Way.

solar corona The Sun's outer atmosphere.

solar cycle A 22-year cycle during which the Sun's magnetic field reverses its polarity twice.

solar day From noontime to the next noontime; for Earth it is 24 hours.

solar eclipse An eclipse during which the Moon blocks the Sun.

solar flare A violent eruption on the Sun's surface.

solar luminosity ($L_\odot$) The total energy emitted by the Sun each second.

solar model A set of equations that describe the internal structure and energy generation of the Sun.

solar nebula The cloud of gas and dust from which the Sun and the rest of the solar system formed.

solar seismology The study of the Sun's interior from observations of vibrations of its surface.

solar system The Sun, planets, their satellites, asteroids, comets, and related objects that orbit the Sun.

solar wind An outward flow of particles (mostly electrons and protons) from the Sun.

solstice Either of two points along the ecliptic at which the Sun reaches its maximum distance north or south of the celestial equator.

south celestial pole The location on the celestial sphere directly above the Earth's south rotation pole.

southern lights (aurora australis) Light radiated by atoms and ions in the Earth's upper atmosphere due to high-energy particles from the Sun; seen mostly in the southern polar regions.

southern highlands Older, cratered terrain in the Martian southern hemisphere.

Southern Wall A huge sheet of galaxies between two voids in the cosmos.

spacetime The concept from special relativity that space and time are both essential in describing the position, motion, and action of any object or event.

special theory of relativity A description of mechanics and electromagnetic theory formulated by Einstein according to which measurements of distance, time, and mass are affected by the observer's motion.

spectral analysis The identification of chemicals by the appearance of their spectra.

spectral lines Dark or bright lines at specific wavelengths in a spectrum.

spectral type A classification of stars according to the appearance of their spectra.

spectrogram The photograph of a spectrum.

spectrograph A device for photographing a spectrum.

spectroscope A device for directly viewing a spectrum.

spectroscopic binary star A double star whose binary nature can be deduced from the periodic Doppler shifting of lines in its spectrum.

spectroscopic parallax A method of determining a star's distance from the Earth by measuring its surface temperature, luminosity, and apparent magnitude.

spectroscopy The study of spectra.

spectrum (*plural* **spectra**) The result of electromagnetic radiation passing through a prism or grating so that different wavelengths are separated.

speed The rate at which an object moves.

spherical aberration An optical property whereby different portions of a spherical lens or spherical, concave mirror have slightly different focal lengths.

spicule A narrow jet of rising gas in the solar chromosphere.

spin (of an electron or proton) A small, well-defined amount of angular momentum possessed by electrons, protons, and other particles.

spiral arms Lanes of interstellar gas, dust, and young stars that wind outward in a plane from the central regions of some galaxies.

spiral density wave A spiral-shaped pressure wave that orbits the disk of a spiral galaxy and induces new star formation.

spiral galaxy A flattened, rotating galaxy with pinwheel-like spiral arms winding outward from the galaxy's nuclear bulge.

spoke A moving dark region of Saturn's rings.

spring tide The greatest daily difference between high tide and low tide, occurring when the Moon is new or full.

stable Lagrange points Locations throughout the solar system where the gravitational forces from the Sun and a planet keep space debris trapped.

standard candle An object whose known luminosity can be used to deduce the distance to a galaxy.

star A self-luminous sphere of gas.

starburst galaxy A galaxy where there is an exceptionally high rate of star formation.

Stefan-Boltzmann law A relationship between the temperature of a blackbody and the rate at which it radiates energy.

stellar evolution The changes in size, luminosity, temperature, and chemical composition that occur as a star ages.

stellar model The result of theoretical calculations that give details of physical conditions inside a star.

stellar parallax The apparent shift in a nearby star's position on the celestial sphere resulting from the Earth's orbit around the Sun.

stellar spectroscopy The study of the properties of stars encoded in their spectra.

stony meteorite A meteorite composed of rock with very little iron; also called a *stone*.

stony-iron meteorite A meteorite composed of roughly equal amounts of rock and iron.

stratosphere The second layer in the Earth's atmosphere, directly above the troposphere.

strong nuclear force The force that binds protons and neutrons together in nuclei.

subgiant A star whose luminosity is between that of main-sequence stars and normal giants of the same spectral type.

summer solstice The point on the ecliptic where the Sun is farthest north of the celestial equator.

Sun The star about which the Earth and other planets revolve.

sunspot A temporary cool region in the solar photosphere created by protruding magnetic fields.

sunspot cycle The semiregular 11-year period with which the number of sunspots fluctuates.

sunspot maximum The time during the solar cycle when the number of sunspots is exceptionally high.

sunspot minimum The time during the solar cycle when the number of sunspots is exceptionally low.

supercluster (of galaxies) A gravitationally bound collection of many clusters of galaxies.

supergiant A star of very high luminosity.

supergranule A large convective cell in the Sun's chromosphere containing many granules.

superior conjunction The configuration when a planet is behind the Sun as seen from Earth.

supermassive black hole A black hole whose mass exceeds 1000 solar masses.

supernova (*plural* **supernovae**) A stellar outburst during which a star suddenly increases its brightness roughly a millionfold.

supernova remnant A nebula left over after a supernova detonates.

superstring A set of theories that hope to describe the nature of spacetime and matter at a more fundamental level than is presently possible.

synchronous rotation The condition when a moon's rotation rate and revolution rate are equal or when a planet's rotation rate equals its moon's revolution rate.

synchrotron radiation The radiation emitted by charged particles moving through a magnetic field; nonthermal radiation.

synodic month (lunar month) The period of revolution of the Moon with respect to the Sun; the length of one cycle of lunar phases.

synodic period The interval between successive occurrences of the same configuration of a planet as seen from Earth.

T Tauri stars Young, variable stars associated with interstellar matter that show erratic changes in luminosity.

tail (of a comet) Gas and dust particles from a comet's nucleus that have been swept away from the comet's nucleus by the radiation pressure of sunlight and impact of the solar wind.

telescope An instrument for viewing remote objects.

temperature (Celsius) Temperature measured on a scale where water freezes at 0° and boils at 100°.

temperature (Fahrenheit) Temperature measured on a scale where water freezes at 32° and boils at 212°.

temperature (Kelvin) Absolute temperature measured in Celsius degree intervals. Water freezes at 273 K and boils at 373 K.

terminator The line dividing day and night on the surface of any body orbiting the Sun; the line of sunset or sunrise.

terrestrial planet Any of the planets Mercury, Venus, Earth, or Mars; a planet with a composition and density similar to that of Earth.

theory *See* **scientific theory.**

thermal energy The energy associated with the motions of atoms or molecules in a substance.

thermal equilibrium A balance between the input and outflow of heat in a system.

thermonuclear fusion A reaction in which the nuclei of atoms are fused together at a high temperature.

thermosphere A layer high in the Earth's atmosphere above the mesosphere.

3-to-2 spin-orbit coupling The rotation of Mercury, which makes three complete rotations on its axis for every two complete orbits around the Sun.

tidal force A gravitational force whose strength and/or direction varies over a body and thus tends to deform the body.

time zone One of 24 divisions of the Earth's surface separated by 15° along lines of constant longitude (with allowances for some political boundaries).

total eclipse A solar eclipse during which the Sun is completely hidden by the Moon, or a lunar eclipse during which the Moon is completely immersed in the Earth's umbra.

trailing-arm spiral A spiral arm pointing away from the direction of rotation, characteristic of all spiral galaxies.

transition (electronic) The change in energy and orbit of an electron around an atom or molecule.

transition zone Region between the Sun's chromosphere and corona where the temperature skyrockets to about 1 million K.

Trojan asteroid One of several asteroids at stable Lagrange points that share Jupiter's orbit about the Sun.

troposphere The lowest level of the Earth's atmosphere.

Tully-Fisher relation A correlation between the width of the 21-centimeter line of a spiral galaxy and its absolute magnitude.

turnoff point The location of the brightest main-sequence stars on the Hertzsprung-Russell diagram of a globular cluster.

21-cm radiation Radio emission from a hydrogen atom caused by the flip of the electron's spin orientation.

twinkling The apparent change in a star's brightness, position, or color due to the motion of gases in the Earth's atmosphere.

Type I Cepheid Population I Cepheid variable star found in the disks of spiral galaxies.

Type Ia supernova A supernova occurring after a white dwarf accretes enough mass from a companion star to exceed the Chandrasekhar limit.

Type II Cepheid Population II Cepheid variable star found in elliptical galaxies and in the halos of disk galaxies that is 1.5 magnitudes dimmer than a Type I Cepheid.

Type II supernova A supernova occurring after a massive star's core is converted to iron.

ultraviolet (UV) radiation Electromagnetic radiation of wavelengths shorter than those of visible light but longer than those of X rays.

umbra The central, completely dark portion of a shadow.

universal constant of gravitation The constant of proportionality in Newton's law of gravitation, usually denoted G.

universal law of gravitation Newton's law of gravitation, which describes how the gravitational force between two bodies depends on their masses and separation.

universe All space along with all the matter and radiation in space.

Van Allen radiation belts Two flattened, doughnut-shaped regions around the Earth where many charged particles (mostly protons and electrons) are trapped by the Earth's magnetic field.

variable star A star whose luminosity varies.

velocity A quantity that specifies both direction and speed of an object.

vernal equinox The point on the ecliptic where the Sun crosses the celestial equator from south to north.

very-long-baseline interferometry (VLBI) A method of connecting widely separated radio telescopes to make observations of very high resolution.

virtual particle A particle and its antiparticle, created simultaneously in pairs and which quickly disappear without a trace.

visual binary star A double star in which the two components can be resolved through a telescope.

void A huge, roughly spherical region of the universe where exceptionally few galaxies are found.

water hole The part of the electromagnetic spectrum at a few thousand megahertz where there is very little background noise from space.

wavelength The distance between two successive peaks in a wave.

waning An adjective that means "decreasing," as in the "waning crescent Moon" or the "waning gibbous Moon."

waxing An adjective that means "increasing," as in the "waxing crescent Moon" or the "waxing gibbous Moon."

weak nuclear force A nuclear interaction involved in certain kinds of radioactive decay.

weight The force with which a body presses down on the surface of the Earth.

white dwarf A low-mass stellar remnant that has exhausted all its thermonuclear fuel and contracted to a size roughly equal to the size of the Earth.

Widmanstätten patterns Crystalline structure seen inside iron meteorites.

Wien's law A relationship between the temperature of a blackbody and the wavelength at which it emits the greatest intensity of radiation.

winter solstice The point on the ecliptic where the Sun is farthest south of the celestial equator.

work Change in an object's energy as a result of a force being applied to it.

wormhole A hypothetical connecting passage between black holes and other places in the universe.

X rays Electromagnetic radiation whose wavelength is between that of ultraviolet light and gamma rays.

X-ray burster A neutron star in a binary star system that accretes mass, undergoes thermonuclear fusion on its surface, and therefore emits short bursts of X rays.

year The sidereal period of revolution of a planet around the Sun.

Zeeman effect A splitting or broadening of spectral lines in the presence of a magnetic field.

zenith The point on the celestial sphere directly overhead.

zero-age main sequence (ZAMS) The positions of stars on the Hertzsprung-Russell diagram that have just begun to fuse hydrogen in their cores.

zodiac A band of 13 constellations around the sky through which the Sun appears to move throughout the year.

zones (on Jupiter) Light-colored bands in Jupiter's cloud cover.

ANSWERS TO COMPUTATION QUESTIONS

Essentials I

2. a. 5.974×10^{24} kg
    b. 1.989999×10^{30} kg
    c. 6.9599×10^5 km
    d. 3.1558×10^7 s
3. a. 8.3 pc = 27.1 ly
    b. 6.52 ly = 2.0 pc
    c. 8450 AU = 1.26×10^{12} km
    d. 2.7×10^3 Mpc = 2.7×10^6 kiloparsecs

Toolbox I-1 3.141×10^9; 3.1831×10^{-9};
27182820000, 0.0000000000367879

Toolbox I-2 $\sim 2.48 \times 10^{13}$ mi, $\sim 3.99 \times 10^{13}$ km, ~ 1.29 pc

Chapter 1

38. 1 more sidereal month than synodic month
48. $R_{\odot} \approx 6.53 \times 10^8$ m

Toolbox 1-1 $\sim 0.53°$. This is the same angle as the Moon makes in the sky; $\sim 0.27°$; $\sim 1.1°$

The Mummy Returns Each day, the Moon is in the same phase as seen everywhere on Earth. For example, when it is full, it is seen as a full Moon everywhere.

Austin Powers II The Moon orbits west to east around the Earth, rather than the east to west depicted in the movie. The motion as seen from the model Earth works out correctly, however, because the model Earth was not rotating. The real Earth rotates west to east about 30 times faster than the Moon orbits the Earth, making the Moon appear to move from west to east each day in our sky. In other words, as seen from the real, rotating Earth, the Moon rises in the east and sets in the west, which is the same motion as seen in the movie.

Chapter 2

13. 5.2 square AU in 1995, 26 square AU in 5 years
14. $a = 100$ AU, maximum distance is almost 200 AU
15. 2.8 yr
16. use $P^2 = a^3$
21. At 2 AU, 1 year $\approx$ 2.8 present years; At 1/2 AU, 1 year $\approx$ .35 present years
22. .01 × present gravitational attraction
23. same acceleration, same length of the year

Toolbox 2-1 Energy increases 9-fold; energy decreases to 1/4; 2000 joules; 0 joules; change its moment of inertia, I, and its angular velocity, Ω.

Toolbox 2-2 Your weight in pounds *or* (Your weight in pounds/2.2)* 9.8 N; same, 9.8 m/s^2 *or* 32 ft/s^2; 9×10^{21}N; it is one quarter the present force.

The Planet of the Apes On a non-rotating space station, people would be floating around. They could stay connected to the floor if they were wearing shoes with strong magnets in them, but they are not wearing such shoes. For a space helmet to do any good it must be hermetically sealed to a spacesuit so that oxygen in the helmet does not escape into the capsule and, thereby, be wasted.

Chapter 3

4. 9 times more
17. 1/25
18. Palomar gathers 10^6 times more light than the human eye.
20. a. 222×
    b. 100×
    c. 36×

Toolbox 3-1 ~ 408 nm; violet; 204 nm; $\sim 3.64 \times 10^{-19}$J

Lara Croft, Tomb Raider The planets don't align on a 5000 year cycle and because they have orbital inclinations, they don't align in a point. Rather, they just cluster together within a few degrees of each other in the sky. You can't use a telescope in a lit room — your eyes would be too dilated. You also wouldn't see all of the planets as disks. Pluto, Neptune, and Uranus would look like fuzzy blobs or points of light. And, you can't see the planets move through the sky in "real time." As seen from Earth, they move too slowly for that.

Chapter 4

10. 7 1/2 times
11. 238 nm
12. 6520 K
13. approaching, 13 km/s
14. receding, 21 km/s
15. about 90,000 km/s
19. $\lambda \approx 500$ nm (same as now), 4 × brighter than now

Toolbox 4-1 ~ 5270K; ~ 9700 nm; infrared

Toolbox 4-2 1/8kg; $\sim .35$ kg; $\sim 17 \frac{1}{2}$

Toolbox 4-3 3×10^4 km/s *or* 0.1c; $\sim 3.3 \times 10^{-4}$; –0.1

Chapter 8

21. 5.0×10^{14} tons

Asteroid Comets are made of ice and rock, whereas asteroids are rock and metal. The comet would have been severely damaged, probably broken apart, and possibly completely pulverized, by the impacts, while changing the orbits of the asteroids only slightly. Because of their angular momentum, the slight deflection of the asteroids might eventually send them Earthward, but not for many decades or longer.

Given the evidence that impacts of large bodies onto the Earth happen very infrequently over our planet's recent history, the development of a space defense system is probably not justified.

Besides the fact that the comet would be in many pieces, the tail would not be streaming behind it, but rather pointing away from the Sun. Also comets are so far away that we can't see them moving across the sky.

Chapter 9

5. next maximum in 2005, next minimum in 2012
16. 1400 kg/m^3
17. 4.8%
18. 500 nm = visible light; 58 nm = ultraviolet; 1.9 nm = x-ray

Toolbox 9-1 ~316,000 years; 6×10^{17} helium atoms; 2.25×10^{14} J

Essentials III

10. 25 times brighter
12. 4.3 pc
14. a. 9.7 pc,
    b. 0.10 arcsec
15. 1585 times brighter

Toolbox III-1 ~4.85×10^{-6} pc; 3.26×10^{-2}; ~2.6 pc or ~8.47 ly; ~237 pc; ~4.22×10^{-3}

Toolbox III-2 ~11.4 pc; ~37.1 ly; +1.74 (Fomalhaut); –1.44 (Sirius A)

Chapter 10

15. ≈ 10$M_\odot$
16. ≈ 10,000–15,000K

Toolbox 10-1 We can ignore M_2 (planet's mass) as tiny compared to the Sun's mass, M_1. Since the Sun's mass in these units is 1, the equation becomes $1 = a^3/P^2$, or $P^2 = a^3$; ~1.56 $M_\odot$; 73 years

Star Wars There should be two shadows, just as you might often see at night when two streetlamps are shining on you.

The shadow is not of the boy, but of Darth Vader, whom he is destined to become.

Chapter 11

26. 2000
27. 200 times longer

Chapter 12

27. about 7470 years ago

Chapter 13

8. 8.9 km, 89 km

Toolbox 13-1 ~1.0×10^{12}kg ; 9×10^9 km; ~2130 $M_\odot$

Black Hole A black hole can be entered from any direction. When falling into a black hole, the tidal force from the hole causes objects to stretch out and eventually come apart.

Essentials IV

Toolbox IV-1 1×10^5 pc; ~6.92; m = 9.83; m – M = 5

Chapter 14

10. 20 times
14. about once every 3750 years

Chapter 15

16. 306 Mpc
17. 7200 km/s

Mission to Mars (a) Since we are in the Milky Way, we cannot see much of it, much less the whole thing as seen from outside it. (b) If the galaxy is an external galaxy, then it is depicted much, much, much larger across than any galaxy that is really visible from the solar system.

Chapter 17

10. 11 billion years

Appendix

Toolbox A-1 5527°C, 9980°F, –270°C, –454°F

INDEX

Note: Page numbers followed by f indicate figures; those followed by t indicate tables; and those followed by b indicate boxed material.

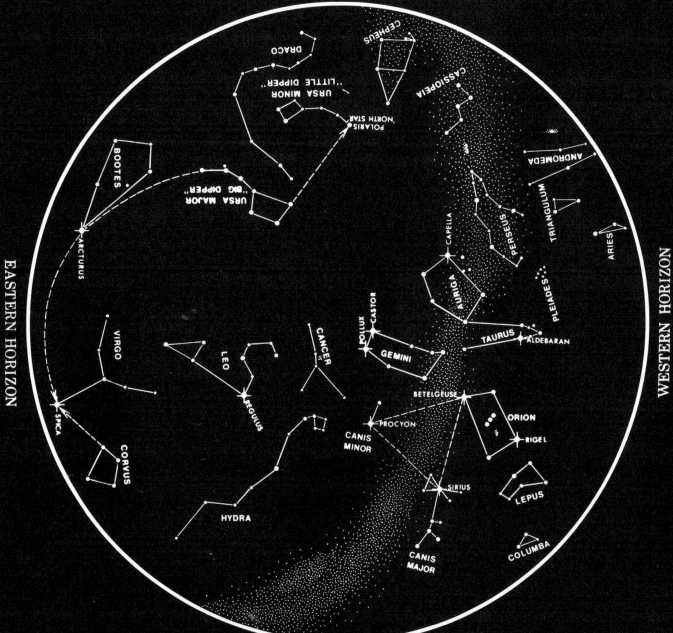

THE NIGHT SKY IN MARCH

Chart time (Local Standard Time):

10 pm...First of March
9 pm...Middle of March
8 pm...Last of March

THE NIGHT SKY IN MAY

Chart time (Daylight Savings Time):

11 pm...First of May
10 pm...Middle of May
9 pm...Last of May